The Quality Yearbook

1997

James W. Cortada
John A. Woods

McGraw-Hill
New York   San Francisco   Washington, D.C.   Auckland   Bogotá
Caracas   Lisbon   London   Madrid   Mexico City   Milan
Montreal   New Delhi   San Juan   Singapore
Sydney   Tokyo   Toronto

Copyright © 1997 by The McGraw-Hill Companies, Inc. All rights reserved. Printed in the United States of America. Except as permitted under the United States Copyright Act of 1976, no part of this publication may be reproduced or distributed in any form or by any means, or stored in a data base or retrieval system, without the prior written permission of the publisher.

1 2 3 4 5 6 7 8 9 0 AGM/AGM 9 0 1 0 9 8 7

ISBN: 0-07-024456-1

The sponsoring editor for this book was Philip Ruppel. Production was managed by John Woods, CWL Publishing Enterprises, Madison, WI. It was designed and composed at Impressions Book and Journal Services, Inc., Madison, WI.

Printed and bound by Quebecor/Martinsburg.

McGraw-Hill books are available at special quantity discounts to use as premiums and sales promotions, or for use in corporate training programs. For more information, please write to the Director of Special Sales, McGraw-Hill, 11 West 19th Street, New York, NY 10011. Or contact your local bookstore.

The Quality Yearbook
1997

Also available from McGraw-Hill

McGraw-Hill Encyclopedia of Quality Terms & Concepts (1995)
    James W. Cortada and John A. Woods
QualiTrends: 7 Quality Secrets that Will Change Your Life (1996)
    John A. Woods and James W. Cortada
TQM in Information Systems Management (1995)
    James W. Cortada
TQM in Sales and Marketing Management (1993)
    James W. Cortada

Contents

Preface	ix
Acknowledgments	xii

PART ONE — **Background for Quality** — 1

The Collapse of Prevailing Wisdom, *H. James Harrington* — 3

Classics in Quality

Some Basics of Statistical Quality Control, *Walter A. Shewhart* — 13

A Generic Concept of Marketing, *Philip Kotler* — 32

Quality Perspectives

The Promise and Shortcomings of Total Quality, *Charles C. Poirier and Steven J. Tokarz* — 47

PART TWO — **Quality by Industry** — 59

MANUFACTURING SECTOR

1996 Manufacturing Outlook: The Dawning of Relationship-Based Competition, *Roger N. Nagel and Napoleon Devia* — 61

Process Redesign and the Successful Turnaround of Shasta Industries, *John Veleris, Robert Harlan, and Sheldon Greenberg* — 69

Achieving Strategic Agility through Economies of Knowledge, *Aleda V. Roth* — 81

Cellular Manufacturing for Small Manufacturers, *Stanley D. Stone* — 91

SERVICES SECTOR

The Value-Driven Bank, *Erick Reidenbach and Terry C. Wilson* — 99

Improving Customer Service in Call Centers, *Dennis S. Holland* — 108

PUBLIC SECTOR

Government

State and Local Governments: Quality in 1996, *Joe Sensenbrenner* — 115

Could a Dose of Deming Transform Government?, *Wayne J. Levin* — 122

Making the Big U-Turn, *The National Performance Review Staff* — 130

Health Care

Hospital Sets New Standard as Closure Approaches: Quality Is Continuous, *Martha L. Dasch* — 138

The Most Important Day, *J. Daniel Beckham* — 146

Reengineering or Rebirth?, *Pamela L. Blyth* — 152

Higher Education

Is Higher Education Ready for the Twenty-First Century?, *Ronald L. Heilmann*	162
TQM in the College Classroom, *Ronald E. Turner*	168
Inside Track to the Future, *Richard Alfred and Patricia Carter*	175
Quality in Business Education, *Niranjan Pati, Dayr Reis, and John Betton*	188

Education K–12

Quality Management in Schools: 1996, *Julie Horine*	194
Total Quality Applied in the Classroom, *Patrick Konopnicki*	205
Quality Excellence in Education, *Franklin P. Schargel*	210
Sharing the Wealth: TQM Spreads from Business to Education, *Robert Manley and John Manley*	218

PART THREE Implementing Quality — 229

QUALITY TRANSFORMATION

Planning

Reinventing Strategic Planning, *John C. Camillus*	231

Leadership

Managing for Tomorrow's Competitiveness Today, *Armand V. Feigenbaum*	243
Power in Organizations: A Look through the TQM Lens, *Paula Phillips Carson, Kerry D. Carson, E. Leon Knight, and C. William Roe*	254
How to Work with, instead of against, Millions of Years of Social Primate Evolution or "How to *Start* Monkeying Around and Be a More Effective Leader", *William Lareau*	267
The Failure of Participatory Management, *Charles Heckscher*	277

Cultural Transformation

Connecting Culture to Organizational Change, *Timothy Galpin*	285
The Soul of the Hog, *Bob Filipczak*	293
Xerox 2000: From Survival to Opportunity, *Richard J. Leo*	301

Voice of the Customer

Maximizing Consumer Quality Begins With Maximizing the Value of Your Brand, *Scott Davis*	315
Customer Loyalty: Playing for Keeps, *Laura Struebing and Anne Calek*	322
Plugging In to Your Customers' Needs, *William Keenan Jr.*	334
Service with a Smile, *John Holten*	340

Training

Measuring ROI: The Fifth Level of Evaluation, *Jack J. Phillips*

Training Success Stories, *Leslie Overmyer-Day and George Benson*	346
	352
Teams and Teamwork	
Teams in the Age of Systems, *Peter R. Scholtes*	360
Moving beyond Team Myths, *John Beck and Neil Yeager*	376
Self-Directed Work Teams: A Guide to Implementation, *Michael W. Piczak and Reuben Z. Hauzer*	384
Systems Thinking/Learning Organizations	
Self-Organization: The Irresistible Future of Organizing, *Margaret J. Wheatley and Myron Kellner-Rogers*	397
Strategic Process Improvement through Organizational Learning, *John H. Grant and Devi R. Gnyawali*	407
Communication	
TQM Can Be DOA Without a Proper Communications Plan, *Kaat Exterbille*	414
The Partnership Facade, *Jim Harris*	419

QUALITY TOOLS AND TECHNIQUES

Process Reengineering

Why Did Reengineering Die?, *Oren Harari*	424
The Missing Piece in Reengineering, *Nicholas F. Horney and Richard Koonce*	431
Benchmarking	
What Benchmarking Books Don't Tell You, *Sarah Lincoln and Art Price*	443
Benchmarking is Not an Instant Hit, *Jim Morgan*	451
Process Management and Measurement	
From Balanced Scorecard to Strategic Gauges: Is Measurement Worth It?, *John H. Lingle and William A. Schiemann*	455
The Effects of SPC on the Target of Process Quality Improvement, *Wen-Hsien Chen*	465
Simplify Your Approach to Performance Measurement, *Philip Ricciardi*	482
When Do I Recalculate My Limits?, *Donald J. Wheeler*	488

FUNCTIONAL PROCESSES

Suppliers and Purchasing

Purchasing and Quality, *Eberhard E. Scheuing*	491
At CAT They're Driving Supplier Integration into the Design Process, *Anne Millen Porter*	501
Five Ways to Improve the Contracting Process, *Pete Hybert*	506

Logistics

Seven Trends of Highly Effective Warehouses, *David R. Olson*

		517
	Build for Speed, *Perry A. Trunick*	522
	Information Technology and Management	
	Getting In the Pink, *Anita Lienert*	528
	Data Warehouses: Build Them for Decision-Making Power, *ohn Teresko*	534
	Accounting/Finance	
	ABC and High Technology: A Story with a Moral, *Frank H. Selto and Dale W. Jasinski*	539
	From Activity-Based Costing to Throughput Accounting, *John B. MacArthur*	547
	Product Development	
	Your Product Development Process Demands Ongoing Improvement, *Preston G. Smith*	555
	Early Supplier Involvement: Leveraging Know-How for Better Product Development, *Francis Bidault and Christina Butler*	567
	Human Resources	
	Do Employee Involvement and TQM Programs Work?, *Susan A. Mohrman, Edward E. Lawler III, and Gerald E. Ledford Jr.*	575
	World Class Suggestion Systems Still Work Well, *John Savageau*	583
	Gainsharing: A Lemon or Lemonade?, *Woodruff Imberman*	591
	STANDARDS AND ASSESSMENTS	
	ISO 9000	
	Now That You're Registered . . ., *Richard C. Randall*	598
	ISO 9000: A Practical Step-By-Step Approach, *Roger S. Benson and Richard W. Sherman*	604
	ISO 14001 Certification: Are You Ready?, *Gregory J. Hale and Caroline G. Hemenway*	612
	Weighing Alternative ISO 9000 Registration: Not for All Companies, *Paul Scicchitano*	620
	Baldrige Criteria/Quality Audits	
	Is the Baldrige Still Meaningful?, *Barbara Ettorre*	625
PART FOUR	**Quality References**	631
	Annotated Bibliography	633
	On-Line Quality Services	668
	Directory of Magazines and Journals	686
	Directory of Business Organizations	704
	Quality Quotes	743
	1996 Baldrige Award Winners	758
	1997 Baldrige Award Criteria	766
	Index	781

Preface

When we started working on this annual anthology and reference in 1993 and published the first edition in January 1994, we hoped we would be delivering articles and information that readers would find useful. Now that we have completed the 1997 edition, we are gratified to know that you and many others like you have found this to be a valuable resource. If you are a previous purchaser and are adding this one to your collection, thank you for your continued support. If you are a first time purchaser, thanks for trying us out. Whether you're new or a repeat customer, continue reading here. We're going to talk about what we think makes this book special and how this edition is different from earlier editions.

The book has two purposes:

1. To provide an *annual documentation* of the most useful ideas, tools, and articles in quality.
2. To serve as an *annual comprehensive reference* for finding information on publications, tools, techniques, and special issues involved in implementing quality management.

We set several goals for ourselves in developing this book. The most important include:

Create a Clearinghouse. Because there is so much coming out in quality, McGraw-Hill and the editors perceived a need for a kind of "clearinghouse" to which people from all different organizations might turn as a starting place to learn about current thinking in the implementation of quality management practices. *The Quality Yearbook* includes the best and most comprehensive selection of such information available anywhere. Its utility is founded on this premise, and its continuing success is a testament to the fact that we have filled this need well.

Create a Book of Lasting Value. Because it is revised annually, *The Quality Yearbook* focuses on information that has come out in the period immediately preceding its publication. While currency is one of its strengths in this fast-changing field, we also intend each edition to have lasting value. You will find articles throughout the book that will continue to be applicable for several years as you go about implementing quality in your organization or helping others to do this. However, a value of each year's new edition is that it gives you updates on all the book's content areas plus selected new topics that reflect the latest developments in the field.

Develop an Authoritative Review. Yet another goal for *The Quality Yearbook* is to contribute to clearer thinking about and implementation of the principles of quality management. This is a broad field. By providing a comprehensive review of the latest discussions of the applications and extensions of quality principles across many categories, we hope we can provide a way to standardize their study.

Organization

As in the previous editions, we have divided the book into four parts:

Part One, Background for Quality, starts with an original article by H. James Harrington titled "The Collapse of Prevailing Wisdom," which looks at some of more basic assumptions about TQM in light of actual practices in different companies. We continue the popular "Classics in Quality" section, this year with an excerpt from Walter A. Shewhart's original *Economic Control of Manufactured Product*, published in 1931. We also reprint Philip Kotler's classic article, "A Generic Concept of Marketing." We think you'll enjoy both of these for their insights and historical value.

Part Two, Quality by Industry, covers TQM as practiced in manufacturing, services, and four different areas of the public sector—government, health care, higher education, and K-12 education.

Part Three, Implementing Quality, presents the best articles and ideas from the past year on the theory and techniques of quality management. We have broken the readings into four categories:

1. Quality Transformation
2. Quality Tools and Techniques
3. Functional Processes
4. Standards and Assessments

Part Three is the longest part of the book and provides a wide spectrum of articles that explain the whys and hows of implementing quality including many different examples from different industries. It includes not only reprints, but original articles prepared for this book by people like Richard Randall in the ISO 9000 area and Eberhard Scheuing on suppliers.

Part Four. Quality References, includes a wide variety of material to help you further your exploration of TQM. Among the content of this part, you will find a comprehensive bibliography of quality articles and books that have come out since in the last edition (this is current through the fall of 1995), a directory of quality magazines and journals. Use it when you are looking for articles on any of the wide variety of subjects taken up in this yearbook. We also include a detailed directory of world wide web sites that deal with quality management issues. And you'll find a large directory of organizations and companies you can contact to learn more about best practices in different industries and functional areas. For reflection and for use in preparing speeches and reports, we continue the section "Quotes on Quality," with another all-new set of quotes. And we've concluded the book with profiles of the four 1996 Baldrige Award winners in this part and a review of the substantical changes in the 1997 Baldrige Award criteria. Be sure to take time to read about the Baldrige Award. It is changing significantly.

What's New in the 1997 Edition?

We've included 80 articles in this edition, about 10 more than in previous years. We've sought to give more consideration to some of the technical aspects of quality management, with articles on specific process management and SPC issues. There is a new section on logistics. And we have continued to expand the amount of material dealing with ISO 9000, including an original article by author and consultant Richard Randall. The ISO section also includes articles on ISO 14001 and good information on becoming ISO certified.

In the 1995 edition in part four on references, we started a new section dealing with on-line quality resources and expanded that in 1996. In the 1997 edition, you'll find a quite complete directory of internet sites, this year rated as to their usefulness. We've also developed a directory of companies and associations you can contact to learn about best practices in different industries and functional areas.

There is lots more as well. Our objective in all this is to provide you with a tool for keeping up with this expanding field. We want these annual volumes to be the first place you look when you have a question about almost any subject in this field. Of course, we welcome your comments on what you like and any suggestions you might have for improvement. You can write us at 3010 Irvington Way, Madison, WI 53713-3414. You can also send e-mail to jwoods@execpc.com.

Consulting Editors

As the yearbook has evolved, we have requested and received the assistance of various consulting editors. Their role has been to help us understand their fields, recommend articles we might include, and, most importantly, to contribute an original article themselves. This year we have had the participation of 10 consulting editors, some who have worked on previous editions and some new. We are very happy to have them associated with this book. You'll find their articles lead off the various sections of parts two and three. The following profiles these individuals. We include their addresses, phone numbers, and E-mail (if available) in case you wish to contact them about their areas of expertise.

- **H. James Harrington**, Special Contributor. Jim Harrington is regarded as on the world's leading figures in quality systems management, with more than 45 years of experience in this field. During his career, he has been involved in developing quality management systems in Europe, North and South America, and Asia. He serves as a Principal with Ernst & Young LLP and is their International Quality Advisor. He is chairman of the prestigious International Academy for Quality. He is an A-level member of the International Organization for Standardization's Technical Committee 176, which wrote the ISO 9000 Quality System Standards. The Harrington/Ishikawa Award was named for him to recognize his support to developing nations in implementing quality systems. He is well-known for several books, including *Business Process Improvement* (McGraw-Hill, 1991) and several others. You can contact Jim at Ernst & Young LLP, 55 Almaden Boulevard, San Jose, CA 95113, (408) 947-6587.
- **Roger Nagel**, Manufacturing. Roger Nagel is Executive Director and CEO of the Iacocca Institute at Lehigh University. He leads 175 professionals in conducting a range of research and developmental programs that facilitate the changes corporations must make to achieve global competitive advantage. Nagel is co-author of the influential book, *Agile Competitors and Virtual Organizations* (Van Nostrand Reinhold, 1995), and a contributor to several academic and industrial journals. He is perhaps most well known for his work on the Iacocca Institute report on a *21st Century Manufacturing Enterprise Strategy*. You can contact Roger Nagel at the Iacocca Institute, Lehigh University, 111 Research Drive, Bethlehem, PA 18015, (610) 758-4086, E-mail: rnn0@lehigh.edu.
- **Napoleon Devia**, Manufacturing. Napoleon Devia is a native of Colombia, South America and is a Senior Iacocca Fellow of the Iacocca Institute and has experience with agility in both academic

and industry settings. At the Institute, Napoleon is spearheading the development and implementation of new organizational concepts, such as the Agile Web of Pennsylvania, the first organization of its kind in the world. Building on this experience, Devia is also helping the development of an Agility Forum in Colombia. He earned a Ph.D. in Chemical Engineering at Lehigh in 1978. You can contact Napoleon at the Iacocca Institute, Lehigh University, 111 Research Drive, Bethlehem, PA 18015, (610) 758-6699, E-mail: NAD3@lehigh.edu.

- **Joe Sensenbrenner**, Government. Joe is president of Sensenbrenner Associates and works in the application of private sector TQM approaches to public-sector service delivery. In 1988 he was recognized as one of the "Ten Most Influential Figures in Quality Improvement." Joe was a three term mayor of Madison, Wisconsin, serving from 1983 to 1989, during which time he pioneered service improvements in virtually every municipal activity. He is the author of articles that have appeared in the *Harvard Business Review*, *Quality Progress*, and *Nation's Business* in the area of implementing TQM in government. He was coordinator of GOAL/QPC's Program on the application of TQM in state and local Government 1989-1994. He frequently speaks and gives seminars around the country. You can contact Joe at Sensenbrenner Associates, 818 Prospect Place, Madison, WI 53703, (608) 251-3100, E-mail: joesense@aol.com.

- **Julie Horine**, Education K-12. Julie is a member of the Department of Educational Leadership at the University of Mississippi where she teaches and conducts research in quality management and organizational improvement. She annually conducts a national research study that examines the deployment of quality practices in educational systems, both in K-12 and higher education. She serves as the lead training facilitator for the Governor's "Strengthening Quality In Schools" initiative in New Mexico. Julie is a five-year Malcolm Baldrige National Quality Award Examiner and has served as a Senior Examiner for the past two years. She is a member of the Baldrige Education Pilot Evaluation Team, which is bringing the Baldrige Award to education. She has served on the Quality Council for the State of Mississippi since 1995. She received her Ph.D. in 1983 from Florida State University. You can reach Julie at Department of Educational Leadership, University of Mississippi, Oxford, MS 38677, (601) 232-5016, E-mail: horine@vm.cc.olemiss.edu.

- **Ronald L. Heilmann**, Higher Education. Ron is a faculty member at the University of Wisconsin, Milwaukee and head of the Center for Quality, Productivity, and Economic Development at that university. He is also a member of the education committee of Milwaukee: First in Quality (MFIQ) and past president of the MFIQ

Executive Committee. Ron is a senior member of the American Society for Quality Control. He is president and founder of the National Educational Quality Initiative, a nonprofit corporation dedicated to improving the quality of the U.S. educational system. You can contact Ron at the Center for Quality and Productivity, University of Wisconsin, PO Box 742, Milwaukee, WI 53201, (414) 229-6259.

- **William Lareau**, Leadership. Bill Lareau is the founder and president of the American Samurai Institute, an organization that provides world-class practices and guidance to business leaders. He is also the Practice Leader in world-class manufacturing and integrated product development of Management Resources, a world-wide consultancy. He is a licensed organizational psychologist and holds a Ph.D. from The Catholic University of America. He is a Senior Fellow of the Center for Competitive Change at the University of Dayton and is an adjunct faculty member at the University of California, Riverside and Newport University. Previously, Dr. Lareau served in executive capacities with ITT and General Dynamics. He is author of numerous books, including *Dancing with the Dinosaur: Learning to Live in the Corporate Jungle* and *American Samurai: A Warrior for the Coming Dark Ages of American Business* (highly recommended for anyone wanting to learn the basics of TQM). You can reach Bill at the American Samurai Institute, 40 Lauramie Creek West, Clarks Hill, IN 47930, (818) 991-2400 (consultancy number), E-mail: dinodoc@aol.com.

- **Scott M. Davis**, Voice of the Customer. Scott is a partner with Kuczmarski & Associates. He lead the firm's focus on brand equity management as a key component of delivering customer value. He has lead projects dealing with the creation of customer-focused innovation strategies with some of the top brands in the United States. Prior to joining K&A, Scott worked for Procter & Gamble, where he coordinated sales, manufacturing, and distribution for packaged soap production in the company's West Coast office. His work has been cited in numerous publications, including *USA Today*, *The Wall Street Journal*, and serves a contributor for *Crain's Small Business* "Business Advisor Column." His work has been published in several marketing-oriented journals. You can contact Scott Davis at Kucsmarski & Associates, 1165 N. Clark Street, Suite 700, Chicago, IL 60610, (312) 988-1533.

- **Eberhard E. Scheuing**, Purchasing. Eb is the NAPM Professor of Purchasing and Supply Leadership at St. John's University in New York. Born and educated in Germany, he received his MBA and Ph.D. degrees from the University of Munich. Strongly involved in purchasing research and education since 1975, Dr. Scheuing is the author of more than 500 articles and 26 books. He is also the

Founder and President of the International Service Quality Association (ISQA), co-editor of *The Service Quality Handbook* (AMACOM 1993), and Co-Chair of international conferences on service quality (QUIS). You can contact Dr. Scheuing at P.O. Box 516, Tivoli, NY 12583, (914) 757-2141.

- **Richard C. Randall**, ISO 9000. Richard Randall is the author of *Randall's Practical Guide to ISO 9000* (published by Addison-Wesley), a book widely acclaimed for its value to organizations with an interest in taking a direct-line to becoming ISO 9000 registered. He was formerly the Southeastern Regional Director for National Quality Assurance (NQA), USA, one of the world's largest and most successful registrars offering more than ninety different scopes of registration. He was also formerly the National Quality Manager for GE Electronic Services, offering instrument calibration and repair services to the nuclear industry, transportation services industry, telecommunication industry, biomedical/pharmaceutical industry, and a variety of manufacturing industries. Randall is a Registered Lead Auditor, qualified through the Institute of Quality Assurance (IQA)-International Register of Certified Auditors (IRCA) program. He is involved in numerous other projects and organizations, all aimed at defining and facilitating ISO-related issues. You can reach Richard Randall at Randall's Practical Resources, 20294 E. Maplewood Place, Denver, CO 80016-1276, (303) 690-0077, Fax: (800) 659-7108, E-mail: rcr9000@aol.com, web site: http://home.earthlink.net/~rpr-online/.

Acknowledgements

This book was a big undertaking for us, and we are indebted to many people who have played a role in its coming together. At McGraw-Hill, Philip Ruppel has given his support to the project from the first edition, and he is an important part of our team.

We want to thank the editors and publishers of the several magazines, journals and books from which we drew the articles that make up the heart of this book. All were very forthcoming in supplying the permissions that allowed us to reprint the pieces you see here.

Jan Kosko at the National Institute of Standards and Technology assisted us by supplying information on the Baldrige award winners for 1996 and criteria changes for 1997. We thank her for her responsiveness to our needs.

Impressions Book and Journal Services, Inc. conscientiously assisted us in being able to deliver this book to you in a timely manner.

A special thanks goes to the reviewers of *The Quality Yearbook* over the years of its existence who have enthusiastically endorsed our project. We also want to thank the editors of *The TQM Magazine* (MCB University Press) for publishing a story on how we develop the yearbook in their December 1996 edition, just as this edition of *The Quality Yearbook* was coming out.

Our wives, Dora Cortada and Nancy Woods, have also been very supportive of our efforts as we shuttle between our houses doing the myriad chores involved in delivering this book to you.

Finally, we want to thank those who have found the 1994, 1995, and 1996 editions of this yearbook to be useful additions to their shelf of quality resources and thank you to those of you who are new buyers this year. You have made this the premier reference book in quality management. We look forward to serving your needs for many years in the future.

Jim Cortada
John Woods

PART ONE

Background for Quality

Total quality management is not new. Elements of the enlightened management practices covered by this term have been around for a very long time. We like to think of TQM as the intelligent aggregation of many different management techniques and approaches that insightful people have talked and written about for centuries. These practices include humane employee policies, problem prevention, customer delight, statistical process control, teamwork, and continuous improvement.

The whole point of TQM is to show how all these practices go together to facilitate the success of all stakeholders in an organization. TQM is not some new and unique approach to management. It is simply a means for helping anyone gain deeper insights into what management work is about and become a better manager of his or her own individual work, team, department, division, or whole organization. TQM is about using the most effective methodologies we know of to get everyone in an organization aligned to effectively and efficiently deliver quality outputs to customers—the exact behavior needed to generate profit. In other words, TQM is not some alternative to current management practices. It is, rather, an examination and refinement of those practices based on this truth: "An organization is a system with processes and has the purpose of serving customers." If you want to manage that system well, you adopt the practices of TQM. Any other approach will compromise your performance.

In time, more and more organizations and their managers will come to know this. In fact, in the future, you can count on this: TQM will go away, but the tools and techniques we call TQM will not. They will have a new name. What will that name be? *Management.* That's right. They will simply be what we will characterize as intelligent managerial action.

TQM Is Not an Ideology

TQM, like all ideas human beings come up with, can turn into an ideology, and it's important to avert that. For example, there is a set of best practices that just seem to go with implementing TQM, such as setting up formal teams across the organization and everyone preparing control charts. This book's lead article, by highly respected consultant H. James Harrington, looks at some of these practices in light of their ability to deliver results. Harrington has found the ideological approach, as is always the case with ideologies, to be wanting. TQM is no formula to be blindly followed. Rather, it is a set of guidelines for thinking through situations from the systems perspective and coming up with actions that make the most sense from that perspective. This approach still requires flexibility and creativity. Jim's article will give you some things to think about in this regard.

The Classics of Quality

In this section of the yearbook, we have sought over the last three editions to bring you seminal writings on TQM. We think you'll find the selections this time especially interesting. We start with the first two chapters from Walter A. Shewhart's book *Economic Control of Quality of Manufactured Product* (this book is available in a special edition from ASQC Quality Press). Shewhart first articulated the principles of statistical process control. These two chapters explain the foundation of his ideas and provide some examples of what he was talking about. If you've never read Shewhart's material before, you'll find it very interesting and surprisingly easy to follow.

We then include Philip Kotler's classic article, "A Generic Concept of Marketing." Kotler is one of the premier business thinkers of the second half of the twentieth century. His textbook *Marketing Management* has been the standard for M.B.A. students for at least thirty years. In this article, Kotler talks about the universality of the marketing concept for guiding mutually beneficial exchanges. The marketing concept is simply another way of saying the organization's purpose is to profitably serve customers. It is what drives all business behavior. Kotler also introduces the idea of stakeholders in this article, although he calls them "publics." If you don't know Kotler's work, this will get you started. This article was published originally in 1972.

Quality Perspectives

Last, we have a section in part 1 called "Quality Perspectives," and we have included a chapter excerpted from a new book, *Avoiding the Pitfalls of Total Quality*. This well-done book provides readers with direction on how to make TQM work and how to avoid making mistakes in its implementation. The excerpt introduces some problems people have and hints at how to prevent such problems. The title of the book is a slight misnomer. The real point is not to avoid the pitfalls of total quality but to avoid the pitfalls that come from believing TQM is a set of mechanical techniques you can impose on an organization and its employees. TQM doesn't work that way. People and organizations are living systems, and if you want them to perform well, you have to manage them as systems.

In reading this excerpt, you'll see that most problems come from misunderstanding the systems view. Indeed, many of the articles in this edition of *The Quality Yearbook* deal with how-to pieces in a variety of contexts. Often these articles discuss both how managers make mistakes in implementing TQM and how they may correct their mistakes. In most cases, the problems come from not fully understanding the systems view and how different actions will reverberate throughout the organization, while successful practice always takes the systems view into account.

We hope you find the articles in part 1 a good foundation for what follows in the rest of this book.

The Collapse of Prevailing Wisdom

H. James Harrington

This article by best-selling author and special guest contributor H. James Harrington reviews the results of the Ernst & Young International Quality Study in the area of best practices. He discovered that it isn't always clear what the best quality-management practices are for any particular company to implement. Those that work at one company might not work at another. As with most things, the way to get the best results ... all depends. This enlightening piece not only affirms the value of quality-management techniques but also suggests that you must proceed with open eyes and realistic expectations.

There is a big difference between wisdom and knowledge. We can define wisdom as a kind of context for our experience. Wisdom shapes what we see and affects our judgments and actions. Conventional or prevailing wisdom comes from past experiences, our education, and our culture. Knowledge comes from information backed up by statistically sound research. Often, as we gather the data and facts that become knowledge, the conventional or prevailing wisdom of the past is shown to be flawed. This article looks at new knowledge gained from the Ernst & Young International Quality Study (IQS) that shows that some our assumptions and beliefs about the application of quality-management practices need to be reconsidered. These practices are not to be applied without a clear sense of their value in different organizations with different problems.

As the results of the statistical analysis began to come in, the idea of a universally beneficial set of best practices proved to be unsound. Many of the practices that we had considered to be basic principles of TQM and the quality movement proved to be ineffective or even detrimental under some conditions. For example:

- Eliminating quality-control inspection
- The use of natural work teams
- Empowerment of the workforce
- Benchmarking
- Not inspecting quality into the product/service

The truth of the matter is that these are not principles; they are conceptual beliefs.

Only Five Real Best Practices

After studying the data for many months, the statisticians could identify only the following five practices as being universal best practices, and even then there is a 5 percent chance that these approaches may not improve your organization's performance.

- Cycle-time analysis
- Process value analysis
- Process simplification
- Strategic planning
- Formal supplier certification programs

Of all the practices we studied, this group of improvement practices showed a beneficial impact on performance, no matter how the organization was currently performing.

Process Improvement Methods

Organizations that made frequent use of practices such as process value analysis, process simplification, and process cycle-time analysis tended to have higher performance than the other organizations. Although the impact was significant on all three performance dimensions—profitability, productivity, and quality—it was strongest for the productivity measure.

Increasing the use of the process improvement practices can be a means to competitive advantage. The techniques are underutilized: organizations are not applying them today with nearly the frequency that the IQS shows to be beneficial. Most organizations say they "occasionally" use these techniques, whereas the best performers say they use these techniques "always or almost always." The benefits of these techniques are becoming more well known, and competitors are adopting them in significant numbers.

Deploying the Strategic Plan

Widespread understanding of the strategic plan by people inside and outside the organization has a broad beneficial impact. The two groups whose understanding showed the strongest impact on performance are middle management (or the medical staff among the hospitals in the study) and customers. Understanding of the plan by suppliers was also generally beneficial.

Most organizations said that their middle management partially understands the strategic plan; increasing that understanding from partial to full is a strategy to gain competitive advantage that positively impacts profit, quality, and productivity. Organizations generally said that customers had little understanding and that suppliers had no understanding of the organization's strategic plan. Increasing customers to full understanding and suppliers to at least a partial understanding also showed widespread benefits.

Supplier Certification Programs

Formal programs for certifying suppliers showed an across-the-board beneficial impact on performance—especially in quality and productivity.

Table 1. **Organizational Classifications**

Performance	Profitability (Return on Assets)	Productivity: (Value-Added per Employee)	Quality (External Customer Satisfaction Index)
Low	Less than 2.0%	Less than $47,000	Low
Medium	2.0% to 6.9%	$47,001 to 74,000	Medium
High	over 6.9%	over $74,000	High

*The productivity figures represent the VAE for 1992. Each year after 1992, an average of 3.1 percent was added to the 1992 value.

Among the IQS participants, certifying vendors is already a standard practice for a large majority (79 percent) of the manufacturers. The practice is rare in banks and hospitals (33 percent of banks and 10 percent of hospitals). The IQS data show such broad benefits of vendor certification programs that we would encourage organizations without such programs to reevaluate whether one may be appropriate for even a portion of their business. The ISO 9000 standards provide an excellent starting point for their supplier certification program.

The Awakening

Imagine our disappointment after spending millions of dollars and writing many books on improvement tools to find that only five of the many improvement tools presently in use are universal best practices. Well, the day was saved when we decided to stratify the data into three groupings.

The analysis team decided to divide the data into three relative performance categories called high, medium, and low performers. Organizations were classified into the three relative performance categories using the criteria shown in table 1.

Statistical analysis of the data related to each of the stratified groups revealed that there were a number of positive and negative practices that had previously been mistakenly considered universal best practices. This analysis also proved that it takes a very different set of activities and beliefs to move a low-performing organization up to the medium-performance level than it does to move a medium-performing organization up to the high-performance level. We also learned that when an organization moves from the medium-performance level to the high-performance level, the organization will need to adopt a very different set of activities and beliefs in order to maintain its high level of performance. Organizations that continue to do the same things that they did to move from the medium- to the high-performance level soon slip back and become medium performers again.

It always amazes me that so many things are obvious once they are called to my attention. It should be obvious that you have to manage an organization

Practice	Performance Level		
	Low	Medium	High
Statistical Process Control	😐	😐	😐
Department-Level Teams	😊	😐	☹
Quality-Related Meetings	😐	😊	😐
Assessing Top Management on Quality	☹	😐	😊
Assessing Mid-Management on Quality	😐	😊	😊
Process Benchmarking	☹	😊	😊
Training	😊	😊	☹
Get Customers' Input on New Product	😊	😊	☹
Evaluating Technology	😐	😊	☹
Measuring Improvement Efforts	😐	😊	😐

😊 = Good   😐 = Bad   ☹ = Ugly

Figure 1. **How different practices impact performance**

very differently if it is on the verge of bankruptcy than you would if the organization is setting the standards for its industry. It is therefore obvious that the organization's approach to improvement should be very different based on its current performance. This conclusion is exactly what the data collected during the study revealed to the study team.

The Good, the Bad, and the Ugly

Based on our statistical analysis of the stratified database, we found that a single practice can have the following three impacts on an organization depending on which performance level the organization finds itself in:

- The Good—The practice has a statistically proven positive impact on the organization's performance.
- The Bad—The practice has no statistically proven impact on the organization's performance. There are probably better ways that the organization should be investing its money.
- The Ugly—The practice has a statistically proven negative impact on the organization's performance.

Figure 1 provides an overview of how some of the different management practices impact the future of the organization's performance based on the current level of performance.

Statistical Process Control
Statistical Process Control (SPC) does not have an impact (either positive or negative) on the performance of the organization in any one of the three performance levels. It was hard for me to accept that the practice of using SPC was not a statistically sound universal best practice, and even harder for me to believe that SPC was not a best practice for at least one of the performance categories. After all, I have spent a significant part of my life praising the merits of and teaching SPC. I know that I have personally worked on processes where the application of SPC has reduced scrap and rework by as much as 80 percent. How could it be anything else but a best practice? But if I believe in statistics (and I do), I have to accept the results that the statistical analysis provides. Perhaps SPC's impact on the whole organization is so small that it is not reflected in the organization's total performance, and as such, we should possibly question whether using SPC is the best way to spend our money. Maybe we should be satisfied that SPC does not have a statistically proven negative impact on the organization's performance. When all is said and done, I still like the tool and use it, but much more selectively.

Department-Level Teams
The data indicate that emphasizing widespread use of teams and other employee-involvement mechanisms is a much more beneficial strategy for low performers than for the other two groups. In particular, department-level teams and cross-functional teams are both strongly associated with improving performance for organizations experiencing quality problems.

Increasing the participation in these practices is less beneficial for the medium-performing organizations. Only two of the practices we studied—department-level teams and problem-solving training—are positively associated with performance for the medium group.

The high-performing organizations show less benefit from these practices, with problem-solving training the only practice showing any positive correlation with performance. Moreover, widespread participation on department-level teams actually reflects a negative impact for the higher group.

We believe that team interaction can be a very effective way for organizations to identify and solve problems—particularly micro-level problems. Low-performing organizations typically have an abundance of such problems and can benefit greatly from the management practices that help identify and solve these problems. The high-performing organizations typically have already solved (or have avoided altogether) many of the micro-problems. For the high-performing organizations, macro-level innovation rather than micro-level problem solving is critical to success. Increasing the team structures seems to contribute little to achieving innovation.

Quality-Related Meetings
You might believe that holding general meetings to communicate quality objectives and organizational performance should have a positive impact on improving the organization's performance. This is not true in all cases. If an

organization does not have the trust of its employees, widespread participation in quality-related meetings can be a waste of time.

- Low-performing organizations—Widespread participation is not helpful at any level of the organization.
- Medium-performing organizations—Widespread participation is generally helpful at all levels of the organization.
- High-performing organizations—These types of meetings are only helpful at the nonmanagement level. Holding them at the middle- and upper-management levels does not provide a positive or negative result and often is a waste of time.

Assessment Criteria

It is generally believed that if an individual is measured on an item, the individual increases his or her effort related to that item. This belief has spurred a trend to tie improvement efforts into a compensation formula for all levels of the organization. The truth of the matter is that organizational performance does not improve in all cases just because individuals or groups are evaluated based on their quality and team performance.

- Low-performing organizations—These organizations benefit most when the nonmanagement employees' evaluations are based on the quality of their work and how well they participate in team activities. At the senior management level, tying compensation to quality and teamwork is not beneficial to the organization and in some cases even lowered productivity and profits.
- Medium-performing organizations—In the medium-performing organizations, tying middle-management compensation into the quality and teamwork of their organization is beneficial to the improvement effort.
- High-performing organizations—In the high-performing organizations, tying executive and middle-management compensation to the overall quality and teamwork within the organization proved to be very beneficial to the organization and is a significant factor in maintaining high performance.

Process Benchmarking

This is a method that has gained considerable favor since Xerox won the Malcolm Baldrige Award. Although theoretically the practice of benchmarking critical processes should help any organization to improve performance, in fact, it does not. Business process benchmarking provides the best ROI when applied to marketing, sales, delivery, and distribution systems.

- Low-performing organizations—The use of benchmarking for marketing and sales can provide a negative impact.
- Medium-performing organizations—In this case, benchmarking is helpful and should be considered.

- High-performing organizations—There is a high positive impact on performance when these organizations use benchmarking. It is a key practice for ensuring that they stay on top.

Training

In the area of training, the IQS examined the average number of hours per employee per year spent in general training and in quality-specific training.

- Low-performing organizations—For the low performers, the findings indicate that more training is related to better performance. This holds true for general and quality-specific training for all levels of employees: senior management, middle management, and nonmanagement employees. Generally, the performance impacts are immediate—and have a particularly strong impact on profit.
- Medium-performing organizations—The benefits of increasing levels of training are much less pronounced for the medium-performer group. Within this group, required training on specific topics for middle managers and nonmanagement employees is the only training seen as beneficial.
- High-performing organizations—There is no benefit from further increasing the training hours.

The IQS also examined the frequency and timing of training "that prepares employees with the skills required to effectively interact with customers." The same patterns described for general training also hold true for this specialized form of training. Customer-relationship training is most helpful to the low-performer group—whether the training is provided when employees are first hired, periodically, or continuously. Indeed, for the lower group, customer-relationship training is critical, as there was a strong negative correlation for low performers who responded that "no such training is provided."

For the medium-performer group, the IQS data show no compelling effect of customer-relationship training on performance. And for the high performers, it is clear that the expense of continuous training is not necessary. A combination of providing customer-relationship training to employees when they are first hired and again when it is urgent seems to work best for this group.

We believe that the need for developing the skill infrastructure is so pervasive in the low-performing organizations that virtually any increase in training can benefit overall performance. Once the base level of skills is in place, though, organizations must invest to maintain the base. They must also become more critical in aligning new training with the particular needs of the organization.

Identifying New Products

The IQS examined the practices that organizations employ to generate ideas for new products and services. It studied the frequency with which organizations use various sources of information, and the relative importance of the contributions of various internal functions.

According to the findings, customer input is important in all three performance groups, but the nature and scope of input broadens and becomes more sophisticated as relative performance rises.

- Low-performing organizations—For the low-performing group, the most important bases for selecting new products and services are straightforward customer requests and customer focus groups. Visiting customers and seeking feedback from current customers also have a positive impact.
- Medium-performing organizations—For the medium group, heeding customer requests is also important, but other sources also come into the picture. Internal market research and survey techniques through the mail and through personal contact are also positively associated with outcome. Suppliers' suggestions are also helpful in identifying ideas for new products and services.
- High-performing organizations—For organizations in the high-performance group, customer feedback is helpful, but external market research is the source for new ideas most positively associated with performance.

We believe that direct customer input is so critical to the low-performance group because a clear and graphic understanding of what customers need and *what they do not need* can be the most effective guiding force for developing products and services and for improving the business processes. High-performing organizations need to maintain and refresh their understanding of customer needs. But in order to reach even higher performance levels, these organizations must look beyond customer requirements *as customers currently perceive them* and anticipate totally new opportunities in the marketplace.

Product Development
Many organizations have had the experience, at one time or another, of correctly identifying customer requirements but then failing to develop products and services that were successful in the marketplace. The IQS examined how frequently various techniques were used to translate customer expectations into the design specifications for a new product or service.

Organizations in all of the IQS industries plan to dramatically increase the use of cross-functional teams that include customers to translate requirements into design specifications. The best practices analysis, however, suggests that this practice is not beneficial across all of the performance levels.

- Low-performing organizations—The findings indicate that the practice that works best in the lower group is using cross-functional teams that include customers. For this group, relying on the development department alone to create the design specifications shows a negative impact on performance.
- Medium-performing organizations—In sharp contrast to the low-performing organizations, for organizations in the medium group,

relying on the development department alone is most beneficial, while no form of cross-functional team shows a benefit.
- High-performing organizations—In the high-level group, cross-functional teams are again beneficial, but only when they exclude the customer.

Again, we see that the core issue is how well the organization understands customer needs. Organizations that do not thoroughly understand all dimensions of the customer's requirements can benefit extensively from having direct customer participation in the development of specifications. However, as organizations gain higher levels of customer understanding, they no longer need to rely on this structure to bring the customer voice into the development process. In fact, as the development effort focuses more and more on innovative new products in high-performing organizations, direct customer participation can limit the development effort if it does not represent a broad enough view of the potential marketplace.

Evaluating the Effect of Using Technology

After new technology has been implemented, organizations often choose to formally evaluate its effects. The IQS asked how frequently organizations evaluate the effect of the technology in areas such as product/service quality, cost of operations, investment requirements, and environmental impact.

- Low-performing organizations—Evaluating the effectiveness of the technologies used does not impact the organization's performance.
- Medium-performing organizations—This group benefits most from the broad use of formal evaluation of technologies.
- High-performing organizations—These organizations actually show a negative impact on performance when they evaluate technologies that they have implemented.

Measuring Process-Improvement Efforts

Many organizations around the world have spent vast sums of money to develop measurement systems to verify the ROI of their improvement processes. The impact of these activities has produced varying results.

- Low-performing organizations—No compelling benefit to measure the effectiveness of their process improvement efforts.
- Medium-performing organizations—The medium group that seems to benefit most from emphasizing measurement of the improvement. Within this group, there is a beneficial correlation with measuring reduced cycle time, reduced cost, less process variation, and fewer customer complaints.
- High-performing organizations—No compelling benefit.

We believe that the benefits of solving problems in low-performing organizations are often so obvious that measuring them is not necessary. The resources needed for such detailed measurement could be better used to identify and solve additional problems.

In contrast, organizations that are already performing at a medium level can benefit greatly from measuring and then using that information to continue to refine the business. But as organizations begin to move from medium- to high-performing levels, further refinement has limited potential for improving the business. New ways of doing business, rather than refinement of the old ways, are often needed to propel these organizations to even higher levels of performance.

Concluding Thoughts

Total quality management and its attendant best practices are conditional. Their successful implementation depends on the nature of our business and our ability to execute. As the results of the International Quality Study show, there are no best practices that organizations can execute blindly and expect great results. But if you intelligently match your organization and its performance level with the practices I've discussed here, I believe you will manage better, and you will see steadily improving results.

CLASSICS IN QUALITY

Some Basics of Statistical Quality Control
Characteristics of a Controlled Quality

Walter A. Shewhart

This article is really the first two chapters of Walter Shewhart's classic book Economic Control of Quality of Manufactured Product, *originally published in 1931, which laid out the basics of statistical process control. Shewhart's work inspired and formed the foundation for the work of W. Edwards Deming. Though Shewhart is often referred to, most of us haven't read his work. The following material will give you a sample of his approach. It is eminently readable, has historical significance, and offers practical insight, all at the same time.*

> When numbers are large, chance is the best warrant for certainty.
> A. S. Eddington,
> *The Nature of the Physical World*

> A situation like this merely means that those details which determine the future in terms of the past may be so deep in the structure that at present we have no immediate experimental knowledge of them and we may for the present be compelled to give a treatment from a statistical point of view based on considerations of probability.
> P. W. Bridgman,
> *The Logic of Modern Physics*

Reprinted with permission from W. A. Shewhart, Ph.D., *Economic Control of Quality of Manufactured Product*, originally published in 1931 by D. Van Nostrand, republished in 1980 by the American Society for Quality Control.

1. What is the Problem of Control?

What is the problem of control of quality of manufactured product? To answer this question, let us put ourselves in the position of a manufacturer turning out millions of the same kind of thing every year. Whether it be lead pencils, chewing gum, bars of soap, telephones, or automobiles, the problem is much the same. He sets up a standard for the quality of a given kind of product. He then tries to make all pieces of product conform with this standard. Here his troubles begin. For him standard quality is a bull's-eye, but like a marksman shooting at a bull's-eye, he often misses. As is the case in everything we do, unknown or chance causes exert their influence. The problem then is: how much may the quality of a product vary and yet be controlled? In other words, how much variation should we leave to chance?

To make a thing the way we want to make it is one popular conception of control. We have been trying to do this for a good many years and we see the fruition of this effort in the marvelous industrial development around us. We are sold on the idea of applying scientific principles. However, a change is coming about in the principles themselves and this change gives us a new concept of control.

A few years ago we were inclined to look forward to the time when a manufacturer would be able to do just what he wanted to do. We shared the enthusiasm of Pope when he said "All chance is but direction thou canst not see", and we looked forward to the time when we would see that direction. In other words, emphasis was laid on the *exactness* of physical laws. Today, however, the emphasis is placed elsewhere as is indicated by the following quotation from a recent issue, July, 1927, of the journal *Engineering*:

> Today the mathematical physicist seems more and more inclined to the opinion that each of the so-called laws of nature is essentially statistical, and that all our equations and theories can do, is to provide us with a series of orbits of varying probabilities.

The breakdown of the orthodox scientific theory which formed the basis of applied science in the past necessitates the introduction of certain new concepts into industrial development. Along with this change must come a revision in our ideas of such things as a controlled product, an economic standard of quality, and the method of detecting lack of control or those variations which should not be left to chance.

Realizing, then, the statistical nature of modern science, it is but logical for the manufacturer to turn his attention to the consideration of available ways and means of handling statistical problems. The necessity for doing this is pointed out in the recent book[1] on the application of statistics in mass production, by Becker, Plaut, and Runge. They say:

---

[1]*Anwendungen der Mathematischen Statistik auf Probleme der Massenfabrikation* (Berlin: Julius Springer, 1927).

It is therefore important to every technician who is dealing with problems of manufacturing control to know the laws of statistics and to be able to apply them correctly to his problems.

Another German writer, K. H. Daeves, in writing on somewhat the same subject says:

Statistical research is a logical method for the control of operations, for the research engineer, the plant superintendent, and the production executive.[2]

The problem of control viewed from this angle is a comparatively new one. In fact, very little has been written on the subject. Progress in modifying our concept of control has been and will be comparatively slow. In the first place, it requires the application of certain modern physical concepts; and in the second place, it requires the application of statistical methods which up to the present time have been for the most part left undisturbed in the journals in which they appeared. This situation is admirably summed up in the January, 1926 issue of *Nature* as follows:

A large amount of work has been done in developing statistical methods on the scientific side, and it is natural for anyone interested in science to hope that all this work may be utilized in commerce and industry. There are signs that such a movement has started, and it would be unfortunate indeed if those responsible in practical affairs fail to take advantage of the improved statistical machinery now available.

2. Nature of Control

Let us consider a very simple example of our inability to do exactly what we want to do and thereby illustrate two characteristics of a controlled product.

Write the letter *a* on a piece of paper. Now make another *a* just like the first one; then another and another until you have a series of *a*'s, *a, a, a, a*, ... You try to make all the *a*'s alike but you don't; you can't. You are willing to accept this as an empirically established fact. But what of it? Let us see just what this means in respect to control. Why can we not do a simple thing like making all the *a*'s just alike? Your answer leads to a generalization which all of us are perhaps willing to accept. It is that there are many causes of variability among the *a*'s: the paper was not smooth, the lead in the pencil was not uniform, and the unavoidable variability in your external surroundings reacted upon you to introduce variations in the *a*'s. But are these the only causes of variability in the *a*'s? Probably not.

---

[2]"The Utilization of Statistics," *Testing* (March 1924).

We accept our human limitations and say that likely there are many other factors. If we could but name all the reasons why we cannot make the *a*'s alike, we would most assuredly have a better understanding of a certain part of nature than we now have. Of course, this conception of what it means to be able to do what we want to do is not new; it does not belong exclusively to any one field of human thought; it is commonly accepted.

The point to be made in this simple illustration is that we are limited in doing what we want to do; that to do what we set out to do, even in so simple a thing as making *a*'s that are alike, requires almost infinite knowledge compared with that which we now possess. It follows, therefore, since we are thus willing to accept as axiomatic that we cannot do what we want to do and cannot hope to understand why we cannot, that we must also accept as axiomatic that a controlled quality will not be a constant quality. Instead, a controlled quality must be a *variable* quality. This is the first characteristic.

> A controlled quality must be a *variable* quality.

But let us go back to the results of the experiment on the *a*'s and we shall find out something more about control. Your *a*'s are different from my *a*'s; there is something about your *a*'s that makes them yours and something about my *a*'s that makes them mine. True, not all of your *a*'s are alike. Neither are all of my *a*'s alike. Each group of *a*'s varies within a certain range and yet each group is distinguishable from the others. This distinguishable and, as it were, constant variability *within limits* is the second characteristic of control.

3. Definition of Control

For our present purpose *a phenomenon will be said to be controlled when, through the use of past experience, we can predict, at least within limits, how the phenomenon may be expected to vary in the future. Here it is understood that prediction within limits means that we can state, at least approximately, the probability that the observed phenomenon will fall within the given limits.*

In this sense the time of the eclipse of the sun is a predictable phenomenon. So also is the distance covered in successive intervals of time by a freely falling body. In fact, the prediction in such cases is extremely precise. It is an entirely different matter, however, to predict the expected length of life of an individual at a given age; the velocity of a molecule at a given instant of time; the breaking strength of a steel wire of known cross section; or numerous other phenomena of like character. In fact, a prediction of the type illustrated by forecasting the time of an eclipse of the sun is almost the exception rather than the rule in scientific and industrial work.

In all forms of prediction an element of chance enters. The specific problem which concerns us at the present moment is the formulation of a scientific basis for prediction, taking into account the element of chance, where, for the purpose of our discussion, *any unknown cause of a phenomenon will be termed a chance cause.*

Scientific Basis for Control

1. Three Important Postulates

What can we say about the future behavior of a phenomenon acting under the influence of unknown or chance causes? I doubt that, in general, we can say anything. For example, let me ask: "What will be the price of your favorite stock thirty years from today?" Are you willing to gamble much on your powers of prediction in such a case? Probably not. However, if I ask: "Suppose you were to toss a penny one hundred times, thirty years from today, what proportion of heads would you expect to find?", your willingness to gamble on your powers of prediction would be of an entirely different order than in the previous case.

The recognized difference between these two situations leads us to make the following simple postulate:

> Postulate 1—All chance systems of causes are not alike in the sense that they enable us to predict the future in terms of the past.

Hence, if we are to be able to predict the quality of product even within limits, we must find some criterion to apply to observed variability in quality to determine whether or not the cause system producing it is such as to make future predictions possible.

Perhaps the natural course to follow is to glean what we can about the workings of unknown chance causes which are generally acknowledged to be controlled in the sense that they permit of prediction within limits. Perhaps no better examples could be considered than length of human life and molecular motion. It might appear that nothing is more uncertain than life itself, unless perhaps it be molecular motion. Yet there is something certain about these uncertainties. In the laws of mortality and distribution of molecular displacement, we find some of the essential characteristics of control within limits.

A. Law of Mortality

The date of death always has seemed to be fixed by chance even though great human effort has been expended in trying to rob chance of this prerogative. We come into this world and from that very instant on are surrounded by causes of death seeking our life. Who knows whether or not death will overtake us within the next year? If it does, what will be the cause? These questions we cannot answer. Some of us are to fall at one time from one cause, others at another time from another cause. In this fight for life we see then the element of uncertainty and the interplay of numerous unknown or chance causes.

However, when we study the effect of these chance causes in producing deaths in large groups of individuals, we find some indication of a controlled condition. We find that this hidden host of causes produce deaths at an average

It might appear that nothing is more uncertain than life itself, unless perhaps it be molecular motion. Yet there is something certain about these uncertainties.

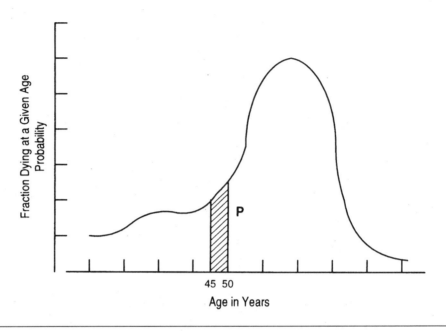

Figure 1. **Law of Mortality—Law of fluctuations controlled within limits**

rate which does not differ much over long periods of time. From such observations we are led to believe that, as we approach the condition of homogeneity of population and surroundings, we approach what is customarily termed a "Law of Mortality" such as indicated schematically in Fig. 1. In other words, we believe that in the limiting case of homogeneity the causes of death function so as to make the probability of dying within given age limits, such as forty-five to fifty, constant. That is, we believe these causes are controlled. In other words, we assume the existence of a kind of statistical equilibrium among the effects of an unknown system of chance causes expressible in the assumption that the probability of dying within a given age limit, under the assumed conditions, is an objective and constant reality.

B. Molecular Motion

Just about a century ago, in 1827 to be exact, an English botanist, Brown, saw something through his microscope that caught his interest. It was motion going on among the suspended particles almost as though they were alive. In a way it resembled the dance of dust particles in sunlight, so familiar to us, but this dance differed from that of the dust particles in important respects,— for example, adjacent particles seen under the microscope did not necessarily move in even approximately the same direction, as do adjacent dust particles suspended in the air.

Watch such motion for several minutes. So long as the temperature remains constant, there is no change. Watch it for hours, the motion remains

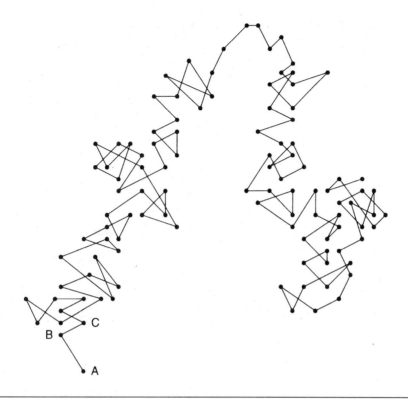

Figure 2. **A close-up of molecular motion appearing absolutely irregular, yet controlled within limits**

characteristically the same. Watch it for days, we see no difference. Even particles suspended in liquids enclosed in quartz crystals for thousands of years show exactly the same kind of motion. Therefore, to the best of our knowledge there is remarkable permanence to this motion. Its characteristics remain constant. Here we certainly find a remarkable degree of constancy exhibited by a chance system of causes.

Suppose we follow the motion of one particle to get a better picture of this constancy. This has been done for us by several investigators, notably Perrin. In such an experiment he noted the position of a particle at the end of equal intervals of time, Fig. 2. He found that the direction of this motion observed in one interval differed in general from that in the next succeeding interval; that the direction of the motion presents what we instinctively call absolute irregularity. Let us ask ourselves certain questions about this motion.

Suppose we fix our attention on the particle at the point A. What made it move to B in the next interval of time? Of course we answer by saying that a particle moves at a given instant in a given direction, say AB, because the resultant force of the molecules hitting it in a plane perpendicular to this direction from the side away from B is greater than that on the side toward B; but at any given instant of time there is no way of telling what molecules are engaged in giving it such motion. We do not even know how many molecules are taking part. Do what we will, so long as the temperature is kept constant, we cannot change this motion in a given system. It cannot be said, for example, when the particle is at the point B that during the next interval of time it will move to C. We can do nothing to control the motion in the matter of displacement or in the matter of the direction of this displacement.

Let us consider either the x or y components of the segments of the paths. Within recent years we find abundant evidence indicating that these displacements appear to be distributed about zero in accord with what is called the normal law.[3]

Such evidence as that provided by the law of mortality and the law of distribution of molecular displacements leads us to assume that there exist in nature phenomena controlled by systems of chance causes such that the probability dy of the magnitude X of a characteristic of some such phenomenon falling within the interval X to $X + dX$ is expressible as a function f of the quantity X and certain parameters represented symbolically in the equation

$$dy = f(X, \lambda_1, \lambda_2, \ldots, \lambda_m) dX, \tag{2}$$

where the λ's denote the parameters. Such a system of causes we shall term *constant* because the probability dy is independent of time. We shall take as our second postulate:

Postulate 2—Constant systems of chance causes do exist in nature.

To say that such systems of causes exist in nature, however, is one thing; to say that such systems of causes exist in a production process is quite another thing. Today we have abundant evidence of the existence of such systems of causes in the production of telephone equipment. The practical situation, however, is that in the majority of cases there are unknown causes of variability in the quality of a product which do not belong to a constant system. This fact was discovered very early in the development of control methods, and these causes were called *assignable*. The question naturally arose as to whether it was possible, in general, to find and eliminate such causes. Less than ten years ago it seemed reasonable to assume that this could be done. Today we have abundant evidence to justify this assumption. We shall, therefore, adopt as our third postulate:

---

[3]That is to say, if x represents the deviation from the mean displacement, zero in this case, the probability dy of x lying within the range x to $x + dx$ is given by

$$dy = \frac{1}{\sigma\sqrt{2\pi}} e^{-\frac{x^2}{2\sigma^2}} dx, \tag{I}$$

where σ is the root mean square deviation.

SOME BASICS OF STATISTICAL QUALITY CONTROL

Postulate 3—Assignable causes of variation may be found and eliminated.

Hence, to secure control, the manufacturer must seek to find and eliminate assignable causes. In practice, however, he has the difficulty of judging from an observed set of data whether or not assignable causes are present. A simple illustration will make this point clear.

2. When do Fluctuations Indicate Trouble?

In many instances the quality of the product is measured by the fraction nonconforming to engineering specifications or, as we say, the fraction defective. Table 1 gives for a period of twelve months the observed fluctuations in this fraction for two kinds of product designated here as Type A and Type B. For each month we have the sample size n, the number defective n_1 and the fraction $p = n_1/n$. We can better visualize the extent of these fluctuations in fraction defective by plotting the data as in Fig. 3-a and Fig. 3-b.

What we need is some yardstick to detect in such variations any evidence of the presence of assignable causes. Can we find such a yardstick? Experience of the kind soon to be considered indicates that we can. It leads us to conclude that it is feasible to establish criteria useful in detecting the presence of assignable causes of variation or, in other words, criteria which when applied to a set of observed values will indicate whether or not it is reasonable to believe that the causes of variability should be left to chance. Such criteria are basic to any method of securing control within limits. Let us, therefore, consider them critically. It is too much to expect that the criteria will be infallible. We are amply rewarded if they appear to work in the majority of cases.

Generally speaking, the criteria are of the nature of limits derived from past experience showing within what range the fluctuations in quality should remain, if they are to be left to chance. For example, when such limits are placed on the fluctuations in the qualities shown in Fig. 3, we find, as shown in Fig. 4, that in one case two points fall outside the limits and in the other case no point falls outside the limits. Upon the basis of the use of such limits, we look for trouble in the form of assignable causes in one case but not in the other. However, the question remains: Should we expect to be able to find and eliminate causes of variability only when deviations fall outside the limits? First, let us see what statistical theory has to say in answer to this question.

Upon the basis of Postulate 3, it follows that we can find and remove causes of variability until the remaining system of causes is constant or until we reach that state where the probability that the deviations in quality remain within any two fixed limits (Fig. 5) is constant. However, this assumption alone does not tell us that there are certain limits within which all observed values of quality should remain provided the causes cannot be found and

> Should we expect to be able to find and eliminate causes of variability only when deviations fall outside the limits?

Table 1. **Fluctuations in Quality of Two Manufactured Products**

Apparatus Type A

Month	Number Inspected n	Number Defective n_1	Fraction Defective $p = \dfrac{n_1}{n}$
Jan	527	4	0.0076
Feb	610	5	0.0082
March	428	5	0.0117
April	400	2	0.0050
May	498	15	0.0301
June	500	3	0.0060
July	395	3	0.0076
Aug	393	2	0.0051
Sept	625	3	0.0048
Oct	465	13	0.0280
Nov	446	5	0.0112
Dec	510	3	0.0059
Average	483.08	5.25	0.0109

Apparatus Type B

Month	Number Inspected n	Number Defective n_1	Fraction Defective $p = \dfrac{n_1}{n}$
Jan	169	1	0.0059
Feb	99	3	0.0303
March	208	1	0.0048
April	196	1	0.0051
May	132	1	0.0076
June	89	1	0.0112
July	167	1	0.0060
Aug	200	1	0.0050
Sept	171	2	0.0117
Oct	122	1	0.0082
Nov	107	3	0.0280
Dec	132	1	0.0076
Average	149.33	1.42	0.0095

SOME BASICS OF STATISTICAL QUALITY CONTROL

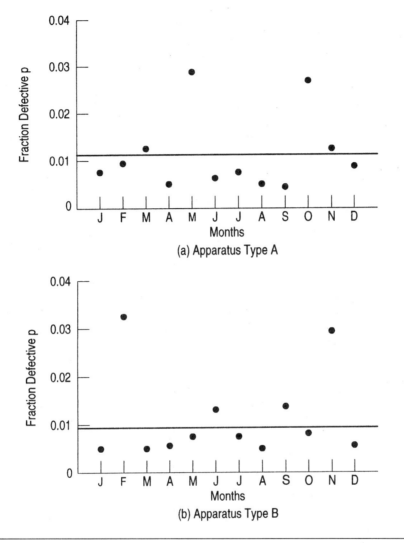

Figure 3. **Should these variations be left to chance?**

eliminated. In fact, as long as the limits are set so that the probability of falling within the limits is less than unity, we may always expect a certain percentage of observations to fall outside the limits even though the system of causes be constant. In other words, the acceptance of this assumption gives us a right to believe that there is an objective state of control within limits but in itself it does not furnish a practical criterion for determining when variations in quality, such as those indicated in Fig. 3, should be left to chance.

Furthermore, we may say that mathematical statistics as such does not give us the desired criterion. What does this situation mean in plain everyday

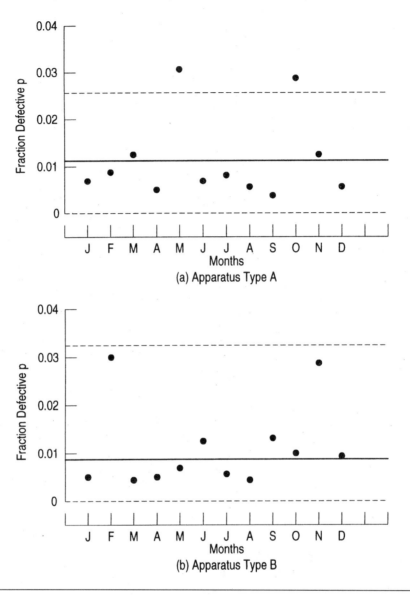

Figure 4. **Should these variations be left to chance?** *a* = No. *b* = Yes.

engineering English? Simply this: such criteria, if they exist, cannot be shown to exist by any theorizing alone, no matter how well equipped the theorist is in respect to probability or statistical theory. We see in this situation the long recognized dividing line between theory and practice. The available statistical machinery referred to by the magazine *Nature* is, as we might expect, not an end in itself but merely a means to an end. In other words, the fact that the

SOME BASICS OF STATISTICAL QUALITY CONTROL

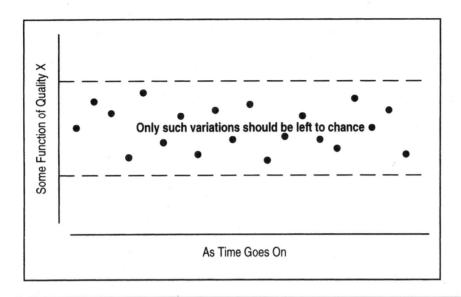

Figure 5. **Judgment plus modern statistical machinery makes possible the establishment of such limits**

criterion which we happen to use has a fine ancestry of highbrow statistical theorems does not justify its use. Such justification must come from empirical evidence that it works. As the practical engineer might say, the proof of the pudding is in the eating. Let us therefore look for the proof.

3. Evidence that Criteria Exist for Detecting Assignable Causes

A. Fig. 6 shows the results of one of the first large scale experiments to determine whether or not indications given by such a criterion applied to quality measured in terms of fraction defective were justified by experience. About thirty typical items used in the telephone plant and produced in lots running into the millions per year were made the basis for this study. As shown in this figure, during 1923–24 these items showed 68 per cent control about a relatively low average of 1.4 per cent defective.[4] However, as the assignable causes, indicated by deviations in the observed monthly fraction defective falling outside of control limits, were found and eliminated, the quality of product approached the state of control as indicated by an increase of from 68 per cent to 84 per cent control by the latter part of 1926. At the same time the quality improved; in 1923–24 the average per cent defective was 1.4 per cent, whereas by 1926 this had been reduced to 0.8 per cent. Here we get some

---
[4] R. L. Jones, "Quality of Telephone Materials," *Bell Telephone Quarterly* (June 1927).

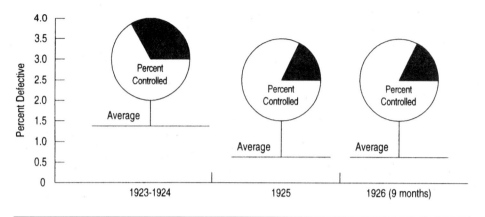

Figure 6. **Evidence of improvement in quality with approach to control**

typical evidence that, in general, as the assignable causes are removed, the variations tend to fall more nearly within the limits as indicated by an increase from 68 per cent to 84 per cent. Such evidence is, of course, one sided. It shows that when points fall outside the limits, experience indicates that we can find assignable causes, but it does not indicate that when points fall within such limits, we cannot find causes of variability. However, this kind of evidence is provided by the following two typical illustrations.

B. In the production of a certain kind of equipment, considerable cost was involved in securing the necessary electrical insulation by means of materials previously used for that purpose. A research program was started to secure a cheaper material. After a long series of preliminary experiments, a tentative substitute was chosen and an extensive series of tests of insulation resistance were made on this material, care being taken to eliminate all known causes of variability. Table 2 gives the results of 204 observations of resistance in megohms taken on as many samples of the proposed substitute material. Reading from top to bottom beginning at the left column and continuing throughout the table gives the order in which the observations were made. The question is: "Should such variations be left to chance?"

No *a priori* reason existed for believing that the measurements forming one portion of this series should be different from those in any other portion. In other words, there was no rational basis for dividing the total set of data into groups of a given number of observations except that it was reasonable to believe that the system of causes might have changed from day to day as a result of changes in such things as atmospheric conditions, observers, and materials. In general, if such changes are to take place, we may readily detect their effect if we divide the total number of observations into comparatively small subgroups. In this particular instance, the size of the subgroup was taken as four and the black dots in Fig. 7-*a* show the successive averages of four observations in the order in which they were taken. The dotted lines are the limits within which experience has shown that these observations should

SOME BASICS OF STATISTICAL QUALITY CONTROL

Table 2. **Electrical Resistance of Insulation in Megohms—Should Such Variations Be Left to Chance?**

5,045	4,635	4,700	4,650	4,640	3,940	4,570	4,560	4,450	4,500	5,075	4,500
4,350	5,100	4,600	4,170	4,335	3,700	4,570	3,075	4,450	4,770	4,925	4,850
4,350	5,450	4,110	4,255	5,000	3,650	4,855	2,965	4,850	5,150	5,075	4,930
3,975	4,635	4,410	4,170	4,615	4,445	4,160	4,080	4,450	4,850	4,925	4,700
4,290	4,720	4,180	4,375	4,215	4,000	4,325	4,080	3,635	4,700	5,250	4,890
4,430	4,810	4,790	4,175	4,275	4,845	4,125	4,425	3,635	5,000	4,915	4,625
4,485	4,565	4,790	4,550	4,275	5,000	4,100	4,300	3,635	5,000	5,600	4,425
4,285	4,410	4,340	4,450	5,000	4,560	4,340	4,430	3,900	5,000	5,075	4,135
3,980	4,065	4,895	2,855	4,615	4,700	4,575	4,840	4,340	4,700	4,450	4,190
3,925	4,565	5,750	2,920	4,735	4,310	3,875	4,840	4,340	4,500	4,215	4,080
3,645	5,190	4,740	4,375	4,215	4,310	4,050	4,310	3,665	4,840	4,325	3,690
3,760	4,725	5,000	4,375	4,700	5,000	4,050	4,185	3,775	5,075	4,665	5,050
3,300	4,640	4,895	4,355	4,700	4,575	4,685	4,570	5,000	5,000	4,615	4,625
3,685	4,640	4,255	4,090	4,700	4,700	4,685	4,700	4,850	4,770	4,615	5,150
3,463	4,895	4,170	5,000	4,700	4,430	4,430	4,440	4,775	4,570	4,500	5,250
5,200	4,790	3,850	4,335	4,095	4,850	4,300	4,850	4,500	4,925	4,765	5,000
5,100	4,845	4,445	5,000	4,095	4,850	4,690	4,125	4,770	4,775	4,500	5,000

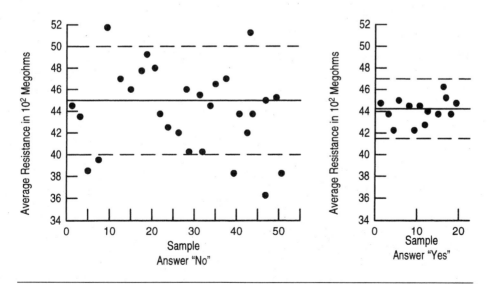

Figure 7. **Should these variations be left to chance?**

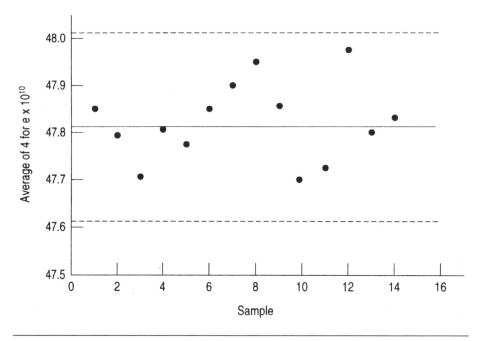

Figure 8. **Variations that should be left to chance—Does the criterion work? "Yes."**

fall, taking into account the size of the sample, provided the variability should be left to chance. Several of the observed values lie outside these limits. This was taken as an indication of the existence of causes of variability which could be found and eliminated.

Further research was instituted at this point to find these causes of variability. Several were found, and after these had been eliminated another series of observed values gave the results indicated in Fig. 7-b. Here we see that all of the points lie within the limits. We assumed, therefore, upon the basis of this test, that it was not feasible for research to go much further in eliminating causes of variability. Because of the importance of this particular experiment, however, considerably more work was done, but it failed to reveal causes of variability. Here then is a typical case where the criterion indicates when variability should be left to chance.

C. Suppose now that we take another illustration where it is reasonable to believe that almost everything humanly possible has been done to remove the assignable causes of variation in a set of data. Perhaps the outstanding series of observations of this type is that given by Millikan in his famous measurement of the charge on an electron. Treating his data in a manner similar to that indicated above, we get the results shown in Fig. 8. All of the points are within the dotted limits. Hence the indication of the test is consistent with the accepted conclusion that those factors which need not be left to chance had been eliminated before this particular set of data were taken.

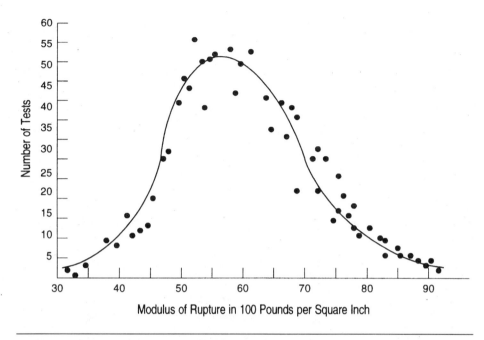

Figure 9. **Variability in modulus of rupture of clear specimens of green Sitka spruce typical of the statistical nature of physical properties**

4. Rôle Played by Statistical Theory

It may appear thus far that mathematical statistics plays a relatively minor rôle in laying a basis for economic control of quality. Such, however, is not the case. In fact, a central concept in engineering work today is that almost every physical property is a *statistical distribution*. In other words, an observed set of data constitutes a sample of the effects of unknown chance causes. It is at once apparent, therefore, that sampling theory should prove a valuable tool in testing engineering hypotheses. Here it is that much of the most recent mathematical theory becomes of value, particularly in analysis involving the use of comparatively small numbers of observations.

Let us consider, for example, some property such as the tensile strength of a material. If our previous assumptions are justified, it follows that, after we have done everything we can to eliminate assignable causes of variation, there will still remain a certain amount of variability exhibiting the state of control. Let us consider an extensive series of data recently published by a member of the Forest Products Laboratories,[5] Fig. 9. Here we have the results of tests for modulus of rupture on 1,304 small test specimens of Sitka spruce, the kind of material used extensively in aeroplane propellers during the War.

> A central concept in engineering work today is that almost every physical property is a *statistical distribution*.

---

[5] J. A. Newlin, *Proceedings of the American Society of Civil Engineers* (September 1926):1436–43.

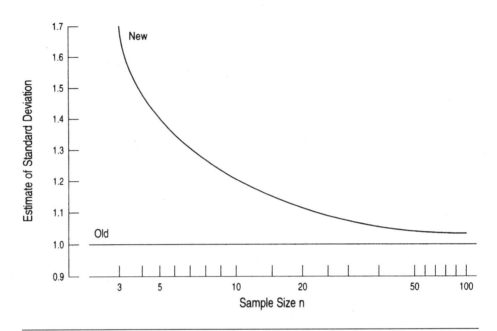

Figure 10. **Correction factors made possible by modern statistical theory are often large—typical illustration**

The wide variability is certainly striking. The curve is an approximation to the distribution function for this particular property representing what is at least approximately a state of control. The importance of going from the sample to the smooth distribution is at once apparent and in this case a comparatively small amount of refinement in statistical machinery is required.

Suppose, however, that instead of more than a thousand measurements we had only a very small number, as is so often the case in engineering work. Our estimation of the variability of the distribution function representing the state of control upon the basis of the information given by the sample would necessarily be quite different from that ordinarily used by engineers, see Fig. 10. This is true even though to begin with we make the same kind of assumption as engineers have been accustomed to make in the past. This we may take as a typical example of the fact that the production engineer finds it to his advantage to keep abreast of the developments in statistical theory. Here we use *new* in the sense that much of the modern statistical theory is new to most engineers.

> This state of control appears to be a kind of limit to which we may expect to go economically in finding and removing causes of variability without changing a major portion of the manufacturing process.

5. Conclusion

Based upon evidence such as already presented, it appears feasible to set up criteria by which to determine when assignable causes of variation in quality have been eliminated so that the product may then be considered to be controlled within limits. This state of control appears to be, in general, a kind of limit to which we may expect to go economically in finding and removing causes of variability without changing a major portion of the manufacturing process as, for example, would be involved in the substitution of new materials or designs.

A Generic Concept of Marketing

Philip Kotler

Philip Kotler is one of the foremost business thinkers in the world. He has been a pioneer in helping managers understand the importance of delivering value to customers, although he has talked about this in the context of marketing. *This article, originally published in 1972, explains how marketing, as the study of creating mutually beneficial transactions, can provide deep insight into all organizational activities. The whole point of TQM is to help the organization effectively and efficiently operationalize its processes to deliver customer value. Kotler discusses how this idea of delivering value is relevant to every organization and helps managers understand how more effectively to serve not just customers but all stakeholders or, as Kotler called them, "publics," a word he uses because the term "stakeholders" was not yet in use. Philip Kotler is S. C. Johnson & Son Distinguished Professor of International Marketing at the Kellogg Graduate School of Business, Northwestern University. His presence there is one reason Northwestern's is the top-rated marketing department in the United States.*

One of the signs of the health of a discipline is its willingness to reexamine its focus, techniques, and goals as the surrounding society changes and new problems require attention. Marketing has shown this aptitude in the past. It was originally founded as a branch of *applied economics* devoted to the study of distribution channels. Later marketing became a *management discipline* devoted to engineering increases in sales. More recently, it has taken on the character of an *applied behavioral science* that is concerned with understanding buyer and seller systems involved in the marketing of goods and services.

The focus of marketing has correspondingly shifted over the years. Marketing evolved through a *commodity focus* (farm products, minerals, manufactured goods, services); an *institutional focus* (producers, wholesalers, retailers, agents); a *functional focus* (buying, selling, promoting, transporting, storing, pricing); a *managerial focus* (analysis, planning, organization, control); and a *social focus* (market efficiency, product quality, and social impact). Each new focus had its advocates and its critics. Marketing emerged each time with a refreshed and expanded self-concept.

Reprinted with permission from the *Journal of Marketing*, vol. 36 (April 1972). Copyright © 1972, American Marketing Association. All rights reserved.

Today marketing is facing a new challenge concerning whether its concepts apply in the nonbusiness as well as the business area. In 1969, this author and Professor Levy advanced the view that *marketing is a relevant discipline for all organizations insofar as all organizations can be said to have customers and products.*[1] This "broadening of the concept of marketing" proposal received much attention, and the 1970 Fall Conference of the American Marketing Association was devoted to this theme.

Critics soon appeared who warned that the broadening concept could divert marketing from its true purposes and dilute its content. One critic did not deny that marketing concepts and tools could be useful in fund raising, museum membership drives, and presidential campaigns, but he felt that these were extracurricular applications of an intrinsical business technology.[2]

Several articles have been published which describe applications of marketing ideas to nonbusiness areas such as health services, population control, recycling of solid wastes, and fund raising.[3] Therefore, the underlying issues should be reexamined to see whether a more genetic concept of marketing can be established. This author concludes that the traditional conception of marketing would relegate this discipline to an increasingly narrow and pedestrian role in a society that is growing increasingly post-industrial. In fact, this article will argue that the broadening proposal's main weakness was not that it went too far but that it did not go far enough.

This article is organized into five parts. The first distinguishes three stages of consciousness regarding the scope of marketing. The second presents an axiomatic treatment of the generic concept of marketing. The third suggests three useful marketing typologies that are implied by the generic concept of marketing. The fourth describes the basic analytical, planning, organization, and control tasks that make up the logic of marketing management. The fifth discusses some interesting questions raised about the generic concept of marketing.

Three Stages of Marketing Consciousness

Three different levels of consciousness can be distinguished regarding the boundaries of marketing. The present framework utilizes Reich's consciousness categories without his specific meanings.[4] The traditional consciousness, that marketing is essentially a business subject, will be called *consciousness one.* Consciousness one is the most widely held view in the mind of practitioners and the public. In the last few years, a marketing *consciousness two* has appeared among some marketers holding that marketing is appropriate for all organizations that have customers. This is the thrust of the original

---

[1] Philip Kotler and Sidney J. Levy, "Broadening the Concept of Marketing," *Journal of Marketing* 33 (January 1969): 10–15.
[2] David Luck, "Broadening the Concept of Marketing—Too Far," *Journal of Marketing* 33 (July 1969): 53–54.
[3] *Journal of Marketing* 35 (July 1971).
[4] Charles A. Reich, *The Greening of America* (New York: Random House, 1970).

broadening proposal and seems to be gaining adherents. Now it can be argued that even consciousness two expresses a limited concept of marketing. One can propose *consciousness three* that holds that marketing is a relevant subject for all organizations in their relations with all their publics, not only customers. The future character of marketing will depend on the particular consciousness that most marketers adopt regarding the nature of their field.

Consciousness One

Consciousness one is the conception that marketing is essentially a business subject. It maintains that marketing is concerned with *sellers, buyers,* and *"economic" products and services.* The sellers offer goods and services, the buyers have purchasing power and other resources, and the objective is an exchange of goods for money or other resources.

The core concept defining marketing consciousness one is that of *market transactions.* A market transaction involves the transfer of ownership or use of an economic good or service from one party to another in return for a payment of some kind. For market transactions to occur in a society, six conditions are necessary: (1) Two or more parties; (2) a scarcity of goods; (3) concept of private property; (4) one party must want a good held by another; (5) the "wanting" party must be able to offer some kind of payment for it; and (6) the "owing" party must be willing to forego the good for the payment. These conditions underlie the notion of a market transaction, or more loosely, economic exchange.

Market transactions can be contrasted with nonmarket transactions. Nonmarket transactions also involve a transfer of resources from one party to another, *but without clear payment by the other.* Giving gifts, paying taxes, receiving free services are all examples of nonmarket transactions. If a housekeeper is paid for domestic services, this is a market transaction; if she is one's wife, this is a nonmarket transaction. Consciousness one marketers pay little or no attention to nonmarket transactions because they lack the element of explicit payment.

Consciousness Two

Consciousness two marketers do not see *payment* as a necessary condition to define the domain of marketing phenomena. Marketing analysis and planning are relevant in all organizations producing products and services for an intended consuming group, whether or not payment is required.

Table 1 lists several nonbusiness organizations and their "products" and "customer groups." All of these products, in principle, can be priced and sold. A price can be charged for museum attendance, safe driving lessons, birth control information, and education. The fact that many of these services are offered "free" should not detract from their character as products. A product is something that has value to someone. Whether a charge is made for its consumption is an incidental rather than essential feature defining value. In fact, most of these social goods are "priced," although often not in the normal fashion. Police services are paid for by taxes, and religious services are paid for by donations.

Table 1. **Some Organizations and Their Products and Customer Groups**

Organization	Product	Customer Group
Museum	Cultural appreciation	General public
National Safety Council	Safer driving	Driving public
Political candidate	Honest government	Voting public
Family Planning Foundation	Birth control	Fertile public
Police department	Safety	General public
Church	Religious experience	Church members
University	Education	Students

Each of these organizations faces marketing problems with respect to its product and customer group. They must study the size and composition of their market and consumer wants, attitudes, and habits. They must design their products to appeal to their target markets. They must develop distribution and communication programs that facilitate "purchase" and satisfaction. They must develop customer feedback systems to ascertain market satisfaction and needs.

Thus consciousness two replaces the core concept of *market transactions* with the broader concept of *organization-client transactions.* Marketing is no longer restricted only to transactions involving parties in a two-way exchange of economic resources. Marketing is a useful perspective for any organization producing products for intended consumption by others. *Marketing consciousness two states that marketing is relevant in all situations where one can identify an organization, a client group, and products broadly defined.*

Consciousness Three

The emergence of a marketing consciousness three is barely visible. Consciousness three marketers do not see why marketing technology should be confined only to an organization's transactions with its client group. An organization—or more properly its management—may engage in marketing activity not only with its customers but also with all other publics in its environment. A management group has to market to the organization's supporters, suppliers, employees, government, the general public, agents, and other key publics. *Marketing consciousness three states that marketing applies to an organization's attempts to relate to all of its publics, not just its consuming public.* Marketing can be used in multiple institutional contexts to effect transactions with multiple targets.

Marketing consciousness three is often expressed in real situations. One often hears a marketer say that his real problem is not *outside marketing* but *inside marketing;* for example, getting others in his organization to accept his ideas. Companies seeking a preferred position with suppliers or dealers see this as a problem of marketing themselves. In addition, companies try to market their viewpoint to congressmen in Washington. These and many other

Marketing consciousness three states that marketing applies to an organization's attempts to relate to all of its publics, not just its consuming public.

examples suggest that marketers see the marketing problem as extending far beyond customer groups.

The concept of defining marketing in terms of *function* rather than *structure* underlies consciousness three. To define a field in terms of function is to see it as a process or set of activities. To define a field in terms of structure is to identify it with some phenomena such as a set of institutions. Bliss pointed out that many sciences are facing this choice.[5] In the field of political science, for example, there are those who adopt a structural view and define political science in terms of political institutions such as legislatures, government agencies, judicial courts, and political parties. There are others who adopt a functional view and define political science as the study of power wherever it is found. The latter political scientists study power in the family, in labor-management relations, and in corporate organizations.

Similarly, marketing can be defined in terms of functional rather than structural considerations. Marketing takes places in a great number of situations, including executive recruiting, political campaigning, church membership drives, and lobbying. Examining the marketing aspects of these situations can yield new insights into the generic nature of marketing. The payoff may be higher than from continued concentration in one type of structural setting, that of business.

> Marketing takes places in a great number of situations, including executive recruiting, political campaigning, church membership drives, and lobbying.

It is generally a mistake to equate a science with a certain phenomenon. For example, the subject of *matter* does not belong exclusively to physics, chemistry, or biology. Rather physics, chemistry, and biology are logical systems that pose different questions about matter. Nor does *human nature* belong exclusively to psychology, sociology, social psychology, or anthropology. These sciences simply raise different questions about the same phenomena. Similarly, traditional business subjects should not be defined by institutional characteristics. This would mean that finance deals with banks, production with factories, and marketing with distribution channels. Yet each of these subjects has a set of core ideas that are applicable in multiple institutional contexts. An important means of achieving progress in a science is to try to increase the generality of its concepts.

Consider the case of a hospital as an institution. A production-minded person will want to know about the locations of the various facilities, the jobs of the various personnel, and in general the arrangement of the elements to produce the product known as health care. A financial-minded person will want to know the hospital's sources and applications of funds and its income and expenses. A marketing-minded person will want to know where the patients come from, why they appeared at this particular hospital, and how they feel about the hospital care and services. Thus the phenomena do not create the questions to be asked; rather the questions are suggested by the disciplined view brought to the phenomena.

---

[5]Perry Bliss, *Marketing Management and the Behavioral Environment* (Englewood Cliffs, N.J.: Prentice-Hall, 1970), 106–8, 119–20.

What then is the disciplinary focus of marketing? The core concept of marketing is the *transaction. A transaction is the exchange of values between two parties.* The things-of-values need not be limited to goods, services, and money; they include other resources such as time, energy, and feelings. Transactions occur not only between buyers and sellers, and organizations and clients, but also between any two parties. A transaction takes place, for example, when a person decides to watch a television program; he is exchanging his time for entertainment. A transaction takes place when a person votes for a particular candidate; he is exchanging his time and support for expectations of better government. A transaction takes place when a person gives money to a charity; he is exchanging money for a good conscience. *Marketing is specifically concerned with how transactions are created, stimulated, facilitated, and valued.* This is the generic concept of marketing.

The Axioms of Marketing

The generic concept of marketing will now be more rigorously developed. Marketing can be viewed as a *category of human action* distinguishable from other categories of human action such as voting, loving, consuming, or fighting. As a category of human action, it has certain characteristics which can be stated in the form of axioms. A sufficient set of axioms about marketing would provide unambiguous criteria about what marketing is, and what it is not. Four axioms, along with corollaries, are proposed in the following section.

> Axiom 1. *Marketing involves two or more social units, each consisting of one or more human actors.*
> Corollary 1.1. The social units may be individuals, groups, organizations, communities, or nations.

Two important things follow from this axiom. First, marketing is not an activity found outside of the human species. Animals, for example, engage in production and consumption, but do not engage in marketing. They do not exchange goods, set up distribution systems, and engage in persuasive activity. Marketing is a peculiarly human activity.

Second, the referent of marketing activity is another social unit. Marketing does not apply when a person is engaged in an activity in reference to a *thing* or *himself.* Eating, driving, and manufacturing are not marketing activities, as they involve the person in an interactive relationship primarily with things. Jogging, sleeping, and daydreaming are not marketing activities, as they involve the person in an interactive relationship primarily with himself. An interesting question does arise as to whether a person can be conceived of marketing something to himself, as when he undertakes effort to change his own behavior. Normally, however, marketing involves actions by a person directed toward one or more other persons.

Axiom 2. *At least one of the social units is seeking a specific response from one or more other units concerning some social object.*

- Corollary 2.1. The social unit seeking the response is called the *marketer*, and the social unit whose response is sought is called the *market*.
- Corollary 2.2. The social object may be a product, service, organization, person, place, or idea.
- Corollary 2.3. The response sought from the market is some behavior toward the social object, usually acceptance but conceivably avoidance. (More specific descriptions of responses sought are purchase, adoption, usage, consumption, or their negatives. Those who do or may respond are called buyers, adopters, users, consumers, clients, or supporters.)
- Corollary 2.4. The marketer is normally aware that he is seeking the specific response.
- Corollary 2.5. The response sought may be expected in the short or long run.
- Corollary 2.6. The response has value to the marketer.
- Corollary 2.7. *Mutual marketing* describes the case where two social units simultaneously seek a response from each other. Mutual marketing is the core situation underlying bargaining relationships.

Marketing consists of actions undertaken by persons to bring about a response in other persons concerning some specific social object. A social object is any entity or artifact found in society, such as a product, service, organization, person, place, or idea. The marketer normally seeks to influence the market to accept this social object. The notion of marketing also covers attempts to influence persons to avoid the object, as in a business effort to discourage excess demand or in a social campaign designed to influence people to stop smoking or overeating.[6] *The marketer is basically trying to shape the level and composition of demand for his product.* The marketer undertakes these influence actions because he values their consequences. The market may also value the consequences, but this is not a necessary condition for defining the occurrence of marketing activity. The marketer is normally conscious that he is attempting to influence a market, but it is also possible to interpret as marketing activity cases where the marketer is not fully conscious of his ends and means.

Axiom 2 implies that "selling" activity rather than "buying" activity is closer to the core meaning of marketing. The merchant who assembles goods for the purpose of selling them is engaging in marketing, insofar as he is seeking a purchase response from others. The buyer who comes into his store and pays the quoted price is engaging in buying, not marketing, in that he does

---

[6]See Philip Kotler and Sidney J. Levy, "Demarketing, Yes, Demarketing," *Harvard Business Review* 49 (November–December, 1971), 71–80.

not seek to produce a specific response in the seller, who has already put the goods up for sale. If the buyer decides to bargain with the seller over the terms, he too is involved in marketing, or if the seller had been reluctant to sell, the buyer has to market himself as an attractive buyer. The terms "buyer" and "seller" are not perfectly indicative of whether one, or both, of the parties are engaged in marketing activity.

> Axiom 3. *The market's response probability is not fixed.*
>> Corollary 3.1. The probability that the market will produce the desired response is called the *market's response probability*.
>> Corollary 3.2. The market's response probability is greater than zero; that is, the market is capable of producing the desired response.
>> Corollary 3.3. The market's response probability is less than one; that is, the market is not internally compelled to produce the desired response.
>> Corollary 3.4. The market's response probability can be altered by marketer actions.

Marketing activity makes sense in the context of a market that is free and capable of yielding the desired response. If the target social unit *cannot respond* to the social object, as in the case of no interest or no resources, it is not a market. If the target social unit *must respond* to the social object, as in the case of addiction or perfect brand loyalty, that unit is a market but there is little need for marketing activity. In cases where the market's response probability is fixed in the short run but variable in the long run, the marketer may undertake marketing activity to prevent or reduce the erosion in the response probability. Normally, marketing activity is most relevant where the market's response probability is less than one and highly influenced by marketer actions.

> Axiom 4. *Marketing is the attempt to produce the desired response by creating and offering values to the market.*
>> Corollary 4.1. The marketer assumes that the market's response will be voluntary.
>> Corollary 4.2. The essential activity of marketing is the creation and offering of value. Value is defined subjectively from the market's point of view.
>> Corollary 4.3. The marketer creates and offers value mainly through configuration, valuation, symbolization, and facilitation. (Configuration is the act of designing the social object. Valuation is concerned with placing terms of exchange on the object. Symbolization is the association of meanings with the object. Facilitation consists of altering the accessibility of the object.)
>> Corollary 4.4. *Effective marketing* means the choice of marketer actions that are calculated to produce the desired response in

The essential activity of marketing is the creation and offering of value. Value is defined subjectively from the market's point of view.

the market. *Efficient marketing* means the choice of *least cost* marketer actions that will produce the desired response.

Marketing is an approach to producing desired responses in another party that lies midway between *coercion* on the one hand and *brainwashing* on the other.

Coercion involves the attempt to produce a response in another by forcing or threatening him with agent-inflicted pain. Agent-inflicted pain should be distinguished from object-inflicted pain in that the latter may be used by a marketer as when he symbolizes something such as cigarettes as potentially harmful to the smoker. The use of agent-inflicted pain is normally not a marketing solution to a response problem. This is not to deny that marketers occasionally resort to arranging a "package of threats" to get or keep a customer. For example, a company may threaten to discontinue purchasing from another company if the latter failed to behave in a certain way. But normally, marketing consists of noncoercive actions to induce a response in another.

Brainwashing lies at the other extreme and involves the attempt to produce a response in another by profoundly altering his basic beliefs and values. Instead of trying to persuade a person to see the social object as serving his existing values and interests, the agent tries to shift the subject's values in the direction of the social object. Brainwashing, fortunately, is a very difficult feat to accomplish. It requires a monopoly of communication channels, operant conditioning, and much patience. Short of pure brainwashing efforts are attempts by various agents to change people's basic values in connection with such issues as racial prejudice, birth control, and private property. Marketing has some useful insights to offer to agents seeking to produce basic changes in people, although its main focus is on creating products and messages attuned to existing attitudes and values. It places more emphasis on preference engineering than attitude conditioning, although the latter is not excluded.

> The marketer is attempting to get value from the market through offering value to it.

The core concern of marketing is that of producing desired responses in free individuals by the judicious creation and offering of values. The marketer is attempting to get value from the market through offering value to it. The marketer's problem is to create attractive values. Value is completely subjective and exists in the eyes of the beholding market. Marketers must understand the market in order to be effective in creating value. This is the essential meaning of the marketing concept.

The marketer seeks to create value in four ways. He can try to design the social object more attractively (configuration); he can put an attractive terms on the social object (valuation); he can add symbolic significance in the social object (symbolization); and he can make it easier for the market to obtain the social object (facilitation). He may use these activities in reverse if he wants the social object to be avoided. These four activities have a rough correspondence to more conventional statements of marketing purpose, such as the use of product, price, promotion, and place to stimulate exchange.

The layman who thinks about marketing often overidentifies it with one or two major component activities, such as facilitation or symbolization. In *scarcity economies*, marketing is often identified with the facilitation function. Marketing is the problem of getting scarce goods to a marketplace. There is little concern with configuration and symbolization. In *affluent economies*, marketing is often identified with the symbolization function. In the popular mind, marketing is seen as the task of encoding persuasive messages to get people to buy more goods. Since most people resent persuasion attempts, marketing has picked up a negative image in the minds of many people. They forget or overlook the marketing work involved in creating values through configuration, valuation, and facilitation. In the future post-industrial society concern over the quality of life becomes paramount, and the public understanding of marketing is likely to undergo further change, hopefully toward an appreciation of all of its functions to create and offer value.

Typologies of Marketing

The new levels of marketing consciousness make it desirable to reexamine traditional classifications of marketing activity. Marketing practitioners normally describe their type of marketing according to the *target market* or *product*. A *target-market classification* of marketing activity consists of consumer marketing, industrial marketing, government marketing, and international marketing.

A *product* classification consists of durable goods marketing, nondurable goods marketing, and service marketing.

With the broadening of marketing, the preceding classification no longer express the full range of marketing application. They pertain to business marketing, which is only one type of marketing. More comprehensive classifications of marketing activity can be formulated according to the *target market, product*, or *marketer*.

Target Market Typology

A *target-market classification* of marketing activity distinguishes the various *publics* toward which an organization can direct its marketing activity. *A public is any group with potential interest and impact on an organization.* Every organization has up to nine distinguishable publics (Figure 1). There are three *input publics* (supporters, employees, suppliers), two *output publics* (agents, consumers), and four *sanctioning publics* (government, competitors, special publics, and general public). The organization is viewed as a resource conversion machine which takes the resources of supporters (e.g., stockholders, directors), employees, and suppliers and converts these into products that go directly to consumers or through agents. The organization's basic input-output activities are subject to the watchful eye of sanctioning publics such as government, competitors, special publics, and the general public. All of these publics are targets for organizational marketing activity because of their potential impact on the resource converting efficiency of the organization. Therefore, a *target-market classification* of marketing activity consists of supporter-directed marketing, employee-directed marketing, supplier-directed marketing, agent-directed marketing, consumer-directed marketing, general

> There are three *input publics* (supporters, employees, suppliers), two *output publics* (agents, consumers), and four *sanctioning publics* (government, competitors, special publics, and general public).

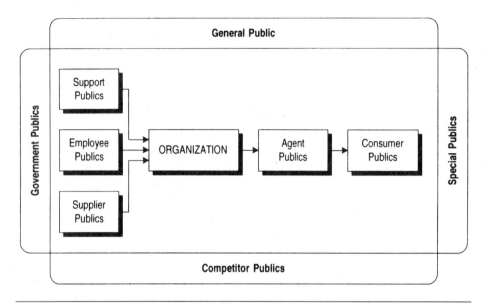

Figure 1. **An organization's publics**

public-directed marketing, special public-directed marketing, government-directed marketing, and competitor-directed marketing.

Product Typology

A typology of marketing activity can also be constructed on the basis of the *product* marketed. Under the broadened concept of marketing, the product is no longer restricted to commercial goods and services. An organization can try to market to a public up to six types of products or social objects. A product classification of marketing consists of goods marketing, service marketing, organization marketing, person marketing, place marketing, and idea marketing.

Goods and service marketing, which made up the whole of traditional marketing, reappear in this classification. In addition, marketers can specialize in the marketing of organizations (e.g., governments, corporations, or universities), persons (e.g., political candidates, celebrities), places (e.g., real estate developments, resort areas, states, cities), and ideas (e.g., family planning, Medicare, antismoking, safe-driving).

Marketer Typology

A typology can also be constructed on the basis of the marketer, that is, the organization that is carrying on the marketing. A first approximation would call for distinguishing between business and nonbusiness organization marketing. Since there are several types of nonbusiness organizations with quite different products and marketing tasks, it would be desirable to build a marketer classification that recognizes the different types of organizations. This leads to the following classifications: Business organization marketing, political organization marketing, social organization marketing, religious organization marketing, cultural organization marketing, and knowledge organization marketing.

Organizations are classified according to their primary or formal character. Political organizations would include political parties, government agencies, trade unions, and cause groups. Social organizations would include service clubs, fraternal organizations, and private welfare agencies. Religious organizations would include churches and evangelical movements. Cultural organizations would include museums, symphonies, and art leagues. Knowledge organizations would include public schools, universities, and research organizations. Some organizations are not easy to classify. Is a nonprofit hospital a business or a social organization? Is an employee credit union a political or a social organization? The purpose of the classification is primarily to guide students of marketing to look for regularities that might characterize the activities of certain basic types of organizations.

In general, the purpose of the three classifications of marketing activity is to facilitate the accumulation of marketing knowledge and its transfer from one marketing domain to another. Thus political and social organizations often engage in marketing ideas, and it is desirable to build up generic knowledge about idea marketing. Similarly, many organizations try to communicate a program to government authorities, and they could benefit from the accumulation of knowledge concerning idea marketing and government-directed marketing.

Basic Tasks of Marketing Management

Virtually all persons and organizations engage in marketing activity at various times. They do not all engage in marketing, however, with equal skill. A distinction can be drawn between *marketing* and *marketing management.* *Marketing* is a descriptive science involving the study of how transactions are created, stimulated, facilitated, and valued. *Marketing management* is a normative science involving the efficient creation and offering of values to stimulate desired transactions. Marketing management is essentially a disciplined view of the task of achieving specific responses in others through the creation and offering of values.

Marketing management is not a set of answers so much as an orderly set of questions by which the marketer determines what is best to do in each situation. Effective marketing consists of intelligently analyzing, planning, organizing, and controlling marketing effort.

The marketer must be skilled at two basic analytical tasks. The first is *market analysis.* He must be able to identify the market, its size and location, needs and wants, perceptions and values. The second analytical skill is *product analysis.* The marketer must determine what products are currently available to the target, and how the target feels about each of them.

Effective marketing also calls for four major planning skills. The first is *product development*, i.e., configuration. The marketer should know where

> Marketing management is not a set of answers so much as an orderly set of questions by which the marketer determines what is best to do in each situation.

to look for appropriate ideas, how to choose and refine the product concept, how to stylize and package the product, and how to test it. The second is *pricing*, i.e., valuation. He must develop an attractive set of terms for the product. The third is *distribution*, i.e., facilitation. The marketer should determine how to get the product into circulation and make it accessible to its target market. The fourth is *promotion*, i.e., symbolization. The marketer must be capable of stimulating market interest in the product.

Effective marketing also requires three organizational skills. The first is *organizational design*. The marketer should understand the advantages and disadvantages of organizing market activity along functional, product, and market lines. The second is *organizational staffing*. He should know how to find, train, and assign effective co-marketers. The third is *organizational motivation*. He must determine how to stimulate the best marketing effort by his staff.

Finally, effective marketing also calls for two control skills. The first is *market results measurement*, whereby the marketer keeps informed of the attitudinal and behavioral responses he is achieving in the marketplace. The second is *marketing cost measurement*, whereby the marketer keeps informed of his costs and efficiency in carrying out his marketing plans.

Some Questions About Generic Marketing

The robustness of the particular conception of marketing advocated in this article will be known in time through testing the ideas in various situations. The question is whether the logic called marketing really helps individuals such as educational administrators, public officials, museum directors, or church leaders to better interpret their problems and construct their strategies. If these ideas are validated in the marketplace, they will be accepted and adopted.

However, academic debate does contribute substantially to the sharpening of the issues and conceptions. Several interesting questions have arisen in the course of efforts by this author to expound the generic concept of marketing. Three of these questions are raised and discussed below.

1. *Isn't generic marketing really using influence as the core concept rather than exchange?*

    It is tempting to think that the three levels of consciousness of marketing move from *market transactions* to *exchange* to *influence* as the succeeding core concepts. The concept of influence undeniably plays an important role in marketing thought. Personal selling and advertising are essentially influence efforts. Product design, pricing, packaging, and distribution planning make extensive use of influence considerations. It would be too general to say, however, that marketing is synonymous with interpersonal, intergroup, or interorganizational influence processes.

    Marketing is a particular way of looking at the problem of

achieving a valued response from a target market. It essentially holds that exchange values must be identified, and the marketing program must be based on these exchange values. Thus the anticigarette marketer analyzes what the market is being asked to give up and what inducements might be offered. The marketer recognizes that every action by a person has an opportunity cost. The marketer attempts to find ways to increase the person's perceived rate of exchange between what he would receive and what he would give up in *freely* adopting that behavior. The marketer is a specialist at understanding human wants and values and knows what it takes for someone to act.

2. *How would one distinguish between marketing and a host of related activities such as lobbying, propagandizing, publicizing, and negotiating?*

   Marketing and other influence activities and tools share some common characteristics as well as exhibit some unique features. Each influence activity has to be examined separately in relation to marketing. *Lobbying,* for example, is one aspect of government-directed marketing. The lobbyist attempts to evoke support from a legislator through offering values to the legislator (e.g., information, votes, friendship, and favors). A lobbyist thinks through the problem of marketing his legislation as carefully as the business marketer thinks through the problem of marketing his product or service. *Propagandizing* is the marketing of a political or social idea to a mass audience. The propagandist attempts to package the ideas in such a way as to constitute values to the target audience in exchange for support. *Publicizing* is the effort to create attention and interest in a target audience. As such it is a tool of marketing. *Negotiation* is a face-to-face mutual marketing process. In general, the broadened concept of marketing underscores the kinship of marketing with a large number of other activities and suggests that marketing is a more endemic process in society than business marketing alone suggests.

3. *Doesn't generic marketing imply that a marketer would be more capable of managing political or charitable campaigns than professionals in these businesses?*

   A distinction should be drawn between marketing as a *logic* and marketing as a *competence.* Anyone who is seeking a response in another would benefit from applying marketing logic to the problem. Thus a company treasurer seeking a loan, a company recruiter seeking a talented executive, a conservationist seeking an antipollution low, would all benefit in conceptualizing their problem in marketing terms. In these instances, they would be donning a marketer's hat although they would not be performing as professional marketers. A professional marketer is someone who (1) regularly works with marketing problems in a specific area and

(2) has a specialized knowledge of this area. The political strategist, to the extent he is effective, is a professional marketer. He has learned how to effectively design, package, price, advertise, and distribute his type of product in his type of market. A professional marketer who suddenly decides to handle political candidates would need to develop competence and knowledge in this area just as he would if he suddenly decided to handle soap or steel. Being a marketer only means that a person has mastered the logic of marketing. To master the particular market requires additional learning and experience.

Conclusion

Generic marketing is a logic available to all organizations facing problems of market response. A distinction should be drawn between applying a marketing point of view to a specific problem and being a marketing professional. Marketing logic alone does not make a marketing professional. The professional also acquires competence, which along with the logic, allows him to interpret his problems and construct his marketing strategies in an effective way.

QUALITY PERSPECTIVES

The Promise and Shortcomings of Total Quality

Charles C. Poirier and Steven J. Tokarz

This article is actually the introductory chapter to a new book that appeared in 1996 titled Avoiding the Pitfalls of Total Quality. *In the excerpt included here, the authors provide an overview of why TQM fails in organizations and introduce a model of the journey organizations take to achieve what the authors call "Big Q."*

Sound the Alarm

Business leaders across the country have sounded the alarm: The United States still trails Japan and other nations in the crucial arena of quality. And according to Xerox's Chairman Paul Allaire, at its present pace, America won't win the quality race.

Quality improvement as a business strategy and an operational discipline can be powerful. It can have great motivational impact on workers and bring new dimensions to customer satisfaction. When the effort includes service as well as product, it can make the difference in keeping customers in highly competitive markets. And if the effort is expanded to improve productivity and costs, quality improvement can dramatically increase profitability.

Yet, America has not fully captured the tremendous power of quality as a strategy. After more than a decade of sporadic intensity, quality efforts have had limited impact on business performance. For all of its promise, this powerful mechanism has failed to achieve the kind of effect it has had in other countries.

Not an Isolated Technique

For most U.S. firms, a quality effort has not led to a competitive advantage because its inherent principles have not been institutionalized within the processes that drive a business. Observes Peter Drucker, noted educator, business writer, and consultant, "It will take a while for American businesses to discover what the Japanese have learned—that quality as an isolated technique has performance limitations" (*Wall Street Journal*, October 2, 1991).

To have the kind of positive impact it is capable of generating, total quality, or total quality management (TQM)—terms that we will use interchangeably—must be a discipline that pervades everything a business does. It must be combined with elements of technology, innovation, cycle time reduction, and enhanced people utilization. And it must contain an unyielding dedication to customer satisfaction, the only road that will lead to long-term profitability. Sales representation, order entry, planning, manufacturing, delivery, communications, distribution, billing, service, and design are all areas of potential satisfaction or dissatisfaction in the eyes of the customer. All of these functions must be conducted in a total quality environment to achieve what global leaders have proven can result. As that environment is created, the effort can then be extended to optimizing the use of resources dedicated to total quality and the building of a holistic improvement process, which we will describe in subsequent chapters.

Quality must be an ingrained characteristic throughout any business, without requiring limitless exhortation and investment. It should be a natural process, fully supported by everyone in the business network, from suppliers to the end consumer. Anything less dooms the organization to living on the edge of customer dissatisfaction and loss of business.

The Promise of Total Quality

Robert D. Buzzell and Bradley T. Gale are the chroniclers of the profit impact on market strategy (PIMS) process. In 1972, they began to study the relationship between strategy and financial results of nearly 3000 business units. They reported, "The 1960s and 1970s brought the realization that market share is a key to a company's growth and profitability. The 1980s have shown just as clearly that one factor above all others—quality—drives market share. When superior quality and large market share are both present, profitability is virtually guaranteed."

The 1990s will show clearly that *three* factors—quality, customer satisfaction, and optimum use of resources—are the elements for success and survival. In this decade, business organizations that are leading the way toward success are showing a consistent pattern. They're expanding the definition of quality to gain more universal acceptance and garner the key management support that is necessary to sustain a quality effort. For them, the concept has evolved from high-quality goods and services to a system of total quality performance throughout the organization.

Total quality has the power to propel a business organization to unprecedented heights of accomplishment. Of the firms that have parlayed TQM into success, Motorola, a standout, used total quality to halt a sliding position in the 1970s that forced the sale of a poor-performing Quasar operation to Matsushita. Moreover, the subsequent turnaround of Quasar by Japanese managers illustrated the power of TQM.

Motorola followed the Matsushita example to gain momentum in the 1980s, eventually becoming a Malcolm Baldrige National Quality Award winner. Today, Motorola is an inspiration as it evangelistically preaches the six-sigma quality gospel that demands that defects per million not exceed 3.4 units.

Milliken and Company used TQM to pursue Japanese textile manufacturers who were using older equipment more effectively. Milliken took the lead and parlayed its devotion to TQM into a Baldrige Award. Then it proceeded to teach other organizations about TQM. Among its accomplishments were: improving on-time delivery from 75 percent in 1984, to 99 percent by 1988; reducing the cost of nonconformance by 60 percent; and winning 41 customer quality awards (Hardie 1990).

Milliken now insists on the same standards of performance from its suppliers that it asks of itself. In one instance, when Milliken told Hoeschst-Celanese (H-C), a Dallas-based chemical group, to improve quality or risk losing its business, the effect was dramatic. Jeff Siebenaller, an official at H-C, reports that after an ineffective 10-year-old quality effort, with TQM they dramatically improved product and process quality and saved the firm $260 million. Coincidentally, H-C was a finalist in the 1988 Baldrige Award competition.

Xerox, another notable advocate of TQM, reports significant accomplishments because of its renewed dedication to TQM: 30 percent reduction in cycle time to bring a new product to market; reduction of defective incoming parts from 8 percent to .05 percent, 78 percent decrease in defects per 100 machines, 27 percent improvement in service response time; 50 percent reduction in unit manufacturing costs; 20 percent improvement in revenues per employee (Kearns 1991).

Within American business, dedication to TQM is happening with enough frequency to reveal a pattern of successful implementation. According to Bill Ginnodo, executive director of the Quality and Productivity Management Association (QPMA), "Those who are leading the way [with TQM] say it is a concept or philosophy for managing operations. It's a set of action-oriented principles, ideas or management practices. It's a way of life, a culture, a mindset. It's a call, a banner for doing things differently. It's a prescription for organizational effectiveness. It's a framework for improvement" (Ginnodo 1991).

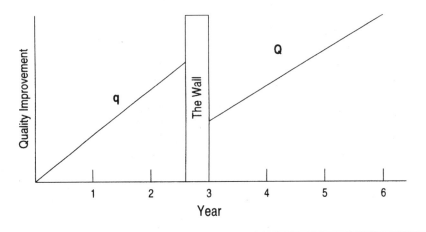

Figure 1. **Making the progression from little q to Big Q**

The Full Meaning of TQM

TQM is the way to succeed in the 1990s, but it is not to be practiced in isolation. It must be a central theme in strategic plans. Many organizations have pursued TQM, but only a few have discovered its full ramifications and achieved success. The business field is littered with those that made a half-hearted, half-intentional, partially understood effort. Most often, the failures spring from a lack of understanding the full meaning of TQM, which creates a pitfall that prevents successful implementation of a total quality effort.

Typically, groups that failed wasted enormous amounts of time and energy seeking a simplistic, cookbook improvement process. What they implemented yielded mediocre-to-poor results because they failed to take the time to build a culture that insisted on high quality standards in all business functions. Rather, they allowed existing cultural features—which typically deny the need for dramatic quality improvement—to prevail and to sidetrack any serious, continuing effort to develop new standards of performance. Others simply shunted quality aside and insisted that new capital investments were the answer to creating future competitive advantages. With so many potential missteps, it is no wonder the scorecard of success is so empty.

Figure 1 is a simplistic diagram of what happens when partial dedication and understanding occur. For about one to three years, the typical organization pursues what is now called "little q," or what people have interpreted as the basics of quality enhancement. Using a tool-driven, or technique-driven approach, for example, implementing statistical process control (SPC) in isolation, these groups made some measurable improvement, only to reach a mental wall that impeded further progress.

At this wall, they were unable to make the jump to "Big Q," or to the fuller quality meaning that embraced the tenets within TQM. A few organizations made progress beyond this wall, usually because a crisis threatened their survival, or they were blessed with enlightened leadership that insisted the organization progress to the higher order of quality initiatives.

Xerox exemplifies this situation. The loss of market share to Japanese competitors, particularly Canon, became a driving force. It took Xerox beyond the inhibiting wall to an ethic that demanded pursuit of the best performance in all features of its big quality effort.

Other organizations mounted an offensive, chased the fuller meaning, and achieved a form of TQM. These firms didn't yield to the wall because they integrated the fuller aspects of quality into their business strategies. The Big Q in those companies means the achievement of the highest possible quality in process, systems, machinery, management, maintenance, service, and delivery to customers inside and outside the organization. These organizations simply did not allow a mental block, a pitfall born of insufficient understanding or the lack of dedication throughout the business, to stall their progress.

IBM Rochester, manufacturer of the AS/400 computing system, is an organization that went right through the wall. It took its customers' interpretation of quality to performance levels that won it the 1990 Baldrige Award. Employees literally drew up new definitions of Big Q that made every person in the organization responsible for satisfying the customer with unquestioned quality standards in every form of job performance. The result was one of the most outstanding total quality team efforts that we have ever encountered.

The Journey to Big Q

Figure 2 depicts a model illustrating the phases many companies progress through as they struggle to improve quality. While it offers a natural progression from a status quo environment to one of Big Q, the journey is fraught with pitfalls. As an explanation of the model unfolds in the following pages, those who have experience in implementing a quality process will likely recognize some of the barriers and obstacles (unexpected pitfalls) encountered along the way. Max Zent, former executive director of productivity and quality for Tenneco, introduced the fundamentals of this model during a 1985 presentation in Toronto to QPMA. This model will be used to set the stage for how quality can or cannot succeed. It will also be used to begin showing the linkage that must exist between productivity, cost, and customer satisfaction if management is to support a long-term quality effort.

The vertical axis of the model lists techniques that progress from caretaking through a series of higher-order changes. When effective improvements have been accomplished, these changes lead to a definite competitive advantage.

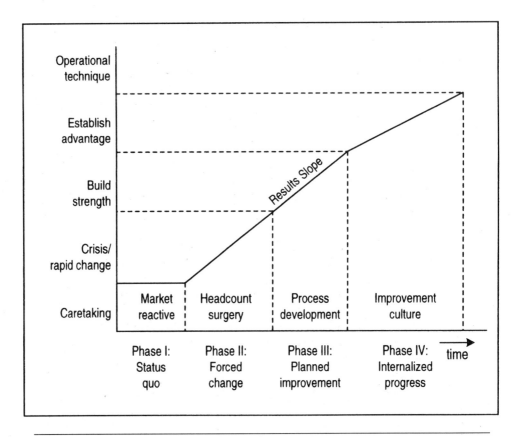

Figure 2. **The relationship between operational technique and development of a total-quality process**

Along the horizontal axis is a series of phases corresponding to developmental changes that take place with strategic alterations. Within the bounded zones are the tactics that are driving the organization during each phase.

Pitfalls in Pursuing the Status Quo—Phase I

Beginning in the first phase, the model depicts an organization that is pursuing the status quo. It's implementing a business plan that calls for annualized improvements, but, in reality, it's caretaking existing levels of performance. Total quality readiness is lowest in this range because no one sees a real need for major change. Results derive from past actions and are driven more by reacting to market conditions than by the dynamics of current managerial forces.

The business will implement minor quality strategies, but these will be more defensive than offensive. The tactics also will be low risk and without innovation. The results line in this phase is flat but it could be slightly up or down. Existing managers perceive performance at a high level, when it is actually static or declining in relationship to disinflation or to more dynamic competitors.

An organization that allows reaction to the market to drive results will typically be found in this sector. Executives will spend an inordinate amount of time searching for excuses for the differences between performance and budget. Managers of most business units will fiercely resist any benchmarking data and will argue that earnings are adequate, even though performance is far from full potential. The reason is that many unit managers do not know the full potential of their organizations. They obstinately avoid comparison with other units or with available market data that would verify that their performance is slowed or declining. Instead, they blame market conditions or differences in circumstances when compared to other, better-performing facilities.

The status quo phase includes low performers and some medium performers. Inadequate-to-marginal return on assets, loss of competitive advantage, and the threat of the unit's collapse can incite management to encourage implementation of a quality strategy to improve the performance of these underachievers. The key word is *encourage*, which implies a lack of strong, serious leadership.

The managers in these underachieving business units will pursue some sort of quasi-improvement strategy in a plodding, frustrating half-attempt to emerge from the status quo. The problem is that such managers produce a continuing string of unachieved budgets. Until the market reacts positively to an overall upsurge in demand, these weak performers become more adept at writing good variance reports than at initiating the real changes necessary. Should the market be lethargic or turn abruptly downward, the more marginal units will close or be otherwise negatively affected.

The status quo is a debilitating experience. Employees tend to divide into two camps: those who want to try something dramatically different and those who want to stand still. It takes a concerted effort by management to pull these factions together and move an organization out of this mode. The obvious pitfall in this sector is the unwillingness by management to face reality; the solution is to force a comparative analysis that dramatically illustrates the gap between current performance and the industry leaders.

> The status quo is a debilitating experience.

Forced Change as a Survival Tactic—Phase II

The tendency in business is to attempt a jump from the status quo to the internalized-progress phase. Senior managers want to move quickly to recap the benefits from quality improvement. They pressure middle management to initiate change, to select an off-the-shelf approach or hire a professional consultant, and to start programs that quickly alter structure, systems, and procedures, as well as boost earnings. But a cultural shift is needed for lasting success, with a revised managerial and operating philosophy.

Only in rare cases, usually at a specific plant or office location, is such a quantum change ever achieved. More likely, a strong-willed manager will hammer through short-term improvement, while the existing cultural aspects remain ingrained. Management gets temporary relief but is left wondering why long-term changes and profit growth were not accomplished. Only a continued aggressive and authoritative supervisory style of management can bring further improvements under these conditions.

When the executive dream of creating an internalized change process is not achieved, the tendency is to abandon the approach and pursue cost reductions. During the 1980s, many large corporations opted to implement one of several quality improvement processes fostered by prominent experts. They did so fully expecting to significantly improve earnings, as well as to make a giant leap forward in quality improvement. The logic was extremely sound, and the approaches had been proven.

Most of these organizations showed short-term achievements, but many found serious problems within their cultures that created an obstacle to successful implementation in the form of an unwillingness to make the necessary internal changes that would make total quality a success. This complication made applying a canned approach to systems that had existed for decades very difficult to achieve. As the effort waned and management withdrew their enthusiastic support, apathetic disappointment crept through these organizations, intensifying the obstruction.

The best improvement philosophies require understanding, commitment, and execution by people who are convinced the inherent principles will achieve their personal as well as their organization's vision of success. These ingredients were often missing in the executives who, in the forced change phase, launched what they hoped would be very successful improvement efforts.

Still pressured for results, senior executives resort to pulling the earnings improvement strings over which they have control. This usually means significant cuts in fixed costs, mainly headcount. Obviously, that's where most of the dollars are budgeted. Downsizing or "rightsizing," as it is called in the politically correct vernacular, can be a favored tactic. In some instances, companies have taken on fat during good times, only to recognize when the cycle turns downward that they're overstaffed and bureaucratic. In these cases, it is time for managerial action and removing the fat, hopefully as quickly and painlessly as possible. Appropriate headcount surgery, if necessary, should be performed to bring cost structures in line within organizations mired in this sector.

Comparative industry data should be used to make certain there are no disadvantages in cutbacks vis-à-vis major competitors. If certain units are at a disadvantage, then survival becomes an issue. The organization must eliminate the gaps in its competitive performance or risk losing the unit. The sensible approach is to determine the skills necessary and work on building in that area, while cutting back in non-value adding sectors. No organization is really ready for this phase. To operate as though the future is at stake runs counter to the thinking of any group beyond low performers. It is simply a favorite management tool that is hard to avoid, particularly for firms in this phase of development.

Planned Improvement Focused on Individual Value—Phase III

Once the appropriate surgery is completed and the comparative analysis has established the performance enhancement that is possible by emulating the leaders, the organization can move more comfortably into the third phase. At this stage, the business unit is ready to build strength. The obvious excesses have been eliminated, but the true efficiencies have not started. Indeed, most post mortems on downsizing efforts have revealed that, beyond a short-term improvement to costs, there are few lasting organizational enhancements. But during this phase, quality, productivity, and value improvement can be initiated.

Management now focuses on the value of each remaining individual. With the organization at the right size, at least for the moment, each person must contribute. Managers must mold teams much as a coach in professional sports works with the players to field the strongest unit to beat all competitors. The results curve may spike upward as easily implemented quality improvements are made. Afterwards, however, it will not slope as dramatically upward as it did in the forced-change period because new processes must be assimilated and because true improvement opportunities, are more deeply imbedded and less accessible. Nonetheless, the goal is to achieve results through management systems and processes that will continue to drive improvement regardless of changes in people.

Process development means that the organization has determined how to attain improvement. Leaders must understand what's involved in the process and know exactly how it will help satisfy the current demands on the firm. They must be ready to move to higher order changes. All unit managers must be able to say, "I know what I need to achieve my objectives, and this process will help me do it." That is an important step in overcoming the illusion that downsizing is the answer for any poor performance.

Now the organization has the beginning of a total quality strategy and sets into motion, in the hands of skilled facilitators, the process that will drive performance. In this phase, eight critical fundamentals will need to be implemented.

1. The CEO clearly establishes the central quality objectives and devotes personal time to visibly reinforce them throughout the organization.
2. The senior managers interpret and articulate the central objectives in terms of mission statements and goals that are meaningful to the units and people under their direction.
3. Resident and department managers translate the mission statements into specific improvement targets.
4. Action plans are prepared to accomplish the targets, not as an exercise to satisfy management but because of the logic of the process.

5. Education and training are accomplished so people know how to execute the process.
6. A communication network is established to discuss, measure, and control implementing the process.
7. Execution teams are set into motion to carry out the intentions of the action plans.
8. Results are tied to the reward system; compensation is firmly linked with implementation.

In the third phase of the progression, the emphasis shifts from conducting business in the usual manner to (a) having the people necessary to accomplish the intentions and objectives of the unit; (b) keeping those individuals directed toward meaningful and rewarding work that contributes directly to an understandable mission; and (c) maximizing the satisfaction of the collective individual efforts for both the person and the organization.

The value of each team member moves toward the optimum, and it becomes imperative that everyone function as part of a coherent, dedicated, and effective unit. As people realize the importance of their individual contribution, the quality of performance will increase and there will be a corresponding increase in productivity and profit.

Managers can then begin to link individual abilities to action teams. These interactions become the vehicle for capitalizing on the opportunities that appear as the unit focuses on how to attain its mission. The teams' achievements will maximize the results of the building-strength phase and lead to optimum use of resources.

The pitfall to be avoided in this sector is the tendency to wait for the rightsizing to instill a survivor mentality that will drive further improvement. That simply will not happen. Overcoming this is accomplished by focusing a team effort on a meaningful quality enhancement effort that stresses the value of each individual contribution.

Internalizing an Improvement Culture—Phase IV

When the organization has learned how to successfully implement an improvement process (and that may involve one or more failures before success), the organization is ready to enter the fourth phase: internalized progress. During this phase, business units build strength to higher levels and establish a distinct market advantage. They do so by internalizing an improvement culture that becomes a way of life. All employees have performance improvement as a constant objective, and the management system continually recognizes and rewards individual and team accomplishment.

At this stage, results in virtually all areas of measurement will be many multiples of what they were during Phase II. A factor we call resource optimization (RO) will start to become a feasible objective. RO is a process that management adopts to focus attention on how to derive the maximum (or optimum) utilization of what quickly becomes scarce resources due to the downsizing and the correlated emphasis on limiting reinvestment funds. (We will discuss this concept in greater detail in the ensuing chapters.) Under the drive for RO, operational techniques and the improvement process merge. No longer must changes in style or approach be discussed. The effort becomes holistic in nature as all effort is focused on improving the quality of job performance throughout the organization.

Teams are constantly looking for improvement opportunities, with an emphasis on making optimal use of all invested resources, particularly key personnel. Action plans are updated and prioritized as an ongoing discipline. Teams are quickly formed to execute on the highest priority issues, and normal work is accomplished as the process continues. Everyone is dedicated to continually improving performance levels to prevent regressing to a status quo environment. With a sense of pride in the capabilities and value of each person permeating the units, no target seems impossible to attain. Optimization is not an ethereal dream, but a potential end result of a dedicated improvement effort. Terms like *six-sigma results* take on real meaning and become a feasible target.

> With a sense of pride in the capabilities and value of each person permeating the units, no target seems impossible to attain.

A Matter of Speed and Results

As we stated earlier, the model offers a natural and logical progression from a status quo environment to one of Big Q. The two fundamental differentiating factors between organizations that have succeeded and those that have not are their results and the speed at which they moved. Companies that avoided most or all of the TQM pitfalls moved with deliberate speed, in a consistent direction, and quickly realized bottom-line results. Companies that became ensnared in obstacles to TQM ultimately floundered, became progressively frustrated, and had little or no profit impact to show for their efforts.

Generally, these unsuccessful companies have either abandoned their efforts or remained on the little q level. The latter is a debilitating setback for two reasons. First, the organization loses valuable time while competitors gain significant advantage via their own improvement initiatives. Second, restarting a quality initiative after a failed attempt is extremely difficult.

PART TWO

Quality by Industry

Although quality-management principles remain the same for all types of organizations, different industries apply these principles in ways particular to their businesses. Part 2 provides a diverse collection of articles on how quality management is making a difference in various industries. We have clustered this collection around *manufacturing, services,* and the *public sector.* In selecting articles for each sector, we sought to strike a balance between principles and applications, while emphasizing the application of these ideas.

Another goal of *The Quality Yearbook* is to garner the assistance of consulting editors to provide their perspectives on applying quality principles in the editors' respective areas. We are fortunate this year to have the participation of several highly respected experts. Each has contributed an original article written especially for the 1997 edition. They include Roger Nagel (executive director of the Iacocca Institute and a leader in the area of agile manufacturing) and Napoleon Devia (senior fellow at the Iacocca Institute) for manufacturing; Joe Sensenbrenner (nationally recognized consultant and former Madison, Wisconsin, mayor) for government; Julie Horine (a leading researcher, writer, and training consultant in applying Baldrige criteria to eduction) for K–12 education; and Ron Heilman (head of the Center for Quality and Productivity at the University of Wisconsin, Milwaukee) for higher education. We commend their articles to you. They provide a wonderful perspective on the value of the quality approach to succeeding in their respective areas.

In the rest of part 2, here's what you'll find:

Manufacturing

We have chosen three articles, two of which are case studies and one a "think piece." The case studies demonstrate how implementing a variety of quality techniques allowed two manufacturing companies who were on the ropes to turn themselves around, becoming highly productive, responsive to customers, and profitable. The think piece by Aleda Roth supplements the article by Nagel and Devia, focusing on how manufacturers must become agile to survive. They do this by turning their plants into "knowledge factories," which use information and their other capabilities to respond quickly to customer needs.

Services

We have chosen two articles that are representative of what's going on in service industries in general. The first article is on how banks need to focus on "value propositions" for customers. The value propositions described are equally relevant for all service organizations. The second article is more how-to in nature, describing how TQM tools and techniques can improve the performance of a customer-service call center.

Public Sector

Organizations of all types are now beginning to appreciate the value of TQM principles for helping them fulfill their mission. For the four areas of the public sector, besides the contributions of our consulting editors, we have selected a variety of pieces that demonstrate how quality management is being applied in these areas. Some articles we especially commend include Wayne Levin's article on how government cannot solve public policy problems without changing the systems from which those problems emerge, and Franklin Schargel's article on how quality practices turned around a vocational high school in Brooklyn, New York.

For additional material on what is going on by industry, please consult the bibliography in part 4. It includes a detailed, annotated list of materials across several industries, of which the articles in part 2 are only a small sample.

Manufacturing Sector

1996 Manufacturing Outlook: The Dawning of Relationship-Based Competition

Roger N. Nagel and Napoleon Devia

This article by consulting editors Roger Nagel, head of the Iacocca Institute, and his associate, Napoleon Devia, explores the new competitive world of manufacturing. This is a world where cooperative relationships with suppliers and customers and even competitors are key. Manufacturers, as the authors say, must move "away from defining themselves as suppliers of products or services" and become "enablers of solutions for customer's needs in the areas of their selected skills." In one sense, that's what customers have always valued, but now, in an intensely competitive world, companies must realize this and make it their operating credo if they are to succeed. This article explains this new trend in depth.

More than ever, manufacturing needs to be aware of what is happening in the world's competitive arena from a strategic standpoint. It is necessary to understand in a systemic way the pressures manufacturing feels in order to anticipate and manage proactively the changes that allow it not only to succeed today but also to strengthen its position for future opportunities.

The strategic outlook is the dawning of relationship-based competition.

In the present outlook, quality, service, and price are assumed to be a given in the competitive arena. They are, indeed, the ticket to enter the competition, but to win, companies have to establish value-based relationships.

More and more companies are understanding that continued economic relationships will be the winning factor in the twenty-first century and are organizing their resources to nurture such relationships.

We have called the resulting value-creating entity the "interprise" to distinguish it from the traditional arms-length hierarchy of organizations that used to deliver goods and services to consumers in the mass-production era.[1]

The coming changes should not be misconstrued to mean that manufacturing has diminished in importance. Our economic society has understood that even though manufacturing will provide fewer jobs as a percentage of the total in the economy, this does not mean that manufacturing is not important, or is disappearing, or should be consigned to the edge of our sphere of interest. Quite on the contrary, manufacturing is becoming more effective, productive, and important. Like agriculture, manufacturing is a key enabler and the tangible support of the new economy. Manufacturing constitutes the foundation for other industries to flourish, and its productivity leads the way toward higher standards of living.

Although American manufacturing has regained competitiveness, there is not yet an undisputed world leader. A look at the top manufacturing companies in the world suggests an even distribution among the United States, Europe, and Asia.[2]

Benefiting from the healthy economy during 1995 and 1996 in the United States, many companies are now profitable and busy—and these profits and work might be a mixed blessing, because renewed activity has blunted the urgent need for strategic changes. Andersen Consulting, Inc.'s, data indicate that only 5 percent of the companies are redefining strategy as part of their transformation efforts. Whether this is a normal fact in the dynamics of change or a lack of foresight raises concerns, because complacency is not the only evil causing stagnation. If companies focus only on catch-up strategies and on using their advantages in innovation or world position as a basis for their leadership, they are risking their future. This danger is real because companies that reposition strategically by following agility principles will surpass companies that do not in ways ever more difficult to imitate in the future.

Some competitors are following a consolidation strategy at present, investing in resources, upgrading their workforce, and preparing infrastructure for the new economy. Profit margins or total revenues are thus not the best indicators of future leadership. Similarly, differences in accounting practices and exchange rates might give a distorted or incomplete view of economic reality.

Manufacturing continues to feel the pressures of globalization as many companies are rushing to China, Eastern Europe, and third-world countries

---

[1]A comprehensive description of the characteristics of the interprise and detailed examples are provided in Kenneth Preiss, Steven L. Goldman, and Roger N. Nagel, *Cooperate to Compete* (New York: Van Nostrand Reinhold, 1996).

[2]See, for example, "A List for the Global '90s," *Industry Week* 245, no. 10 (1996): 12, for a list of the top 1,000 manufacturing companies in the world.

not only to explore markets and the building of manufacturing facilities there, but also to leverage their human resources for worldwide operations. Even though some companies are struggling with quality levels in foreign locations, downsizing side effects at domestic plants, and other cross-cultural and policy issues, global manufacturing will continue to expand as a means to have access to global markets as well as to balance exchange rates and regional economic fluctuations.

Competition has intensified on all sides. The deregulation in the energy, transportation, communications, and finance industries on one hand and the merging of physical products with services and information on the other hand are together forcing more and more companies to operate in what were formerly exclusive territories. Thus, long-distance communication carriers and local communications companies are competing for the same customers and invading each other's turf. Automobile makers are selling financial services, and software producers, once considered service companies, are now manufacturing products and offering personal finance services.

It is important to grasp firmly this brief description of the most relevant drivers in the competitive environment—those changes that affect manufacturing in companies. In this intense, information-filled environment, manufacturing is striving to manage an increased technology offering, to maintain a somewhat higher demand for products, and to accede to additional demands for organizational changes.

The greatest of these change drivers is relationships. Creating new ways to build relationships has made and is making a most profound change on strategy in manufacturing operations.

The principles of the emerging competitive environment are still the same as those described in earlier work,[3] but the understanding of their main characteristics is becoming more clear and widespread.

Specifically, we review here broad areas in which substantial progress has been observed. The observations in these areas converge in one fundamental conclusion: to obtain outstanding economic performance, skillful leadership of value-based relationships across all operational levels is becoming the key determinant.

> Creating new ways to build relationships has made and is making a most profound change on strategy in manufacturing operations.

A New Mind-Set: Growth through Product and Service Extensions

Companies have realized the existing opportunities for growth by looking at available resources with a new vision or mind-set. They are pursuing such opportunities more aggressively. A new mind-set in which growth is again

---

[3]For a detailed description of agility see Steven L. Goldman, Kenneth Preiss, and Roger N. Nagel, *Agile Competitors and Virtual Organizations* (New York: Van Nostrand Reinhold, 1994). You can also visit the Agility forum Home page at http://agilityforum.org.

the main component has begun to erase the downsizing craze, and organizations are reshaping their own people and information resources to create profitable new businesses. Therefore, those who have this new vision and turn it into action can attain better company performance than would otherwise be possible.

As a consequence of this new mind-set, the manufacturing/service industry classification boundaries are blurring, as indicated by the consolidation of the Fortune 500 list.[4] This consolidation means much more than the fall of a classification scheme that has been useful for the last forty years to represent business reality. It points out the radical transformation that is occurring in the business itself.

> The manufacturing/service industry classification boundaries are blurring, as indicated by the consolidation of the Fortune 500 list.

By leveraging their people, information, and other resources, companies once considered manufacturing archetypes, such as GE, are obtaining sizable revenues from activities currently seen as services. More and more manufacturing companies are discovering additional sources of income in services and information. Amdahl and Unisys are selling not only computers but also information services, and Oracle is selling not only databases but consulting services. Automotive companies have enlarged their leasing operations and expanded into other services. Nissan and a few others are now selling not only brand cars but also luxury transportation services that include a "loaner car" while the customer's car is being serviced. Software companies that began as service companies have moved to manufactured products as they sell packaged software and will probably move back to service as they sell processing on specially configured software applications over the Internet in the near future.

The new vision is changing the way companies define their operations and allows them to re-create themselves for the future. It has intensified competition in some traditional niches but most importantly is permitting the creation of totally new markets, with new combinations of goods and services.

Collaborative Production, Virtual Infrastructures and Environments

Enhancing competitiveness through cooperation has advanced both internally and externally in organizations that have manufacturing operations. Besides using the Internet for product information dissemination and some business transactions, intranets and other information and communication technologies are now being used to form virtual teams that achieve time compression of most manufacturing operations.

Beyond the highly publicized Boeing 777 case of design network, Hughes has recently put in place a network of 20,000 workstations linked in order to

---

[4]Thomas A. Stewart tells an interesting story about the reasons for making this decision. See "A New 500 For the New Economy," *Fortune*, May 15, 1995.

share product-development data. But intranets are not solely defined by videoconferencing and electronic data interchange. Communications tools are now being used to set in concurrent motion the flow of needed resources to create profits for the organization—wherever in the world, and whenever an opportunity develops. Manufacturing is not only about enterprise-wide resource planning, but also about planet-wide opportunity realization. The Agile Web Inc. in Pennsylvania, although still very young, is actively exploring the challenges of this new form of organization. Interest in intercompany cooperation and partnership agreements is growing at double-digit rates, although companies are still struggling to increase their success rate in strategic alliances and virtual relationships. Although still in embryonic form, manufacturing is discovering how to use the new virtual infrastructure to support business relationships that increase effectiveness in new product and business development.

Smart Physical Products

Manufacturing is leveraging people skills and information not only at the organizational level but also at product and process levels. A greater percentage of digital and communication technologies are being included in what will be the next generation of physical products: smart products.

For instance, electronically controlled engines and suspensions, self-diagnosis, and global positioning devices make up a complete information system that will change a Mack truck, and other similar vehicles, into platforms for continued relationships over a lifetime of transportation solutions. Or to consider another manifestation, smart cards, widely promoted in the 1996 Olympic Games by Motorola and few other companies, are the continuation of a series of products leading to "smart everything."

Manufacturing is also incorporating and benefiting from information technology built into industrial process machines of all types. Advances in sensors, wireless communications, lasers, micro devices, and microprocessors are modifying not only the operation and maintenance of industrial machines but also the relationships between suppliers and customers over the lifetime of processing solutions. Real-time machine performance transmitted to a central computerized control center now allows manufacturers to plan and to execute machine maintenance based on the actual physical needs and not on statistically determined times or service hours. Maintenance is becoming automated or guided in most cases, with parts supplied as needed within the required time frame. The capital costs of equipment might not have decreased appreciably, but price/performance or price/power has improved substantially, allowing manufacturing to continue lowering costs while coping with the challenging demands of customization.

Motorola and Mack Truck are certainly not alone. Deutsche Telecom's management of the information system that collects engine data from electronic sensors located in the engines of Lufthansa aircraft and sends it via satellite to the Lufthansa control center for maintenance scheduling illustrates the above observations. Otis Elevator is using its Otisline, an integrated customer and service information system, not only to make information available to all company functions but also to show potential customers the actual field data of product performance and service response data. With regard to Otis, reliability and instant access to the service information upon demand is a strong starting point for their business relationships with customers. Scitex makes it possible for customers' systems around the world to interact with Scitex's maintenance center, and for customer personnel to interact with sales and other company units through its global computer network. These are just a few examples that illustrate the principles along which the new competitive environment is now being shaped.

In addition, American manufacturing is very active in transferring defense-based technology to civilian applications and vice-versa. This tendency toward cross-fertilization is also exhibited by the current active technology exchange between Eastern Europe and the Western world.

The challenges of commercial transportation in the next decade might help to alleviate the enormous impact of the transition from defense to commercial markets for many manufacturing operations. Resource reallocation within the economy is not easy or painless, but many organizations are making substantial progress at capturing a sizable piece of the action in the commercial markets.

More technology and knowledge resources are now available as a result of these changes. Manufacturing is synthesizing new products and services at record-breaking rates. The nature of the new economic success is forcing a shift from the internal world of manufacturing operations, where efficiency was the king, to the external world, where other types of constraints limit the rate at which society embraces their offerings. Smart products will push this balance even more to the right, and challenges and opportunities are therefore expected to demand higher levels of strategic performance.

In a fashion similar to the route by which machines freed humans from exhausting and slow mechanical tasks, smart products are now releasing humans from repetitive logical tasks, allowing people to concentrate on more creative, productive, and enjoyable activities. However, these achievements yield their fullest benefits only with the dynamic coupling of the interactive economic relationships along all stakeholders in the interprise.

Manufacturing organizations are using the increased technological capabilities to position themselves in networked environments in which smart products are used as platforms for continued economic relationships.

Resource Integration through People and Technology

Manufacturing is using advances in logistics, information technology, concurrent engineering, and cooperation more extensively in order to bring about the interprise and shorten the concept-to-cash cycle times in new product and new business development cycles. Although progress realized in these areas might allow organizations to gain some advantages over competitors still in mass-production mode, the pursuit of higher improvements is forcing manufacturing to make decisions that, in practice, configure the interprise.

Companies involved in manufacturing are now locating resources at customers' plants more frequently as the number of special projects using virtual cross-factory teams and product-development teams becomes larger. Connecting people and information systems is reducing inventories, linking business dynamically, and enabling the interprise to seek higher levels of economic performance proactively.

Motorola is linked to its suppliers by its schedule sharing method, in which shared information systems allows them to manage inventory levels and deliveries in a coupled dynamic fashion.

Ford and its J. Walter Thompson advertising agency found a way to organize electronic interactions between consumers and the people who design their cars. This on-line information system enables designers to understand the needs and desires of their consumers, and consumers in turn get an improved and affordable product.

World-Class Supplier to World-Class Enabler

Many companies are still playing catch-up and are still running toward the goal of becoming world-class suppliers. Others, though, have started to look at themselves and to think of themselves differently, in a way that enables them to discover and proactively upgrade their target markets. Such companies are moving away from defining themselves as suppliers of products or services, to become enablers of solutions for customer's needs in the areas of their selected skills. In order for this change to bring about results, it must be not a semantic exercise, but a real transformation effort across the entire organization. Important lessons in many key aspects of this transformation effort are illustrated in the experience of Ross Operating Valves of Livonia, Georgia. Seeking increased margins in a very competitive environment, Ross Operating Valves created the *RossFlex Process*, a technology-enabled, structurally different organization of the available resources of knowledge, skills, and relationships within the interprise. The results were dramatic increases in orders, customers, profits, product line, relationships, and interprise capabilities. The whole perception of the world changed for Landis and Gear when they shifted their mind-set from a supplier of control systems to a enabler of its customers to control climate efficiently in shopping centers. The Steelcase effort to position itself in the marketplace not as a manufacturer of desks, chairs, and credenzas, but as a company that enabled its customers to create effective working environments for their employees illustrates progress in this direction.[5]

At the Iacocca Institute, we see more awareness of these trends and the imperatives to change. However, there is still a lot of work to do in order to achieve safer routes to strategic transitions. Some companies are exploring new ways on their own; some are helping to explore tools and game plans to introduce changes; but many are still just waiting for them. More cooperation

> Many companies are moving away from defining themselves as suppliers of products or services, to become enablers of solutions for customer's needs in the areas of their selected skills.

---

[5]For example, see Steelcase's ad in the *Wall Street Journal*, 22 February 1996, section A7.

among industry and academia is needed to do exploratory work and find the right tools more quickly.

Manufacturing companies are struggling with the need to manage the "touchy-feely" stuff—that is, the "people issues" implicit in the new economy. There is not enough available operational knowledge of how to go about them or how to measure them. For instance, how should a company tackle motivation and compensation in the new economy? How should it deal with the intellectual property and trust issues in empowered-teams environments? How can a company organize the needed skills, knowledge, and rapid upgrades on a continuing basis? How should a company appropriately respond to societal pressures of job creation, quality education, environmental protection, and diversity in the new global economy?

As stated in the beginning, manufacturing needs now more than ever to be aware and to understand what is going on in the competitive environment. Internal focus is no longer enough. Running plants efficiently is no longer enough. Manufacturing has to be an enabler of an organization's strategic actions. Reducing waste, costs, and complaints is no longer enough. Producing good-quality products faster, better, and cheaper is no longer enough. Without losing focus on their manufacturing operations, manufacturing leaders have to establish multiple but coordinated customer and supplier contacts, participate actively in customer opportunity teams, package their organization's skills and knowledge in new ways to protect and enhance value, and help to gather the key information needed to engineer the value-based and affordable new products and services that will evolve into new businesses. Manufacturers have to link their operations to their suppliers and customers to bring about the interprise, not only to manufacture current products but also to contribute to the discovery process of opportunities for creating the future business.

We see manufacturing winners positioning themselves in the design, engineering, and product realization processes of the next generation of physical products, in which information-rich and communication-capable devices will replace the predesigned flexibility in products of the present to make the customization of total solutions a daily reality and the main source of revenues for modern corporations.

The message is clear and simple in manufacturing: Change operations strategically in sync with the competitive environment, using technology quickly to enable individualized economic relationships globally. If you do not do it, someone else will do it, isolating you from the future source of revenues: the continued economic relationship in the interprise.

Process Redesign and the Successful Turnaround of Shasta Industries

John Veleris, Robert Harlan, and Sheldon Greenberg

Shasta Industries is a manufacturer of recreation vehicles. The company had a lot of manufacturing problems that affected quality and raised costs. This article describes how Shasta transformed its manufacturing processes by implementing quality management techniques to, among other things, raise productivity 50 percent.

In January of 1995, Grant Thornton LLP was asked to assist in a productivity improvement effort by Shasta Industries, a $50 million recreational vehicle manufacturer. Shasta is more than 50 years old and remains one of the oldest names in the recreation vehicle (RV) industry. One its two manufacturing plants is located in Elkhart, Indiana where many other recreational vehicle manufacturers and their suppliers are concentrated.

The company had been operating in the red since 1993. Early in the 1990s, Shasta had begun to watch its market share and margins shrink. By 1993, the company was operating in the red. It was time to take action. In response, management began to focus on sales by enhancing product design and overhauling the entire sales process. These changes, along with a favorable economic climate in 1994, resulted in a significant upward sales trend. Unfortunately, manufacturing was unable to gear up sufficiently enough to meet the increasing demand. The result was a six-month backlog. Manufacturing couldn't increase production due to the high labor and total quality costs related to the manufacturing operations.

Breaking Down the Problem

The challenge facing the company was comprised of two distinct components, each having major strategic and operational implications. These components included:

- How to increase the profitability of manufacturing operations and improve product quality and,

- How to increase and sustain manufacturing capacity in order to respond to growing demand.

Both challenges needed to be addressed within six months. That was the company's real window of opportunity. If the company waited more than six months, customers who had unsatisfied demand or requests would defect to competition and take with them more marketshare.

A situation assessment and problem analysis identified a number of primary causes contributing to the problem. The Cause-Effect diagram in Figure 1 outlines the main contributing causes to poor plant productivity at the start of this project. The change-resisting corporate culture had inhibited previous corrective actions, and necessitated both employee involvement and disciplined plant-wide communication mechanisms.

One major business environment constraint was the low employment and chronic unavailability of skilled labor, which is endemic to the Elkhart County labor market. According to management, relocation was not an option. The company was paying wages more than 30 percent lower than those offered by the competition. Benchmarking demonstrated that plant labor productivity was also at 50 percent of that of the competition. Furthermore, availability of skilled labor extending to supervisory ranks was almost nonexistent, even at competitive compensation levels.

In essence, the company was hiring unskilled individuals who, when they learned certain skills, would defect to Shasta's competitors. As a result, Shasta's entire labor force was constantly riding the learning curve. Labor turnover was approaching 250 percent, with 90 percent of the employees having a tenure of less than a year.

Project Organization and Planning

The real challenge was not only to identify and enact a solution, but to do so in a phased-in, 6-month approach that would optimize available internal resources and ensure long-term sustainment. This necessitated the development of an internal capability for continuous manufacturing process improvement and employee training programs. Together with Shasta's President Bill Snook, Grant Thornton LLP helped develop a carefully crafted plan to deliver quick results and to provide knowledge transfer.

The project core team (Figure 2) provided project management and knowledge transfer skills to help initiate change, develop an internal mechanism that would sustain that change, and guide Shasta through the most crucial steps. Using this process, the company would hire a manufacturing engineer who would provide implementation assistance. This manufacturing engineer would also select three employees to assume the responsibility of carrying out a training plan. The manufacturing engineer, the trainers, and other selected middle management and supervisory personnel were under the project management and training of Grant Thornton LLP.

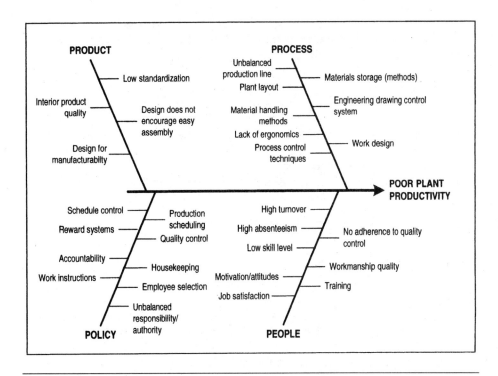

Figure 1. **Poor plant productivity contributing causes**

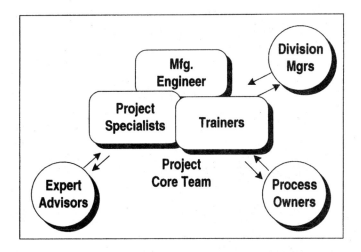

Figure 2. **Project organization**

The turnaround strategy focused on four critical success factors. First, job satisfaction had to improve in order to reverse employee retention problems and extend the average employee tenure. Stabilizing turnover rates would provide needed time for all other training and productivity improvement programs to work. Second, the learning curve would be shortened through rapid, focused training and "quick work efficiency hitters related to re-work avoidance." Next, productivity would be further improved through process and work methods redesign. Finally, a performance incentive plan would pass a portion of the productivity gains to plant employees to ensure higher compensation was distributed equitably. This strategy would also allow for a gradual build-up of the critical mass required for a second shift. A communication plan was also developed to promote teamwork and employee involvement in the implementation of the strategy. It was also developed to ensure employee acceptance and support.

An Innovative Approach

To support the strategy objectives, a comprehensive approached was designed. Specifically, this new approach was tailored to do the following:

- Assist in changing the current organizational culture;
- Provide project management assistance and technical expertise in industrial and manufacturing methods engineering;
- Assist in job analysis, work-redesign, and production line balancing;
- Develop a labor training program and a performance incentive plan; and
- Identify and prioritize short-term productivity improvement projects.

The approach involved four phases (Figure 3). The objective of phase one was to analyze the current manufacturing state and to establish an improvement baseline. Phase two aimed at short-term (2 months) work performance improvement through work redesign. The development of performance measures and job descriptions would provide the foundation for labor training programs. Phase three sought to improve the entire manufacturing process over a longer-term horizon (6 months) by improving the flow of work and materials, production line balancing, and appropriate performance incentive systems. Finally, phase four set sights on the development, adjustment, and systematic delivery of the training program. This last phase was scheduled to begin three months from the beginning of the project.

Job analysis focused on manufacturing workstations that presented the highest workmanship quality problems and, subsequently, required re-work. Understanding work activities that contributed to re-work was the first step in identifying "quick productivity hitters." Primary causes for each workstation re-work activity were then identified. In most cases, Paretto's law triumphed: 80 percent of the re-work was generated by errors in two dozen

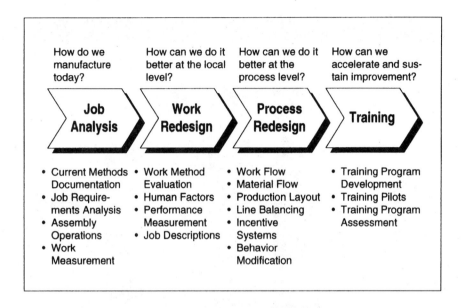

Figure 3. **Project approach**

assembly and installation activities upstream. Most of the errors could be avoided through clear work instructions and on-the-job, quick training.

Documentation and analysis of manufacturing activities and tasks were prioritized based on their relative contribution to value and quality. Because of the high frequency of new product designs and the multitude of product models, proper development and maintenance of work instructions was of paramount importance. The goal was to help workers do the job right, and to reduce mounting frustration caused by not knowing how to do things the right way. But there was a problem. The company lacked a system that would help them reduce the cost of quality, and would promote a sense of accomplishment among the hourly employees.

To leverage job analysis with the development of work instruction and job description systems, a database (Figure 4) was created that mapped all work elements (manufacturing activities and tasks). The database tracked the sequence of the work elements, their time duration, man-hours required for completion, and workstation involved, as well as other pertinent information. This activity and job analysis database saved a lot of analysis time and proved invaluable in re-balancing the production line, and in developing a training program.

Work redesign was a natural extension of the job analysis. It focused on human factors and was based on the premise that work performance can be improved through increased job satisfaction.

Work redesign offered the opportunity to interview employees and re-enforce the message that their voice mattered, and that job satisfaction was a key factor in improving labor productivity.

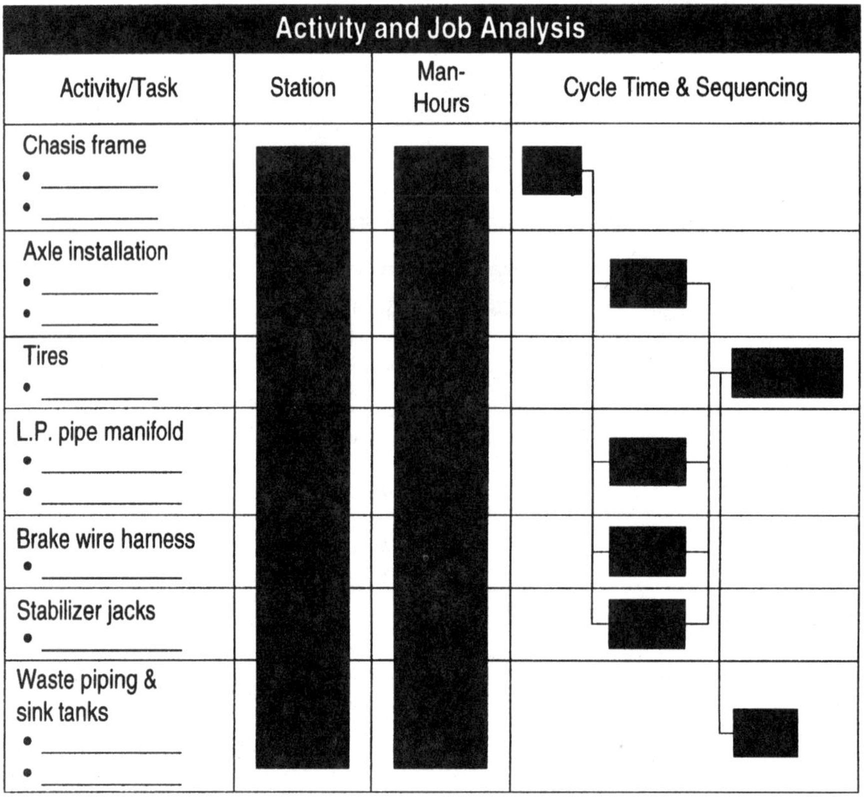

Activity and Job Analysis			
Activity/Task	Station	Man-Hours	Cycle Time & Sequencing
Chasis frame • _____ • _____			
Axle installation • _____ • _____			
Tires • _____			
L.P. pipe manifold • _____ • _____			
Brake wire harness • _____			
Stabilizer jacks • _____			
Waste piping & sink tanks • _____ • _____			

Figure 4. **Activity and job analysis**

Labor and management had to work together to identify work redesign opportunities that, along with initial training, would help employees become more productive and able to share the gains equitably through performance incentives.

New recognition policies and rewards—problem-solving meetings, employee of the month awards, etc.—were also activated. An employee suggestion system was even created. The suggestion system was initially used to solicit all productivity improvement-related ideas. The response was excellent and, along with the employee interviews, more than 200 improvement opportunities were identified. Improvement opportunities were mapped in the activity and job analysis database for evaluation. Next, they were classified into work redesign- or process redesign-related areas. The opportunity evaluation resulted in dozens of short-term and longer-term action items that were incorporated into the project plan for scheduling, resource allocation, and appropriate follow through.

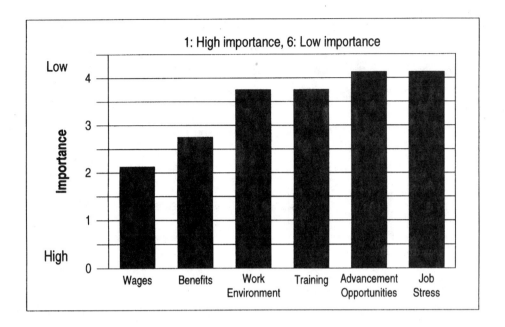

Figure 5. **Key job-satisfying factors**

A survey of the hourly workforce was conducted to determine key job satisfying factors and the extent to which these factors could improve job satisfaction (Figure 5). The results of that survey verified that wages and benefits were of primary importance, although work environment, training advancement, and job stress were also cited as key satisfiers. Work analysis focused on improving the work environment and reducing job stress. These two factors were improved through better ergonomics and work methods, work station layout, and a cleaner plant environment, which facilitated easy movement.

Process Redesign

Process redesign focused on improving the work and material flow, redesigning and balancing the production lines, improving the overall plant layout, and inducing behavior modification through schedule control and performance incentive rewards. Process redesign also involved redistributing employee accountability and responsibility in conjunction with job enlargement and job enhancement.

Initially, the manufacturing production lines were organized around 13 to 15 workstations each requiring three to six employees. Quality accountability was collective and highly diffused with a final quality inspection station at the end of the line. Most vehicles would exit the production line with significant quality deficiencies. These deficiencies would then be corrected either in the inspection station or, most frequently, in the parking lot. Line quality

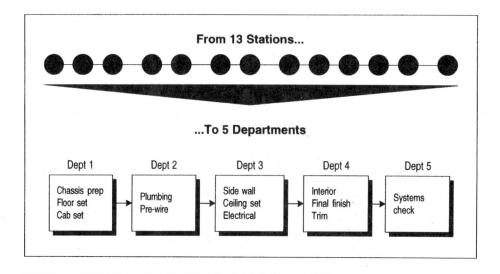

Figure 6. **Manufacturing process redesign**

inspectors were responsible for identifying and keeping track of specification compliance problems, but they had no authority.

With a minor exception, the employees knew only how to perform a narrow set of activities and tasks at the workstation level. The narrow skill set required by each workstation facilitated quicker learning, but it also had some serious drawbacks. Specifically, absenteeism and turnover created significant manning "holes" in several workstations on a daily basis. The result was bottlenecks that delayed the entire line. There were five employees who could fill such holes, but even that number was insufficient. Pressure to "get the product out the door" hadn't allowed anyone the luxury of systematic job rotation. The foreman performed less supervision and more "fire-fighting." He even performed line work.

Our first priority was to group existing stations into five distinct departments (Figure 6) by grouping similar activities. Each department would require a wider skill set, but would reduce the impact of an absent employee on overall department performance. A wider skill set per employee also offered work variety through horizontal job enlargement. The jobs of the foremen were enriched to a supervisory level by performing more planning, organizing, and control. Each department would be responsible for its own workmanship and quality, and its foreman would be held accountable accordingly. Embedding and maintaining production performance and quality at the local level was a critical success factor for the upcoming incentive system. To ensure that quality wasn't sacrificed for productivity, a new quality policy was developed that gave line employees more accountability for their work. The line inspectors had more authority to halt production when

> To ensure that quality wasn't sacrificed for productivity, a new quality policy was developed that gave line employees more accountability for their work.

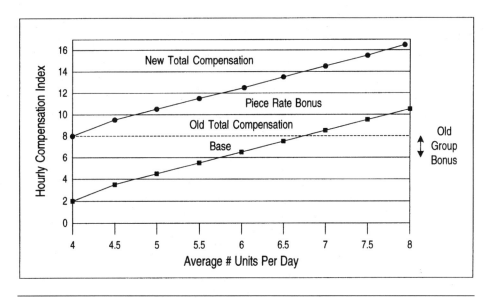

Figure 7. **Piece-rate incentive system**

quality was sub-standard, and it allowed sales personnel a final product review on a sampling audit basis.

On a fundamental level, the internal focus on quality was changed so that every department was performing as either a vendor or a customer of another department.

An incentive system was designed to successfully couple individual and department responsibility with performance economic rewards. The employees were solicited for ideas that would make such a system equitable and implementable. Several variations of incentive systems were evaluated. A variation of a piece-rate group system surfaced as the best alternative (Figure 7). The system would reward each department for its own performance. Previously, the company was paying hourly employees additional compensation at 25 percent of the wages. This flat, extra compensation was euphemistically called a "group bonus" to remind everyone that it was intended only as temporary means to help the employees become more productive, and it could be revoked at the discretion of management!

The challenge of implementing the new incentive system was to improve productivity by about 10 percent so that the new base would match the old total compensation. It was thought that such productivity improvement could be achieved through the early stages of work redesign and training. As productivity continued to improve, the new total compensation would surpass the older fixed compensation ceiling, which would offer employees a proportional piece-rate bonus increase. The system was designed to offer employees a total compensation potential equal to—if not better than—the best-of-class in the industry and best-of-breed in the area.

For successful implementation to occur, some initial efficiencies needed to be achieved. For starters, the company had to abandon the old group bonus. Furthermore, we needed to address the way each plant had unique problems and different tenure levels, which made implementation of the system particularly difficult. Eventually, we achieved excellent results because management and labor made a sincere effort to making the system work, with the understanding that it would take several months of continuous adjustment.

Training

The development and delivery of a self-sustaining training program proved extremely successful. Critical success factors included the fact that the program was designed to deliver 20 percent of the knowledge required to learn 80 percent of the related tasks, and/or avoid 80 percent of the problems encountered in manufacturing.

Designated trainers were sent initially through a rigorous "train-the-trainer" program that included training methods, measurement and evaluation, communication and feedback, and program administration. Consequently, the trainers assisted in the development of the program itself, conducted planning for training sessions, and ultimately continued the administration of the program. The accumulated knowledge gained in the activity and task analysis, as well as work redesign, was used initially to develop an employee training program comprised of modules at three levels:

- Orientation training (two-hour training);
- Common quality problem avoidance (half a day on-the-job training); and
- Manufacturing/Assembly (scheduled on an "as needed" basis and master/apprentice concept).

The Manufacturing/Assembly training module was designed for each department, and included assembly and industry guidelines, instructions on how to read engineering drawings, process diagrams with step-by-step assembly instructions, and "tip sheets" on how to perform tasks or overcome problems.

Implementation Challenges

The multi-dimensionality of the problem raised several serious challenges during implementation of the project approach. The most important challenges were:

- Project management and coordination because of the multitude of critical project activities, the short time frame for bringing about significant change, and the fear of business disruption by overburdening internal resources. A careful balance had to be maintained between spreading team responsibilities and existing workloads for all individuals involved.
- Handling concerns about piece-rate implementation. Management was concerned with the possibility that line employees

wouldn't have the skills required to be successful under the new, aggressive work standards and the piece-rate system. Furthermore, doing away with the old "group bonus" compounded these concerns because the hourly employees had been conditioned to taking it for granted. Unsuccessful implementation of the incentive system could result in a mass labor exit. This risk was minimized by using a two-pronged approach. First, the ability of the hourly employees to meet the new work standards was tested through schedule control. This technique encompassed laying out all the tasks required for a specific area in the assembly line in Gantt chart format. Supervisors were then responsible for ensuring that line employees performed tasks according to the schedule. Next, the incentive system was piloted in a department at one plant, and was introduced gradually to the rest. The same approach was followed in the second plant.

- Overcoming the established corporate culture. Resistance to change required extra hand-holding and communication. It took a significant amount of management commitment to coach and involve the hourly employees through the process. Short-term results had to be achieved, and communicated properly in order to satisfy the need for instant gratification. New symbols and messages had to be constructed and delivered systematically—involving top management in the orientation training sessions, regular status meetings, etc.

Results

During the project, several key performance measures were developed to track progress and ensure continuous adjustments on a proactive basis. These measures included voluntary turnover, avoidable and unavoidable absenteeism, labor cost as a percent of value produced (at the wholesale price level), overtime as a percent of regular time, hours of training delivered, and productivity as a ratio of value produced per labor hour (productivity index set at "1" at the beginning of the project). Absenteeism proved an excellent leading indicator of voluntary turnover.

These performance measures were tracked on a weekly basis, and were incorporated in the weekly management and supervisory status meetings. The measures allowed management to monitor trends with regard to their relationship with implementation actions, and to make appropriate adjustments when necessary.

Over the 35 weeks following kickoff of the project, productivity increased in both plants by an average of 50 percent.

Labor as a percentage of the value of the units produced was also reduced significantly for one of the two plants. For the other, it dropped by an average of approximately 2.5 percent, which translated to about a 16 percent improvement.

More importantly, the workforce at Shasta Industries has undergone a cultural shift. Line employees and managers have gone from a seat-of-the-pants mentality approach to fighting fires to systematic planning, controlling, and managing operational issues for continuous improvement.

Important lessons were learned. Complex business problems require multi-dimensional and innovative approaches, management determination and commitment to solving the problem, and a spirit of genuine collaboration between labor, management and, where needed, business advisors and outside experts.

Sheldon Greenberg is responsible for the manufacturing/distribution operations and productivity consulting practice at Grant Thornton LLP's Chicago office.

Robert Harlan is a senior consultant in the manufacturing/distribution operations and productivity consulting practice at Grant Thornton LLP's Chicago office.

John Veleris is a management consultant with 17 years of experience in manufacturing.

Achieving Strategic Agility through Economies of Knowledge

Aleda V. Roth

Complementing the article by Nagle and Devia, Roth argues for the importance of agility and the creation of what she calls **knowledge factories**, *"an organic, accelerated learning organization that produces knowledge as a key by-product." She insists that this is not merely a neat idea but a business imperative in a time when change is constant, competition relentless, and information universal, requiring companies to be able to respond quickly to customer groups as small as one buyer. Such notions are summarized using the term* **agility**. *This article will help you understand the importance of agility in manufacturing today.*

Manufacturers are in the midst significant transformations in history. The kaleidoscopic nature of the global economy has created an environment of unprecedented hyper-complexity and dynamic change. So much so that the rules of the competitive game are shifting faster than managers can react to them. Moreover, these dynamics bode continuous sea changes in work and sources of value-added. Looking to past solutions for rationing resources and controlling costs is not an option. The drivers of change indicate one point: Successful manufacturers tomorrow will not look much like those of today.

Manufacturing pacesetters are seeking new competitive weapons. Prominent in the search is strategic agility—the capability to produce the right products at the right place at the right time at the right price. By definition, strategic agility is achievable only with competitive strength in a combined set of generic capabilities, namely quality, delivery, flexibility, and price leadership. Managers then have options. They can use one or more capabilities to preempt or imitate fast global competitors. Armed with multiple capabilities, manufacturers will be better prepared for the changes ahead.

Acquiring strategic agility is no easy task. It requires "economies of knowledge" through accelerated enterprise-wide learning. Economics of knowledge means that the firm is able to use its business acumen, combined with skilled people and experience with advanced technologies, to create an

---

This article is reprinted with permission from *Strategy & Leadership* (formerly *Planning Review*), March–April 1996. Copyright © 1996 by The International Society for Strategic Management. All rights reserved.

organization that consistently identifies, assimilates, and exploits new knowledge more efficiently and effectively than the competition.

Organizational Knowledge Drivers

The quantum need for strategic agility stems from chaotic markets and technological forces. Global markets in constant flux have become "virtual." At dizzying speeds, customers, suppliers, and competitors swap boundaries, niches, and roles. It is not uncommon for competitors to serve dual roles as customers or suppliers as long as it is profitable to do so. The plethora of emergent technologies and customer information bases only fuel the infinite ways of dynamically dicing up the marketplace and add to the multiplicity of forms that competition can take.

> Global markets in constant flux have become "virtual."

The global stage is set for opportunistic firms to engulf their ill-prepared prey with no warning. Consider the drivers of organizational knowledge: customers, information, speed-to-markets, and technological choices.

- We live in an age of customer choice, and today's savvy customers want it all! Worldwide, customers' access to product information and databases enables them to make more informed choices. Global customers are demanding uniformity in product quality and service standards. One only has to look at the telecommunications and electronics industry to understand this perspective.
- Information is becoming a global commodity and alone provides little strategic value-added. Access to the Internet and other on-line services provide information to the masses. Computer-based commerce, such as electronic data interchange, speeds up order fulfillment cycles. Because information is widely available, competencies in synthesis and analysis boost the strategic arsenal.
- Speed-to-market is the rallying cry of world class manufacturers. An estimated one-third to two-thirds of a firm's annual sales are derived from products developed within the past two years. On the downside, the same processes that enable rapid new product introduction also increase product proliferation and process complexity.
- The array of technological choices available is growing exponentially. The ability of individuals, organizations, and even society at large to exploit their potential has not kept pace with new technology. As technological advancement impinges upon every aspect of daily work, the expertise, spirit, and creativity of people—not capital and equipment—are the limiting resources. The rapidity of change has outstripped traditional methods and definitions of production. As a result, managers must turn upside down the tenets they held in the past about how to organize work, how to share work, and how to strategize. Yet much of what managers know today and the decisions they make are based upon

industrial, mass production logic. This is a major source of entropy, siphoning valuable resources and misdirecting talent.

Mass-production logic has become a corporate millstone, weighing down its potential for learning and slowing its progress toward strategic agility. Removing the millstone requires "out-of-the box" thinking to capitalize on knowledge creation. World class manufacturers have succeeded by thoroughly "un-learning" old production rules and contextualizing their know-how with postindustrial, information logic.

To turn our organizations upside down and to capitalize on knowledge, managers must broadly address three questions:

- What are the beliefs concerning the competitive landscape and the differentiating priorities?
- How are work processes modified, changed, and evaluated?
- How are the sources of value-added changing?

> Mass-production logic has become a corporate millstone, weighing down its potential for learning and slowing its progress toward strategic agility.

Evolving Management Perspectives

Management debate around these questions can benefit by understanding the distinctions among various manufacturing epochs—those strategic moments in which top management perspectives of competitive capabilities shift radically and require revolutionary approaches to change. The trajectory of the dominant paradigms of production logic and best practices can be understood by using a strategic map to look back to production's historical roots and to help recognize its major disconnects between today's and tomorrow's passages. (See Exhibit 1.)

Each epoch highlights the time-phased best practices of the globally competitive manufacturers in industrialized regions. Notably, the outstanding performers, defined here as "world class," are approaching the 21st century priorities today. The less competitive laggards remain vested in the past, where the fundamental premises of management date back to the industrial age of Henry Ford and Alfred Sloan.

Clearly, within each manufacturing epoch, managers must see things differently in order to manage differently. To illustrate, consider how long it took U.S. automotive manufacturers to recognize and respond to the structural change in their markets caused by Japanese competition. Why didn't Roger Smith quickly "do something" different rather than let GM's market share erode? Only fifteen years ago did a handful of world class manufacturers embark upon a journey called lean production, now considered a best practice. Shattering old axioms, total quality management and just-in-time logic are changing management mindsets entirely.

Industry leaders are now making a call to arms for a new production logic, called agile manufacturing. They hope to use their manufacturing capabilities to deluge the marketplace with a wider variety of higher value products.

Beyond that is strategic agility—the ability to leverage enterprise-wide operations to turn on a dime, providing the right product at the right price,

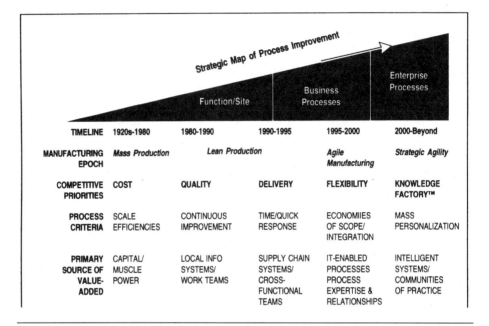

Exhibit 1. **Evolving management perspectives**
Source: Aleda V. Roth, "Neo-Operational Strategy: Linking Capabilities-Based Competition with Technology," in *Technology Handbook,* ed. G. Gainer (New York: McGraw-Hill, 1996).

anywhere. In other words, the pacesetters will transcend the functional manufacturing boundaries to develop fluid, enterprise-wide operations in order to be globally agile.

Competitive Priorities

The decision calculus for manufacturing investment choices is typically associated with competitive priorities—the manufacturing-related capabilities that win new customers and retain them. In a competitive world there is little room for waste. An emerging body of research suggests that the dominant set of competitive priorities, especially within the industrialized countries, is tied to manufacturing epochs. The strategic importance of investing in epoch-based capabilities lies in its contribution to building sustainable competitive advantage and superior profitability.

Using mass-production logic, investments are made by trading-off competitive capabilities, e.g., quality or cost or flexibility. Following a product life cycle, manufacturing priorities shifted as volume increased, such as from quality to low cost. Lean production logic differs. Quality, delivery, and costs are viewed as complementary capabilities, not to be traded-off for one another. Within agile manufacturing, flexibility—the ability to make product variety with high quality and reliable deliveries at a competitive price—is of paramount importance.

Strategic agility requires a metamorphosis from the organization as mechanistic "working machine" to the "Knowledge Factory®"—an organic, accelerated learning organization that produces shared knowledge as a key by-product. The conceptual analogy is the way that a breeder reactor generates more fuel than it consumes. As operations evolve from physical production of products to the manufacture of invisible, intangible assets such as services, knowledge, and ideas, work becomes largely cognitive, organic, and virtual.

Thus with each manufacturing passage, past investments must frequently be treated as sunk costs as technologies, knowledge, and skills are dramatically redefined. Moreover, a firm's relative position on the strategic map determines its potential to leverage generic capabilities through operations processes at the site, the business unit, or the enterprise level. For example, manufacturing priorities of quality and delivery are typically implemented at the plant. Here production benefits from more localized process redesign projects that build upon its manufacturing process knowledge. By the same token, integration of business unit processes is necessary to realize flexibility. Attaining strategic agility means spanning national boundaries in order to leverage a portfolio of global assets. To do so, firms must rationalize resources and processes on an enterprise-wide basis.

> Strategic agility requires a metamorphosis from the organization as mechanistic "working machine" to the "Knowledge Factory.®"

Process Performance Criteria

Evaluation of operations process performance varies by epoch as seen in Exhibit 1. Efficiencies from economies of scale, continuous improvement, and quick response are the outcomes of mass and lean manufacturing, respectively. With agile manufacturing, the business unit profits from organizational connectivity. Economies of integration, which are enabled by technology, will enhance manufacturing's ability to work in tandem with other functions, customers, and suppliers.

With strategic agility, process performance is based upon the relative degree of mass personalization—the ability to provide products and services based upon each individual customer's specific characteristics, attributes, needs, and wants. Mass personalization is demarcated from producing for the needs of "clusters" of customers within defined market segments. Take, for instance, the personalized production of Panasonic brand bicycles of Japan's National Bicycle Industrial Company. Each bike is manufactured to fit the customer's exact height, weight, and color preference. Operations must know customers very well and become less narcissistic in their approaches. Evaluation must be based on asking: "What can we do for you?" in contrast to "How well are we satisfying you?"

Sources of Value-Added

The sources of value-added change whenever a dominant paradigm passes into another epoch. Under mass production, value-added was achieved by applying capital and muscle power. In the lean era, quality, local information systems such as just-in-time, and natural workgroups and teams provided significant returns when properly deployed. As the focus shifts to information technology (IT) enabled processes, value-added comes from excellence in supply chain management and cross-functional teams. Tearing down functional silos is key to quick response. Agility means that design engineering and customer services interface through business process management.

For strategic agility, value is derived from leveraging enterprise assets worldwide. Recent research on 1,300 manufacturing business units across North America, Europe, and Japan calls for neo-operations strategy where superior value is derived from knowledge-based competencies, e.g., combining IT and state-of-the art manufacturing processes with technologically competent workers and intelligent organizations. Without question, revolutionary—not evolutionary—changes in management mindsets and practices are necessary. Managers must shift their thinking from technology-based competencies to knowledge-based portfolios—enterprise-wide intelligent systems and knowledge workers joined in "communities of practice."

> Managers must shift their thinking from technology-based competencies to knowledge-based portfolios.

Value is created by knowledge workers (who include managers, engineers, technicians, and information workers and specialists) employing increasingly powerful, versatile, and user-friendly advanced technologies to enhance their productivity. The development of communities of practice—a network of knowledge workers and workgroups who carry out the actual work, whether or not they formally fit into the proper organizational boxes—enabled by advanced technology, will be an essential means of sharing expertise.

Take for instance, Oticon, a Danish manufacturer of hearing aids. At Oticon there are no "fixed" jobs. President Lars Kolind is quoted as saying: "In my experience, it takes a revolution to change a traditional, machine-like organization into one that has a high level of flexibility and the ability to advance knowledge." Within two years, Oticon evolved into a so-called "spaghetti organization," which Kolind describes as a chaotic tangle of interrelationships and interactions. Oticon's knowledge workers bring clusters of skills to the job; they bring the intelligence behind the computers, telecommunications, and automated systems; and they dynamically reconfigure themselves into "virtual" communities of practice. People co-locate seamlessly between projects and tasks as required. Seamlessly, they convene, address a problem, and disperse as needed. The bottom line results are increased profits, productivity, and innovation.

A Progression of Knowledge-Based Competencies

As strategic thinkers, managers must understand how generic factors critical to success arise. By definition, these factors are among their portfolios of current competitive capabilities and those needed in the future. Empirical evidence supports a new theory of competitive progression. (See Exhibit 2). Accordingly, generic capabilities are developed sequentially and are interdependent over time, regardless of intended priorities. Roth's theory (1996)

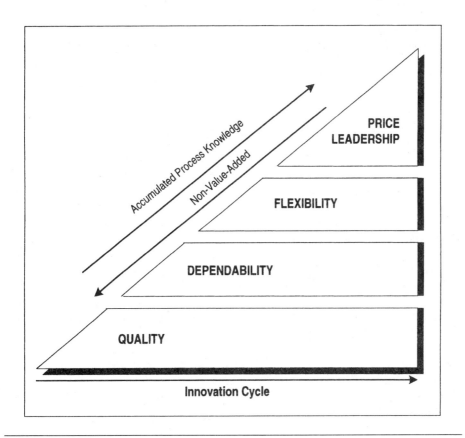

Exhibit 2. **A progression of knowledge-based competencies**

explains why generic capabilities observed within manufacturing firms accumulate in a progression—from quality and then to delivery and then to flexibility and then to price leadership during an innovation cycle.

Quality is first, not only because it is the ante for competing, but because it increases process predictability and increases organizational learning. Mastery of total quality management (TQM) as a new business science and deployment of its principles provides a foundation for organizational learning. As people are permitted to practice quality in daily work, they gain new social skills along with improved functional and technical expertise. TQM requires them to come together, communicate, dialogue, and share expertise. A quality umbrella can ready the organization for ongoing change.

Delivery reliability expands local process capabilities. Fast and reliable delivery is possible only after a company can control its product quality. Delivery builds upon product quality by expanding the need to share technical information and process knowledge cross-functionally and along the supply

> A quality umbrella can ready the organization for ongoing change.

chain. For reliable delivery, manufacturers must be able to provide the same level of support and service worldwide. Increased predictability of distribution processes is essential. Because products and services usually have some "local" contents (i.e., addressing market-specific, environmental, safety, energy, and cultural issues), understanding distribution processes augments the firm's knowledge portfolio.

Flexibility shatters organizational boundaries. Quality and delivery pave the way for flexibility due to improvements in subsystem predictability and know-how. Now manufacturers must come to grips with larger, unwieldy business unit processes. Business process redesign and innovations take out more organizational slack. Proverbial fire walls at the boundaries are destroyed in order to get closer to suppliers, partners, and especially customers. Flexibility requires an inordinate closeness to customers. Intimate customer knowledge provides opportunities for unprecedented cross-organizational process integration. Unfortunately, few companies deliberately build capabilities on long-term customer relationships.

Price leadership is the final step in the progression. Price leadership means that the firm has significant flexibility in pricing because of lower manufacturing and distribution costs. It occurs only when product and process innovations reduce costs over a product family life cycle and controllable costs are sustainable, or declining. For global players, total price leadership is all about leveraging the entire enterprise and its network of suppliers, customers, partners, and facility locations. Price leadership is not a one-time event but requires a clear understanding of enterprise-wide processes underlying quality, delivery, and flexibility.

Barriers to Growth in the Knowledge Era

Interestingly, the competitive progression mirrors the macro evolutionary pattern depicted on the strategic map (Exhibit 1). Having higher levels of combinative generic capabilities is the essence of strategic agility, and the most cost-effective path of accumulating generic capabilities follows from competitive progression theory. Competitive progression theory is the conceptual linchpin between manufacturing epochs and the evolution of the enterprise as a whole.

Combined competitive capabilities are broadly associated with the changing organizational architecture, technological progress, and roles of people—both employees and customers. For example, within a Knowledge Factory, the customer becomes a collaborator in the production process. Customers, supported by intelligent technologies like virtual reality software, participate in the design and production of their own goods and services.

The primary barriers to progress toward strategic agility are a lack of skilled human resources and management expertise to deploy these resources. Technology will not be the issue because all firms will have access to comparable technology. The question will be: How do we build the thoughtware, and how do we gain the organizational knowledge?

> Within a Knowledge Factory, the customer becomes a collaborator in the production process.

The answer is intuitively simple, but not easy to implement. Firms must work through the competitive progression faster and, thereby, accelerate the number of innovation cycles. The Knowledge Factory® model shows that this is possible by establishing deliberate strategies and a predisposition for learning.

People Competencies. Managers must realize that knowledge is a socially constructed phenomena. Learning requires collaboration and sharing of insights. Xerox Parc, for example provides convenient places where people can get together routinely. John Seely Brown of Xerox calls it the "distributed coffee pot" phenomenon. Communities of practice are self-reinforcing at problem solving. Employees need to work together, build, and cross-fertilize the enterprise knowledge bases. Huge knowledge gaps exist in most organizations as companies using industrial-age logic continue to segregate workers. Accelerated learning requires deliberate opportunities for face-to-face communication, collaboration, and periodic co-location.

Management Competencies. The second key to achieving organizational knowledge is through expanded management competencies. The industrial logic that brought successful companies to the 1980s was built upon models of functional excellence. Today there is a generation of managers who are really "incompetent." They remain vested in the past. Blind to the sea changes ahead, many are complacent in their own professional niches and individual comfort zones; others are uncomfortable with the new environment; and yet others, fearing the tumult of downsizing, are playing out an end-game. Ironically, for some of these managers, who excelled by 1980s standards, the financial signals today still indicate that everything is OK. Yet in the turbulence of the 21st century, major changes will come quickly to the unprepared who are waiting for their proverbial "ship to come in" rather than swimming out to meet it.

The lack of preparation for the challenges ahead creates a drag on institutional systems, either by pigeonholing the organization into traditional methods or managing by fads. Barriers to developing new managerial competencies are exacerbated by the rapidity of change, which leaves little time to react, and by the reinforcement of old behaviors by the short-term orientation of management compensation systems.

In the 1980s, the "best" managers were identified by their functional expertise such as finance, sales, or production, and their ability to "command and control" their operations. (See Exhibit 3.)

Today, managers need different, broader competencies—business unit process expertise, at the least. They must understand and value team-based, cross-functional management. And as firms move to the 21st century knowledge era, the best of breed managers will need enterprise process expertise. They must be able to "see 'round the world." Because the bodies of scientific know-how and technological progress are evolving radically over time, development of economies of knowledge are critical to managerial prowess. Consequently, the investments in management renewal must be an ongoing process.

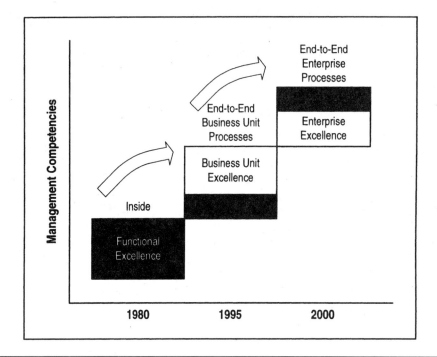

Exhibit 3. **The evolution of management competencies**

References

Craig A, Giffi, "The New Manufacturing Strategy," *Harvard Business Review* (May/June 1994): 154–55.

Polly LaBarre, "The Dis-Organization of Oticon," *Industry Week* (18 July 1994): 23–26.

M. J. Piore and C. F. Sabel, *The Second Industrial Divide: Possibilities for Prosperity* (New York: Basic Books, 1984).

Aleda V. Roth and Craig A. Giffi, "Winning in Global Markets: Neo-Operations Strategies in U.S. and Japanese Manufacturing." *Operations Management Review* 10:4 (1995).

Aleda V. Roth, Ann S. Marucheck, Alex Kemp, and D. Trimble, "The Knowledge Factory for Accelerated Learning Practices," *Planning Review* (May/June 1994): 26–33, 46.

John Seely-Brown, "Research that Reinvents the Corporation," in *The Learning Imperative*, ed. R. Howard, (Boston: Harvard Business Review, 1993).

Aledo V. Roth is an associate professor at the Kenan-Flagler Business School of the University of North Carolina at Chapel Hill. Her co-authored books include Competing in World Class Manufacturing, World Class Banking: Benchmarking the Market Leaders, *and* The Management of Continuous Improvement: Cases in Health Administration.

Cellular Manufacturing for Small Manufacturers
A Practical Approach

Stanley D. Stone

This article documents the transition to cellular manufacturing at Electronic Hardware Corp., a supplier to the aerospace industry. This company was on the ropes, and management saw the move to manufacturing cells as key to the company's survival. EHC revitalized itself by its actions, which included TQM techniques such as kanban scheduling, SPC, and other quality-management techniques that go hand in hand with the cellular-manufacturing approach.

Electronic Hardware Corp. (EHC) is a small manufacturer of value-added plastic components consisting primarily of control knobs for the aerospace, industrial and consumer markets. Like many other companies located on Long Island, EHC has struggled to survive defense cutbacks and the lingering recession affecting the New York area.

EHC is a 33-year-old privately owned company employing approximately 90 people in a union environment. EHC utilizes its 20,000-sq.-ft. facility to produce approximately $6 million in sales. The company was founded in 1962 as a supplier to the burgeoning defense industry and continued to grow and prosper until the mid-1980s.

As defense spending decreased, so did profits. By the early 1990s, the company had downsized and began to struggle for survival. Cash flow was poor, accounts payable stretched beyond 100 days, suppliers were beginning to cut off raw materials and the company had maxed out its credit. EHC needed to improve cash-flow and fast! The company had to make radical changes to the way it manufactured products and handled inventory. The following is the actual step-by-step process which began in 1993 and is credited for turning the company around.

The Cell Project Team

Selecting the team. The team should consist of those individuals who will be ultimately responsible for the implementation of the project. Functions represented should include top management, manufacturing, engineering, quality and material management/production control. A group consisting of five to eight people is usually appropriate to achieve the proper group dynamics (sub-teams can be formed as necessary to tackle specific tasks). At EHC, the primary team consisted of seven people: president, director of manufacturing, engineering manager, quality manager, two manufacturing managers and the production control manager.

Developing a cellular manufacturing environment will require a firm commitment from upper management and everyone on the project team. It is a radical re-thinking of how the company conducts business, and the team may encounter many obstacles including: resistance to change, the re-allocation of resources, language/communication barriers, and re-training and cross-training of employees, supervisors and (most importantly) managers. Existing paradigms will be dispelled as the team learns to "think outside the box."

> Developing a cellular manufacturing environment is a radical re-thinking of how the company conducts business.

At EHC the transition to cellular manufacturing was the top priority—it had to be. The team believed that the cell approach would dramatically improve efficiency and lower manufacturing costs.

The primary team met weekly after-hours for three-hour "skull" (brainstorming) sessions. Additionally, the team held monthly status meetings on Saturdays. No one on the team had any prior experience implementing cellular manufacturing or kanban, so self-education became the first order of business. The team allocated the first 30 minutes of each meeting to self-training. This included a group study of videotapes, books and articles that were available on the subject and periodic mentoring from a local consultant.

Secondary teams were commissioned as needed to work on special projects such as the factory layout, flowcard design, training and kanban.

EHC had little or no budget to finance this undertaking, so the cell team had to be very creative. Improvising became a way of life in order to provide the resources necessary to proceed. Existing machines and tooling were cleverly reconfigured to operate in the cells.

Sales and Product Analysis

The key step for any company considering cellular manufacturing is performing a sales/product analysis. Since many companies produce diverse product lines it is necessary to determine which products will produce the greatest economic benefit if converted to cellular manufacturing. Typically these are products produced in a repetitive or continuous flow fashion.

The analysis determines which manufacturing processes the products have in common. Products are sorted into "family groups" by common characteristics such as the manufacturing routings and bills of materials. This step can be fairly easy if the company uses intelligent part numbers or assigns manufacturing categories to part numbers.

Once product family groups have been identified, the pareto technique can be used to identify which product families will provide the "biggest bang for the buck," that is, the products that run at the highest volume, have the most processes and components in common, and generally seem to be good candidates (sufficient volume) for cell manufacturing.

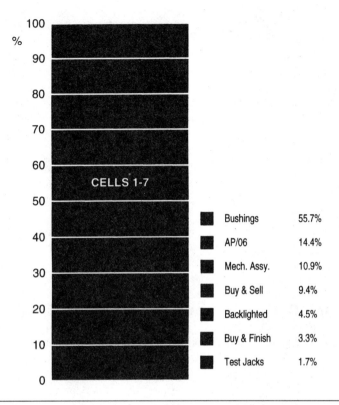

Figure 1. **Weekly shipments by dollars, January–June 1993**

The analysis of EHC products (see Figure 1) revealed significant links between many products which were considered otherwise discrete. The sales analysis revealed that 60 percent (sales value) of the products produced had very similar attributes. A common denominator between these products (knobs) was they used a similar component (a bushing) and basically followed many of the same manufacturing steps.

Therefore, all knobs with bushings were now considered to be part of the same family—the bushing family. Because of high volume in both sales and units produced, the team agreed to pursue the bushing product family for the first cell project since it offered the greatest economical benefit.

Cell Design

The bushing family shared many of the same processes and used similar types of equipment. Even though products in the family were comparable, they were not identical, so additional processes and equipment needed to be considered for the cells.

> The most desirable approach to cellular manufacturing is to build the entire product—start to finish—in the same cell.

The most desirable approach to cellular manufacturing is to build the entire product—start to finish—in the same cell (this was the approach EHC would take). If the entire product cannot be built in one cell, the approach should be to assemble a series of cells each building a logical subassembly and feeding a next-higher-assembly cell.

Determine the product family equipment needs. To accomplish this, a process matrix is developed to match each process to the equipment required to perform the process. By using the matrix, common and unique equipment can be identified for each product to be run in the cell.

Next, determine how many cells will be required (or can be built with the existing equipment) to manufacture the entire family group. To do this, two things are needed: the projected capacity of a cell and an inventory of the existing equipment.

Capacity analysis. At EHC, a detailed time study of each process used by the bushing family group was performed. In this case between six and 10 operations were required to manufacture a knob. Since a cell is intended to function as a continuous flow process, the slowest operation (constraint) will determine the throughput of the cell. (The troop cannot march faster than the slowest soldier.) Once the slowest process is determined, it is easy to estimate the capacity (rough cut) of a cell. With the approximate capacity established, the number of cells required and their configurations can be determined. To improve the capacity of a cell, begin with increasing throughput at the constraining process—if possible.

Equipment inventory. Once the equipment has been identified, an inventory must be taken to determine which assets are available for the cell(s). EHC had no formal asset inventory, so the company took this opportunity to photograph each piece of equipment and affix a numbered inventory tag to it. A specification sheet was completed for each item inventoried to identify attributes such as physical size, power, water and air requirements, as well as the general condition of the equipment. The photograph was attached to the specification sheet and filed in a three-ring binder. This information proved to be indispensable during the cell design and layout stages. If equipment needs to be purchased (and some may), this is a good time to prepare the budget or get creative.

At EHC it was calculated that seven cells were required to meet the demand for the bushing family. Each cell would be staffed with two operators. All seven cells were designed to carry out all the manufacturing steps necessary to build a complete product—in this case, from a molded plastic shell to a completed knob packed and ready for immediate shipment to the customer.

Cell Implementation

Project management. A project of this magnitude requires conscientious project management. This was accomplished at EHC by defining the critical action steps (milestones), planned completion dates and the individuals who would be responsible. All this information was entered on a Gantt chart which was reviewed at the weekly cell meetings.

The prototype. At EHC, a prototype cell was designed and assembled for use as a "test bed" to prove out the cell concept and to expose any potential problems that had not been anticipated. The cell was arranged in a "C" configuration which would support two operators, one working from the inside and another working on the outside. The throughout and the quality of the knobs produced on the prototype cell exceeded the expectations of the cell team and met with high marks by the cell operators and the customer.

Once the prototype cell was in operation, the team was able to obtain valuable feedback from the operators. Cell operators offered numerous suggestions for improving the process and the ergonomic factors affecting cell performance. Many of these suggestions were incorporated into the design of the remaining six cells.

Plant layout. To accommodate the cells, much of the facility had to be rearranged. We constructed a mock layout and entered it into AutoCAD. Using the software we were able to move entire departments around to come up with the best layout. The actual physical moves were conducted after hours and on weekends so production would not be disrupted.

Set-up time reductions. As the cells came on line, the team realized that changes to the existing tooling fixtures and molds were necessary to achieve rapid tool changes. Fast tool changes were important because our intention was to make the cells flexible and replace large batch production with small lot production driven by kanban. The initial set-up time was nearly one hour and has been continuously reduced to approximately 12 minutes.

Quality. Statistical process control was designed into the cell. Each cell was supplied a kit containing all of the necessary quality tools such as calipers, go/no-go gauges, thread gauges and depth gauges. Operators were taught how to use these tools and, most importantly, empowered to stop production if a process went out of control. This was difficult at first because the previous philosophy had been *never stop production!*

A database was developed to record actual operational performance and used to set reasonable expectancies (R.E.s) for production at each cell. It was decided that a bonus system would be used to reward the operators for meeting or beating the R.E. To achieve the bonus, the operators must work as a team and balance the line completing as many parts as possible. The bonus system at EHC has been successful and operators have earned an average of six percent over their base pay.

Kanban

Kanban. Kanban is a Japanese word which literally means card. In the Toyota production system, this "card" is used to initiate the "pull system" for manufacturing scheduling. Kanban/pull systems are used as a scheduling technique and to eliminate the "waste" of overproduction. At EHC we implemented a kanban system using color-coded bins instead of cards. The system is used to supply the cells with component parts as needed to manufacture knobs. These component parts include molded plastic shells produced in-house as well as parts purchased from suppliers.

EHC uses a three-bin kanban system; two bins are located at the point of use and one bin is located at the supplier (internal or external). The kanban system uses yellow bins which have sample parts attached to them for fast part recognition by the operators.

As soon as a bin is emptied (each bin contains enough parts for one shift) it is returned to the internal supplier and a reserve bin is taken and returned to the point of use. The empty bin signals the supplier to make new parts and once the bin has been filled, the supplier stops production of this item (no waste). The supplier has 24 hours to complete the transaction. The second bin at the point of use is extra—it is kept as emergency buffer stock—in case of machine breakdown.

All primary suppliers were brought on board before the program was started and agreed to support the Just-in-Time (JIT)/kanban system. When a bin containing a purchased part is emptied, it is brought to the receiving department. Next, the operator faxes a kanban request form to the supplier. The supplier immediately ships replacement parts to EHC using a next-day delivery service. Once the replacement parts arrive, they are placed into the empty bin and returned to the point of use. The operator who ordered the parts will now inspect them and put them into the kanban system. Note: The operator performs both the purchasing and incoming inspection functions for this transaction.

Visual Factory

At EHC, the seven new cells now manufacture products accounting for approximately 60 percent of total dollars shipped. Because of the financial impact, downtime and missed schedules can create a very serious problem.

In the beginning, measuring cell downtime and throughput was difficult because of the lack of systems and timely information. Because the computer system at EHC was "state of the ark" (not "state of the art") technology, it took a minimum of 24 hours to get feedback on cell performance. If problems occurred (and they did), information was received too late to take corrective action.

Resolution

A red warning light was mounted above each cell to be used to indicate that the cell was experiencing a quality problem, production problem or delay (including repairs or machine set-ups). The cell operators were empowered to stop production if the process generated defects. The warning light now provides instant feedback to the manufacturing and quality managers who take immediate action to resolve any problems.

To provide "real-time" information on the throughout performance of each cell, inexpensive dry-erase boards were mounted on the cells. These boards are used to display the hourly performance of each cell. Schedules are set based on the reasonable expectancy established for the particular product running. The schedule is cumulative over the entire eight-hour shift and the operators enter their actual production for each two-hour time segment. If a cell is on or ahead of schedule, the operators enter the production number using a green marker. If the schedule is not achieved, the operator uses a red marker.

All cells and schedule boards face the same direction so factory managers and floor supervisors can obtain a "status at a glance" and offer help if needed. The schedule boards worked so well in the cells that this concept was extended to many of the other non-cell operations. The results have been excellent—the operators know what is expected of them and management knows what is happening.

Prior to the cell project, various size and color bins were used to contain the work-in-process (WIP). This was very confusing and unproductive. WIP (especially rework) was lost for weeks at a time! The team instituted a new system which standardized the size of the bins used and established a color code for the bins. A three-color system was adopted: yellow, blue and red. Yellow bins were designated for kanban use only. Blue bins were used only for standard parts not on the kanban. Red bins were designated for rework or customer returns only. At a glance, the mix of work on the EHC factory floor can now be assessed and rework can no longer hide!

Education

Training. Training should begin at the start of the project and never stop! Within six months all seven cells at EHC had been placed into operation. Training became a major focus for the cell team. Operators who previously had only one simple task (such as working a drill press) were now responsible for several operations (such as drilling, taping, reaming and statistical process control). To complicate matters, most of the operators spoke little or no English.

An English as a Second Language (ESL) training program was instituted—and continues today. The operators were educated on JIT, kanban and statistical process control (SPC). An SPC system was put into place and the operators were furnished with calipers, depth gauges, and an assortment of go/no-go gauges. All cell operators were trained how to use these instruments.

Moving On

Repeat as necessary. Once all of the previous steps have been completed, repeat them as necessary for all the remaining product families.

Measurable Benefits

The implementation of cellular manufacturing, kanban and the simple visual factory techniques has had an astonishing impact on EHC. Here are some of the measurable benefits accrued to EHC in 1994, the first year following implementation:

- Work-in-process was reduced 55 percent.
- Manufacturing cycle time was reduced from five weeks to less than one week.
- Quality (based on customer returns) was reduced from 4 percent to 1.5 percent.
- Production efficiency was improved from 39 to more than 60 pieces per manufacturing hour.
- Sales increased by 36 percent.
- The morale and attitude of the operators increased.

The most important benefit was the positive cash-flow which was reinvested and used to pay down creditors. In 1994, EHC showed a profit for the first time since 1987!

In 1995, EHC has continued to make improvements. Small molding machines are being tested in cells and will eventually eliminate the inventory of all plastic components—including kanban. EHC's financial situation is dramatically improved. And thanks to the gainsharing program implemented in January, the employees are now sharing in the success that they have created.

Stan Stone, CPIM, is director of manufacturing operations for Electronic Hardware Corp. in Farmingdale, New York. He has 22 years of production control, systems development, and operations management experience in large and small companies.

Services Sector

The Value-Driven Bank

Erick Reidenbach and Terry C. Wilson

Banks, like any service organization, prosper to the degree they efficiently deliver value to their customers. This article talks about the idea of customer value and presents six "value propositions" that banks and any service organization should use to drive strategic business decisions and actions. Though the article is about banking, read it in relation to your own business; it is equally relevant.

Perhaps not since the Depression era has the plight of the banking industry been so unsettled. The invasive competitive intrusions from nontraditional competitors; a regulatory policy that is antiquated and arguably biased; a declining share of financial assets that has reached a low of 25% (down from 71% in 1960);[1] and bank closures, mergers, acquisitions, and consolidations that are sure to eliminate as much as 30% of banking jobs mark this industry as one in decline.[2] Is this decline inexorable? Is it immutable? The answer is No. For banking to survive and prosper in the ensuing years will require a Herculean effort to turn the industry around, one bank at a time. The prescription for this turnaround is found in a new philosophy of business—one based on the idea of injecting value into the customer relationship.

In Search of a New Philosophy of Business

Every business has a philosophy of how it operates. This philosophy is either explicit or implicit, but it nonetheless directs the organization's relationship with its markets and its customers. The degree of fit the organization enjoys with its customers is dependent on the congruency of the philosophy of business with the operating reality of the market. The philosophy of business creates this fit by dictating a bank's structural configuration and relationships,

> The degree of fit the organization enjoys with its customers is dependent on the congruency of the philosophy of business with the operating reality of the market.

Reprinted with permission from *The Bankers Magazine*, September–October 1995. Copyright © 1995 by Warren, Gorham & Lamont. All rights reserved.

its communication and power systems, and its reward systems. All of these systems, in turn, align the bank's relationship and its approach to its customers, its markets, and its competitors.

The current business philosophies of most banks do not fit with current operating realities. These philosophies can be described as a hybrid—an amalgam of business philosophies (itself composed of remnants of a bankrupt production mentality), blended with partial tenets of a sales philosophy, and sprinkled lightly—and we mean very lightly—with elements of a customer orientation. Blended together, these ingredients constitute a strategic recipe for failure—one that is sure to hasten the decline of many banks.

This hybrid business philosophy is not capable of dealing with the current operating realities facing banking. The very assumptions embedded within these antiquated business philosophies and their operationalizations have created the problems plaguing the banking industry. For example, the production philosophy assumes a condition in which the demand for banking products and services exceeds the supply. This is patently obsolete given the ever-increasing number of both traditional and nontraditional competitors. Equally ineffectual is a sales philosophy that says that market share or asset size can be increased by increasing the selling effort. While this may provide a short-term fix, it ignores the most compelling operating reality of all. Customers, and the way they relate to banks, have changed.

To solve these problems—to implement a customer value-driven philosophy of business—requires a new paradigm or business model. This article describes this new value paradigm and how it can be implemented within the banking industry. The discussion of the value paradigm begins with an examination of the concept of value and of what are called "value propositions," which form the underpinnings of the new paradigm. In conclusion, what is required to make this value paradigm operational is considered.

Customer Value: A Business Paradigm for the 1990s

Although the idea of creating customer value is not new, it has, however, been forgotten or, at least, relegated in importance. In the midst of all the new business fads (e.g., TQM, reengineering, customer satisfaction), the idea of delivering outstanding customer value was lost. Fortunately, it has been rediscovered.

Value can be defined in a number of ways depending on what type of value is pertinent. For example, there is the notion of value added and there is intrinsic value, incremental value, value trade-off, additive value, or compositional value. All these types of value are important and have relevance, depending on the industry in which they are used. For the purpose of banking we are talking about a perceived value and accordingly define this value as the ratio of perceived customer benefits to the relative price the customer pays to receive those benefits, a value-in-exchange.

Customer Value = Benefits/Price of Benefits

The greater the benefits perceived by the customer relative to the price the

THE VALUE-DRIVEN BANK

	Relative Quality		
Relative Price	**Low**	**Average**	**High**
High	Inferior Value	Poor Value	Average Value
Average	Poor Value	Average Value	Better Value
Low	Average Value	Better Value	Superior Value

Exhibit 1. **Strategic value options**

customer pays, the greater the perceived value. In its simplest configuration, using service quality as the global benefit and the price the customer pays to receive this quality, a number of value options become apparent. Exhibit 1 illustrates these potential value options.

Banks that are offering their customers average quality at an average price can only hope to leverage an average value option. Average value will not stem customer migration to alternative financial service providers offering greater value. Banks offering less than average quality at an average price are competing using a poor value position, an untenable long-term position. It is only those banks that can offer high quality banking services at a low relative price that will find themselves in a superior value position, one they can leverage for superior performance.

There are two other critical points to be derived from Exhibit 1. First, value is relative. It is relative to the perceived quality of services and products offered by financial service competitors, and it is relative to the perceived price the customer pays for those services and products. This is an important point, because many bankers tend to discount the importance of competitive offerings, focusing only on the quality of their own product/service mixes.

Second, value is perceptual in nature. It is absolutely unimportant and irrelevant what the bank management thinks of its current value offering. What is important is how the customer, both actual and potential, perceives the value offering relative to other offerings. Value is in the eyes of the customer.

Value is in the eyes of the customer.

Value Propositions

There are a number of value propositions that form the underpinnings for this new value paradigm. These propositions, if properly implemented, will allow banks to transform their current philosophy into a value-based philosophy, the first step to leveraging superior performance.

Value Proposition 1

> The existence and perception of value are necessary conditions in any exchange situation.

The existence and perception of value are necessary conditions in any exchange situation. This is basic economics, and being basic, is worth remembering. Exchange entities will agree to transact if, and only if, each entity perceives the existence of value in the outcome of the transaction. Customers will build a relationship with your bank if, and only if, they perceive value in the relationship. Conversely, there are a number of different potential relationship opportunities that banks may choose to forgo because there exists no value for the bank to engage in the relationship. Many banks are beginning to understand that some market opportunities are not potentially profitable, a basic measure of value to the bank. Consequently, to be all things to all customers is not prudent.

Value Proposition 2

Some banks will be better able to identify, create, and maintain value opportunities than other banks. Value Proposition 2 is the driving force behind success or failure in the marketplace. It is an undeniable fact of competition that some banks will be better than others at the essential tasks of value competition. Those banks that master the skills and strategies of value creation will be able to leverage this competence into greater performance and less vulnerability to competitive intrusion. Those banks that continue to do business under the old philosophies or paradigms will find themselves fodder for the merger/acquisition mill.

Value Proposition 3

Value opportunities are product/market specific. What is valued by an elderly couple looking for a private banking relationship will not be the same as what is valued by a young couple seeking a mortgage product. What is valued by an aggressive investor will not be the same thing valued by a conservative saver. To ignore the value success requirements within diverse product/markets creates a homogeneous value offering, one that will have little if any appeal to different customers.

Each product/market has embedded within it certain success criteria for value creation. Each product/market represents a value arena in which the bank competes with its value offering against other banks, car companies, retailers, software manufacturers, and brokerage houses. To survive and prosper in these value arenas requires a sophistication in marketing planning, enabling the institution to identify value opportunities more clearly, to create value more closely aligned with customer demands, and maintain those value opportunities provided within each individual product/market more effectively than competition. This process of identification, creation, and maintenance of value requires a level and skill for market planning that is not well established in most banks. Unfortunately, without it, no bank can expect to establish a differential value advantage in any product/market.

Value Proposition 4

The bank that can identify, create, and maintain a differential value advantage will be in a stronger position to leverage that value advantage into superior performance. Absolute emphasis on customer satisfaction is misplaced. Customer satisfaction is not correlated with profitability. Customer retention is. Customer retention is directly related to customer satisfaction, which is an evaluation of the relative value provided the customer by the bank. The "customer value-customer satisfaction-customer retention-profitability-shareholder value" linkage is a crucial one.

> Customer satisfaction is not correlated with profitability. Customer retention is.

The difference between customer value and customer satisfaction is an important distinction, as the following example shows. A customer may choose a specific mortgage option from a particular bank, believing that that mortgage fulfilled his or her purpose. The benefits of the mortgage were well worth the price of the mortgage. This is value. However, when comparing mortgage options across banks, that same customer may find better options elsewhere. This comparison would lead to a condition of dissatisfaction for our hypothetical customer and an increased likelihood of defection. By providing the greatest mortgage value, your bank stacks the comparison process in its favor and eliminates the potential for dissatisfaction.

How much profit is created by a customer who values a relationship with an organization? Reicheld and Sasser estimate that reducing customer defections by 5% can increase profits between 25% and 85%, depending on the industry.[3] Kmart calculates a loyal customer to be worth about $75,000,[4] and Sewell Cadillac in Dallas estimates a loyal customer to be worth about $300,000.[5] How much is a customer who values his or her banking relationship worth to a bank?

Consider what we know about the economics of customer retention. In *Customer Service Renaissance: Lessons From the Banking Wars*,[6] it was pointed out that a customer who finds greater value elsewhere typically takes with him or her on average a three-product relationship and a deposit balance of $23,000. Moreover, according to Luke Helms, former CEO of Seafirst and now with Bank of America, customer acquisition costs run about three to five times the costs of customer retention.

A May 1990 report by the American Banking Association further demonstrates the impact on bank profitability of not providing valued service to bank customers.[7] The example involves a bank with 200,000 customers. Ten percent of their customer base switches banks every year for various reasons. Of this customer migration, 21% leave because they are receiving value-denigrating service levels. For each lost customer, this bank projects a negative impact of $121 gross profits/year. When this is contrasted with a customer acquisition cost of $150 per customer, it becomes possible to calculate the annual cost of value-denigrating service. If 20,000 of their customers are migrating to other financial services institutions (0.10 × 200,000) and 21% of this migration is due to poor service, a full 4,200 customers are defecting because they are not receiving adequate value in their banking relationship. At a cost of $121 in gross profit per customer this adds up to $508,000 in loss profits. Added to this is the cost associated with the acquisition of 4,200 new customers to replace the defections. This accounts for an additional $630,000 (4,200 × $150) or a total bottom-line impact of lost profits of $1,138,200. This is the cost of value-denigrating service to this bank. The customer value-customer retention linkage has a direct and significant impact on bank profitability. Value is the cement that links customers to profitability and shareholder value?

Value Proposition 5

What constitutes value today will not necessarily define value tomorrow.

What constitutes value today will not necessarily define value tomorrow. Current bank-delivered value is not consonant with the expectations and needs of contemporary banking markets. Customers, both retail and commercial, big and small, are finding better value elsewhere. Customer definitions of value have changed, but the industry's perception of what constitutes value has not. To punctuate this point one needs only to look at what is happening with respect to the bank credit card business.

Credit cards have been historically, one of the most, if not the most, profitable products for banks, earning about a 5% return on assets across the industry. Enter AT&T with their Universal Card. They took advantage of a complacent industry by linking sophisticated marketing with a positive image for outstanding service and operational excellence to create a unique and unsurpassed value offering. The Universal Card rocketed to success based on what AT&T correctly perceived to be a shift in value definitions unheeded by traditional competitors. Currently, it ranks second in the industry with more than 20,000,000 cardholders who account for more than $11.1 billion in receivables.[8] Customer definitions of value change and the bank that can anticipate these changes will be the bank that scores.

Value Proposition 6

The bank that can identify, create, and maintain a differential value advantage will be less vulnerable to competitive pressures. The bank that can provide its customers with superior value will be less susceptible to competitive intrusion. Our studies indicate that the bank that can create the greatest perceived value is also the bank with the fewest customer defections. It is the bank that can leverage the dynamics of the customer value-shareholder value linkage to its optimal level. This is a core competency that your bank must develop.

Becoming a Value-Driven Bank

How does a bank that finds itself operating under the conventional production/sales philosophy become a value-driven bank? How can a bank take advantage of the customer value-shareholder value linkage?

There are no quick-fix solutions, no nostrums for change, no magical incantations. There is only commitment and dedication to the idea of identifying, creating, and maintaining customer value. We offer the following brief description of the key components of this change process, a process we developed working with a number of other clients, in and out of the financial services industry.

Critical Components and Processes

The change to a value-driven bank begins with the development of a value infrastructure. The three critical components of the value infrastructure are vision, mission, and culture. The bank's vision is an articulation of a desired end state for the bank. Vision tells why the bank exists. Mission tells what the organization does in support of that vision. Culture is the pattern of beliefs and values that give rise to both the vision and the mission, but perhaps more importantly supports these two critical infrastructure components. Without a coherent mix of components within this infrastructure, no organization can hope to compete on a value basis.

If a bank is laboring under a production philosophy, one that focuses on standardization of product/service mixes and cost reduction and treats all customers the same, its infrastructure and market relationship will be shaped under these assumptions. No matter how hard it tries to compete on a value basis it will not be successful. The mental filter of management and employees will be focused on the identification, creation, and maintenance of opportunities that are congruent with the tenets of a production mentality, not on a value basis. The bank that wants to compete as a top value deliverer must do more than simply talk about value—it must eat, sleep, and breathe value.

Value opportunities arise from two basic sources: the environment and customers. Environmental opportunities are sourced in the interplay of technology, competition, the cultural and social fabric of an increasingly global market system, the economy, the legal/political/regulatory environment, and resource supply. The constantly changing nature of these interactive environments creates opportunities for identifying, creating, and servicing customer value, but only for the bank that has an operating value infrastructure.

Customers must be coproducers or codesigners of the very value production system itself. This means that both producers and consumers must know a great deal about each other. This in turn requires that the bank become a knowledge system and cease to view itself as a factory.

Both of these critical environmental components require sensing capabilities that many banks currently do not possess. These environmental components are the source of value opportunities that can be sensed only by the bank that has a value infrastructure in place.

The organization component requires the fitting together of several critical processes and subcomponents that include employees, performance standards, innovation, and networks. Employee concerns involve such factors as effective selection and recruitment, training, teamwork, and retention. Performance standards determine both the quality and the pace of effort your bank puts out. Innovation is fundamental to your bank's success. We hear much of learning organizations. These are organizations that continually learn, unlearn, and then learn again. They are innovative engines. Finally, networks form an organizational web of interconnecting relationships capitalizing on technology transfers and process expertise.

Organizational Value Alignment

While a bank may excel at one or more of these components or processes, the key to organizational success lies in the bank's ability to align, or fit together, forging these critical components into a meaningful whole that is greater than the sum of the individual parts. Only those organizations that have optimally aligned vision, mission, and culture (infrastructure) with organizational components and process (employees, innovation, performance measures, and networks) and ultimately with customers, will be able to achieve a differential value advantage. These are the organizations that can truly be called learning organizations.

Banks have demonstrated a singular inability to align these components and processes. Bank claims of a customer orientation are not aligned with customer needs for access. Bank operating hours are set not with customers in mind but rather are determined by cost considerations. Increased demands on loan officers to sell are not supported by incentive systems designed to reward selling effort. Centralization of holding company decisions retards a branch's ability to provide timely and responsive service to its customers. Several bank clients have offered product/services motivated by a potential cost savings to the bank only to discover that there is little, if any, market for the product/service. The increased costs of marketing a product/service to a segment that is uninterested in the offering wipes out much of the potential cost savings. These are all symptoms of misalignment.

This alignment process is not static but dynamic. Its dynamism takes the form of a continual adjustment process assuring that corporate vision and marketplace reality are connected. In this adjustment process lies the ability of the bank to continually upgrade the value it delivers to its customers.

Cross-functional teams are essential in this adjustment and value upgrading effort. Many financial service products and services depend on some sort of process being completed to deliver the product or service. For example, the mortgage origination process involves a number of related steps in order to be completed. A client bank identified 107 individual steps in this process, from the point at which the customer approaches the bank to closing. If the bank were able to complete each one of these 107 steps with 99% accuracy (i.e., no errors), then only 34 times out of 100 would they be able to complete the entire process without mistakes.

A bank that embraced a value philosophy would view this process as an excellent opportunity to employ a cross-functional team charged with two objectives:

- Identify each point in the process where redundancies and gaps exist. This would enable the bank to provide faster and more responsive mortgage services to its customers; and

- Drive out all costs not associated with the delivery of these benefits.

An error rate of 66% in the origination process means slower, less responsive service to the bank's customers and the additional costs associated with rework, clearing exceptions and disclosure errors, and chasing down customers to complete the work. Mapping the process and focusing on value creation provides an opportunity for the bank to increase its service while at the same time reducing costs and passing these savings on to customers in the form of lower fees. Banks that do not embrace a value philosophy would be less vigilant of, if not blind to, this opportunity for creating value.

Conclusion

Value is what is driving customers today. Southwest Airlines, Taco Bell, Lands End, Mazda, and Walmart are the models that commercial banking must emulate. The old factory metaphor of the production philosophy or the Willie Loman ideal on which the sales philosophy is built cannot solve the current problems of banking.

When Wayne Gretzky was asked what made him so great, he responded, "I'm great because I skate to where the puck will be, not where it is." If banking is to reclaim its stature as an industry, it too must skate to where the puck will be—creating customer value.

Notes

1. George G. Kaufman and Larry R. Mote. "Is Banking a Declining Industry? A Historical Perspective," *Economic Perspectives* (Federal Reserve Bank of Chicago, May–June 1994), 7.

2. Kelly Holland, "Blood on the Marble Floors," *Business Week* (Feb. 27, 1995):98.

3. Frederick F. Reicheld and Earl W. Sasser, "Zero Defections Quality Comes to Services," *Harvard Business Review* (September–October 1990).

4. Conversation with Kmart executives at Kmart Leadership Conference. West Virginia University, May 1994.

5. James L. Heskett, Thomas O. Jones, Gary W. Loveman, W. Earl Sasser, and Leonard A. Schlesinger, "Putting the Service-Profit Chain to Work," *Harvard Business Review* (March–April 1994):164.

6. M. Ray Grubbs and R. Eric Reidenbach, *The Customer Service Renaissance; Lessons From the Banking Wars* (Chicago: Probus, 1991), 8.

7. May 1990 Report, American Banking Association.

8. Michael Treacy and Fred Wiersma. *The Discipline of Market Leaders* (Reading, Mass.: Addison-Wesley, 1994), 64.

R. Eric Reidenbach and Terry C. Wilson are principals in the Valtec Group, Hattiesburg, Mississippi, and coauthors of The Value Driven Bank, *Probus Publishing.*

Improving Customer Service in Call Centers

Dennis S. Holland

This article spells out how to use TQM tools and techniques to improve the operations and processes of any customer-service call center. The author emphasizes the use of process management, empowerment, and continuous improvement to help the center perform better and be more responsive to customers.

The bond that ties customers to companies is no longer a handshake but rather a telephone call.

Over the last decade, the playing field on which companies compete for customer relationships has changed dramatically. Where once customer loyalties were won through face-to-face interactions, they are now being defined by telephone transactions. In short, the bond that ties customers to companies is no longer a handshake but rather a telephone call.

Businesses striving to cut costs, streamline operations, and enhance customer services by implementing call centers must be aware that just as they once worked hard to gain and retain customers through high quality face-to-face customer service programs, they must now strive to optimize the quality of relationships created and managed over the telephone. And just as a combination of competence and trust serves as the basis for high-quality face-to-face relationships, so too are these the critical elements for success in call centers.

This is especially true given the recent proliferation of technologies such as Interactive Voice Response (IVR) and Computer Telephony Integration (CTI). IVR systems that provide automated responses to routine customer inquiries free agents from this workload, but leave them with the task of responding to more complex questions. And, while CTI provides agents access to the computer applications they need to answer these questions, it also broadens the range of inquiries to which they must respond. The results are an agent population pushed, by both customer and company, to consistently do more, faster, better; and a need for call center management to ensure that these ever-increasing expectations are met—on a consistent and continuous basis.

Reprinted with permission of the author. Copyright © 1996 by Dennis Holland. All rights reserved. The article orginally appeared in *The Quality Observer*, April 1996. Dennis Holland is director of marketing for Teknekron Infoswitch, a worldwide provider of customer call-center solutions. For information on Teknekron products and services, call 1-800-835-6357.

By focusing on quality as the competitive differentiator in the call center marketplace, managers now have the means to respond to these spiraling expectations by enhancing the competencies of their agents and fostering trust in their customer relationships. This will turn the telephone transaction into a formidable weapon for their company's long-term competitive benefit.

Convergence Leads Emergence of Quality

The emergence of quality as a competitive advantage did not happen overnight, however. In fact, the practical implementation of quality within the call center is only now possible because of the convergence of business processes, management philosophies, and new technologies. It is this convergence that, for the first time, creates the opportunity for call center managers to turn quality into a system that can be understood, measured, and improved upon—an approach not easily or readily available in face-to-face relationships. Prior to this convergence, call center managers had few methods and tools available to them for the purpose of ensuring and measuring the quality of their customer relationships. And, those tools they did possess seldom provided a fair and representative picture of either the agent's or the call center's performance. For example, even the most visionary of managers relied on isolated approaches, such as real-time monitoring, to capture performance data, simply because it was the only tool available. Unfortunately, the benefits of such approaches were limited, at best, and in many cases created an environment where managers were forced to act more like guards than coaches.

Furthermore, without a fair sampling of an agent's performance on a consistent basis and in a variety of customer interactions, managers could not accurately determine what, if any, training was required. The result was disgruntled agents, dissatisfied managers, and a perception of total quality in the call center as an unattainable goal.

Today, the quality convergence presents new opportunities to move beyond the uncertainty and disappointment associated with such quality initiatives of the past. To leverage the quality convergence to your competitive advantage, let's explore its components—beginning with the process by which to establish and measure quality in the call center, as depicted in the above graphic.

Let the Process Begin

The first step in the quality process is data collection. In this step, every effort must be taken to ensure that the performance data collected is relevant, accurate, and above all, fair. Many call center managers accomplish this by capturing both the "human" element of the calling experience (e.g., the quality of the call) together with the statistical outcomes associated with the agents productivity (e.g., average call times, abandonment rates, orders taken, etc.).

The primary tool used to collect the human element is service observation. As opposed to real-time monitoring, contemporary service observation tools allow for the automatic, random sampling of agent performance over a period of time that is meaningful to the call center. Service observation provides the mechanism to analyze the quality of the caller's experience; determine how agents perform against quality criteria established by the call center; and identify "best practices" on the part of agents that can be replicated or "cloned" across all agents.

The quality data collected through service observation can and must be balanced with productivity measurements to present a true picture of agent performance. Productivity statistics can be easily acquired through a variety of sources available in the call center—from the statistics captured by the automatic call distributor, voice response unit, scheduling applications, mainframe, etc. The bottom line is that this step in the process must be as accurate, as complete, and as fair as possible to achieve the magnitude of potential gains.

The second step of the process, performance analysis, allows for the evaluation of both the quality and productivity data in a manner that is objective and consistent across all agents and agent groups. Using the data collected, trends are analyzed, star performers identified, and areas for improvement pinpointed. As a result, development needs can be quickly assessed, and training programs can be targeted and/or modified to effectively meet those needs.

The final step of the process is the incorporation of the quality and productivity analysis into the development planning process. This dynamic approach to development planning will serve to continually modify the criteria against which performance data is collected, progress is measured, and so on—establishing a continuous cycle of improvement based upon the changing needs of both the call center and its customers. This is illustrated in Figure 1.

Reinforce Process with Relationship Philosophies

It is not enough, however, to simply implement processes. Many companies have tried and failed as they focused only on "process for the sake of process," while not devoting the time, attention, and patience required for these new and oftentimes bold processes to take hold and thrive in their environments. Therefore the philosophies—the second component of the quality convergence—that businesses and call center management exhibit, must be in line with a total commitment to quality as competitive differentiator. Reinforcing this commitment, from the top down, will serve to create a call center environment in which all team members are empowered to consistently excel in their delivery of service to the customer.

The questions, "Why is my philosophy toward quality important?" and "Am I prepared to let quality flourish in my call center?" are typical of those that many managers ask themselves at this point. To determine an answer to these and similar questions requires analysis in two dimensions: 1) What philosophies do I prescribe to in the implementation of quality in my call center?

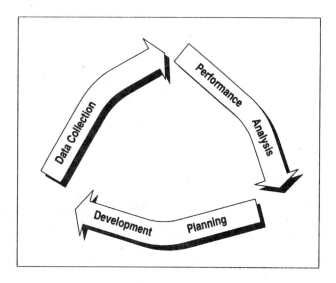

Figure 1.

and 2) What roles do my agents assume in response to these philosophies? By analyzing these two dimensions, call center managers can better understand if they, and the tools they use, are in fact enablers of quality in the call center or inhibitors. The model in Figure 2 illustrates the relationships between management philosophies and the tools they use to reinforce those philosophies, and the resulting role of the call center agent.

Exploring the lower left corner of the model reveals that a management philosophy based on enforcing quality in the call center is typically reinforced with such tools as real-time monitoring. Leaning over the shoulder or "plugging in" to a call often creates a feeling, on the part of the agent, of invasion and, many times, intimidation. A resulting response of compliance—"I'll do it because you make me do it"—is therefore the most that call center managers can expect of their agents, making this an environment where the introduction and incorporation of a quality focus and quality processes is doomed to failure.

On the next level of the model, management philosophies that encourage a focus on quality will likely be supported by such tools as service observation and some level of performance feedback. The less intrusive methods of automated service observation and the more open approaches to communicating the results of sessions promote a response of active participation in quality programs by agents.

The ultimate goal is found at the third level where management promotes a philosophy of empowerment in the call center. Empowerment philosophies assume that an individual agent has the competence to perform his or her job well and can be trusted to do so. This philosophy toward quality is the enabler to a true, continuous quality process such as described earlier and is one that

> Empowerment philosophies assume that an individual agent has the competence to perform his or her job well and can be trusted to do so.

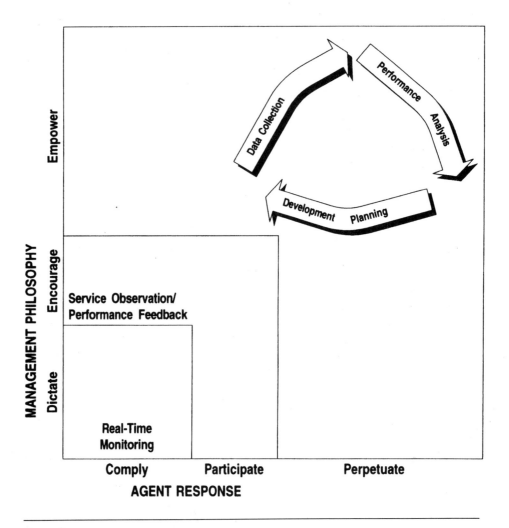

Figure 2.

engenders an agent role that is dedicated to perpetuating quality both in the call center and in the customer relationship. The agent competence and management trust that thrives under a philosophy of empowerment will quickly and easily translate into a customer relationship that is equally based on a combination of competence and trust—once again, the basis for any profitable relationship managed via the phone.

Make Technology Work for You

Achieving the maximum value from quality processes and philosophies in the call center is highly dependent upon the intelligent application of the third

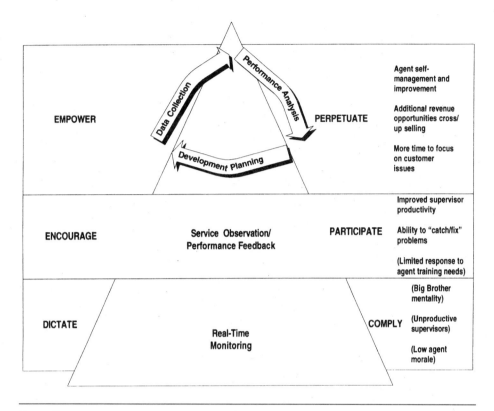

Figure 3.

and final component of the quality convergence, technology. Advancements in technology now allow for the automated scheduling, recording, and playback of service observation sessions, including both voice and screen data. In addition, such service observation technologies can be integrated with performance management tools based on measures of both quality and productivity. These technologies can be easily customized to meet a call center's specific data collection and performance analysis needs, and can be integrated into either existing or reengineered development planning practices. Furthermore, the technology to support such quality processes has proven to be practical as well, with many leading call centers experiencing pay-back periods of as short as six months.

Using technology to enable the quality process will establish the benchmark to which the competition will aspire. Creating an environment in which this process can be perpetuated will set the standard of excellence by which your agents will flourish. And just imagine the impact this alone will have on your business and on your customer relationships.

Bottom Line—Use It or Lose It

With all this talk of a convergence on quality, some final questions must undoubtedly be surfacing—questions such as "What's the bottom line for me and my company?" and "If I just continue with what I'm already doing, will I not eventually reach the same place?" Figure 3 illustrates the total quality picture and its effect on the competitiveness of today's call centers. The key is in determining quickly (i.e., faster than your competition) which benefits will position you not only to achieve but sustain competitive advantage, as well as which risks you're not willing to take on behalf of your customer relationships, and ultimately, your company's future.

In today's competitive markets, often the only thing between you and your customers' loyalty is a phone call. The question is, "Will you provide your customers with the reason to call someone else, or will you exploit the power of quality as a competitive weapon to ensure that the customer calls your number for a long time to come?" If the latter was your answer, you've already begun the battle for competitive advantage.

Public Sector

GOVERNMENT

State and Local Governments: Quality in 1996

Joe Sensenbrenner

In this article, consulting editor Joe Sensenbrenner, former mayor of Madison, Wisconsin, and nationally recognized consultant to governments in implementing TQM in the public sector, tells us that TQM and its techniques at the state and local government level are not making as much progress as many would like these days. He explains some of the reasons for this and some of the initiatives still taking place.

Introduction

It was hard to find an upbeat internal quality practitioner in state or local government in 1996. This is all the more significant because one of the most widespread characteristics of these activists is congenital optimism.

The outlook was, however, understandable for several reasons. Most practitioners were not involved in the major planning activities precipitated by the impending block grant devolution of responsibilities signed into law but not actually implemented by most states in 1996 (see "Government Levels and Relationships"). Also, many of the state budgets passed and taking effect during the year required reductions in resources and personnel across programs and agencies. Participating in these wrenching changes can reduce credibility and divert resources from improvement activities. Last, there was a growing sense that what for several years had been a successful set of theories and practices had recently reached the limit of its success in the current environment, while at the same time it was apparent much more needed to be done (see "Containing Systems and Community").

These and related themes may be viewed from three perspectives: (1) inside government: losing internal momentum as new relationships with a major partner possible, (2) between governments: major structural relationships between levels of government will change as will most of the major government programs, and (3) containing systems: new focus and inquiry into governments' environment-community.

Inside Government

Metrics

Activity in ongoing programs was influenced as the election results of 1994 took hold, resulting in heightened scrutiny of government activity. Across the country, many quality practitioners were involved in developing and deploying measurements of performance. They fall into two broad categories. The first involves setting customer-related measurable goals and measuring such outcomes. Prominent among them are Texas and Arizona Performance Budgeting; Minnesota Milestones; the Oregon, Florida, and Utah benchmarks; and Iowa's efforts to develop return-on-investment analysis and performance budgeting pilot programs.

They are characterized by a strong emphasis on measurement of program outcomes as well as on effectiveness of delivery. Most have created a direct linkage to the budget-development process and embody a regular reporting of results to stakeholders.

The second category aims to get better information on the actual cost of parts of government service delivery. Leading efforts include Indianapolis's activity-based costing of its services and programs and Texas's strategic budgeting, which marries strategic planning, performance measurement, and priority-based budgeting.

Many of these efforts are driven by an exploration of contracting out or a competitive process that could precede such a move. Thus, although measurement and feedback processes and customer-driven measurements were basic features of 1950s quality, the drivers for these activities in 1996 tended to have ideological perspectives and political agendas unrelated to TQM.

Keith Smith of the California Department of Transportation states, "The two key changes that will start the rebuilding of government's credibility are to report to stakeholders on program outcomes and to develop accurate methods to cost services. Only then can fact-based decisions be made as to which programs to support and whether the public or private sector is best equipped to deliver services."

The two key changes that will start the rebuilding of government's credibility are to report to stakeholders on program outcomes and to develop accurate methods to cost services.

Unions

There is one public sector milestone that may offer a paradoxical opportunity not evident in the private sector: The high degree of unionism in public employment.

In 1994, public-sector union membership stood at 38.7 percent of the total public labor force, a level of union density higher than ever attained in the private sector. The greatest union density in the private sector was 36 percent, reached in 1956, an era when the prevailing pattern of labor-management relationships was ritualized adversarialism. To note that labor relations in the mid-1950s was disconnected from customer concerns is understatement. In 1994, union density in the private sector stood at 10.9 percent.

And there is ample empirical evidence to suggest that unionized industries have far greater success and achieve longer-lasting results when implementing a quality initiative. (See The Mutual Gains Enterprise, Kochan and Osterman, for example.) Somewhat paradoxically, the data show that strong unions are an important positive variable in measuring the success of quality efforts.

In the public sector, however, there is not only a strong union presence but also a marked difference in the attitude of these unions toward quality and the need to change the way government does business.

Virtually all major public-sector unions have embraced the need for improving the work methods and processes of government, the need to redesign government. The American Federation of State, County, and Municipal Employees (AFSCME) passed resolutions to this effect at both its 1994 and 1996 International Conventions. The Service Employees International Union (SEIU), American Federation of Teachers (AFT), and American Federation of Government Employees (AFGE) have taken similar stances. Formal arrangements embodying these principles have been in place to guide quality efforts in Milwaukee, Minneapolis, and the state of Ohio for more than five years now.

None of these unions has abandoned their traditional role in advocating basic rights and benefits for members. Unions continue to pursue their interests in these areas, often with vigor. What they have come to understand, however, is that the principles of the quality movement are not dissimilar to the principles of the labor movement. They have also discovered that a quality workplace improves conditions for employees, as well as providing a measure of stronger employment security.

The current issues of devolution and block grants, along with ever-tightening state and local government budgets, are likely to increase tension and employment security anxieties in the workplace. These tensions, in turn, will result in increased union attention to matters such as layoffs, retraining, and "bumping" (a worker's right to remaining jobs if his or her current position is eliminated). This environment necessarily limits the energy and flexibility that union leaders will bring to work-process improvement, at least if they wish to maintain credibility with the members who, in essence, have hired the union to represent them. This has certainly been the reality in New York and other jurisdictions facing sharp budget cutbacks.

Yet these same conditions also present windows of opportunity for the development of strong union-management partnerships in the public sector, partnerships focused on enhancing the quality of government services.

Paul Goldberg, executive director of the 39,000-member AFSCME state employee union in Ohio, observes that "If we look at the data, it is clear that the public sector has the potential for developing an amazing level and depth

> These unions have also discovered that a quality workplace improves conditions for employees, as well as providing a measure of stronger employment security.

of successful quality initiatives. We have: (1) the need to improve government; (2) workers who will benefit from the creation of high quality workplaces; (3) taxpayers who want greater value for their tax dollars; and (4) unions in our most highly-unionized sector which are prepared to partner with management to achieve these results. The only thing that seems to be missing is management recognition of the situation, and the willingness of both parties to set aside adversarial instincts long enough to determine the true value of partnership."

Private-sector unions and managers seem to have missed the opportunity to change in time to avoid plummeting fortunes. Public-sector employees and managers now face many of the same unsettling factors that emerged in private marketplaces forty years ago. As was the case for their private-sector counterparts, there are no guarantees—job or otherwise—for public employees. Competition, lifelong learning, workplace change, a willingness to take personal responsibility, and a willingness to take risks are all critical elements of public employment—now, and well into the twenty-first century.

What is also critical, however, is the window of opportunity for genuine partnership between unions and management in the public sector. It will be tragic if we allow that window to slip shut because we weren't prepared, or we didn't know what to do, or simply because we never did it that way before.

ASQC
An organizational update: In their third year, the Public Sector Network of the American Society for Quality Control surpassed 1,700 members and was organized into the fifteen regions of the ASQC structure. Counselors are available in each region to assist people interested in public-sector quality on a volunteer basis.

Government Levels and Relationships
Last year, it appeared that the long-awaited and little-understood change in financing large areas of federal program responsibility, commonly referred to as block grants, would occur. It didn't until late in the year. This prevented state legislatures from considering their options during regular spring legislative sessions. The presidential and related state elections then foreclosed substantive consideration of these matters by the 1995–96 state legislatures.

Two things, however, were rather clear as far as quality practitioners were concerned. The first is that they were rarely involved. In state after state, the usual suspects of top department heads and their familiar consulting partners put together the broad outlines (and often detailed specifics) of state plans. Whether there will be major successes or unanticipated difficulties remains to be seen.

The lack of involvement of traditional quality practitioners raises troubling questions. Why have they not been successful in the terms of the organization's decision makers? Perhaps they have been successful with something other than what the decision makers now deem as the core competency now needed. Do state quality leaders (or, for that matter, does traditional

TQM) have in their toolboxes what is needed? Have the practitioners continued their learning?

The second thing that is clear is that the set of activities related to reorganizing or redesigning myriad existing programs will be where the action is for state and local governments for the next several years. Most of the plans reported in the press related to opportunities created by block grants have proposed better coordination of existing programs and relationships. This is a recipe for failure. These efforts can achieve only the level of success permitted by the knowledge gained through the process used and the resulting interactions of the systems (see "Containing Systems and Community") and must, therefore, respond to the following questions:

> The set of activities related to reorganizing or redesigning myriad existing programs will be where the action is for state and local governments for the next several years.

1. *Is the public being invited to participate?* Are the users of services being consulted on the design of the programs that affect them? Or is it the familiar scenario of six white males gathering in a conference room to divide the spoils according to the customary political calculus of whose electoral interest will be served and whose constituencies will be protected?

   Only through an inclusive process can we find how things really work and also build broad public credibility for the programs we develop. We must invite the people who are the intended beneficiaries to be co-planners. We need to hear what drives them and why they make the decisions they do. Once known, this information can be used to design programs, processes, and incentives that correspond with the way things really work and how people really behave.

   It may seem chaotic and unpredictable, but this approach has the best chance of identifying the right goals, asking the right questions, developing public support, and keeping governments from being blindsided.

2. *Will the planning be conducted to take into account all parts of the system?* Or will it be piece by piece and program by program? A piecemeal approach to the needs covered by block grants is a recipe for disaster. If you cut funds for medical assistance, you also have to be able to predict how that will affect work habits, day-care needs, nursing-home costs, shelter demands, and other factors. Otherwise, you may solve one problem but defeat several other goals.

3. *Will systems be flight-tested in pilot projects?* Because nothing happens in isolation, it is reckless for states even to think of experimenting with millions of lives without first gathering solid information about how all parts of the new system will interact. Citizens should be asking for hard facts, developed in city and county pilot programs, about how affected clients will actually make decisions and respond to new rules.

4. *Is the system designed to correct itself?* States are moving into uncharted territory. The programs they design must have built-in mechanisms for data collecting, monitoring, and alteration if they are not to collapse in the face of sudden new realities.

   Public officials sometimes fancy themselves expert predictors—but the life of society is unpredictable.

The new federal budget means a sea change for the states. The big questions should be asked and answered now—before the deluge.

Containing Systems and Community

As expertise and learning have accumulated in areas of process improvement, it has become increasingly noticeable that the impact of isolated excellence has been disappointing. The often modest levels of success reported in the press for private-sector quality initiatives may well be mirrored in the public sector.

Inquiry into this often perplexing record has led many of the pioneers of the quality movement to look for answers in the developing understanding of systems theory and its implications for governance.

Although included as one of the four aspects of his "system of profound knowledge,"[1] the practical implications of Dr. Deming's concept of "appreciation of a system" had not until recently been much explored by the quality movement. The popularization of the core concepts by Peter Senge in his *The Fifth Discipline* brought widespread attention to this dimension of organizations, social institutions, and, indeed, our universe. This interest has been formally recognized by several of the leading quality organizations in ways that may shape our lives beyond the workplace.

The Association for Quality and Participation

In 1996, the Association for Quality and Participation undertook a substantial "broadening of our view to the communities that are home to our workplaces." They integrated their organization with the World Center for Community Excellence founded in 1991 in Erie, Pennsylvania.

The World Center had the mission "To promote, encourage, guide and catalyze people in communities throughout the world to work together to continuously improve themselves, their organizations and their communities." Now the publisher of the *Journal for Quality and Participation*, with 65 local chapters and a national training program, will disseminate literature on starting community quality councils, guidelines for sustaining mature chapters, best practice information regarding structures and programming, as well as training materials to all its membership.

Kellogg Foundation

This line of inquiry has caught the attention of foundations. Perhaps the most ambitious is the project funded in 1996 by the W. K. Kellogg Foundation to "create an infrastructure for community-wide learning in Jackson, Michigan, to impact improvement of the community as a whole system, and to engage in a 'Learning Exchange' with three other communities engaged in large-scale change ... through social system and leadership development." The other

---

[1] The elements of Deming's system are knowledge about variation, psychology, a theory of knowledge, and appreciation for a system.

communities are Greenwood, South Carolina; Madison, Wisconsin; and Grand Rapids, Michigan.

Each of these communities has a considerable experience with quality as applied to the processes and people in business and industry. Each has an established community-wide quality organization that wishes to learn more about large-scale engagement and change. And each has developed a different basis for organizing activities, including a community college (Jackson), a foundation and chamber of commerce (Greenwood), and a not-for-profit independent membership organization (Madison). The three-year study led by David and Carole Schwinn in Jackson, Michigan, should be watched as the learning and improvement strategies can help us understand alternative community change processes.

The Deming Institute

The W. Edwards Deming Institute is an association of persons who wish to carry on the vision of Dr. Deming in four areas including "Network and Support." To pursue this area, they have developed a partnership project with a local community "to collaboratively work to explore and address local needs and issues." Nineteen communities in twelve states submitted proposals in 1996 for this initial experiment to "work to understand the system of issues behind community needs and help in implementing a system of responses to those community needs." The community selected in August was Takoma, Washington. Those wishing additional information can contact Mr. Britt Hall at the Center of Quality and Innovation at Waukesha County Technical College, (414) 691-5461.

Summary

The question of public quality impact and success will depend on the same basic factors affecting of our colleagues in the private sector: because there is no such thing as a job guarantee, the future belongs to those who are ready to continue to be lifelong learners and take risks to create the future.

Could a Dose of Deming Transform Government?

Wayne J. Levin

This article explores a very important point about government and politics: We can't solve problems that are inherent in the system without improving the system. The author explores the experience of Canada and its fruitless attempts to reduce budget deficits. The approach Canada took parallels that of the United States, and the results are the same. In both cases, political and economic self-interest has proven to be an intractable roadblock. Levin provides some modest suggestions about what to do about all this. This is a thoughtful and thought-provoking piece.

Consider, if you will, that for many of us:

- Our major public and private institutions in both Canada and the US have lost relevance and there appears to be growing participation in the underground economy.
- In both Canada and the US our government and public institutions are in serious deepening debt. The severity of the problem has increased to such a level that for many, government is itself becoming irrelevant.
- Both collectively and individually, government is failing and increasingly unable to do enough good for those whom they govern.

New thinking is needed—Clearly the theories upon which past efforts were built are no longer capable of achieving the outcomes for which they were employed. There is no reason to believe that the reapplication of methods based on this thinking, even with renewed vigor, will lead to improvement. New thinking, a more useful theory, is required to realize a profound improvement in the performance of government in Canada.

Perhaps profound new thinking will work . . . Fundamentally, the reason why Japanese industry has achieved such world class success is that the theory in which they based their actions were profoundly different and much more useful than what Western industry held dear at the time (and largely still does).

Reprinted with permission from the *Journal for Quality and Participation*, January–February 1996. Copyright © 1996 by the Association for Quality and Participation. All rights reserved.

The Japanese were introduced to this theory in the early 1950s by Dr. W. Edwards Deming. To this day, Deming remains revered in Japan for his contribution over the years; his portrait, along with that of the founder and the current chairman, hangs in the lobby of Toyota's head office. The most coveted prize in industry is named for Dr. Deming and the winners' list reads like a *who's who* of Japanese industry.

Could we apply Deming to the political process? The application of the Deming philosophy in the domain of political science, though equally applicable as in management science, is uncharted territory. In this article, I will try to demonstrate both the need for and the benefit to be derived from the application of Deming's theory in politics.

A word of warning before proceeding . . . Most of the practices advocated by Deming were resisted by management. Yet, in time and as a result of the pioneering work of some industry leaders, these practices and the theory in which they are based, are slowly gaining acceptance. No doubt then that the analysis and illustrations which will be provided here will provoke some uneasiness. But I argue that the theory is useful, for it offers a new understanding of the deficiencies of the existing system of government as it prescribes remedies.

Applying a Deming Analysis to the Canadian Political System

Our political leaders, like most Canadians, do not appreciate that measures such as those which follow are outcomes; simple barometric gauges, mere symptoms of an increasingly ineffective system of government:

- The level of unemployment . . .
- The high school dropout rate . . .
- The number of those living below the poverty line . . .
- The number of individuals and corporations who pay little or no tax . . .
- The number of those who require medical attention . . .
- Fiscal deficits.

Focus on the system, not its outputs—Our concern should not be for improving any of these metrics directly. Merely modifying the primary variables associated with each of these outcomes will not yield improvement in the long term. Rather, the concern should be to optimize the system of government in Canada since these outcomes are its products.

Any one or more of the measures mentioned above would do to illustrate my point, but I will concentrate on fiscal deficits for brevity's sake.

The system of fiscal deficits—The Mulroney government established a goal in 1984, at the commencement of their term of office, to eliminate the deficit. They were, with benefit of doubt, sincere in their desire to achieve this goal. During Mr. Mulroney's reign each of his 8 budgets included a 5-year plan to achieve this goal. Despite assertions to the contrary, no appreciable or sustainable reduction has been realized in the Federal deficit. A prolonged period of economic expansion, which would otherwise constitute a remarkable opportunity to reverse the long trend of fiscal deficits, has been squandered.

What was the problem? The Government had a goal but no method. By no method I mean that there was no sound theoretical basis for the actions taken. The basic approach to reduce the deficit has been to close the gap between its two primary variables: revenues and costs.

The cost cutting schemes tried include:

- Restraint initiatives . . .
- Privatization . . .
- Civil service payroll freezes and days off without pay . . .
- Cuts to transfer payments (either in real terms or a reduction in the rate of increase).

These restraint initiatives have been the medical equivalent to attempting to cure chronic diarrhea by suturing the offending orifice. Systems theory tells us that since the human body is a system, this approach will lead to grave consequences. The same can be said of government. Since all methods based on the theory that cost cutting will improve fiscal performance have failed, we need to modify or develop a new theory.

Revenues increased substantially over the years yet not sufficiently to offset the cost of government. Of late, more increases have been put into effect including expanding the tax base to include taxing health benefits. Adjustments of this sort, while being substantial in and of themselves, pale in comparison to the magnitude of the deficit.

These cost-cutting and revenue-generating methods are unsound for another critical reason—while they attempt to account for the primary variables, they fail to sufficiently account for the interaction between the primary variables as well as the ripple effects that propagate to their subsidiaries.

The government as casino . . . Rather than responding to its failures, our political leaders have expanded the application of their supposition that simply raising revenues contributes to fiscal health. Lotteries exemplify this. It began innocently enough with a national lottery to fund the '76 Olympic games in Montreal. Later it advanced to provincial and inter-provincial lotteries of various sorts to fund activities that might otherwise be funded from the government treasury. When pressure for more revenues increased, lotteries advanced to other forms of gambling—with the advent of *Sportselect*, to outright betting. Today, in an effort to increase revenues we now have casinos opening up in various locations throughout Canada. Revenues increase, but there are interactions with other variables causing an increase in costs as well. With more jurisdictions turning to casino gambling as a source of revenues, there will be less tourist benefit. Casinos inevitably and invariably attract organized crime, drugs, and prostitution. No doubt then there will be increased burden on law enforcement agencies and the judicial system. The full extent of the undesirable consequences on the social well-being of the communities involved will never be known.

This example also illustrates how baseless actions can submerge our values. Many Canadians view gambling as a form of stealing. No one twenty years ago could have predicted government development of a gambling industry. But gambling per se is not the source of concern; that governments find salvation in the money it generates is. Revenue generating schemes such as lotteries, and in particular gambling, encourage the worst in us while taking advantage of some basic human weaknesses.

The problem: no constancy of purpose—The deficit example described above (and the examples to follow) illustrate that the principal problems with government are that it lacks sufficient knowledge and is without relevant purpose. Absent knowledge or a clear purpose, government pursues a capricious collection of confused and conflicting ends. With some exceptions, the means that are selected to achieve ends are at best dubious, usually without theoretical foundation, and often take advantage of the weaknesses and/or ignorance of the governed.

To the detriment of the citizenry of this country, politics is more about fostering favorable political self-image than it is about achieving equity, improving the human condition, and maximizing the potential of all Canadians. Like a game of *broken telephone,* the original guiding principles of government have been lost. Like making an endless series of photocopies of photocopies, we may not know the appearance of the original, but we know that the latest edition is no facsimile.

The root of the problem: the system has no purpose—Without constancy of purpose, without a consistent and enduring aim, we cannot view our Canada as a system, an interdependent network of communities, cities, provinces and territories that work together to accomplish our common aim. This is at the root of our problems. Without a system-wide aim each component of the system wanders, expanding according to its own desires and purposes causing the whole system to suffer.

Sub-optimizing economic development . . . Federal, provincial and many municipal governments have departments of economic development. Some activity of this sort may be necessary at all levels of government but because there is no aim to our system of government, frequently there is no integration of efforts such as these. In addition to this duplication of effort, without an aim system components have a propensity to work at cross purposes.

- Ontario's Ministry of Transportation sells its mailing list to generate revenues of $14 million facilitating Ontario's Ministry of the Environment to have need to find landfills to dispose of the resulting junk-mail refuse.
- The Federal government is responsible for refugees. Until recently, they did not allow refugees to work while their claim was being processed, leaving provincial governments to bear the cost of maintaining them through social assistance.

The list of inter- and intra-government conflicts that result from a lack of appreciation for a system goes on and on.

> Without a system-wide aim each component of the system wanders, expanding according to its own desires and purposes causing the whole system to suffer.

Optimizing the System

Our political leaders need to undertake two very ambitious jobs, both based on systems thinking, in order to put Canada back on the rails of prosperity. The first is to foster industry of substantially increased added-value. Through legislation, regulation and policy, government can encourage the development of high-value industry without spending substantial amounts of taxpayer's money. New, high-value industries revitalize the economy and provide an enduring contribution to the social and economic well-being for all and favorably impacts fiscal performance. The second, our main subject here, is to optimize the system of government.

Optimization means achieving the most favorable or desirable state for all concerned—achieving the greatest good for the greatest number of people. It means everyone wins. Optimization, then, is the antithesis of self-interest which is win-lose and ultimately, *we-all-lose*. Though not selfish, optimization does not mean being selfless either. Pursuing optimization means recognizing that only by serving the interest of others, will your interests be served.

> Pursuing optimization means recognizing that only by serving the interest of others, will your interests be served.

When self-interest rules any member of a system, it ultimately destroys the system . . . All system components must cooperate to achieve optimization. A car is a system. It has many critical components in which the failure of any one will render the system sub-optimal. A loose fan belt may cause the alternator to fail to charge the battery sufficiently. The car may run for a while but its performance is sub-optimal and its longevity is curtailed. The failure of a $100 fuel pump would completely incapacitate even the most expensive car.

Following self-interest is endemic to government . . . In their quest to raise their popularity, at the expense of the governing party, opposition parties engage in activities that threaten the viability of the system.

Self-interest in opposition is matched by self-interest in government itself. Decisions on where to locate centers of government activity, including prisons, service contracts, and administrative offices, are frequently either genuinely or are allegedly influenced by one or more of a variety of self-interests. Whether it is a genuine allegation or not is irrelevant since the mere presence of this attitude is indicative of self-interest destroying the system. Expensive and elitist special interest groups and professional lobbying institutions sometimes employ former senior ranking civil servants and retired political leaders to permeate self-interest through the boundaries of the system of government. Thus, self-interest has been refined and elevated to new levels of science, sophistication and "professionalism."

The promotion of one self-interest *ipso facto* means that the interests of others are thwarted. Sometimes that interest is a shared interest. Take for example the pressure put on the government of California to weaken policies designed to protect the environment. In this particular case certain interests have, so far, successfully impeded the development of a high-value industry with the potential to produce tens if not hundreds of thousands of jobs and improve the environment for us all.

Self-interest is easy to overcome if the system adheres to a relevant aim . . . If the system has relevant purpose, and if that purpose is shared by its components and those that it serves, the system's aim can act as an antidote to self-interest.

As an aim for our system of government I propose the following:

To foster equity, develop the human condition, and maximize human potential by developing and harnessing the skills, creative talents, and intellectual proficiencies in Canada for the benefit of all Canadians.

With such an aim those who seek government intervention would then have to indicate how their proposal is consistent with the system's aim and how, therefore, it would contribute to system optimization.

Aim provides focus and purpose . . . Though essential, the aim alone, no matter how good, is not sufficient to yield optimization. Improvements must be made to the system proper. Improving outcomes from a system's perspective is a new theory, a sound supposition serving as a basis for action that is quite distinct from existing political doctrine.

Optimizing government operations—Consider, for example, the Goods and Services Tax (GST). It is an unpopular tax that was introduced by the Conservative government at the commencement of 1991. The Liberal Party, after obtaining power by promising to replace the GST, promptly established a committee, after it was elected, to investigate GST alternatives. In addition to creating the illusion of a promise fulfilled, its expert testimony lends credence to whatever course of action is selected. Either one alone satisfies the need to project a favorable self-image; neither one leads to optimization. The reasons for this are clear:

1. The GST is not a bona fide attempt at system optimization. Its replacement, therefore, is not likely to be either.
2. The use of expert witnesses to generate alternatives will not be fruitful because they represent the same knowledge that produced the GST.

We should not be surprised that all these experts had to offer were mere modifications of the existing tax.

Tax law incorporates no natural elements, does not mimic anything in nature, nor does it adhere to any universal laws. Yet is has achieved doctrine status. A huge industry employing thousands of people has emerged growing in proportion to the complexity of the tax system. With all this activity, technology and sophistication, an alien could be forgiven for concluding that taxation was centered on divine law. To improve the tax system, and to improve the effectiveness of government generally, we need to look for a more useful pool of knowledge.

Government deficits, the level of unemployment, high-school dropout rates and other like social metrics are related. Any action directed at improving one of these metrics, without appreciation for a system, is likely to have a detrimental effect on at least one of the others. The corollary is that a system's approach at improving any one of these metrics will have a positive impact on one or more of the others. The system of government is integrally linked to all social outcomes, therefore, government is in the best position to affect the kind of positive social change that comes from system optimization.

The Need for New Knowledge

To know that we know what we know and that we do not know what we do not know, that is true knowledge.

Confucius

Getting outside of your system, your box—New knowledge is needed but new knowledge is difficult to obtain because, as Deming frequently pointed out, a system can not know itself; it may know well the activities that it is engaged in but this does not mean that it knows itself. By not knowing itself, it will not appreciate its predispositions and deficiencies, and will therefore not be motivated to improve.

The human body, again since it is a system, offers several examples of this including the following: the reader may recall the first time he or she ever heard a recording of his or her voice: it sounded unfamiliar. This is because the system generating the voice is predisposed to hearing it a certain way; a recording is independent of any such preconception. Singing instructors and sport coaches are sources of outside knowledge for those seeking improvement in these areas. That Deming's theory was (and to some extent still is) rejected because of its counter-intuitiveness demonstrates just how hard it is for outside knowledge to penetrate even educated minds.

Knowledge from outside of the system does not necessarily mean outside of Canada. Many of the solutions mentioned earlier are well precedented in the West. Our system of taxation is more or less the same as those in other industrialized countries. Our GST was inspired and adopted from other Western industrialized countries, notably New Zealand. Though it comes from outside Canada, the thinking that created the GST is not outside of the thinking that created prior forms of taxation. But as the solutions are well precedented, so are its failures. Almost all industrialized Western countries are similarly plagued.

Advisors are frequently used to educate political leaders but they too are not necessarily knowledge from outside the system. The Federal finance committee seeking to replace the GST found no new knowledge from their bevy of experts. They learned a lot of details, gained a few new insights, but garnering information is not acquiring knowledge, it only engenders the illusion of knowledge.

The metric chosen determines the measurement . . . We see a similar dilemma in science. The selection of a measurement instrument already limits what can be learned because the choice of metric carries with it an implicit assumption of the essentials of what is being measured. At one time, for example, it was believed that there was a proportional relationship between brain size and intelligence. With this thinking a caliper was developed to measure the cranium so as to determine intelligence. We know this action to be without foundation because the theory that intelligence is relative to brain size does not explain the existence of numerous stupid people with large heads.[1]

Some Closing Thoughts

Do our political leaders appreciate the need for outside knowledge? If they did they would be taking a radically different approach on how they govern. Engaging in practices that are largely anti-systemic and not theoretically based has squandered resources and brought about fiscal debt without sufficient enduring economic and social benefit. Oddly, in Canada, the usual response to a failed regional economic development program is another regional economic development program.

Regrettably, the same limited thinking is still employed and it is for this reason that I doubt that we'll see improvement in government any time soon.

In the case of government, criticisms, such as those raised here, are sometimes responded to by stating that there is a system of *checks and balances* that eventually leads to the best approach. Any reasonable evaluation of such arguments on the basis of Deming's theory clearly exposes substantial shortcomings. Theory, generally, and the *Plan-Do-Study-Act* (PDSA) cycle, a methodology for advancing theory, imposes outside knowledge for those who seek a better way. Under the PDSA regimen, a failure triggers either the need to modify the theory employed, to improve its effectiveness and ability to predict, or search for a completely new one. If this practice alone were habitual in government, we would not be experiencing fiscal, economic and social malaise today.

Dr. Deming's system of profound knowledge is a solid theoretical framework to develop and test new approaches. It provides both a worthy critique of past efforts and an explanation for their failures while prescribing new remedies to be tried and the means by which to test their effectiveness. Since Canadians and Americans are fortunate to live in a democracy, those of us who appreciate the gains to be realized through application of Deming's methods, some of which were expounded in this article, should exert their conviction publicly.

Wayne Levin is a professional engineer with a master's degree in Management Science who specializes in statistical methods and the application of Dr. W. Edwards Deming's System of Profound Knowledge.

---

[1] Since the theory does not explain these failures, it needs to be modified or abandoned. It seems ludicrous, doesn't it, that we ever considered such a possibility.

Making the Big U-Turn

The National Performance Review Staff

This article is an interim report on some of the progress of Vice President Al Gore's reinventing government initiative. It especially notes how a focus on customer satisfaction has been a driving force of these efforts and points out that the Social Security Administration has the best telephone customer service of any organization in the country. You can learn more about the National Performance Review on the Web at this site: http://www.npr.gov/homepage/view1/html.

Why would a company like Ford that already has a reputation for quality go to the trouble and million-dollar expense of publicizing specific promises about what service their customers can expect? Why promise to tell you the status of your car "within one minute of your inquiry"? Why not just try to do it and keep the one-minute goal quiet? Why let the customers in on your measures of success, or failure?

Ford's doing it this year for the same reason the reinvented U.S. government started doing it two years ago. To build confidence among its customers—in Ford's case, confidence that a good experience in the service department is no fluke. It's like Babe Ruth pointing out where the next homer's going. You have to promise and then deliver. That builds confidence that the whole organization is designed and managed to deliver the results the customer wants.

If Ford's customers don't have confidence in the company, Ford is in trouble. But if the government's customers lose confidence, all of America is in trouble. Because the government is—or should be—nothing more than Americans working together to solve problems that can't be solved any other way. Like educating our next generation to compete in the world market. Or protecting our borders. Or making our streets safe. If the American people don't even believe our government can answer the phone and give quick, courteous service, how can they believe it can keep America great and free?

We have to restore confidence that we can all work effectively together through self-government. And the government has to build confidence just like Ford—or any good company—does: with each and every customer.

For a while there, the government was getting away from us. It stopped being the government of the people. It was marching to different drummers—special interests, Washington professionals, well-meaning people with good

---

Reprinted with permission from *Quality Progress*, March 1996.

intentions—on a path that seemed to be headed away from the tax-paying customers of government.

In March 1993, President Bill Clinton asked Vice President Al Gore to lead the National Performance Review.[1] The Clinton-Gore reinvention initiative mapped out a dramatic change in direction, a big U-turn, to head government back to the people.[2,3,4] All of this basic work on customer service is the force behind the big U-turn.[5] It all begins with listening to what people want.

Just Listen to the People

Whether you work in the Ford Service Department or the Social Security Administration, customers all want the same thing. They all want to be listened to. They all want quality service.

They all care about:

- Getting things done right
- Getting things done quickly, or at least knowing how long it will take
- Dealing with knowledgeable, reliable people
- Resolution in one contact, or one point of contact until resolution
- Knowing where to turn if a problem shows up
- Choices on how and where to get services
- Readily available information
- Clarity in forms, publications, process descriptions, advice, and correspondence
- Courteous treatment from people who are friendly, respectful, trusting, and willing to listen
- Readily accessible, clean, and safe facilities
- Names and phone numbers so they can call directly to get questions answered

Whether you work in the Ford Service Department or the Social Security Administration, customers all want the same thing.

But even knowing that customers care about all these things is not enough. You have to keep asking customers what they want, and listening to what they say. You have to find out what matters most. Companies that don't ask enough questions get it wrong. Remember when Ford decided its customers wanted Edsels? Or when Coca-Cola decided it was high time for a new taste?

Government agencies are not telepathic either. Take the Internal Revenue Service (IRS). It firmly believed that good customer service meant mailing you your tax forms right after New Year's Eve. Until it asked, that is. Then the agency found out that what people really want is as little contact with the IRS as possible and a quick refund.

So next spring, about 20 million taxpayers in 50 states who use the 1040-EZ form can forget the form and file on their touch-tone phone. It's quicker (about six minutes) and easier than EZ (the phone system does the math). It

BIG-TIME PROMISES — STANDARDS THAT TOUCH MILLIONS

The list below is a sampling of customer service standards and approaches contained in "Putting Customers First '95: Standards for Serving the American People":

Consumer Product Safety Commission. Hot line available 24 hours a day for customer complaints and details of the latest product recalls.

Department of Agriculture. If we get things wrong in the Rural Business and Cooperative Development Service, we'll apologize and make it right.

Internal Revenue Service. Tax refunds due on complete and accurate paper returns in 40 days; 21 days for electronic returns.

Veterans Canteen Service. Food court service in three minutes; unconditionally guaranteed customer service.

National Archives and Records Administration. Within 15 minutes of walking in, you'll have either the information or the help you need.

National Park Service. Great Smokey Mountains visitor center open every day but Christmas.

Occupational Safety and Health Administration. Inspectors will be respectful and helpful, and focus on the most serious hazards.

Bureau of Labor Statistics. Data any way you want it: from a live person, by recorded message, fax, microfiche, diskette, tape, Internet, or telecommunications device for the deaf.

Department of Commerce/STAT-USA. CD-ROM orders shipped in one day or the CD is free.

Minerals Management Service. Easy access to us in our office or by phone, or we'll meet you at a more convenient location you request.

Environmental Protection Agency. In our voluntary programs, publicly recognize the achievements of business.

U.S. Coast Guard. Search and rescue on demand, 24 hours a day, seven days a week.

Education Department. Special education customers will be seen within 10 minutes.

Forest Service. Our offices and visitor centers will be open at times convenient to you.

Social Security Administration. New and replacement cards mailed within five days; we'll tell you the Social Security number in one day if it's urgent.

FedWorld. On-line transactions complete in seven seconds.

U.S. Mint. Orders taken 24 hours a day, seven days per week.

requires no contact with any person at the IRS. And you'll get your refund within 21 days (rather than the 40 it takes if you file a paper form).

See? Government is listening to you. Government is coming back to the people.

The IRS, Social Security Administration, and Postal Service were the first agencies to publish customer service standards two years ago—the courageous pioneers. But it's not just them. In 1994, 150 agencies laid it on the line—putting their promises down in black and white for all to see.[6]

The 1995 edition contains the customer service standards of 214 federal agencies. New additions include the Securities and Exchange Commission, with its 24-hour hot line and one-day response, and the Peace Corps—always eager to recruit—which promises to send out information on job openings within one day. The Architectural and Transportation Barriers Compliance Board and the National Endowment for the Humanities are examples of the many smaller agencies publishing standards for the first time. And major departments—such as Transportation, Labor, Agriculture, Veterans Affairs, and Commerce—added standards on customer service areas not covered earlier.

When we released the standards in 1994, we said we were just getting started, that we have a long way to go. It's still true that much remains to be done.

We have to reorganize to be responsive where we serve customers; we've been organized for top-down control. We have to train employees to deliver results to customers; they've been trained to follow what Gore calls "mind-numbing rules." We have to have systems designed to please customers; up to now, we've had systems that were designed to please bosses, headquarters, and management committees.

These things are changing. The customer service teams that put the standards together have a fire in the belly that comes from knowing that what they do can truly change government. Together, they are pounding away at the old ways of doing things. They are putting together new, customer-driven systems. And they are building on the strongest possible base—the desire of federal employees to serve America, to do what they signed up to do. This desire, coupled with the discretion to make decisions to serve customers, is a potent force for change.

Agency leaders are also engaged. Many spent big chunks of their time this past year outside Washington listening to customers and frontline employees, seeing what it is like to work where government touches Americans. They are finding out firsthand what needs fixing.

Making all the necessary changes will take time. These are things that go to the very core of government. So rather than make a promise we cannot keep—a cardinal sin in customer service—some of the customer service standards are pretty modest compared to the best in business. But this approach leads to improvement, too.

In 1994, the Department of Veterans Affairs (VA) promised veterans no more than a 30-minute wait to see a benefits counselor. VA knew 30 minutes wasn't good enough, but thought that was the best it could do. During 1995,

The customer service teams that put the standards together have a fire in the belly that comes from knowing that what they do can truly change government.

> **FORD PROMISES**
>
> The following are Ford's dealer-to-customer standards:
>
> - Appointments available within one day of customer's requested service day
> - Write-up begins within four minutes of arrival
> - Service needs continuously identified, accurately recorded on Repair Order, and verified with customer
> - Vehicle serviced right on first visit
> - Service status provided within one minute of inquiry
> - Vehicle ready at agreed-upon time
> - Thorough explanation of work done, coverage, and charges

the 30-minute standard focused VA's work on shorter waiting times. The new promise is that veterans will wait no more than 20 minutes. For tomorrow, VA is working on changes to virtually eliminate the wait.

Across the board, standards are getting better. Agencies are staying open longer hours to serve you. They're promising to keep listening to their customers and keep improving their standards. Some even pledge in writing to admit when they're wrong and to try to make it right.

You seldom have to stop by a government office anymore. Dozens of agencies are offering their customers more modern service choices: telephone, fax, the Internet. For example, many Social Security, welfare, and veterans payments now go out electronically: safe, secure, and easy. Scores of agencies provide information and interactive services over the World Wide Web, the latest, most powerful, and most customer-friendly service station on the information superhighway. [A good place to start is http://www.whitehouse.gov.] The Federal Emergency Management Agency's web site (http:www.fema.gov) offers 5,000 pages, words, and pictures of the latest emergency information. The National Cancer Institute's web site (http://www.nci.nih.gov) has a vast but easy-to-use store of information on preventing, detecting, and treating the dreaded disease. The U.S. Business Advisor (http://www-far.npr.gov/VDOB) gives one-stop electronic access to rules from every agency of the federal government that regulates businesses, large or small. Electronic government is on the way.

A Revolution

It's not just more and better standards. It is an all-encompassing customer service program. It's an entirely new way of governing. It's the president and vice president putting customers first.

It started in 1993 with a recommendation from Vice President Gore's National Performance Review team and the president's Executive Order that set all federal agencies on a customer service campaign. The president ordered agencies to survey their customers to see what kind of service people want and whether they are getting it. To get ideas from frontline workers who deal with customers day to day. To give customers choices and easy access, and a way to complain and get problems fixed. He set the goal for the government to deliver service equal to the best in business.

This year, the president reinforced his order to put customers first. He reminded agency heads to keep pushing for improved service and told them to measure their performance and report the results to customers. The agencies will soon pass out report cards that show how they are doing. The IRS, for example, will post its results in this year's tax booklet.

President Clinton's new directive leaves no doubt that the goal is a revolution in how government does business, top to bottom, so that customers are the focus. Customer service standards and measures are to be part of strategic plans, training programs, personnel systems, and anything else that ought to be changed to advance the cause.

It Works

Publishing standards is risky business. Everybody knows the minute you blow it. But you do it anyway if you care more about improving service than saving face.

Look at the Postal Service. It was one of the first brave hearts, publishing crystal-clear standards back in 1993—like local delivery overnight, absolutely, positively. The first year's results? A public relations nightmare. Performance in Washington and New York fell in the 50% range—the system even lost Gore's Christmas cards. Chicago came in at 66%, with some mail found burning under a bridge.

Was that any worse than before the Post Office set standards and measured performance? Nobody knows.

Has it gotten better since the agency set standards and measured performance? You bet! Washington, New York, Chicago: now, all performing above 80%. Nationally, on-time delivery is up from 74% to 87%.

When you measure performance, you get performance. And making the whole process public helps a lot.

The Social Security Administration (SSA) made big public promises in 1993 about delivering quick, courteous, and competent toll-free telephone service. This year, *Business Week* reported that an independent survey of the country's best over-the-telephone customer service ranked the new improved Social Security Administration top in the nation.[7] The SSA beat companies like L.L. Bean, Federal Express, and Disney. The SSA was slow answering the phone, but so good at getting problems solved quickly and courteously that the agency still got the highest score overall. The SSA is the best in the business. And to stay the best, it is training 3,300 more operators this year so it can get to all the calls quicker.

Grand successes like that are made up of millions of individual successes like this one described in a letter from Maurice Dopp of San Francisco (September 23, 1994):

> An independent survey of the country's best over-the-telephone customer service ranked the new, improved Social Security Administration top in the nation.

"A month or so ago, I called the SSA with a query. A [recorded] voice said that the telephone shop was now open from 7 a.m. to 7 p.m., and that if I wished to hang on, someone would be with me in an estimated six minutes. In six minutes someone helped me: "Oh, you need to fill out form XX. Would you like to come in, or would you like to do it over the phone? What day can someone call you? What time?" On the appointed day, at the appointed hour, my phone rang and a friendly voice asked me questions. The next day the mail arrived with the form filled out, ready for my signature, and with a postage-paid return envelope. No private business that I can recall has ever given me service this good! Cheers to all concerned!"

That's the kind of government America can be proud of.

For the Convenience of Customers

More than any other time in the history of America, the government is listening to what the people want, and delivering. It's all here in this book.

Just like last year, the service standards in this year's book are organized for the convenience of the customers. Not alphabetically by agency, so that government employees can find their goals; they did the work, and they know who their customers are. But organized by customer group, so that, for instance, beneficiaries can look in one section to find out what kind of service they can expect from a government agency, whether it is Agriculture, Health and Human Services, Housing and Urban Development, Labor, Treasury, or whatever. The same for business owners, local law enforcement agencies, travelers and tourists, and so on. If standards apply to more than one group, we repeated them so that customers don't have to hunt around. If you really want to look things up by agencies, there is an index in the back. The organization of the book is symbolic of the reinvented government. It puts customers first.

Notes

1. Report of the National Performance Review, "From Red Tape to Results: Creating a Government That Works Better and Costs Less," (Washington, D.C. Government Printing Office, 1993). The report can also be found on the Internet's World Wide Web at http://www.npr.gov/homepage/view1.html.

2. Brad Stratton, "Reinventing Government: An American Imperative," *Quality Progress*, October 1993, 5.

3. Brad Stratton, "How the Federal Government Is Reinventing Itself," *Quality Progress*, December 1993, 21–34.

4. Karen Bemowski, "Trailblazers in Reinventing Government," *Quality Progress*, December 1993, 37–42.

5. "Clinton Champions Customer Service Actions," *Quality Progress*, June 1995, 17–18.

6. Report of the National Performance Review, "Putting Customers First: Standards for Serving the American People," (Washington, D.C.: Government Printing Office, 1994). The report can also be found on the Internet's World Wide Web at http://www.npr.gov/homepage/view1.html.

7. "Top Providers of Telephone Customer Service," *Business Week*, May 29, 1995, 6. This was a small chart in the magazine's Up Front section. The source for the chart was the "World Class Benchmarks" study done by Dalbar Financial Services Inc., Boston, Mass.

The National Performance Review is located at 750 17th St. NW, Box 101, Washington, DC 20006, (202) 632-0150, fax (202) 632-0390. Its Internet World Wide Web home page is http://www.npr.gov.

Public Sector
HEALTH CARE

Hospital Sets New Standard as Closure Approaches: Quality Is Continuous

Martha L. Dasch

This article documents how a navy hospital in Orlando, Florida, went about revamping its same-day surgery unit processes to be simpler and more responsive to patient needs and thus improve the quality of its care.

Continuous improvement in a medical facility can occur at any time, even when it is scheduled to close its doors. This is what happened at Naval Hospital Orlando in Florida. Although its staff was aware of the hospital's impending closure (the hospital was notified officially on July 1, 1993, that it had to close), it saw a need to improve and decided to do what was necessary to make it a better, more efficient medical facility. Specifically, a self-directed team was interested in improving its same-day surgery (SDS) unit. The improvement plan that it developed faced some rough spots, but in the end, not only did SDS procedures dramatically improve, an ambulatory procedures unit under the SDS unit was established as well, making the hospital more efficient and patient friendly.

Reprinted with permission from *Quality Progress*, October 1995. Copyright © 1995, American Society for Quality Control. All rights reserved.

Naval Hospital Orlando

Naval Hospital Orlando, built in 1982, was a four-story, 153-bed facility with five operating rooms (OR). It provided health care for the Orlando Naval Training Center's military members and their families. The hospital had an outpatient clinic that covered most specialties, and it provided ancillary services to support inpatient and outpatient needs. When it was built, however, it was not constructed with an SDS unit. Because of an increasing need for a unit of this kind, the hospital converted 2,200 square feet of the third-floor surgical ward into an SDS unit in 1990. The 10-bed unit combined pre-operative teaching, staging, recovery, and administrative services all in one area. The surgery was performed in the OR on the second floor.

SDS Satisfaction Survey

In 1993, the SDS unit conducted a satisfaction survey of its internal and external customers. Patients were asked to rate and comment on areas such as the care they received, time it took to check in, and staff helpfulness. The staff received a different survey that asked them to rate and comment on such items as the timely receipt of patient information, patient notification, and the check-in process. The results revealed several areas that could be improved. First, the results indicated that patients were dissatisfied with the disorganized pre-admission process. While it only took two hours for some patients to complete the process, it took others up to two days. Because the unit did not specify a time for completing the pre-admission work and did not schedule appointments for patients, long delays occurred, particularly in anesthesia. If patients became impatient with the long delays, they took their pre-admission packages home. When they returned to the hospital, information inevitably was lost or out of order, forcing the SDS administration to reorganize the paperwork, resulting in further delays. In addition, pre-admission was not located in a single area. Elderly patients, sick patients, and those with disabilities had to visit three different floors to complete the process.

Second, because patients did not have appointments, the OR unit was not able to schedule its staff appropriately to conduct pre-operative sessions. Before a surgical procedure, patients are provided with information explaining the procedure and what they can anticipate prior to, during, and after the surgery. Patients are also given a list of instructions about how long to fast before surgery, how to physically prepare before coming to the hospital, and when to arrive. Patients with additional questions or concerns can be reassured by the staff prior to the day of the surgery.

Third, because there was not a dedicated SDS staff, but rather a staff that rotated from the surgical ward, there was a lack of commitment and poor follow-through in the SDS unit.

Fourth, each of the seven clinical services that referred patients to the SDS used different pre-admission packages. Pre-admission packages consist of the doctor's orders: the patient's medical history: authorization for surgery: ancillary service information, such as lab, X-ray, and electrocardiogram (EKG) test results; progress notes; nursing assessment forms; and admission authorizations. The variety of pre-admission packages created confusion, and errors resulted when patients were prepared for the SDS unit.

Fifth, the SDS unit received late or incomplete pre-admission packages from clinical services, which caused delays or surgery cancellations.

Sixth, the SDS unit's physical layout was stressful for patients. The pre-admission and recovery areas were side by side: prospective patients who were participating in the pre-operative sessions could observe post-surgical patients coming from surgery.

SDS Improvement Team

In June 1993, a group of concerned staff members got together to review the results from the SDS satisfaction survey. This multidisciplinary group included a surgeon; the associate director of surgical nursing; an OR nurse; a nurse anesthetist; and representatives from patient administration, the laboratory, EKG, and the nursing staff. Scheduling clerks from the seven clinical services that used SDS also participated.

The group's meetings weren't highly productive at first. Comments such as, "You don't understand how our service does this" or "This won't work" showed a lack of understanding and trust. But, as the meetings continued and the different departments of the hospital explained their particular needs, the group agreed that there had to be a better way. The group came up with a plan to chart patient flow from the day surgery was deemed necessary to the day of the surgery. Figure 1 illustrates the many steps a patient had to complete during the pre-admission process.

The initial plan was to analyze the flowcharts of the seven clinical services and then develop a single admission procedure that all could follow. It was hoped that this would decrease the pre-admission time for patients.

The group analyzed the flowcharts, used Ishikawa diagrams, and brainstormed in order to develop a list of improvement opportunities. About six months later, the group prepared to implement several of its improvement solutions. While some could be implemented immediately, others were considered long-term improvements. The improvements for immediate trial included the following:

- Schedule pre-admission appointment times for patients to eliminate "walk-ins."
- Combine pre-operative teaching, laboratory blood tests, and admission authorization to decrease the number of stops patients must make (see Figure 2).
- Implement a single pre-admission package and checklist for use by all seven clinical services to reduce confusion and increase efficiency.
- Centralize the processing, storage, and retrieval of pre-admission packages by moving responsibility for completion to the SDS unit.

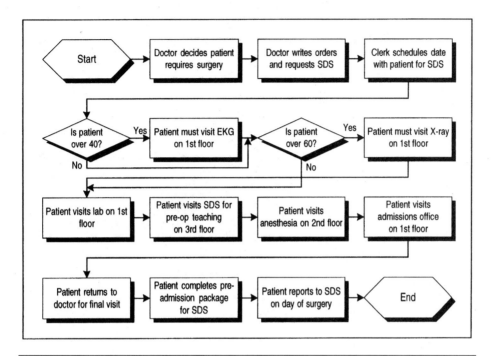

Figure 1. **Pre-admission process for SDS (before improvements)**

Improvements Sparkle Then Fizzle

At first, success seemed certain. After polling the hospital staff, results indicated that the implemented changes had reduced patient pre-admission time and had increased staff satisfaction. However, no time line had been established for long-term improvements, and the group stopped meeting. With no meetings, communication channels began to close. Instead of fine-tuning improvements when questions arose, old methods were reinstated.

In early 1994, the group realized what was happening and began meeting again to adjust the short-term improvement plan and develop a long term improvement plan. After carefully evaluating the problems that were occurring, the group decided on the following:

- Assign permanent staff to the SDS unit to facilitate cross-training in administrative, laboratory, and EKG duties.
- Relocate the SDS unit to the second floor to allow for greater space and collocation of the unit with the OR and anesthesia staff.
- Provide the capability for patients to complete all pre-admission work in one place, with the exception of X-rays.

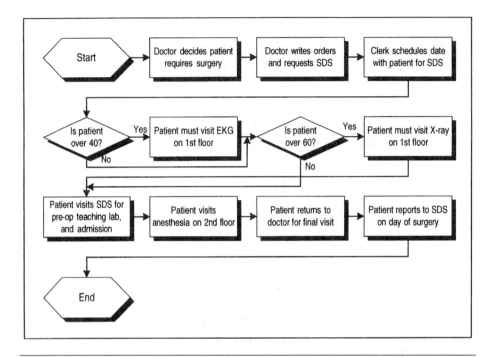

Figure 2. **Pre-admission process for SDS (after first improvement change)**

Downsizing Closes Obstetrical Ward, but Presents SDS Opportunity

As part of the Base Realignment and Closure Commission process, the hospital closed its obstetrical ward in June 1994. This closure enabled the improvement group to implement its long-term improvement efforts. The SDS unit moved to the second-floor obstetrical and nursery ward. The unit now occupied 4,000 square feet and took advantage of the 12 beds, two delivery rooms, and four labor rooms. The nursery area was converted into an administrative pre-operative area where all pre-admission work except X-rays could take place. It also separated the pre-operative patients from patient staging and recovery. Most of the pre-admission procedures now were taking place in one area on one floor, greatly improving the pre-admission process for patients (see Figure 3).

The second-floor location was more convenient because the administrative pre-operative area now was located across the hall from the OR and anesthesia staff. In addition, permanent staff was assigned to the SDS unit and cross-trained so that personnel could handle SDS, pre-operative teaching, and admission responsibilities.

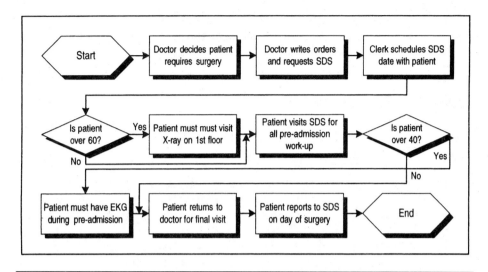

Figure 3. **Pre-admission process for SDS (after final improvement change)**

The increased space motivated the group to expand on their improvement vision. With the additional space now available, why not create an ambulatory procedures unit (APU) under the SDS unit? Nurses in the APU unit could be trained to give intravenous medication, known as intravenous conscious sedation (IVCS), and assist in the performance of endoscopies (the insertion of a lighted tube down the throat of a patient for diagnostic purposes) and other minor procedures. This would free anesthesiologists from having to be present for every minor procedure and allow them more time for the major OR surgeries. Patients could be treated on an outpatient basis in the APU or be brought in from the inpatient nursing department. Centralization of these capabilities would enable providers to do more procedures more efficiently with dedicated staff.

On June 24, 1994, the SDS/APU unit began initial operations. The labor and delivery rooms were easily adapted to perform ambulatory procedures under IVCS, such as endoscopies, formerly performed in the surgery clinic area. Also performed in this area were uncomplicated ear, nose, and throat procedures formerly performed in the OR. Many clinical areas came to the APU with suggestions of procedures that could be performed in the new space.

A number of efficiencies resulted from combining the SDS and APU units:

- The anesthesia staff was used more effectively.
- All minor-procedure and endoscopy equipment was located in the APU area, which expedited its use and assignment.
- Nurses' professional capabilities were enhanced because they were cross-trained in IVCS skills. Credentials for clinical competency in IVCS were instituted.
- The main OR was made more available for complicated cases.

- The APU's capacity for procedures increased due to quicker turnaround time and the ability to stage and recover patients outside the procedure room.

Lessons to Be Learned

The group at Naval Hospital Orlando achieved improvements by trial and error. The lessons it learned may be of value to others wishing to make similar improvements in a health care facility:

- Be persistent. Don't give up when it seems as though nothing will happen. In this self-directed team, there was no just-in-time training or quality advisor, which led to some initial frustration and floundering.
- Don't stop meeting after you think you have the improvements in place. The group stopped meeting in late 1993 after implementing initial improvements, and it provided no communication channel for clinical services that had changes that needed to be made. Again, a quality advisor may have advised the group to continue to meet in order to perform all steps in the plan-do-check-act (PDCA) cycle. Once the group made improvements in the "do" phase, they needed to monitor and "check" to ensure that the improvements produced the gains anticipated.
- Ensure that the improvements are continued after initial implementation so there is no regression. The organization must take steps to institutionalize changes. Since the change in procedures had not been institutionalized, when clinical services ran out of the new pre-admission checklists, they reverted to their old forms and methods. Again, PDCA would have assisted in institutionalizing this change.
- Continue to expand your vision and your opportunities to improve. When you begin to reach your goals, look around for other opportunities. The group experienced a significant breakthrough in developing the APU as part of the SDS improvement effort.
- Improvement efforts should not be a short-term investment; they need top-level support and multidisciplinary collaboration with the intent to improve patient care. They are time-intensive and can be frustrating, but they are ultimately worth the investment.

> Don't stop meeting after you think you have the improvements in place.

Continuous Improvement Lasts Forever

Improvements are worthwhile in any organization, regardless of the time it has left. Because of the changes that were made at Naval Hospital Orlando, patients' quality of care and employees' quality of work life improved, even if only for a while. Although the Naval Hospital Orlando has closed and ceases to function, the SDS unit stands as a benchmark for others to emulate. The members of the improvement team will take their planning, team, and communication skills to other facilities. Physically the doors have shut, but in the minds of those who made the change, the improvement will last forever. Continue improvements as long as you have an organization to improve.

Martha L. Dasch is a medical services corps officer in the U.S. Navy, stationed at the Armed Forces Medical Intelligence Center at Ft. Detrick in Frederick, Maryland. She received a master's degree in business administration from Hood College in Frederick, Maryland.

The Most Important Day

J. Daniel Beckham

What is the most important day to which the author refers? It is the day devoted to understanding customers and their needs as any health-care facility tries to figure out how to reduce costs. Beckham points out that if health-care professionals don't take customer value into consideration, all their efforts will be for nought. He then explains a procedure for effectively exploring who the health-care facility's customers are and what they value.

Cost-cutting campaigns are put in place. Expenses are reduced. Management declares victory. Theoretically, the organization now is positioned to compete more effectively on price because its costs are down.

Unfortunately, it's rarely that simple. More often than not, the real result is lasting demoralization of the workforce, degradation of customer satisfaction, and a slow reabsorption of expenses. Like a wrung-out sponge, the organization quickly soaks in new expenses until it's once again waterlogged.

In pursuit of competitiveness, pressure to reduce costs has caused every industry, including healthcare, to re-examine itself. Often, this self-examination leads to a flurry of expense reduction. Although some of these cost-reduction campaigns are introduced within the context of quality improvement or re-engineering, they often quickly come unglued from these philosophical underpinnings.

In the rush to meet expense-reduction targets and deadlines, managers are incented with bonuses and employees scurry to be survivors. Noble commitments to quality and customer satisfaction are rarely translated to performance goals. What gets measured gets done, and it's dollar savings that get measured.

The Expense-to-Value Ratio

I've said it before and I'll say it again: A chimpanzee with a calculator can reduce expenses. Volume derived from market advantage, however, is much harder to come by and harder to recover once lost. Of course, expenses and volume are two sides of the same coin, inextricably linked with one another. Impact one and you can dramatically impact the other. Theoretically, you can reduce expenses to zero. But long before you get there, you may have reduced the organization and its services to something the customer won't bother with.

Reprinted with permission from *Healthcare Forum Journal*, May–June 1996. Copyright © 1996 by *Healthcare Forum Journal*. All rights reserved.

Maintaining the ratio of expenses-to-customer value is the essence of a sustainable business. Get either side of the ratio too far out of whack and you're heading for disaster. Driven largely by financial concerns in the Seventies, for example, GM decided to use the same chassis, power drives, and very similar body components across all of its divisions. The result was Pontiacs that looked like Buicks. Cadillac even marketed a cosmetically altered Chevy under its marque. Quality was a minor consideration.

The line of cars was an accountant's dream. The potential for standardization and economies of scale was immense. Of course, it was a grand plan that almost destroyed the company. Customers quickly turned to the variety and quality offered by the Japanese.

Unfortunately, most expense-reduction initiatives aren't nearly as coherent as GM's. They are willy-nilly and piecemeal. Teams are formed and chase costs with little or no consideration for how expenses in one area interact with expenses in others. Or how much impact a little expense reduction here may have on customer satisfaction somewhere else in the organization.

Such suboptimizing rarely reflects failure of the teams or individuals who bring it about. Instead it reflects the failure of leadership to see and communicate things whole.

The Only Customer

Reducing cost structures is fundamentally important to competitiveness. So important that it should be undertaken only after leadership has dedicated itself to a serious understanding around how the work of the organization relates to the needs of the customer. Much pain, frustration, and misspent energy can be saved if leaders pause before launching expense reductions and take some time to envision what the organization really does and how it creates value in the marketplace.

What makes the counterproductive flailing and lack of lasting effect in so many expense-reduction programs especially troubling is that they could have been so easily avoided if leaders had dedicated a single day to being more disciplined. Every management team that undertakes a major expense-reduction program ought to set aside a day dedicated to thinking about the work of their organization from the customer's perspective.

The first agenda item for such a gathering is this question: Who's the customer? This simple query will surely set off a torrent of debate. Some will argue that it is the employer. Others will assert that surely it is the doctor. And of course some will suggest, perhaps timidly, that it is the patient or the consumer. Others will wonder if it isn't all of these.

It is an important debate. Not because the answer is uncertain but because the debate underscores how far many healthcare organizations have strayed from their purpose.

The only customer is the ultimate customer: those people who are the ultimate consumers of the unique value-added capabilities of the organization. For hospitals, physicians, health plans, and insurers, that customer is someone who, whether healthy or ill, seeks to be assured of access to a caregiver. Everybody else is an intermediary or a link in the value creation process.

A Dangerous Concept

One of the most dangerous concepts to work its way into the lexicon of organizations lately is that of the "internal customer." There are no internal customers. All customers live outside the organization unless they are in their role of consumer (a hospitalized nurse, for example). The notion of an internal customer degrades the primacy of the real ultimate customer.

In one large medical center, I watched as members of the nursing staff, well inoculated in the concept of the internal customer, asserted that their needs as customers ought to take primacy over those of patients. The issue was parking spots. The nurses insisted their parking spots ought to be closer than those of patients and families. As many a hospital parking clerk will attest, this same line of thinking has been adopted by many physicians as well.

Debate this question if you want, but when push comes to shove in the marketplace, there's only one customer who matters. And that's the one who ultimately selects and uses the product. As long as consumers continue to choose their doctors and their health plans, they remain the ultimate customer.

Are they likely to lose or give up that choice in the future? I wouldn't count on it. Anyone who thinks that the role of customer will slip from the hands of consumers needs to take a closer look at the demographics of the communities he or she serves.

Let's look at these customers whom a lot of experts would have us believe can be shifted like sheep from one corral into another. Consider, for example, those folks who already comprise 40 percent of the patient load in most hospitals and consume 70 percent of the healthcare resources: the "docile" elderly. Take a look at the American Association of Retired People. It has 28 million members and more than 5,000 state and local chapters. It is larger than any organization in the United States other than the Catholic Church. It is twice the size of the AFL-CIO. If it were a private company, its annual cash flow would put it near the top of the Fortune 500. It includes one in every four registered voters.

Here's another insight on the "malleable" elderly. They have a voter participation rate of 60 percent. The average for the non-elderly population is 30 percent. Today, they constitute 12 percent of the population. In 10 years, that number will surge to 40 percent. When people ask me, "What kind of care will the elderly get in the year 2000?" I respond by saying, "Whatever kind they want."

The Market Buys Value

It's become popular to suggest that all the healthcare market cares about today is price. That it is price—and only price—that shapes purchase behavior in healthcare. But no market and no customer buys on price alone.

Markets buy on value. And value is part of an equation with price only serving as the denominator. The numerator is comprised of benefits relevant to the customer: quality, peace of mind, reliability, access, breadth of service, compassion. The market always buys on value. And the market is inevitably—sometimes slowly, but inevitably—wise.

I've spent 20 years involved in some fashion with marketing, a year of that as chair of the American Marketing Association, and I've never seen the market get it wrong. It bought Hondas when it could have bought Buicks for less. It bought Sony when it could have bought RCA for less. It bought Seiko when it could have bought Timex for less.

Economists and most business consultants have a pretty narrow definition of what it means to be "value added." Value-added activities are those that directly create benefits or utility for the customer. Everything else is "overhead."

Some of the more hard-nosed observers of the healthcare industry contend that the only activities that are value added are those that directly touch patients or keep them healthy; if you don't, you're overhead. In healthcare, that leaves a lot of overhead. The organizations that will thrive in the health field in the future are those that can most dramatically impact both the numerator and the denominator of the value equation.

Your Mother and the Doctor

Having said all this about who the real customer is for healthcare organizations and what they want, let me suggest that there are some folks who should be sitting in a room with the management team during its one-day meeting—doctors. No one has more influence over the customer than the doctor.

The consumer isn't a set of actual tables. She's your mother. When she buys healthcare, she's purchasing what Jeff Goldsmith has described as "the most intimate of all human services." Someday, she may buy those services when her life is balanced on the precipice. And at that moment, her most precious relationship is not going to be with a health plan executive, a clerk, or a hospital administrator. It will be with a doctor.

> The consumer isn't a set of actual tables. She's your mother.

Every day, the most powerful force in the healthcare marketplace, the ultimate customer, sits down in the waiting rooms of doctors. No one in healthcare is in a better position to impact both the numerator and the denominator of the value ratio.

You've heard it before: 80 percent of the cost and quality in American healthcare flows out of the tip of a physician's pen. And no one puts more hands on patients or is in a position to more directly contribute to their well-being. It goes without saying, I hope, that also present in the room should be the other value-added players in American healthcare: nurses.

Get to Work

Now that everyone's in the room and there's been some earnest discussion of the customer and the importance of value, it's time to get to work:

1. Hand everybody three oversized index cards and ask him or her to write down (preferably using a marking pen and large, legible writing) what value customers are seeking when they visit the doctor's office, or the hospital, or whatever your healthcare organization happens to be.

   You may get some complaints like, "I can't get this all on three cards; I've got five values." (Stand firm. Such narrowing disciplines thinking and forces tough choices.) Or, "This is really hard." (Be sympathetic. It *is* hard.)

2. Tape these cards to the wall. Ask the group to consolidate the cards that are redundant or related to each other. Create a "header card" that succinctly describes the key customer value conveyed by each cluster of cards. (Even though you gave everybody only three cards, you could end up with five or six clusters.) Drop off the outlier cards that don't relate.

   If everyone truly took a customer perspective, then on the wall before you should be a set of cards that describes why you exist as an organization. On the header cards, you may find words like "diagnosis," "intervention," "recovery," "maintenance," or "prevention." Those are highly important words. They describe your core customer processes. Some of the clusters will have more cards in them than the rest. Those are your highest priority core processes.

3. Now have the group take each of the high-priority core processes and ask (again from the customer's perspective): "What has to happen to deliver the value of that process?" "What are the big macro-steps taken to accomplish the process?" Ask everyone to try to boil it down to between five and seven steps for each core process and write each step on index cards.

4. Then for each core process, put the macro-step cards on the wall. Have the group cluster redundant and related cards. This time, a connection should be established between the clusters. One cluster should lead to the next (or it may branch off into a separate track). Draw arrows on cards and place them between each cluster. Repeat this approach for each core process. The result should be a map that relates all the macro-steps for each core customer process.

5. Now ask the group to rate each macro-step in each core process diagram on a scale of one to ten based on the level of "value added" it creates for the customer. Depending on the size of the group and the amount of time available, you may want to dedicate a team to each of the core customer processes.

By the end of the day, the group should emerge with something it probably didn't have at the beginning of the day: an understanding of the basic work of the organization from the perspective of the customer and a sense of how much value this work creates for the customer.

Mapping Your Way to the Customer

As basic as it may be, this simple mapping exercise can foster an enlightened, productive, and sustainable approach to expense reduction because it creates a framework for deploying cost-reduction initiatives, generates a sensitivity for the interrelatedness of such efforts, and provides direction regarding which initiatives are likely to have the greatest impact (negative or positive) on customer value.

Such a map also establishes customer need and the processes that serve it at the top of the hierarchy of the many processes that make up the work of the organization. Finally, and perhaps most importantly, such high-level scrutiny of the organization helps pull expense reduction out of functional silos and into a systematic approach.

The object of creating the maps is not re-engineering (although it may help such an effort); it is understanding. Such understanding of the work of the organization is an appropriate first step for much that management must do, including strategic planning, organization design, marketing, quality improvement, and, of course, cost reduction.

Here are the basic rules for an effective single-day mapping exercise:

- Look at everything from the customer perspective.
- Keep it simple enough to explain to an eight-year-old.
- Place time limits on each step.
- Remember that the object is understanding, not improvement or re-engineering.
- Don't be constrained by flow-charting rules or tools.
- Map things as they are, not as they should be.
- Assume all the knowledge you need is in the room.
- Keep a high-level perspective.

This is a short article. Intentionally so. The approach that is outlined here is intentionally short as well. It is simple. But the deliberations and the thinking it embodies will be hard. It may tie your stomach in a knot. Keep at it. This may prove to be one of the more important days in the life of your organization.

The process described above was developed and applied at Central DuPage Health System in Winfield, Illinois, as part of that organization's Transformation Initiative.

J. Daniel Beckham is president of The Beckham Company of Whitefish Bay, Wisconsin, a healthcare consulting firm focused on strategy and integration. He can be reached at (414) 963-8935.

Reengineering or Rebirth?

Pamela L. Blyth

Health-care facilities of all sorts are seeking changes that make them more efficient and responsive to patients as well as many other stakeholders. This article profiles two such facilities and how they reengineered themselves, with a special emphasis implementing self-directed work teams. The article includes sound advice for any such facility seeking to make such changes. And although the term TQM is not mentioned, what these facilities were trying to do is entirely consistent with quality-management principles.

Reengineering, rightsizing, downsizing—whatever you call it, it is revolutionizing the way business is done in health care. As the industry continues its turbulent transition into the 21st century, no single department or facet of service delivery is left untouched by the process, and the managers of facility and support services have felt their share of the trauma.

For some, the trauma has been as simple and intense as being told to reduce staff to levels that are—to put it bluntly—ridiculous for the level of service they are expected to continue providing. For others, it has meant a quick and often fairly unethical (depending on your point of view) loss of jobs to outside contractors who make grand promises of having the expertise to "do more with less." And for some, it has meant just giving up: a retreat into other industries that offer a more sensible approach to service delivery and business management.

But are any of these approaches right . . . or necessary? Some of the managers who have survived will answer with a resounding "no!" These adventurous souls have fought against the odds to find solutions that meet the traditional characteristics of successful change and compromise. They have carved out unique and customized "win-win" situations for themselves, the employees for whom they feel responsible, the patients served by their facilities, and the business interests that now increasingly control—with ever less intervention from the clinical side—the delivery of care.

Don't misunderstand. There are no supermanagers with easy solutions; the solutions area always challenging and often painful. But they can guide others through the complex evolution to a new paradigm of service delivery and management. And the payoffs are amazing.

Several of these managers talked with *HFM* about their current organizational plans and the strategies they used to move from more traditional approaches to the way they are successfully organized today. Because of the dramatically different appearance that their organizations assume after this process, their stories are almost more about rebirth than reengineering. Most importantly, all have survived—and all feel the better for having experienced the process.

Case 1: The John M. and Sally B. Thornton Hospital, University of California San Diego Medical Center

This hospital is unique because, before it was built, the staff was able to develop an organizational strategy that was unorthodox for health care facilities. In fact, key faculty physicians worked side-by-side with staff to plan and prepare the facility for its opening in July 1993.

Although the UCSD Medical Center had long focused on quality health care, research and the education of doctors, it decided that an alternative organizational design for this new facility would be needed for this new hospital. The planning team began to examine existing patient-focused care programs in other hospitals. When they didn't find what they were looking for in other health care models, they turned to industry.

Although they remained committed to the concept of patient-focused care, it was in industry that they found the organizational model for an innovative and effective approach to the delivery of health care services: work teams. At Thornton Hospital, the concept of work teams eventually developed into *self-directed* work teams, an approach that has been successful hospitalwide.

Core Values

Shawn Bolouki, M.Sc., M.B.A., director of support services, was involved in the planning process for the new facility almost from the beginning. It quickly became clear that in order to support an entirely reengineered process with reengineered roles, a brand new team structure would be required.

Employee teams had to be empowered to make critical decisions if the patients' needs were to be met efficiently and effectively. But ensuring that appropriate decisions would always be made was a challenge.

In the end, reengineering the traditional organizational concept around *core values* provided the shared vision to keep everyone on track throughout the process. These core values were eventually identified as customer service, teamwork, integrity, caring, leadership, self-directed learning and crosstraining.

Self-Directed Work Teams

For Thornton, a self-directed work team (SDWT) is a highly trained group of six to eight individuals who are fully responsible for creating a well-defined segment of work, or even an entire work process. That segment can be either a final product, such as a completed insurance claim or an inpatient stay, or an intermediate service or procedure, such as those provided by an operating room team or a radiology team. An SDWT represents the conceptual opposite

> Although they remained committed to the concept of patient-focused care, it was in industry that they found the organizational model for an innovative and effective approach to the delivery of health care services: work teams.

What Are the Benefits of the Self-Directed Work Team Model?

The benefits to Thornton Hospital, including cost reductions and savings, are numerous. SDWTs have demonstrated their ability to manage more efficiently and increase the quality of service to their internal customers. Benefits include:

- **Reduced layers of management.** Teams are composed of highly skilled technical staff who, thanks to the absence of a supervisor, can share in decision making on everything from technical issues to administrative tasks.
- **Improved communication.** Better communication results from more opportunities to communicate directly instead of through multiple layers.
- **Increased productivity.** Efficiency of workers at Thornton increased to the point that they were able to absorb an additional 38,000-square-foot building without increasing the size of the staff.
- **Improved coordination of work.** Group decisions and ownership of responsibility have increased follow-through and timely completion of tasks.
- **Improved relationship with the union.** Increased collaboration with the union has resulted in an increase in trust and mutual respect because of the hospital's view of all employees as business partners.
- **Enhanced customer service and satisfaction.** Patients communicate directly with service providers while the nursing staff remains focused on clinical care. Team members use business cards and respond directly to requests from patients and other staff members, which has resulted in a 95 percent patient satisfaction rate.
- **Cross-trained staff.** Team behavior and core values encourage both formal and informal cross-training opportunities, because the competence of each individual affects the team's total performance.
- **Higher morale and job satisfaction.** All team members are "stakeholders" who see the value of their personal contributions; their support for each other results in peer evaluations that are aimed at improving work performance (instead of merely criticizing it) and a greater sense of accomplishment and contribution.
- **Higher quality.** The team effort has encouraged the use of all available talent, resulting in better, more effective solutions and improved service delivery.
- **Reduced cost.** Team development of 14 energy conservation interventions resulted in a reduction of $40,000 in energy costs for Thornton.
- **Costs savings.** The team is developing a model to determine "hard" and "soft" cost savings in comparison to the traditional engineering department.

> ### WHAT DOES A WORK TEAM DO AT THORNTON?
>
> - Identify its goals and objectives.
> - Plan and adopt a budget.
> - Monitor expenditures for suppliers/labor; recommend/make changes
> - Schedule work and maintain timekeeping records
> - Develop/implement training plans
> - Screen and interview job applicants
> - Conduct weekly staff meetings to gain agreement on assignments
> - Persuade new team members to adopt the team's goals/objectives
> - Set standards for performance
> - Self-correct team and individual performance through feedback
> - Write policies and procedures
> - Prepare for the JCAHO review.

of an assembly line, where each person assumes responsibility for a narrow technical function. The teams support the reengineered work processes and represent a dramatic, radical change to work.

Bolouki is quick to point out that an SDWT is not unmanaged; it is simply managed differently. At Thornton Hospital, there is a formal leadership structure consisting of leaders, coordinators and shift leaders. In addition, a more innovative leadership structure supports the decision-making activity of the team; it includes team leaders, coaches and shared governance committees.

Each team has both a leader and a coach. They either volunteer or are elected by the team to lead, coach, teach and support the team by acting as a conduit of information from the facility's leadership. Each serves for one year, and there is no additional pay for this position. Team leaders are mentored by coaches who attend ongoing training in team effectiveness. A new incentive program that grants both individual and team bonuses has been implemented as well.

Organization of the Teams

The 300 or so employees of Thornton are organized into 23 teams with a team leader and coach for each team. Thornton's leadership and staff have developed what might be called an "organization allergy" to organization charts. Although a chart exists, Bolouki emphasizes that the way the names are listed neither reflects any type of priority nor represents the true relationship between the teams.

For example, a typical medical/surgical team is composed of nurses, clinical care partners (hospital assistants), environmental care partners (responsible for food service and cleaning) and unit secretaries. The respiratory care practitioner and pharmacist are connected to that team, and they chart information on each patient in the same electronic file provide in-services to the

Thornton's leadership and staff have developed what might be called an "organization allergy" to organization charts.

> **THORNTON HOSPITAL'S GUIDING PRINCIPLES**
>
> - **Customer service.** Everything is organized to bring service directly to the patient and doctor, and services that might not normally be a part of the traditional functional unit have been brought into each team.
> - **Information services.** As much as possible, information systems are available where staff actually do their work so that they have ready access.
> - **No sacred cows.** The fact that something has always been done a certain way no longer matters if another method is more appropriate.
> - **Leadership.** Leadership is expected from everyone. "That's not my job" has been eliminated from the culture.

med/surgical staff, and participate in day-to-day decision making and delivery of care.

Facilities Services Teams

Bolouki previously had experimented in another facility with "cycle cleaning teams," which were renovation teams that coordinate complete annual renovation of patient rooms and other specifically targeted areas of the facility. Staff members with varied specialties from facilities engineering worked with clinical engineering and environmental services to complete the team's work. Although Bolouki met with these teams and provided resources, they were self-directed and totally responsible for developing their schedule and procedure for these cycle-cleaning assignments.

There was much conflict regarding roles and assignments in the first month. Gradually, with Bolouki's assistance, the teams were able to work effectively together. In fact, they successfully renovated 52 rooms under one budget with a mere two-day delay from the targeted completion date!

This first concept of a self-directed work team provided the model for the first Thornton facilities engineering team, which was the first team hired for the new Thornton Hospital. The team's job was to work directly with the contractor to complete the construction and readiness process for the facility. They did this largely on their own, with some assistance, coaching and mentoring from Bolouki.

The Transition to Team Behavior

The way in which the facilities engineering team at Thornton has evolved has interested project participants and observers. In its present form, it is structured and operates differently from its pre-operations design, thus providing a clear example of continuous process improvement in action.

Essentially, a team of watch engineers, maintenance mechanics, electricians, a plumber, a locksmith/carpenter, a painter, an HVAC mechanic and one administrative assistant provide support to the entire hospital. They do not have a coordinator or supervisor, and they elect a team leader and a coach to lead the team. The team is also supported by an internal team structure that includes an education and training committee, a hiring committee, and a budget committee, as well as specific individuals or volunteers who are responsible for safety, scheduling and timekeeping.

To improve operations, the facilities engineering team has a central customer service desk that allows them to respond quickly to requests for service for all support services departments throughout the hospital; a customer need call only one main number. The desk is staffed by one person on a five-day basis during day hours, but all staff are cross-trained in customer service and can take trouble calls. Customers continue to be fascinated that a request for temperature control can be made with only one phone call and that, using the automated building system and the $2.5 million computer system set up to control the environment, the temperature can be adjusted without anyone having to enter their room.

The facilities engineering team is now completing a comprehensive cross-training program that will allow anyone on the team to respond to the customer or quickly involve the necessary staff members in the request.

Future Plans for the Model
This model is earmarked for continuation thanks to its overwhelming success. Future plans include fine-tuning the self-directed work team approach to further reduce the number of job classifications. Bolouki says that leadership and staff now find it difficult to imagine working in an organization where layers of management exist on top of staff. The model also has been expanded to other support service areas, such as telecommunications, nutrition and environmental services, all now in various stages of development toward the self-directed work team model.

Recommendations
Structuring a nontraditional organizational model during times of change is, to say the least, risky. But the Thornton leadership team has been the recipient of a high level of trust by both the chief executive of UCSD health care and the Thornton operational leader. They continue to scrutinize the financial operations and achievements of the Thornton facility. Bolouki feels that support from the CEO and top leadership is essential.

The implementation of SDWTs is a deliberate and planned process—it does not happen by itself. It must be thoroughly planned, well implemented and given time to develop. Staff training is essential, and employees in supervisory roles should understand that they are considered key players in planning and training. They should then be given the opportunity to become part of the new model. Their value as resources and contributing players should never be underestimated. Open and frequent communication should be fostered and supported both emotionally and operationally. And everyone must understand the core values and guiding principles that drive behavior, including that of the vendors.

Bolouki also says that the key to success of the SDWT approach is in the empowerment of every staff member to address issues and correct performance problems themselves with little or no feedback from leadership. And

The implementation of SDWTs is a deliberate and planned process—it does not happen by itself. It must be thoroughly planned, well implemented and given time to develop.

finally, adherence to the core team values is what ensures that the decisions made will be appropriate. The Thornton leadership originally hired the team members based on those values, and the facilities engineering team hired its own team members using the same criteria; both groups did a good job.

Case 2: St. Charles Medical Center, Bend, Ore.

This 180-bed, 330,000-square-foot facility serves as a regional referral center for 140,000 people across 25,000 square miles of central and eastern Oregon.

Unlike Bolouki at Thornton Hospital, Michael B. Severns, director of facility services, didn't have the opportunity to design a facilities services department as part of the creation of a new hospital. His story is a prime example of how existing departments can survive the difficult transition from a traditional organizational structure to a truly self-directed work team model.

Motivation for the Change

In December 1993, executives decided to reorganize St. Charles into operational clusters in order to reduce operational costs, as well as improve service delivery. Eventually, the facility moved from a structure consisting of executive, senior, middle and supervisory management levels to a much more streamlined approach. The organization now looks very different. As a result of the transition to "cluster management," 54 traditional departments (with 54 directors) were eliminated and "reborn" as 16 clusters, each run by a leader/manager.

The 16 leader managers now comprise the Leadership Council that governs the day-to-day operations of the hospital. The Executive Council (senior management) now focuses its efforts and expertise outwardly on such issues as business development and strategic planning.

Operational Clusters

Each cluster is composed of four to seven functionally oriented teams. Supervisory positions have been eliminated because the teams are entirely self-directed with assistance from a team leader. These teams have clearly defined goals, which are to:

- Increase productivity
- Be action-oriented
- Be customer-driven
- Streamline operations

Severns was named the leader/manager of the facility services cluster, which now includes facility grounds, biomedical services, plant operations, environmental services, material services and linen services.

The Transition Process

Responsibility for the success of this new model rested squarely on the shoulders of all managers, and it was clear from the start that a team effort was required. To foster an environment in which team-building could occur, the intent to reorganize was communicated in a forthright manner, openly and honestly, from the beginning. Several town hall–style meetings were held on each shift to communicate and discuss the impending changes, as well as the rationale behind the decision to reorganize.

All existing management positions were eliminated, and everyone was given an opportunity to apply for the cluster leader/manager positions. The same was true for the team leader positions. Final selections were based on previous work performance and history, people skills, and management skills. Severns notes that any employee not selected for a leadership role was offered the opportunity to integrate into the new model or be given help and resources to look for another job.

Characteristics of the Team

Team leaders are empowered to coordinate day-to-day activities of the team within the budget parameters set by the leader/managers and the executive council. Although budgets are ultimately the responsibility of the leader/manager, all budgets are created, developed and implemented by the teams in a process coordinated by the team leaders.

Although individual team members have specific responsibilities, all members are cross-trained as much as possible to ensure effective service delivery. For example, environmental services project personnel are cross-trained to make building and equipment rounds. They help monitor the physical plant equipment, such as boilers and chillers, which enables the plant operations team to create more flexible schedules.

As at Thornton, the hospital planned and implemented most of the training inhouse. Training focused on techniques for managing change, people-centered team building, developing and implementing interaction agreements, and communication skills.

Staff Evaluation

The team leader positions in this model are considered permanent. They are evaluated both on their individual performance and the performance of the team. Monitoring the performance of each team is accomplished through:

- Weekly meetings with the leader/manager and the team leader
- Bimonthly meetings with the leader/manager and the entire team
- Ongoing communication reports that document continuous improvements and the measurements used to confirm them
- An annual report to the hospital's Performance Excellence Process Committee

All staff evaluations, including those of the leader/manager and the team leader, consist of self-evaluation, peer evaluation, and evaluation of demonstrated leadership, which includes a review of goals and objectives.

Reengineering the Technology

As the facility services cluster moved into this new self-directed work team model, one of the first logical activities to emerge was a staff analysis of the current use of technology. Using internal and external resources, Severns coordinated a needs analysis that focused on evaluating methods, technologies and systems to improve the operation and efficiency of the building while removing environmentally unsafe materials, such as PCB ballasts.

Again, the process was a team approach using the talents of all players. By benchmarking against other facilities, Severns and the team were able to identify cost savings opportunities that could be realized through redesigning building systems. In many cases, replacing outdated equipment with high-efficiency alternatives offered a more effective use of technological resources.

Severns and his team obtained approval to develop a partnership among the facility services' cluster teams, the HVAC performance contractor and the local utility company. Once management reviewed and approved the details of the plan and its potential impact on cost containment, it simply became a matter of partnering with the utility to implement the details.

Because the company realized the substantial benefit of power saving, it shared the cost of the program, thus reducing the facility's financial investment. In fact, St. Charles earned an energy rebate of more than $500,000. Other benefits included better defined control over temperatures for specific areas, improved availability and delivery of hot water, and increased customer response efficiencies.

Cost Savings

By adopting a "green" policy, the facilities team reduced utility consumption by 40 percent over two-and-a-half years while making the hospital environment a safer and more comfortable place for staff, patients and visitors.

Increased team productivity and technological upgrades have made possible a significant reduction in outside contracts and services, which has translated into additional dollar savings.

Increased team productivity and technological upgrades have made possible a significant reduction in outside contracts and services, which has translated into additional dollar savings. One particularly noteworthy accomplishment was reduced nitrogen costs. By adapting a process used in the wine-making industry that takes compressed air, passes it through a membrane, and separates the gas into nearly pure nitrogen, St. Charles became the first hospital in the country to produce its own supply on-site.

In all, the facility team at St. Charles increased the total building size by 30 percent while decreasing facility operating costs by an overall 15 percent—and all with no increase in positions!

The Impact of Reengineering

Severns has made presentations to the community to show the hospital's success as well as how others can benefit from similar projects. And the U.S. Environmental Protection Agency presented the facility with a commendation for the scope and effectiveness of the project.

These successes have permanently changed the way the St. Charles evaluates and uses facility management technology. A commitment has been made to incorporate energy efficient design features, as well as equipment and fixtures, in all future construction and renovation. All architects and engineers who work for St. Charles in the future will have to design not only the buildings themselves but also the systems they use. And the systems will be user friendly and complement the healing environment.

Recommendations

Severns encourages benchmarking and data trending for teams that want to keep improving their systems. The better the use of technology, the more efficient the operation. That includes such simple things as using e-mail capabilities of local area networks to reduce the paperwork in scheduling tasks, appointments and meetings. In fact, he encourages the use of any electronic media that will facilitate document sharing, work processing or the use of spreadsheet applications.

But the most important element of successful reengineering, according to Severns, is the serious commitment to training and mentoring all managers and staff. Learning more effective ways of communicating, interacting and problem-solving are key components to building a successful reengineering team.

He encourages staff to question every aspect of operations, from the frequency of preventive maintenance measures to the method of scheduling staff. Fine-tuning operational systems should become a continuous and self-perpetuating activity that results in automatic quality improvements and cost savings.

The facilities manager of the '80s typically handled plant operations, biomedical electronics, grounds, renovation and construction projects, and often environmental services and security. The manager of the '90s and beyond will also have to coordinate some of the following:

- Technology, including data communications and network management
- Materials management, including vendor relations and partnerships, visual management and inventory systems
- Planning and development, including new building designs and healing health care environments
- Research and development projects
- New business opportunities

According to the facility managers who are successfully "weathering the storm," one thing is certain: the days when one brilliant manager had total responsibility for generating all decisions and solutions are over—it's no longer an efficient way to deliver service.

Empowering high-performance teams is essential if services are to be efficiently delivered in tomorrow's health care environment. Staff talents and problem solving abilities must be used to the max. Creating an environment that not only helps staff make this paradigm shift, but supports this new mode of teamwork, is emerging as the real challenge for today's health care leaders.

Pamela L. Blyth is president of Pamela L. Blyth & Associates, Durham, North Carolina.

Public Sector
HIGHER EDUCATION

Is Higher Education Ready for the Twenty-First Century?

Ronald L. Heilmann

This article by consulting editor in higher education Ron Heilmann examines the important issue of whether higher education is ready for the information age. Right now, most universities are clearly operating under nineteenth-century industrial-age assumptions, not coming to grips in meaningful ways with the changes wrought by modern communication and information technology or the demands of the new workplace. Improving and reforming the system so that it can deliver services more efficiently and effectively is at the heart of this issue. This article raises important questions that higher education must address over the next several years.

As the millennium approaches, views pertaining to nearly every facet of society, indeed our world and beyond, are being advanced, discussed, debated, and accepted or rejected with increasing frequency. Although this can take on the character of a popular new game, the opportunity for serious, future-focused thought and dialogue should not be passed by. This is especially true if the subjects of such thoughts and dialogue are those facets of our world that have great potential contributions to the quality of life, as does higher education.

The intent of this article is to stimulate thought and dialogue about higher education and its role in the twenty-first century. More questions will be raised than answers provided. It is hoped that the process of thinking and talking about the questions now will sow the seeds for an early initiation of those changes that higher education will have to make to be ready for the twenty-first century.

Two of the future-focused views pertaining to higher education appearing during the past year are particularly thought provoking. The first is a study titled "Transforming Higher Education: A Vision for Learning in the twenty-first Century" written by Michael G. Dolence and Donald M. Norris and published by the Society for College and University Planning (SCUP). The second is an address given at the 1996 annual meeting of the American Assembly of Collegiate Schools of Business (AACSB) by its outgoing (1995-96) president, Scott S. Cowen. Dr. Cowen, dean and Albert J. Weatherhead III professor of management at the Weatherhead School of Management at Case Western Reserve University, titled his address "Are We Going to Shape the Future or Is the Future Going to Shape Us?" Each view is independent and offers a very different perspective.

The SCUP study starts from the premise that society is undergoing a fundamental transformation from the Industrial Age to the Information Age. The study then postulates that "Those who realign their practices most effectively to Information Age standards will reap substantial benefits. Those who do not will be replaced or diminished by more nimble competitors." Thus the core issue of the study is an assessment of whether higher education has done, is doing, or seems prepared to do enough to meet Information Age standards.

The assessment process provides a point-by-point comparison of a set of descriptors offered as the Industrial Age (present) higher-education paradigm with a set of descriptors constituting the Information Age (near future) paradigm. Both sets of descriptors are listed in table 1. For purposes of this article, I will elaborate on only a few of the comparisons—enough to provide a sense of the case being made by the authors.

The initial comparison is of the present "teaching franchise" and the future "learning franchise." The teaching franchise is described as the current system of credentialed faculty in accredited institutions awarding course credits to and conferring degrees upon class-attending, tuition-paying students. This system is contrasted with information and knowledge bases, scholarly exchange networks, or other learning mechanisms readily available to anyone choosing to access them and who has the resources to compensate the provider.

Other critical differences between the paradigms that directly impact the *learning* process include passive versus active learning; set time for learning versus individualized time and pace; time out for education versus just-in-time learning; and separate learning systems versus unbundled learning experiences. Each of these elements of the paradigms relate to the manner in which students choose to learn. The traditional classroom will be one of options, but technology will make many more options available; thus higher education can anticipate competition.

Several other elements relate to organizational facets of how educational services are or will be delivered. From this perspective, the current model is labeled the "factory model." This label derives from the fact that the present organizational structure is designed for the efficient "mass production" of teaching. However, in an environment in which *customers* are demanding

Table 1. **Higher Education Paradigms**

Industrial Age	**Information Age**
Teaching Franchise	*Learning Franchise*
Provider-driven, set time for learning	Learner-driven, individualized learning (including time for and pace of)
Primarily passive learning	Primarily active learning
Information infrastructure as support tool	Information infrastructure as the fundamental instrument of transformation
Individual technologies	Technology synergies
Time out for education	Just-in-time learning, fusion of learning and work
Continuing education	Perpetual learning
Separate learning systems	Fused learning systems
Traditional courses, degrees, and academic calendars	Unbundled learning experiences based on learner needs
Teaching and certification of mastery are combined	Learning and certification of mastery are related, yet separable issues
Front-end, lump-sum payment based on length of academic process	Point-of-access payment for exchange of intellectual property based on value added
Collections of fragmented, narrow, and proprietary systems	Seamless, integrated, comprehensive, and open systems
Bureaucratic systems	Self-informing, self-correcting systems
Rigid, predesigned processes	Families of transactions customizable to the needs of learners, faculty, and staff
Technology push	Learner vision pull

Source: Adapted from "Transforming Higher Education: A Vision for Learning in the 21st Century"

"mass customization," new organizational structures are likely to appear because, once again, technology will make new options available. The SCUP report refers to these new technology-based systems as "knowledge networks," which enable "network scholarship," or "network learning."

Inherent in the choices available to learners will be responsibilities. The SCUP report states bluntly that "Learners in the Information Age will not be able to afford the luxury of on-again, off-again learning." It will be the learner's responsibility to stay "connected" and pursue learning throughout his or her lifetime. It will be the learner's responsibility to learn how to navigate the knowledge networks to achieve specific learning outcomes.

The choice within higher education is the extent to which to be a participant in the new network-based systems or a provider of related services. Because the future envisioned within the SCUP report is not that distant, *current leaders* throughout higher education are likely to be making many of these necessary decisions. Therefore, assessing the state of academic leadership seems appropriate. Dean Cowen's address at the 1996 AACSB annual meeting did just that—focused on academic leadership. And though his remarks were made to deans of business schools, they are more generally applicable.

He acknowledged an array of forces currently having an impact on education by stating, "In recent years we have witnessed significant change in management education at the undergraduate, graduate and executive education levels. The catalyst for these changes are the macro-environmental trends facing all organizations—globalization, technology, diversity and rising expectations of consumers. Added to this list is the reality that higher education is under close scrutiny by the public and increasingly viewed in a critical light. Whether we like it or not, the public often believes that higher education is not appropriately responsive to its stakeholders, is not committed to being productive, and is falling behind in producing the types of graduates needed to lead organizations and society in the twenty-first century."

Dean Cowen asserts that "academic leadership is the single most important issue facing higher education . . . as we approach the twenty-first century." And it is the "exercise of effective leadership" that will enable making the changes necessary to regain the confidence of the public.

Cowen challenges current leaders to undertake self-assessment regarding the provision of effective leadership. He accomplishes this by asking a series of questions:

1. Are we providing the necessary intellectual leadership for executives and organizations as we approach the twenty-first century?
2. Are we operating our schools in a way that promotes overall high quality, continuous improvement and assurance to those we serve that we practice what we preach?

To give perspective to this question, he suggests considering the following statement:

Assume you were told the following characteristics of a hypothetical organization in a hypothetical industry:

- The services provided are driven by what the organization wants to do.
- Customer service/responsiveness are anathema to the organization's culture.
- A majority of the organization's costs are fixed and committed.
- A significant percentage of its workforce has lifetime employment contracts without incentive compensation or systematic performance processes.
- The culture often values process more than results.
- The organization's key human resources can spend at least 20 percent of its time on activities external to the organization, including working for a competitor.

- The leaders of the organization often lack the knowledge and skills needed to lead and manage an effective organization.

It is clear that such characteristics are not likely to be associated with an organization promoting high quality and continuous improvement. And it is no accident that these characteristics are those some may attach to higher education.

Additional questions that flow from question 2 and the hypothetical scenario include

- Are we willing to assume the personal risk . . . to make the kinds of radical infrastructure and cultural changes that are needed in many of our organizations?
- Are we ready to tackle the issue of tenure?
- Are we ready to empower others to foster entrepreneurial behavior in our institutions?
- Are we willing to become learning organizations?

He then inquires:

3. Are we and our schools truly committed to fulfilling a mission based on learning and outcome assessment?

Again, additional questions flow from this and include

- What would my school look like—in terms of programs, processes, performance indicators—if learning were its primary mission?
- Would we teach classes on Monday, Wednesday and Friday in fifty-minute periods?
- Would courses be designed as fixed seven- or fifteen-week blocks?
- Would our students have to be physically present on campus?
- Are course grades necessary?
- Should students solve cases written by others or should they become case writers?
- How should we measure the quality and impact of our research?

Cowen's fourth and last central question is

4. Are we properly preparing the next generation of faculty and academic administrators in the context of the future, or are we clinging to the traditions of the past?

Here he is clearly taking a longer view, recognizing that bringing about the changes implied by many of his questions will fall to the next generation of leadership. All of the questions in the synopsis of the SCUP report and those quoted from Dean Cowen are designed to force critical introspection by professionals in higher education. There are no right or wrong answers. On the other hand, we must accept that maintaining the status quo is not an acceptable answer. Nor is, as stated in the SCUP report, merely superimposing "a silicon veneer onto outmoded Industrial Age systems, techniques and organizational cultures."

In 1989 Peter Drucker stated, "Today, we are in the early stages of a social and technological revolution that should drastically and irrevocably change the meaning of education and the art of teaching in our society. *In fact, education will have to change more in the next three decades than it has since the modern school was created by the appearance of the printed book over 300 years ago*" [Emphasis added.]

Now that the approaching millennium is motivating others to become future focused, attention is being given to what those changes foreseen by Drucker will be. It is hoped that this brief summary of some of the current thinking about higher education's future will motivate more of us whose careers are linked to higher education to think about how we are going to prepare for that future and contribute to it.

References

More on Dean Cowen's views can be found in his coauthored book: *Innovation in Professional Education: Steps on a Journey from Teaching to Learning* by Richard E. Boyatzis, Scott S. Cowen, David A. Kolb, and Associates, published by Jossey-Bass, Inc.

TQM in the College Classroom

Ronald E. Turner

This article by a university economics professor provides fifteen quality concepts translated into specific actions that college faculty and administrators can use to improve the quality of education. These actions take into account the needs of customers—students—and how to use their feedback for improvement.

Adopting the philosophy of total quality management (TQM) in the classroom could revolutionize how education is pursued. Unfortunately, I have found that most college faculty members perceive it as an alien business philosophy, especially when terms such as "customer" and "value" are used. Since incorporating TQM into my classroom, faculty members have discussed their concerns with me regarding the use of TQM in education. The following 15 ideas address those concerns and should put to rest many of the misconceptions educators have about TQM. The underlying philosophical implications TQM can have for education will help resistant faculty see why TQM works not only in business, but in the classroom as well.

Embracing a Customer Focus Doesn't Mean Giving Students All A's

The idea of treating students as customers is probably the most controversial for educators because of the implied shift in power. Customer focus means that a customer's needs should be met, and success is determined by how well those needs are met. Faculty, therefore, would define their success by how well they meet their students' needs. This means asking students, "What do you need in order to call your experience here a success?" The predictable answer, "straight A's and no work," worries some instructors.

While this statement does answer the question, the answer does not meet the students' needs. A school that gives automatic A's and has no standards will have no value. Kaoru Ishikawa, a leading Japanese theorist regarding quality, warned that, "Yes, [the] consumer is king. But there are too many kings who are blind."[1] Embracing a customer focus, therefore, does not mean giving out straight A's and abandoning standards.

A customer focus in the classroom means that instructors should measure success by how well students are learning and how worthwhile the students find their class experiences. Just because students sit in the class does not mean that they are getting something out of it. In addition, customer focus suggests that instructors become open to student feedback because it is the most important feedback that they can receive regarding quality.

If Students Fail, the System Has Failed

When students fail in a non-TQM system, typically they are blamed for not working hard enough or for not being adequately prepared when they started the course. If the student is treated as a customer, however, failure will be interpreted as meaning that the student's needs were not met. Thus, the focus shifts from telling students to work harder to making the system work for students.

Routinely, I used to fail 25% to 30% of the students in my introductory economics course. Clearly, the needs of these students were not being met. After exploring the reasons why so many students were failing, I concluded that many of them lacked the math skills to read graphs and interpret simple equations.

The first week of class, I started giving a math pre-test and warned students to drop the course and take an algebra course if they couldn't answer all of the math questions correctly by the end of the first week. The first year I did this, 25% of the students dropped the course in the first week. My failure rate correspondingly dropped to 5%.

It took two years of giving the pre-test before the word got out that students should understand algebra before taking economics. I no longer get the high dropout rate in the first week, and the failure rate of the economics course has remained very low. Enrollment in algebra classes has increased significantly.

Faculty Members are Customers of Those Who Teach Students Prerequisite Courses

Instructors who teach prerequisite courses are in the role of supplier to those who teach courses that have prerequisites. After adopting TQM, I started asking instructors in the business department which economics concepts they would like me to focus on in order to make their jobs easier. I was surprised to discover that the marketing course, for instance, relies heavily on the concept of demand in order to teach price theory. Discussion must occur between instructors about material that will be used in other courses downstream. I have requested that the marketing instructors start giving me feedback on what students remember about economics. This will help guide improvement efforts in my class.

Students Are Better Off Quitting Than Flunking Out

When I encounter students who don't come to class because they have no interest in the material, I advise them to seek other alternatives. Some students complain that the course is required for their programs, but they disagree with the requirement. When they tell me this, I encourage them to discuss with their department chairs either changing the requirement or finding an option that better fits their needs. Some students admit that they really shouldn't even be in school. They are better served by quitting than flunking out. Failing will result in an F on their transcripts, whereas quitting will result in a W (withdrawal). Future options for these students will be more open with a W than with an F.

As part of my end-of-the-year student feedback, I ask students if they agree that my course should be a requirement. The results are passed on to the department and the next group of students. My reasoning for sharing these results with students is that typically the reputation of a course shapes beginning expectations. By giving students feedback from the previous class, I hope to offset false expectations.

Treating Students as Customers Means Allowing Students to Choose Not to Come to Class

Instructors should not take on the role of parent or boss. Instructors are supplying a service. If students don't come to class, that is their choice, just as they can choose not to go to a concert for which they have bought a ticket. On the other hand, when students miss a concert, they don't expect the band to give them a private performance. Instead of being judgmental about students who miss class, instructors should ask themselves, "Why do students prefer skipping my classes?" Getting rid of attendance requirements is like lowering the water level in a stream. It will reveal the underlying boulders that hinder passage down the stream. Instructors need to recognize that if the only way they can get students to attend their classes is through threats, then there must not be much of value taking place in the classroom. Some instructors worry that this gives students too much credit for being able to recognize value. However, students are adults ranging in age from 18 to 70; they don't need an instructor to hold their hand. They can tell when they are learning something. They also can tell when a class is a waste of time.

Students Can Sometimes Learn the Material Without Coming to Class

Some of my students choose to skip classes and only come on the days of tests. Frequently, I discover that some of these students can get A's and B's by reviewing classmates' notes and reading the appropriate materials.

Some instructors become infuriated by this kind of behavior. But, if students can master the material without coming to class, that's okay. Their underlying needs have been met and standards have not been abandoned.

Completing the Syllabus Is Not a Measure of Success

At the 1993 GOAL Conference in Boston, MA, a speaker from the University of Mississippi told a story about the Midland, England, bus system. Passengers of the local bus line started calling the local newspaper to complain about bus drivers who were failing to stop to pick up passengers. Apparently the drivers would grin and wave as they drove by the bus stops. When the bus company was contacted by the newspaper, the company explained that the drivers had discovered they couldn't keep on schedule if they had to stop for the passengers.

Unfortunately, too many instructors drive their syllabuses in the same fashion. They consider it more important to complete the syllabus than to take any students along for the ride.

New and Tenured Instructors Should Visit Each Other's Classrooms

Whose classrooms should be visited: new instructors' or old timers'? The traditional non-TQM viewpoint is that administrators and tenured instructors should visit the new instructors to determine if they are good enough to retain. A classroom visit, however, should be a learning experience rather than a fault-finding mission.

By visiting the classrooms of instructors who are known to run the best classes, both new and tenured faculty members can learn new methods to use in their own classrooms.

Therefore, rather than inspecting what new instructors are doing, administrators and other faculty can visit to discover what can be learned from them.

Instructors Need to Work On Improving the Educational Process

Time spent on continuous improvement efforts should consume 5% to 10% of the work week. For instructors, this will mean taking three to five minutes out of a 50-minute lecture for continuous improvement purposes. Gathering feedback from students or brainstorming with students as a means of improving the learning process are two methods that can be used.

Time spent on continuous improvement efforts should consume 5% to 10% of the work week.

Once a week, I ask students to complete forms with the following feedback questions:

1. Was the class today worthwhile?
2. Any comments/suggestions for improving this class?
3. Did you do any reading/homework for this class?
4. Any comments/suggestions regarding the reading?
5. Any questions you want answered?
6. How many hours of studying have you done for this course since the last class?

These feedback forms reveal helpful information. For example, I showed a film that took an entire class period. The next class, I conducted a discussion about the film. The discussion proved to be a dud. The students' feedback suggested I stop the film every 10 minutes to discuss that 10-minute segment. I tried this the next time I showed a film, and the response was enthusiastic. Later, a communications specialist told me that films should be interrupted every seven to 12 minutes for discussion.

Taking time to collect student feedback hasn't cost me instruction time. Rather, the suggestions I get save me time by improving my teaching effectiveness. I spend less time going back over materials that students didn't understand the first time.

Feedback from Students Can't Be Used against Instructors

I don't have trouble handing out student feedback forms when a class goes well, but I am resistant to handing them out when a class doesn't seem to go well. If these feedback forms were to be reviewed by the academic dean, that resistance would be compounded. Data collection should be used for continuous improvement, not as a way to inspect instructors.

Interestingly enough, I have discovered that even on days when students have lacked enthusiasm, they still gave positive ratings. On several occasions, when students have reported that the class wasn't worthwhile, they have said it was not because of me, but because they were feeling sick or tired or hadn't done their assignments.

Get Rid of Performance Appraisals

W. Edwards Deming was correct when he advocated ending performance appraisals.[2] The annual faculty evaluation used at my school focuses on the past, rating people in different areas as being satisfactory or needing improvement. For the most part, these evaluations are used to get rid of "bad apples."

Evaluations should look to the future. Discussions should concentrate on what the administration can do to help people do a better job. Instructors should be asked about long-range ambitions so that these can be melded into the future focus of the organization. Tools that will assist faculty to become better teachers should be offered.

For faculty members who have not yet proven themselves in the classroom, feedback should be provided that lets them, and administrators, know if they are performing to some minimum standard. New instructors need to be monitored and discussions aimed at improving current performance. An ineffective teacher should not be left alone until the end of the year; that individual must be helped from day one.

This is a very different attitude from assuming that teachers are born good or bad. It means that instead of seeking reasons for their dismissal, the administrator should seek ways to help them improve.

> It means that instead of seeking reasons for their dismissal, the administrator should seek ways to help them improve.

No Matter How Good the Test, Luck Will Be Involved

Students learn at an early age that testing is part luck. Because of time constraints, tests can't cover everything. If students are lucky, they will have studied the right things, even though their overall knowledge might be limited. If they are unlucky, their test scores will reflect those areas that they didn't study as intensely, and will give a too-low measure of their overall mastery of the subject. This is true regardless of the testing format.

This does not suggest that student test scores can't be an indicator of study efforts and underlying ability. But—and this is critical—test scores will always be flawed because luck will have played a role in them.

I have changed my test policies to reflect the role of luck in two ways. First, students are given the right to take a test again if they believe the test has underestimated their knowledge of the subject. About 5% of students take me up on this. Second, instead of giving three preliminary tests and a final exam, I now give 10 to 12 tests a semester. When there are fewer tests, the impact on a student who is having a lucky or unlucky day will have more significance.

When Luck Is Involved, Test Results Should Come Out in a Bell-Shaped Curve

Many educators rely on grading by a curve. I have seen final exam scores ranging from 90 to 30, with 55 to 60 being the average. Instructors who look for bell-shaped curves assume the curve reflects underlying ability, much like an IQ test, and they interpret bell shapes as being indicative of a good test.

The bell curve, however, is just as reflective of the outcomes of rolling two dice. In fact, the more luck there is in a process, the bigger the range between low and high scores and the more likely a bell-shaped distribution will occur.[3] Typically, I introduce statistical process control to teachers by showing them the control limits for real-life test results. They are always surprised to see that when average grades are low, the control limits are farther apart, hence indicating that more luck is involved.

Beyond this, if faculty perceive students as falling into two distinct groups—those who study and those who don't—then a bimodal distribution should instead occur, reflecting this underlying difference. If all students are studying, then a bell-shaped curve should result, but the difference between high and low should be small.

Faculty Members Should Attempt to Bring Their Teaching Styles into Alignment

Lack of consistency in teaching due to different teaching styles, textbooks, and testing procedures makes it difficult for instructors to learn from others' experience. If several instructors teaching the same course cannot come to an agreement about how to teach the course, then it will be more difficult to learn from one another.

When there is more than one person teaching a particular course, it is advantageous for all instructors to reach consensus on course content, books, tests, and other materials. For many educators, this is seen as an attack on their individuality.

It is critical that educators understand that consensus means no one is being forced to adopt standards, books, and techniques of which they disapprove. Consensus means each person has a voice in the decision. A discussion aimed at reaching consensus will enable instructors to realize their philosophical differences and will help the group start forging a common vision of what education is all about. There should be no peer pressure or coercion from administrators.

Adjunct Faculty Members Should Be Sought for the Long Term

Schools should not use adjunct faculty who serve only a semester or two and then move on. The school should be focused on finding faculty members who will stay for the long haul and are committed to continuous improvement of their teaching processes.

> The school should be focused on finding faculty members who will stay for the long haul

Adjunct instructors who only teach for a semester or two won't have time to improve and develop their course curricula, search out better textbooks, or discover the learning styles of students attending the school. Moreover, adjunct instructors who come and go will not be in a position to work with other faculty members. They will not share in the learning of others and won't be able to share their own experiences.

TQM in the Classroom Can Transform the Educational Process

While total quality concepts appear focused on business, they are extremely beneficial when used in the classroom. They should not be perceived as threatening. Students are called customers, but they are still students; classes may be skipped, but the course still exists. Applying TQM in the classroom means that instructors consider how their students can effectively learn the material they have to offer. TQM is a way of thinking. Even if all 15 concepts are not adopted, they will promote the fundamental discussions that are needed among educators if a true transformation of the educational process is to take place.

Notes

1. Kaoru Ishikawa, *What Is Total Quality Control? The Japanese Way*, translated by David J. Lu (Englewood Cliffs, N.J.: Prentice Hall, 1985).
2. W. Edwards Deming, *Out of the Crisis*, (Cambridge, Mass.: Massachusetts Institute of Technology Center for Advanced Engineering Study, 1986).
3. Ronald E. Turner, "What Is the Probability That Your Employees Will Succeed?" *Quality Progress* (June 1995).

Ronald E. Turner is an instructor at Eastern Maine Technical College in Bangor, Maine. He received a master's degree in economics from the University of Maine in Orono. Turner is a member of ASQC.

Inside Track to the Future

Richard Alfred and Patricia Carter

This insightful piece by two community college activists explains what community colleges must do to remain competitive in the future with alternative sources of education. The authors explain the importance of being agile (though they don't use that word), which will require rethinking all the traditional assumptions of community-college education. The authors discuss and describe the three Cs that everyone involved in community colleges must focus on: competitors, customers, and culture.

A looming issue on many community college campuses is the need to bring outdated administrative structures and systems into line with new visions about what a college should be and how it should deliver education. Discussions with college leaders about organizational change invariably turn to strategies that can be used to redesign the institution. The conventional wisdom asserts that colleges which fail to change will lose ground to competitors who are eager to capture new markets. "Losing ground" is easy to understand—it is a loss of market share followed by a decline in enrollment and operating revenue and, ultimately, in competitiveness.

Despite the pervasiveness of this conventional wisdom, we are only now beginning to discover that our colleges are falling behind. Burdened with high fixed costs, outdated programs and management systems, and hierarchical administrative structures, they are caught in a vortex of change that threatens to undermine their competitiveness. Tinkering will not be enough. The only way out will be to develop structures and leadership that create new forms of value for students and, in so doing, enable community colleges to get to the future first. The purpose of this article is to describe some of the changes that community colleges will need to make to compete effectively. Our intent is to view change as a variable process that can move quickly or slowly depending on the interaction of three C's: *competitors, customers* and organizational *cultures*. We start by looking at the competitive arena in which our colleges are now playing.

> Burdened with high fixed costs, outdated programs and management systems, and hierarchical administrative structures, they are caught in a vortex of change that threatens to undermine their competitiveness.

Challenges from New Competitors

Competitors are fast at work reshaping the postsecondary education market. Five in particular are having a significant effect on educational design and delivery, yet have not been met with adequate responses from community

---

Reprinted with permission from *Community College Journal*, February–March 1996. Copyright © 1996 American Association of Community Colleges. All rights reserved.

colleges. They include: 1) companies and corporations providing on-site programs for current and future workers, 2) corporate giants in the communications industry with a capability for distance delivery into homes, workplaces, shopping centers, and areas where people congregate, 3) supplementary education providers, such as private tutoring companies, which use proven techniques to produce positive learning outcomes in students, 4) K–12 schools partnering with business and industry to prepare work-ready youth, and 5) temporary service agencies using training programs to prepare flexible workers for many different jobs.

What are these competitors doing, or could they do, to challenge the preeminence of community colleges? They are creating value in ways that surpass our colleges. Value is nothing more or less than the advantages of education provided by an institution that makes it more or less desirable. It comes in many forms, some of which are summarized in Figure 1. Community college faculty and administrators are most likely to recognize value in the form of cost, which makes education more affordable to students. Proprietary schools, however, can also deliver value through cost by streamlining courses and curricula to move students into the market more quickly, thereby reducing foregone earnings. Value can be realized as convenience which makes access easy by bringing courses and services directly to the customer.

Potentially, telephone and cable companies are in a strong position to make education convenient for students through distance delivery. Value can also be created in the form of great programs and services which attract students because of their distinctive design and delivery. Take, for example, the skill of corporations with high-powered training programs (e.g., Motorola and General Electric)—first in identifying employee needs, next in high-quality program design, and then in fast program development. These capabilities transform otherwise pedestrian training programs into niche programs that can be delivered outside of company walls and bring the corporation that developed them into the business of education. Value is also achieved through customer intimacy—operating practices which distinguish one institution from another because they reach out and identify beneficiary needs and find ways to help them achieve important goals. Some private colleges have succeeded in competition with community colleges through student intake procedures (i.e., admissions and financial aid) which anticipate student needs and provide a direct link (in the form of job guarantees) between education and work.

Finally, value can be created in the form of maverick ideas which provide a competitive advantage to organizations which invent totally different ways of delivering education and surprise and outstrip traditional institutions. K–12 schools partnering with business and industry to introduce children in the primary grades to technology and work skills could potentially reshape community college enrollment patterns by preparing students directly for work in corporations. These students would move from high school to full-time employment, eventually appearing on college campuses as part-time learners already engaged in careers, not as full-time students seeking careers.

FORMS OF VALUE IN POSTSECONDARY EDUCATION

Core processes that...	Maintain efficient delivery systems and control costs	Provide accessible programs, courses, and services and market them skillfully	Generate new ideas, translate them into programs and services	Provide solutions/ Help students succeed in programs, courses	Invent totally different ways of doing things that surprise and outstrip competitors
Organizational structure that...	Has central authority and a finite level of empowerment	Uses operating units to determine student preferences for courses, program	Decentralizes authority and acts in a flexible, loosely-knit, and ever-changing way to achieve	Encourages operating units to reach out in determining student needs and finding important goals	Frees leaders at all levels in the organization to identify processes and delivery systems that depart from current practice.
Management systems that...	Maintain standard operating procedures and unit costs	Constantly identify client needs and strategies for delivering accessible programs and services	Reward creative and innovative ideas of individuals and put them into practice immediately	Allocate resources to providing customer service and maintaining customer loyalty	Value different ways of doing things and underwrite risk in development
College culture that...	Acts predictably and believes that quality, low-cost operations can meet the needs of most clients	Views students and off-campus groups as primary customers and assigns value to institutional responsiveness to changing needs	Experiments and thinks outside "current" practice at all levels in the organization	Is flexible and thinks that customers should receive maximum assistance in achieving specific goals	Permits freedom for maverick managers to pursue ideas and see them through to implementation

Building tomorrow's delivery systems

Shoring up today's programs

Figure 1.

Value, whatever its form, will be the vehicle that determines success in tomorrow's market. Organizations which succeed will find ways to deliver programs, courses, and services better and more economically than other organizations. Motorola, for example, possesses a range of resources that could yield a competitive advantage in the delivery of postsecondary education. Using the interplay of values described in Figure 1, Motorola could choose to partner with a telecommunications firm and deliver education at reasonable cost to customers in homes, workplaces, or any convenient location. It could offer truly distinctive courses and curricula by bringing together the best work of countless teachers and authors on the information highway. It could establish its own form of credit for these courses and circumvent the traditional

Value, whatever its form, will be the vehicle that determines success in tomorrow's market.

accrediting apparatus by connecting credit with important outcomes (e.g., jobs, job skills, and advanced technology training). The academic semester and credit hour would become irrelevant and so would when, where, and how a course is taught. Success would be determined by the number of students enrolled and the speed at which they move through courses and into jobs. Most important would be value created for the customer through a combination of cost, convenience and quality.

What is astonishing about competitors like Motorola is their focus on the "customer" and their capacity to reinvent organizational culture to serve new and existing markets. Research has shown that five factors contribute to their success:

- Well-developed competencies that serve as launch points for new programs and services.
- An ability to fundamentally renew or revitalize by periodically changing the services they deliver and delivering new ones.
- A focus on "operational excellence" which in some way distinguishes the programs and services offered by the organization from those offered by competitors.
- A flattened decentralized organization that enables staff to identify and respond quickly to changing needs.
- A dedication to "customers" which enables the organization to "live" the customer's problems; it understands that the best program or service isn't the best value if the customer is unable to use the program or service effectively. (Senge, 1990; Prahalad and Harnel, 1990; Schein, 1993; Treacy and Wiersema, 1995)

Competitor organizations have ways of creating and delivering value that set them apart from community colleges. They focus on learning required to make transformational changes—changes in basic assumptions needed to succeed in today's fast-moving, often turbulent market.

Tinkering or Fundamental Change?

If competitors are seeking and finding new ways to deliver value, what are community colleges doing to keep pace? The answer is "tinkering" within the organization and "leveraging" resources. Tinkering involves minor adjustments with programs, services, and management systems as a response to change while leveraging involves increasing possibilities for accomplishing goals through incremental resources. Although both are legitimate and important activities, they have more to do with shoring up today's programs and services than with building tomorrow's delivery systems. On the value chart (Figure 1), they would fall on the left side under "cost" and "convenience" while high-performing competitors would concentrate more of their efforts on the right side—building "distinctive programs and services," "customer intimacy," and "maverick ideas."

Tinkering is not sufficient to provide a competitive advantage in the future. Although it worked in earlier years when advantages of cost and access put community colleges ahead of traditional competitors, the situation today is very different. A quick look inside most of our colleges will show that senior administrators are spending more and more of their time evaluating the behavior of competitors. The scene is familiar: the weekly president's cabinet meeting attended by the vice presidents or deans. The college's educational delivery system is divided into traditional and corporate services and continuing education programs serving diverse markets. Each is feeling pressure from strong competitors and has raised questions about the performance of the college:

- Why does it take only minutes and minimal paperwork to register at a neighboring college, but twice that time and annoying paperwork to register at our college?
- Why do private consulting firms get lucrative contracts for workforce training from business and industry by putting together programs in half the time and at three-quarters of the cost that it takes us to do it?
- Why does a regional proprietary school have detailed information about each student's job placement and employment record, but our college lacks even the most basic information about what happens to students after college?
- Why can workers taking contract education courses in the corporate services division of our institution get superior service for a 10-hour class, but evening students enrolling in a regular credit course can't get any advice?
- Why can a neighboring community college charge considerably less in-district tuition and fees for a three-credit course than our college?

Implicit in these questions are a number of beliefs that amount to tinkering as a strategy for coping with rapid change. The first is that access and low cost will be sufficient to maintain an advantage over competitors. In other words, students will continue to pay for education that is lower in cost and more accessible than comparable offerings at other institutions. After all, cost and convenience are critical factors in enrollment decisions. The second is that the investment needed to inform and prepare faculty and staff for massive change will not provide a full return because the turmoil associated with resistance to change could divert the institution from its fundamental goals. Indeed, there may even be instances—most notably, when downsizing and outsourcing strategies are used to control costs and improve quality—where intense feelings expressed by staff could lead to a negative image with the public. A final belief is that it will be possible to hold or capture major segments of the market by getting more resources for existing programs and services—more and better equipment, more staff, more marketing, and more technology. Adding resources is a way of preserving the base of the institution—its capacity to respond—while avoiding the use of harsh measures to adjust to change.

The extensive work we have carried out in developing strategic plans and effectiveness models with community colleges shows that these assumptions are deeply flawed. They ignore the depth of change that will be required to compete for the future. We meet many leaders who tell us about the good things their colleges are doing to serve the community and today's market. But serving today's market certainly does not equal market leadership tomorrow. Think about two sets of questions:

Today

Which customers do you serve today?

Through what delivery systems do you reach students today?

Who are your competitors today?

What programs and services make you unique today?

What is the basis for your competitive advantage today?

In the Future

Which customers will you serve tomorrow?

Through what delivery systems will you reach students tomorrow?

Who will be your competitors tomorrow?

What programs and services will make you unique tomorrow?

What will be the basis for your competitive advantage tomorrow?

If faculty and administrators don't have reasonably detailed answers to the "future" questions and if the answers are not significantly different from the "today" questions, there is little chance that their colleges will be leaders in tomorrow's market. Competition in today's market takes place mostly with established competitors such as four-year colleges and proprietary schools. Competition in tomorrow's market will take place outside the boundaries of postsecondary education as we know them. Moreover, while benchmarking the successful practices of today's competitors is sufficient to gain or maintain a competitive advantage, tomorrow it will be the minimum price of market entry. Keeping score of competitors and your college's advantages or disadvantages is not the same thing as inventing new advantages.

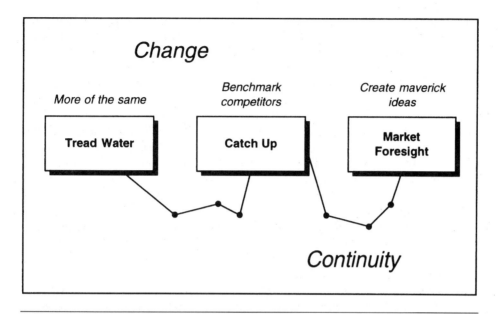

Figure 2.

The market a community college dominates today is likely to change substantially over the next 10 years. Community colleges will need to fundamentally change their core processes and administrative structures to compete in the future. To do this, they will need to challenge their own orthodoxies—to get beyond established ways of designing and delivering education by making quantum leaps.

Strategy, Structure, and Leadership

Fundamental change must be driven by a point of view about the future of the market. How do we want this market to be shaped in five to 10 years? What must we do to ensure that the market evolves in a way that is maximally advantageous to us? What processes and capabilities should we be building now to occupy the market high ground in the future? These questions point to three drivers of change that any college must consider if it wants to get to the future first: strategy, structure and leadership.

Strategy. Since most community colleges have not developed a long-term strategy for the future, leaders' first task is to forge strategy that will help define the future. Figure 2 depicts different approaches to strategy. At the most basic level, strategy can amount to treading water—fine-tuning programs and services to meet changing needs. Essentially, the goal is to preserve existing ways of doing things based on a belief that value in the form of low cost and convenience will continue to attract students. As competitors begin to bite into market share by providing new forms of value, a different strategy will need to be pursued.

> Major changes in programs, services, and delivery systems will be needed to overtake competitors.

Major changes in programs, services, and delivery systems will be needed to overtake competitors. One method for doing this is to assess customer needs and benchmark the successful practices of competitors—what we call "catching up." But as our colleges catch up, they also fall behind. Customers and competitors never wait for institutions to catch up—they continually reshape the market through new needs and new services. A strategy to "get ahead of the market" by anticipating future needs and creating new approaches to education may be the best answer. We call this strategy market foresight—it involves changing the shape of the market through developing entirely different ways of delivering programs and services. Colleges would not compete with rivals, but would challenge themselves by looking at opportunities on the horizon and creating maverick approaches to delivering programs and services.

Transformation Community College—a fictitious suburban college in the midwest—has developed a delivery system that reaches beyond its competitors. In 1992, the college's position seemed unassailable. With an enrollment of 12,000 credit students, it had recorded its tenth straight year of enrollment growth and looked forward to continuing growth. Its strategy was simple: keep costs low compared with competitors and make programs and services relevant and accessible to a wide variety of learners. Curriculum review was used to identify obvious problems with programs and new courses and services were added to promote enrollment growth. This water-treading strategy worked beautifully as long as revenue flows were steady and competitors remained predictable in their approach to the market.

The first sign of trouble came in 1994 when fall enrollment dropped 5 percent. Major declines were recorded in career and technical programs and in part-time enrollment. A quick scan of program markets and part-time student needs and preferences revealed some startling results.

- Equipment inventories in most technical programs were obsolete, leading to substandard instruction.
- A gap existed between changing employer skill requirements and the skills learned in career and technical programs.
- A new wave of competitors were working directly with employers to develop distance delivery instruction to prepare workers for jobs in the regional labor market.
- Part-time students' expectations had shifted to include guarantees of a job and lifetime placement as factors in enrollment decisions.

The college responded with a catch-up strategy to bring programs and services into line with changing market conditions. It surveyed employers to determine changes in worker skill requirements and current and prospective students about course taking and scheduling preferences and service needs. It also benchmarked the best practices of regional competitors. A clear message jumped from the data: course offerings in career and technical programs would have to be radically altered and a new delivery system developed for student services. College leaders moved to eliminate programs with obvious problems and reallocate resources for course and curriculum development. A 33-credit general education requirement including computer literacy was established; special course sequences in business skills and cooperative education were developed in career programs; and a firm mandate for planning, curriculum review, and resource development was given to program advisory committees. Student services were restructured using a customer service model designed by consultants in the human services industry.

These changes seemed to help at first, but Transformation's competitors had gotten so far out in front that the effect was minimal. Workforce requirements and student preferences continued to change at a rapid pace and faculty and staff could no longer keep up. A more aggressive strategy would be needed to maintain market share. The executive team latched on to the idea of market foresight and committed itself to putting the college at the forefront of the market by 2000. A "change team" was assembled to rethink the college's direction and to challenge fundamental assumptions. The team was broken down into task groups and each group was given an assignment. One group considered in detail the economic threats to the college and the opportunities afforded by the information highway. Another group studied new ways of designing and delivering programs that could be used to change the shape of the market. The remaining groups developed a view of market needs that was substantially different from current views, explored new opportunities on the horizon, and considered how to devote more institutional resources to building competencies and developing opportunities.

Transformation's new strategy can be captured in three words: entrepreneurialism, speed, and a focus on outcomes. This strategy is based on the college's success in restructuring the total system of educational delivery by moving to an enterprise system. Rather than employing permanent faculty and staff who over time would create obstacles to change, the college chose to purchase instructional services and contract for academic support from faculty and staff organized into educational enterprises. Faculty, for example, are clustered into teaching enterprises to offer specially crafted courses and curricula to academic divisions. These enterprises are both small and large, and their focus is on a discipline, field of interest, type of student served, or pedagogical style. Each teaching enterprise sets its own guidelines with regard to teaching loads, techniques, and other policies. The focus is on getting results with students using innovative instructional techniques.

Faculty working in enterprises are given nonprobationary appointments or contracted over short periods depending on the preference of administrators. This change gives the college greater freedom to set performance standards for instruction and the opportunity to shift the focus from teaching to learning.

Structure and Leadership

Transformation College emerged from the process as a faster, more entrepreneurial organization. It had a more creative view of its market and it was primed to take risks in the design and delivery of education. This view was held not only by a few executives or innovative instructors, but by most staff. Indeed, those who participated in the process thought it contributed as much to leadership development as it did to strategy development.

To create the future as Transformation College did requires structure as well as strategy. Why do we talk about structure in addition to strategy? The traditional hierarchy does not work effectively in fast-moving markets. The hierarchy has enabled community college leaders and top managers to feel that they are in control by giving them three core responsibilities: to be the institution's chief strategist, its structural architect, and the developer and manager of its information and control systems. However, it is clear that the management model that follows from this doctrine—today's pyramid supported by fragmented departments and service units—cannot change quickly enough to deliver competitive results.

> The traditional hierarchy does not work effectively in fast-moving markets.

From atop the hierarchy, presidents and deans look down on order and uniformity-a neat configuration of tasks and activities parceled out among academic and administrative divisions. As their label implies, divisions divide. The divisional model fragments institutional resources. It creates vertical communication channels that insulate academic departments and service units and prevent them from sharing their strengths with one another. Consequently, the whole of the institution is often less than the sum of its parts. Furthermore, the divisional structure has little built-in capability for renewal—for discarding old ideas and assumptions as they become obsolete. In other words, for all their growth, community colleges have become inflexible, slow to innovate, and resistant to change.

To address these problems, community colleges will need to consider new designs for management and leadership. Structural solutions such as skunk works, strategic alliances, and downsizing are interesting concepts that may deliver early results, but they do not remove cultural impediments to renewal; they only side-step them. The challenge is: How do community colleges reinvent their organization and the postsecondary education market at the same time? Tomorrow's market will be horizontally integrated with institutions competing and collaborating without central coordination. No agency or coordinating board will set rules for competition or tell institutions what to do. And when community colleges lose their proprietary advantages, speed—the capacity to change quickly to meet or get ahead of the market—will be what matters most. To succeed, our colleges will need to provide value equal to or better than competitors while working faster to maintain market share.

> Tomorrow's market will be horizontally integrated with institutions competing and collaborating without central coordination.

What new organizational designs can we expect to see in the future? It is likely that "adhocracy" will become a dominant theme of leadership and management! Leaders will work to create organizations which are fluid, dynamic, and temporary. Institutions—and units within institutions—will change shape according to the demands of the market. A premium will be placed on decentralized structures that delegate decision-making responsibility to faculty and staff and reward entrepreneurial skills, risk-taking, and a commitment to experimentation and innovation. Streamlined units that stress teamwork and speed will be a centerpiece of the "new" management in community colleges. Leadership will be transformational and it will be provided by individuals who enjoy imagining directions and envisioning the future. Tomorrow's leaders will move beyond conventional strategic planning, where endless analysis precedes action. They will focus on the core purposes of the institution and find ways to achieve ambitious goals.

If leaders are to be successful in creating a new kind of organization, they will need to consider some or all of the following actions:

- Identifying enduring core values that provide a sense of critical purpose for the college.
- Determining what should and should not change about the college; distinguishing between core values that provide continuity and practices that must change to maintain organizational vitality.
- Developing an organizational structure that increases speed by creating a small college "soul" in a big college body:
  — Flattening and decentralizing the organization.
  — Emphasizing teamwork.
  — Cultivating an "owner mindset" in faculty and staff.
- Creating stretch by striving for big gains without an idea of exactly how to achieve them:
  — Seeking to exceed rather than achieve a goal.
  — Not establishing a horizon for organizational performance.
  — Using teams comprised of instructors and staff to develop strategy.
- Breaking down boundaries and walls to innovation based on organizational structure and function; creating a boundaryless organization by reconsidering the hierarchical and departmental structure of the college.
- Deepening the commitment of faculty and staff to the values, goals, and operating strategies of the institution by adopting an open-book approach to management:
  — Ensuring that all faculty and staff learn and understand the college's "financials," along with all other numbers that are critical for tracking organizational performance.
  — Ensuring that faculty and staff learn that part of their job is to move performance indicators in the right direction.
  — Ensuring that faculty and staff have a direct stake in the college's success by sharing resources and information.
- Preventing budgets from minimalizing college and staff performance by creating goals that:
  a. stretch the organization through faculty and staff effort and
  b. work outside traditional constraints because they are not attached directly to the operating budget.
- Developing systems for planning that immerse faculty and staff and use external sources of information to chart the future of the institution.

- Estimating the "organizational will" for change within the institution—a measure of the extent to which faculty and staff at all levels in the organization are interested in and committed to change.
- Determine the capacity for collective action in carrying out the process of change.

Questions for the Future

We conclude with an assignment* for community college leaders: Pick 10 to 15 faculty and administrators in your institution. Ask them one simple question: What one educational innovation would you like to try that would put this college ahead of the market? Do not tell them what you mean by "educational innovation" or "ahead of the market"—let them use their own definitions. Give them a week or a month, but insist on an answer that fits on one page.

When all of the results are in, perform several analyses:

- First, when they talk about the market, do they include traditional competitors or all competitors? That is, how comprehensive is their view of competitors? Institutions cannot plan effectively for the future if they have a shortsighted view of competitors.
- Second, determine whether their concept of educational innovation is sufficiently broad and encompassing. Have they escaped the myopia of current practice so they understand how competitor behavior and customer needs are reshaping the delivery of education?
- Third, do they identify innovations that will have real impact? Without such a focus, a community college can pursue a whole range of competing ideas, spend lots of money on competing activities, and never get to the future first.
- Fourth, would their answers surprise competitors? Are they competitively unique?
- Finally, can an action plan be distilled from their answers? Do they know what they will do differently this year, next year, and so on? Many colleges have a vague concept of the long term, and a lot of specifics and budget pressure in the near term—but nothing in between. The linking actions that get institutions from the short term to the long term are missing.

Many leaders think of the short and long term as separate agendas, but in reality, they are interwoven. Our colleges will not get to the long term in one big jump. The goal should be to understand what objectives are possible this year that will have enormous implications for the future. Can institutions point to the five or six innovations—such as partnerships, or experiments with beneficiaries—that hold great portent for the future? These innovations should attract a disproportionate amount of leaders' attention. This attention is where leaders can add value that distinguishes our colleges from their competitors.

Note

1. Adapted from G. Hamel and C. K. Prahalad, *Competing for the Future* (Boston, Mass.: Harvard Business School Press, 1994).

References

Alfred, Richard. *Transforming Community Colleges to Compete for the Future.* Ann Arbor, Mich.: Unpublished paper, 1995.

Alfred, Richard and Scott Rosevear. "Organizational Structure, Management and Leadership for the Future." In *Handbook on College and University Management.* Maryville, Mo.: Prescott Publishers (in press).

Alfred, Richard and Patricia Carter. "Changing Managerial Imperatives." *New Directions for Community Colleges,* no. 84, 21(4). San Francisco: Jossey-Bass Publishers, 1994.

Case, John. "The Open-Book Revolution." *Inc.,* June 1995, 26–43.

Hamel, Gary and C. K. Prahalad. *Competing for the Future: Breakthrough Strategies for Seizing Control of Your Industry and Creating the Markets of Tomorrow.* Boston: Harvard Business School Press, 1994.

Hamel, Gary and C. K. Prahalad. "Competing for the Future." *Harvard Business Review* 94(4) (July–August 1994):122–28.

Schein, E. H. "How Can Organizations Learn Faster? The Challenge of Entering the Green Room." *Sloan Management Review* (Winter 1993):85–92.

Senge, Peter. *The Fifth Discipline.* New York: Doubleday, 1990.

Treacy, Michael and Fred Wiersema. *The Discipline of Market Leaders: Choose Your Customers, Narrow Your Focus, Dominate Your Market.* Reading, Mass.: Addison-Wesley, 1995.

Richard Alfred is an associate professor at the University of Michigan's Center for the Study of Higher and Postsecondary Education, and is the director of the Consortium for Community College Development. Patricia Carter is the executive director for the Consortium for Community College Development.

Quality in Business Education
A Global Perspective

Niranjan Pati, Dayr Reis, and John Betton

This article makes some radical suggestions about the importance of creating global standards for higher business education, recommending, for example, "building validity measures into outcome assessments that make grades independent of instructors, institutions, and the country in which the business school is located." This may not happen, but this article provokes some thinking on such issues, based on the importance of improving business education to prepare students for competing globally.

Higher educational institutions play a crucial role in the new global economy. The competition for global advantage hinges in part on whether a nation has equipped its available human resources with adequate expertise to satisfy the demands placed on them by a diverse group of global customers. A preamble to competitiveness in the melting pot of a global economy is a strong national commitment to aggressively, actively, and continuously integrate quality initiatives with higher educational systems. The fifth-ranked educational goal of the National Council on Educational Standard and Testing (N.C.T.E., 1992) report states, "by the year 2000, every adult American will be literate and will possess the knowledge and skills necessary to compete in a global economy and exercise the rights and responsibilities of citizenship." This illustrates the national resolve of the United States to be a major player in the global economy.

Educational institutions have often been misinterpreted as "knowledge factories." The rhetoric of managing higher education establishments like businesses is a familiar strategy in the United States—a strategy that is supported by lawmakers but contested by the academic community. The debate on whether to characterize the students as "customers" and the faculty as "knowledge-factory workers" seems never ending.

The "value-added" concept in education is a complex idea due to the interchangeability of its constituents depending upon the context. For example, students can be characterized as customers in knowledge gathering,

---

Reprinted with permission from *The Quality Observer*, June 1996. Copyright © 1996 by *The Quality Observer*. All rights reserved.

products in service delivery, and providers in tuition payment. Similar analogies apply to faculty members, administrators, the community at large, legislators, products, employers, etc. There is no rationale that could establish a unique classification process for providers, products, and customers in higher education. The definition of these entities is interchangeable in an economy that is increasingly orienting itself to encompass global providers, global products, and global customers. This leads to the focus, discussed in the next section, on attaining quality in undergraduate business education where educators play roles as providers and customers in addition to others.

Global Benchmarking of Business Education

The higher education community often becomes engrossed in the short-sighted discussion of how to utilize taxpayers' dollars economically, to be accountable to its stakeholders, to satisfy its student customers, etc. (Wagner, 1989) However, rarely is there discussion about the way in which the educational services are provided to the constituents served.

In the name of academic freedom much damage has been done to the process of education. The curricula required to enrich students' experience at universities are developed to satisfy the intellectual curiosity of faculty in the name of flexibility and diversity in course design. In fact, we contend education is a process that should be standardized globally. How educational materials are delivered to students must be an issue related to academic freedom but the content of this material should not be. This calls for "best-in-practice" global benchmarking of the educational process. Benchmarking, perfected at Xerox, provides a tool to improve business education as a process when business schools are compared against the best current practices. Since an educational process is primarily curriculum driven and faculty steered, global benchmarking of the curriculum, faculty, and outcomes is required to maintain credible standards of business education.

> In the name of academic freedom much damage has been done to the process of education.

Curriculum

Business students, in a real-world situation, are expected to transact business that involves interaction with two or more parties in the areas of their expertise upon completion of their education. Therefore, a business curriculum should ideally comprise: 1) general education, 2) common body of knowledge, and 3) elective specialty components.

The general education aspect of business education includes: knowing the language and culture of business environments; developing writing and oral communication skills; building numerical and mathematical reasoning; and developing critical thinking ability. In the formative years of higher education in the United States, the general education aspect included mastering Greek, Latin, and Mathematics (National Center for Education Statistics, 1993). Gradually, the curricular emphasis shifted to more of a technical orientation, at the expense of ancient studies and the classics. In business education, which is increasingly incorporating a global dimension, learning one's own and foreign languages, as well as one's own and foreign cultures, is an ingredient for success in conducting business.

The aspect of numerical literacy that includes numerical and mathematical reasoning is also important for conducting successful business transactions. The critical thinking skills that are part of business education aim to help decision makers generate alternatives.

The common body of knowledge in business must explore, analyze, and tie together the strands of business characteristics that run across the specialized areas. Typically, the common body of knowledge covers the fundamental principles and premises of business operations and the aspects that bring together the principles of different areas to help decision makers make better and more informed decisions.

The "choice" or elective component in a business curriculum provides the tools specific to an area of specialization. This might include specific course-work that is concerned with the changes in global business practices.

While the general education courses and electives need to be specific in terms of geography, language, and discipline, the common body of knowledge component should be independent of language, functional areas, and the country in which the material is delivered. Therefore, the curriculum for the common body of knowledge component is suitable for global benchmarking, whereas the general education and the elective specialty can be benchmarked within a specific country or within a large geographic area (e.g., Central Europe, Pacific Rim, Latin American Cone).

A vital issue pertinent to benchmarking of the above curricular components is the criteria with which the components should be benchmarked. We observe that in the community of business academics there is a dangerous shift toward making business schools a training ground for a few major businesses. It is important for us as educators to understand that business education is not the same as in-house training. The real quality of business education does not lie in efficiently delivering "how to" but rests on its ability to let students understand "why" in order to keep their long-term professional curiosity alive. Therefore, criteria for benchmarking in this context should precisely incorporate the conceptual aspects of a business curriculum that can be translated into other countries, contexts, etc.

Outcome Assessment

Outcome assessment in business education, ideally, must evaluate the overall knowledge enhancement that occurred in the course of business studentship. The usual assessment measure of assigning grades for coursework in the United States has validity problems. The grades assigned in a course in one institutions would not translate into a similar grade if the esteem level of institution differ. Furthermore, course expectations in different institutions (and in the extreme case, within the same institution) may differ, causing grade variation. The above situation raises questions as to the validity of letter and numerical grades. The problem of grade variation becomes even more difficult in transnational comparisons between higher education systems, some of which do not use grades. More often than not, businesses prefer employing graduates with superior academic performance in the hope that they will also perform well on the job. Due to lack of uniformity in grades across business schools, such expectations are rarely realized, raising important questions about the validity of grades. Questions of validity also arise due to lack of agreement on curricular materials which vary among instructors, schools, and countries. It is important to understand that a benchmark for assessing students' performance assumes the assessment procedure to be valid.

Therefore, we contend that a uniform educational assessment standard will essentially reduce the problem of establishing validity measures. Especially in the case of the common body of knowledge component of business education, global outcome assessment procedures must be established. A preferred method for building validity measures into the outcome assessment is to make the grades independent of instructors, institutions, and the country in which the business school is located. This calls for external administration of tests and blind evaluation of achievement outside the student's local affiliation. This method of assessment would render the evaluation valid.

Though this external evaluation procedure sounds very simple, the coordination and monitoring of participating business schools are crucial to successful implementation of the process. Undesirable suspicion and opposition from the academic community in the name of "academic freedom" may hinder the realization of the process. Similar methods of evaluation can be adopted for the enculturation and location of the business institutions (i.e., typically the tests for this component should be common for a country).

For undergraduate business education, four common tests in the freshman, sophomore, junior, and senior years will ensure the quality of the outgoing students. Because of the validity of the external examination and evaluation processes, employers will feel confident using academic performance as a substitute measure of on-the-job performance. In fact, the external evaluation system is in vogue in many Pacific Rim countries (U.S. Department of Education, 1993) regarded as major emerging competitive business forces.

The external evaluation system should not be confused with the certification measures of professional proficiency attained on the job. In order to benchmark the outcomes (i.e., the proficiency level of students) common ground is required. We strongly believe that the existence of a common method of assessment would help to establish a valid benchmark. Standardized tests used for establishing entrance standards—such as SAT, GMAT, GRE, etc.—do not reflect the specific skills acquired by students in schools or measure the kinds of knowledge, skills, and abilities required in a competitive global economy.

Quality of the Faculty

Another cornerstone of business education is the quality of the faculty. The quality of the faculty is usually measured along three dimensions:

Teaching

In the changing economic environment of the current decade, businesses are constantly looking for business graduates able to think globally and act locally. An example of such a paradigm shift is the fact that the American Assembly of Collegiate Schools of Business (AACSB) requires business schools to incorporate a global dimension in the business-related courses taught as part of their accredited curriculum. Therefore, one of the contemporary quality dimensions in teaching is whether the present faculty is equipped to handle the challenges brought in by this new era of heightened competitiveness. A recent survey of U.S. business school deans (McGrath and Hargrove, 1992) concluded that 48 percent of their faculty members are adequately or well prepared to provide a global perspective to the business courses they are teaching. This is not particularly encouraging considering the fact that AACSB introduced this global dimension almost 20 years ago. Therefore, one of the benchmarking criteria used to assess the quality of business school faculty members should be the level of expertise they demonstrate in teaching global dimensions within their specific disciplines.

Utilizing student evaluations as a measure of faculty quality is a common practice in North American business schools. However, this process is flawed. Instructors can be discouraged from making courses more intellectually challenging for the students because of the perception that this will result in less favorable student evaluations. The argument of making instructors accountable to students for their learning becomes difficult unless this accountability can be adequately defined in terms of strategic learning objectives or specific outcomes. But these measures are rarely utilized by business schools. An external evaluation system, adjusted for input quality, would adequately reflect the level of expertise gained in the class. This indicator might serve as an unbiased global measure of faculty quality which can be used subsequently for benchmarking purposes.

Research

Advancing the boundaries of theoretical knowledge and/or conducting research on applied business topics are two primary research areas the faculty of a business school must demonstrate. The hallmark of a quality school in this research dimension is a balance between the theoretical and applied research. The benchmarking of faculty research in a business school setting must be peer-based. The quality and quantity of the scholarly and applied research production of a business school faculty would provide a good benchmark for the research quality internal to the institution. Citation indexes also provide a popular quantitative measure for the quality of a publication (i.e., a good quality publication is the one that is most often cited). Benchmarks for research quality can be subsequently established, which might include the scholarly papers produced per FTE, etc.

Service

Business school faculty must continuously engage in the stimulating exercise of reaching out to the greater community in which the business school is located. Establishing meaningful relationships with business for the purpose of reality checking should continue as a measure of faculty. The quality benchmark for the service component of a business school would also include gauging the perceptions of the community at large using focus group interviews, perception questionnaires, etc.

Issues and Challenges

The issue of setting benchmarks for quality in business education is complicated due to the recent change in demographics experienced by institutions of higher education. The gradual shift from the traditional school-age population to an older generation of students, particularly apparent in business schools, poses a challenge because of the differences in the expectations of a diverse student population. Catering to the taste and quality expectations of a culturally diverse student community is another challenge faced by business schools. Legal inclusion of the student group protected by the Americans with Disabilities Act also introduces an additional dimension of diversity. It is a challenging task for any educator, business educators in particular, to maintain a consistent quality of education when the process of education needs also to recognize the sometimes competing needs of heterogeneous groups. Thus, a benchmarking system for business schools that accommodates special characteristics of the current student population is necessary to ensure the level of quality delivered at each business school, in terms of the expectations of its stakeholders and institutional missions.

Suggested References for Further Reading

McGrath, L. C. and C. L. Hargrove. "Internationalizing the Business Curriculum: A Status Report." Paper presented at the 11th Annual Eastern Michigan University Conference on Languages and Communication for World Business and the Professions, Ypsilanti, Michigan (March 1992).

National Center for Education Statistics. *120 years of American Education: A Statistical Portrait.* Edited by Thomas D. Snyder, Washington, D.C.: 1993.

The National Council on Education Standard and Testing. *Raising Standards for American Education.* Washington, D.C.: 1992.

U.S. Department of Education. *Standards for the 21st Century: Opening Statements of Ministers at the Asia-Pacific Economic Co-operation Education Ministerial.* Washington, D.C.: 1992.

Wagner, Robert. *Accountability in Education.* London: Routledge, 1989.

Public Sector

EDUCATION K–12

Quality Management in Schools: 1996

Julie Horine

This article by Julie Horine, consulting editor in K–12, highlights national Baldrige initiatives that are providing a comprehensive framework for assessing quality excellence in K–12 schools. The article summarizes the results of the education pilot of the Malcolm Baldrige National Quality Award Criteria and discusses lessons learned from pilot schools as well as future implications.

Although an increasing number of American schools are actively implementing quality management principles in an effort to more effectively and efficiently meet the needs of students and stakeholders,[1] research suggests that quality results still appear to be isolated and not interconnected.[2] Noticeably lacking is a common definition of what constitutes "quality excellence." Lack of agreement among educators on how to define quality excellence produces a wide variation of responses to the question "How does a school system determine if it is delivering ever-improving educational services to students or improving overall school performance?" Without a standard of quality excellence and a diagnostic tool that enables a school system to assess itself against that standard, a school cannot determine how well it is doing, how it compares to other schools, or how to improve.

The Malcolm Baldrige National Quality Award "Education Pilot" criteria,[3] piloted nationally in 1995 and formally evaluated in 1996, may provide K–12 educators with both a nationally recognized standard of quality excellence and a diagnostic assessment tool. As schools face the tough challenges of limited resources, public demands for accountability, and state mandates, the criteria offer schools a practical, comprehensive approach for pursuing performance improvement. This article summarizes the Baldrige Award program, why the Baldrige Award Criteria are needed in education, the Baldrige

Education Pilot framework, results of the 1995 pilot, progress from several state and individual school efforts, and future implications of the Baldrige criteria in education.

What Is the Baldrige Award?

In an effort to stimulate American companies to improve quality and productivity, the Malcolm Baldrige National Quality Award was signed into public law in 1987 by President Ronald Regan. Since 1988, the Baldrige criteria have served as the basis for selecting twenty-four national award-winning organizations that demonstrate business excellence and quality achievement. Award winners serve as models of excellence and help promote the sharing of best practices across sectors. Although over one million copies of the criteria have been disseminated, the most prevalent use of the criteria is *not* for award purposes, but as an internal diagnostic tool to improve organizational performance.

> the most prevalent use of the criteria is *not* for award purposes, but as an internal diagnostic tool to improve organizational performance.

Why Baldrige in Education?

National initiatives such as Goals 2000 reflect a growing national consensus to improve and strengthen education. As a result of these initiatives and national interest in establishing a Baldrige Award category for education, the Baldrige criteria were adapted and extended in 1995 to education and health care on a pilot basis. No awards were presented in the 1995 pilot, however, and each applicant received written feedback including strengths and areas for improvement regarding its performance management system. A program evaluation was conducted in 1996 to address the many issues associated with extending the award eligibility to education. Importantly, the Baldrige education pilot criteria offer education a framework for defining quality excellence and a diagnostic tool to systematically improve educational services and organizational effectiveness.

The Baldrige Education Pilot Framework

The education pilot criteria are built upon a set of eleven core values and concepts that provide the foundation for developing and integrating twenty-eight results-oriented item requirements. The core values (see figure 1) are embedded in the seven category item requirements (see figure 2) and reflected in the framework shown in figure 3. The category components work together as a system to achieve the goals of the organization.

In order for a school to determine its effectiveness in meeting and making progress against the criteria requirements, the school must evaluate its quality-management practices using three evaluation dimensions: approach, deployment, and results. These three dimensions examine (1) the effectiveness of the school's approaches, (2) the extent to which the approaches are deployed throughout the school, and (3) the results achieved from the approaches. Together, the criteria and evaluation dimensions form a diagnostic

> **Core Values and Concepts**
>
> The 1995 MBNQA Education Pilot Criteria are built upon a set of 11 core values that provide the foundation for developing and integrating all requirements.
>
> 1. Learning-Centered Education
> 2. Leadership
> 3. Continuous Improvement and Organizational Learning
> 4. Faculty and Staff Participation and Development
> 5. Partnership Development
> 6. Design Quality and Prevention
> 7. Management by Fact
> 8. Long-Range View of the Future
> 9. Public Responsibility and Citizenship
> 10. Fast Response
> 11. Results Orientation

Figure 1. **Core values**

Only six of the twenty-eight items ask for results; all other items focus on the approach and deployment dimensions.

tool for assessing the strengths and weaknesses of a school's quality-management practices. Only six of the twenty-eight items ask for results; all other items focus on the approach and deployment dimensions.

The Pilot Process and Results

The 1995 education pilot generated nineteen applications, nine from K–12 and ten from higher education. Applicants submitted a seventy-page application to the Baldrige Award office that addressed the twenty-eight items of the education pilot criteria (see figure 2). Teams of six nationally trained Baldrige evaluators independently reviewed the applications, listing the strengths and areas for improvement and assigning a score based upon the scoring guidelines. Three school systems, one from K–12 and two from higher education, were selected for site visits. All applicants received a written feedback report.

Although the relatively small number of education applications (nineteen) tends to exaggerate the percentages, some observations may be made from the relative score distribution between the education and business applications. Figure 4 shows the differences at the different ends of the scoring ranges. Forty-two percent of the education pilot applications were in the 0–250 point range, while only 6 percent of the business applications were in this range. At the other end of the scoring range, 36 percent of the business applications scored above 550, while no education applications scored above 550. These differences suggest a difference in the level of maturity between applicants in these two sectors.

Categories	Items
1.0 **Leadership** (90 pts)	1.1 Senior Administration Leadership 1.2 Leadership System and Organization 1.3 Public Responsibility and Citizenship
2.0 **Information and Analysis** (75 pts)	2.1 Management of Information and Data 2.2 Comparisons and Benchmarking 2.3 Analysis and Use of School-Level Data
3.0 **Strategic and Operational Planning** (75 pts)	3.1 Stategy Development 3.2 Strategy Deployment
4.0 **Human Resource Development and Management** (150 pts)	4.1 Human Resource Planning and Evaluation 4.2 Faculty and Staff Work Systems 4.3 Faculty and Staff Development 4.4 Faculty and Staff Well-Being and Satisfaction
5.0 **Educational and Business Process Management** (150 pts)	5.1 Education Design 5.2 Education Delivery 5.3 Education Support Service Design/Delivery 5.4 Research, Scholarship, and Service 5.5 Enrollment Management 5.6 Business Operations Management
6.0 **School Performance Results** (230 pts)	6.1 Student Performance Results 6.2 School Education Climate Improvement Results 6.3 Research, Scholarship, and Service Results 6.4 School Business Performance Results
7.0 **Student Focus and Student and Stakeholder Satisfaction** (230 pts)	7.1 Current Student Needs and Expectations 7.2 Future Student Needs and Expectations 7.3 Stakeholder Relationship Management 7.4 Student and Stakeholder Satisfaction Determination 7.5 Student and Stakeholder Satisfaction Results 7.6 Student and Stakeholder Satisfaction Comparison

Figure 2. **Baldrige criteria for education**

The majority of the 1995 applications in the education pilot program (seventeen) were within the lower three scoring bands. Most of the education applicants could be characterized as having the following attributes:

- beginnings of systematic approaches to the primary purposes of each item, nearly stages of transitioning from reacting to problems to a general improvement orientation
- major gaps existing in deployment that could inhibit progress in achieving the primary purposes of the item

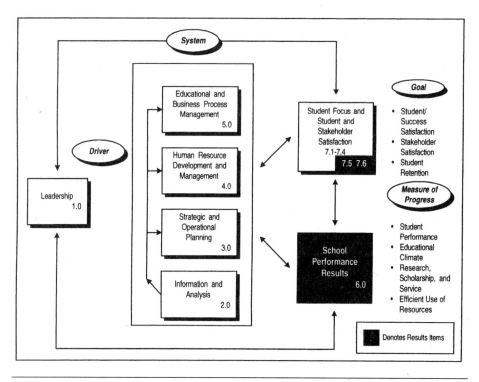

Figure 3. **Education pilot criteria framework, dynamic relationships**

- early trends are being established, but many results are not yet evident.

In the education pilot, there were no actual winners selected. However, previous national Baldrige Award–Winning companies agree that quality improvement is a continuous process and that it takes years to actually become a Baldrige Award winner and serve as a national role model. Both of the 1995 business Baldrige winners, Armstrong World Industries' Building Products Operation and Corning Telecommunications Products Division, started their quality journeys in the 1980s. The results of the Baldrige education pilot suggest that the pilot applicants overall are not yet at the same level as the business applicants, even though they are making good efforts and progress, and may be among the best in their class.

Lessons Learned from the Pilot

Several of the education pilot applicants discussed their experiences during the 1996 Quest for Excellence VIII Conference in Washington, D.C.[4] The following comments regarding the application process and feedback reports are taken from three schools that wrote an application, Pinellas County School

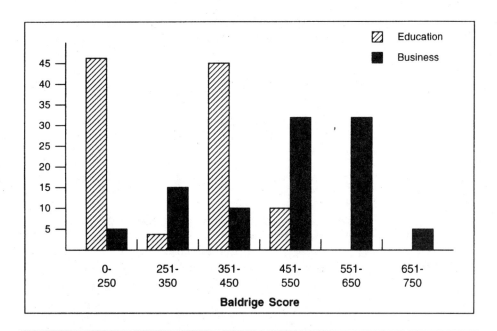

Figure 4. **Education and business applicants by Baldrige score**
Source: 1995 Baldrige education applicants (19) and business applicants (47).

District, Belmont University, and Babson College, and one school that received a site visit, Pearl River School District. Representatives from each of these schools made formal presentations at the conference.

What did we learn from the self-assessment process?

The application process created tremendous awareness and generated action. We learned that we collect a multitude of uncoordinated 'stuff' and that we have a lack of comparative/benchmark data. We learned that we have numerous activities, but few processes, and that sometimes we lose sight of the student.

—*Babson College*

The Baldrige process is a valuable tool for helping us drive student achievement. We see the Baldrige as one small but very integral piece of that in terms of Pinellas County School's quality transformation. First you have to create a culture. You just don't jump in the middle of things and say let's go do a Baldrige. In order to support an empowered learner, it involves employing the community, looking at the school as the smallest unit of change, how that impacts the classroom, and what happens at the process level in terms of the teaching and learning that impacts student achievement.

—*Pinellas County School District*

The application process is useless without taking the feedback and applying it. Take the time, it is worth it to your organization.

—*Belmont University*

How did we use the feedback report?

To us, the feedback report is the most important part of the process. The feedback allows an organization to roll the recommendations into the strategic planning process. It provides valuable jewels that can really help you to grow and improve continuously.

—*Pearl River School District*

We see the Baldrige process as a key component in creating a quality culture. We did not do quality because it was fun. We did it because we absolutely have no choice, but to impact student achievement. We will use the feedback to look at the relationships between all of the parts of the organizations that must be pulling together and optimizing in order for that to happen. The information that we received from the Baldrige process will certainly help us drive student achievement.

—*Pinellas County School District*

The Baldrige is such a reflective process. For most people, the feedback should not be much of a surprise. We got what we expected. Our feedback report has validated and has probably given us that little extra boost to make sure that we keep focused and realize that we have a long way to go yet.

—*Babson College*

The feedback is really the key to the whole Baldrige process. We have constantly focused on our approach. We have been able to look at our deployment strategies based on our feedback report. But we found out that we need to become more linked, more integrated as a system.

—*Northwest Missouri State University*

How does the Baldrige compare to accrediting agencies?

In general, in comparing accrediting agencies and the Baldrige, accrediting agencies tend to focus on programs and not necessarily understand or make us demonstrate that what we have is a system that is all interrelated. If we cannot demonstrate that all the individual areas somehow comprise a system for managing the institution and a way to see things, then we are not truly able to understand how we can improve and do better.

—*Babson College*

Individual District and Statewide Baldrige Efforts

Some individual school and statewide partnership initiatives are actively using the Baldrige education pilot criteria to drive school improvement. Several of the more mature efforts include Pinellas County School's Superintendent's Quality Challenge and Minnesota's Partners for Quality Education. Each of these initiatives is summarized below.

Pinellas County Schools

As part of its Total Quality Schools initiative, Pinellas County Schools, a large Florida urban district of approximately 100,000 students in 138 schools, uses a Baldrige-based self-assessment process at both the district and individual school levels to guide its improvement efforts.[5] At the district level, the Baldrige criteria guide the annual strategic-planning process, while at the school unit level, the criteria guides annual improvement plans.

The Superintendent's Quality Challenge (SQC), patterned after the national Baldrige criteria, was introduced in 1994 to provide an internal process to evaluate progress and to guide continuous improvement efforts. At the school-unit level, the voluntary self-assessment process is still relatively new. Thirty-five schools are voluntarily using the challenge criteria to conduct self-assessments as a basis for completing annual improvement plans mandated through state reform legislation, Blueprint 2000.

According to training specialist Chris Harwell, "quality is a way of life at Pinellas County Schools. You could walk into any school and people would know what you are talking about." To support school assessment efforts, support groups composed of certified examiners offer technical assistance and provide feedback on written assessments. In 1996, ninety-three business and school volunteers, ranging from principals, teachers, and guidance counselors to community business leaders, attended a three-day training session to become certified examiner/consultants. Examiners assist individual schools and departments in understanding and using the Baldrige self-assessment as input to their annual school improvement plans.

Pinellas County Schools is becoming a role model for other schools as evidenced by its receipt of the state of Florida's Governor's Sterling Award in 1993, an award process patterned after the Malcolm Baldrige National Quality

Quality is a way of life at Pinellas County Schools. You could walk into any school and people would know what you are talking about.

Award. In addition, the district's Quality Academy, which has trained hundreds of teachers, parents, and other community residents in quality principles within Pinellas County, is now assisting other schools and states, such as North Carolina, with their quality efforts.

Minnesota's Partners for Quality Education

In 1996, 105 school sites received quality training as part of Minnesota's Partners for Quality Education Initiative.[6] Thirty-eight school sites worked on completing Baldrige self-assessments, while sixteen schools worked on improvements based on assessment feedback reports. The initiative, based upon the Malcolm Baldrige National Quality Award criteria, provides a voluntary process through which schools and districts can improve the quality of their management, measurement, and learning systems to meet the needs of students, parents, employers, and the community. The initiative is administered through the Minnesota Academic Excellence Foundation, a nonprofit, public/private partnership initiated in 1990.

Presently, Minnesota's quality initiative is affecting over half of the state's student population, roughly 1,500 schools and 380 districts, because many of the participating school sites represent the larger, metropolitan schools. School site teams include administrators, school board members, faculty, support staff, students, and parents.

One of the initiative's critical success factors is matching schools with business partners. Results of the initiative's first impact study revealed that schools with active business partners made the most progress. Findings suggest that schools need assistance and encouragement to understand and stay committed to the continuous journey of quality improvement.

Program results show that the level of understanding of quality concepts has increased markedly—84 percent of participating school sites report moderate to high understanding of quality concepts compared to 23 percent when they first began the two-year program. Similarly, 61 percent of the sites report they are now practicing these concepts, compared to 18 percent when they first began the partnership.

Although data suggest that Minnesota's Partners for Quality Education initiative is working, it is taking longer than expected to achieve its goals. Impact findings suggest that completing the self-assessment is "hard work" and that incentives to improve results are not supported by the present education funding system. One monetary incentive, however, may provide schools with the resources necessary to finance process improvements. The School Improvement Performance Contracts Pilot Program, recently established by the Minnesota Legislature, provides schools with grants up to $400,000 to increase student achievement and to improve key processes such as instruction, curriculum, assessment, and staff development.

Future Implications

Educators and community members involved in Baldrige-based education initiatives believe that the criteria may offer the necessary framework to help schools deliver ever-improving value to students and other stakeholders, and improve overall school performance. The decision to proceed with a full-scale Baldrige national award program for education, however, depends upon many factors, such as funding and legislation. During 1996, as the pilot program was being evaluated, an education task force began efforts to raise $20 million in private-sector funds to support an education award program. Ultimately, legislation will be required to expand the existing Baldrige award program's eligibility categories. The existing award program is presently limited to for-profit organizations (manufacturing, service, and small business) under Public Law 100-107. The future of a full-scale award program for education and health care will depend largely on a continuing support from these two sectors and the prospects for long-term funding.

The future impact of a Baldrige Award in education may result in providing education with benefits similar to those gained by business. These benefits would include (1) helping schools improve performance practices through an integrated, results-oriented set of performance requirements; (2) facilitating communications and sharing of best practices information within and among schools of all types, based upon a common understanding of key performance requirements; (3) fostering the development of partnerships involving schools, businesses, human-service agencies, and other organizations; and (4) providing a working tool for improving school performance, planning, training, and organizational assessment. Such factors could have a significant impact on the future of American education.

For More Information on the Education Pilot

You may receive a free copy of the Education Pilot Criteria by contacting

Malcolm Baldrige National Quality Award Office
National Institute of Standards and Technology
Route 270 and Quince Orchard Road
Administration Building, Room A537
Gaithersburg, Maryland 20899-2036
Phone: 301-975-2036
Fax: 301-948-3716

Resource Names

Minnesota Academic Excellence Foundation

Partners for Quality Education
Zona Sharp-Burk, Executive Director
971 Capitol Square Building
550 Cedar Street
St. Paul, MN 55101
Phone: 516-297-1875
Pinellas County Schools

Superintendent's Quality Challenge
James C. Shipley, Executive Director, Quality Academy
301—4 Street SW
Largo, FL 34649-2942
Phone: 813-588-6295
Pearl River School District

Quality in Public Education
John DiNatale, Assistant Superintendent
275 East Central Avenue
Pearl River, NY 10965
Phone: 914-588-6295

Notes

1. Anne Calek, "Fifth Quality in Education Listing," *Quality Progress*, 28, no. 9 (September 1995):29–77.

2. Julie E. Horine, William A. Hailey, and Robert O. Edmister, "Quality Management Practices and Business Partnerships in America's Schools," *Planning and Changing*, 25, no. 3–4 (Fall–Winter 1994):233–40.

3. United States Department of Commerce, *Malcolm Baldrige National Quality Award 1995 Education Pilot Criteria*.

4. Quest for Excellence VIII Conference, The Official Conference of the Malcolm Baldrige National Quality Award, February 5–7, 1996, Washington Hilton and Towers, Washington, D.C.

5. Pinellas County Schools, Largo, Florida, *Education Pilot Application*, September 1, 1995.

6. Minnesota Academic Excellence Foundation, *Partners for Quality Education Initiative*, April 1996.

Acknowledgments

The author expresses her appreciation to the following individuals for material used in this article: Sue Rohan, Malcolm Baldrige National Quality Award office; James Shipley, Pinellas County Schools; and Zona Sharp-Burk, Minnesota Academic Excellence Foundation.

Total Quality Applied in the Classroom

Patrick Konopnicki

Here you'll read about the experiences of Virginia Beach Public Schools in adopting and applying TQM practices in the classroom. Teachers and students have begun to embrace these principles as a better way to facilitate learning and to improve performance.

For some time educators have been urged to prepare student graduates for the high-performance work place of tomorrow. Schools across the country-like many businesses, are under pressure to produce results. Communities are clamoring to get more for their tax dollars.

As Lloyd Dobyns, the former TV news correspondent turned quality guru, said at a recent Community Quality Day in Virginia Beach: "There is a greater vision where everyone can benefit from doing one thing: adopting a quality system of management."

For education, the greater vision may lie with the use of total quality management strategies in K–12 classrooms and administration. While school system leaders have expressed growing interest in applying quality processes and theory, few have been successful at reaching the ultimate customer—students in our classrooms.

In our school district, the second largest in Virginia with 76,000 students and 82 schools, TQM strategies are being applied in the classroom with positive results.

With changing demographics and a decline in the number of graduates attending four-year colleges, the district began to apply TQM principles to help better prepare our next graduates.

Initial Approach

Four years ago, our school district first considered TQM in the classroom as a method of continuously improving instruction, first in the areas of school-to-work transition and technical and career education. As a first step, a small group of technical and career teachers were trained in TQM at a two-day workshop. Subsequently other teachers and administrators were encouraged to attend staff development on TQM awareness provided by city government in Virginia Beach.

Reprinted with permission from *The School Administrator*. Copyright © 1996 by *The School Administrator*. All rights reserved.

Teachers found TQM interesting at first, but they were not sure how it would fit into the classroom. Technical and career education teachers figured this would be another fad that soon would pass. To respond to these concerns, a voluntary group of teachers, administrators, and business people began meeting monthly beginning in August 1992 to share quality strategies in education. As a result, the national office of Vocational Industrial Clubs of America selected Virginia Beach as a test site for its TQM curriculum.

Quality teaching practices began to spread to instruction when the school district's Office of Instructional Support Services adopted the mission of "continuously improving instruction." TQM paved the way for educators in leadership positions to spread the word that instructional practices should not be static and always can be improved.

Quality Deployment

After hours of introducing team training, facilitation skills, and TQM tools, the old classroom practices of "chalk and talk" and "drill and kill" started to fade in technical and career education. As more teachers became involved in training, growing numbers of academic teachers wondered how TQM would work in their classrooms.

The breakthrough for academic teachers, especially at the elementary level, happened after reading *Future Force* by Elaine McClanahan and Carolyn Wicks. This book served as a valuable staff development catalyst that immediately empowered teachers with tools and strategies to make their students lifelong learners and problem-solvers.

The Virginia Beach school system did two things to increase TQM's impact upon instruction. First, the district created a position of TQM trainer in the Office of Technical and Career Education. This ensured continuity of training for both classroom teachers and administrators, says Debbie Gentry, the district's coordinator of TQM/industry standards.

"Many of the teachers have discovered that they already were using elements of total quality management in their classroom. The training provided a vehicle for refining their use," she notes.

Second, the school district has given recognition to the teachers and teams of staff who have improved instruction by using the five cornerstones of quality endorsed by the superintendent:

- Continuous improvement,
- Customer focus,
- Teamwork,
- Empowerment, and
- Data-driven decisions.

Each month a Superintendent's Quality Award is given to an employee or student who demonstrates these cornerstones to improve teaching and learning. This recognition has been ongoing since December 1992. Meritorious projects include:

- a fourth-grade classroom TQM implementation using force field analysis, flow charts, cause-and-effect diagrams, issue bins, and run charts;
- a statistical process control field trip by manufacturing and statistics students to the Ford Motor Company's assembly plant in Norfolk;
- implementing TQM in a middle school social studies class; and
- benchmarking fourth-grade reading and math progress at Williams Elementary School.

All classroom teachers have access to TQM training, but students may choose to have more intensive training by enrolling in a TQM pilot course offered at Salem High School in Virginia Beach. Students have embraced the team-oriented teaching style of instructor John Ledgerwood because they are empowered to make decisions about the way the class is taught.

The highly interactive, 90-hour TQM course asks students to examine the elements of the continuous improvement process. Students apply these elements to solving both personal and organizational problems. Ownership, team building, and decision-making activities provide students with opportunities to improve their education and attain future career goals. Most emerge from the class with a working knowledge of quality on both a personal and leadership level that will prepare them for a competitive global workplace. Students with at least a B average can receive three hours of college credit.

Some students make a special trip every day from their home school to take the TQM class at Salem High.

On an administrative level, the Strawbridge and Williams elementary schools in Virginia Beach are training entire faculties and staffs as a part of their schools' strategic planning process. Both school principals have requested that all teachers receive a minimum of 14 hours of TQM training. It is not unusual to see teachers using the tools of total quality—nominal group voting, course mission statements, plus/delta charts, cause-and-effect diagrams, affinity diagrams, force field analysis, and other charts to collect data and measure progress with their empowered students.

Dianne Kinnison, a third grade Strawbridge Elementary School teacher, helped her students develop a class mission statement: "We are the Strawbridge Sharks and the totally cool members of the exciting Seahorse Class. We will have a happy, safe, and clean learning environment that's a radical place to be. We will make learning fun by using successful teamwork and intelligent thinking. We will help each other stay focused so we can all learn together."

Early Results

Qualitative and quantitative data are being collected to assess the value of introducing TQM philosophy and strategies to classrooms. Ernie Sawyer, a Salem High School assistant principal, says, "The TQM initiative has created a new enthusiasm for instruction by teachers and a new enthusiasm for learning by students. It's really making a difference."

One parent of a student enrolled in Salem High School's TQM course noted on a parent survey the positive impact the class had had on her daughter. "Since its beginning, my daughter has been running strong and in the right direction. She has demonstrated a wonderful assertiveness I had not noticed before. More power to her!"

Students in Elaine Johnson's fourth grade class at Williams Elementary School demonstrated the most promising results by scoring 28 points higher than the school average on the Iowa Test of Basic Skills. Johnson, a 20-year veteran teacher, says, "I just challenged my students to have the best year ever. I highly encourage all principals to take a look at TQM. While hard data is not in yet, nearly all teachers trained who have tried the approach could tell you participation has increased and teaching is more fun."

Definitive results will be forthcoming next year since the National Education Association has chosen the Virginia Beach school system as Virginia's Learning Lab project site. Learning Lab projects around the country are designed to create change and innovations in education. Our project will research the use and measurement of TQM strategies in the classroom at two elementary schools, one middle school, and one high school. The study will determine the impact of TQM on academic excellence.

Contributing Partners

One exciting development of the system's progress has been the formation of the Virginia Beach Quality Alliance, a liaison among the school division, city, business, and military communities. While the mission of the alliance is to provide a sharing of quality materials, services, resources, and expertise, a primary feature is to certify Virginia Beach students in TQM.

In partnership with classroom teachers, the alliance has developed a competency-based TQM certification for students, the only one of its kind in the country. Alliance members serve as mentors for students who elect to go through the certification process. At a recent executive meeting of the alliance, three corporate board members pledged to fund three $500 scholarships for students who completed the certification process.

Timothy R. Jenney, superintendent of Virginia Beach schools, sees TQM certification for students as "a contribution to the local corporate culture that will boost economic growth and development."

One corporate partner, Reza Hashampour, president and CEO of IMPAQ Computers Inc., says he was startled to discover public schools in Virginia Beach teaching TQM. "I worked with Fortune 500 companies who were just starting to implement quality approaches," he noted. Hashampour's company sponsored one of the student scholarships.

The certification process has been endorsed by the following quality-focused businesses: Motorola; the National Alliance of Business; Dobyns, author of the book *Quality or Else*; General Motors; Ford Motor Company; and

National VICA. Endorsements are pending from Arthur Anderson and the American Society for Quality Control.

To date, seven students have undergone the rigorous TQM certification process, and 40 more students now are seeking certification. The first TQM-certified graduate has been hired to conduct company TQM training for new store employees at Harris Teeter, a major supermarket chain.

Our experience makes it apparent that public schools can empower students to succeed. Virginia Beach's quality focus is creating partnerships to ensure that students succeed academically, personally, and professionally within a thriving quality community.

Patrick Konopnicki is Director of Technical and Career Education, Virginia Beach Public Schools, Virginia Beach, Virginia.

Quality Excellence in Education

Franklin P. Schargel

This is the story of a vocational high school in Brooklyn, New York, that has a heavily minority student body that suffered all the traditional ills of similar schools—high dropout, skipping classes, poor support by parents, and lack of educational results. After one administrator discovered TQM, the school began a slow process of implementing quality practices across the board. In doing this over a period of four years, this school has completely turned itself around and become a benchmark institution for vocational high schools. This inspiring story dramatically affirms the value of TQM in education.

Today's employers continually complain that young adults are poorly prepared for the working world. This "work force of the future" not only lacks basic literacy skills, but it is also found to be deficient in social skills, attitudes, dress, punctuality, and many other work-related skills valued by the business community. The student body must, of course, accept some of the responsibility for these shortcomings. However, the real culprit in this unfortunate situation is America's inefficient and out-dated system of education.

America stands poised to enter the 21st century equipped with school calendars developed in the 17th century; teaching methods developed in the 17th and 18th centuries; and classroom layouts developed in the 19th century. American education is like the man in the boat furiously rowing to the future while the boat is firmly tied to the rooted paradigms of the past. We cannot succeed as a nation until we untie the boat.

America cannot compete against low-wage, low-skill countries. And, if we are to compete against countries with advanced economies (see matrix shown at right), we must improve our education by integrating quality practices and processes into the fabric of our schools so they become an integral part of everything a school does. It no longer makes sense for quality management techniques to cascade down from industry to the schools—instead it should percolate up from the schools to industry.

The problem is that our schools have failed to keep pace with the changing demands of America and the global marketplace and have become a drag on our economy. Sustained economic growth can only be achieved through a

> America stands poised to enter the 21st century equipped with school calendars developed in the 17th century.

Reprinted with permission of the author. Originally published in *The Quality Observer*, June 1996. Copyright © 1996 by Franklin P. Schargel. All rights reserved.

high-skilled world-class work force—a work force made up of individuals who are equal to or better than the best workers in the world.

Some people believe that if we devote enough energy and resources to the problem we will halt the decline in education. However, this argument simply cannot be supported in the reality of the '90s.

In the coming years, education will face serious competition for declining government resources. Society can no longer support the idea that pouring more money into our schools will work, given the current structure of the schools. We will not solve our problems by spending more or spending less, by creating new public bureaucracies, or by "privatizing" existing schools. Taxpayers are demanding a greater return on their investment. After 10 years of educational reform and $60 billion in new funding, tests scores are stagnant and dropout rates are higher than they were in 1980. Critics of the schools insist on lengthening the school year by extending the school day or by making students and teachers work harder. The "More-Longer-Harder" strategy of educational reform doesn't work. (Dr. Deming would have called it "tampering.") The way to achieve educational transformation is through a systemic change in the way our schools are organized and run, and the ways in which teaching and learning take place.

As a result of applying quality management techniques, George Westinghouse Vocational and Technical High School, an inner-city school, has been successful in reducing dropouts, increasing post-secondary school attendance, increasing parental involvement, designing and implementing business and school partnerships, and developing a quality improvement curriculum.

George Westinghouse Vocational and Technical High School, located in the heart of downtown Brooklyn, is in many ways a typical urban inner-city school. Although the school is open to all city residents, most of the students live in the inner-city neighborhoods of Brooklyn. Of the approximately 1,800 students currently enrolled:

- 70 percent are African-Americans
- 26 percent are Latino
- 3 percent are Asian
- .5 percent are white

Many of the students come from single-parent, low-income families—62.1 percent live below the poverty line. Most graduates will be the first in their families to obtain a high school diploma.

Westinghouse has problems typical of many inner-city schools. It has an aging faculty; a high student-attrition rate; students who entered with poor reading and math skills; and students with a lack of motivation, low self-esteem, and a history of failure.

In the fall of 1990, an assistant principal at Westinghouse attended a seminar. He returned to the school convinced that utilizing Total Quality Management (TQM) to change the instructional process would make a significant difference in the academic results of the school. The principal of Westinghouse agreed and pledged his support. For the next four months, they immersed themselves in reading all they could about quality. Not having many

The "More-Longer-Harder" strategy of educational reform doesn't work. (Dr. Deming would have called it "tampering.")

educational models, they decided to benchmark the nation's leading enlightened industrial firms with successful TQM programs in place. These companies generously shared their expertise by training the assistant principal in TQM tools and techniques; and in December 1990, the school administrator agreed to assume the responsibility of quality coordinator along with his other duties.

In January 1991, the concept of TQM was introduced to a highly unionized, skeptical staff. The skepticism stemmed from the proverbial notion that quality management is just another fad, or "flavor of the month," that will soon pass. The assistant principal felt that if he couldn't convince the staff to accept the process, then it was doomed to fail. On January 30, 1991, the principal stood before the faculty and stated: "As long as I am principal, this school will use TQM techniques and tools to address the challenges which we face." As a result, the staff wrote a mission statement for the school: "The purpose of George Westinghouse Vocational and Technical High School is to provide quality vocational, technical, and academic educational programs that will maximize each student's full potential in today's changing technological society and prepare students to meet the challenges of our rapidly changing world. In an era of intense international competition, each student will be prepared to meet the demands of the world of work, pursue post-secondary school education, and address life's challenges."

The first step toward making this mission a reality was building a TQM foundation by establishing credibility with the faculty. The assistant principal developed and delivered several TQM training workshops. This training familiarized participants with the quality philosophy, as well as the actions, tools, and techniques that would help them achieve a higher quality product and reduce chances of failure. The faculty pinpointed 23 areas of concern that needed to be addressed. They prioritized them, and it was agreed upon to address one new obstacle a month while continuing to deal with the previous month's concerns. Since January, a cross-functional steering committee has directed its energies toward removing the root causes of each error factor.

One year later, January 1992, we began to introduce these techniques to the parents. In January 1993, the process was introduced to the students.

Short-Term Results

As a result of our efforts in promoting quality concepts during the past four and one-half years, we have seen many significant changes and improvements. Here are some examples:

Student Dropout Rate
When the process began our dropout the rate was 12.9 percent. It is now 2.1 percent. To put this in perspective, New York City, as a whole, has an average dropout rate of 17.2 percent.

Post-Secondary School Attendance
Although most of our graduates are the first in their families to finish high school, more than 72 percent of them will pursue post-secondary education.

Apprentice Training Program

Teachers in our vocational and technical departments have redesigned our 9th-grade program. Like most high schools, Westinghouse has a higher dropout rate for the entering class than any other. Our Apprentice Training Program pairs an entering freshman with a senior mentor. For 10 weeks freshmen are assigned to seniors in shop classes. The 9th grader works side by side with the 12th grader who guides the younger student through class experiments. Freshmen pick up 10th and 11th-year skills, and the seniors reinforce their own learning and assume leadership responsibilities. Teachers say there is less boredom in classes, thus less disruption and a lot more focus on work. This program provides one-on-one education at no cost to the school, taxpayers, or Board of Education.

As a result of the pilot year, 28 freshmen in the Apprentice Training Program received grades of 85 percent or better. Of an equal number of those not in the program only 14 received grades of 85 percent or better. Also attendance of those freshmen in the Apprentice Training Program was higher. The success of piloting the Apprentice Training Program resulted in expanding it to include all freshmen after the initial experiment.

Quality Improvement Class

We established a Quality Improvement (QI) Leadership Class for students so that we could begin involving students in the quality process. The class, composed of between 25–30 students from various grades, meets five days a week and informally serves as the Student Quality Team. The team meets monthly with the principal and tries to locate ways to improve overall student performance at the school. We also developed Quality workshops for students and some have taken Quality Training at National Westminster Bank USA and the IBM Executive Training Facility at Palisades, N.Y.

Summer Quality Academy

Raising student achievement is also one of our goals for the Summer Quality Academy. Two hundred randomly selected incoming 9th-grade students received training in the application of TQM processes before the start of the 1992–1993 school year. The students were taught:

- "How to do it right the first time"
- critical thinking and decision making
- active listening
- test-taking
- team-building
- time management skills

Classroom Cutting

One of the staff's concerns was class cutting. By using TQM techniques and tools, we have been able to reduce cutting by 39.9 percent in a six-week period. Another major ongoing concern is class failure. On January 30, 1991, we identified 151 students who had failed every class. Using the tools and techniques provided by TQM, we were able to reduce the number of students who failed every class from 151 to just 11 by June 30, 1991—a 92 percent improvement.

> Using the tools and techniques provided by TQM, we were able to reduce the number of students who failed every class from 151 to just 11.

Student Recognition and Reward

During the past school year, each subject area or department has recognized an outstanding departmental achiever as the Quality Student of the Month. The student is recognized by having his or her name and picture placed on a bulletin board.

The Quality Students of Westinghouse, those attaining an 85 percent and above average, are listed on a bulletin board in the hallway outside the principal's office.

A Quality Student Recognition Ceremony is held each term to honor our high achievers and their parents. Parents receive a congratulatory letter from the principal at the conclusion of each term thanking them for contributing to their youngster's success.

Faculty Recognition and Reward

We have started a staff recognition and reward program. A Quality Staff Selection Committee, representing a cross-section of the staff, reviews staff and student nominations and chooses a Quality Staff Member of the Month. Those selected receive a plaque and a small check, as well as the honor of having their names announced over the public address system. In addition, their names are placed on our "Quality Staff" bulletin board for all students, staff, parents, and guests to see.

Staff Steering Committee

As we progressed, staff members began to see concrete results (e.g., improved school tone, reduced cutting, greater student achievement). A cross-functional Staff Steering Committee was formed by volunteers to address, monitor, and institutionalize our quality efforts.

Tearing Down Departmental Walls

Inter-departmental meetings have been held on a regular basis. Members of the English and Social Studies departments have held joint meetings to discuss how to coordinate learning programs and how to implement writing across the curriculum.

The three trade departments have held joint meetings to coordinate the ordering of supplies, as well as to handle instructional problems that they share.

School Contract

Our Quality Staff Steering Committee, during the 1991–1992 school year, wanted to improve student achievement by raising student attendance, reducing student tardiness and cutting, and developing a consistent school-wide policy to promote student achievement.

As a means of achieving these goals, committee members drafted a student contract that describes requirements to succeed in subject classes. They began by meeting once a month after school but soon requested weekly meetings. They worked for four months hammering out the wording of a proposed school contract. Their only payment was two dozen doughnuts and the opportunity for input into restructuring the school's learning environment. Members of the committee presented the proposed contract at monthly faculty conferences to obtain further input, comments, and suggestions. In June, after months of fine tuning, the contract was approved by 75 percent of the faculty. The contract was implemented in September 1992.

Union Grievances
As a result of continued involvement with our staff and their union, grievances fell from 29 in 1993 to three in 1994.

Parental Involvement
We felt that without parental support, our process would fail. Since parents are considered internal customers at our school, we felt we needed to survey them so that we could best satisfy their needs. A questionnaire was developed and completed by parents at an evening orientation for incoming students at the first Parents-Teachers' Association meeting. The questionnaire sought to determine the best day and time to conduct PTA meetings and the topics our parents wished to discuss. In addition, a personal profile of our parents' income level, educational attainment, and family composition was developed.

The survey results demonstrated that our parents wished to select meeting topics and invite guest speakers. When the parents were brought into the decision-making process, they found their voice and our PTA was transformed. In spite of their decision to more than triple their dues, membership increased dramatically to more than 200 paid members. Monthly attendance at meetings followed suit. For the first time in years, the PTA adopted formal bylaws, and leadership positions were contested at the annual election of officers.

The parents asked for and received training in negotiating with teenagers, conflict resolution, evaluating and selecting colleges, etc. As the school year progressed, parents increasingly internalized the idea of shared responsibility and they became enthusiastic, creative, and proactive. In response to the social tensions created after the Rodney King verdict, our parents requested and organized a Family Night on June 1, 1992. The event's theme was that we are one family—the Westinghouse High School family—all working to create a caring environment for learning. More than 175 parents, students, and staff shared ethnic food, singing, and dialogue. The overwhelmingly positive response to Family Night persuaded us to continue the event as an annual function.

Satisfying Our External Customers
TQM Training
Our external customers in the business community—such as Colgate-Palmolive, Digital Equipment, IBM, Motorola, the Marriott Corporation, NYNEX, and Xerox—have been generous in sharing their TQM expertise.

Business Partnerships

The Ricoh Corporation, a leading Japanese manufacturer of photocopiers and facsimile machines, visited our school and was impressed by our application of the TQM process. Ricoh hosted 60 of our students on a tour of its facilities and followed that up with a partnership offer for our students to repair Ricoh products.

The school has raised more than $2,000,000 from industry in new or additional programs and services.

National Westminster Bank USA and the Westinghouse Electric Corporation have established multi-year "Continuous Improvement" scholarships for our most improved student graduates.

Business Advisory Councils

School-Business Advisory Councils have been established in the electronics, woodworking, and optical areas. Another is being developed with the jewelry industry. Council members—from business, industry, and the faculty—meet regularly with the principal to suggest ways to upgrade our programs and make our students more employable. The school has been visited by numerous business leaders who have donated equipment and supplies. The school has received more than 150 computers, donated by the manufacturer, to be used to further the TQM effort.

Partnerships with Colleges

Polytechnic University and New York City Technical College have agreed to run coordinated programs with Westinghouse High School. Project Care allows students to take courses at the colleges while they are still attending Westinghouse. The 2 + 2 Tech Prep Program is another college preparatory program that the school has developed with New York City Technical College. The National Center for Research in Vocational Education (NCRVE) has said that our program should be the "benchmark" for the nation's Tech-Prep programs. Polytechnic University students now tutor our students in Scholastic Aptitude Test preparation.

State Grants

The school received several funding grants, including a New York Working Grant, that assured the school $143,000 a year for three years to establish a year-round, on-site employment office (i.e., Career Development and Employment Center). Westinghouse was one of only six New York City high schools, and the only Vocational and Technical school, to receive this grant.

Media Attention

The school has been featured in four international videos, at least 10 books, and roughly 40 articles—including *Business Week*, *Fortune*, *Management Review*, and *The New York Times*.

Conclusion

What we are achieving has been done without any funding from our Board of Education or the city of New York. We have worked with a minority population in a hostile learning environment. If we can make it happen in New

York City, it can happen anywhere! Like most organizations in the midst of quality improvement, we can best be described as a work in progress. We have taken a few steps in the process of problem solving and improvement, team building, strategic planning, and leadership—and those will continue. Perhaps the most important thing we've learned is that we are on the right track because the process works and it is here to stay. We recognize that TQM is a proven process being used by industry worldwide to fundamentally change the methods of doing business, to make the best use of its resources, and to realize the greatest human potential.

We are not finished! We never will be! The road ahead is longer than the one we've traveled. If we are to have a skilled and dedicated work force, then the idea of quality must be integrated into the day-to-day operations of the schools. What we have done is to provide a beginning that any school, teacher, or parent can undertake today using existing resources.

The re-emergence of any nation's businesses has to be predicated on the success or failure of its public educational system. If the educational system fails to deliver qualified graduates as workers, then the business community is doomed to fail.

Franklin Schargel is an education consultant helping schools improve through using TQM tools and techniques.

Sharing the Wealth: TQM Spreads from Business to Education

Robert Manley and John Manley

Deming's fourteen points are just as relevant to education as to business. This article documents how one school district used the fourteen points to transform its approach to education in administration and in the classroom.

The American educational environment is poised for change, as it was at the turn of the 20th century when Frederick W. Taylor's scientific management method was applied to the American school system.[1] The scientific management method is based on the premise that product specifications define product performance standards. In education, this translates into an emphasis on testing students at the end of a process. The same comparison can be made between an approach such as management by objectives, which uses specific objectives as the standard of comparative measurement, and outcome-based education, which compares desired outcomes against actual outcomes. In contrast, W. Edwards Deming's philosophy of quality improvement emphasizes the *processes* by which results are produced, rather than the end results. It is this philosophy that is finding its way into education today.

There is some concern, however, that the difficulties that have plagued the American business community in accepting, understanding, and implementing total quality management (TQM) may also cripple efforts in education.[2] The business and educational communities have trouble with the fact that TQM cannot be applied in a cookbook fashion—it can only be applied on a case-by-case basis. Examining the experiences of one school district that implemented a quality program, however, may help other districts with the same mission.

Defining Total Quality Education

The West Babylon School District in Long Island, NY, began its quest for TQM by formally defining the educational philosophy of continuous improvement and quality. Thus, total quality management became known to the district as total quality education (TQE), and business terms such as customers

> Deming's philosophy of quality improvement emphasizes the *processes* by which results are produced, rather than the end results.

Reprinted with permission from *Quality Progress*. Copyright © 1996 by the American Society for Quality Control. All rights reserved.

and mass inspection became ideas and concepts related to education. The name change shifted attention away from preconceived notions of TQM and focused attention on the district's commitment to providing quality education, encouraging lifelong learning, and enabling students to take pride in their work. This definition describes TQE as a student-oriented instructional philosophy of continuous improvement that involves the following:

- Commitment of the administration and the board of education
- Participation by a critical mass of stakeholders
- Using statistical tools for analysis
- Continuous review of processes
- Exercising strong quality leadership
- Providing a training and retraining program
- Using portfolio performance and other hands-on expositions for student evaluations

In 1991, West Babylon's superintendent and board of education committed themselves to integrating quality into their administration based on Deming's 14 points.[3] The superintendent attended training programs taught by Deming and J. M. Juran. Training tapes were purchased from the Public Broadcasting System and Massachusetts Institute of Technology for use in seminars for executives and supervisors. Principals, department supervisors, and the superintendent trained the remaining staff in TQE. In addition, site visits were made to a Long Island engineering firm to see how a business implemented TQM.[4] The district was then prepared to apply Deming's 14 points to education.

The 14 Points of TQE

1. *Create constancy of purpose.* At a General Motors-sponsored seminar in 1993. Deming was asked who the customer is in education. He answered, "We go overboard on words. Society should be the beneficiary. We don't have customers in education—don't forget your horse sense."[5] The student is the product of the schools. Society benefits from educated youths who believe that community service and lifelong learning are personal responsibilities.

    To create constancy of purpose, the following mission statement was developed by the superintendent's management team, the board of education, instructional and support staff, parents, and community members. The plan-do-check-act cycle allows the statement to evolve as the community grows and changes.

    > We, the West Babylon School Community, declare our commitment to provide educational experiences of quality that enable all our students to learn, share, and succeed. Trustee and employee contributions are vital in our efforts to have students become lifelong learners

who take pride in their work and in service to others. We, therefore, endeavor to continually improve the quality of our work.

2. *Adopt the new philosophy.* In West Babylon, employees began adopting TQE by meeting in cross-functional teams, or quality circles, to set priorities for staff training. Quality circles are defined as "intensive, small-scale teams or units focused on improving the . . . process by getting input from those who do the actual work."[6] Throughout the district, the use of quality circles and statistical data is an integral part of analyzing the work that is accomplished. In addition, the instructional staff measures what students apply in a simulation or laboratory setting, while teachers remediate what students have not yet learned to apply, thus creating a continuous learning environment.

Several changes in the district illustrate the benefits of adopting the new philosophy. For example, demand for school lunches rose significantly when lunch services personnel began monitoring students' desires for specific foods, such as pizza with multiple toppings. Second, attendance at after-school tutorials tripled when parents and teachers encouraged students to perceive them as additional assistance for success—not as punishment. Finally, in the elementary computer centers, staffing needs were reduced with the help of grandparent volunteers who assist library media specialists. By actively seeking parental and community assistance, parental support and guidance increases, as does community concern for education.

3. *Cease dependence on mass inspection.* Support teams of transportation, lunch, clerical, and building maintenance workers meet throughout the school year to reduce variation in financial services, training, and materials, and to increase satisfaction among students, teachers, parents, and administrators. Nine cross-functional quality teams made up of teachers from seven schools meet regularly to discuss educational goals. These cross-functional teams include teachers from two grade levels and support staff from different buildings. The teams analyze the work flow and learning processes of students. Unit, chapter, and midterm exams are de-emphasized: instead, the application of concepts, portfolio work, individual performance, and integrated or thematic projects dominate the evaluation of learning.

The following examples demonstrate how end-of-process testing was reduced.

- At the elementary schools, entire grades engage in thematic learning processes, such as re-creating an American colonial village in the school gymnasium.
- At the middle school, students select a single theme to explore, such as the future of news.

- At the high school, five academic departments now use hands-on portfolio work for 35% of the year-end New York State Regents exams.

4. *End the practice of awarding on price tag alone.* District regulations require that all municipal purchases go to the lowest responsible bidder. Since responsible suppliers would not provide shoddy or faulty products, those that deliver poor-quality products are removed from the list of bidders. In addition, orders are placed early in the spring or fall from a single supplier to reduce variation. Applying this concept, the custodial quality circle identified a need for higher quality brushes, mops, and cleaning fluids. The result has been longer-lasting equipment and greater quality service at no additional net cost.

5. *Improve constantly and forever the system of production and service.* Every person and every department must commit to constant improvement. In West Babylon, school-based quality circles work together to improve the system of delivering services. The continuous improvement of services is illustrated by the work of the fifth- and sixth-grade team, which is composed of a fifth-grade teacher from each of the five elementary schools, two sixth-grade teachers, a supervisor from the grade-school level, and the supervisor of special education for grades six through eight. The team used a fishbone brainstorming technique (see Figure 1) to identify issues that needed attention, such as the common perception among teachers that students did not successfully meet the challenge of engaged learning time (the time spent covering curriculum subjects) at grade six—too much performance variance was exhibited across the grade. After dividing the work of each fifth-grade class into 30-minute blocks, it was discovered that some classes had as much as 3.5 hours more engaged learning time per week than the average fifth-grade class. The team analyzed the variables that pull students out of class during instructional time and recommended several changes that added a total of 43.5 hours of instruction per week for the entire grade.

6. *Institute training and retraining.* A quality circle helps the human resource division select training programs to increase the job skills of the support staff. During the 1994–95 school year, in response to a request from one quality circle, 86 paraprofessionals received training in assertive discipline practices to help them manage students more consistently. Instructional staff members also receive training and retraining in new classroom technologies, as well as in the research findings of their individual disciplines. In addition, all employees receive at least two days of training annually to help improve the quality of their work, and teachers now attend half-day training sessions in the use of multimedia in the classroom.

7. *Institute leadership and discover barriers to pride in work.* West Babylon instructional support supervisors believe that to lead is to help others do their jobs better. Supervisors seek to continuously train employees.

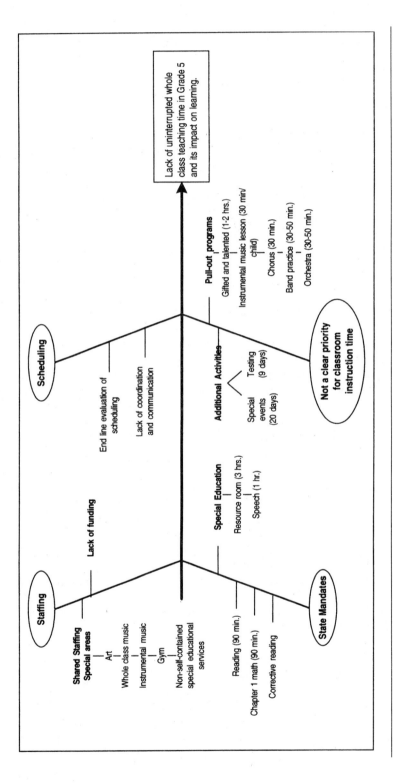

Figure 1. **Fifth-grade cross-functional team analysis**

Teachers, supervisors, principals, and other certified staff receive training in synergetic or cooperative supervision and work in cooperative teams to improve instruction and learning.

All 35 supervisors meet with human resource personnel five times a year for advanced training in cooperative supervision. An instructor from the University of Tennessee spends one morning a year with the supervisors to enhance their techniques and share the latest research with them. This exchange results in two or three new practices each year. For instance, the social studies chairperson now allows teachers to write their own key improvement indicators after a supervisory review.

8. *Drive out fear.* Suggesting new ideas is risky; people fear punitive assignments or other forms of discrimination or harassment. To counter this, the district encourages applying new methods, creating new solutions, and implementing innovative experiments. This has been accomplished, in part, by quality circles that encourage interaction and open discussion. All quality circles are made up of volunteers who meet at least three times a year. In the past, teachers and administrators complained about the number of meetings they were required to attend. The use of quality circles, however, has led to far fewer complaints since quality circles replace other committees, are voluntary, and achieve results quickly. For example, quality teams at the primary-school level asked for parent training programs. They identified the need, wrote the program with parent volunteers, and provided training. Now, parent training occurs at the preschool and kindergarten level.

9. *Break down barriers between staff areas.* Deming said people can work superbly in their respective departments, but if their goals conflict, they can ruin a company.[7] Quality circles improve cross-communication and encourage brainstorming among members of the district. The superintendent's quality council (SQC), which includes senior executives from personnel and administration, curriculum and student services, testing, finance, transportation and lunch programs, and elected members of the community, ensures that local quality teams do not act in conflict with one another. Community groups and union leaders receive all board of education policies prior to implementation. Nine cross-functional, 30 departmental, and seven school-based quality teams ensure that the district's mission and vision are achieved. A wellness committee, composed of managers, supervisors, and union personnel, evaluates recognition practices and helps determine recognition programs.

10. *Eliminate slogans and targets for the work force.* It is necessary to provide the means to the desired ends. For instance, the transportation staff prevents waste and reduces accidents by managing its own quota-free environment. Administrators and teachers focus on improving student learning without using elitist or confusing slogans. Instead, students are encouraged to master the learning expectations of the curriculum and are tutored according to their individual progress. Report cards in grades

> Community groups and union leaders receive all board of education policies prior to implementation.

Language Arts	1st	2nd	3rd	4th
Uses clear oral language				
Demonstrates good listening skills				
Self-selects writing topics and ideas				
Engages in and sustains writing activities				
Writes effectively for a variety of purposes				
Uses a variety of vocabulary and sentence structures				
Self-edits and revises for clarity and meaning				
Uses punctuation appropriately				
Uses capital and lower-case letters appropriately				
Uses legible handwriting				

Key
E = Evident—consistently demonstrated
OE = Often evident—working toward consistency
SE = Sometimes evident
NE = Not evident
NA = Not assessed at this time

Figure 2. **Sample from third-grade student progress report**

K-3 were redesigned by the staff with parental input to reflect this cultural change (see Figure 2).

As a result of the continuous effort to encourage learning, the number of middle school students attending summer school has decreased 50%. Also, due to a quality circle analysis of school activity participation, ninth graders now attend a five-week high school orientation program. As a result, ninth-grade participation in school activities has risen from 66% to 92%.

11. *Eliminate numerical quotas.* The behaviors of hard work, pride, service, and growth, should be emphasized instead. For instance, school bus transportation schedules and stop locations are now planned and altered to meet the community's needs. The goal is to support active learning endeavors, not increase the number of students transported. Teachers and administrators reject numerical results as a measurement of success;

failure rates are not predetermined by a bell-shaped curve. Instead, student achievement, participation in learning and service activities, and parental support are measurements of quality. Students are evaluated by demonstrating their knowledge and judged by their performance. As a result, cost and service improvements are identified by individuals or quality circles and data analysis.

Removing quotas encourages workers to apply new ideas to old tasks. For years, the district's payroll checks were printed off-site. Under their own initiative, the clerical staff compared the cost of off-site check printing to the cost of purchasing a software package, computers, and a local-area network. They recommended a switch to on-site check printing. The additional responsibility could have been a barrier to the change, but the staff's involvement in the decision led to their voluntary commitment to the project. Now, the $200,000-a-year savings is being used to hire additional classroom teachers.

> Student achievement, participation in learning and service activities, and parental support are measurements of quality.

12. *Remove barriers to pride of workmanship.* Inspection identifies defects but does not explain how to prevent them. Intrinsically, workers want to produce products without defects. Leadership and continuous system improvement make this possible. To achieve pride in work, supervisors must listen to staff members, and employees must listen to the recipients of their work (the community). Often, these groups do not have the same knowledge base or purpose of work. Communicating the purpose of work and achieving respect among district employees and the community is one of the greatest challenges school leaders face.

    In West Babylon, both the support and instructional staff focus on training needs, material improvements, and process and procedural controls. Teachers and administrators meet with students to learn why they make errors, and supervisors and teachers examine the barriers to continuous learning. There are no quick fixes—experience suggests looking for patterns in the system that inhibit learning. Process improvement is slow at the classroom level since each classroom is a social system in itself. Nevertheless, it is clear that successful leaders are produced by persistence and careful analysis. West Babylon's TQE effort has resulted in a dropout rate of less than 2%; completion of a year of algebra by all ninth graders; participation by 85% of secondary students in after-school sports, music, and club activities; and an average student attendance rate of 96%.

13. *Institute a vigorous program of education and reeducation.* Workers must continuously acquire knowledge to be of service to the community. West Babylon support staff receive regular training in computer technology and advancement skills. Administrators and teachers attend programs on research in their fields or other areas of interest. In addition, instructional staff members participate in a yearly training program in an area identified by quality circles, such as multimedia technology for the classroom. In 1994, the entire West Babylon special education staff

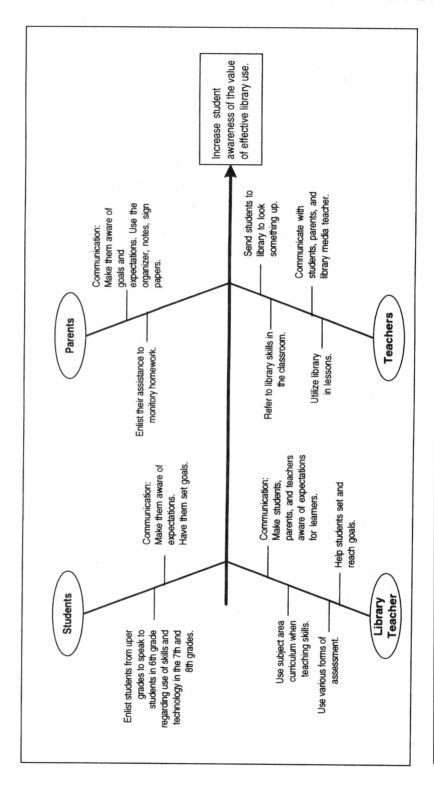

Figure 3. **How to improve student competency within the library**

began learning new ways to serve students with special needs. As a result, 80% of the special education staff chooses to participate in voluntary after-school programs.
14. *Take action to accomplish the transformation.* This can be done using the plan-do-check-act cycle—study a process, decide how to improve it, develop and implement a plan, and observe the effects. Each semester, employees follow the requirements set by the SQC to identify one to three goals, and a plan is written to implement, measure, analyze, and adjust the steps to accomplish these goals. Statistical tools such as Pareto charts, fish-bone diagrams, histograms, and scatter diagrams are used to structure and solidify the process and help visualize a situation.

The social studies department in the high school used a fish-bone brainstorming exercise to examine obstacles to student achievement. The team determined that the greatest barrier to learning was the predominant teaching style of lecturing and note taking. Now, training in different teaching techniques such as cooperative learning and thematic teaching is provided. In addition, a quality team examining library skills determined a need for students to realize the value of research skills. The fish-bone diagram shown in Figure 3 demonstrates the actions implemented to improve student competency with library skills.

If TQE is to work in a school district, continuous support from the superintendent, board of education, staff, and parents is a must. Constant provisions for training and implementing quality improvements demand resources from the entire district. Time must be allotted for planning, experimenting, implementing, checking, analyzing, sharing, and learning from academic experiences. Schools must adhere to the pursuit of quality and continuous improvement.

As the 21st century approaches, greater district and school autonomy will create an educational system that provides quality, improves instruction, and generates growth in learning. This is expressed in the paradox of American education as noted by Decker Walker: "The distinctive American talent for invention and adaptability leads, in its school system, to a series of novelties that burn brightly but fade quickly . . . if Demingism were to spread its influence from business to schools, the gridlock might begin to break much more quickly than in the past, for it would release the innate creativity of schools as institutions."[8]

> Statistical tools such as Pareto charts, fish-bone diagrams, histograms, and scatter diagrams are used to structure and solidify the process and help visualize a situation.

Notes

1. Maurice Holt, "The Educational Consequences of W. Edwards Deming." *Phi Delta Kappan,* January 1993.
2. Herbert W. Hoover Jr., "What Went Wrong in U.S. Business's Attempt to Rescue Its Competitiveness," *Quality Progress,* July 1995.
3. W. Edwards Deming, *Out of the Crisis* (Cambridge, Mass.: Massachusetts Institute of Technology Center for Advanced Engineering Study, 1986).
4. Mary Walton, *The Deming Management Method* (New York: Perigee Books, Putnam Publishing Group, 1986).

5. Maurice Holt, "Deming on Education: A View from the Seminar," *Phi Delta Kappan*, December 1993.

6. Ruth E, Thaler-Carter, "Measuring the Tangible and Far Reaching Effects of TQM," *School Foodservice and Nutrition*, February 1995.

7. Walton, *The Deming Management Method*.

8. Holt, "The Educational Consequences of W. Edwards Deming."

Robert Manley is superintendent of the West Babylon School District in West Babylon, New York. He has a doctorate in educational administration from St. John's University in New York City. Manley is a member of ASQC.

John Manley is an assistant professor of finance at Savannah State College in Georgia. He has a doctorate in finance from Rutgers University in Newark, New Jersey.

PART THREE

Implementing Quality

Part 3 presents a large collection of articles, almost all published in 1996, that deal with the strategic and tactical issues surrounding the implementation of total quality management. Reflecting what organizations have learned over the past several years, the material in this part is organized into four logical groupings: *Quality Transformation, Quality Tools and Techniques, Functional Processes,* and *Standards and Assessments*. Within each of these groups, you will find practical advice and techniques relevant to all types of organizations.

Quality Transformation

Quality management is **not** just a set of special techniques to add to your management repertoire. Quality management requires a long-term commitment at the top and a disciplined, informed transformation of company culture. Managers **must** focus on supporting and developing the people who carry out the interrelated processes that characterize organizational work. This section includes an array of articles on *planning, leadership, cultural transformation, voice of the customer, training, teams and teamwork, systems thinking/ learning organizations,* and *communication*. Our selection is designed to give you a clear sense of the attitudes and practices that can help you make the transition to a culture that supports continuous improvement of processes and quality for customers. Building a culture based on the values of quality management is not easy. It takes long-term commitment and discipline. But it's worth it, because it creates an organization that is vibrant and supportive of employees, that continually becomes more efficient, and that will have a sustainable competitive advantage.

In this section, we are happy to have two original articles by William Lareau (in *leadership*) and Scott Davis (in *voice of the customer*). Lareau is author of *American Samurai* (1991), one of the best books on TQM principles and practices to have come out in the past several years. Davis is a highly regarded marketing consultant. We are pleased with all the articles included, but some you may especially want to note include Bob Filipczak's "The Soul of the Hog," a review of the Harley-Davidson company culture; Peter R. Scholtes's "Teams in the Age of Systems," which includes an especially valuable review of different types of teams and when to use them; and Margaret Wheatley and Myron Kellner-Rogers' article "Self-Organization: The Irresistible Future of Organizing." This thoughtful piece argues strongly for why understanding organizations as systems makes sense and the implications of such an approach for succeeding today.

Quality Tools and Techniques

In the 1997 edition of *The Quality Yearbook,* we continue to emphasize process management and include three subsections, *process reengineering, benchmarking,* and *process management measurement.* These include a variety of practical articles to assist you in thinking about these issues such as, Oren Harari's "Why Did Reengineering Die?" John H. Lingle and William A. Schiemann's "From Balanced Scorecard to Strategic Gauges: Is Measurement Worth It?" Nicholas Horney and Richard Koonce's "The Missing Piece in Reengineering" and Philip Ricciardi's "Simplify Your Approach to Performance Measurement."

Functional Processes

Important work is under way in many industries at the functional level (for example, accounting, information management, and human resources), making it possible to begin documenting successes and procedures. We have noted this and chosen to highlight six areas: *suppliers and purchasing, logistics, information technology and management, accounting and finance, product development,* and *human resources.* The subsections on logistics and product development are new to this edition. Our guideline here was thoughtful practicality, and these articles describe how quality principles can help shape and improve practices in all these areas.

Standards and Assessments

We conclude part 3 with three articles that look at the issues surrounding ISO 9000 and the Baldrige Award. We are pleased to have two original articles in the ISO section. Richard Randall's "Now that You're Registered . . ." examines what happens once a company is registered—is that an end or another part of the journey? We also include three other articles on issues related to ISO 9000. The articles section of the yearbook concludes with a thoughtful look at where the Baldrige Award now stands.

We again refer you to the detailed bibliography in part 4. You'll find an extensive listing of current articles and books covering all the topics taken up in part 3.

Quality Transformation
PLANNING

Reinventing Strategic Planning

John C. Camillus

Strategic planning, like other activities in an organization, is a process. This suggests that the process can be analyzed and that methods can be developed to make the process more effective. This article suggests a specific methodology for undertaking the strategic-planning process. This process systematically examines bipolar organizational choices in a way that allows organizations to effectively match their capabilities to satisfy customers with what customers value so as to avoid decisions that by less disciplined planning end in market failure.

If there is any consensus on the nature of the management challenge today, it is that business environments are more dynamic, competitive, and unpredictable than ever before and that businesses have to be more flexible, organic, and innovative in how they structure themselves. Different environmental characteristics and different organizational forms require new and different ways of defining business strategy.

New concepts of strategy will inevitably require or be created by new and different processes. With both the content of strategy and the process of strategy development requiring change, it follows that there is a need to reinvent "strategic planning" in organizations.

This article is reprinted from *Strategy & Leadership* (formerly *Planning Review*), May–June 1996, with permission of the Strategic Leadership Forum (formerly The Planning Forum), the International Society for Strategic Management. Copyright © 1996 by the Strategic Leadership Forum. All rights reserved.

This need is reflected in the turmoil that is evident in both the practice and theory of strategic planning. Gerstner of IBM, Eaton of Chrysler, and Gates of Microsoft, in different contexts and for different reasons, all have questioned the value of a strategic "vision." Mintzberg has written the obviously premature obituary and projected the resurrection of strategic planning.[1] Strategy has been freshly described through the lenses of "hyper-competition," resources/core competence/capabilities, population ecology, time/speed, chaos/complexity, and issue management. Sifting through these multitudinous and individually persuasive perspectives to create an integrated, useful, and manageable approach requires that we take a step back and remind ourselves of why a statement of strategy is needed and what the nature is of the issues that strategic planning processes address.

Functions of Strategy Statements

Strategy statements at the very least need to:

- Provide direction, preferably in an evocative and inspirational fashion that ensures consistency over time, especially as reflected in resource allocation decisions.
- Offer a framework for action that ensures consistency across managers.

These minimalist objectives suggest that an effective strategy statement should (1) flow from or create a shared and enduring vision of the future of the organization, (2) inform capital budgeting and human resource development decisions, (3) guide managerial actions and provide benchmarks of performance, and (4) direct management attention to emerging strategic issues.

Dimensions of Strategic Choice

At the core of strategy is a framework of fundamental alternatives. If we can identify the dimensions of this framework, we can describe the foundation of a strategy. These dimensions are unique to each business and constitute a simple, understandable, powerful, and effective way to define an organization's strategic profile.

The power of this approach to articulating strategy can be assessed by the ability of the approach to serve as a basis for meeting the four objectives of strategy statements and by its applicability in vastly different organizations. This approach has been adopted in dozens of organizations; the following examples are drawn from actual strategy development efforts in very diverse industries.

A Health-Care Business

An organization in the health-care business identified five key dimensions of choice as the basis for profiling its strategy. The dimensions were:

- Type of client the organization would serve
- Source of revenues

- Basis for diversification
- Orientation toward growth
- Kind of technology relevant to the organization

These five dimensions, illustrated in Exhibit 1—Example A, can readily define alternative, significantly different "visions" of the organization. At the left extreme, we have an organization that chooses to serve clients of limited means by accessing support from philanthropic foundations, engages in new activities only if they have been proven successful by other organizations, emphasizes the status quo, and employs unsophisticated technologies. At the other extreme, we have an organization that serves clients who have the ability to choose between alternative providers, relies on fees from clients or insurance, is an innovative leader in the business, is oriented toward aggressive territorial and product-line expansion, and employs advanced technologies.

The left extreme of these dimensions defines an organization similar to the typical visiting nurses association. The organization is characterized by altruistic motives, support by the United Way, and the offering of basic services such as home-health care. The other extreme describes an organization such as Humana, where the leadership and growth aspirations position it at the frontier of technology with innovations such as the artificial heart, while simultaneously engaging in nationwide product-line/service expansion.

The five key choices describe extraordinarily different but logically consistent concepts of business or visions—different values, different aspirations, and different competencies. The driver of the strategic position of an individual health-care business is the values of its board of directors in deciding which clients to serve. The logical consistency of the organization's strategic profile can be readily checked by assessing the fit of the choices on subsidiary dimensions with the selected position on the driver.

A Credit Union

If an organization decides to change its competitive position, i.e., its profile along its strategic dimensions, the implications in terms of new competencies, strategic programs, and measures of performance are readily identifiable. This example describes a federal credit union that is faced with the strategic decision of whether to respond to its members' demands for the equivalent of checking services (known as "share drafts") and electronic banking. Currently, members can only buy "shares" through payroll deductions and obtain loans by applying in person. This organization's current strategic profile occupies the left extreme of the dimensions of strategic choice in Exhibit 1—Example B. To move to an alternative profile, new technology would have to be acquired and managed; new personnel with appropriate capabilities would have to be recruited; the membership field would need to be broadened to acquire the critical mass to economically support the investments in technology and personnel; merging with another credit union may be desirable to achieve the necessary critical mass; and a growth orientation may be required to support continued investments in technology and to support the culture necessary to respond to competitors' responses.

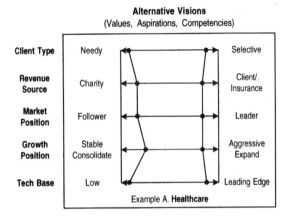
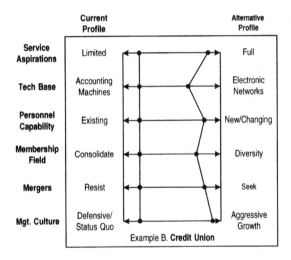
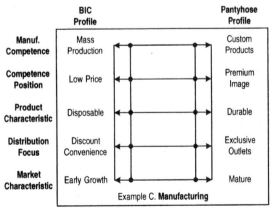

Exhibit 1. **Dimensions of strategic choice**

The gaps between the current profile and the alternatives being considered define the strategic action programs that would need to be implemented. The gaps also indicate the performance milestones that must be developed to assess implementation effectiveness.

Each profile will have one fundamental strategic choice that determines the appropriate position on subsidiary dimensions. In this example, the driving choice is the service aspirations of the management. With a different top management team, aspirations or values may find expression in a defensive, status quo orientation rather than an aggressive, growth orientation (which in Example B is the last dimension). If this management culture is unchangeable, then it would become the driver and, in effect, would rule out the consideration of the growth-oriented profile as an alternative.

Manufacturing

The dimensions-of-choice approach can effectively add qualitative, strategic criteria to the financial assessments of investment opportunities. This example has been tailored to make it relevant to the well-known and unsuccessful decision by BIC Corporation to add pantyhose to its existing product lines of disposable ballpoint pens, cigarette lighters, and razors. BIC's strategic profile occupied the left extreme of the dimensions in Exhibit 1—Example C. BIC's distinctive, core competence in precision mass production and in low-cost assembly of small metal and plastic components served as the driver of its strategy and led it to great success. The low cost arising from high volume and technical competence enabled BIC to make and sell its disposable products at price points as much as 80 percent below its competition. The disposable character of BIC's ballpoint pens, lighters, and razors reinforced the high volume of sales and also enhanced the convenience of use of its products in comparison to fountain pens, refillable lighters, and traditional razors. The distribution channels fitted ideally with the character of its products. The focus on growth markets supported the volume characteristics consistent with BIC's mass production competence. Not surprisingly, the logical consistency of BIC's strategy led to enormous success.

BIC's decision to add pantyhose to its product lines was prompted by financial analyses that indicated positive returns when a French source offered the opportunity to BIC. The venture was a striking failure and would probably have been readily identified as a non-starter if the dimensions of strategic choice had been employed to qualitatively assess the fit of this opportunity with BIC's successful, well-implemented existing strategic profile. In the first place, the product had nothing to do with BIC's manufacturing competence. Second, the product, far from being 80 percent lower in price than existing competition, was actually priced higher than the No Nonsense and L'Eggs brands. Third, pantyhose are disposed of when they run. In the context of pantyhose, "disposability" is likely to be viewed as an annoying lack of durability rather than as a convenience. Fourth, high-priced pantyhose are distributed through department stores and specialty stores, not through supermarkets and convenience stores. Fifth, BIC deviated from its practice of

> BIC's decision to add pantyhose to its product lines was a striking failure and would probably have been readily identified as a non-starter if the dimensions of strategic choice had been employed.

focusing on growth markets and entered a domain that was saturated and dominated by significant existing players.

The lack of strategic fit is strikingly evident from the perspective of the dimensions of strategic choice. Qualitative assessments such as this could raise relevant and significant questions regarding the validity of the assumptions that underlie the necessary financial analysis.

The ability of the dimensions-of-strategic-choice approach to focus managerial attention on issues of strategic significance can be illustrated through any one of the preceding examples. Developments relating to the driver among the dimensions are likely to be of strategic significance. For example, in the case of the credit union, federal regulations affecting the type and combination of services that may be offered would affect the strategic profile in a fundamental way. On the other hand, developments relating to the subsidiary dimensions are more likely to affect action plans and deadlines and may not require a radical rethinking of strategy.

These selected examples illustrate how the dimensions-of-strategic-choice approach can be useful in articulating an organization's strategy:

- It can readily and evocatively identify an organization's vision—its enduring and distinct values, aspirations, and competencies.
- It highlights the qualitative assessments that can lead to better informed resource allocation decisions.
- It identifies the strategic programs and related, performance measures by vectoring the existing strategy to the desired strategic profile.
- It makes strategic issue identification more efficient and effective by focusing attention on relevant sources of change and assessing the strategic or operational nature of the impact of developments.

This approach has been employed in a wide variety of organizations. It has been applied in service and manufacturing industries; in businesses with annual revenues ranging from a few million to several billion dollars; in not-for-profit organizations; in highly regulated organizations such as utilities; and in highly competitive, bottom-line oriented businesses. In all of these contexts it has been viewed by management as a more enduring, evocative, action-oriented, communicable, and understandable way of articulating strategy than the traditional statements that focus on product-market-technology combinations, that define the orientation of the functional areas, or that describe an industrial-organization-economics-based relationship between the organization and its industry. Especially in the context of ephemeral products, transient markets, dynamic technologies, and blurred industry boundaries, the characteristics of the dimensions-of-strategic-choice approach merit consideration.

Process Design

Traditional processes employing SWOT analyses necessarily have to assume a strategy, usually the existing strategy, in order to ascertain what is a strength, weakness, opportunity, or threat. Inevitably, therefore, traditional strategic planning processes are oriented toward incremental changes in existing strategies. The premise that change is more rapid, significant, and unpredictable than ever before enhances the need for fundamental, *de novo* development of strategies. If the process cannot be anchored in an existing strategy, the alternative anchor or starting point is the strategic issues that can be identified. Issue generation techniques, usually variations of the classical brainstorming approach, can identify a universe of possibilities, potential, concerns, and aspirations that are the raw material of strategy creation.

In designing a strategic planning process that moves from strategic issues to the dimensions of strategic choice, it is important to bear in mind that strategic issues possess characteristics similar to the class of problems that are technically known as wicked. Wicked problems cannot be effectively resolved using the classic, linear problem-solving process. Wicked problems and strategic issues possess the following characteristics:

1. The problem is not independent of the solution.
2. Multiple stakeholders with different and sometimes conflicting priorities exist.
3. Unexpected contingencies and hazards will inevitably be encountered.
4. Underlying factors giving rise to a wicked problem are complex and intertwined, and cause and effect are difficult to distinguish.
5. No perfect or obviously right answer can be found.
6. Wicked problems and emerging strategic issues in a dynamic world have few, if any, precedents that can help in responding to them.

It is possible to design several alternative processes that can carry an organization from "strategic issues" to the "dimensions of strategic choice" and that incorporate the characteristics necessary to deal effectively with the wicked characteristics of strategic issues. One simple and effective approach that has been widely applied is described in Exhibit 2.

The first step is issue generation. Groups that engage in this activity must be designed to be diverse in both functional areas and levels of responsibility, while minimizing direct reporting relationships between members. From a wicked problem standpoint, this step in the process must provide for inclusion of relevant stakeholders and must emphasize divergent thinking.

The second step is one of convergence. The top management team could categorize and prioritize the issues. The classic criticality/urgency matrix can be effectively employed in this step. Exhibit 3 illustrates the application of this matrix in the context of an actual organization. Issues that are categorized as both highly critical and highly urgent are the focus of subsequent steps in the process. Although this is a convergent process, in case of doubt, issues should be assigned to the "high-high" category.

The third step in the wicked-problem-solving process involves clustering the issues into groups. While this appears to be a simple step, it is important

> It is possible to design several alternative processes that can carry an organization from "strategic issues" to the "dimensions of strategic choice".

Step	Techniques/Variables	Characteristics
Generating Issues	Brainstorming, nominal group techniques, and scenarios Hierarchy avoidance	Inclusiveness Divergent thinking
Prioritizing Issues	Top management involvement Criticality/urgency matrix	Covergence
Clustering Issues	Mind mapping, affinity diagrams, story boarding Multiple alternatives "Theories of the business" Paradigm shifts	Experimentation Surfacing assumptions Modularizing, but with interconnections
Generating Alternatives	Task forces Plenary sessions Overlapping membership "Ubiquitous" planners	Divergence Experimentation
Defining Dimensions	Top management involvement Brainstorming "Driver" identifcation 7+/- 2 dimensions Analytic hierarchy process	Experimentation Surfacing assumptions
Defining Strategic Profiles	Values/core capability drivers Reconstituted task forces Analytic hierarchy process	Convergence Actions —> strategy

Exhibit 2. **Strategy development and a wicked problem-solving process**

in that the criteria for forming the clusters and the connections between the clusters in effect reveal the implicit theory of the business. For instance, if competitive issues are linked to technology issues, a very different view appears than if competitive issues are linked to regulatory or legislative developments. It is important to develop and debate alternative ways of clustering issues in order to surface underlying assumptions and debate alternative theories of the business. The clusters respond to the wickedness of strategic issues by creating coherent and manageable groups of issues. However, the linkages between the clusters must be recognized in order to integrate the analyses that follow.

The fourth step is to assign the clusters to teams to generate alternative responses to the issues. This step is very significant in that it deviates from the classic strategic planning processes by requiring the development of alternative actions without the context of a strategy. It is a creative, divergent

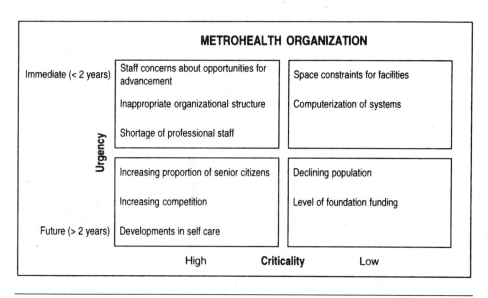

Exhibit 3. **Criticality-urgency matrix**

step where the focus is on what the range of alternative actions is rather than what actions support a given strategy. Lindblom and Hayes have discussed the logic underlying this inductive approach to strategy formulation in their classic articles.[2,3]

Exhibit 4 provides a stylized representation of the alternatives generated in response to a single issue in the Metrohealth organization. The teams must be constituted so as to support new approaches. The team leader cannot be the executive responsible for the organizational area closest to the issues being considered. The CFO, for example, should not be on the team looking at new financing options, but may be part of another team responsible for marketing or technology-related issues. Second- or third-level finance executives may, however, be part of the team looking at finance issues in order to serve as resource persons. Throughout this step, communication between the teams is essential and should be ensured through mechanisms such as plenary sessions, overlapping memberships, and "planners" who attend all team meetings.

The fifth step leads to defining the dimensions of strategic choice. In this step, a small group of senior executives takes up, issue by issue, the alternatives generated by the teams. Working with the first issue, they brainstorm to identify the dimensions along which each action alternative differs from the other alternatives. They repeat this process for every issue until no new dimensions are identified. Experience suggests that it is rare to encounter more than 12 or 13 dimensions. The dimensions can then be prioritized using techniques such as the "analytic hierarchy process" to arrive at the most

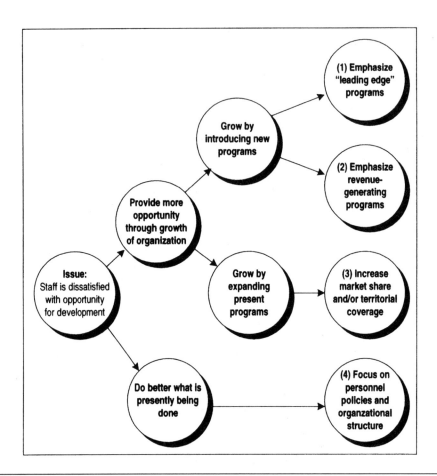

Exhibit 4. **Alternative responses to a strategic issue**

significant and evocative ones. Exhibit 5 reports the dimensions developed in the Metrohealth organization by considering the alternatives illustrated in Exhibit 4.

Selecting the driver that ensures logical consistency in creating strategic profiles is an important decision. The driver usually is the most enduring element among all the dimensions, reflecting the values, aspirations, or core competencies/capabilities of the organization. It usually characterizes an aspect of the organization rather than its environment. It also has the most obvious impact on what positions are feasible along the other dimensions.

The comprehensiveness of the dimensions selected must also be assessed before moving to the next step. One or more of the dimensions selected should affect every functional area or department in the organization, should reflect the major stakeholder groups, and should relate to every cluster of issues that was analyzed. It is desirable, however, to bear in mind the long-established limit to human comprehension of $7 +/- 2$ items.

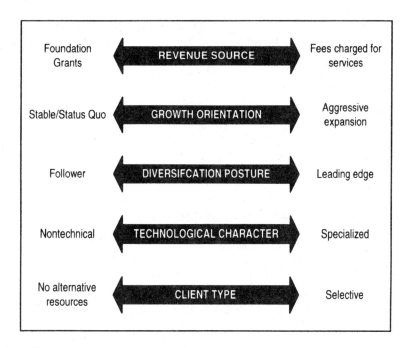

Exhibit 5. **Dimensions of strategic choice**

The sixth step involves the creation of possible strategic profiles. These profiles readily flow from the previous step. Strategic profiles are developed by identifying distinct positions along the driver and linking each of them to the appropriate, logically consistent positions along the subsidiary dimensions. These strategic profiles can then be applied to the alternatives generated for each issue to identify the one or limited few alternatives that fit with the profiles. The strategic profiles and related sets of actions can then be debated by top management to arrive at the profile that constitutes the vision of the organization and the action programs necessary to realize this vision. Multiple strategic profiles should be developed and debated in keeping with the recommended approach to handling wicked problems in which no single optimal answer can be objectively developed. The strategic profiles and related action programs can be created by a single senior management group or by teams of executives drawn from those who generated alternative responses to the issues. These new teams would be formed such that each reconstituted new team would include at least one representative from each of the previous teams that generated action alternatives.

One lesson stands out from the dozens of organizations that have gone through such a process. Communication is critically important. Not only should the selected strategic profile be communicated, but progress through the process should be shared, especially with those who had been involved in the initial issue generation step.

The dimensions-of-strategic-choice approach to defining strategy—both content and process—is designed to respond to today's turbulent environments and changing organizational characteristics. It describes the organization's strategy in an enduring fashion and yet provides the needed flexibility. The process responds to the "wickedness" of strategic issues, and yet, it is easy to communicate, remember, and use. The combination of process and content brings the people in the organization together to a shared vision and a framework that integrates resource allocation, critical success factors, strategic programs, and issue management.

Notes

1. Henry Mintzberg, *The Rise and Fall of Strategic Planning* (New York: Free Press, 1994).

2. C. E. Lindblom, "The Science of 'Muddling Through,'" *Public Administration Review* (Spring 1959):79–88.

3. Robert H. Hayes, "Strategic Planning—Forward in Reverse," *Harvard Business Review* (November–December 1985):111–19.

John C. Camillus is the Donald R. Beall Professor of Strategic Management in the Katz Graduate School of Business, University of Pittsburgh.

Quality Transformation
LEADERSHIP

Managing for Tomorrow's Competitiveness Today

Armand V. Feigenbaum

In this article, quality pioneer Armand Feigenbaum states in his own terms the importance of leadership in implementing quality-management principles and practices for any company that wants to remain competitive. He suggests that total quality leadership will be the requirement for success. He defines what this is in terms of five different components and why it is necessary. Feigenbaum remains an influential thinker on quality issues, and his thoughts deserve careful consideration by all managers.

Because of this delayed learning factor in large complex systems, it is not so paradoxical that it is often more productive to project the long term future of management and quality for the American economy over a period of decades than it is to estimate business conditions for the next two or three years. For example, we first reported our experience in successfully implementing quality-based competitive leadership in 1951 through publication of our initial *Total Quality* book. Our strategy and its execution were straightforward:

- Companies that satisfy the quality requirements of their customers better than competitors will have far better sales growth and profitability.
- Quality and cost are a sum not a difference—partners not adversaries—and companies that improve their quality significantly will also reduce their costs significantly.

Reprinted with permission from the *Journal for Quality and Participation*, March 1996. Copyright © 1996 by the Association for Quality and Participation. All rights reserved.

We emphasized that quality was likely to become the most powerful competitive force for U.S. companies for the second half of the 20th century (1950 to 2000) just as production had been for the first half (1900 to 1950).

We were right. . . . Systems and processes for total quality strength have reordered the priorities of U.S. companies and have become a hugely important factor in restoring America's international competitive leadership. Moreover, these systems and processes have become a major factor in businesses throughout Europe, Asia, Latin America, the Middle East and Micronesia.

Adapting to the conditions for 21st century success—However, as we enter quality's second half century, there is now seismic change in the customer's value expectations:

- In the U.S. and global marketplace . . .
- In the economic conditions now facing American manufacturing and service organizations . . .
- In the fundamentally different management approaches to human leadership needed to serve these demands.

The new requirement is how to position *quality-based competition* for implementation of the actions that fit these demands. Our strategy and its execution are again straightforward:

- Companies must achieve complete customer satisfaction in their key markets—not that customers are just satisfied but that they're enthusiastic about what you're doing.

This means companies must produce or deliver products and services which are:

- Affordable, for today's increasingly economically constrained consumer and business buyer . . .
- Essentially perfect for him and her . . .
- Determined by this user.

One of the single largest company costs is delivering customer satisfaction. Reducing the costs of failure in doing this provide companies both funding to support quality value enhancement as well as major additions to net operating income itself.

Quality leadership defined—This strategy and its execution characterizes organizations whose competitive strength is now being built to directly respond to the marketplace, cost and human leadership changes that are included here. We are coming to think of this organization as the *Quality Leadership Company*™ which:

- Sees its basic objective as the continuous acceleration of value for customer, for investors, and for employees by delivering consistently improving business results no matter what the business conditions are—good times or bad.

- Emphasizes that market driven means that its quality is what its customer—not what the company—says it is.

This, for example, is a major differentiator among strong versus highly challenged service organizations today. It remains very difficult for some service providers in healthcare or in education or in government to accept the principle that service quality is the *call* of the service receiver—not of the service provider.

- Leads with a fundamentally new competitive combination of passion, of discipline and of populism (you can feel it in what it does) reflected in a bias for improvement action (not delay) and an emphasis on widespread communication.

This connects the company with the *here-and-now* business environment which is powered by: global markets, customer development, employee attitudes, supplier trends and investor and public expectations.

- Recognize that its sustained growth demands the combination of constantly increasing customer satisfaction, operating cost leadership in its industry and markets, human resource effectiveness throughout the organization, and close integration with its supplier base—all four, all the time.
- Bring back the enthusiasm (what I call the fizz) to work in the organization by the deep commitment to fundamental business improvement that comes from the knowledge, skills and attitudes about solving problems democratically, about the value of teamwork.
- Implement these results relentlessly by strengthening the processes, tools, resources and strategies that help every man and woman in the company think and act and decide on how he or she both individually and as part of a team can help to provide complete customer satisfaction.

It is only this type of relentless implementation and support of quality leadership that will enable all of your employees to meet today's accelerating values and how to accomplish this in ways that help their companies successfully face the huge new competitive cost pressures that exist at all levels of business today.

Why is *Total Quality Leadership*™ the *new* requirement for success? Why and how we must now move toward this new objective of *Total Quality Leadership*™ focuses on quality that's:

- Consistently affordable . . .
- Essentially perfect . . .
- Determined by the user.*

> It is only relentless implementation and support of quality leadership that will enable all of your employees to meet today's accelerating values.

---

*The costs of failure in delivering this complete customer satisfaction is progressively reduced through individual and group improvement efforts. These both fund quality value enhancement as well as provide major additions to net operating income itself.

My take on this is shaped by my own and General Systems' experience from projects in the United States and throughout the world (this experience is supported by the quantitative information maintained in General Systems' Research Division's competitive trend database). What I see through this window on the world makes three trends very clear to me.

1. The future of quality-based competition will be staggeringly different and far more demanding than the past which, while difficult enough, we're going to remember primarily as a warm-up practice for what lies ahead.

There are two reasons for this rapid and vast change:

- The rapidly expanding customer satisfaction demands for much greater purchase value from today's very savvy customer as well as business buyer . . .
- The growing cost pressure forced by the rapidly increasing cost and quality efficiency in the many new Japan's—not only in Asia but also other parts of the world—that both U.S. companies and the Japanese themselves will face.

Taken together these demands are forcing a rapid further acceleration in the traction of American quality improvement programs over and beyond the present rate—a demand that some U.S. companies which have successfully passed through the hard times of the past few years are finding it difficult to accept and to face.

2. Somewhat paradoxically, today's much improved business situation and the full plate of quality projects already working through many organizations, makes it an even more challenging management and quality professional job to bring about acceptance and implementation of this acceleration as compared to a few years ago when every man and woman in the organization came personally face-to-face with how tough the quality demands were.

The challenges ahead require very specific objectives and timetables and the recognition that an organization's culture for quality improvement is far more than a philosophy supported by speeches and motivational seminars. It's instead the collective result of the organization's actions for improvement which can be advanced only by a clear and relentless process that drives these actions derived from customer determined quality improvement as fully in good times as in bad.

3. The companies which have emerged as today's pacesetters are those that have been achieving these results because they are led and managed with a new kind of *competitive discipline* (I use the term without apology) that has emerged from the painful economic crucible in which we in the U.S. have been immersed.

This required competitive discipline is based on building world-class competitive strength in the three fundamental areas that are the keys to leadership in the brutally competitive global markets of the next several years. These are:

- Creating constantly increasing product and service value for customers at any price level . . .
- Getting the operating cost leadership that makes this economically possible for companies . . .
- Bringing back the fizz—the enthusiasm—to people's jobs in the organization that drives all this.

Seismic Shifts in the Marketplace

This brings me to what I earlier described as the seismic changes that have now created the conditions that make this kind of strength necessary for competitiveness and to a discussion of the three most important of these changes—beginning with the fundamental shift in customer value expectations in the US and global marketplace.

The seismic shift in customer's valuations of products and services—for many years General Systems has been surveying customer buying patterns in the major US and international markets.

Our survey data show that nine out of ten buyers now make quality their primary purchasing standard as compared with ten years ago when it was three to four out of ten. But the even more fundamental shift is that these buyers now consistently express their wants for affordable quality—quality and price but also time saving and service—not simply as a trade-off between quality and price.

What's driving the consumer change? That this is so is a statement of a very large economic and social shift for both consumers and business buyers in the American marketplace. It's driven for consumers by the fact that U.S. wage earner family income is no higher in constant dollars than it was twenty years ago and this has zoomed cost in purchasing skyward.

What's driving the business consumer change? For business buyers, the change has been driven by pressures over the past few years which have demanded:

- Reliable, predictable operation of the equipment . . .
- Services purchased with no equipment backup . . .
- Little or no tolerance for the time or cost of any failures.

Moreover, these buyers (both consumer and business) have now become convinced by the widespread awards, the media and accompanying advertising attention to quality that any company can provide essentially perfect quality if managed to do so—their belief is that it ought to be a given in any consumer or business purchase transaction.

The measure of quality has also shifted—The result is that market quality leadership today isn't merely measured in such traditional terms as quality

> Our survey data show that nine out of ten buyers now make quality their primary purchasing standard as compared with ten years ago when it was three to four out of ten.

defects—zero or otherwise. It is instead measured in terms of the total customer perception of quality. This means that the quality of steel or plastic or the service the customer receives is an important part but *just a part* of the complete support, billing accuracy, delivery reliability package he or she expects they are buying.

The 21st century winners—The companies that understand this recognize that in today's markets competitive leadership depends upon accelerating the increase in the things done right—both small and large—that buyers want, not merely reducing the things done wrong.

A second chance, no way! In many consumer products, one out of every three resolved customer complaints nonetheless shatters the loyalty of that customer and the prospects for his or her retention.

> In many consumer products, one out of every three resolved customer complaints shatters the loyalty of that customer and the prospects for his or her retention.

This huge shift of customer quality perception is the reason for the disparity in their quality satisfaction measurements that increasingly concern some companies in the results of their market quality surveys.

But we're improving! And a great deal! Organizations in these industries and markets continue to state to the media and their potential or old customers that their data shows that their quality has greatly improved—meaning in defect reduction. But our survey data shows that buyers in the same markets are saying that it hasn't improved—meaning in complete customer satisfaction—and they reduce their buying from these companies accordingly.

Increasing customer expectations have gathered speed . . . Our surveys show that it isn't always because buyers are saying that products have gotten worse. What they are saying is that they haven't gotten better fast enough for the price to satisfy today's increasingly knowledgeable user.

> Buyers are saying that products haven't gotten better fast enough for the price to satisfy today's increasingly knowledgeable user.

The huge difference between a completely satisfied customer—affordable quality, essentially perfect product or service, determined by this customer himself or herself—and a customer who just thinks his product is okay, is a five to seven times greater likelihood of customer retention and of return sales.

The fundamental shift for businesses and service providers—Successful market alignment of a company's customer quality leadership program now requires focusing its customer satisfaction strategy on objectives and results which consistently lead and exceed customer requirements for enhanced quality value, not merely react to them on a defect reduction basis which was the traditional approach to quality improvement. This value enhancement objective is a fundamentally different business quality goal to drive the company's processes from product and service design and development to customer satisfaction measurement than this traditional approach.

The new requirements for management—For many companies this means that they must transform their management orientation and their quality systems, as a fundamental key to meet their sales and marketshare growth, as well as their customer retention goals. For example, in durable household products, when buyers like the product today they tell six other possible buyers; when they don't like it they tell twenty-two.

> When buyers like the product today they tell six other possible buyers; when they don't like it they tell twenty-two.

That's the huge leverage that customer satisfaction has on consumer product sales. It is this emphasis that is the basis for our *Total Quality Leadership*™ technology and systems installations in major manufacturing and service companies today.

The seismic shift in economic pressures on business—The second change I want to discuss is the huge and new economic pressure on companies that results from the business forces I've been reporting.

The competitive scissors . . . It can be visualized as a competitive scissors whose sharp jaws now are closing in on many companies. One jaw is the strong upward pressure on company cost created by the increases American companies continue to have in spite of their best cost containment efforts; the other jaw comes from the severe downward buyer pressure on price.

What makes this particularly demanding today is that after all these years of cost accounting, many companies still don't know what things really cost. Thus they have no real foundation for effective cost reduction. This has been why it has taken companies so long to recognize that the cost of delivering customer satisfaction to buyers is their single largest cost except for material purchases.

Reducing the cost of failure . . . In the companies in which total quality has been led and managed right their total quality initiatives have established the work and teamwork processes which, by making quality better for customers, have greatly reduced the costs of failure in delivering this customer satisfaction.

These very large and positive reductions in cost have become a centerpoint of the competitive strength of the *operating cost leadership* these strong companies have for addressing the affordability issues I've been discussing.

These major reductions in failure costs flow both to support quality value enhancement improvement programs as well as additions to bottom-line net operating income in an explicit way that can be reported to shareholders and investors.

Why slash and burn costs cutting has failed . . . There is an enormous difference between this kind of competitive leadership and those companies whose approach has been based on slash-and-burn cost reduction programs which simply shift quality problems from one department to another.

Corporate anorexia . . . Experience over the past few years makes it crystal clear that cost reduction, unless synchronized with specific improvements in the company's way of working and in serving customers, is like weight reduction without a change in lifestyle—it doesn't stick. It just leads to anorexic companies—more cost reduction, and more difficulty in dealing with customer market leadership, and on and on and on.

A New Type of Management Is Required

These two changes I've been discussing toward the shift in customer value expectations and toward the huge new economic pressures brings me to the third change—the fundamentally different management approaches to human leadership needed to serve these demands.

> Cost reduction, unless synchronized with specific improvements in the company's way of working and in serving customers, is like weight reduction without a change in lifestyle—it doesn't stick.

False doctrine: drive the bosses' ideas into the hands of workers . . . The new management is based upon the rejection of a false management doctrine that dominated many organizations throughout the 1980s and the early 1990s.

The false doctrine was that good management and successful improvement means getting the ideas out of the boss's head into the hands of the workers. Implementation of this doctrine has been characterized by the kind of inward looking restructuring which looks at your own corporate navel while too often growing out of touch with rapid changes in new markets, new employee attitudes and new management approaches.

This false and ineffective doctrine was summed up in the recent two panel cartoon where, in the left panel, the management group is shown around a business table with the caption saying "We've just to restructure and downsize around here," while in the right panel, labeled one-year later, the same group is shown around the same table with the caption, "Now that we're restructured, what do we really do?"

Disconnected organizations—I call the companies that have attempted this kind of approach to business improvement *disconnected organizations*. They are increasingly disconnected from:

- Their customers—Their disconnected approach emphasized top-down planning isolated from the rapid changes in their customers' buying preferences.
- Their employees—Their human resource programs have been keyed to the fireworks motivational seminars, seasoned by regular doses of management speechmaking. When you, as an employee, returned to your job to use what you've heard, you continued to be faced with the same ambiguous management processes that you can't influence, and with trying to thread your way through organizations that are still a group of separate departmental islands without any bridges among them.
- Their suppliers—Their approach to suppliers has been with baseball bats, not integration.

What does good management look like? Today we know that good management in fact means:

- Using the knowledge, skills and attitudes of every man and woman in the organization . . .
- Encouraging and supporting the freedom to innovate . . .
- Supporting solving problems democratically . . .
- Valuing the sense of teamwork that the great majority of the men and women who work already bring to their job because of the basic traditions of American life.

Successful improvement in many of the companies which are America's competitive leaders today is built around a powerful marriage of encouragement of these human strengths together with strong management leadership.

The *death* of total quality has been greatly exaggerated! We are frequently asked by the press and magazines to explain "Why the durability of total quality"—again when led and managed right—is the single most important competitive effectiveness influence on American companies.

Total quality is durable because it is the competitive connector—it connects successful companies with their employees, their customers and their suppliers and investors in these demanding times. It is key to the change in the quality leadership now evolving in the new manufacturing and services pacesetters of America.

What are the most important aspects of the *new* quality leadership? At General Systems we refer to the most important of these aspects or actions as fingerprints of quality leadership.

- **Senior management's ability to lead the quality effort**—The first fingerprint throughout all of these organizations—there are a total of five fingerprints—is senior management leadership ability to be part of creating the organization's quality improvement playbook and be its quarterback during the scrimmages on the field rather than simply being a spectator in the stands.

It is leadership that creates an organization-wide atmosphere of superior performance which recognizes that the pursuit of excellence is the most powerful emotional motivator in any organization. And it is leadership with the sureness and lightness of senior management touch that understands that a good total quality management system, like a good stomach, works best when you scarcely know it's there.

- **Senior management's ability to focus on firm improvement goals**—The second fingerprint is a relentless focus on identifying the firm goals required for full achievement of the necessary improvement for the organization.

This is a very different type of leadership centered on goals and the supporting execution steps which are based upon analysis and planning by men and women from the bottom to the top of the organization—the people who really understand the details of what has to be accomplished—and not by the traditional central executive office bureaucracy.

- **Senior management's ability to create and support an empowering atmosphere**—The third fingerprint is the kind of empowerment that I prefer to think of as organization flexibility. The emphasis is not merely on increasing employee responsibility step by step at his or her workplace but upon a far more basic approach.

This type of empowerment builds among all employees a foundation of openness, of trust and of several—way communication. It creates an environment for what I think of as individual job entrepreneurship—it encourages men and women to develop their own forms of teamwork and their own personal ownership of competitive improvement.

It is leadership with the sureness and lightness of senior management touch that understands that a good total quality management system, like a good stomach, works best when you scarcely know it's there.

There is always a better way and the men and women who know this are the men and women closest to the work itself. A several percent quality improvement achievement in jobs throughout an organization compounds at a remarkable rate throughout that organization and provides an enormous increase in organization wide performance.

- **Senior management's ability and self-discipline to completely satisfy the customer**—The fourth fingerprint of the competitive leader is in his or her emphasis (as a focus for sales growth) that their product or service must have complete customer satisfaction in their selected key target markets.

Complete customer satisfaction is obtained when customers are not just satisfied but that they're enthusiastic about what the company is doing. If the product or services misses that mark, the leader's self-discipline requires that the company change and improve its customer value drivers and to do it fast.

The leader's discipline continues to emphasize that even when complete customer satisfaction is achieved the company must continue to listen as carefully as before to the customer so that when the customer changes, so does the company.

- **Senior management's ability to relentlessly strengthen individual and group leadership capacity**—The fifth fingerprint of the competitive leader is that it implements these results by relentlessly strengthening the processes and the tools and the resources and the strategies that help every man and woman think, decide and act on how he or she, both individually and as part of a team, can support the company's quality efforts:

  - In ways that help to provide the improved user quality that meets today's accelerating values . . .
  - In realistic ways that help their institutions successfully face the huge new economic pressures that exist at all levels of institutional life today.

Summing Up

When these fingerprints of action are taken together and quantified in our competitive trend database, the data indicates that the effectiveness of their customer quality delivery systems are three to four times greater than that of their competitors.

It is this type of leadership, which is results—oriented and has bias-for-action centered by a quality discipline, that is altering the face of the management of quality improvement in America.

It is this type of leadership that represents a fundamental ingredient in the re-invention of American management. This re-invented leadership takes the customer value shifts, the cost pressures and the human leadership change abilities and molds them into the dominant force for improvement in quality's second half century. And it's a major factor in the evolution of the total quality field itself.

If I use the framework employed through several editions of my *Total Quality Control* book with which those of you who use the book may be familiar, I would say that this evolution moved beyond the first steps toward quality control for products and services, has integrated the step toward total quality management of the organization's infrastructure and is now moving toward total quality leadership which provides the competitive results connector.

Total quality leadership, I have found, provides the energy necessary to achieve the objective of creating and maintaining complete customer satisfaction, which is:

- Affordable, for today's increasingly economically constrained consumer and business buyer;
- Essentially perfect for him and her, and
- Determined by this user.

A total quality leadership company should expect of itself, the following:

- Sales growth accelerated by complete customer satisfaction;
- Operating cost leadership powered by processes that have been failure cost reduced;
- Human resource effectiveness characterized by the return of the fizz—the enthusiasm—to work.

Business leadership is being reordered as we move into the business future in favor of the companies that successfully achieve this complete customer satisfaction leadership goal. The question leaders must ask of themselves is this: "Is my leadership helping my organization to accelerate its total quality leadership as we move into quality's second half century?"

Dr. Armand V. Feigenbaum is president of General Systems Company, Inc., an international engineering firm located in Pittsfield, Massachusetts. Dr. Feigenbaum is originator of the term and Concept of total quality control and is the author of the classic and still contemporary book on it: Total Quality Control.

Power in Organizations: A Look through the TQM Lens

Paula Phillips Carson, Kerry D. Carson, E. Leon Knight, and C. William Roe

The effective use of power is crucial to leading effectively. Social scientists have been studying what power is and how we use it to facilitate performance. This article reviews that research and provides some specific insights, especially into how to use social power bases to facilitate individual and team performance. This article synthesizes many ideas that have been around for a long time and provides a TQM slant on them.

In a total quality environment, the empowered employee alters the traditional supervisor-subordinate relationship. Instead of passively executing orders, empowered employees assume both the responsibility and authority necessary to anticipate and respond to workplace problems.[1] Because total quality management (TQM) is a relatively new philosophy, managers are often perplexed about how to supervise empowered employees. While managers may understand that empowerment does not equate with the total abdication of their authority or accountability, they may not know how best to exercise their power in this new environment. Attempts have been made to theoretically model[2] and measure the process of empowerment,[3] but current research has yet to investigate the relationship between supervisory power and employee empowerment.

Because of its intuitive appeal and practicability, J. R. P. French and B. Raven's conceptualization of social power is the framework most widely accepted by researchers investigating leader influence.[4] These two social psychologists identified five distinct power bases:

- **Coercive power.** The employee's perception that the manager can mediate punishment.
- **Legitimate power.** The employee's perception that the manager has the authority to prescribe behavior.
- **Reward power.** The employee's perception that the manager can mediate rewards.

Reprinted with permission from *Quality Progress*, November 1995. Copyright © 1995 by the American Society for Quality Control. All rights reserved.

Table 1. **Effect of Power Bases on Employees**

Power Base	Satisfaction with Supervision	Job Satisfaction	Performance
Expert	Strong positive effect	Strong positive effect	Strong positive effect
Referent	Strong positive effect	Weak positive effect	Weak positive effect
Reward	No effect	No effect	Moderate positive effect
Legitimate	No effect	No effect	No effect
Coercive	Strong negative effect	Moderate negative effect	No effect

- **Referent power.** The employee's identification with and attraction to the manager.
- **Expert power.** The employee's perception that the manager possesses unique or rare knowledge or skills.

Managers exercise various types of social power to facilitate employee execution of requests and instructions. How employees respond to these requests and instructions depends on the type of power exercised by the manager. The optimal reaction a manager would like to receive from an employee in a TQM environment is commitment—the employee internalizes and enthusiastically embraces the manager's request. A less desirable outcome is compliance—the employee conforms to the request but with minimal effort and support. A third possible outcome, though undesirable, is resistance—the employee opposes the request either overtly or through subtle means, such as making excuses or delaying.[5] To ensure that commitment rather than compliance or resistance occurs, a manager must carefully consider which social power base to exercise.

> The optimal reaction a manager would like to receive from an employee in a TQM environment is commitment.

The Social Power Bases: From Potentially Defective to Effective

Research on the use of French and Raven's five social powers has demonstrated that the power bases yield different outcomes depending on their desirability.[6] This is demonstrated in Table 1, which summarizes a meta-analysis of the relationships among the social power bases and employee satisfaction with supervision, job satisfaction, and performance.[7]

Table 1 reveals that coercive power is most likely to elicit dysfunctional consequences because it diminishes employee satisfaction. Similarly, reliance on legitimate power is deficient because it lacks the capacity to stimulate employees to improve their attitudes or performance. And while the proper exercise of reward power may improve performance, the use of referent and expert power has been empirically determined to be most effective.

These findings present a challenge to organizations because the least effective power bases—coercive, legitimate, and reward—are those most likely to be used by managers. In fact, these three power bases are collectively referred to as "positional" forms of influence, so called because managers inherit these powers when they take a supervisory job. Alternatively, referent and expert powers are "personal" forms of influence, meaning that they must be cultivated and strengthened through interpersonal interactions and relationships with employees.

By closely examining these five power bases—from the potentially defective (coercive) to the most effective (expert)—managers can draw conclusions as to how they might effectively exercise power in the enlightened context of TQM.

Coercive Power

Management's reliance upon coercion has diminished over time because of legal constraints, the potential for retaliation, and a greater understanding of employees' physical and psychological needs.[8] Furthermore, managers are recognizing that the use of coercive power is detrimental to supervisor-employee relationships in total quality environments. Using it diminishes trust and creates fear. According to W. Edwards Deming, organizations should attempt to drive out rather than induce fear.[9] Barriers such as authoritarian managers who control by threats and punishment heighten conflict and stand in the way of quality enhancement.

Employees must feel secure if quality is to improve. They must be able to ask questions, report quality problems, and take a firm position on necessary improvements without concern of reprisal. Theodore Lowe, vice president of quality improvement/corporate planning for Venture Industries, observed that, because of fear, teamwork has been destroyed in many businesses, and a positive bond between managers and workers has been prevented. "Fear robs an organization of its potential; it is the major reason for the difference between the full potential of the people and the current achievement of the organization."[10]

"Fear robs an organization of its potential; it is the major reason for the difference between the full potential of the people and the current achievement of the organization."

Even though many managers have come to view the use of coercive power as dysfunctional or, at best inappropriate, recent economic factors have increased its use as a means of survival. And, contrary to fears that coercion will consistently undermine individual performance, researchers have not found a stable relationship between these two variables.[11] There may be several explanations for this unexpected finding.

First, employees may fear that if their performance declines, the coercion will continue. Consequently, their negative reactions are manifested through reduced satisfaction (see Table 1), which enhances the possibility of organizational withdrawal.[12] Second, rather than directly influencing performance, coercion may produce more aggressive reactions, such as attempts to retaliate. As Philip Crosby recognizes, "People just want their rights until you try to trample them. Then they want revenge."[13] Third, coercive power may indirectly reduce "citizenship behaviors"—activities that employees engage in, such as mentoring, that are essential to effective organizational functioning but are not explicitly included in the job description.[14]

Practical Suggestions for Maintaining, Augmenting, and Utilizing Power Bases

Coercive Power

Guidelines for preserving the power base
- Remind employees that the authority to use coercion is inherent in managerial positions.
- Don't use coercion if unable to follow through on a threat.
- Punish immediately when an employee's behaviors are detrimental to quality performance or when they jeopardize the safety or productivity of others.

Effective use of the power base
- Punish only when expectations and the penalties for not meeting them are understood.
- Punish only when employees have the ability to improve quality performance.
- Ensure the punishment is fitting to the infraction and punish in private rather than in public.
- Convince staff that coercion may be necessary during economic crises or emergencies.

Ineffective use of the power base
- Showing an employee who is boss, making an example out of an employee, or encouraging an employee to quit.
- Using coercion for personal rather than organizational benefit.
- Punishing for low quantity when quality may suffer if quantity is increased.

Dysfunctional consequences of ineffective use
- Employee trust is diminished if the manager is seen as uncaring and autocratic.
- Legal liability is created based on potential claims of discrimination and wrongful termination.
- Employee motivation and morale are diminished and fear is heightened.

Legitimate Power

Guidelines for preserving the power base
- Understand and regularly exercise the authority inherent in the position.
- Ensure employees understand that requests are to be executed.
- Confirm that requests were executed, and take corrective action when they are not.

Practical Suggestions for Maintaining, Augmenting, and Utilizing Power Bases (continued)

Effective use of the power base
- Augment legitimate power usage with reward and coercion, when appropriate.
- Make requests politely but firmly, avoiding the appearance of begging or being apologetic.
- Make requests directly rather than through intermediaries.

Ineffective use of the power base
- Exceeding the boundaries of authority inherent in a position.
- Making unethical or impossible requests.
- Failing to allow employees to offer constructive alternatives to a request.

Dysfunctional consequences of ineffective use
- Employees may ignore requests, even if only to test the limits of the manager's authority.
- Employees may attempt to undermine or sabotage requests.
- Employees may go "over the head" of the legitimate power-holder.

Reward Power

Guidelines for preserving the power base
- Remind employees that authority to dispense rewards is inherent in managerial positions.
- Ensure that rewards offered can be delivered.
- Constantly try to secure access to more resources and rewards that may be dispensed to employees.

Effective use of the power base
- Offer rewards that are desirable to employees.
- Clearly identify criteria for achieving rewards.
- Use intangible as well as tangible and monetary rewards.

Ineffective use of the power base
- Rewarding quantity rather than both quantity and quality.
- Failing to dispense promised rewards when employees comply with requests.
- Using rewards in a manipulative manner.

Dysfunctional consequences of ineffective use
- Trust is undermined and referent power is diminished.
- Withholding a reward may be perceived as coercive, or the reward becomes expected and loses its reinforcement potential.
- Extrinsic rewards may undermine intrinsic motivation.

Practical Suggestions for Maintaining, Augmenting, and Utilizing Power Bases (continued)

Referent Power

Guidelines for preserving the power base
- Eliminate status differentials between managers and employees.
- Be a positive role model for employees.
- Build trusting interpersonal relationships between managers and employees.

Effective use of the power base
- Indicate that a request is personally important to the power-holder.
- Demonstrate interest in and respect for employees.
- Coach employees in the proper techniques for fulfilling a request.

Ineffective use of the power base
- Making inappropriate requests that will jeopardize positive relationships.
- Chastising employees for mimicking inappropriate behaviors of the manager.
- Relying on referent power for personal gain or benefit.

Dysfunctional consequences of ineffective use
- Employees may feel exploited if requests exceed the limits of the relationship.
- Employees may be torn between a need to please the manager and a need to work toward organizational objectives.
- Relationships between managers and employees may become strained if employees' personal and developmental needs are not considered.

Expert Power

Guidelines for preserving the power base
- Maintain and augment specialized knowledge through professional development and educational activities.
- Ensure employees are aware of the unique information possessed by the powerholder.
- Be confident and decisive when sure of a recommendation.

Effective use of the power base
- Provide logical, rational evidence that a request will yield functional outcomes.
- Share relevant information with employees and allow for participation.
- Do not act arrogant or condescending when sharing information.

> *Practical Suggestions for Maintaining, Augmenting, and Utilizing Power Bases (continued)*
>
> Ineffective use of the power base
> - Withholding information to maintain perceptions of expertise.
> - Dispensing information when unsure if the information is correct.
> - Fabricating or embellishing expertise.
>
> Dysfunctional consequences of ineffective use
> - Attraction to and trust in the manager may be undermined decreasing referent power.
> - The environment for empowerment may be undermined because employees are unlikely to be asked to participate in decision making.
> - Employees may lose their desire for continuous learning and development.

A fourth explanation may be that during tough economic times, employees understand that workers who are not aligned with the organization's quality culture threaten the very existence of the company. Norman Blake, a turnaround expert and chief executive officer of USF & G, explains the nonsignificant relationship between coercion and performance with his lifeboat theory: "The boat can't hold everyone, and not everyone is contributing to the boat's forward motion." Therefore, ineffective hands are thrown overboard, and employees come to learn this is their potential fate.[15]

Legitimate Power

Legitimate power is conferred upon individuals occupying positions of leadership,[16] resulting in the institutionalization of formal authority. Because legitimacy depends solely upon one's position in the organizational hierarchy, requests that are made using this power are likely to result in compliance but not commitment. Crosby suggests that reliance on legitimate power is like reliance on organizational policy manuals. "It is very difficult to get even motivated [employees] to read the procedures—let alone follow them. Having a large book of policy and practices never saved any company from disaster."[17]

Just as policy manuals need to be supported by managerial communication and role modeling, legitimate power needs to be bolstered by one of the other social power bases if commitment is desired. In fact, the primary advantage of legitimate power may be that it involves control over resources, ultimately giving rise to reward power. An appropriate reward can bolster an employee's attraction to the leader, strengthening referent power. Illegitimately using legitimate power, as when a leader makes requests beyond the scope of his or her authority, can decrease referent power.

Reward Power

Reward and recognition are integral components of the quality philosophies of Deming, Crosby, and J. M. Juran. Indeed, the results shown in Table 1 indicate that reward power is positively related to performance. Such a relationship, however, may be realized only when rewards are given for high-quality performance; that is, when the desired behavior, and not some other factor, is rewarded.

Another requirement for the effective use of reward power is knowing what rewards are valued by employees. Aleta Holub, manager of quality assurance for the First National Bank of Chicago, discovered the importance of this knowledge. At first, the bank offered incentives to employees based on what it thought they would like. Holub, for example, wanted the opportunity to have lunch with the president, so lunch in the executive dining room was offered as one of the incentives to improve performance. Unfortunately, not all employees considered this an incentive. In fact, many found this type of event quite stressful. "We had been good about going to our customers to ask what was important to them, and we realized that it was just as crucial to go to our employees to ask what was important to them."[18]

In some companies, managers have offered rewards that not only are valued by employees but also benefit the organization. Consider a perk offered by Harley-Davidson. Any employee who purchases a Harley-Davidson motorcycle can receive a 20 percent rebate. Front-office and assembly-line employees get the chance to own, drive, and assess the quality of the bikes produced by their company. While many consider the reimbursement to be the best part, the company also profits from the feedback provided by employees who have firsthand experience with the product.[19]

While reward power can be used to improve individual and organizational performance, over-reliance on this power base can have adverse consequences for both the company and its customers. For example, ethics may be sacrificed to attain expected performance levels. Under threat of termination, Sears Roebuck chairman Edward A. Brennan instituted a compensation program aimed at encouraging employees to boost sales. Two years later, Sears' Tire and Auto Centers were accused of overcharging customers an average of $235 for unnecessary repairs. Consumer affairs officials blamed the ethical problem on unreasonably high quotas, commission-based compensation, and attractive incentives for top sellers.[20]

TQM cannot succeed in an environment that rewards only quantity. When the reward structure focuses exclusively on "how much" instead of "how well," flirtations with impropriety are hardly surprising. A classic case occurred at a GM light-truck plant. A manager was caught with a device in his office that periodically speeded up the assembly line beyond the rate designated in union contracts. Confronted with the evidence, he pointed out that the company's production specifications and reward system were based on the line running at maximum allowable speed 100% of the time. He was simply trying to make up for inevitable downtime.[21]

> While reward power can be used to improve individual and organizational performance, over-reliance on this power base can have adverse consequences for both the company and its customers.

Reward power creates another problem. When employees anticipate extrinsic reinforcements, the result may be overjustification, in which employees begin focusing on earning rewards and lose interest in their work. Thus, intrinsic motivation is inadvertently undermined.[22] To prevent overjustification, nonmonetary awards should be included in the reward structure of TQM organizations. Sam Walton, founder of Wal-Mart, emphasized that a paycheck and a stock option will buy one kind of loyalty, "but nothing else can quite substitute for a few well-chosen, well-timed, sincere words of praise. They are absolutely free and worth a fortune."[23]

Relatedly, if an individual comes to expect a reward each time a task is performed, the reward may be perceived as a right rather than as a discretionary bonus. Withholding the reward may then be interpreted as a coercive act.[24] Finally, a reward may be perceived as a bribe,[25] which can diminish trust and lead the employee to label the manager as manipulative.[26] Nevertheless, when used appropriately, reward power can be an effective motivator of quality performance.

Referent Power

Referent power allows an employee to identify with the manager. This identification positively influences performance outcomes, but supervision is necessary for satisfaction (see Table 1). When relationships are based on this type of power, employees tend to imitate the behaviors and adopt the attitudes of supervisors, even though they are unaware that such power exists. This identification reduces dysfunctional workplace conflicts and facilitates quality production.

Although referent power is slow to develop, the process can be hastened by engaging in symbolic actions that decrease status differentials. Altering traditional terminology is one mechanism for reducing this differential. For example, in TQM environments, the labels "coaches," "facilitators," and "associates" are used instead of "bosses," "supervisors," and "subordinates." Reciprocal trust between the manager and the employee also strengthens referent power. Juran observes. "The manager must trust the work force enough to be willing to make the delegation, and the work force must have enough confidence in the managers to be willing to accept the responsibility."[27] Crosby notes an additional benefit of being trusted: "Trustworthy people acquire no worthy enemies as they go about their labors."[28]

Expert Power

The use of expert power seems most conducive to improving work-related attitudes and performance.

Extant research demonstrates that the use of expert power seems most conducive to improving work-related attitudes and performance (see Table 1). But to maintain expert power, employees must trust that the powerholder is dispensing all relevant information honestly and completely and that the knowledge the expert powerholder is sharing is technically correct.

Unfortunately, expert power is most susceptible to erosion because it is based entirely upon employee perceptions of a manager's credentials. As leaders share with others the specialized knowledge and unique skills they posses, dependency upon the powerholder is reduced and, thus, expert power is diminished. In turn, this may decrease the level of referent power attributed to an expert powerholder.

Some expert powerholders have gone to extremes to protect their power by exaggerating their position's responsibilities and obliterating alternative sources of information.[29] In a TQM environment, such tactics can be detrimental. If managers try to maintain expertise by withholding information from employees, poor quality performance will result.[30] In a quality management culture, sharing technical information is of paramount importance. It is a manager's responsibility to help employees get the information they need, as opposed to unilaterally possessing information and dispensing it only when deemed appropriate. Furthermore, managers in a TQM environment are expected to help employees arrive at solutions rather than simply dictating solutions to be implemented. "Management has to go to the employee base and convince people that they are valuable, that their ideas are valuable, and that management will listen to their ideas," says Jim Litts of Johnson & Johnson.[31]

When managers invite their employees to participate in decision making, they find that their expertise is questioned and challenged. Instead of perceiving this as a threat, however, total quality managers ought to encourage empowered employees to disagree. Such two-way communication is likely to result in the generation of more creative solutions to problems. According to Crosby, "There is absolutely nothing more demotivating or demeaning to [employees] than to have to go to meetings where the assigned role is to be a faithful listener."[32]

Total quality managers must recognize the expertise of others, but it is also important for them to posses the unique knowledge required to effectively perform their roles. While withholding expert information from employees is an inappropriate technique for maintaining this power base, there are certainly other, less subversive methods. These include keeping oneself informed and continually upgrading one's own expertise.

It is a manager's responsibility to help employees get the information they need, as opposed to unilaterally possessing information and dispensing it only when deemed appropriate.

Using Social Power Bases for TQM Purposes

Legitimate power can elicit compliance with managerial mandates; therefore, manager requests and instructions based on this power source should be polite, rational, clear, and confident.[33] Requests that are outside the bounds of legitimate power (that is, one's authority) should be avoided. Reward power has a moderately positive effect on performance, and thus can be used to encourage quality initiatives. Rewards should be desirable, achievable, and appropriate to the action. Nonmonetary rewards can also enhance quality performance. Using coercive power to influence individuals should be avoided, rather, the use of positive discipline ensures that employees understand expectations, are warned before sanctions occur, and are punished fairly and consistently. The frequent necessity for coercion may indicate that a manager's expectations and managerial style are not consistent with TQM expectations and therefore should be reexamined.

Because expert power is susceptible to erosion, efforts should be undertaken to protect this power base by continually upgrading one's skills. Protection, however, should not include being reluctant to share expertise with employees. Sharing helps employees meet the organization's goals and provides a good basis for the development of referent power. Referent power, which is based on trust, can be increased by treating empowered employees fairly and being sensitive to their individual needs. Employees will accept information and knowledge from managers only if they trust them.

In some situations, the selection of power bases may be dictated by environmental forces. Coercion may be necessary in harsh economic times, despite the negative effect it has on satisfaction. Also, it might be appropriate when emergencies arise, or when there is a need to act quickly. Business settings may influence a power base choice. For example, sales managers may find reward power to be most effective, while safety directors may place greater reliance on coercion, using such techniques as issuing citations for infractions. Project team coordinators may have to rely on personal power bases if positional power is unavailable.

Bertrand Russell proclaimed that power is "the most fundamental concept in social science."[34] Similarly, Roger Dahl declared, "The concept of power is as ancient and ubiquitous as any that social science can boast."[35] Early social theorists such as Plato, Aristotle, and Machiavelli have struggled with the issue of power. Total quality managers may be struggling with the issue of power as well. By reviewing the characteristics, uses, and consequences of the five social power bases, managers will have a better idea of how to effectively exercise the power they have inherited.

Notes

1. Arthur R. Tenner and Irving J. DeToro, *Total Quality Management: Three Steps to Continuous Improvement* (Reading, Mass.: Addison Wesley, 1992).
2. Jay A. Conger and Rabindra N. Kanungo, "The Empowerment Process: Integrating Theory and Practice," *Academy of Management Review* 13, no. 3 (1988) 471–82.
3. Bob E. Hayes, "How to Measure Empowerment," *Quality Progress*, February 1994, 41–46.
4. John R. P. French and Bertrand Raven, "The Bases of Social Power," in *Studies in Social Power* (Ann Arbor, Mich.: University of Michigan, 1959).
5. Gary A. Yuki, *Leadership in Organizations* (Englewood Cliffs, N.J.: Prentice-Hall, 1989).
6. French and Raven, "The Bases of Social Power."
7. Paula Phillips Carson, Kerry D. Carson, and C. William Roe, "Social Power Bases: A Meta-Analytic Examination of Interrelationships and Outcomes," *Journal of Applied Social Psychology* 23, no. 14 (1993): 1150–69.
8. Daniel Katz and Robert L. Khan, *The Social Psychology of Organization*, 2d ed. (New York: Wiley, 1978).

9. W. Edwards Deming, *Out of the Crisis* (Cambridge, Mass.: Massachusetts Institute of Technology, Center for Advanced Engineering Study, 1986).

10. Samuel C. Certo, *Profiles in Quality: Blueprints for Action from 50 Leading Companies* (Boston, Mass.: Allyn & Bacon, 1991).

11. Philips Carson et al., "Social Power Bases: A Meta-Analytic Examination of Interrelationships and Outcomes."

12. Peter W. Hom, Roger W. Griffeth, and Paula Phillips Carson, "Turnover of Personnel," in *Handbook of Public Personnel Management* (New York: Marcel Dekker, 1995).

13. Philip B. Crosby, *Quality Is Free: The Art of Making Quality Certain* (New York: McGraw-Hill, 1979).

14. Dennis W. Organ, "The Motivational Basis of Organizational Citizenship Behavior," in *Research in Organizational Behavior* (Greenwich, Conn.: JAI Press, 1990).

15. Brian Bremner, Mark Ivey, and Ronald Grover, "Tough Time, Tough Bosses," *Business Week* 25 November 1991, 174–80.

16. Bernard Bass, *Stogdill's Handbook of Leadership* (New York: Free Press, 1981).

17. Philip B. Crosby, *Quality without Tears: The Art of Hassle-Free Management* (New York: McGraw-Hill, 1984).

18. Certo, "Profiles in Quality: Blueprints for Action from 50 Leading Companies."

19. H. Allen, "Sweet Ride to Success: America's Premier Motorcycle Manufacturer Owns the Road in Employee and Customer Loyalty," *Corporate Report Wisconsin* 9, no. 4 (1993): 8.

20. Kevin Kelley and Erie Schine, "How Did Sears Blow This Gasket?" *Business Week*, 29 June 1992, 38.

21. K. Labich, "The New Crisis in Business Ethics," *Fortune*, 20 April 1992, 167–76.

22. Daniel J. Bem, "Self-Perception Theory," in *Advances in Experimental Social Psychology*, vol. 6, (New York: Academic Press, 1972).

23. John Huey, "Sam Walton in His Own Words," *Fortune*, 29 June 1992, 98–106.

24. Faye Crosby, "A Model of Egoistical Relative Deprivation," *Psychological Review* 83(1976): 85–113.

25. Jerald G. Bachman, Clagett G. Smith, and Jonathan A. Slesinger, "Control, Performance, and Satisfaction. An Analysis of Structural and Individual Effects," *Journal of Personality and Social Psychology* 4(1966): 227–36.

26. Harry Levinson, "Psychoanalytic Theory in Organizational Behavior," in *Handbook of Organizational Behavior* (Englewood Cliffs, Prentice Hall, 1987).

27. J. M. Juran, *Juran on Leadership for Quality: An Executive Handbook* (New York: Free Press, 1989).

28. Crosby, *Quality Is Free: The Art of Making Quality Certain*, 126.

29. D. Hickson, C. Hinings, C. Lee, R. Schneck, and J. Pennings, "A Strategic Contingencies Theory of Intra-Organizational Power," *Administrative Science Quarterly* 16(1971): 216–29.

30. James W. Dean Jr. and James R. Evans, *Total Quality: Management, Organization, and Strategy* (Minneapolis–St. Paul: West, 1994).

31. Dan Ciampa, *Total Quality: A User's Guide for Implementation* (Reading, Mass.: Addison-Wesley, 1991).

32. Crosby, *Quality without Tears: The Art of Hassle-Free Management*.

33. Yukl, *Leadership in Organizations*.

34. Bertrand Russell, *Power: A New Social Analysis* (New York: W. W. Norton, 1938).

35. Roger A. Dahl, "The Concept of Power," *Behavioral Science*, 2(1957): 201–18.

Bibliography

Labich, Kenneth. "The New Crisis in Business Ethics." *Fortune*, 20 April 1992, 167–76.

Salibian, C. E. "Foster Empowerment, Profile Follows." *Rochester Business Journal* 9, no. 45 (25 February 1994): 7.

Webber, Ross A. *Management: Basic Elements of Managing Organizations.* Homewood, Ill.: Irwin, 1975.

Paula Phillips Carson is an associate professor of management at the University of Southwestern Louisiana in Lafayette.

Kerry David Carson is an associate professor of management at the University of Southwestern Louisiana in Lafayette.

E. Leon Knight Jr. is a professor of marketing at Texas A&M University in Corpus Christi.

C. William Roe is a professor of management at the University of Southwestern Louisiana in Lafayette.

How to Work with, instead of against, Millions of Years of Social Primate Evolution or "How to *Start* Monkeying Around and Be a More Effective Leader"

William Lareau

Bill Lareau is the author of one of the best books on TQM: American Samurai *(Warner Books, 1991). In this article, excerpted from a work in progress, Lareau reviews some basic primate attributes that help explain our behavior in organizations. He suggests that leaders who understand these attributes are far more likely to be successful. Why? Because they will take actions that are consistent with their attributes. Lareau has a down-to-earth style that makes this article not only interesting but fun to read.*

A hidden mechanism is attacking productivity in most businesses. This insidious destroyer of communications, productivity, and profits is social dynamics. These dynamics are the hardwired, natural social behaviors that human beings always use when they interact with others. If left to operate naturally and unchanneled, group dynamics can wreak havoc in a organization—and they usually do. It's not that group dynamics are bad—on the contrary, they have tremendous value for family, team, and human race survival. The problem is that group dynamics weren't designed to operate in modern organizations. For group dynamics to benefit an organization, the organization must change to accommodate them; the reverse is impossible. Group dynamics have a life of their own that cannot long be denied.

Human group behavior is driven by universal rules that almost all people follow even if they don't "know" them. These rules evolved for one purpose only: to increase the probability of survival of small groups of subhuman primates facing relatively constant and known hazards in primitive environments until the primates could reproduce and raise the next generation of offspring. In that situation, the social dynamics that evolved are unmatched in their effectiveness. We, today's human beings, are proof that these social dynamics have worked effectively for over six million years.

The problem is that, in a mere blink of the evolutionary eye, our minds evolved the ability to reason. The result is that, in the short span of two hundred thousand years, a vast machine- and computer-based civilization has sprung from primitive, hide-wearing hunter-gatherers. Yet all human beings still interact with each other and other groups with social and group dynamics almost identical to those that ensured the success of primitive hunter-gatherers. In essence, human thinking abilities evolved at light speed, while our social skills have not much changed. The employees working in business today must do much more than breed, seek food, and survive, but their social behaviors have not evolved to match these new, rapidly changing demands.

The consequence is the typical confusion, conflicts, suboptimization, mistakes, and stress that are the hallmark of traditional organizations. Without direct, planned intervention in a focused manner, this sorry tale is a certainty—it's human nature. Once we understand exactly how human beings are programmed by evolution to behave, it's possible to take advantage of these innate tendencies and use them to both the advantage of business and the happiness of the people who work there. Let's take a look at a few of the many social dynamics that can impede or benefit any business organization.

Group Formation and Group Size

Human beings are primates, very close relatives to chimpanzees and gorillas and somewhat more distant cousins to baboons and various monkeys. A given of all primates is that they form small groups. This happens naturally, whether you like it or not and whether it's helpful or not. Group formation happens because being part of a group had survival value for all of our ancestors over the last five million to seven million years. Primates that "went it alone" were killed off or starved to death, and the genetic imperative toward solitary behavior was thus eliminated from the gene pool. Those who survived and had offspring passed along genes with a facility for, and a dependence on, group endeavor. As a consequence, all primates form small, cohesive groups and enjoy working and living in them more than working alone or working in very large groups.

This group formation can be seen in every primate type, whether its members are bowling-league members, gorillas, chimpanzees, or baboons. Each gathering consists of many subgroups. In a chimpanzee troop, the nursing mothers form one group, the adolescent males another, the subdominant males another, and the young females yet another. Groups of "friends" will form others. The members of these subgroups spend much more time with each other than with other members of the larger group. This structure is no different than the sections of a department, the departments in an organization, or groups of friends in a college dormitory.

The ideal number for a group is typically said to be "seven, plus or minus two." Why seven, more or less? Partly because, given the organization of our visual and auditory senses, we cannot communicate effectively with more than about eight other individuals without yelling and straining to hear. And we cannot get close enough to observe the subtle nonverbal cues that are so much a part of the primate social landscape.

The absolutely critical insight here is that it doesn't matter whether some higher authority believes a group or department is functioning as a closely knit team with fifteen to one hundred members. Social primates will break this larger group into many smaller groups because it is almost impossible to have a single, focused group of more than nine or ten people. Over time, these smaller, informal groups will come to have differing views of the organization, the world, and business priorities. Unless steps are taken to compel these informal groups to adopt a similar view of the world, they will develop different procedures for doing the same work and, typically, increasingly negative views about the other groups (such as the conflicts between engineering and manufacturing in most traditional organizations).

What does this mean for work groups? The bottom line is that work groups, teams, committees, and task forces should be kept to a maximum of about nine people. Groups larger than this will break into smaller groups, and like it or not, these subgroups will begin to compete with each other. This means that instead of an executive committee of fifteen, it's wise to broaden the authority of several of the members and decrease the overall size of the team to no more than nine.

The same holds true for "intact work groups" or "natural work groups." Suppose there are twenty workers on an assembly line. This is too many to form a single team and expect them to develop close-knit ownership over their processes. The "social primate approach" is to divide the twenty people into two or three teams, each team "owning" a more limited portion of the assembly line. This approach requires management to do more work to set up and lead more teams, but there's no alternative. The twenty workers, left to themselves, will create their own small, informal teams anyway. The only decision that management has to make is, will the teams be the ones that make sense to how the process should be managed, or will the teams be catch-as-catch-can informal groups whose interests may conflict with business objectives? The popular and successful "cell" concept of modern pull systems is a fortuitous tip of the hat to this small-team concept.

Each work group should have its own information center (such as kaizen boards) to foster a sense of group pride and ownership of its area. The key to getting the various work teams to work as part of a larger, cohesive department, rather than focus on competition with other groups (which is natural primate behavior) is communication and measurement. Each team must be required, through the reporting, analysis, and improvement of key measures, to get better at how it serves both the upstream and downstream processes

The twenty workers, left to themselves, will create their own small, informal teams anyway.

and teams. In this way, each team can work as a tightly knit unit and compete with other teams in how well it looks good by doing the right things that serve the greater good of the entire process.

Group Norms and Changing Behaviors

Every group has norms for behavior—guidelines for what's acceptable and what's not. For most groups, these norms are never spelled out, but everyone knows them; people act differently in a bar than they do in a church. Being social primates, people are expert in quickly assessing group norms for situations within their own culture (and are surprisingly good at estimating norms for new situations after a very short time). Norms govern speech patterns, deference to authority, how much candor is tolerated, personal space limits, the level of emotionality that's allowed, and so on.

A critical insight about norms is that they are the rules that a group actually follows, not rules that it should follow or that would be good for it to follow. Many business organizations have rules, regulations, mission and vision statements, policies, and procedures. These legalistic guidelines are not norms *unless* they are truly followed. Many managers mistakenly believe (or perhaps hope) that policies, procedures, and their own exhortations are norms: guidelines for behavior that are followed. In a world-class organization, this assumption is probably true. In organizations that are successful over the long term, most groups have a remarkably similar sets of norms that overlap with both the organization's norms and the norms of individuals.

Employees will not display behaviors that "put the customer first" simply because this message is broadcast in speeches and memos. Even employees who have extensive collections of "quality program" coffee cups and wallet cards understand and follow the actual, operating norms, not what's on the coffee cup. Employees know what to do if, at the end of the month, a shipment should be held for quality problems but is needed to make the monthly sales plan. They know from past experience whether management will quickly switch from speaking about quality to looking for scapegoats if they hold the shipment. So smart employees nod their heads about quality when they hear the speeches but handle the problem shipments in the manner that satisfies the actual norms they've seen. In traditional organizations, this means, "get the shipments out any way you can." They know from past experience in identical and related areas that you get in a lot less trouble if you obey the actual norms, rather than the official, spoken norms. Remember, for millions of years, primates have survived by recognizing what works for survival and what gets you killed. Just because there are no more saber-toothed tigers around anymore doesn't mean that people have stopped being perceptive about the rules for survival, be they life threatening or career ending. Our ability to perceive norms accurately is hardwired into us.

Need for Leadership

Insights are sometimes found in strange places. The next time you visit the zoo, check out the butts of the baboons. Yes, I said, "baboon butts." You'll notice that some of the baboons have very red and apparently swollen buttocks. These red areas are called ischeal tuberosities. They are genetically hardwired signs of authority and leadership in baboons. The tuberosities appear in exaggerated form only as a baboon becomes more dominant. Being dominant causes the tuberosities to grow larger. If the dominant baboon is deposed from power, the redness fades. The newly dominant baboon then quickly develops larger, redder tuberosities.

The baboon style of leadership has great survival value for baboons. The dominant baboon's authority does not allow a lot of fighting and conflicts. The structure he or she provides allows for smooth, day-to-day operation of the troop. This is critical because baboons in the wild have few resources for infighting; the entire troop's energy must be devoted to gathering food and protecting the troop. And because the leader is strong, experienced, and confident, he or she remains in power until an exceptional replacement grows up and successfully challenges him or her. This style of leadership works because one dominant baboon will know what to do as a leader as well as any other. Two million years ago and today, few things change in the environment of baboons over the course of a few generations that would require radically different leadership skills.

All primates have this same sort of dominance scheme, although not all show such obvious body-part changes. The need for leadership is genetically engineered into all primates. The good news is that such a need makes it easy for a good leader to assert authority: primates are ready and willing and want to follow. The bad news is that primates will embrace bad leadership rather than none; we can't help ourselves. The insights here are several: (1) any human organization will readily accept and form many more levels of leadership that it rationally needs in order to satisfy the "I feel secure with my own leader" mentality of small groups and (2) followers will put up with a lot of incompetence and harassment before they will take steps against a leader (such as complaining to higher management). It's in our genes to tolerate leaders—it's not in our genes to select good leaders for today's competitive marketplace.

A further primate insight is valuable for today's leaders. Long-term studies of chimpanzee behavior (especially by Franz De Waals) have demonstrated that successful chimpanzee leaders do not rely on raw power or status. Instead, they cultivate favor with all of the small groups within the chimpanzee troop. They allow themselves to be pawed over by the toddler chimpanzee children in the presence of the mothers. They stick up for the weaker members of one group in the presence of the other groups. They drive apart meetings of the up-and-coming, would-be dominants (think of them as conspiratorial vice presidents of the troop) and then reconcile with each one individually by food sharing and grooming. The message is "Work with me and we'll all prosper; work against me and you'll suffer."

Leader chimpanzees, our closest relatives, are compelled by innate influences to practice this type of relationship building because it has survival value. So are we. Yet, in our huge, reason-built organizations, leaders all too

It's in our genes to tolerate leaders—it's not in our genes to select good leaders for today's competitive marketplace.

often exert authority by memo, title, or raw power. And we wonder why it's not effective with the people who do the work? Our leaders could learn from the chimpanzees: concentrate on building personal relationships up and down the organizations, let people see you as a person (remember "management by walking around"), and take the time to stand up publicly for the lower-level members of the organization who stick their necks out to do the right thing. And, because we're still social primates, leaders should watch out for those scheming vice presidents and keep them apart (now is it clear why so many vice presidents get two-year assignments to factories in small towns in the Midwest?)

Changing Behavior

Managing change is a key concern of all business leaders. They are faced with redirecting workforces that are caught up in a maelstrom of economic and technical changes. Successful change requires establishing a new set of organizational norms in place of the norms that don't work anymore. Given that driving rapid change is the single most critical leadership element these days, you would think that most business managers would understand the basics of human (or primate) behavior change. Sadly, this is not the case.

> The only way to change norms is to alter the consequences to behaviors that people observe and perform.

The only way to change norms is to alter the consequences to behaviors that people observe and perform. If behavioral consequences are seen to be meaningful and consistent over a period of weeks and months, only then will norms start to change. Speeches, slogans, posters, vision-wallet cards, and the like only set up an expectation—one that's not often met in traditional organizations. "Talking the talk" is only hot air if behavioral consequences are not observed and felt. Managers must "walk the talk" after they "talk the talk," and they must walk it for many months before employees will believe that things are different from the dozens of other "talks" they have heard over the years.

A few key points about behavior change are important here. The first is to appreciate that any behavior consists of four elements. They are the physical *action* that accompanies the behavior, the *thought* that takes place while the behavior is occurring, the *emotional* response that accompanies the behavior (happiness, sadness, hope, despair, etc.), and the *physiological* response that takes place (sweating, heart rate, blood pressure, etc.). The only element that anyone can control in a situation is action. The other three elements tend to become conditioned responses that accompany a specific action.

This conditioning has incredible survival value because it enables a person to act instinctively with the first whiff of danger. For example, slamming on the brakes to avoid hitting a suddenly noticed stopped vehicle in the road ahead is a behavior that consists of all four elements. The *action* is stepping the brake and tightening the grip on the steering wheel for the possible impact. The almost automatic *thought* is "Oh, no, how I am going to blame this on the other driver?" The *physiological response* is increases in galvanic skin response, heart rate, and blood pressure. The *emotional response* is anger, despair, guilt, or hostility, depending upon the nature and past history of the individual.

Most behaviors in most situations we encounter at home, at play, or at work are highly conditioned. In almost every case, the action part of the behavior follows the highly physiological, emotional, and thought components that have accompanied the actions in the past. For example, you hear that another extended staff meeting is scheduled, and you think about how much work you have and how the meeting is a waste of time. Perhaps you feel depressed or angry. Yet you must go to the meeting. When you're there, you're emotionally, physiologically, and cognitively ready to be resentful, hostile, and bored. The only part of the behavior you can really control, having been to dozens of such meetings, is the act of showing up for the meeting. Once you show up, because you have no real choice, you are conditioned to be bored and to expect—and act, feel, and think as if you expect—the worst.

After a very few experiences, these types of behaviors become highly conditioned habits that consist of all four dimensions. These behaviors become conditioned because they work for us on some level (or we think they work). Once this happens to any behavior, in any situation, it becomes extremely difficult to change the behavior. This is one reason why it is so difficult to get managers to adopt a new quality program, have a spouse stop smoking, or get engineers get out on the shop floor with the workers occasionally. No matter how bad the action appears to us, the accompanying thoughts, emotions, and physiological states are providing the person with some comfort.

The quickest and most reliable way to change a behavior is to change the action component. You can't force someone to change their thoughts unless the consequences of their actions change. If you can change the action, you have a shot at changing the other three dimensions. Yet how does management in most organizations attempt to change behavior? You got it: they try to change thoughts or make emotional appeals. This can't work. Employees may listen and may want to believe they should change their rational expectations, but then they are compelled by reality to perform the old action and the accompanying "old" thought, physiological, and emotional components come right along and provide comfort. If you want to change a behavior, you must begin to compel small changes in action *in advance of changing thoughts or emotions.*

For example, if you wish to encourage a group of uncooperative, seemingly uninterested, and perhaps hostile employees to participate in a group discussion, it usually doesn't help very much to appeal to their thoughts. They've heard such appeals before and then saw no real changes. They're not stupid; they know what the real norms are. What you must do is get them to perform, in small ways, the action components of team work behaviors, even if they don't like it or believe it will work or even recognize they're doing it. Once the action components begin to change, the other components will begin to fall into place. Thus the first step is often to have the group brainstorm ways to improve some part of their process. As part of the brainstorming, each person is required to write down his or her observations, one at a time, on a

> The quickest and most reliable way to change a behavior is to change the action component. Yet how does management in most organizations attempt to change behavior? You got it: they try to change thoughts or make emotional appeals.

piece of paper and tack it up on the wall. It is key to this technique to have each individual write out and then walk up and tape the paper to the wall. What you are doing is forcing the person to change an action from a nonparticipatory one (sitting and grousing) to a participatory one (being part of the activity, even though he or she may still complain).

Cognitive Dissonance

It doesn't seem like much, but these simple changes in actions are an incredible start. You see, people have a hard time doing one thing and then thinking something that's contrary to what they've done. It's called cognitive dissonance. People want (and need to have) their self-image and behavior match what they think they are, what they expect, and what they observe. Over the ages, this has helped keep people's behavior consistent and predictable. Consistent and predictable behavior has always had high survival value because what's lethal and what's safe never used to change much over the course of a couple of generations. Only now, in the modern world, has our environment begun to change faster than our innate comfort level with accepting changes.

> People have a hard time doing one thing and then thinking something that's contrary to what they've done.

So business is faced with daily changes, but people are not comfortable with daily changes. People interpret anything that happens, familiar or new, in light of what they already know and expect. Because they already have a repertoire of learned behaviors (actions with accompanying physiological, emotional, and thought components), people respond almost automatically with an established behavior that reinforces what they believe is happening (or want to happen). Most often, as in the "hitting the brakes" example, this instant response works well, as it has for social primates over the last several million years. The key is that even the largest behavior change starts with small changes in actions that are consistent, observable, and performed by each person. If this pressure on actions is kept up for many months, in some small way, on a daily basis, there is hope that true behavior change can occur. Of course, the consequence of the new action must provide people with satisfying thoughts and emotional and physiological responses. In other words, people have to get something (such as power, fun, freedom, or belongingness need satisfaction) from performing the new action.

Propinquity and Attraction

Many people believe that opposites attract. This is generally not the case in terms of group identity, friendships, and marriages. People are comfortable with and seek out others who share similar views, opinions, attitudes, and even physical characteristics. This phenomenon is called propinquity: we seek out, value, accept, and identify with people who are similar to us in any dimension. The more of our dimensions they share, the more we will tend to like them and seek them out. This had tremendous survival value for early primates because it reinforced group cohesion and loyalty. In changing times, this very same imperative can stop people from seeking out different views and opinions, especially when they come from different departments.

The most powerful dimension is physical location. All other things being equal, human beings feel closer to those who are near them in terms of distance. Thus, each of us will tend to identify with our neighbors more than with people six blocks away (even if we don't know them personally). At work, we will develop bonds and feelings of loyalty to those who are seated near us, regardless of organizational or reporting membership. This instinct is strong because it is central to maintaining small-group cohesion. Any of our ancestors without this instinct probably ended up being eaten by a tiger because he or she didn't stay close enough to a group.

At the same time that physical location is so important, other dimensions can reinforce propinquity. Height, weight, race, religious preference, political beliefs, manner of dress, sports team preference, recreational activity, type of car driven, and so on, are factors that can influence propinquity. The most visible factors are always the most influential because we are primarily visual creatures. Just consider the number of dimensions that separate management from other employees in a traditional organization: physical comfort (light, airy offices versus dirty, dark factories), dress (suits, dress shirts, and ties for management versus casual wear for others), parking areas (reserved parking versus the south forty), and entrances to the facility (front office lobby versus factory back door). The list could be expanded almost infinitely to include company social functions, amount of information available on what's happening, benefits, bonus programs, and so on. Is it any wonder that management is seen as an "outside" group by most employees? How could they feel otherwise, when every possible signal screams, "Management is not like us."

The key to using propinquity as a positive force is to reduce the various distances between all levels of management and employees. Elimination of coat-and-tie dress codes boldly demonstrates that the barriers are not as great (company uniforms for everyone with first names on them are an ideal solution that also provides a tremendous economic benefit for employees). Placement of management offices near production and service areas is another way to reduce distances, both physical and in the similarity of information that each group perceives.

All other things being equal, human beings feel closer to those who are near them in terms of distance.

The Bottom Line

The bottom line is that if you want to be an effective leader of human beings in today's rapidly changing business world, you'd better start "monkeying around" right away, at least in terms of developing a better understanding of primate behavior. While you're learning, here are some things you can do right away:

- People will form small groups, so lead them to form ones that are meaningful to their work processes.
- Groups and departments will compete with each other because of propinquity; you can't stop it. Ensure that the competition is always win-win by establishing a series of measurements that enables any or all areas to win by excelling in doing things that help

other areas (such as never allowing a downstream process to run dry because you miss a product cycle).
- People want to be led and will respond well to good leadership. Select supervisors at every level based upon their concern for people and their understanding and feel for working with people.
- People will stick with old behaviors unless you can get people to change their actions in small ways. Introduce planned, long-term changes in steps of small, required changes in actions rather than as large, visible emotionally or thought-based programs. For example, don't kick off a massive Toyota Production System installation. Instead, start with five-minute morning updates for each work group.
- There are social and status castes in all primate groups; they can't be stopped. Lessen their effect on your organization by reducing the distances between all types and levels of employees. You can start with dress codes, office location, information availability, and parking.

Good luck in your search for knowledge and success. May you develop huge ischeal tuberosities (at least figuratively) and the rewards they bring though your mastery of primate leadership. I look forward to grooming with you someday.

The Failure of Participatory Management

Charles Heckscher

TQM suggests teamwork, empowerment, and meaningful participation in the organization by employees at all levels. Unfortunately, organizations that seek to implement teams and empowered employees often fail because they do this within the context of the assumptions of the bureaucracy, which run counter to such actions. The author suggests that for organizations and their leaders to be successful, they must appreciate that "Increasing autonomy—leaving people alone—leads the wrong way; increasing interdependence and interaction across walls leads the right way." For a manager, this article, like the others in this section, provides insight into how to bring out the best in employees to their and the organization's benefit.

Participatory management has been one of the longest-running management trends of the post-'50s era. While fads have come and mostly gone—job enrichment, quality of work life, quality circles, autonomous teams, and more recently TQM and reengineering—one underlying theme has steadily picked up momentum: increasing the involvement of lower levels in decision making.

The reason is clear: Bureaucracy has reached its limits. The simplicity and power of top-down, rule-based administration created competitive advantage in the past, but it blocks the responsiveness and continuous innovation that are keys today. That is why "teamwork" and "empowerment" are seen almost everywhere as the road to success.

But does that road really lead to success? For five years I have interviewed middle managers in companies that are downsizing and restructuring; I've talked to more than 250 managers in 14 large organizations, including divisions of General Motors Corp., AT&T Co., Pitney Bowes Inc., Honeywell Inc., and DuPont. One thing I can say is that from their viewpoint participatory management has generally *not* accomplished much. It is rarely successful in breaking the walls of bureaucracy.

In fact, downsizing and restructuring as they are normally practiced have the opposite effect: As middle managers have consistently told me, these changes *increase bureaucracy and increase organizational politics.* Their experience is that the organizations they work in have become more rule-bound

> Downsizing and restructuring as middle managers have consistently told me, *increase bureaucracy and increase organizational politics.*

Reprinted with permission of the Conference Board from *Across the Board*, November–December 1995. Copyright © 1995 by the Conference Board. All rights reserved.

and narrowly focused rather than less, and less entrepreneurial rather than more.

There is a huge gap between the views of the middle and the top. Top managers consistently believe they are breaking down old cultures and creating a new environment of collaboration and innovation. They talk of education, increased information-sharing, systems of teamwork, and communication. In most cases little of this has actually penetrated to the middle layers. The rhetoric and the programs talk of one thing, but the reality in the heart of the organization is another.

Hence, as a growing series of studies has found, the enormous efforts at restructuring are yielding mostly disappointing results. While they usually produce some short-term cost reductions, executives looking for fundamental increases in organizational effectiveness aren't finding them.

Restructuring has great potential, but many current approaches only undermine the long-term strength of organizations. Instead of transforming bureaucracy, they unwittingly reinforce it. I have seen just a few cases that avoid the bureaucracy trap.

The Empowerment Double Play

Part of the problem is that there are two sharply different meanings of "teamwork" or "empowerment": One of them creates something new, but the other just reinforces bureaucracy. By confusing the two, many organizations set themselves on the wrong path.

The basic problem with bureaucracy is that it operates by dividing work up into small pieces and building walls between the pieces. That is the foundation for the control system: Each person is supposed to focus on a particular set of programs, skills, and objectives. When an individual completes his piece, he throws it "over the wall" to the next step in the process.

Instead of tackling this problem, the most popular version of "empowerment" simply gives individuals bigger pieces: It gets rid of micromanaging superiors and unnecessary rules, and it allows people to do the job they are assigned without interference. "Delayering" is often seen as a way to get rid of micromanagement and restore some real power to the front lines.

That sounds good, but it's not new: Alfred Sloan recommended that back in the 1920s when he was creating the classic bureaucratic structure of General Motors. Letting people do their jobs is nothing more than a principle of good bureaucratic management. Sure, organizations tend to build up unnecessary layers and rules, and it is wise to prune them periodically. But this approach increases the autonomy of different parts of the organization rather than helping them to work together. It therefore *strengthens* the walls that block systemic innovation and responsiveness.

Another version goes a step further, but only a step. I am referring to "autonomous team" systems that get rid of an immediate supervisory level and create a team from those who would normally be a bunch of subordinates. This has the advantage of reducing layers and, to some extent, increasing flexibility within the group—people can cover for each other and share information more directly. But this kind of team, being permanent, quickly builds its *own* walls around itself. The members begin to share an identity that leads them to resist breaking up and recombining, and they don't generally deal well with "outsiders." I have come across many cases of autonomous teams fighting to protect their turf and unity against change efforts. While this kind of teamwork increases immediate flexibility, it again raises the barriers to larger-scale responsiveness.

The kind of empowerment that really breaks the bureaucratic mold is something different: Instead of increasing autonomy, it increases the ability to *work together effectively across walls*. These walls exist between levels, blocking open communication, and they exist between functions, blocking effective cooperation on complex problems. The organizations that are starting to leap beyond bureaucracy are the ones who have built a capacity to pull people together quickly, from different parts of the system, with different skills and knowledge, into effective *temporary* teams working on a task. This capacity to quickly bring together diverse expertise into an effective unit is *the* characteristic that marks a quantum leap beyond bureaucracy and that will create competitive advantage for the next economic phase.

Very few companies in my experience can do this well. Though many have been trying cross-functional, task-focused teams, the middle managers I have spoken to widely regard them as ineffective and distracting. Empowerment programs that expand *autonomy* make the situation worse, not better.

Restructuring Pitfalls

Restructuring not only fails to create the new kind of participation just described; it has also undermined the *old* forms of participation that kept good bureaucracies working reasonably effectively in the past.

In traditionally effective organizations, a key form of participation was through the direct supervisor. That was the position responsible for listening, for supporting, and for representing people to the organization. Studies have always shown that most people are pleased with their supervisors, and that this is their most important relationship in the corporation. Again, new programs to create "participatory" relations between managers and their subordinates may be helpful in systematizing past practice, but they do not bring anything fundamentally new.

But at the same time that these programs try to build those "coaching" relations, the larger restructuring process undermines them. Delayering widely increases the span of control and reduces direct supervision. That may mean that people control their own jobs more but are *less* able to bring up concerns and issues to someone they trust. Focus groups and skip-level meetings do not substitute for the personal relations that used to give people a sense that they had someone to go to in a pinch.

The old organization also had another crucial form of "teamwork." Middle managers were never the narrowly job-focused, rule-following automatons they were made out to be. Fortunately for the companies they worked for, they have always stretched their job limits by building communication *networks* throughout the system, cutting across official lines.

These management networks were informal, based on personal trust and hit-or-miss contacts. You might go through your first orientation session with Joe from engineering. A few years later, when you're handed a project that needs more engineering talent than you can muster from your own people, you call Joe up and ask for help. He might give you some tips; he might even lend you one of his people for a few weeks. And, of course, you'd do the same for him.

Now, any bureaucratic organization that didn't have these informal networks was doomed to fail. If middle managers really had to go through every channel and follow every rule to get what they needed, the whole system would quickly come to a grinding halt. It was because people *found ways to cross the walls* that bureaucracies functioned at all.

What happens in a downsizing or restructuring? One manager summed it up nicely: "The changes did two things: They destroyed networks, and they destroyed career paths. Now, more and more, people don't trust each other."

The first thing is that the networks are disrupted: Joe is gone. The second thing is that the wider basis for trust is destroyed. The main reason Joe was willing to help you was that he knew you were both in this for the long haul, and that when he came back to you for a favor you'd still be around. If he was lucky, you might even have gotten promoted and be able to help him more extensively. Now his replacement, if there is one, doesn't know any of that. He's got enough problems of his own; he can't spare time or people. Why should he help you out?

Things are even worse if, as in many companies, there is increasing short-term pressure to meet objectives. What that means is that the *formal* requirements of the organization are being rewarded more than ever, but the *informal* teamwork that gave it life is pushed further under the surface. It becomes still harder to go out of your way to help someone in the expectation that it will all work out for the best in the end.

In the absence of trust, people feel more isolated, more lonely, more vulnerable; they rely more on formal rules than before, and they are quicker to protect their turf. All of that is the opposite of what is intended, of course. But in the absence of trust, no amount of training, of participatory programs, or of communication makes much of a difference.

The Loyalty Trap

There is more: Restructuring not only disrupts middle managers' traditional forms of involvement and collaboration; it pushes them into a defensive mentality that blocks new forms.

Nearly every company I have been involved with has been trying to get managers more involved in the business. They have elaborate programs to distribute strategic information on a far wider scale than ever before, and they have set up all kinds of systems to encourage critical feedback and discussion. The logic is that it is not enough anymore for people to work together on particular tasks: They need to feel an ownership in the whole and to understand its relevance to their work. It is a noble objective that is seldom realized—because managers *flee* from this level of participation.

Despite all the programs, two widespread problems appear. First, middle managers do not understand the business; they have little or no conception of the competitive environment and the strategic positioning of the firm. They understand their own piece extremely well, but not more. The whole set of changes threatening them makes very little sense.

Most of the top-level managers to whom I report this find this hard to believe. "We tell them all the time; we tell them everything," they say—and they're largely right. The problem is not usually that middle managers haven't *been told* about the business; it's just that they haven't *heard* it.

Second, though they have deep private doubts about their leaders' competence, they refuse to express these doubts in public. They almost refuse to admit them to themselves: The same person will commonly in one breath criticize top management and in the next say that he has complete faith in them. Though many companies are going all out to provide opportunities for dialogue, middle managers are reluctant to use them fully.

This is an irrational pattern but a typical one. It's partly due to fear of retaliation—but only partly. A more basic source is, surprisingly, the loyalty that managers continue to feel towards their companies. Despite the downsizings, despite the violations of long-term expectations, most of the people I have talked to care deeply about their organizations and want to hold on. But they are *trapped* by their own loyalty.

Let me try to interpret what they are feeling. The deal in loyalty is that if you do your job well the organization will take care of you—you do your part and the higher-ups will do theirs. Now a bunch of your superiors are coming down and saying to you, "We want you to know what's happening. We want you to know the information we have. We want you to know how difficult and confusing the environment is. And we want you to tell us honestly what we're doing wrong and what we can do better." What middle managers hear is: *They don't know what they're doing.*

That's not a message the middle managers want to hear; they actively resist hearing it. When their superiors come out with another batch of business information, it basically means that more layoffs and disruptions are on the way. They can't do anything with the knowledge, and it just makes them anxious. "I think what we all struggle with," one manager said, "is how much we can influence the business—because we know the business can influence us."

So most don't want to hear too much or criticize too much; they don't want to come face to face with the fact that the situation is no longer in control and no one quite knows what is going on. Even the private watercooler conversations are contradictory: On the one hand they say the top is

> The problem is not usually that middle managers haven't *been told* about the business; it's just that they haven't *heard* it.

making all sorts of stupid decisions that don't make any sense; on the other they say (or hope) that the top probably does know what it's doing.

The situation is, in short, one of *loyalty without trust*. People hang onto their attachment to the company as a kind of security blanket in a world gone mad, but they no longer have the network of trusting relationships so crucial in the past. They cope by withdrawing into a narrow world, putting their heads down, getting by, waiting it out:

"I have no doubt that it's going to get worse, but I have no doubt that we will recognize it and turn it around," a manager told me. "The only thing we can do is just wait. We have a good organization here, when we get directed to do it, we do it. I can't help but think that the pendulum will swing back. We will just have to wait it out, because there is really nobody to talk to."

Overcoming the Barriers

All of this adds up to a pattern in which no matter how hard you try, you stay stuck within bureaucracy. Downsizing and restructuring produce quick benefits because they cut costs and temporarily improve the alignment between structure and strategy. But they don't increase the basic flexibility of the system; in fact, they reduce it by *increasing* the barriers between people. So most organizations end up putting people back on (though they may hide it by changing the categories) or restructuring over and over to try to keep up with change.

As I've said, few in my study have escaped the trap. The ones that have are the ones that have changed the *most*. They have created a kind of teamwork that is not just a modification of bureaucracy, but that truly has brought down the walls. Most managers have served on multiple teams—some lasting for a few hours, some for a few years, many of them spontaneously created from the middle, all focused on solving problems.

I have never found this kind of management through an entire corporation: It has appeared so far only at the level of plants and divisions. In my own study one of the successes is a General Motors plant, another a Honeywell division; and recently I have been impressed by AT&T's Network Services Division. I have also heard convincing case studies of similar units at General Electric Co., IBM Corp., and a few others. But it clearly remains rare.

In these organizations the managers I have interviewed are not putting their heads down and avoiding reality; they are working together to master it. They are not looking for more *autonomy* but are enjoying a sense of *interactivity*.

"The middle managers really make this place run," one manager said. "There's a strong desire among all of us to work the problem. We work together, we're friends, we have a lot of kinship in getting a job done."

How have they done it? One element is simply to be clear about what you're doing—about the difference between "empowerment" *within* bureaucracy and breaking bureaucracy. Increasing autonomy—leaving people alone—leads the *wrong* way; increasing interdependence and interaction across walls leads the right way.

At least two other things seem crucial:

A Focus on Purpose

The new environment does increase the mobility and independence of managers. They can no longer rely on personal contacts and informal networks as a basis for trust, and they need to pay attention to their own needs rather than counting on the company to take care of them. The problem is how to get these more individualistic employees to work together. A major answer seems to be to a focus on the organization's *purpose*. The most successful organizations in my sample have created a rich and thoroughly shared picture of what they are trying to accomplish together in the next few years.

The purpose, in this sense, is different from most definitions of "vision," "goals," or "mission."

- On one side, "visions" tend to be simple, eternal statements of the organization's identity that can be put on a wall: Mazda Inc.'s, to take one classic example, is "Everlasting effort for everlasting cooperation." That supports traditional loyalty, but it doesn't create a sense of working on a *problem* together.
- On the other side, many organizations heavily emphasize this year's goals or numbers. That gives a visible target but is too short-term to sustain a sense of shared commitment and enthusiasm; it quickly degenerates into "just hitting the numbers."

A purpose is something in the middle: a complex statement of the organization's challenges and objectives *over a three-to-five-year time frame*. It usually can't hang on a wall, because it is far too complicated. The limited time frame keeps the purpose concrete, away from airy abstractions, but it's long enough to provide a challenge worth *caring* about.

People who understand the purpose can talk knowledgeably about competitive positioning, comparative advantages, and strategic objectives. This has been the clearest difference in my sample: In the troubled companies middle managers cannot talk about those issues; in the dynamic ones they can, and that sense of purpose pulls them together.

Telling the Truth

All corporate leaders say they want to communicate the reality of change, but most are held back by their end of the loyalty bargain. They are supposed to protect "their people"; they are almost unconsciously reluctant to admit that they can't do that, and they worry that their people won't be able to deal with the full reality of what's happening. So they pull their punches. They try to reassure people: "We may not be able to guarantee you a job, but we'll do everything we can to take care of you." They also tend to hold back the full picture, doling out reality a bit at a time, allowing people to believe that things may return to "normal."

The dynamic companies don't do that. They tell the full picture of competitive problems, and they make clear that promises of security are no longer possible. For most managers, including lifelong loyalists, this has the effect of a bracing dose of truth: The world *makes sense*, even if it isn't entirely pleasant.

I don't mean that the solution is to be cruel: Hauling people out of buildings under armed guard, or refusing to do anything to help those laid off, may be ways of getting the truth across, but they don't contribute a sense of shared purpose. The problem is to replace the paternalistic ethic of *caring for* people with a professional ethic of *mutual commitment*.

Companies with a sense of honesty and purpose have high levels of participation. They don't do it with participative programs, but they create an environment in which understanding and criticizing the business direction are important to people's daily jobs. In that situation people work together rather than fragmenting into isolated pieces.

Establishing New Relationships

"Participation" is one of those words that can mean almost anything. In this case it can mean at least two opposite things: reinforcing bureaucracy or overcoming it. Managers and others have no difficulty in "participating" within the traditional framework: That is, they want to be left alone to do their jobs, to get out from under unnecessary rules; and they are delighted to discuss with superiors how their jobs can be done better. These kinds of "participation" date back at least to the 1920s.

Flexible organizations, however, require that people participate beyond their jobs: Everyone needs to understand and contribute creatively to the overall purpose of the organization. That kind of participation profoundly threatens traditional relationships.

Surprisingly, this is the crux of the change in expectations. Layoffs and downsizings do not shatter the contract of loyalty. Managers are able to rationalize these for a long time as responses to crisis, and to settle into a waiting posture. They don't fundamentally give up the expectation that, if they can just get through this period, things will return to normal.

Being asked to get involved in strategic purpose is a far more serious threat, because it establishes new relationships. If middle managers (and others) have real involvement in issues beyond their jobs, it means that the whole system of bureaucratic coordination needs to be reconstructed. It means that middle managers have to be treated far more as independent agents than before.

It means that the uncertainty and threats of the "outside" world have to become a daily reality for everyone, not just the top.

Charles Heckscher chairs the Labor Studies and Employment Relations Department at Rutgers University. He is author of White-Collar Blues: Management Loyalties in an Age of Corporate Restructuring (Basic).

Quality Transformation

CULTURAL TRANSFORMATION

Connecting Culture to Organizational Change

Timothy Galpin

Productive organizational change requires a culture that supports that change. Everyone knows that quality initiatives that aren't supported by a quality culture are bound to fail or yield mediocre results at best. This article, although it does not directly address the relation between culture and TQM, does provide real insight into how culture encourages or discourages successful change efforts. It enumerates ten components of an organizational culture and shows how these components act as a "screen" for changes, allowing some through and blocking others. Recognizing this, you can manage this screen to facilitate positive changes.

Achieving and sustaining the goals of organizational change is difficult at best—and for many organizations seemingly impossible. Most change initiatives fall short of reaching their goals where "the rubber hits the road"—during implementation and follow-up.

Effective implementation requires that changes in operations, systems, procedures and the like be clearly connected to an organization's culture. Making this connection not only enables effective implementation, but also embeds change in the day-to-day life of the organization, sustaining desired effects such as lower costs, increased revenue, improved customer service, fewer errors or quicker processes. Creating a connection between culture and change requires passing proposed changes through a "cultural screen" to identify how best to implement and embed changes in an organization.

Defining Organizational Culture

Components of culture can be isolated, yet no one component alone fully describes the culture of an organization. Organizational culture is a mosaic of interrelated elements. As these individual elements interact each work day, they collectively create the organization's culture. For example, a manager who continually gives orders to subordinates can be called authoritarian. But a single manager's leadership style does not create an authoritarian culture. In fact, many other managers in the same organization may frequently solicit input from employees, collectively creating a participatory culture.

There are 10 components that together establish an operational description of organizational culture:

- Rules and policies
- Goals and measurement
- Customs and norms
- Training
- Ceremonies and events
- Management behaviors
- Rewards and recognition
- Communications
- Physical environment
- Organizational structure

Identifying discernible components of culture allows organizations to determine which tangible elements can be managed to help implement and sustain change. But just as none of the 10 components alone defines culture, none of them can individually support desired changes.

Managing Culture during Change

The primary motive for managing culture during change is to implement *and* sustain *changes.*

The primary motive for managing culture during change is to *implement* and *sustain* changes. Too often, executives and managers struggle when implementing changes because they don't understand how to make them important to employees.

For example, a major retailer identified a decline in customer service as a major obstacle to revenue growth. Service wasn't bad, but an analysis of the competition and customers' wants indicated that competitors were moving well ahead in customer service. Management tried to improve the situation by declaring that excellence in customer service was important and that poor service "would not be tolerated." The company also spent a significant amount of money, and management and employee time, on training employees in new service behaviors.

Unfortunately, those two actions were not enough to bring about the desired changes in behaviors and make them stick. Although there was a brief increase in service ratings, once the training was completed, the ratings quickly fell off.

This example illustrates how managing only one or two aspects of an organization's culture—in this case, training and communications—fails to effectively implement or sustain change. How could the company have done better? It should have embedded service excellence in the culture of the organization by managing as many of the 10 components as possible. Deciding which components to use to implement and sustain change requires sending the desired changes through a "cultural screen." Doing so will identify *all* of the elements of organizational culture that can be leveraged to successfully implement and sustain change.

Applying a Cultural Screen

All 10 components can be applied to the change process as a cultural screen. Evaluating a desired change through the elements of this screen allows organizations to identify the cultural components relevant to implementing and sustaining desired changes. Although all 10 of the cultural components will not always be needed to bring about particular changes, all changes should be screened to identify the cultural aspects that a company can leverage.

For example, a large manufacturing company wanted to implement a new purchasing process to lower costs and improve tracking accuracy. At first glance, implementation did not appear to be all that complex, since the new process was easy to understand and "looked good on paper." Yet, when presented with the new purchasing process, the employees involved were not eager to make the required changes. When the change was explored in more depth from an implementation perspective, the company learned that the old process was still being reinforced by the old training, old goals and measurement, old management behaviors and the like.

When the company applied the cultural screen to the proposed process changes, it found that seven of the cultural components would help to make the process changes stick and would reinforce the new purchasing process. The seven components were identified as setting new goals and measurement, developing new rewards and recognition, establishing new training, supplementing communications, redefining rules and policies, changing management behaviors, and establishing new customs and norms. Once applied, these components helped the company implement the new process within 60 days and achieve its goals of lower costs and improved tracking.

Once a desired change has been put through the cultural screen and the components of the screen that will help implement the change chosen, specific actions should be identified for each component. Some examples for each of the 10 cultural components are described in the chart of Implementation Actions for Cultural Components.

Developing and Implementing an Action Plan

After identifying implementation actions for each cultural component that will reinforce the desired changes, the company should develop an action

IMPLEMENTATION ACTIONS FOR CULTURAL COMPONENTS

Rules and Policies
- Eliminate rules and policies that will hinder performance of new methods and procedures.
- Create new rules and policies that reinforce desired ways of operating.
- Develop and document new standard operating procedures.

Goals and Measurement
- Develop goals and measurements that reinforce desired changes.
- Make goals specific to operations. For example, establish procedural goals and measures for employees conducting the process that is to be changed, rather than financial goals that are a by-product of changing the process and that employees cannot easily relate to their actions.

Customs and Norms
- Eliminate old customs and norms that reinforce the old ways of doing things and replace them with new customs and norms that reinforce the new ways. For example, replace written memos to convey information through the organization with face-to-face weekly meetings of managers and their teams.

Training
- Eliminate training that reinforces the old way of operating and replace it with training that reinforces the new.
- Deliver training "just-in-time" so people can apply it immediately.
- Develop experiential training that provides real-time, hands-on experience with new processes and procedures.

Ceremonies and Events
- Establish ceremonies and events that reinforce new ways of doing things, such as awards ceremonies and recognition events for teams and employees who achieve goals or successfully implement changes.

Management Behaviors
- Develop goals and measurements that reinforce the desired behaviors.
- Provide training that focuses on the new behaviors.
- Publicly recognize and reward managers who change, by linking promotion and pay rewards to the desired behaviors.
- Penalize managers who do not change behaviors. For example, do not give promotions or pay increases on bonuses to managers who do not demonstrate the desired behaviors.

> *Implementation Actions for Cultural Components (continued)*
>
> **Rewards and Recognition**
> - Eliminate rewards and recognition that reinforce old methods and procedures, replace them with new rewards and recognition that reinforce the desired ways of operating.
> - Make rewards specific to the change goals that have been set.
>
> **Communications**
> - Eliminate communication that reinforces the old way of operating; replace it with communication that reinforces the new.
> - Deliver communication in new ways to show commitment to change. Use multiple channels to deliver consistent messages before, during and after changes are made.
> - Make communications two-way by soliciting regular feedback from management and employees about the changes being made.
>
> **Physical Environment**
> - Establish a physical environment that reinforces the changes. Relocate management and employees who will need to work together to make changes successful. Use "virtual offices" to encourage people to work outside the office with customers and telecommunications to connect people who need to interact from a distance.
>
> **Organizational Structure**
> - Establish an organizational structure that will reinforce operational changes. For example, set up client service teams, eliminate management layers, centralize or decentralize work as needed, combine overlapping divisions.

plan. This plan should focus on successfully leveraging each cultural component to execute and sustain the desired changes. The action plan should also include the people involved, time frames and resources needed.

Effective implementation requires adherence to all facets of the action plan—resources, people involved, and timing. Failure to stick to the plan will send strong messages to the organization that management is not serious about changing, and the commitment needed to implement and sustain change will be lost.

For example, a large chemical company wanted to change its recruiting, performance management and salary administration processes by transferring most of those activities from HR professionals to managers and employees. The company first identified the activities to be performed by managers and employees, then ran these changes through the cultural screen.

The screening identified eight cultural components to use to expedite making and sustaining the changes: rules and policies, goals and measurements, customs and norms, training, ceremonies and events, management behaviors, rewards and recognition, and communications. The company developed action plans for each of the eight components. When the action plans were implemented, however, training schedules slipped, senior managers did not show up for rewards presentations, events were canceled and little of the planned communication occurred. These events showed managers and employees alike that the company was not serious about making the changes. As a result, implementation went poorly, and the desired changes were short-lived.

Measuring the Impact

After implementation has begun, it is important to measure the results of the changes. First, the company must be able to determine when goals have been achieved. Is absenteeism down, customer service improved, cost reduced or teamwork enhanced? Reliable measurement is needed to answer these questions credibly without guesswork.

Measurement also provides a way to track progress. By determining what to measure and how to measure it, the organization can clearly see whether it is on track to achieving its change goals. Accomplishing goals shouldn't be a surprise. Periodic measurement allows people to see progress toward the goals. Managers and employees alike can observe themselves achieving implementation milestones. Attaining intermediate landmarks helps people stay motivated and keeps them committed to achieving the final goals.

Measurements used should build a complete picture of the effectiveness of changes through a combination of quantitative and qualitative measures. Behavioral observations of employees and management are one measure of effectiveness that goes beyond financial data. Behaviors can be counted, described and summarized to draw conclusions about the changes made, such as improved customer service, enhanced teamwork or better communication.

Management and employee feedback can also help determine the impact of changes made. Managers and employees are often the best sources for learning about shifts in perceptions about job satisfaction, what changes are or are not working well, and what impact the changes are having on customers. Organizational innovation can be indicated by the time it takes to develop new products and services or the number of new products developed in a year. As with any measurement, fundamentals such as assigning accountability for information gathering, establishing methods of data collection, setting the frequency of collection, using sound data analysis techniques, and creating useful reporting formats apply.

Effective Measures

These guidelines can help establish effective measures of change:

- **Set specific, numeric expectations.** Any activity can be measured in time, units, money, or customer satisfaction. Using this strategy allows people in the organization to begin "measuring the immeasurable."

- **Keep it simple.** Keeping it simple helps people understand the measurement used. It also helps people stay focused as they work toward the goals of the change effort.

- **Be creative.** Creativity unlocks new methods of measurement. Don't fall for the lure of using the "same old measurement for the same old issues."

- **Involve managers and employees in designing their own measures.** Managers and employees are often a big help in designing credible and creative approaches to measurement. Involving them in designing measures also encourages their commitment and buy-in to making change happen.

- **Determine the frequency of measurement.** Define how frequently measurement will take place. Will it be hourly, daily or weekly? Measuring too often will put unneeded pressure on people. Measuring too infrequently will cause people to forget about the change.

- **Establish responsibility for keeping score.** Measurement is nothing more than score keeping. By regularly keeping score, people can assess their progress against a given standard and have a point of comparison against themselves, other teams and/or competitors.

Determine who will be responsible for keeping score. Rotating the responsibility will help people stay aware of the score and prevent one person from "burning out" on tracking the progress. For individual goals and measurement, assign each person the responsibility of keeping his or her own score.

Creative Measures

There are no limits to the way measurement can "look." Here are a few examples to help spark some ideas for measuring both abstract factors such as customer service, teamwork, and communication skills and concrete ones such as cutting expenses.

- **Use self-rating.** Self-rating offers several benefits. First, it establishes buy-in to the measurement. Second, it motivates people—the change becomes important to them. Third, it creates an awareness of the factors being rated, and awareness is the first step toward change and improvement. Many managers find that employees' self-ratings often are more critical than the managers' appraisals.

- **Measure discrete behaviors.** Measuring discrete behaviors can be as simple as determining whether bank tellers regularly say "thank you" or tracking the amount of time a manager mingles with the loading dock team each day.

- **Make measurement visual.** People like to see what is happening around them. Providing visual measurements is a good way to keep people motivated and interested in their performance. One company developed a novel approach to cutting spending with a visual measurement system that increased employee awareness of expenses.

At the beginning of each month, the company distributed each department's budget to department heads as "budget bucks." Employees were asked to staple the "money" to expense vouchers and use it to "purchase" supplies from the company stockroom. Department heads submitted the money with requests for checks to pay invoices. Through this technique, the company hoped to reduce expenses as a percentage of sales by 3 percent. As people became more aware of cash flow, expenses actually fell by more than 5 percent.

Continuing Management of Cultural Components

To help ensure that change goals are achieved once implementation and measurement have begun, it is essential to continue managing the cultural components to reinforce and embed change in day-to-day operations. For example, while changing its stock replenishment procedures, a major retailer learned from management and employee feedback that communication needed to be improved between merchandise buyers and the stores. The company reinforced communication by moving buyers into stores so they could customize inventory for local markets.

Conversely, "taking one's eye off the ball" once changes have begun can cause implementation efforts to slip. Change that is not continually managed will not yield sustained results. One company's efforts to increase sales by changing its sales process failed, because the company ignored the need for continued management. The company conducted excellent training for all employees involved; provided frequent communication through multiple channels; and realigned its goals measurement, and rewards and recognition toward the desired changes.

But once the new sales process was kicked off and an initial increase of 10 percentage points across all regions was achieved, management did not continue to manage the process through ongoing measures and rewards. As a result, the goals achieved within the first two months were short-lived, and after six months, sales volume returned to previous levels.

Many organizations ignore cultural change because it appears so difficult to manage. Instead, they focus on the more "tangible" targets of change: operations, equipment, systems and procedures. But achieving and sustaining the goals of organizational change efforts require a connection between culture and operational changes.

Applying a cultural screen will help facilitate the implementation of desired changes. An organization committed to changing its operations, systems, procedures and the like must tirelessly apply as many of the 10 cultural levers in the screen as possible.

Timothy Galpin, Ph.D., is an organizational development specialist and a principal in the consulting division of Pritchett & Associates, a Dallas firm specializing in organizational change and merger integration.

The Soul of the Hog

Bob Filipczak

Almost everyone admires the Harley-Davidson turnaround that TQM played a central role in. This article describes the Harley culture, based on five values: tell the truth, be fair, keep your promises, respect the individual, and encourage intellectual curiosity. The author tells how these values manifest themselves at this company. Although Harley-Davidson's workforce is heavily unionized, its employees know customer value is vital, and that knowledge is an important motivator in their performance.

It was his first day on the job, two years ago, and Ken Sutton was at a vending machine getting a cup of coffee. Behind him, a machinist he had never met asked, "you're new here, aren't you?"

Sutton allowed that he was.

"Well, let me buy you a cup of coffee," the machinist offered.

It was a simple gesture, but two years later, as vice president and general manager at Harley-Davidson's power train plant in suburban Milwaukee, Sutton remembers what it was like to be welcomed to the family. He learned quickly that "relationships" were an important—maybe the most important—component of Harley-Davidson's corporate culture.

Relationships? isn't that a bit warm and fuzzy for a work force full of hard-core bikers building power trains for motorcycles? Here's a factory full of tough-looking, tattooed and (except for the women) bearded workers. They're producing the most infamous icon of go-to-hell individualism in America, and it turns out they're all worked up about relationships?

Well, yes, says Margaret Crawford, corporate director of training and employee development. "We are very much a relationship-based company." Building strong working relationships—between co-workers, between unions and management, between supervisors and the supervised, between executives and machinists—is a core concern at Harley-Davidson.

That doesn't mean you'll hear phrases like "I hear what you're saying" or "Thanks for sharing" from Harley workers. They are a little more *direct* than that. Harley-Davidson's idea of a healthy working relationship is embedded in five formal values that fit easily on a 3- by 5-inch laminated card and constitute a code of behavior for everyone. Most workers at the power train plant seem to know them verbatim:

> Here's a bunch of hard-core bikers producing the most infamous icon of go-to-hell individualism in America, and it turns out they're all worked up about relationships?

Reprinted with permission from the February 1996 issue of *Training* magazine. Copyright © 1996 by Lakewood Publications, Minneapolis, Minnesota. All rights reserved. Not for resale.

- Tell the truth.
- Be fair.
- Keep your promises.
- Respect the individual.
- Encourage intellectual curiosity.

Oil Puddles

You can't get so much as a shoeshine anymore without receiving a card engraved with somebody's mission or vision or values, of course, and all of this would be neither here nor there except that Harley-Davidson is something of a special case.

Not to put too fine a point on it, Harley was at death's door in the early 1980s. The company had a reputation for poor quality, it had lost most of its market share to the Japanese, it was beset with debt, and it was generally going the way of many American manufacturing businesses that are no longer with us. The period from 1969 to 1981 is still referred to as the "AMF years," when Harley was owned by American Machine & Foundry. The history of the AMF years, and the subsequent resurrection of the company, is documented in the book *Well Made in America* by Peter C. Reid.

Reid doesn't attribute Harley's troubles solely to AMF as many observers have, and in fact argues that much of Harley's current success is due to investments made by AMF. But while the bike's reputation for reliability was far from unsullied even before AMF bought the company—bikers took perverse joy in pointing to any oil puddle in the street and speculating that a Harley must have been parked there—the quality of Harley-Davidson's products indisputably suffered during the AMF years.

Reid describes how AMF began to turn this around during the last years of its stewardship with a triad of quality techniques learned from the Japanese: just-in-time inventory (JIT), statistical process control (SPC) and employee involvement. Even so, when AMF put the company up for sale in 1981, it looked to be a death sentence for Harley. Only a last-minute leveraged buyout by 13 of Harley's top managers saved the business. Company veterans still talk about how current CEO Rich Teerlink scrambled from investor to investor on Dec. 31, just hours before Harley would be forced to declare Chapter 11 bankruptcy.

The rest of the tale is now famous as a classic American turnaround, a Cinderella story. Harley survived constant setbacks to reclaim its market share, eliminate its long-term debt, and regain the respect of its customers. It did so by producing bikes of the highest quality and reliability. In 1994, the company had $1.5 billion in sales and a return on equity of 27.5 percent. It owns about 56 percent of the U.S. super-heavyweight motorcycle market—and the only reason its share isn't higher is because the company can't produce enough motorcycles to meet demand. Depending on which bike you order, you'll wait six to 18 months to get a new Harley.

Part of the reason for Harley's success is, of course, a uniquely fanatical brand of customer loyalty. Even during the AMF years, Harley riders kept the faith—though it was sorely tested—and kept buying the company's motorcycles. Employees like to remind people that their company logo may be the only corporate symbol that customers actually tattoo on their bodies. It's also a testimony to the Harley mystique that more than half of all employees own a bike. Most of the executives, including Teerlink, are Harley riders. Workers know that the machine they are currently building might turn out to be their own . . . or the CEO's.

Living Quality

Because Harley-Davidson was one of the first victories of the American quality movement, and perhaps the most dramatic, the company has become a poster child for total quality management (TQM). But you don't hear many people at Harley talking about quality anymore; it's so ingrained in the culture that it has become a way of life.

Many employees have been around long enough to remember the AMF years, however, and there is a palpable sense that nobody wants to take his eyes off the ball ever again. Lest anyone forget that poor quality almost destroyed the company, the other side of the laminated values card presents a list of "issues" or continuing concerns:

- Quality
- Participation
- Productivity
- Flexibility
- Cash flow

The one TQM phrase that seems to have survived in the vocabulary of employees is "continuous improvement." It has come to mean not just process improvement on the factory floor, but a more personal improvement of the relationships between people. That's where the "values" kick in. And when people agree to tell the truth, be fair to each other, keep promises, respect each other and encourage curiosity, there is nothing necessarily "soft" about their ensuing interpersonal dealings. As Curt Kapugia, senior quality engineer, puts it, "I could hate your guts, but I still have to respect you as an individual at work so that we can work together."

Here is how some of those values work in practice.

> The one TQM phrase that seems to have survived in the vocabulary of employees is "continuous improvement."

Truth

The five values were initially hashed out in the late '80s as part of a bigger plan called the Business Process, an umbrella concept that Teerlink has been pushing for the last four years as a way to educate employees about all aspects of the business, including its strategic direction. Executives held discussions with employees at every level about what it meant to "tell the truth" at Harley. Did it mean everybody must tell the whole truth and nothing but the truth all the time? It was generally conceded that sometimes managers couldn't tell everyone all they knew. But if a manager or anyone else had information he couldn't release, it was unacceptable to dissemble. He had to look the person straight in the eye and tell her that he couldn't give her the straight scoop—and, when possible, *why* he couldn't tell her what he knew.

Manufacturing engineer Dave DiJulio tells the story of an improvement that he and some other engineers wanted to make to the arrangement of a workstation. It was a change that the operators might not like, and the engineers were wondering how they were going to finesse it. Finally someone pointed out that "telling the truth" was supposed to be standard operating procedure. So they decided to lay it on the line: Tell the workers exactly what they were planning and elicit ideas on how to do it better.

"Instead of jamming it down their throats," says DiJulio, "we went through it, got ideas, and actually didn't implement quite what we were talking about." This seemed remarkable to DiJulio, who joined Harley just two years ago. Now, he says, he's used to it. "That's kind of the way it is around here. Somebody will have a concept, we'll go through it, and by the time it gets implemented it doesn't look anything like [the original idea]."

Fairness

There are two unions at the power train plant, and the cooperation they practice with management may be unparalleled anywhere else in America. General manager Sutton remembers that one of the primary concerns when he started at the company was productivity. There was talk of opening a new plant. Then, as now, most manufacturing companies tended to build new plants in the South, where unions don't have as strong a presence.

Instead, the leadership of the company went directly to Harley's unions and presented them with the problem of expanding production. The theme Sutton kept hearing from veteran managers was that the unions had stuck by Harley when times were tough, to the point that machinists brought in their own drills because the company couldn't afford to buy new equipment. Management's attitude, he says, was: "We're not going to try to go get the fast dollar or the low labor wage someplace else. We're going to stick with the people who stuck with us." In the end, the company decided to expand production in the existing facility instead of building a new plant.

Bob Klebar is president of Lodge 78 of the International Association of Machinists and Aerospace Workers (IAM), one of Harley's unions. His inclination to view management as a partner instead of an adversary was born when the 13 Harley executives bought the company from AMF in 1981. At that point, says Klebar, the managers legitimately could have asked for a new contract from the unions. They stuck with the negotiated contract instead.

Management's good-faith dealings with the unions have continued. Harley-Davidson has a history of in-sourcing, which means that it tries to bring as much work as possible into the plant to forestall any layoffs. And Klebar's union has considerable control over what kind of work is outsourced to other companies. By choosing the work that his group does in-house, he bolsters the job security of his people. Contrary to the practices of most companies, "When times are good, we'd outsource it. But if things get tougher, we'd look at those [projects] and maybe bring them in-house to keep the employment," says Klebar.

Largely as a consequence of management policies and actions like these, the machinists and tool and die makers Klebar represents are willing to take terms like "customer focus" seriously. "Instead of focusing on what we think we have a right to," Klebar says, his people work closely with the production department, and everyone concentrates on the final customer. "It really binds us together and makes us go in the right direction."

The union sometimes will even censure its own workers for shoddy performance, Klebar says: "Instead of protecting someone who's not doing the work, we take him aside and say, 'Hey, we really need your help to make this successful, and from what we see, you're not pulling your weight.'"

That willingness among workers to police themselves may become more crucial as Harley's power train plant begins a move toward self-managed teams this year. As training manager Darlene Rindo describes it, the plan is to create "semiautonomous work groups." These teams won't be completely self-managing, but team members will set their own schedules and be cross-trained. The change will be gradual, Rindo says; for one thing, it would violate the company values to dump a lot of new responsibilities on employees all at once. Management and the unions are cooperating in the creation of these work groups.

Respect

Harley's employee-involvement commitment, started in the late '70s, contains an element of respect for the individual. But this value becomes most apparent when employees talk about the company's process for consensus decision-making.

You'd think that a bunch of Harley workers would have as much use for facilitated, consensus-driven meetings as they would for Hondas. After a few conversations with employees, however, you come to understand that these meetings are anything but calm affairs. People tick off multiple examples of meetings that became forums for workers to express dissenting opinions to executives. In meetings with CEO Teerlink, says Sutton, "I've seen machine operators do what in other companies would be considered 'talking back to the president.' I've also seen that he didn't flinch. He listened."

This brand of respect goes both ways. Executives confront employees in straightforward terms, too.

A training course called "Meetings Harley Style" teaches employees how to run meetings and how consensus decision-making is supposed to work. According to training manager Rindo, the basic idea is that while not everyone in the group will agree with a decision, they have to be able to go out and support it on the factory floor. That means never moving on to another agenda item without reaching closure. But there is a stylistic element to all this that the course covers as well, Rindo says. "It's very much teaching the Harley culture at the same time, meaning that when people lay it straight—and people at Harley lay it straight a lot—it's OK."

"I've seen machine operators do what in other companies would be considered 'talking back to the president.' I've also seen that he didn't flinch. He listened."

"We get a lot of resistance in meetings," says Sutton, "and something we call 'push back,' where if you don't like something, it's your obligation—not just your right—to state what you don't like, what you think is wrong, and get that out in the open."

There is one caveat to this open meeting style, Sutton says: No foul language.

No cussing? At Harley-Davidson? That's right. Foul language has nothing to do with respecting other people, says Sutton, so it isn't allowed at the meetings.

Curiosity

The final statement in the declaration of values is "encourage intellectual curiosity." Harley-Davidson's Learning Center is a facility dedicated to lifelong learning. Its primary role is to serve employees who want to keep their skills current. In some cases, this means remedial training in basic skills, but mostly the center is a place to come with requests for specific job-training courses.

John Boyd works as a powder-coat operator at the power train plant, powder-coating being essentially a painting process. As Boyd explains it, all you have to do is come up with a course you want taught, enlist other employees who have similar interests, and appeal to the Learning Center. It will organize the course.

Harley also sponsors a Leadership Institute, a separate program set up by Marquette University in Milwaukee in which employees can take college courses for credit on the company's premises and get a baccalaureate degree in leadership at the end of four years.

The Leadership Institute also offers a course on values, in which employees are encouraged to take a hard look at their own values in an effort to see where they match Harley's declared values. Corporate training director Crawford, who teaches the course, says it isn't just a feel-good exercise but a very serious examination.

Originally the values course started at the top of the organization. Later Crawford decided that this kind of soul-searching would benefit everyone. CEO Teerlink is himself certified to teach the values course, and co-presented with Crawford when they took the vice presidents through the program.

A striking example of someone whose intellectual curiosity was encouraged is training coordinator Gail Rosenthal, a member of United Paperworkers International Union Local 7209. A quick look at her career path is a lesson in lifelong learning. During her 26 years at Harley, she has been in the electronics department, worked in golf cart assembly, progressed to snowmobile assembly (both during the AMF years, when Harley manufactured a variety of products), was the first woman engine tester and the first woman engine repair mechanic. Now she's a trainer who teaches a course in interpersonal skills and loves it. Rosenthal was selected for the job because interpersonal skills can be a dicey subject for bikers. "[Management] felt people on the floor would trust someone 'who's been there,' " she says.

There's so much training in progress at Harley, in many different areas, that it has become Rindo's mandate to keep track of what's going on in all the departments and collate everything into an integrated plan. For example, says Rindo, there's safety training, environmental training, apprenticeships, training programs developed at the power train plant, programs developed at the corporate level, the Leadership Institute, values training, product-knowledge training, and all the programs going on at the Learning Center. She's got her work cut out for her.

Stakeholders

This is a word that comes up often in conversations with Harley workers. The corporate values aren't intended solely for interactions between co-workers but are meant to include all the stakeholders of the company. That means workers, managers, unions, the community, the environment, the government and, most importantly, the extended family of Harley-Davidson: its zealous customers. Employees are encouraged to attend some of the rallies the company sponsors (major annual events include gatherings in Sturgis, SD, and Daytona, FL) to talk to customers and find out what they want. And like its employees, Harley-Davidson's customers aren't shy about expressing their opinions.

> The corporate values aren't intended solely for interactions between co-workers but are meant to include all the stakeholders of the company.

That may help account for the surprising lack of complacency at Harley. Nobody is smug about that 18-month waiting list for a new bike; everyone we talked to agonized about keeping customers waiting that long. And no one seems to take the company's present success for granted. Talk around the factory is that if they don't constantly change, constantly improve, constantly keep an eye on the competition, they could all quickly find themselves in the unemployment line. "What's our competition doing?" asks DiJulio rhetorically. "Our competition is doing a lot of things to try and get our market share. Where we have a problem is we can't make enough bikes."

One quick recession, Boyd remarks, and that 18-month list of orders could vanish overnight. When times are tight, worries powder-coat technician Jimmy Kilbourn, the toys are the first thing to go.

Another surprise at Harley is that so little of the talk you hear about the corporate values is self-congratulatory. Rather, everybody wants to be the first to admit that they aren't fully living up to the ideals. Does every decision achieve a perfect balance of all stakeholder interests? No. Does everyone tell the truth every time? No. Are they always respectful of individuals? No. Do they always keep promises? Boyd comments that one manager promised a monthly party as an incentive for more productivity. Boyd hasn't seen any parties yet.

Engineer Kapugia produces a notebook and displays a list of concerns that employees have raised with him. It's a pretty long list, but he doesn't want to forget even one of these issues. "All you've got to do is forget somebody," says Kapugia, "and before you know it, that somebody is going to say to somebody else, 'Hey, don't bother talking to him. He never gets back to you.'"

Bob Filipczak is staff editor of Training *Magazine.*

Xerox 2000: From Survival to Opportunity

Richard J. Leo

Xerox had a real wake-up call in the early 1980s and had to respond to the Japanese challenge by taking the quality-management approach. The company's management now understands that the initiatives that the company took to turn itself around are no longer adequate. The company has developed a model for efficiently meeting customer needs into the next century. This article explains this model and how it is being implemented throughout the company. The Xerox approach, based on a careful review of how to effectively deploy quality management principles and practices throughout the company, can be useful to other companies also engaged in transforming themselves.

For all the changes that had taken place at Xerox Corporation in the past decade, one thing remained sacred: Leadership Through Quality, the total quality strategy launched by the company in 1983. It's credited with saving Xerox from extinction. It changed the Xerox culture by giving Xerox employees a working knowledge of total quality and the tools and techniques to achieve it.

Using this ambitious quality strategy, Xerox not only survived, it prospered. In the past decade, the company has regained market share it had lost to the Japanese and received more than 20 awards worldwide for its quality achievements. More recently, Xerox's share price rose from a low of $29 in 1990 to a high of $145 in 1995.

Leadership Through Quality is, in a sense, the bible of the Xerox quality movement. Nothing, however, stays the same in a complex, highly competitive market, not even the company bible, whose principles have been embraced company-wide. Observers might have been surprised when Xerox announced plans to update Leadership Through Quality, but they shouldn't have been. Even quality strategies that have worked well must be improved to meet current and future business needs.

That was the message Paul Allaire, Xerox chairman and corporate executive officer (CEO), gave senior managers in February 1994 as he outlined the

---

Reprinted with permission of the author from *Quality Progress*, March 1996. Copyright © 1996 by Xerox Corporation. All rights reserved.

strategy for Xerox 2000, an initiative designed to assure Xerox continued success into the next century and beyond. The heart of Xerox 2000 is the new Leadership Through Quality.

"The story of how Xerox turned quality around in the face of some extremely tough competition has really become one of the major success stories in the annals of business history," Allaire said. "But we know there's a vast difference between feeling proud about this and feeling smug about it. The seeds of decline lie in complacency; if we're not continuously improving, we're falling behind."

According to Allaire, the competition isn't standing still, and Xerox customers expect continuous improvement in products and services. "We are facing a crisis of opportunity. On one hand, we see attractive markets and have superior technology. On the other hand, we won't be able to take advantage of this situation unless we can overcome cumbersome, functionally driven bureaucracy and use our quality process to become more productive."

Company Self-Analysis Leads to Change

Allaire had based his conclusions on a thorough organizational self-analysis conducted by the Xerox corporate office the previous year. The self-analysis assessed what the company had learned from the process of applying for numerous quality awards; evaluated Xerox successes in market share, growth, and return on assets; and examined the quality tools Xerox was using, including QuickJIT and Xerox's policy deployment process called Managing for Results.

The corporate office found that, over the years, some things had not changed. The Xerox quality policy, which outlines the company's customer-first philosophy, was still valid. Other initiatives had also proven valuable, including customer focus, managing for results, problem solving, benchmarking, employee and supplier involvement, and management by fact (the company's use of statistical and management tools to evaluate business results, make decisions, and introduce process improvements).

Despite its progress, Xerox faced several challenges. Within the company, senior managers and quality experts saw shortcomings in the company's quality approach and were struggling with ways to overcome them. While senior managers saw the power of quality practices and knew they could use them to achieve Xerox's business goals, the company suffered from concept clutter. Xerox had a large number of quality tools and processes, but its people weren't using them consistently or effectively. Quality experts wanted to help senior management use Xerox quality improvement to achieve business results, but they couldn't find a way to integrate the array of practices and tools into the business management process that senior managers were using.

The company's self-analysis revealed more specific challenges as well. In addition to improving process discipline and the use of quality tools, Xerox needed to increase clock speed to get ahead of the competition. Clock speed referred to more than just turning out products; the company needed to be faster at making decisions, implementing them, and measuring their impact.

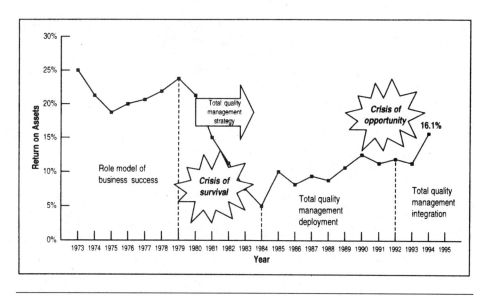

Figure 1. **Moving from survival to opportunity**

Allaire has been quick to point out that Leadership Through Quality is still sound. In fact, Allaire and the Xerox corporate office have used the term "reaffirmation strategy" when presenting the new approach to Leadership Through Quality, emphasizing that Xerox stands firmly behind the basic plan it implemented more than a decade ago. But Xerox had to make some adjustments for the future that would broaden Leadership Through Quality from a strategic approach to a methodology that would integrate total quality management (TQM) into business planning and daily operations.

Xerox needed to make two fundamental changes that would affect Leadership Through Quality; it needed to integrate quality into all aspects of business operations and use quality to deliver improved business results. In particular, Xerox needed to focus on two critical objectives: profitable revenue growth and world-class productivity.

Linking quality to productivity improvement was not a part of the initial strategy. When Xerox leaders first developed Leadership Through Quality, they purposely did not tie the program to productivity measures but to the pursuit of quality through teamwork and process improvement. They reasoned that team-based quality improvements would inevitably lead to better business results, which proved to be true.

Once Xerox began to focus on quality in the 1980s, there was a corresponding improvement in business results as a response to the crisis of survival faced by Xerox. Allaire cautioned, however, that gradual improvements won't be enough to remain competitive as the year 2000 approaches. He said it's time for quantum improvements; Xerox is now facing a crisis of opportunity (see Figure 1). Xerox needs to set aggressive productivity and revenue

Xerox needed to focus on two critical objectives: profitable revenue growth and world-class productivity.

goals and use its quality process to achieve them. Updating Leadership Through Quality was the key to achieving these business goals.

The changes are subtle but important. By integrating business results objectives with the self-assessment of Xerox TQM practices and the Managing for Results process, the Xerox 2000. Leadership Through Quality strategy doesn't just tell Xerox employees what they should try to accomplish, it shows them how to do it.

Management Model Illustrates What Is Needed to Run the Company

The new methodology was designed to build on the original foundation of the Xerox quality movement and consolidate everything Xerox has learned since. Considering the large number of quality initiatives Xerox has used to improve its worldwide operations, this was not an easy task.

Xerox uses more than 60 specific process improvement initiatives, employee involvement programs, and quality tools to manage the various parts of the business. It also uses a range of business policies and approaches, quality intensification efforts, and benchmarking. These elements, plus input from Xerox employees and leaders, have been integrated into Xerox 2000. But Xerox had to do more than collect all of the lessons learned in one place. It had to organize them so they could be easily understood and used by people at all levels of the company.

Xerox's management clarified the company's vision, focused on essential objectives, and organized and deployed its resources accordingly. But how does a company communicate such a far-reaching, seemingly complex methodology to all of its employees? How would Xerox, which spent four years training 100,000 employees worldwide in the principles of Leadership Through Quality, signal the new direction?

It began simply with a diagram. Xerox created the Xerox Management Model to illustrate the concepts of Xerox 2000 Leadership Through Quality. The model shows in detail how Xerox has restructured its approach to quality. Simply put, it shows all of the quality practices and tools Xerox uses and describes how they will be used in various parts of the business. With this logical framework, Xerox employees are able to communicate and implement quality improvement more effectively on a daily basis.

> Experience has taught that, other than major advances in technology, there is no better way to improve productivity than to select the right tool to accomplish a job.

Experience has taught that, other than major advances in technology, there is no better way to improve productivity than to select the right tool to accomplish a job. Employees can't produce the desired results with a tool if they don't know what it's for or where to find it. Picture a toolbox without separate compartments: Hammers, screwdrivers, wrenches, and pliers are tossed together with no rhyme or reason, and the nails, screws, nuts, and bolts are rattling around at the bottom. Most have experienced the frustration of trying to get a job done quickly when the toolbox is a mess.

The Xerox Management Model is a toolbox for all Xerox employees, and it holds everything needed to run the company. It's been described as a holistic

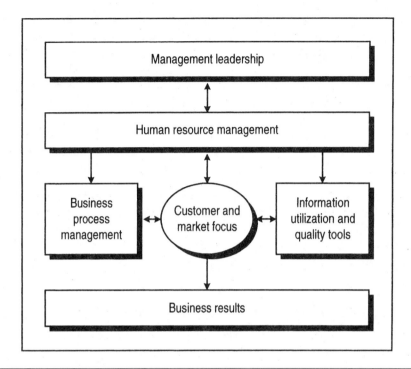

Figure 2. **Xerox management model**

management model because it addresses every aspect of Xerox work: planning, creating, leading, managing, changing, organizing, communicating, learning, and rewarding. Even with so much inside, it's not a grab bag; all of these tools have been neatly placed into separate, but interrelated, business process compartments.

The model shows six groups of basic business practices within Xerox (see Figure 2). All of the quality initiatives, statistical tools, and management practices used at Xerox fall into one of these categories. Each category has a desired state or a vision. The five interrelated groups of practices work together to deliver the sixth category: business results. The categories and their desired states are:

- Management leadership. Xerox management displays a customer focus, exhibits role model behavior, establishes clear long-term goals and annual objectives, establishes strategic boundaries, and provides an empowered environment to achieve world-class productivity and business results.
- Human resource management. Xerox management leads, motivates, develops, and empowers people to realize their full potential. All employees are personally responsible for continuously learning and acquiring competencies required to achieve business

objectives and to continuously improve productivity for customers and Xerox.
- Business process management. Business processes are designed to be customer-driven, cross-functional, and value-based. They create knowledge, eliminate waste, and abandon unproductive work, yielding world-class productivity and higher perceived service levels for customers.
- Information utilization and quality tools. Fact-based management is led by line management. It is achieved through accurate and timely information and by the disciplined application and widespread use of quality tools.
- Customer and market focus. Current, past, and potential customers define the business. Xerox recognizes and creates markets by identifying patterns of customer requirements. By anticipating and fully satisfying those requirements through the creation of customer value, Xerox achieves its business results.
- Business results. Business results are determined by how well Xerox performs in the first five categories.

Establishing objectives is one thing, but there has to be something more specific to back up the good intentions. All of the categories contain practices that define the categories in more detail and deliver the actual business results. Currently, there are more than 35 measurements and practices (see Figure 3) within the model that deliver seven business results. For example, business process management has the following six practices:

- Business process ownership and documentation
- Process breakthrough
- Continuous improvement
- Process measures
- Management system
- Technological excellence

Each practice, in turn, has its own desired state along with processes and core measures to gauge progress against the goal. For example, the desired state for process breakthrough encourages innovative use of business processes to attain business results. Xerox teams and organizations use process reengineering techniques to create business processes that radically change the way employees work and achieve breakthrough gains in productivity.

Processes and core measures for process breakthroughs include a reference to the strategy contract, which is an agreement between the corporate office and other Xerox organizations as to the key deliverables that will be achieved in the next three to five years. In-process measures and the value of final outputs show breakthrough improvement in terms of productivity and customer value. Investments are made for process reengineering as part of the strategy contract. Two- to tenfold improvements are achieved in targeted processes.

- Vision and strategic direction
- Managing for results
- Behaviors and quality values
- Fact-based management
- Empowerment
- Communication
- External responsibility
- Selection and recruitment
- People development
- Management development
- Reward and recognition
- Employee involvement
- Work environment
- Valuing diversity
- Business process ownership and documentation
- Process breakthrough
- Continuous improvement
- Process measures
- Management system
- Technological excellence
- Customer first
- Customer requirements
- Customer data base
- Market segments
- Customer communications
- Customer query and complaint management
- Customer satisfaction and loyalty
- Customer relationship management
- Customer commitment
- Benchmarking
- General quality tools
- Statistical and management tools
- Specialized and advanced tools
- Information management
- Quality and productivity network

Business results
- Customer satisfaction
- Employee motivation and satisfaction
- Market share
- Return on assets
- Productivity
- Profitable revenue growth
- Balance sheet and cash flow strength

Figure 3. **Nonfinancial measurements for continuous improvements**

Xerox 2000 Leadership Through Quality does more than tell Xerox employees about the company's business goals, it gives them the tools to pursue these goals and track their progress. Business priorities are defined each year by top management and translated into specific business activities for each level of the corporation down to the individual employee through the company's Managing for Results process.

Xerox Achieves Goals with Its Managing for Results Model

Managing for Results is a Xerox management model process and a major contributor to the success of the Xerox 2000 Leadership Through Quality strategy. By providing a structure for business plans, self-evaluation, and corrections, it makes the difference between merely setting ambitious goals and actually achieving them.

The Managing for Results process has three basic steps: setting direction, deploying direction, and managing direction (see Figure 4). Setting direction begins at the corporate level with the CEO. The fundamental Xerox values and priorities influence the three- to five-year goals and strategies that the corporate office establishes. The corporate office then translates these long-term goals into corporate annual objectives. At the business division level, direction-setting activities include developing strategy contracts and annual plans. Throughout the direction-setting phase, the corporate office and the division level use two-way communication. They discuss goals to ensure that they are in agreement with corporate strategy, determine process capability, identify required enablers, and establish in-process measures for evaluating progress.

During the direction deployment phase, the goals are communicated to the remaining levels of the corporation. The division objectives are translated by every level of management into individual objectives at the process level. When the deployment phase is complete, all employees understand how their individual objectives are tied to their manager's objectives, the division objectives, and the corporate objectives. Two way communication at every level ensures that the objectives are achievable.

In the direction management phase, every manager evaluates plans and processes established in the deployment phase and implements them, continually checking progress and comparing outcomes to the goals. During this phase, managers and their teams use business assessment, management-by-fact information, and statistical tools to assess their performance and analyze reasons for shortfalls.

As part of Managing for Results, the business-assessment process serves as the report card for progress at different levels of the corporation. All management levels perform business assessment to identify gaps between the current state and desired state for each of the Xerox management model practices. These gaps determine the annual objectives established for the following year, and action plans are set to close the gaps.

> All management levels perform business assessment to identify gaps between the current state and desired state for each of the Xerox management model practices.

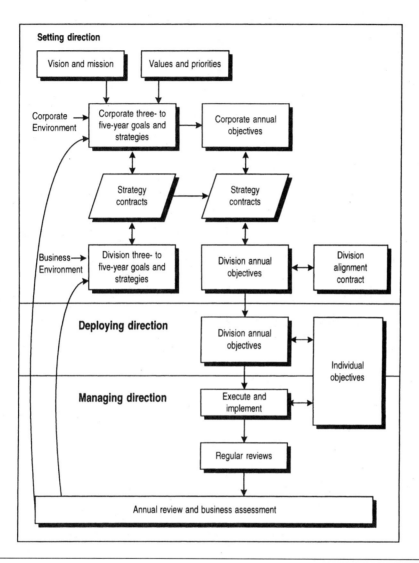

Figure 4. **Managing for results**

Groups that perform business assessment undergo both a self-examination and a validation conducted by individuals outside the group. For the self-assessment, groups use a storyboard to chart progress for each of the 35 practices and seven results objectives (see Figure 4). The storyboards sum up detailed information at a glance by targeting five key areas: performance, strengths, improvement areas, contributing performance factors, and actions. Each action is defined through the storyboard headings of what, who, how, when, and status. This ensures that accountability is defined for each action and progress is measured (see Figure 5).

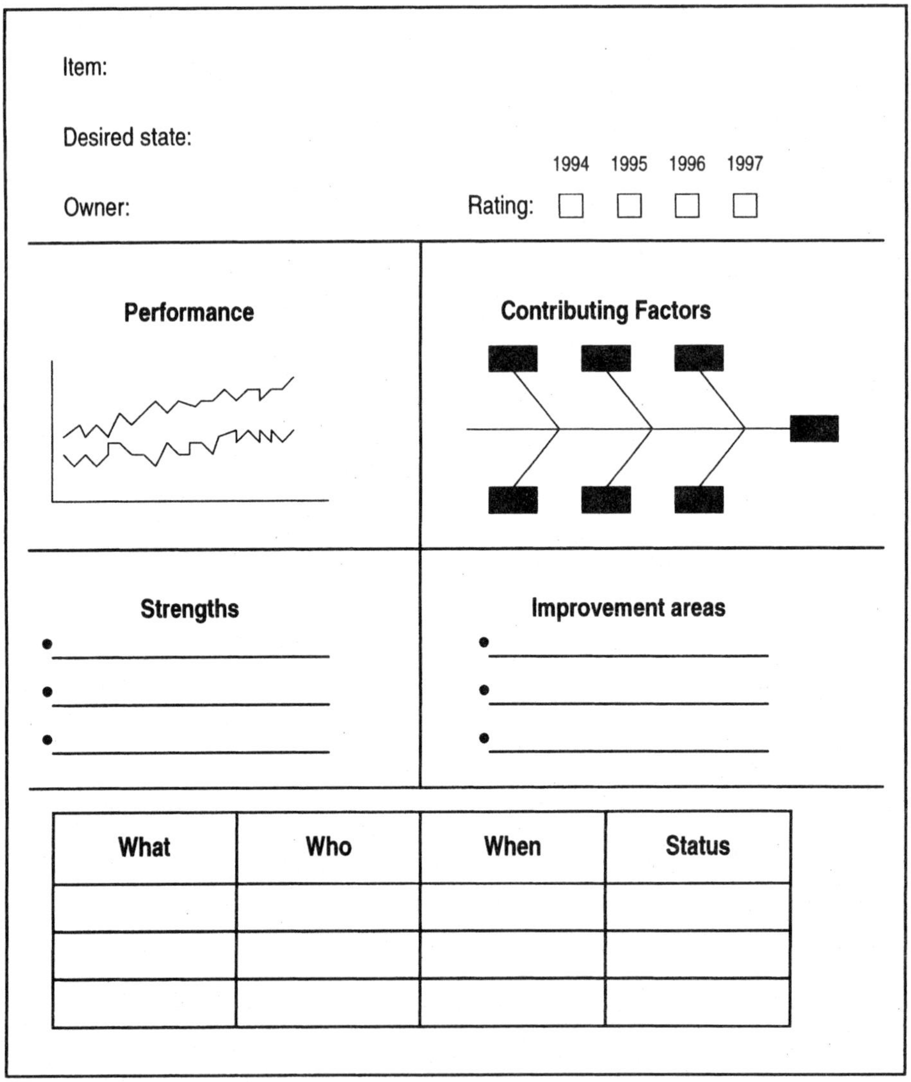

Figure 5. **Self-assessment storyboard**

After the self-assessment, managers and others involved in the business assessment evaluate all of the Xerox management model practices and results on a seven-point scale. They then determine the importance of each practice for achieving objectives and select the vital few practices most in need of improvement at the process level. They define the performance gaps and analyze their root causes using management-by-fact information and statistical tools.

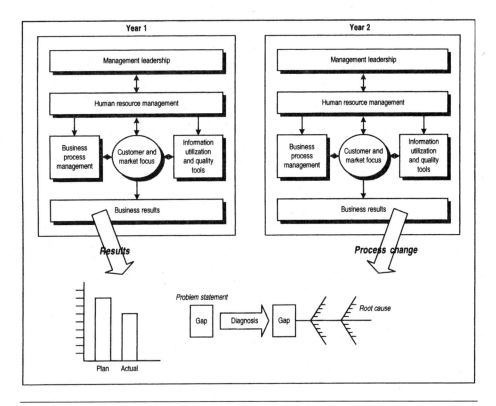

Figure 6. **Management by fact**

In addition to performing an assessment of the 35 practices, Xerox groups use management by fact to diagnose annual business results, analyze the root causes for shortfalls in performance, and develop and prioritize countermeasures. They then use the Managing for Results process to deploy the corrective actions needed to close the gaps (see Figure 6).

An optional but frequently used part of the Xerox quality method is business assessment certification. It serves as a second opinion to a group's self-assessment. During certification, a group that has completed its self-assessment invites a validation team to inspect its findings. The team members could be senior managers from other Xerox divisions or individuals from outside the company. The certification team reviews the self-assessment results and might request evidence to support self-assessment ratings and storyboard data. It then provides feedback and suggestions to the group (see Figure 7).

Looking Into the Company's Mirror

The structure of the management model mirrors Xerox's company structure. Management leadership is at the top of the diagram, showing that management commitment and leadership guide all other activities. The arrow leading

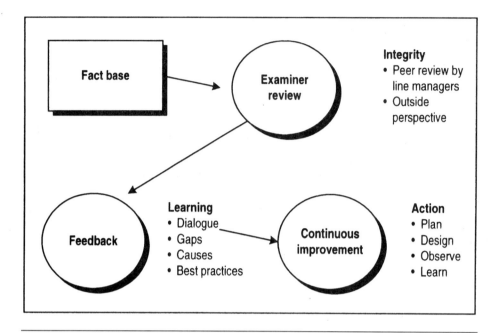

Figure 7. **Business assessment certification**

to and from management leadership emphasizes the importance of open and honest communication with—and feedback from—employees.

The remaining components are arranged around customer and market focus: the focal point of the model. Allaire explained, "Everything we do begins and ends with the customer. Customers define our business and are the center of our universe."

The management leadership section is complemented by the human resource management, business process management, and information utilization and quality tools sections. These sections surround and directly affect all Xerox efforts, and business results are the outcome of the relationships of all the components. Arrows between the sections represent Xerox's internal and external customers and the markets that Xerox serves, and some of the arrows are bidirectional to show the two-way communication between Xerox and its customers.

The model offers many benefits to Xerox. It provides a framework that integrates, connects, and simplifies all of the quality practices and tools that Xerox employees use. It facilitates organizational learning and renewal by indicating which business initiatives should be pursued and which should be eliminated. If process improvement can be tied to one of the 35 practices in the model, it will strengthen Xerox performance. Allaire has explained, "If it can't be linked to one of the practices, we probably shouldn't be doing it."

The model also helps to expose the duplication of efforts. If there are several initiatives directed at process improvements for a specific business

> "Everything we do begins and ends with the customer. Customers define our business and are the center of our universe."

practice, it's probably a waste of resources. The model improves communication by giving management and employees a simpler language for talking about continuous quality improvement. It's flexible; depending on business needs, Xerox can add or delete elements of the model in the future. The model is all-inclusive, covering virtually every aspect of the business.

Allaire expects Xerox employees to use the Xerox management model to define processes, individual roles, and responsibilities at regular operations reviews; for coaching and training employees; in management meetings; when introducing new business initiatives; for identifying internal benchmarks and best practices; and as the foundation for business assessments and certification.

The Future of Xerox's Management Model

As with the original Leadership Through Quality initiative, Xerox will implement this updated quality strategy gradually with the ultimate goal of reaching every Xerox employee.

Deployment began in 1994. Senior managers already have used the model for their business assessment and have selected a set of vital few practices for process improvements. Currently, managers are being trained in the use of the model, as they will be the agents of change in launching this new strategy. They will communicate information about the model to their employees and provide leadership and resources for Xerox employees to implement the methodology and conduct their own assessments. Frontline managers will coach empowered teams to make process improvements in daily work practices so that business results are produced. Employees will manage their own work processes; they are to use the model as a gauge of how well they perform their jobs. Therefore, the model can be used consistently by everyone, from individual employees to top management.

By using the management model in combination with computer technology, customer-driven changes will be introduced on a companywide basis, resulting in global information sharing, best-practice identification, and new process deployment at speeds previously not achievable. It's an ambitious plan, but it has the benefit of a strong foundation. The Leadership Through Quality strategy implemented in the 1980s proved that quality improvement initiatives can produce significant business results.

In the past several years, Xerox has steadily improved in virtually every important measure: customer satisfaction, productivity, market share, revenues, and return on assets. In 1991, the year Xerox reshaped its corporate structure to increase its customer focus and boost productivity, its document processing revenues climbed 2% over the previous year to $13.8 billion. The following year they rose another 6% to $14.7 billion, and the return on assets was 14.7%, the highest rate in more than a decade. In 1994, Xerox achieved even more substantial improvements, increasing document processing revenues by 6% to $15.1 billion. Return on assets rose 3.5 points over 1993 levels to 16.1%, exceeding the company's goal of 15%.

The confidence Xerox has gained from past successes enables it to look to the future with increasingly ambitious business goals and quality objectives. It might seem incongruous that, at a time when many companies are focusing on downsizing, Xerox has chosen the theme, "The time to grow is now." But using the Xerox 2000 Leadership Through Quality strategy as its guide, Xerox is preparing its people to grow the company's business results, productivity, and quality performance.

Richard J. Leo is the vice president and general manager at Xerox Quality Services in Rochester, New York.

Quality Transformation

VOICE OF THE CUSTOMER

Maximizing Consumer Quality Begins With Maximizing the Value of Your Brand

Scott Davis

Successful management means creating a mutually beneficial relationship between a company and the customers it serves. When this happens, customers will be satisfied with their purchases, and organizations will earn the profit that allows them to stay in business and grow. In this original article written for the yearbook, Davis explains how to effectively manage product brands as assets to create and perpetuate long-term customer satisfaction and loyalty and company profitability.

What is a brand? Is it a name, a slogan, an identity, a logo, or an image? Yes, a brand is all of these and much more. It is a level of familiarity, trust, consistency, and quality upon which a consumer can depend use after use and year after year to make purchasing decisions. For a company, a brand is also an asset that requires consistent focus and action to maximize its potential and shareholder value. In truth, a brand is a multifaceted combination of image, product quality, service delivery, and asset valuation.

When thinking about five familiar product categories—soap, gum, tires, cameras, and razors—several brand leaders emerge: Ivory, Wrigley's, Goodyear, Kodak, and Gillette, respectively. As most marketers know, these products have been brand leaders and preferred by consumers since 1926 and before. For seventy years, these companies have all defined and managed their brands across three dimensions: unified name, look, and brand feel; consistent brand performance and quality; and maximization of the brand as an asset.

At a recent conference on new product development for consumer packaged goods companies, over two hundred middle- and senior-management executives were queried: *"How many of you believe that you ultimately control the brand decision consumers make at the retail grocery shelf?"* Surprisingly, less than one dozen managers raised their hands. Oftentimes, companies do not look at their brands as the truest and purest way of helping consumers cut through the product clutter to make confident, quality-based purchase decisions.

Despite the few strong brand-based companies already mentioned, the lack of influence over a customer's brand decision becomes further amplified as competition increases. Competition now comes from traditional category competitors, private labelers, technological advancements, and the overall proliferation of new products. In combination, the highly competitive landscape provides a murky product and service environment that directly challenges the advancement of product quality and often attacks brand loyalty. Unless a better approach is taken, companies are ultimately placing the decision on which to brand to purchase in consumers' hands. This often results in customers' making brand purchase decisions based on the lowest common denominator, price.

Adopting what I call a Brand Asset Management approach can help companies secure the long-term positioning, quality, and service expectations of their brand. Managing the brand as an asset ensures a level of quality that demanding consumers require while choosing among the plethora of brands. This approach also provides a means to better understand where brand dollars should be spent to maximize overall consumer satisfaction and sales.

Company and Consumer Research Show That the State of the Brand Is Uncertain

My company's research has found that even within this increasingly complex "brand world," nearly 75 percent of the consumers polled state that they make brand choices based on several key brand decision drivers. These include the following:

Primary Brand Decision Drivers

1. Quality and reliability
2. Performance experience
3. Familiarity with the brand and its benefits

Secondary Brand Decision Drivers

1. Availability and convenience of accessing the brand
2. Price/value relationship

Other Drivers
1. Advertising
2. Service provided
3. Friend's advice/experience

It's important to note that over 60 percent of consumers truly resist making brand decisions away from their brand of choice. This is often because their choice brands represent and maintain a level of quality and fulfill an anticipated level of product and service satisfaction.

What is disconcerting, however, is that these same consumers can easily state four brand disloyalty triggers that may result in short-term decisions away from the loyal, choice brand. These disloyalty triggers are

1. Peer recommendations to try a different brand
2. New product offerings competing directly with their brand
3. Perceived shift in the brand's price/value relationship
4. Strong advertising from the competition.

Obviously, these triggers highlight the fragile nature of the brand and the lack of control that most companies feel over the world surrounding their brand.

In addition to our consumer research, several studies with companies managing over $100 billion worth of brands across consumer and industrial industries indicate several broad conclusions. In general, these companies did not understand how to effectively manage their brands to maximize consumer satisfaction and overall shareholder wealth. Key results from this study include the following:

Key Brand Asset Management Study Findings
- There is no universally agreed-upon definition of a brand.
- Brand value is defined relative to awareness and recall, not to the potential strategic contributions and benefits the brand can make to a company.
- Only 35 percent of the companies surveyed felt that they conducted enough market-based research prior to setting brand strategy.
- Few companies leverage any type of reward system tied to long-term brand performance, which results in many short-term actions and decisions.
- The primary reasons that there is limited long-term strategic brand focus are management's continued short-term or quarterly focus, profitability requirements, and lack of understanding of the brand's true potential.
- However, 73 percent of companies surveyed elevated the responsibility for the long-term success of the brand to a vice-president level or higher, indicating an increased awareness of the brand's importance.

- Last, the study uncovered what brand managers perceive to be the biggest threats to the long-term success of the brand assets. Overall, companies revealed that the most critical threat to the brand was a lack of internal understanding of what the brand "stands for." In addition, they felt that there was inadequate funding, not enough research, and increasing private-label competition.

Maximizing Consumer Loyalty and Satisfaction and Returns for the Company Calls for Adopting a Brand Asset Management Approach

Mismanagement of a brand jeopardizes its long-term value, consumer satisfaction, and loyalty. Brand Asset Management will help companies regain control of their brands as well as their consumer base. There are three basic tenets behind successfully implementing a Brand Asset Management mindset:

1. Management must be willing to invest to truly understand their brand today in order to make better brand decisions tomorrow.
2. Management must clearly articulate the long-term strategy of their organization to determine what role the brand should play in helping to achieve that strategy.
3. Management needs to be willing to activate and support a long-term, brand-based strategy, regardless of short-term pressures.

If companies adopt and agree on the outlined tenets of a Brand Asset Management approach, they can expect the following benefits:

1. The maximization of consumer satisfaction, support, and loyalty to the brand
2. The protection of the brand vis-à-vis competitive threats and consumer disloyalty triggers
3. The growth of the brand in terms of increased volume, market share, margins, and so on.
4. Improved returns on all investments made in the brand
5. Increased company dedication, support, and loyalty to the brand

Developing a consumer-based Brand Asset Management strategy includes following three steps:

Step 1: Developing a brand vision and brand picture

Step 2: Uncovering brand-building platforms and crafting a five-year brand asset management strategy

Step 3: Activating the five-year brand asset management strategy

Let's look at these steps in more detail.

Step 1: Developing a Brand Picture

Step 1 integrates five critical inputs, all aimed at providing the organization with a clear picture and understanding of where the brand is today vis-à-vis consumers and relative to the competition; a definition of which untapped opportunities the brand can leverage to grow the brand tomorrow; and a vision for the brand that forces management to articulate what the brand is expected to "do" for the organization over the next five years, relative to brand value and brand revenue and profit contributions. The inputs to developing this brand picture include the following.

Developing a Five-Year Brand Vision

Management must be able to clearly state the company's overall five-year vision before the brand discussion can take place. This vision must include financial and strategic objectives. Brand financial and valuation goals can then be set based upon the following inputs: historical growth of the brand, growth of the brand's competitive category, growth of competitive brands, or the expected returns of the brand's new product pipeline.

Building a Consumer Model

Building a consumer model entails clearly defining how the consumer acts within the brand's competitive category. The model also needs to include how purchase decisions are made, all category-driving purchase decision criteria, key decision influencers, and how consumers rate purchase attributes relative to the different category players.

Understanding the Brand Persona

Understanding the brand persona is a critical step to truly understanding how to create an exceptional strategy for the brand. The persona captures the images the brand creates in the consumer's mind, the identity it owns, and the true advantages it possesses over the competition. Each of these criteria needs to be truly understood.

Developing a Brand Contract

Developing a brand contract entails having the brand's consumer define the rules, bylaws, promises, and contract stipulations that are connected to the brand. This, in effect, becomes a company's contract with the consumer. Consequently, when consumers perceive the contract is broken, it may drive them to make purchase decisions away from the brand. The brand contract becomes an even more powerful tool when it includes an understanding of the top two or three competitors' contracts.

Determining Brand Extendability Opportunities

Once a company understands the world in which the brand competes from a consumer and brand perspective, managers can conduct targeted unmet needs, wants, and opportunities research and determine what opportunities exist for the brand. These opportunities identify what the brand currently does not satisfy in addition to what is not being satisfied by top competitors. The easiest way to categorize extendability opportunities is to consider the following five dimensions: new products, communications, pricing and positioning, channels, and internal communications.

Step 2: Crafting a Five-Year Brand Asset Management Strategy

Once a company clearly understands and agrees on the five key inputs to the brand picture, opportunities that strengthen and leverage the brand over the long run become clear. If a company applies rigorous analysis to the brand picture findings outlined in step 1, it will uncover several brand-building platforms. The goal then is to develop brand-based strategies tied to each platform that include clear goals and objectives, expected investments, returns, the timing of each, and a game plan that outlines the executional strategy. The inputs to crafting a five-year brand asset management strategy include the following.

Uncovering Brand-Building Platforms

After completing step 1, several outputs emerge that can serve as the platforms for crafting alternative Brand Asset Management strategies to be tested with consumers. Specific outputs that a company needs to assess and analyze include

- key purchase criteria within the category
- perceptions of the brand relative to competitive brands in the identified purchase criteria
- the brand's contract relative to competitive brands
- the major unmet needs and wants of the category.

Developing Brand-Based Strategies

Once consumers have helped prioritize alternative brand-building strategies, companies can determine the specifics attributes affiliated with each:

Brand-Based New Product Strategy—Define the goals and roles new products should play in support of the brand and identify potential needs-based categories to pursue with new products.

Brand-Based Communications Strategy—Articulate the key positioning, messages, and vehicles required to support the brand over the next five years.

Brand-Based Pricing and Positioning Strategy—Determine the brand's position and craft pricing strategies that align with the brand contract and help capitalize on extendability opportunities.

Brand-Based Channel Strategy—Possibly realign the channel strategy to more accurately reflect consumer desires based upon how consumers buy the brand and what influences them.

Brand-Based Internal Communications Strategy—Because internal communications are the primary source of brand strategy failure, it is important for the company to determine a system for communication that includes how the information should be communicated and how often. The strategy must be created prior to launching the Brand Asset Management strategy.

Crafting the Plan

Once all inputs are agreed upon and the marketplace has confirmed the direction, the company needs to craft a plan that details goals, how the goals will be met, who is responsible, and what investments and returns are expected. In addition, management should establish metrics that help determine the Brand Asset Management plan's success. Management should also establish rewards that recognize the success of participants. All this should be based on a set of management promises or norms that demonstrate the company's long-term vision and commitment to maximizing consumer satisfaction and loyalty.

Step 3: Activating the Five-Year Brand Asset Management Strategy

Although many companies successfully activate several areas within traditional brand management, the changing face of the consumer and vision of a long-term horizon should help companies adopt a new approach for success. Leveraging a Brand Asset Management approach will help companies achieve what should be a primary objective for the brand—maximizing its long-term value, both from a consumer and a dollar return perspective.

Scott Davis is a partner heading up the Products and Brand Asset Management Practice at Kuczmarski & Associates, a new products and services management consulting firm. The strategy explained in this article was developed by his firm. Contact him at (312) 988-1500, 1165 North Clark Street, Chicago, Illinois 60610.

Customer Loyalty: Playing for Keeps

Laura Struebing and Anne Calek

Are you aware that a customer retained for life can be worth a million dollars in business? This person isn't necessarily going to spend a million dollars, but the compounded effect of that retained business in terms of lowered costs and word-of-mouth advertising can be a very large number. This article tells more about this point and provides many specific ideas, using examples from real companies, for how organizations can retain customers, a vital aspect of growth and survival today.

Why are your customers saying they are satisfied when 15% to 40% of them are defecting from your company each year?[1] It may be because customer satisfaction is not enough anymore. Most customers now have high standards for routine transactions. They expect to be treated in a courteous and professional manner, and they expect their needs to be met and problems solved. Since customers *expect* to be satisfied, a satisfied customer is just as likely to defect as one who is dissatisfied. With the array of choices that customers have, companies must go beyond satisfaction to win customers' loyalty.

Completely satisfied customers are much more loyal than satisfied customers; any drop in total satisfaction results in a major drop in loyalty.[2] A. Blanton Godfrey, chairman and corporate executive officer of Juran Institute, Inc., said, "In a Banc One study, people who rated the bank as outstanding were four times less likely to leave the bank than those who ranked it satisfactory, neutral, or unsatisfactory. These people were also five times more likely to buy another financial service than those who ranked it satisfactory, neutral, or unsatisfactory."

Godfrey explained that the concept of customer satisfaction has left many promises unfulfilled. "Most believe customer satisfaction is tied to market profitability and market share. Although many companies have high levels of customer satisfaction, less than 2% of companies see measurable bottom-line improvement due to customer satisfaction alone. Therefore, companies need to move beyond customer satisfaction toward customer loyalty."

Reprinted with permission from *Quality Progress*, February 1996. Copyright © 1996 by the American Society for Quality Control. All rights reserved.

According to Godfrey, having loyal customers—those who spend a high percentage of their allotted money on the products and services that the company offers—pays off. Loyal customers provide higher profits, repeat business, higher market share, and referrals. So, as loyalty goes up, operating costs go down. "For example, if 100% of sales go to a few customers rather than 1% each to thousands of customers, things such as billing costs, distribution costs, and market relationship costs all go down."

Laura Gregg, technical trainer and writer for Wizard Textware, explained that there is a five-to-one net revenue advantage to having loyal customers. If companies increase their customer retention by 2%, this is equivalent to cutting their operating costs by 10%.

Sheila Kessler, principal of Competitive Edge and author of *Total Quality Service*, said loyal customers are critical to a company's profit because they are a continuous source of income, which leads to high levels of profit. For example, if a company multiplies the profit the customer generates for the company each year by the lifespan of the customer and then multiplies that number by every customer that the company gains through word-of-mouth advertising, it can figure out what one lifetime customer is worth. "One lifetime customer of Minute Maid is worth $1 million and one Delta Airlines' lifetime customer generates $1.5 billion."[3]

She added, "Organizations should seek lifetime customers because it costs five to seven times more to find new customers than to retain customers you already have."

According to Godfrey, in 1990, it cost MBNA bank more than $50 to acquire a new credit card customer during the first year and it didn't start making any money until the second year. Even though most of the profit is with the old customers, 98% of MBNA bank's marketing costs were being spent on getting new customers and almost no money was spent on keeping old customers. When MBNA bank started to ask its customers why they left and fixed their problems on the spot, it retained 50% of those who tried to leave and became one of the most profitable banks in the country.[4]

A lifetime customer is also worth all of the sales a company makes from all of the people he or she refers to the company. For example, Kessler said, "A Minute Maid orange juice customer provides $26,000 by word-of-mouth advertising alone."

According to Gregg, companies should strive for a relationship with customers in which their customers are so excited about their product that they are willing to talk positively about it and the company to other people. "These customers are so loyal to your product that they are willing to put their own reputations on the line."

But What Do Customers Want?

According to Tom Connellan, co-author of *Sustaining Knock Your Socks Off Service*, companies need to know their customers and what they want. This, however, might not be obvious because a customer's experience consists of an outcome (what the customer gets) and a process (how the customer gets it). Although customers might initially come to a company for an outcome—what it can do, create, produce, or deliver—the process by which that outcome is delivered is at least as important.

Connellan said two customers can get the same outcome—postage stamps, for example—but the process that each went through might not have been the same. One might have received the stamps from a machine and the other from a clerk at the post office. Some might wonder why they both didn't get their stamps from a conveniently located machine to avoid the hassle of waiting in line at the post office. This is because a machine can't call a customer by name or ask how he or she is doing, which is important to many people. Companies need to be concerned with the whole experience of receiving their product, not just the product itself.

Connellan said that to satisfy and retain customers, companies must pay as much attention to the process as they do to the expected outcome, because a customer might be satisfied with the outcome but not with the process, or vice versa. For example, a person might get his or her driver's license renewed—the outcome that is expected—but the process might be such a hassle that the customer feels dissatisfied. On the other hand, the process could be outstanding—the facility is clean and attractive, the employees are friendly, and the line moves quickly—but the customer might not get the outcome he or she expects—actually receiving a renewed license.

Gregg said, "Both customers who receive what they want but are hassled in the process and customers who enjoy the process but receive results that do not meet their expectations are at risk of leaving the company. Since most companies spend a fair amount of their improvement efforts on the results side, if they start focusing on the process side, they will very quickly move into customer loyalty."

> Companies need to be concerned with the whole experience of receiving their product, not just the product itself.

Are Companies Giving Their Customers What They Want?

To deliver good customer service, companies must find out what factors drive customer loyalty. Godfrey warned that, many times, when companies measure customer satisfaction they measure what someone inside the company thinks the customer wants. "Don't assume what customers want; ask them. Otherwise you will spend a lot of money improving things that are not important to the customer."

According to Kessler, companies need to do research and measure customer retention to understand what their customers value. "If you are not measuring customer retention, you're probably not managing it. Three things companies should measure include customer satisfaction, why customers left your company, and the attrition rate of customers. You can't have customer loyalty without knowing what these measures are."

Customer satisfaction measures include both hard and soft measures. Hard measures are behaviors that are measured by cash register receipts, market share, revenues, and profits and include customer retention levels and

CUSTOMER LOYALTY: PLAYING FOR KEEPS

number of referrals from other customers. Hard measures show what customers are actually buying.

Soft measures center on customer perception. They include surveys, focus groups, interviews, and observations. They help determine what customers think of a service or product relative to the competition. But it is important to remember that what people say and what they do are often different. For example, an individual might be delighted with the taste of the extra raisins and nuts offered in a high-priced, well-advertised cereal but still end up buying the generic brand.[5]

To make sure companies are measuring what they want, measurement tools, such as surveys and focus groups, need to be matched with company objectives. If a company wants qualitative data, such as what questions to ask in surveys or a feel for what is important to customers, it should use a personal approach, such as focus groups and interviews. If it wants to obtain quantitative data or find out what a large number of people think of the company, then written surveys should be used (see the sidebar "Does and Don't of Conducting a Survey" p. 326).

Companies should conduct focus groups or interviews so that the right questions are asked in written surveys, and to understand why customers make purchase decisions and what is needed to delight them. Most people have a hard time articulating their specific needs because they don't spend much time thinking about the nuances of a company's service or product, but a trained interviewer can help customers clearly define the specific needs they want that service or product to meet.[6]

Talking to Satisfied Customers Is the Easy Part

Gregg said, "In addition to talking to satisfied customers, you need to seek out your dissatisfied customers and the people who left. They will provide very valuable information and you might have the opportunity to get them back by being willing to engage in dialogue with them."

She said companies should make it easy for customers to complain because criticism from customers is more valuable than praise. "Complaints are priceless insight into your business. If customers don't tell you that they are upset, they might never come back and you won't know why. If you know what the problem is, you can take care of the problem, take steps to ensure it doesn't happen again, and improve based on their complaints. Therefore, maintain regular communication with your customers so you can quickly resolve their product and service complaints."

"Complaints less insight ir business."

James Rooney, general manager at the Process Safety Institute and vice president of ASQC's section affairs, added that when trying to satisfy customers, a company shouldn't replace a product without also asking questions, such as how the product broke and how it was being used when it broke. "To build loyalty, you need to find out what happened and why it happened so you can use that information to improve the product."

A "no hassles" service guarantee is one way that companies can quickly learn about customers' changing expectations, so they can adapt their processes to satisfy or even delight customers more often. A no hassles guarantee

DOS AND DON'TS OF CONDUCTING A SURVEY

For survey success, use the following tips:

- Do pilot your survey with representative customers to make sure that you have included the key questions and that the questions are clearly and neutrally worded.

- Do find ways to motivate people to complete a mailed survey, such as including an incentive for completing it or telling customers how you will use the information to enhance your service to them.

- Do ask for qualifying information so you can determine whether the respondents are frequent customers or those who have never used your services.

- Do ask questions about your competition and include competitors' customers in your sample.

- Do leave room for comments.

- Do leave space for customers to write their names and phone numbers, but make it optional.

- Do ask customers to rank services according to importance and performance so you can determine where your resources should be allocated.

- Do conduct surveys to measure both what customers want and their attitudes about what they think they are getting.

- Don't guess what your customers want. They will be willing to tell you what is important to them, especially if you make it easy for them to do so. For example, create a short questionnaire that customers can complete when they pay.

- Don't overuse or abuse surveys by frequently sending out a superficial set of written questions to a sampling of customers.

- Don't only ask questions that you think are important; find out what the customers think are important through focus groups and interviews.

- Don't force customers to answer your survey. The whole idea behind asking them questions is to create a place where people are going to be happy to do business. If you pester them, they're not going to be happy.

is better than a "no questions asked" guarantee; no questions asked means that the company has no opportunity to ask what went wrong so it can learn from the response.[7]

Solve Customers Problems with Flair

Respect for customers is especially important when trying to solve their problems. Companies should strive to make amends with targeted customers when, inevitably, something goes wrong. Kessler said, "Since one lifetime customer could be worth $1 million, companies should try to decide in the customer's, not the company's, favor."

Gregg said that while the customer is not always right, the customer must always be made to feel that he or she is right or at least feel that the company is concerned about his or her welfare and about the resolution of the problem. "The whole issue focuses on care and concern. Sometimes giving a customer a chance to vent or just the fact that you asked what he or she wants is all it takes to build loyalty. But remember, just because you ask your customers what their wants, needs, and expectations are does not mean you have to meet them all."

According to Carl Sewell's book *Customers for Life*, when something goes wrong, a company should acknowledge its error, fix it immediately, and apologize. If a customer says there is a problem, 99% of the time there really is; companies shouldn't let the other 1% be a reason to mistreat everyone else. The problem should be fixed while the customer is still there, and, having decided to give the customer what he or she wants, the company should give in completely. Haggling over the amount will cost the company all the goodwill it was trying to gain.[8]

According to Connellan, a customer will be more loyal and more likely to pass on favorable comments about the company to others if he or she receives an elegant solution to a problem than one who never had a problem at all. When a mistake happens, the company should recover with such flair that the customer remembers not the error but the treatment after the mistake. This is what will transform an angry customer into a loyal one.

A Technical Assistance Research Program study found that 63% of all dissatisfied customers will never do business with an offending company again. But if a company resolves the customers' problems, 90% of those dissatisfied customers will remain loyal to the supplier.[9]

In fact, a national survey of 1,179 department store customers by Burke Customer Satisfaction Associates, a large customer satisfaction research company, found that the customer who complains is most likely to be one of the store's best and most satisfied customers. The survey found that this is because the complainers are those who care enough to seek remedy and expect that their problems will be resolved to their satisfaction.[10]

Carey Watson, senior vice president of marketing for Burdiness department store in Miami, FL, regards complaints as a positive thing, especially from frequent shoppers, who are traditionally a store's strongest supporters.

"When customers are loyal to your store, they care about you. So when you disappoint them, it's kind of a shock. When they complain, they are saying, 'Hey, it's unusual but you've let me down and I just wanted to let you know.'"

Surprise and Delight Customers in Moments of Truth

Connellan said that since as many as 40% of customers are at risk of going to a competitor, now more than ever, everyone in the company needs to be committed to amazing and delighting their customers with unexpected quality.

According to Jan Carlzon, CEO of Scandinavian Airlines System, hundreds of small "moments of truth"—all of the times that a customer comes into contact with a company, including billing, advertising, telephone calls, reservations, complaints, and using the service or product—make up the customer's perception of that company. In fact, customers can generalize about an entire organization based on one moment of truth.[11]

Connellan said that to delight customers, companies must have preprogrammed responses for the five or six things that might go wrong to ruin a customer's moment of truth with the company. "For example, Walt Disney World has a vice president in charge of its parking lots, who tries to eliminate any parking-related issues before they become a problem.

"Why is Disney World so concerned about its parking lots? Because the parking lot is the customer's first and last moment of truth when attending the Disney theme park. If something goes wrong here, such as the customer not being able to find his or her car in a 75,000-car lot, the customer's entire Disney theme park experience might be ruined."

Kessler said that with 11,000 magazines, 2,000 types of beer, and 150,000 types of pain relievers on the market, customers have a lot of choices. Therefore, companies need to go above and beyond to stand out. They need to undersell and overdeliver. For example, the attrition rate for cellular phone users is 43%. Cellular phone companies are using a system that overpromises and underdelivers, which encourages people to switch: When their customers' bills are too high, they quickly leave and look for a better deal.

She said data base marketing is one way to delight customers. Some companies are now using a data base to note customers' personal preferences, shopping habits, names, and addresses so they can treat their customers as individuals and speak to them on a personal level. They can pinpoint offers for them and demonstrate sensitivity about individual customers' wants and desires. This is important because generally the more personal the service, the more loyal the customer is.

For example, said Kessler, the Ritz-Carlton has a frequent-guest system. When a guest requests something special at one hotel location, the next day any other Ritz-Carlton hotel can call up that guest's record and note that special request. "Once when I stayed at a Ritz-Carlton, the hotel's maid recorded that I brought my own satin pillow cases when I stayed with them, and the next time I stayed at a Ritz-Carlton, to my surprise and delight, the hotel had satin pillow cases on my pillows."

While small improvements to delight the customer can set a company apart from the rest, companies still need to be sure that what they are doing adds value for the customer. For example, sweeping the parking lot every hour will probably guarantee that a company has the cleanest parking lot around, but customers likely won't do business with the company because of it.

Another way to delight customers is through mass customization. Today, small shops are not the only businesses offering customized services. Increasingly, large companies are now customizing service to their customers. According to Kessler, this mass customization is becoming so popular that people are now expecting and demanding it. Customers are saying of products and services, "I want it exactly when I want it, how I want it."

> This mass customization is becoming so popular that people are now expecting and demanding it.

Kessler cited many examples of companies that now offer mass customization. Levi Strauss gives customers the option of paying $10 more for jeans that are custom-fit and sent directly to their homes. American Airlines expedites meal service on flights for those who want their meals sooner. Companies such as Federal Express and United Parcel Service put signatures on file so customers don't have to be at home when their packages are delivered.

Give Loyal Customers the Respect They Deserve

Loyal customers need and deserve acknowledgment, and it is critical for companies to show respect for these customers. People like individual attention and like to be remembered, so companies need to go out of their way to provide extra help or attention to their customers. For example, according to Kessler, Delta Airlines employees will tell frequent customers, "I notice that you fly with us a lot." This small effort shows respect and individual attention.

She said that since a company's best customers generate higher profits, to develop customer loyalty, they should be rewarded and given the best value so they can also enjoy the benefits of the value they create for the company. Even though loyal customers are more willing to pay a higher price for a product, they shouldn't have to make up for the high cost of acquiring new customers. Instead, loyal customers should be pampered, get price breaks, and get something in return for their continued loyalty, because they have earned it.

Another way to reward loyal customers is to find ways to help make their lives easier and more enjoyable. For example, some companies get involved in customers' lifestyles, organize events around their interests, and help them meet each other through clubs and newsletters. Saturn is one company that gets involved with its customers. According to Kessler, it provides its customers with car maintenance clinics, car clubs, a newsletter, and a toll-free phone number. It also sponsored a Saturn Homecoming in June 1994, which was attended by 42,000 customers.[12]

Get Customers Involved

"Since 80% of successful new products and services come from customer ideas, you should ask customers what new products and services they would like," Gregg said.

For example, said Godfrey, when Ford built its 1994 Mustang, it invited 200 loyal customers, which included members and presidents of Mustang clubs, to be a part of the design team. These customers provided critiques and new design ideas through fax and e-mail. "Basically they told the design team what made a Mustang a Mustang."

He said these customers were not only reviewing the product but also test driving Cameros and Firebirds so they could tell Ford what they liked and disliked about the competitors' cars. Ford also asked drivers of these competitors' cars for their opinions.

According to Godfrey, the input from Ford's customers was so important that this group overruled the design team twice. The first design included from wheel drive and a four-cylinder engine. "The 200 members went directly to their Mustang clubs and collected 30,000 letters written to Ford saying that Ford could make a four-cylinder, front-wheel-drive sports car but they were not allowed to use the word Mustang. According to these customers, a Mustang is a V-8 with rear-wheel drive."

Originally, there was not going to be a convertible in the line, said Godfrey. But the customers were saying. "The coupe is probably what I will end up with, but there'd better be a convertible in the showroom or I won't be in the showroom; Mustangs have convertibles."

With help from these customers, the 1994 Mustang was brought to market in 25% less time with 30% fewer dollars spent than any comparable development program in Ford's recent history. The entire year's production was sold out by March 1994.[13]

Frontline Employees Are the Company

Bob Hayes, director of research at Medical Consultants and author of *Measuring Customer Satisfaction*, said, "Frontline employee attitudes are related to customer satisfaction. If employees are not happy, this will be reflected to the customers."

If a cashier is rude or a salesperson isn't helpful, the customer will think negatively about the entire company. Aside from price and product issues, as far as customers are concerned, frontline employees and the processes they use *are* the company. Therefore, how they interact with customers has a direct impact on customer expectations. If a customer has a bad experience securing a mortgage, returning damaged goods, or buying a product, he or she will probably have a bad impression of the company.[14]

"Companies need to recognize that frontline employees shape and create the customer's impression of them as a company and can make the difference in building or undermining customer relationships," Gregg said. "Many customers can be lost after only one purchase because they had a bad experience with an untrained or uninterested employee."

To extend the best service to customers, the people on the front line must have the authority, resources, and training to resolve problems and satisfy customers. Many well-meaning company policies can rob employees of the ability to do what might be in the best interest of both the customer and the employee. For example, by not allowing an employee to refund a customer's money on a small-ticket item, a company might lose a customer who spends hundreds or thousands of dollars with the company annually.

According to Gregg, companies need to train, empower, and then reward employees for frontline problem solving. "Federal Express empowers employees to make decisions and solve problems right away as they see fit. They are allowed to take the appropriate risks to keep the customer's business and they are not penalized for the occasional error because Federal Express believes that well-intentioned efforts are just as important as successes."

In fact, Abt Associates' studies of 50,000 employees have found that the degree to which an organization's frontline employees believe they have the capability, tools, and organizational support to provide for customer needs mirrors the degree of customer satisfaction. The more empowered the employees, the higher the level of customer satisfaction.[15]

This research also showed that employees' job satisfaction is directly related to customer satisfaction. Job satisfaction stems from employees' assessment of how much the company values them and treats them with respect and dignity. When managers produce high levels of job satisfaction, employees in turn produce high levels of customer satisfaction. An effective organization has capable, satisfied, and dedicated employees.[16]

In the United States, there is 30% turnover in all frontline jobs.[17] Companies can help avoid this by hiring the kinds of employees who have the potential to learn and acquire the skills and competencies that customers require, and who are capable of delivering the behaviors that customers want.

Kessler said, "Companies must manage retention, because high employee retention equals high retention of customers. One way companies can manage retention is to pay a lot of attention to who they are hiring."

What are the criteria for hiring frontline employees? Gregg said, "As they say in retail sales, success is location, location, location; with frontline employees it is personality, personality, personality. While you can teach just about anyone what they need to know about the products and services and the procedures for dealing with customers, you cannot develop really solid interpersonal skills, which are needed for directly dealing with a customer."

Kessler said, "Another reason to hire the right people is that if attrition is high, the company might not have knowledgeable employees because the new people will always be learning about the company's products. Customers want knowledgeable employees to deal with, but it is hard to have this if there is constant turnover."

It is annoying for a customer who expects a company's employees to be knowledgeable to have someone say, "Sorry, I'm new here. I don't know." Even if the customer sympathizes with the employee, he or she still isn't receiving the level of service that was expected.

Another way to reduce attrition is to pay employees more generously. The extra money offered will allow companies to hire more talented people who

When managers produce high levels of job satisfaction, employees in turn produce high levels of customer satisfaction.

will generally make fewer mistakes, need less supervision, and be more willing to do whatever it takes to make the customer happy.

Studies have found that outstanding organizations tend to pay frontline service people above-average wages for their respective industries. In most organizations in which service is mediocre, employees tend to be paid at the low end of the scale and their pay is not based on how well they perform.[18]

According to a survey by the Society of Incentive Travel Executives (SITE) Foundation, while cash is a popular incentive, so are training, stock options, trips, recognition at company meetings, and merchandise. All of these carry significant recognition value.[19]

Overall, people who believe they are doing a good job and providing a valuable service to customers tend to enjoy what they are doing, are more productive, stay with the organization longer, have lower rates of absenteeism, and have higher satisfaction with themselves, their colleagues, and their employer. Therefore, organizations that have satisfied people serving their customers typically have more satisfied customers.[20]

Gimmicks Alone Won't Produce Customer Loyalty

Frequent buyer programs reward people for doing business with a company and show them that they are valued customers. Frequent flyer miles, punch cards (e.g., buy nine cups of coffee and the 10th is free), or reward certificates are as valuable as money to customers. Businesses can also partner with other companies to provide purchase incentives to a targeted audience. These alliances can increase sales and encourage brand loyalty, while saving on promotional costs.

Gregg said, "Whatever reward system you choose, it should be easy to use and learn and each purchase should increase the value of the previous one."

Incentive programs should be so appealing that customers will not want to leave because it will cost them too much. For example, Sprint has a new theme, "It pays to stay." The company has now made it profitable for customers to stay with Sprint by adding a 10% rebate payable to loyal customers at the end of the year. Previously, there was no barrier to exit, so customers could take an incentive from AT&T or MCI to switch, and then come back to Sprint later without having lost anything.[21]

Companies must remember, though, that customer loyalty programs are only as good as the entity offering them. Customers will only remain loyal if they are completely satisfied, which is why seemingly loyal airline customers defect when they exhaust their frequent flyer miles. Everything else in terms of the customer experience and basic product and service must be equal before frequent buyer awards from the company will keep customers coming back.

Kessler said many people think of customer service as providing frequent flyer programs and free gidgets and gadgets. Companies cannot, however, give away free items to apologize for not getting the basics of quality right the first time. Before offering frequent buyer programs, the basics of quality must be right first.

> Everything else in terms of the customer experience and basic product and service must be equal before frequent buyer awards from the company will keep customers coming back.

Only when a company has incorporated quality into the basic elements of its products or services, has a recovery process for counteracting bad experiences, and excels in meeting customers' rising expectations can it move from customer satisfaction to the level of customer loyalty. Today's customers have a lot of choices; they expect to be satisfied, and companies must continually find ways to give them something extra to win their loyalty and keep them as lifetime customers.

Notes

1. Michael W. Lowenstein, *Customer Retention: An Integrated Process for Keeping Your Best Customers* (Milwaukee, Wis.: ASQC Quality Press, 1995).
2. Thomas O. Jones and W. Earl Sasser Jr., "Why Satisfied Customers Defect," *Harvard Business Review*, November–December 1995, 88–99.
3. For more information on calculating how much a loyal customer is worth, see Wilton Woods, "After All You've Done for Your Customers, Why Are They Still Not Happy?" *Fortune* 11 December 1995, 182.
4. For more information, see Frederick F. Reichheld and W. Earl Sasser, "Zero Defections, Quality Comes to Services," *Harvard Business Review*, September–October 1990.
5. Sheila Kessler, *Total Quality Service* (Milwaukee, Wis.: ASQC Quality Press, 1995).
6. Ibid.
7. James Bredin, "Keeping Customers," *Industry Week*, October 2, 1995, 62.
8. Carl Sewell and Paul B. Brown, *Customers for Life: How to Turn That One-Time Buyer into a Lifetime Customer* (New York: Doubleday, 1990).
9. Kessler, *Total Quality Service*, 74.
10. For more information on this survey, contact Burke Customer Satisfaction Associates, 805 Central Ave., Cincinnati, OH 45202, (513) 684-7659, fax (513) 684-7717.
11. Jan Carlzon, *Moments of Truth* (Cambridge, Mass.: Ballinger, 1987).
12. For more information on Saturn, see David A. Aaker, "Building a Brand: The Saturn Story," *California Management Review*, Winter 1994, 114–33, or Karen Bernowski, "To Boldly Go Where So Many Have Gone Before," *Quality Progress*, February 1995, 29–33.
13. "The New Mustang," *The Quality Minutes*, video, Juran Institute and Center for Video Education, vol. 1, no. 1.
14. Jeffrey J. Zomitsky, "Frontline Facts," in *Leveraging Frontline Capability* (New York: The Conference Board, 1995). 13–15.
15. Wendell J. Knox, "Strengthening Relationships for Better Service," in *Leveraging Frontline Capability* (New York: The Conference Board, 1995). 9–11.
16. Ibid.
17. Zomitsky, "Frontline Facts."
18. Thomas K. Connellan and Ron Zemke, *Sustaining Knock Your Socks Off Service* (New York: AMACOM, 1993).
19. Ibid.
20. Ibid.
21. Tom Harvey, "Service Quality: The Culprit and the Cure," *Bank Marketing* 1 June 1995.

Laura Streubing and Anne Calek are editors at Quality Progress *Magazine.*

Plugging In to Your Customers' Needs

William Keenan Jr.

Surveys are an important part of listening to the voice of the customer. This article profiles the use of surveys in three different companies, shows what the companies do with the information they glean, and provides some brief advice on preparing effective customer surveys.

Please, stop with the clichés.

"Our company is customer-focused."

Enough already. Have mercy.

"We're committed to our customer."

All right. Is it money you want? Just don't say . . .

"At our company, the customer always comes first."

Ahhhhh. Give me a straitjacket—42 long!

We've all heard the clichés about customer service so often they've become trite—and agonizing. Heck, shouldn't the notion of being customer focused be standard-operating procedure? A given?

And yet, for some reason too many companies feel compelled to tell the world in their advertising and slogans about their commitment to the customer. Invariably, it is those companies boasting like peacocks who really have a fuzzy idea about their customers' true needs and demands. Their concept of a customer satisfaction survey is measuring how many units they sold this month versus last month—not determining whether those customers will reorder next month.

For every business crying wolf about customer satisfaction, though, there are hundreds of Fortune 500 companies and many small and mid-size companies that have some sort of customer satisfaction initiative underway. Such companies as Dow Chemical, Eastman Chemical, IBM, Ralston Purina, 3M, and Xerox are surveying customers and collecting data on satisfaction levels. The intention? To use that data to identify and resolve obstacles for better customer relations, to formulate sales and marketing strategies, and to retool sales and service skills.

Too often, customer satisfaction initiatives fail, or provide misleading results, because they ask the wrong questions of the wrong people and produce

> Too often, customer satisfaction initiatives fail, or provide misleading results, because they ask the wrong questions of the wrong people and produce data that is used in the wrong way.

Reprinted with permission from *Sales and Marketing Management*, January 1996. Copyright © 1996 by Bill Communications. All rights reserved.

data that is used in the wrong way. The problem is compounded when companies fail to inform the customers they've surveyed of how their input has changed the way the company does business.

On the other hand, the companies that have successfully used customer surveys have learned two key lessons: how you measure customer satisfaction can often be as important as what you measure; and, you must be ready to adapt and adjust your surveying methodology to ensure that "listening to the voice of the customer" is more than just a cliché.

Survey as Sales Tool

The Xerox Corporation, which has been surveying customers since the late 1970s, might be considered the granddaddy of customer satisfaction surveyors. And still, the company has had to adjust its methods to improve the survey's impact.

In the mid-1980s, Xerox sent mail questionnaires to 40,000 customers per month—broadly targeting decision makers, operators, and administrators at customer companies. But according to Peter Garcia, manager of customer satisfaction for the copier giant, it found that even with a 13 percent response rate, those responses came from operators and administrators much more than they came from decision makers. As a result, "We didn't know what the decision maker thought of us," Garcia says.

Over time Xerox has standardized its rating scales and the core elements of its questionnaires (always leaving room for more specialized questions depending on the market). It adjusted the survey language to better represent customer needs, and focused all of its survey work on two measures. "We wanted to identify the gap between where we were and one hundred percent satisfaction," says Garcia. "And we wanted to be benchmarked against our key competitors."

Today Xerox uses phone surveys and goes after decision makers exclusively, conducting about 10,000 surveys per month. It has also gone beyond measuring customer satisfaction to gauging customer loyalty as well. "We began to find that people who said they were 'very satisfied' rather than simply 'satisfied,' were six times more likely to buy from us again," Garcia says. As a result, Xerox continues to measure satisfaction, but also asks whether the customer would recommend Xerox to an associate and whether she would buy from Xerox again. It also tracks survey data by three classes of customer: first-time buyers, replacement or upgrade buyers, and additional equipment buyers, giving the company a finer read on who purchases what, when, and why.

Xerox has also implemented a "Customer Relationship Assessment" survey for its top customers. The goal, again, is to go beyond satisfaction to look at customer loyalty. These are face-to-face surveys conducted by the account management teams of about 500 global, national, or "named" accounts—essentially Xerox's best customers.

The process involves the customer directly in the survey process. The Xerox account team meets with top people at their key accounts to determine

> "We began to find that people who said they were 'very satisfied' rather than simply 'satisfied,' were six times more likely to buy from us again."

> ## CUSTOMER SURVEY MISTAKES TO AVOID
>
> Even the best-intentioned of customer surveys can go wrong, says Bill Alper of the Hay Group, a management consulting firm.
>
> One way is in whom you survey. "Are you surveying decision makers, or are you surveying users?" Alper asks. "Some companies might not want to survey the decision maker. Others may want to survey two or three different people within a customer organization. Some companies send surveys without knowing who is the decision maker. And in other cases, the target customer may receive the survey but pass it off to someone else to complete." All of these miscues can have an effect on the quality of the data a survey generates.
>
> Other factors that could influence the validity and usefulness of customer satisfaction data are the questions that are asked (are they too simple, too complex, or simply too many in number?), how often customers are surveyed (beware of the burnout factor), and the format by which customer satisfaction data is obtained (face-to-face is best, but mail is much less expensive).
>
> Stanley Brown at Coopers & Lybrand's Centre for Excellence in Customer Satisfaction warns that even a high customer satisfaction rating is no guarantee that you're doing the job. "A customer may be satisfied but have low expectations," Brown says. "Another supplier may come along and meet its needs better."
>
> Brown's advice for developing a customer satisfaction survey is to work backwards. "Start by asking what are the customer's priorities and develop questions that deal with those priorities," he says.
>
> "And stay focused on what's going to drive customer retention—not just satisfaction."

jointly the content of the assessment survey, its frequency, and who within the customer organization should receive it. "It may involve ten different people in the account," says Garcia. "We let them decide."

The key feature of the process, though, is that once a Xerox sales rep presents the results to a customer, the two try to devise ways to solve any problem the survey uncovered. They set criteria and tasks for both Xerox personnel and those on the customer side to perform, as well as dates of completion and periods to review progress. "It's more than a measure of customer satisfaction," says Garcia. "It's a relationship-building tool. It forces action for improvement both within our organization and the customer organization. It becomes a key element in the account management process."

Garcia expects this sort of relationship-building approach to customer satisfaction surveys to expand at Xerox, engaging more customers in closer dialogue and giving customers an even stronger voice within Xerox. "We've discovered that the relationship is just as important as the core product in most situations," says Garcia. "Customers are assessing us on how interested we are in them and how well we're listening to their needs. More and more we're dealing in markets of one."

The Closer the Better

While conducting a good survey can become an almost scientific process, no study is foolproof. With written surveys there's the nonresponse bias, and with telephone surveys there's the bias created when a respondent says anything to hurry off the phone.

"There are biases to every survey methodology," says Bill Barnes, director of customer satisfaction for the Eastman Chemical Company. For that reason, Eastman Chemical, based in Kingsport, Tennessee, has chosen to conduct all of its own surveys and not rely on an outside firm for assistance. The belief: the company can stay closer to its customers' true feelings that way.

To Barnes, the success of a customer satisfaction survey relies on convincing customers that there's value for them to participate. The Eastman survey, which has been ongoing since 1986, asks customers to rate the importance of several key buying criteria—things like manufacturing support, packaging, sales expertise, and responsiveness—and evaluate both Eastman's performance in these areas and the performance of Eastman's closest competitor.

Eastman salespeople hand-deliver the survey questionnaire to a select group of customers in each of Eastman's business units. By delivering the questionnaire in person, salespeople can explain the importance of the survey as a way of improving Eastman's ongoing relationship with the customer. (Salespeople are assured that the survey document won't be used to judge their performance.) Customers are asked to complete the survey immediately and to mail it back to Eastman's headquarters.

Salespeople review the survey results with their customers on follow-up sales calls. "We want to show customers that the survey will be a working tool to build our relationship," Barnes says. "Our business is based on relationships. We call on customers weekly. Some of our salespeople have only one customer. Why wouldn't a customer be willing to tell us how we were performing?"

Eastman uses the data gleaned from its surveys to drive organizational improvements, and to see if previous changes have improved customer satisfaction levels. To that end, each question on the survey is assigned to a department within Eastman, which is then responsible for analyzing the data and determining what corrective action is needed. "If the question is related to product quality, manufacturing takes ownership of the question," says Barnes. "If it's related to order entry, customer service takes ownership."

Internal ownership also allows Eastman to make continuous improvements in the survey process itself, says Barnes. "We've adjusted the level of detail we're looking for in specific questions to make it more consistent throughout," he says. "We've had business units and the sales organization work more closely together to determine which customers to survey and which contacts within a customer organization to survey. And we've changed the way we ask customers about the importance of key buying criteria." It now asks customers to rate importance from this perspective: "Would you give more business to suppliers who went from average to excellent in the performance of this key buying criteria?"

Inside or Out?

Like Eastman Chemical, Dow Chemical Company initially used an internally developed customer satisfaction questionnaire that was hand-delivered to customers by Dow salespeople. It asked customers to evaluate Dow's performance in a variety of product, service, and operational areas and compare it to the performance of Dow's best competitor.

The problem, says Dick Sosville, vice president, sales and marketing for Dow North America, was that "the survey was not user-friendly. It was time-consuming for the customer to fill out. It was written with no sense of feeling for how the data could be used. And it failed to make any effort to find out what customers were interested in." David Fischer, group manager of strategic research services for Dow, adds another problem—hand-delivery of the surveys. "A customer is not in the right mind-set to be critical and objective when a salesperson hands over a customer satisfaction survey and says, 'I'll wait while you fill it out.'"

Dow resolved its doubts about survey design and methodology by going to a third-party vendor. It now has an outside consultant conduct phone surveys of customers. "We wanted to get the company away from an emphasis on the process of measurement and on to using the results of that process. Now we can focus on using the data to find things we can work on to position our strategy to understand and meet the needs of the customer," Sosville says.

Dow still maintains strict control over the survey process by getting directly involved at the survey-development stage. "We first ask customers what's important to them," says Fischer. "We go out to some customers and reach others through focus groups to find out what's important to them."

Then the questions are validated internally by having them reviewed by the departments within Dow that would be affected by specific questions and would be using the data from those questions. Dow then pilots the questionnaire with a small group of customers before surveying its larger customer base.

The goal of the survey process, Sosville says, is to make Dow more market driven in each of its businesses. "Nearly eighty-five percent of our people never have direct contact with a customer, but all have a hand in serving the customer. We use the survey to teach them to think about the customer first."

For Dow, the voice of the customer is something quite literal. The vendor it uses, Audits and Surveys World, records customer responses both to specific, quantitative questions ("How do you rate Dow on a scale of one to five for on-time delivery?") and to more open-ended, qualitative questions ("Tell us about your chemical recycling efforts and how Dow might be able to help you there"), and codifies the responses by key words and phrases. Customer comments can then be played back to give Dow employees—in particular, that 85 percent who have no direct customer contact—the true "voice of the customer."

"That way," says Sosville, "employees get a better feel for who our customers are and what their needs are than they would from any set of charts and graphs."

Service with a Smile

John Holten

This article explains how Norwest Mortgage, a national mortgage company, has used customer surveys to enhance customer service, increase its business, and keep its current customers. The company has developed a sophisticated survey program and provides incentives to its employees for excellence in customer service. These efforts have contributed to record growth in the company, affirming once more that profit and customer satisfaction are intimately related.

What do you get when you send 400 employees of Norwest Mortgage, Inc., on a four-day cruise? No, this is not a riddle. It's a business proposition. So instead of a punchline, the correct answer is, a return on investment in the form of outstanding customer service.

For the past four years, Des Moines–based Norwest Mortgage has offered a sunny getaway to all employees in any branch office that meets its target for customer service. Like any good incentive program, the offer spurs employees to put more emphasis on a specific aspect of performance.

In the first year, one branch office reached the customer-service target, earning a getaway to Boca Raton, Florida, for six employees. This past year, 31 of Norwest Mortgage's more than 700 offices nationwide were rewarded for stellar achievement in customer satisfaction.

To make room for 815 people, including one free guest per employee, Norwest chartered an entire ship for a voyage to the Bahamas, Key West, Florida, and back to Miami. Norwest insists that the trip has paid for itself many times over. Pete Wissinger, executive vice president and managing director of retail operations for Norwest Mortgage, explains: "We were interviewing a branch office today for a videotape. They were saying, 'We've gone out and told every Realtor that we earned a cruise by providing great customer service. They all want to do business with us. And everyone in town wants to work for us.' The branch offices are using the incentive to gain a competitive advantage through service and as a recruiting tool."

Although just a fraction of all branch offices win the trip, the payoff for Norwest is that customer service has improved dramatically companywide (see Figure 1). And, as Wissinger says, Norwest believes that better service will strengthen its position as the nation's largest retail lender.

"We've gone out and told every Realtor that we earned a cruise by providing great customer service. They all want to do business with us."

Reprinted with permission from *Mortgage Banking*, May 1996. Copyright © 1996. All rights reserved.

Customer Satisfaction Boom

Broad efforts to improve service, including a travel incentive for branch offices that score high on a customer survey, have succeeded in raising customer satisfaction at Norwest Mortgage.

Year	Qualifying Branch Offices	Company-wide Satisfaction Rating
1992	1	81%
1993	12	84%
1994	8	88%
1995	31	90%

Figure 1.

At the heart of the incentive program and improved service lies a customer survey and technology—the tools used to measure customer satisfaction. Overcoming the subjective nature of customer satisfaction, Norwest Mortgage uses automated survey processing to quantify, measure and compare performance by its branch offices.

Norwest Mortgage originally started surveying in 1990 to get feedback from borrowers. "We wanted to know what our customers [thought] of our service and whether they would refer others to Norwest Mortgage," Wissinger says.

Loan Production Administration, part of Norwest's retail operations, began by mailing surveys once a month to 20 percent of recent borrowers. A local company manually tallied the returned surveys in a process with a three-month lag time. Norwest's internal growth and acquisitions further strained the system. As loan volume rose, so did the volume of surveys. In 1990, when Norwest originated $8.8 billion in home loans, the company was sending out about 1,500 surveys a month. By 1992, when loan originations had grown to $21 billion, Norwest was mailing 3,500 surveys a month. Meanwhile, Norwest declared a commitment to unbeatable customer service.

"One of our strategic imperatives is to win with every customer," Wissinger says. "We feel that, going forward, customer service is what will differentiate us. We want to be a premier service organization."

The commitment to service reflects a few facts of life. First, the life expectancy for home loans is shrinking. "People don't have a loan for 30 years," Wissinger says. "They move up or refinance."

Second, Realtors have long memories. "Most of our business is relationship selling," Wissinger says. Realtors will build a relationship with the lender that keeps their clients happy.

Third, the mortgage company is not alone. As part of a larger corporation with the parent being a bank, Norwest Mortgage and other Norwest operations give and receive business referrals. "We have a lot of cross-selling in place," Wissinger says. "When we provide good service, it definitely is beneficial for other parts of Norwest and makes them feel confident in referring business to us."

With the focus on winning with every customer, sampling was no longer an adequate tool to measure customer satisfaction in the mortgage lending operation. At the time, polling all new borrowers meant sending out 17,500 questionnaires a month. In 1992, under Wissinger's direction, Norwest chose to automate processing with the help of National Computer Systems (NCS), a provider of information collection products, processing services and management systems. "We looked for a company like NCS that was large enough and had the technology to survey 100 percent of our borrowers," Wissinger says. "We also wanted a more professional-looking survey."

NCS designed a new questionnaire that asks borrowers to mark their answers in fill-in bubbles. Nine questions offer a satisfaction scale of one to five. One of those questions asks whether customers would refer others to Norwest Mortgage. The form also invites additional comments. Norwest transmits the name and home address of recent borrowers to Minneapolis-based NCS, which addresses and mails surveys twice a month. In 1995, with loan originations reaching a record $33.9 billion, Norwest mailed about 28,000 questionnaires a month. Forty to 50 percent of borrowers complete the survey and seal it in a return envelope addressed to the Norwest Mortgage's Survey Processing Center.

The actual survey asks borrowers to respond and rate the service they received in the following areas:

- Overall, I am satisfied with the service I received from Norwest Mortgage.
- The employees at Norwest Mortgage provided correct, clear and understandable information to assist me in choosing the loan program best suited for my needs.
- Norwest Mortgage employees adequately explained what information was required of me to obtain my loan.
- Norwest Mortgage promptly advised me of any additional information needed and explained why.
- Norwest Mortgage employees promptly and accurately informed me about the costs of my loan.
- I was kept updated in a timely manner by Norwest Mortgage employees on the status of my loan.
- Norwest Mortgage employees promptly returned my phone calls throughout the loan process.
- Norwest Mortgage employees clearly explained the details I needed to know to close my loan (i.e., time, place, cash required to close, etc.).
- I would refer others to Norwest Mortgage.

Processing the Responses

The Norwest Mortgage Survey Processing Center consists of one full-time employee and an NCS OpScan 5 scanning system. The employee can load as many as 300 surveys at a time, punch the "go" button, then work on inputting written comments for five minutes while the scanner automatically feeds and "reads" the stack.

The OpScan 5, a mid-range desktop model with a 28-by-13-inch footprint, performs optical mark reading. The same technology that brought fame to the No. 2 pencil, it simply identifies which bubbles or boxes have been filled. Now highly refined, it reads with discretion. You don't have to cover every speck of white to register an answer. And you don't have to fret over a poor job of erasing because a special feature distinguishes between a mark and a smudge. Programmed to look for marks in specific locations, the scanner requires a custom-designed document that NCS prints under tight tolerance. "The scanner has been great," Wissinger says, adding that it has made internal processing fast, accurate and cost-effective.

Precision in printing is vital with scannable documents. If the bubbles aren't printed where the programming says they should be, the scanner will miss or misinterpret the answers. For that reason, photocopies of scannable documents shouldn't be used to collect data.

Reports for each branch location show how borrowers rated the office on each question. Branch managers and regional managers use the reports to track progress, detect trends and identify training opportunities. Now Norwest processes 5,000 to 8,000 surveys per month and has a better management tool.

NCS also customizes surveys for Norwest. In 1995, when Norwest Mortgage merged with Riverside, California-based Directors Mortgage Loan Corp., NCS sent survey forms to customers of Directors that were printed with Directors' name. Changing Directors' name to Norwest Mortgage was completed April 1, 1996. In the interim, those locations went by the name Directors/Norwest. Norwest continued to use the Directors name because it was well-known and respected in California, where Norwest had little name recognition, Wissinger says.

Likewise, customized surveys acknowledge Norwest's joint-venture partners, for example, if the borrower closes with the Towne Square joint venture with Wells Fargo Bank. And the survey results are shared with the joint venture's staff, including Julie Piepho, president of Towne Square and a Norwest senior vice president and division manager.

With the sophisticated new system, Norwest can measure customer satisfaction, make credible comparisons and provide quick feedback to branch offices—everything needed to run an incentive program.

"We didn't automate with plans to start an incentive program," Wissinger says, "but Norwest was looking for other ways to incent employees, and we saw that this could work." So the company started rewarding employees, based on survey results.

One Crucial Question

Earning a trip depends on one question: Overall, how satisfied are you with the service of Norwest Mortgage? Branches that meet a minimum profit requirement and receive a positive satisfaction rating on 95 percent of surveys can start packing sunscreen.

During the cruise, employees receive awards from Mark Oman, president and CEO of Norwest Mortgage, and participate in a Service Stars conference, where employees share their secrets to outstanding customer service. "We get together as one big Norwest group and talk shop," Wissinger says.

The branch manager earns a trip to the national sales conference when the branch earns a 4 or 5 on 90 percent of survey forms. "Our branch locations are paying attention to customer service," Wissinger says. "Winning the cruise is added incentive. They also take pride in having high customer satisfaction."

Overall satisfaction (a score of 4 or 5) has increased companywide from 81 percent in 1992 to 90 percent in 1995. "That is an important success story for us," Wissinger says. "We had a record-volume year and raised overall customer service."

Adding technology to improve cycle time on loan approvals has helped increase satisfaction. In 1994, Norwest started to track the percentage of loans that met its goal of 15 days or less of processing time—from application through closing. That first year, one-third of new loans were completed within 15 days. In 1995, processing time on two-thirds of loans was in 15 days or less. In the first quarter of 1996, 80 percent met the goal, with an average processing time of 13 days.

About 70 percent of Norwest loan applications now are taken on a laptop computer either in a real estate office, during an open house, at a proposed building site or at other convenient locations for borrowers, Wissinger says.

Norwest also has developed a system to automate credit analysis. Using thousands of previous loan approvals as a model, the system learns which factors are routine and which should be flagged for closer study. Combining artificial intelligence and a laptop computer, which can upload applications directly to a mainframe, Norwest has transformed what was once a 60- to 90-day process to as little as a day. "We've invested a lot to improve the drivers of customer service," Wissinger says. "Our goal for 1996 is to approve 50 percent of all loans within 48 hours."

Refining the Program

The survey and incentive program have required some minimal refinements. For example, to meet the needs of the growing Hispanic population, the questionnaire went bilingual, with English on one side and Spanish on the other.

Norwest also recently hired a manager to communicate more survey information to branch locations. Because employees companywide have raised the standard for service, branch offices will have to do even better to qualify for future trips. This year, they will need to score 4 or 5 on 96 percent of questionnaires. Also, Norwest will start rewarding superior service. Branch offices will qualify by earning a 5 on 88 percent of returned survey forms. After seeing a steady rise in service, Wissinger believes that dreamy destinations will motivate employees to take customer service to the highest level. "A 5," he says, "that's a 'wow' in our standards."

John Holten, a former business reporter, is the owner of Holten Communications in Minneapolis.

Quality Transformation
TRAINING

Measuring ROI: The Fifth Level of Evaluation

Jack J. Phillips

Training is important, everyone says, but is there a way to measure the return on investment in training? The answer is that such a measurement is still somewhat subjective, but methods are being developed to actually measure the monetary payback from investing in training. This article explains how to do that.

For too long, the training and development process has escaped the scrutiny of accountability. While expenditures have grown, many training departments have not taken the extra step to show the payoff of their efforts—particularly the more elaborate, comprehensive, and expensive programs. Tools and techniques to measure return on investment (ROI) are now available, and the process has become reliable and acceptable. Measuring the return on training investment should be a requirement in most organizations, at least for some programs.

Before dismissing the idea of measuring your company's return on training investment, consider the rewards of such an evaluation. An example of how valuable training can be is evident in the literacy training program at Magnavox Electronics Systems Company, West Coast Division, in Torrance, California.

Many literacy programs are undertaken because they are needed, but little attention is paid to the program's economic returns. Magnavox wanted to know if this type of training returned an economic dividend. After its 18-week literacy program was initiated, which covered verbal and math skills, it was shown to have a significant payoff for the company. The benefits of the program (reduced scrap, rework, and increased productivity) were converted to dollar values. The program yielded a cost-benefit ratio of 8.4:1 and a 741 percent return on training investment! And while these numbers are impressive, the program designers consider them conservative.

In every corner of the training and development field, the pressure to measure the return on investment is increasing. At some time or another, virtually every organization will face this important issue. Many progressive organizations are taking a logical and methodical approach to developing ROI for a limited number of programs, using a sampling basis. Six trends have increased the interest in, and use of, ROI measurement in training and development:

- Training and development budgets are continuing to grow, which creates more pressure for accountability.

- Training and development are linked to competitive strategies, which make them important areas for measuring program contribution.

- Many programs have failed to deliver what was expected; consequently, program sponsors have requested ROI calculations/justifications.

- The concern for accountability in all functions in an organization is increasing; thus, the training and development function becomes one of many support efforts under scrutiny.

- Top executives in a large number of organizations now require ROI information.

- To justify their contribution, trainers have increased their interest in the ROI process.

> The program yielded a cost-benefit ratio of 8.4:1 and a 741 percent return on training investment!

A New Model for Evaluation Levels

Developing ROI for training requires a key modification of a classic model. The four-level framework developed by Donald Kirkpatrick in 1959 does not focus directly on the ROI issue. As shown in figure 1 Kirkpatrick defines Level 4 evaluation as the results linked to training. These results could take the form of reduced absenteeism and turnover, quality improvement, productivity, or even cost reduction. But this level of evaluation does not require a specific monetary value (cost savings) to be determined. To obtain a true ROI evaluation, the monetary benefits of the program should be compared to the cost of implementation in order to value the investment. In effect, this process moves evaluation to the next level—Level 5 in our revised Kirkpatrick model. Thus, the fifth level of evaluation is developed by collecting Level 4 data, converting the data to monetary values, and comparing them to the cost of the program to represent the return on training investment.

In practice, many organizations are taking evaluation to this new level for a few selected courses, often using some form of sampling. When the ROI formula is developed, evaluation is conducted at all five levels.

Level	Questions
1 Reaction and planned action	• What are participants' reactions to the program, and what do they plan to do with the material?
2 Learning	• What skills, knowledge, or attitudes have changed and by how much?
3 Job applications	• Did the participants apply what they learned on the job?
4 Business results	• Did the on-the-job application produce measurable results?
5 Return on investment	• Did the monetary value of the results exceed the cost for the programs?

Figure 1. **Five levels of evaluation**

For example, in the Magnavox case. Level 1 evaluation—reaction of employees—was measured by post-course surveys. Level 2 learning was measured by TABE (Test of Adult Basic Education) scores before and after training. At Level 3, changes in the behavior of employees were measured by daily efficiency ratings. At Level 4, business results were measured through improvements in productivity and reductions in scrap and rework. Finally, at Level 5, ROI was calculated by converting productivity and quality improvements to monetary values and comparing these to the full program costs to yield an ROI value.

The ROI Process

Calculating ROI requires a process model, as depicted in figure 2 below. The various elements of evaluation (design instruments, levels, and purposes) form the specific data collection plans. A variety of data collection tools, ranging from questionnaires and surveys to monitoring on-the-job performance, are available to trainers.

Once data is collected, the next step of the ROI analysis begins with deliberate attempts to isolate the effects of training on the data items. At least ten strategies have been used to accomplish this:

- use of controls
- trend line analysis (time series)
- forecasting methods

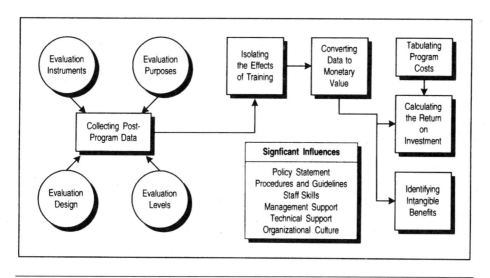

Figure 2. **ROI process in HRD**

- participant estimates of training impact
- supervisor estimates of training impact
- management estimates of training impact
- customer input
- expert estimates of training impact
- subordinate input on training impact
- calculations/estimations of the impact of other factors.

The next step is to convert collected data to monetary values. This requires a direct conversion of hard data, such as quantity, quality, cost, or time, which is an easy task for some programs such as technical training. For "soft" data, the task is more difficult, although a variety of techniques are used to place values on the improvements. Among the techniques used are

- historical costs
- supervisor estimation
- management estimation
- expert opinion
- participant estimation
- external studies.

The next step is to calculate the costs for the program. Although there has always been a need to capture training costs, the need is amplified with more attention on accountability and the ROI calculation. The ROI formula is the annual net program benefits divided by program costs, where the net benefits are the monetary value of the benefits minus the costs of the program. The ROI formula is as follows:

$$\text{ROI (\%)} = \frac{\text{Benefits} - \text{Costs} \times 100}{\text{Costs}}$$

This model also recognizes that there should be intangible benefits that will be presented along with the ROI calculation.

ROI Strategies/Best Practices

Although it is difficult to uncover companies' precise strategies, a recent search identified over 2,000 organizations that could be contacted in an effort to determine the nature and status of the ROI process. Although this was not a carefully designed research project, it presented a review of the efforts of many progressive organizations. Several common strategies began to emerge that can be considered best practices for calculating an ROI in training and development. A review of such best practices is as follows:

Set targets for each evaluation level. Recognize the complexity of the evaluation levels described earlier. Some organizations attempt to manage the process by setting targets for each level. A target is the percentage of training programs measured at that level. For example, at Level 4, where it is difficult to measure, organizations have a low level of activity—usually less than 20 percent. Level 5 evaluation—ROI—is even less likely to occur—usually around five percent, reflecting the complexity of a process that commands significant resources and budgets.

The process of establishing evaluation targets has two important advantages. First, it provides measurable objectives for the training staff to clearly measure progress for all programs or any segment of the process. Second, adopting targets focuses more attention on the accountability process, communicating a strong message to the training staff about the commitment to measurement and evaluation.

Evaluate at the micro level. Training measurement and evaluation usually focuses on an individual program or a few tightly integrated courses. The ROI process is more effective when assessing the direct payoff of an individual program. Attempting to evaluate a group of courses conducted over long periods of time is quite difficult. The cause and effect relationship becomes more confusing and complex. Also, it is inappropriate to attempt to evaluate an entire function such as quality training or technical training. For this reason, evaluation must be a micro-level activity.

Use sampling for ROI calculations. Determining the desired level of ROI calculations is an important issue. There is no prescribed formula and the number depends on many variables, including

- staff expertise on evaluation
- nature and type of training programs
- resources that can be allocated to the process
- support from management for training and development
- organization's commitment to measurement and evaluation.

Other variables specific to the organization may enter the process. Most organizations settle on evaluating one or two sessions of their most popular programs. For example, the federal government's Office of Personnel Management has developed an ROI calculation for one of its most popular courses—Introduction to Supervision. Still others may select a program from each of their major training segments. In a large bank with six training academies, a program is selected from each academy, each year, for an ROI calculation.

For organizations implementing the ROI concept for the first time, only one course should be selected for a calculation as part of the ROI learning curve. In the final analysis, the selection of programs for ROI calculations should yield a level of sampling where top management is comfortable in its assessment of the training and development function.

Jack J. Phillips is president of Performance Resources Organization (P.O. Box 1969, Murfreesboro, TN 37133-1969; 615/896-7694).

Training Success Stories

Leslie Overmyer-Day and George Benson

Just as its title suggests, this article provides an overview of the success in training in various companies as documented by the American Society for Training and Development. The best practices of different companies profiled, such as Aetna Life & Casualty, Andersen Consulting, and Boeing, provide hints for what others can do and affirm the payoff of training in improved individual and organizational performance.

Are companies really using technology to deliver training? If so, how? Which techniques are successful in developing knowledge-based training for experts? How can my training group get involved in our company's reengineering effort—before people come to us for training?

These are hard, but important, questions for training professionals. The search for new techniques is a constant activity. Competition, downsizing, and reorganization continue to affect organizations, including training departments. As they do more with less, practitioners seek ideas for managing change while still providing high-quality instruction.

Here are some success stories from companies that are members of the American Society for Training and Development Benchmarking Forum—their best and most successful training practices, as well as the methods they use to identify such practices.

In 1995, the Benchmarking Forum selected 14 "best or successful practices" from its 48 members. The selections are wide-ranging—from electronic training delivery to successful reorganization of a training department. Most fall into the new-practice category; many reflect current trends in the training field.

Aetna Life & Casualty: Redesigning the Education Department

Most training professionals know that they need to link their departments and programs more closely with the businesses they support, especially in such highly competitive and rapidly changing industries as health care. The Forum recognized the education department at Aetna Health Plans for its redesign of the department and for the training staff's contribution to the organization's overall reengineering effort.

In the past, Aetna's education department consisted of managers who supervised teams of training specialists assigned to major internal-customer groups. Because staff members focused on one area, they tended to develop a deep knowledge of that area without gaining more global understanding of the business. This customer alignment also needed to be adjusted each time that the business reorganized. So, the department embarked on a large-scale redesign of the home-office training unit.

Each staff member selected a team on which to participate based on one of four conceptual models. The teams had six weeks to develop a proposal that expanded the conceptual model into a working design. Each team presented its ideas at an all-staff meeting; management selected the winning ideas to create a new structure for the department.

The redesigned department—which includes innovations proposed by all of the teams—is flatter and now run by a smaller team of managers rather than many single manager-led teams. It's structured around business needs, in line with Aetna's business plan. The new structure emphasizes the development of multiskilled educators instead of training specialists assigned to a specific business group. It also includes the following elements:

- a 360-degree-feedback review process
- an internal network of coaches
- a resource-allocation team
- a commitment to the ongoing development of staff members
- project managers with increased responsibility and customer exposure.

The new structure has met Aetna's expectations: strategic positioning for organizational changes; increased skill levels of staff members in the education department; and more dispersed leadership among training staff members, who have shown an increased sense of ownership for their work. In addition, productivity and quality levels are higher. There has also been an increase in information sharing and employee-development opportunities. The cost of the redesign was minimal, and educational services weren't interrupted.

The education department has also changed the way it does business. For example, it formed a partnership with the customer administration department during the planning stages of the reengineering process. Educators worked with business process specialists to conduct a needs analysis and then developed the training. The training team also worked with human resources to create and implement a change-management plan. These partnerships enabled the training and education department to proactively assess training needs upfront during a business change, which minimized the need to revise training materials. The education department also gained visibility and set a precedent for future partnerships.

Aetna's Health Plans Training and Education department's best practices clearly demonstrate the success of developing flexible, multiskilled educators, that training is a key factor in successful business transitions, and that the training department must be involved closely with the business groups it supports.

Andersen Worldwide: Concept Mapping and Pattern Matching

Training professionals frequently seek better methods for clarifying training needs upfront and evaluating training outcomes later. Andersen's Center for Professional Education uses concept mapping and pattern matching to clarify expected employee competencies and evaluate the effectiveness of training designed to develop the competencies. In this context, the purpose of concept mapping is to paint a clear picture of training expectations and priorities by explicitly integrating multiple stakeholder views. After the training is complete, the expectations are compared to the outcomes.

For many trainers, it's a standard practice to ask stakeholders to identify and prioritize their expectations through brainstorming and negotiation. What's unique about concept mapping and pattern matching are the sophisticated statistical techniques that enable the integration of input from different people. The result is an easy-to-understand map that shows the relative importance of each competency area. This map is the foundation that guides the training design, development, implementation, and evaluation. Andersen practitioners have used this method to compare how well the learning and performance outcomes of training are meeting the initial expectations. Through the use of concept mapping and pattern matching. Andersen has advanced the increasingly popular notion of using these outcome measures to assess "return on expectations," or ROE.

The Boeing Company: Knowledge Modeling

Most training professionals find it challenging to convey complex information and promote knowledge-based skills. Traditionally, training staffs have used job-task analyses to develop course content. But task analyses are inadequate for jobs or tasks with heavy cognitive components—in other words, tasks that experts think about a lot before they act.

The Forum recognized Boeing for meeting the unique training needs of its many knowledge-based employees through techniques borrowed from the field of artificial intelligence. These techniques shortened the nine-month learning curve for new users of complex computer-aided drafting and modeling software, known as CAD-CAM. Boeing now uses the Knowledge Analysis and Design Support methodology (KADS) to identify and transfer the "thinking" of expert CAD-CAM users. KADS makes nonobservable thinking processes explicit.

Boeing's KADS analysts interviewed and observed expert users in order to identify their thinking processes for solving problems, dealing with uncertainty, and minimizing risks. Then, the analysts worked with course developers to identify learning objectives and to integrate the expert CAD-CAM practices into a training curriculum. Though KADS was originally developed to build automated knowledge-based systems, Boeing has also used it effectively in instructional design, reference documentation, and process improvement.

In a field comparison of 70 engineers trained in the expert CAD-CAM practices and 30 untrained engineers, the benefits of the new approach became evident. All of the trained engineers were able to perform a construction task; less than half of the untrained engineers could. In addition, the slowest time for the task among the trained engineers was equal to the fastest time among the untrained engineers. The conservative estimate of return on investment for one program was 4,000 percent per year—based on less productive time lost among new users, less inefficiency among current users, reduced errors in data, and less demand on computing resources due to inefficient data.

Digital Equipment Corporation: Technology Solutions

The role of advanced technology and alternative delivery systems are of growing interest to training professionals. In this spirit, the Benchmarking Forum recognized Digital Equipment for several practices that are among the leading-edge applications of communications technology found in training. These practices provide a window on the future and highlight the potential of the Internet and other computer networks for training and performance support.

Two of Digital's best practices demonstrate the benefits of using the World Wide Web to deliver training. One, Digital currently maintains its curriculum catalog on the Web for instant access to employees located anywhere in the world. This on-line system has increased course enrollment because of the easy access, search capability, and user-friendly graphical interface.

In addition, the hypertext format and graphics capability of the Web enables Digital to provide the following:

- sequential curriculum maps
- pre-course skill checks
- links to contacts within the training department for course information, the course-enrollment system, and education resources outside of Digital.

Digital also uses the Web to deliver a basic finance course for managers. In the past, internal instructors taught the course, which includes such topics as interpreting a balance sheet and using ratios. But lately, Digital has had fewer finance employees available to instruct. The Web enables Digital to offer a self-paced, computer-based course that employees can access at their convenience from any PC system, eliminating travel expenses and time away from work.

The course consists of eight modules, including pre-module self-assessments, exercises, review questions, documentation, and real-life examples from Digital's annual report. The course's reference guides provide just-in-time, just-what-you-need resources. Web delivery also offers the unique benefit of being able to update the course instantly from a single site without having to recall diskettes.

Digital estimates that the cost of course conversion from paper-based to electronic will be recouped through the elimination of paper, reproduction, postage, and instructor expenses. Cost-control efforts also led to the development of an alternative method for delivering hands-on training to technicians who service Digital's complex computer products. Like many companies, Digital has reduced expenditures for capital equipment and travel in recent years, creating instead a centralized location for electronic delivery—a remote-access lab with various hardware and software accessible in real time from sites around the world. Users at remote sites also receive either lecture material or training videos. This approach provides technicians with hands-on experience working with complex computer products and network environments. In fact, the training system uses the actual technology they work with, as well as built-in simulations and exercises. A full-time administrator is available online to monitor sessions and answer trainees' questions. As network technologies, the Internet, and video-broadcasting techniques continue to advance, it will be easier, faster, and more cost-effective to implement these approaches.

> In fact, the training system uses the actual technology they work with, as well as built-in simulations and exercises.

Digital's most versatile technology application was developed by the learning-and-development arm of its customer-services department, which serves 20,000 employees who install, maintain, and service thousands of Digital's and other suppliers' products worldwide. With so many products, technicians often have to provide service to a customer without having complete or up-to-date information for diagnosis and repair. So, the department developed the Multivendor Customer-Service Learning Utility to provide access to and retrieval of training and documentation for technicians whenever and wherever they need it.

The MCS Learning Utility is an electronic-delivery system with Web access through high-speed modems and multiple servers. Using a "search engine," course maps, and modular outlines, technicians can select training courses and documentation and then download them to a laptop. The system also provides handbooks, reference guides, schematic diagrams, and simulations. With the MCS Learning Utility, Digital anticipates improvements in technicians' productivity and skills, as well as reduced training costs.

Tektronix Education Consortium: Partnering

Shrinking resources are a common phenomenon in many training departments. Several technology-based businesses in the Portland, Oregon area have been working together to broaden educational opportunities for their employees. Tektronix joined with Sequent, Intel, Wacker-Siltronic, Hewlett-Packard, Automated Data Processing, and Mentor Graphics to seek inter-company opportunities for meeting shared training needs. The companies' training managers meet monthly to discuss ways to pool their resources and to expand the availability and content of their courses cost-effectively.

This consortium led to the creation of the Regional Workforce Training Center, which secured several state grants to promote Portland's economic development. The center has also improved educational opportunities for employees of the consortium companies. Because the consortium shares internal courses and combines resources for supplier-provided training, Tektronix has been able to expand its course offerings and cancel fewer classes due to underenrollment.

Texas Instruments: Automated Evaluation

Training professionals are frequently asked to document the long-term outcomes of their training programs. But it's difficult to obtain feedback from participants after they've returned to their jobs. Due to cost and time constraints, this step is often skipped. So, Texas Instruments developed an automated e-mail survey system to obtain feedback from participants for course managers.

In the past at TI, the evaluation of training transfer was a nonstandardized process that was both time- and labor-intensive. Now, the automated system enables course managers to register courses online for evaluation. The system sends an e-mail message to all participants 90 days after they complete a course asking them to fill out a short questionnaire. The seven questions, which can be customized, garner information about skills transfer to on-the-job performance. The responses are recorded online, and the results are compiled in a database that course managers can reference.

The Automated Evaluation Project has increased the use of evaluations, reduced cycle time for data collection, and provided a standard measure for evaluating the transfer of skills and behavior to the job. Texas Instruments estimates that it will experience significant savings in labor costs to conduct evaluations. It has already seen improvements in both the quantity and quality of participants' feedback.

Special-Interest Practices

The Forum also selected several practices that reflect HR-management issues. Four focus on assessing employees' skill levels and the skill requirements for jobs and the efforts to close skill "gaps": differences between the required and actual skill levels. All of the practices aim to improve performance of the workforce.

Allstate Insurance Company: Task and Skill Profiles

In 1989, Allstate reorganized several processing centers around the country into three data-management centers. Task and skill profiles were used to assess the skills of the employees in those locations to match their existing skills to the skills required in the new environment and to develop learning agreements to bring them up to the required levels. This process resulted in the closure of hundreds of skill gaps and enabled every employee in those three centers to transition to their new roles with minimal impact on productivity. This success led to the implementation (70 percent complete) of a skills management system and an employee skills and training database, for all 60,000 Allstate employees and agency staff.

All employees, including managers, prepare a "learning agreement" to list their future training needs by priority, based on their job responsibilities. Employees can take courses based on the learning agreements to close any skill gaps. This approach integrates task and skill profiles for comprehensive performance management. Allstate credits this process with significantly increasing employees' understanding of their jobs, reducing training expense through targeted training, increasing employee satisfaction with training, and improving customer-focused performance.

All employees, including managers, prepare a "learning agreement" to list their future training needs by priority, based on their job responsibilities.

Digital Equipment Corporation: Individual Assessment Standards and Guidelines

Digital Equipment has developed standards for managing individual competency data in order to ensure that it has current, accurate, and useful information while protecting employees from misuse of the data. The standards were created because various approaches to assessment were being used across the country, approaches based on different assumptions or knowledge about what constituted correct practice.

A worldwide Assessment Standard was created by the Development and Learning organization, working with business management, legal, and HR functions. The standard is supported by guidelines and resources for managers to ensure that the standard is properly applied. Digital has also formed a Competency, Assessment, and Measurement Group that provides consulting to Digital's business units for a fee. Digital anticipates that the benefits will include hiring people with appropriate skills, identifying training needs, planning appropriate development activities for employees, and certifying employees.

Sprint: The LINK Performance-Management System

In 1990, the Sprint Corporation was made up of business units with different HR systems for performance planning and assessment, and no common culture or language for communicating Sprint's vision, values, and mission to employees. Now, Sprint uses the LINK Performance Management System with a common language for integrating business objectives, employee-development plans, 360-degree-evaluation instruments, and educational courses.

The system includes seven Sprint core dimensions and 29 subdimensions that share a language and value system for describing Sprint's work and culture. The Sprint dimensions are defined through job analyses, employee and executive interviews, and external industry data. They are the foundation of all training, performance management, selection, career development, and assessment within Sprint.

The LINK system operates continuously in a cascading format. Before the new calendar year, executives announce their business plans. Then, four to six key business objectives are identified for each associate aligning the Sprint dimensions and subdimensions to support the accomplishment of the objectives. Each associate creates an individual development plan with input from a 360-degree instrument. The LINK process also involves two interim reviews, one annual performance appraisal, and a yearly salary adjustment. This aspect is unique in that compensation is based on whether an employee attains his or her objectives as well as how.

Texas Instruments: Job-Role Profiles

Texas Instruments' information-technology group developed job-role profiles and an integrated skills-management system in order to ensure rapid response to changing business demands and to facilitate a reorganization from a hierarchical organization to a team-based one. The goal was to analyze and document current job requirements, project future needs, and provide training on skill gaps. The group identified 43 job functions within the information-technology environment and developed seven to eight critical skill requirements for each function—including skills in communication and teamwork, as well as job-specific technical skills. The group compiled the jobs and skills in one source distributed to all employees.

The approach focuses equally on business and personal needs in developing education strategies. The job and skill profiles are used in the following areas:

- self-assessments
- career development
- skill inventories
- project assignments
- curriculum development.

Each job's skill requirements are tied to The Education Center's courses. If an employee recognizes a deficiency, he or she knows which course to take. Previously, skill assessments were conducted informally and ad-hoc. Now, a computer database integrates job-role profiles, project requirements, employee profiles, and skills training.

Some of the practices mentioned in this article, such as Digital's use of the World Wide Web to distribute its curriculum catalog, are early examples of what are likely to become common practices in the near future. Overall, the best and successful practices identified by ASTD's Benchmarking Forum reflect increasingly applied training trends. They're also examples of how successful training professionals can accommodate organizations' ever-changing training needs.

Leslie Overmyer-Day is a senior research analyst at Amerind, Inc., and a consultant to the ASTD Benchmarking Forum.

Quality Transformation

TEAMS AND TEAMWORK

Teams in the Age of Systems

Peter R. Scholtes

In this article, Peter Scholtes discusses the importance of teams from the systems view of organizations—a view that reminds us that work gets done as interdependent processes, which teams are often best to facilitate. This article is especially valuable for the table it includes, which classifies seven types of teams, when to use them, and other valuable information.

> We trained hard—but it seemed that every time we were beginning to form up into teams, we would be reorganized. I was to learn later in life we tend to meet any new situation by reorganizing, and a wonderful method it can be for creating the illusion of progress while producing confusion, inefficiency, and demoralization.
> —*Petronius, 66 A.D.*

This delightful observation from Petronius, a Roman satirist, tells us.[1]

- Teams aren't new. They have been around for a long time.
- It takes a while for a group of individuals to form into teams.
- From time immemorial, teams have been undone by managerial decisions that disregard how the workers in the workplace will be affected.
- For a long time, people have learned to settle for the illusion of progress.

Reprinted with permission from *Quality Progress*, December 1995. Copyright © 1995 by the American Society for Quality Control. All rights reserved.

I add to Petronius' observations two of my own:

- Many managers mistakenly believe that "adopting a new philosophy" (W. Edwards Deming's point No. 2) involves establishing a proliferation of teams.[2] "Teams" and "quality" are not synonymous. Too many teams or the wrong kind of teams interfere with adopting a quality philosophy and give teams a bad name. Teams themselves can become an illusion of progress.
- Teams are often disconnected. They usually exist in isolation from each other, working on projects that individually might be worthwhile but collectively lead nowhere. Sometimes teams even work at cross purposes. For example, a cheese company was having a problem consistently producing top-grade commercial cheese. When the cheese was below grade, it had to be sold to the makers of animal feed. This company ended up with two teams: a production-based team, which sought to remove the causes of below-grade cheese, and a marketing-based team, which sought new markets for below-grade cheese.

How can organizations avoid a proliferation of disconnected teams? They need to adopt a new philosophy toward teams. This philosophy is built on two prerequisites:

1. *Teams must support customers, systems, and improvement.* The business world is changing from an era focused on individuals, hierarchy, and control to one focused on customers, systems, and improvement. The team philosophy must reflect this new focus.
2. *Teams must not only support the system, they must also be a system.* Just as the net value of a team should exceed the sum of its individual members (this is called synergy), the net value of all teams in an organization should exceed the sum of the individual teams. Synergy and interdependence should exist not only within teams but across teams. There needs to be a system of teams—an interacting network of efforts that operate in sync toward common purposes. This requires a different mode of leadership by management.

The Basics of an Organization

Every organization—whether it is a business or a Friday-night poker group—consists of three basic elements: *a group of people* who are pursuing a *common purpose* using *systems, processes,* and *methods.* Conventional management methods focus on the element of people: If managers want to successfully pursue a purpose, they introduce interventions to the employees. Managers "motivate" employees, reorganize them, put them into teams, hold them accountable, empower them, and offer them promises of rewards or threats of punishment. This has been the prevailing approach to running a business since the mid-1800s.

The quality movement, begun in Japan in 1950, represents a shift in focus from people to systems, processes, and methods. The philosophy of management taught by Deming, Kaoru Ishikawa, and J. M. Juran emphasizes that virtually all problems are caused by inadequacies in systems.[3] Therefore, all of the motivated, teamed-up, empowered, accountable, self-directed, "incentivized" employees in an organization cannot compensate for the organization's dysfunctional systems and processes. Unless managers let go of their obsession with employees as the cause and the cure of poor quality, they will never discover what quality is all about.

> All of the motivated, teamed-up, empowered, accountable, self-directed, "incentivized" employees in an organization cannot compensate for the organization's dysfunctional systems and processes.

The shift in management paradigms that began in 1950 is not a shift from a hierarchical, individual focus to a team focus. Rather, it is a shift from a hierarchical, individual focus to a systems focus. In the hierarchical paradigm, teams are seen as an extension of management, used to assist and support its needs. In a systems paradigm, teams are participants in and stewards of the system, serving customers' needs.

Teams in the New Era

The new perspective on teams is shaped by two concepts that are inseparably linked to each other:

1. *The customer-in mentality.* The customer-in mentality is more than just a frame of mind that is centered on delighting the customer. It embraces a set of values and priorities that are translated into routines and daily behaviors. Customers dominate management's thinking, values, planning, and decision making. The customer in mentality affects how teams work, how teams define what is considered a good job, how their priorities are identified, how they measure progress and much more.

   The notion of customer-in, which is adapted from the ideas of Ishikawa and Noriaki Kano (a Japanese teacher of quality at Tokyo Science University), is best understood when contrasted with its opposite, a product-out mentality (see Tables 1 and 2).[4] The tendency in this country—especially after World War II—has been a product-out mentality.

2. *Gemba.* Gemba is a Japanese expression that has no equivalent English word. It is formed from two Chinese symbols : "gem," which means "important work," and "ba," which means "a place of action." So, gemba is the work flow within an organization that directly contributes to the products and services that customers receive. In other words, gemba is the assembly of critical resources and the flow of work that contribute directly to those efforts adding value to the customer.

Every organization consists of the gemba and those who support the gemba (see Figure 1). While an organization consists of many systems, only those systems directly related to a work flow that adds value to the customer is gemba. Gemba is the part of the organization that does the work about which customers care the most. The gemba's measure of success is customer delight. The success of the rest of the organization depends on how well it serves the gemba.

Product-Out Mentality	Customer-In Mentality
• Our company defines what a "good job" is. • We design products or services that please us and then convince the customer of their value. • Marketing consists mostly of sales promotion: "How can we get more customers to buy this?"	• Our customers define what a "good job" is. • We learn from the customers what they value and design products and services to meet and exceed those expectations. • Marketing consists mostly of customer research: "What do customers need that they are not getting? What are they getting that they do not need?"

Table 1. **Differences in Mentalities**

Table 3 shows what is and isn't gemba in an organization. The purpose of this distinction between gemba and non-gemba is not to establish a new hierarchy of importance among employees or a new internal pecking order. The purpose is to define the organization's systems and identify which functions should systematically serve others.

Gemba teams are groups that collaborate (within and between each other) for the smooth functioning of the gemba systems and for the delight of the external customers. All other teams should interact with the gemba teams: when a team is commissioned, it should describe its purpose, its mission, and how its work is relevant to the needs of the gemba and external customers. The needs of the gemba should shape the focus of the teams, not vice versa. In essence, gemba teams are the customers of every other team in the organization.

The Bigger Picture

While the concepts of customer-in and gemba give an important new context to the purpose of teams, it is also important to look at the elements of systems and teamwork (i.e., an environment of collaboration). As depicted in Figure 2 there is an interdependence between purpose, systems, and teamwork:

Some Attitudes of a Product-Out Mentality	Some Attitudes of a Customer-In Mentality
• Arrogance: "We are the experts. We know better than others."	• Humility. "We must learn continually from our customers."
• A not-invented-here reaction: "If an idea is worthwhile, we have already thought of it. If we didn't think of it, it isn't worthwhile."	• Altruism and flexibility: "The customers will help us identify ideas that are good for them. We must be nimble in our response, wherever the idea originated."
• Win-lose: Self-absorption and complacent isolation.	• Win-win: Thoughtfulness, empathy, responsiveness, and modesty.

Table 2. **Differences in Attitudes between the Mentalities**

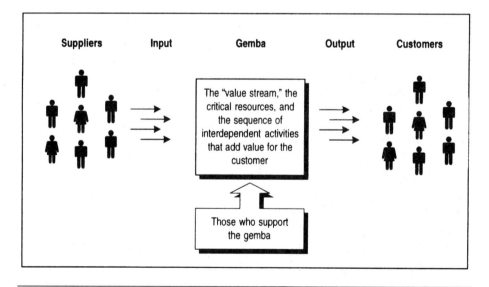

Figure 1. **The gemba**

The Gemba	Not Gemba but Provide Services to the Gemba
• Product or service design • Product development activities • Service development activities • Potential customer contact and sales • Product delivery • Service delivery • Instructional and other after-delivery services for the customer • Routine customer maintenance services	• Most management services • Customer research • Human resources • Plant or facilities repair and internal maintenance • Payroll and other financial services • Purchasing • Administrative services • Training • Budgeting • Management information services

Table 3. **What Is and Isn't Gemba in an Organization**

- *Purpose is necessary to create a system.* If a person is asked to clean off a table without knowing the purpose, he or she won't know the appropriate method or system to use. The appropriate approach for cleaning the table for the purpose of eating is different from cleaning it off for the purpose of performing surgery. As Deming taught, without a purpose, there is no system.[5]

- *Purpose is necessary to create a team.* Purpose gives definition, focus, and direction to a team. The purpose will help management decide whether a team is necessary and, if so, which capabilities or resources are needed. Without a purpose, a team is just a room full of people trying to figure out why they are there. This team won't last very long. As mentioned earlier, a team's purpose should be the external customers' needs, the work of the gemba, or a task supporting the gemba.

   Sometimes it is possible to infer a team's purpose by observing what it does and how it spends its time. Using this approach, one would conclude, for instance, that the purpose of a major league baseball team is to argue over money.

- *Systems and teams require each other.* A team with a purpose and no method will flounder and fail for the lack of a means to achieve its end. But having a system, process, or method for doing work is insufficient if there are no working relationships between those who are pursuing the purpose.

> Purpose gives definition, focus, and direction to a team. The purpose will help management decide whether a team is necessary and, if so, which capabilities or resources are needed.

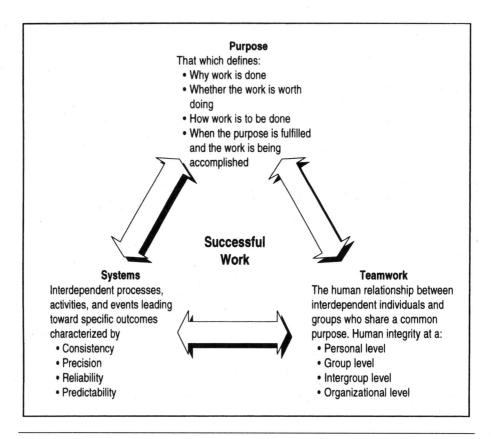

Figure 2. **The interdependence between purpose, systems, and teamwork**

The interdependent dynamics of purpose, systems, and teamwork that exist at the team level also exist at the organizational level (see Figure 3). When the larger organization lacks one of these dynamics, the team's work will be much more difficult. There should be consistency and alignment between the team's sense of purpose, systems, and teamwork and the larger organization's sense of those dynamics.

Organizations, too, exist in a larger context, such as a community, society, nation, or planet. Misalignments of purpose, systems, and a sense of community between organizations and the larger context will create dissonance and disequilibrium. Companies are slowly becoming aware of such misalignments, as evidenced by emerging studies in how organizations can start living with the planet rather than seeking to control it.

An organization's systems and interdependent interactions are complex. Thus, it is challenge for a team to succeed in the midst of these possibly conflicting dynamics. The challenge to leaders is to lead systems and not merely to proliferate teams. So far, a few concepts that can help managers better lead systems and teams have been explored:

Teams should be an expression of and vehicle for a systems view of the organization.

TEAMS IN THE AGE OF SYSTEMS

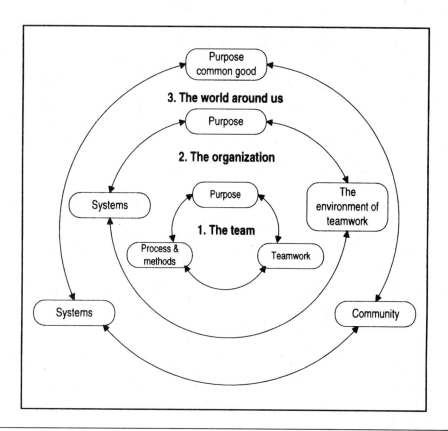

Figure 3. **Teams and systems**

- Teams should be an expression of and vehicle for a systems view of the organization.
- Teams should adopt a customer-in mentality.
- Teams should be either doing or supporting gemba work.
- No team acts in isolation. Teams are micro systems within macro systems. A team will benefit or suffer from the degree of consistency and alignment between its area of limited focus and the larger systems surrounding and interacting with it.
- The starting point for all teams is, "What is this team's purpose?" If the purpose is inconstant or unclear, there will be no team.

Using these concepts, the following examines some of the popular approaches to teams.

The Types of Teams

I decided to develop a catalog of types of teams and spent several months asking various people from various organizations, "What kinds of teams do you have?" In addition to interviewing people about their teams. I conducted a search of current books and articles on teams.

The results of the interviews and literary search are contained in Table 4. For each of the seven different types of teams listed. I have provided a brief description: identified whether it could be a gemba team: described its purpose, strengths, and vulnerabilities; and indicated when to use it and what strategies, methods, and tools might be needed. This table, of course, only briefly covers the seven types of teams. These teams are described in more detail in many of the books referred to in the sidebar "For More Information."

Two teams that are not in Table 4 (starting on page 370) are high-performance teams and self-directed teams. In regard to high-performance teams, I kept searching for what was unique about them. I could determine only one difference: Some teams are called high-performance teams, while others are not. It seems that "high performance" is simply a trendy phrase.

In regard to self-directed teams, leadership can be viewed as a role in which one person performs several functions or it can be seen as a system of functions that can be shared by several people. There is no necessary advantage to being self-directed. Self-directed teams are just as capable of genius or stupidity as teams directed in any other manner.

Systems of Teams and Networks of Efforts

As stated previously, too often there are teams that individually do worthwhile work but collectively go nowhere. This is usually indicative of an absence of internal strategies and systems that link teams and their individual missions into a larger collaborative purpose.

Kano urges people to go "an inch wide and a mile deep" rather than "an inch deep and a mile wide."[6] Rather than working on many priorities superficially (too many teams working in diverse directions), an organization should coordinate its efforts around fewer priorities.

Beginning with a carefully selected priority—one that benefits customers and the organization's long-term survival and growth—management should ask, "What will it take to successfully accomplish that priority?" The first round of answers is likely to be a list of subordinate goals. For each, management should repeat the what-will-it-take question, perhaps continuing through further sequences of subordinate goals, until it starts listing actions.

Figure 4 shows a tree diagram, one of the seven management tools, which can not only help in this process, but also serve as a way of thinking. The following should be considered when using the tree diagram:

- Using the tree diagram goes beyond management by objectives (MBO). MBO consists of goals (usually too many) with no method. MBO is not leadership. It is the abdication of leadership.
- Not all efforts must be undertaken by a team. The architectural adage "form follows function" applies here. The use of a team is a choice of form and, as such, should be determined by the function or work to be done.

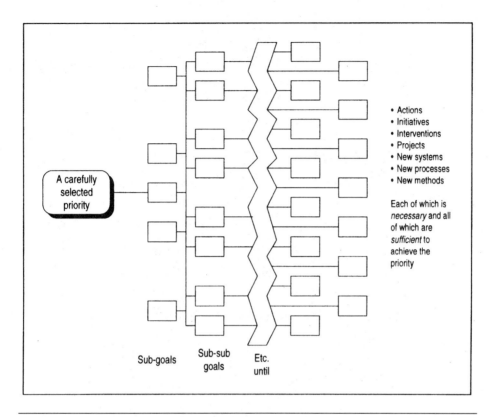

Figure 4. **The tree diagram**

- The work of each individual or group must be linked to all other efforts in a network of collaboration. Everyone must be doing his or her share, each pulling in the same direction.
- This system of work should be overseen by managers or specially selected teams of managers who offer support and challenge, provide links to line management and key internal resources, and periodically review the separate efforts. These leaders conduct the "check" step of the plan-do-check-act cycle and lead by asking good questions.

Although some of the necessary and sufficient actions identified in the tree diagram might be undertaken by one person with a simple stroke of the pen, many will probably involve complicated projects that will require an assignment to an already existing team or the commissioning of a special ad hoc team. The following questions can help an organization decide whether a team is needed and, if so, what type of team would work best:

1. *Is a team needed?* Before deciding which type of team to use, an organization first has to determine whether a team is actually needed. This

Table 4. **Types of Teams**

Type of team	Description	Purpose	When to use it	Strengths	Vulnerabilities	Strategies/methods/tools	Comments
Natural work group (self-directed or not) This can be a gemba team	People who work together every day; same office, same machine, same location, or same process. Sometimes are developed into quality circles	• To study and improve the process, to receive and/or communicate vital information, and to participate in planning • To establish and maintain the current standardized methods	• When the issues need a local perspective • When the focus is limited to this group's process, customers, and everyday work life • Should be constant and in every organization	• Creates routine monitoring, feedback, study, and quick response systems • Creates local ownership and pride • Improves awareness and attentiveness	• Overemphasizing part of the system to the detriment of the whole system • Tampering with the system • Chasing after symptoms of deeper systemic causes	• Key process indicators • Feedback loops from internal and external customers • Standardization • PDCA cycle* • Communication links with line management	Leaderless groups will need to identify the functions of leadership and the methods by which these functions will be carried out without a designated leader. These functions, too, should be standardized, best-known methods.
Business team (self-directed or not) Ordinarily these are gemba teams.	Usually a cross-functional team that oversees a specific product line or customer segment. Depending on the business it "owns," might include such functions as sales, production, marketing, and customer service	• To maintain and improve an entire coordinated system of customer-focused activities from start (marketing orders) to finish (delivery, installation, and service) • To rapidly adjust to changing customer needs	Should exist in every organization, at least as a vehicle for periodic review of the organization's performance regarding particular customers, products, and services	• Creates a systems-based flow and interaction within a conventional, hierarchical organization • Creates more rapid and nimble responsiveness to market changes	• Stretching people too thin; a small marketing staff might be required to be part of too many business teams • Having authority diminished might cause resistance from conventional managers (e.g., a plant manager) • Losing sight of larger systems	• Market and customer research • Quality function deployment • Production and service planning • Key process indicators • Monitoring of ongoing customer satisfaction	A major challenge will be for business teams to successfully shift from an autonomous-function mentality to an interdependent systems mentality. It requires that each member adapt a team mentality and the whole group create an environment of teamwork.

TEAMS IN THE AGE OF SYSTEMS

Team	Description	Purpose	Frequency	Outputs	Pitfalls	Skills	Notes
Management teams: Executive team — This team should support the gemba but is not a gemba team.	A group of managers who are peers and the person to whom they commonly report. The managing director, chief executive officer, president, etc., and his or her reports	• To lead that part of the organization over which they have control • To give direction and focus to the organization	Should be a constant presence and have regular meetings	• Builds consensus and common purpose at the top • Creates support for each other	• Developing groupthink: an unquestioned acceptance of the correctness of its beliefs • Introducing managerial fads and "viruses"	• Skills in gathering and analyzing data • Inquiry and review • An obsession with pleasing the customer • Systems and statistical thinking • The seven management tools • The methodology of improvement • Change strategies and methods • Stability strategies and methods • Consensus-building group processes	• All management and business teams need profound knowledge.** • People in leadership roles need a clear sense of purpose, values, mission, and priority and must be masters at communicating these. • Too often what is called a management team is a collection of independents who listen to each other's pronouncements. • These groups need planned and facilitated-meetings; the managers need to learn group skills. • These hierarchically based teams must develop processes of coordination with systems-based teams.
Management teams: Linchpin teams — These are not gemba teams, but they support the gemba (commitments here apply to any middle-management team).	A cascading network of teams, starting with the executive team, in which each manager is a member of a team led by his or her boss and leads a team consisting of his or her direct reports	• To create a leadership network so that constancy and consistency is maintained throughout the organization • To deploy information and collect input rapidly	Should probably be constant, meeting regularly (The frequency of "regularly" will be shaped by the work and needs of the organization, but the *minimum* frequency would ordinarily be monthly.)	• Maintains ready avenues for communication throughout the organization • Helps maintain constancy of purpose • Helps plan cycles for communicating emerging priorities and eliciting input	• Being arrogant the "not invented here" syndrome • Short-term thinking • Avoiding conflicts and risks • Being not a team, but an aggregate of independent functionally autonomous leaders; (e.g., a council of sovereigns) • Not working collectively for the good of the whole system • Insufficient planning for, and management of, the meeting process		

Table 4. (continued) Types of Teams

Type of team	Description	Purpose	When to use it	Strengths	Vulnerabilities	Strategies/methods/tools	Comments
New product and service design teams (skunk works or conventional) These are usually gemba teams.	Usually a cross-functional group assigned to redesign all or part of a product or service. "Conventional" means the team works within the ordinary context and environment of the organization. "Skunk works" means a radical departure from how and where such work is ordinarily done.	• To develop: New markets New products New services New applications or capabilities to be offered to customers • To create and maintain a market edge	As one innovation is being introduced, another one or more should be planned.	Maintains the vitality and continued well-being of the organization	• Developing groupthink • Creating innovations that please the innovators but disregard the needs of the customers	• Market research • Continuous closeness to the customer • Quality function deployment • Understanding the processes and methods of creativity • Understanding the processes and methods of data-based studies of important characteristics and indicators • PDCA cycle is a reflex and a constant	Customers generally can't tell you directly what innovations to create. But a close understanding of the customers or users and their experiences, concerns, frustrations, etc., can identify needs that the innovators can then seek to satisfy.

TEAMS IN THE AGE OF SYSTEMS

Team type	Description	Purpose	Frequency	Results	Pitfalls	Methods	Notes
Process redesign or systems re-engineering teams. These can be gemba teams.	Similar to new product and new service teams, except that they deal with the internal operations that create and deliver the product or service	New methods, processes, or systems for design, development, delivery, and/or service	Continuously but selectively used as part of a larger strategy (probably an annual strategy) to pursue carefully selected organizational priorities	• Results in dramatic reductions in cycle times, waste, and staffing needs • Allows the organization to do more with higher quality and at less expense	• Tampering; not basing changes on data, but on guesswork • Disregarding the needs of the internal and external customers	• PDCA cycle • Standardization • Systems analysis • Statistical thinking • Measurement systems • Participative methods	Redesigners must keep in close contact with those who will bear the brunt of their innovations.
Improvement project teams. These teams ordinarily provide a service to the gemba.	A natural work group or cross-functional team whose responsibility is to achieve some needed improvements on an existing process; an ad-hoc assignment	To improve an existing system or process for the design and delivery of a product or service. These improvement projects are usually more specific and have a narrower focus.	Don't overdo these: To successfully undertake these efforts requires considerable support from the entire organization.	These projects can be combined and integrated into a system of efforts aimed at achieving needed improvements. When successful, these projects eliminate the causes of problems. Problems don't recur.	• Using overly complicated methods (overkill) • Cutting corners without an understanding of what is lost (underkill) • Not applying the specific, proven, hard-nosed methods; reverting to old-time problem-solving methods	• Improvement methodology in general • Standardization • Seven-step method • Methods for communicating with, educating, and involving others who work with the process under study	Much of the quality improvement literature has been written with this approach in mind. Don't create a proliferation of these projects and teams. These are easily started but hard to sustain.

Plan-do-check-act cycle
*"Profound knowledge" is W. Edwards Deming's final legacy. See *The New Economics for Industry, Government, Education* (Cambridge, Mass.: Massachusetts Institute of Technology, 1994).

determination involves examining the nature and purpose of the work to be undertaken.
2. *Is the issue simple or complex?* Generally, the more complex the issue, the more it must be broken down into its component parts, each possibly needing an individual or subgroup to do the work.
3. *Does the issue require no special expertise or multiple experts from multiple disciplines?* While not all areas of expertise need to be represented on a team (some can be occasional consultants to the effort), the level of expertise needed depends on the issue. In general, the more complex the issue, the more expertise needed.
4. *Can the issue be dealt with quickly (days) or will it take a long time (months)?* With a prolonged effort, there are more likely to be transitions and turnovers. More members can help reduce disruption and ensure continuity.
5. *Does the work of the team involve issues that are single-function oriented, or does it cut across functions?* Cross-functional work usually needs a cross-functional team.
6. *Is the issue controversial?* It is usually better to put some dissenters on the team to help with the work rather than having them snipe at the work after it is completed. Besides, their dissent might well be based on legitimate considerations.
7. *Will implementing the project's recommendations be easy or complex?* Those who are needed to implement a solution will understand it better and commit themselves more to the solution if they helped develop it.

Tend What Is Sown

Organizations should develop a healthy skepticism about teams. Teams should not continue to exist if their usefulness and necessity can't be demonstrated. Teams should be part of a coordinated effort—a system of teams pursuing high-priority needs. Teams should be able to describe how they contribute directly to meeting external customers needs or how they contribute to those who are in the work flow that leads directly to external customers.

It's not hard to establish a lot of teams, just as it's not hard to plant a lot of seeds in a garden. The hard work of gardening and setting up teams is lending to them, nurturing them, supporting them, and preparing to process what they produce. Such direction, focus, challenge, support, and caretaking must come from leaders. It is part of the leader's new job.

Notes
1. Petronius. *Satyricon.* 66 A.D.
2. W. Edwards Deming. *Out of the Crisis* (Cambridge, Mass.: Massachusetts Institute of Technology, Center for Advanced Engineering Study, 1986).
3. Deming, *Out of the Crisis:* Kaoru Ishikawa, *Guide to Quality Control* (Tokyo: Asian Productivity Organization, 1982); and J. M. Juran, editor, and Frank M. Gryna, associate editor, *Juran's Quality Control Handbook,* 4th ed. (New York: McGraw-Hill, 1988).

4. Ishikawa, *Guide to Quality Control*, and Noriaki Kano, "Profit and Growth Through Quality," seminar sponsored by the Madison Area Quality Improvement Network, August 12–13, 1993, Madison, Wis.

5. W. Edwards Deming, *The New Economics for Industry, Government, Education* (Cambridge, Mass.: Massachusetts Institute of Technology, Center for Advanced Engineering Study, 1993).

6. Kano, "Profit and Growth Through Quality."

Peter R. Scholtes is the principal of Scholtes Seminars & Consulting in Madison, Wisconsin. He received a master's degree in education from Boston University in Massachusetts. Scholtes is a member of ASQC.

Moving beyond Team Myths

John Beck and Neil Yeager

In this article, the authors remind us that teams are made up of individuals and require leadership for both the individuals and the team as a whole. The authors propose a simple model to understand various leadership modes appropriate to optimizing individual and team performance. From this model, they provide guidelines for helping team leaders make teamwork work.

The director of organization development for a medium-sized electronics company had been working overtime to eliminate some barriers that existed between divisions. In his determination to unleash the organization's full potential, he established cross-functional teams designed to create a new kind of organization that would thrive on consensus. So far, the all-out effort has produced endless meetings but few results. The managers involved are praying that "this too shall pass" so they can get back to their "real" jobs.

In another case, a partner of a large accounting firm has encouraged her managers to improve teamwork by empowering the staff. "How can we empower anyone when we have no real power ourselves?" asks one manager. "We're merely the conduit for pushing work down the organization. Teamwork here means getting him to do this and her to do that."

The president of a major chemical company wants his senior vice-presidents to facilitate team meetings that involve employees in transforming the company. Why, he wonders, are they so unwilling to take the leap of faith?

"We're too busy putting out fires," says one.

These are common scenarios across organizations. Executives, in their attempts to keep up with the dramatic changes of the past decade, have grasped for new ways to manage their organizations. Teamwork has become essential in this brave new world, but the conventional wisdom fails to deal with the complexities of creating and sustaining a high-performing team.

Myths versus Realities

Conventional wisdom about teamwork is based on several assumptions that can prevent a team from reaching its full potential. The common myths are as follows:

Style 3	Style 2
• supporting people with their responsibilities by helping them think through problems • asking questions to help others analyze and solve problems • giving people recognition for seeking or accepting support	• involving people in your responsibilities, making decisions based on their input • seeking information for analyzing and solving problems • giving people recognition for making contributions to problem solving
Style 4	Style 1
• giving people responsibility; letting them make decisions on their own • maintaining limited communication through briefings and updates • giving people recognition for accepting responsibilities	• taking responsibility; making decisions on your own • giving information about what, how, and why to do something • giving people recognition for following directions

Figure 1. **Common leadership styles. A snapshot of the leadership styles many managers use.**

- The importance of working together as a team has replaced the importance of individual contributors.
- Putting high-performing individuals together in a group automatically creates a high-performing team.
- It takes a team a long time to be up and running.
- A team has to work through conflicts before it can be productive.
- Decision making by consensus is the best way to make a team work effectively.
- Team accountability means that everyone is responsible for everything.
- There are no leaders or followers on teams; everyone is equal.

To replace these myths, managers need a new reality-based understanding of what it takes to make teams work. One reality is that teams need leaders. Another is that leaders must know how to be effective in today's team-based environments. The Leader's Window is a model that guides leaders to accomplish those two realities.

Before examining the Leader's Window, it's important to understand the four leadership styles used by many managers, shown in figure 1. These styles aren't new. They've been part of every leadership theory since people started writing about the subject. But the way the styles are applied using the Leader's Window provides a new approach to successful teamwork.

Each leadership style offers a potential blessing and curse: It's a blessing when the style matches the conditions; it's a curse when there's a mismatch between the leader's actions and what followers need in a given situation. The labels in figure 2, "Applying the Leadership Styles," show how each leadership style can be effective if used at the appropriate times, or ineffective if used at the inappropriate times.

Our research in talking with people about what makes a "best" leader has taught us that to be effective, team leaders require some unconventional wisdom. First, they must recognize that a team is made up of individuals and the group dynamics that arise when individuals come together. Leaders must also realize that individual assignments and group efforts are not polar opposites, even though conventional wisdom says that one is either an individual contributor or a team player. Many trainers are fond of saying that there's no "i" or "u" in the word "team." The implication is that teamwork is a "we" exercise.

> Leaders must realize that individual assignments and group efforts are not polar opposites.

The result of that kind of thinking is the mistaken notion that teamwork happens only when all team members are together in a room. The reality: Not every team requires a carefully orchestrated group effort—like a basketball or football team. Some teams—such as golf, tennis, and track—call for individual excellence and limited group interaction. Most business teams require both.

To lead a successful team, a manager must know how to orchestrate effective group dynamics and how to translate a group effort into individual accountabilities. We offer new ideas for leading groups and individuals, and the Leader's Window—a new way to integrate individual and group leadership.

> To lead a successful team, a manager must know how to orchestrate effective group dynamics and how to translate a group effort into individual accountabilities.

Leading Groups

Leaders of groups must understand the predictable stages of team development: forming, storming, norming, and performing—a model developed by Bruce Tuckman in 1965, based on T-groups and other leaderless groups.

It may be comforting to believe that storming is natural, that storms will turn into productive norms, and that the endpoint will be performance. But Tuckman's model provides a distorted view of the realities of teamwork.

Most people think that forming means just meeting your teammates. Team orientation requires a lot more. It includes clarifying the team's mission, defining goals and roles, and establishing procedures for getting the work done. At the outset, a team has to form around a clear purpose and then focus on the best ways to accomplish that purpose.

Another distortion is that storming is inevitable. Actually, many teams are able to focus and become productive without storms. They come together, clarify what they have to do, and get on with the job of doing it. Storming can or might occur, but it doesn't have to. Of course, if you believe it's supposed to happen, it probably will.

It's also a distortion that storming automatically turns into norming. Some team leaders still think that dwelling on conflicts is the best way to

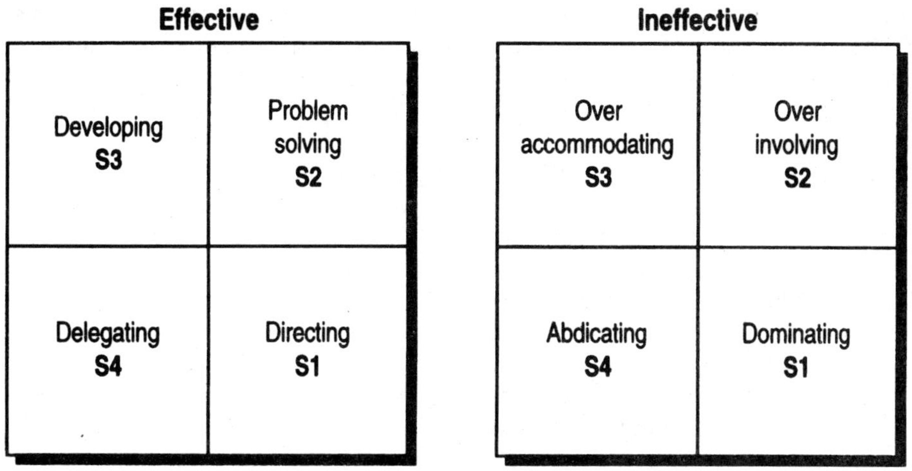

MOVING BEYOND TEAM MYTHS

Effective

Developing S3	Problem solving S2
Delegating S4	Directing S1

Ineffective

Over accommodating S3	Over involving S2
Abdicating S4	Dominating S1

Figure 2. **Applying the leadership styles**

prepare a team to perform. In reality, storms occur when a team isn't focused: the mission and goals are ambiguous, roles are confusing, and operating procedures are dysfunctional. These problems don't go away by themselves or resolve through arguing. They require that team members answer these questions: What are we expected to do? How are we going to coordinate our efforts to get it done?

A final distortion is that performing is the endpoint of group development. In our experience, teams often move beyond performing and level off. They become complacent, burnt out, or defensive. Such symptoms arise when a team goes on autopilot once it hits the performing stage. If leaders don't assume that performing is the endpoint, they can counter any leveling-off by refocusing and revitalizing the group.

To overcome these distortions, we've created a new business-oriented, reality-based model of group development: forming, focusing, performing, and leveling. We've also identified the leadership styles for effective group dynamics. It's as simple as 1-2-3.

A full team orientation requires both forming and focusing. In forming, team members need style 1, directing, from their leader to find out why they're on the team and where they're going. They need to know their mission, goals, and management's expectations. In focusing, a leader should use style 2, problem solving, to involve the team in "owning" the mission and in determining how it will achieve it. Team members need to know their roles and responsibilities, norms for communicating, decision-making procedures, and coordinating mechanisms.

If a team is formed and focused properly, it can move on to the performing stage without conflicts or power struggles. If a team doesn't form around a clear purpose and focus on how to accomplish it, storms are likely to occur.

> Storms occur when a team isn't focused: the mission and goals are ambiguous, roles are confusing, and operating procedures are dysfunctional.

But it's important to view storms as a symptom of being out of focus, not as an inevitable stage of development. If conflicts do arise, a leader should use style 2 (which involves making decisions based on team input) to clarify what the team is expected to do and to organize ways to get it done.

When a team moves on to the performing stage, the members are ready for action. That involves distributing responsibilities among team members so that each one knows what he or she must do. The most important part of performing doesn't happen in a meeting room; it happens in the workplace where individuals have to deliver on commitments to the team. For that to occur, a leader should shift to style 3 (developing) to give team members the level of responsibility they can handle and to ensure that they receive the support they need from other members.

To keep the team in the performing mode, a leader should use style 3 (which involves listening and supporting team members' decisions) when the team comes together as a whole. The members should meet to share progress reports, update changes, identify and solve problems, and capitalize on opportunities. Without such meetings, team members may redirect their focus back to day-to-day assignments, and team productivity may level off.

To make team meetings effective, a leader should avoid the trap of over-accommodating—trying to get the group to make all of the decisions. The way to avoid that trap is to distribute responsibility so that individual team members are empowered to make decisions, with the team's input. This is very different from group decision making. The result is shared leadership with a strong bias toward action.

When a decision must be made, a leader can use style 2 (problem solving) to get the team's input to his or her decision, or style 3 (developing) to guide the team member who has agreed to make the decision. The designated decision-maker should listen to other team members and then decide. If there's consensus, the decision is easy. If not, the decision-maker should be empowered to make the decision. If every decision has to wait for total buy-in, a team will get bogged down in "groupthink" or in what one of our clients calls "consensus-run-amok."

To summarize, a leader should use style 1 (directing) and style 2 (problem solving) to take charge of the forming and focusing stages that launch a group. When a team reaches the performing stage, the leader should shift to style 3 (developing) to empower team members to take action with team support. To keep a team in the performing mode and avoid leveling, a leader should bring the members together at regular intervals, using style 3 to help them make decisions and style 2 when the leader must decide.

Leading Individuals

Another myth about teamwork is that it diminishes the importance of individual contributions. This couldn't be more untrue. Individual members must be empowered in order for the team as a whole to take action. So, team leaders need to know what it means to empower people.

> ### FROM MYTHS TO MUSTS
>
> To move beyond the myths of teamwork and achieve team success, hold to these truisms:
>
> - Teams need leaders.
> - Though teamwork is important, it doesn't end the need for individual contributions.
> - With proper leadership, teams can be up and running quickly.
> - Focused teams avoid conflicts that can slow them down.
> - Teams work best when it's clear who's empowered to make which decisions.
> - Accountability resides with individual team members, not the team.
> - Effective team leaders enable all team members to lead and support each other.

If a leader just lets go, the result isn't empowerment but abdication. Leaders are guilty of abdication when they don't communicate clear expectations, when they leave team members alone too much, when they don't provide support when team members need it, and when they waffle when it's their turn to make tough decisions. Unfortunately, that's what is happening in many organizations under the banner of empowerment. You can't just let go and pray. Real empowerment means providing clear directions, delegating responsibility, being available to give support when it's needed, and being prepared to make timely decisions when necessary.

In terms of the four leadership styles, the sequence is 1-4-3-2: style 1 (directing) to clarify expectations, style 4 (delegating) to hand off meaningful assignments, style 3 (developing) to stay in touch and help team members make their own decisions, and—if necessary—style 2 (problem solving) to make tough decisions based on team members' recommendations.

According to our research, the most effective leaders practice this 1-4-3-2 empowerment sequence. Yet, that approach is counter to conventional wisdom. In the past, most managers were trained to pick the right leadership style for a given situation. But in most situations, all four styles are needed.

Conventional wisdom also holds that the best way to develop employees is to move from style 1 to style 2 to style 3 to style 4, gradually letting out the kite string of responsibility. That approach has worked only on occasion, even when the business world was slower and more predictable. In this age of rapid change, it doesn't stand a chance. In today's downsized organizations, leaders have to delegate more than ever. The best leaders use the 1-4-3-2 sequence. That doesn't mean that they treat everyone the same.

When team members take on new assignments, the most effective leaders emphasize the beginning and the end (styles 1 and 2) of the 1-4-3-2 sequence:

- 1: clear and complete directions to team members

Real empowerment means providing clear directions, delegating responsibility, being available to give support when it's needed, and being prepared to make timely decisions when necessary.

- 4: limited time for team members to work alone
- 3: rare occasions for team members to make decisions with support
- 2: a lot of identifying and correcting problems.

When team members are handling familiar assignments, the most effective leaders emphasize the middle (styles 4 and 3) of the 1-4-3-2 sequence:

- 1: limited directions to team members
- 4: a lot of time for team members to work alone
- 3: numerous opportunities for team members to make decisions with support
- 2: rare occasions when problem solving is necessary.

> Many teams break down because their decisions aren't translated into individual accountabilities.

Many teams break down because their decisions aren't translated into individual accountabilities. Each team member needs clear expectations. When team members can't get the support they need or timely decision making, they get stuck and aren't likely to accomplish their goals. The 1-4-3-2 leadership approach is a team leader's guide for getting team members to take action.

Making Teamwork Work

Now, how do you put all of this together? Remember: A team is made up of individual team members plus the group dynamics that occur when they come together. The key to successful teamwork is using the power of the group to focus and coordinate individual actions. Team meetings are useful at the outset for forming and focusing the group on actions. Meetings also provide an efficient structure for the communication needed to sustain the performing stage and to avoid leveling off into complacency. But in between these meetings, the work has to be done by individuals who are accountable for their own responsibilities and supported in their efforts.

To accomplish that, leaders must integrate the 1-2-3 sequence for leading groups with the 1-4-3-2 sequence for leading individuals. The result is the Leader's Window—a simple four-phase template for getting teams to perform.

Phase I: Team Orientation

A leader can manage a team's forming stage by using style 1 (directing) to explain the team's mission and goals. Using style 2 (problem solving), a leader can focus the team by involving members in determining the best ways to accomplish the mission and goals. Everyone must be clear about his or her roles.

Phase 2: Individual Assignments

Using style 1 and style 4 (delegating), a leader can ensure that each team member knows what he or she is empowered to do. What are the deliverables? When are they due? How do team members plan to accomplish their tasks? What do they need from the leader, other team members, and people outside the team to get their work done?

Phase 3: The Work

A leader can use style 4 to give team members the freedom to accomplish their assignments. Using style 3, a leader can check in from time to time, offer encouragement, and be responsive when the team needs support.

Phase 4: Team Problem Solving

A leader should bring together team members at regular intervals to keep them informed of the team's progress, to identify and solve problems, and to coordinate each member's efforts. When it's necessary to make decisions, a leader can use style 2 to obtain the team's input to his or her decision, or style 3 to help team members receive support for decisions they've agreed to make. A leader should always encourage his or her team to seek ways to work smarter.

Tapping the power of the group to focus individuals can help your organization's teams perform better, struggle less, and produce the results you always thought they could.

John Beck and Neil Yeager are members of the Charter Oak Consulting Group, an OD firm based in East Berlin, Connecticut. This article is based on their book, The Leader's Window: Mastering the Four Styles of Leadership To Build High-Performing Teams (John Wiley and Sons, 1994.)

Self-Directed Work Teams: A Guide to Implementation

Michael W. Piczak and Reuben Z. Hauzer

This article is a kind of mini-primer to implementing self-directed work teams. It includes information on what makes them work and how they fail. It is especially valuable for its information about trying to get SDWTs going in a unionized organization. The article includes a box with a case study from one manufacturing company, Boart Longyear, on how they instituted SDWTs and the results.

It seems virtually impossible to pick up a current edition of a business publication without finding an article about self-directed work teams (SDWTs). SDWTs, also called semiautonomous work groups or self-managing teams, have been implemented in an increasing number of companies with varying degrees of success. Such companies include Chevron Chemical, Coca-Cola, Federal Express, General Electric, General Motors' Saturn Division, Motorola, Procter & Gamble, and Xerox. The idea appears to be catching on outside the United States as well, with such Canadian companies as Babcock & Wilcox, Boart Longyear, Campbell Soups, Dofasco Steel, Honeywell, Northern Telecom, and Steelcase Equipment reporting various stages of implementation.[1]

The list of U.S. and Canadian companies adopting teams covers a broad spectrum of industries. Self-managing teams can be found in petrochemical, electronics, consumer products, heavy manufacturing, and pharmaceutical companies.

SDWTs are often viewed as an extension of or a vehicle for the continuous improvement process, which may have begun with statistical process control training. Teams are also seen as a logical structure for the implementation of total quality management, ISO 9000, or lean production techniques. Some firms see SDWTs as part of world-class manufacturing, which can include such techniques as total predictive maintenance, just-in-time manufacturing, and employee involvement. The business literature consistently refers to gains attributable to teams in the areas of quality, productivity, flexibility, commitment, and customer satisfaction.

Reprinted with permission from *Quality Progress*, May 1996. Copyright © 1996 by the American Society for quality Control. All rights reserved.

This article provides a road map for the successful implementation of SDWTs. The points raised here are based on a review of the literature, field visits to a number of organizations experimenting with the team concept, an examination of and participation in SDWT Internet forums, and hands-on experience in implementing teams with several organizations.[2]

Also set out in this article are key success factors that must be managed in the course of SDWT implementation. Recommendations are provided on what should be done, how it should be done, and by whom. In addition, some caveats—or key failure factors—are issued so that these practices can be avoided or, at least, anticipated and managed.

SDWTs Defined

It is helpful to begin the discussion armed with a working definition of SDWTs. Briefly stated, an SDWT is a highly trained group of employees (generally 6 to 15 volunteer or assigned members) that is fully responsible for turning out a well-defined segment of finished work (in this case, finished in the sense of processing responsibility).[3]

SDWTs are distinguished from most other types of teams in that they often have more resources at their disposal, a broader range of cross-functional skills, much greater decision-making authority, and better access to information. Such teams plan, set priorities, organize, coordinate with others, assess the state of processes, and take corrective action. They take on responsibilities typically assigned to frontline supervisors; in this way, the job of supervisor can evolve into a coaching or mentoring role, or the supervisor can be redeployed into some other position. Organizations might wish to consider eliminating this level of management, but the timing of such a change should be planned carefully.

> SDWTs plan, set priorities, organize, coordinate with others, assess the state of processes, and take corrective action.

Unlike earlier employee structures, such as quality circles or people involvement programs, SDWTs are not organizational appendages that require employees to engage in activities after work or apart from their daily activities. Self-directed teams do for themselves what they think needs to be done and, in this way, engage in a new and expanded set of activities.

SDWTs represent a different way of working whereby individuals behave as though their sphere of activity were their own business. In this sense, SDWTs take a step backward in time to an era when there was a greater degree of individual ownership in work. At the same time, SDWTs represent a step forward for organizations that are trying to instill a sense of ownership or entrepreneurial spirit in the work force. SDWTs require a change in the way work is organized and executed. Team members should have genuine input into substantive decisions that affect their work lives. If employees perceive their participation as being limited to marginal decisions, or worse, that management is simply engaging them in a program of the month, they will regard teams as a naked attempt to get them to work harder. In reality, SDWTs represent a new paradigm of organizational life for both team members and those who interact with them.

The transition to team-oriented work can be wrenching for all concerned. New expectations are created for members of the team as well as for members of management. Support staff must be prepared to deal with individuals for whom work has become a new and, possibly, intimidating experience.

SDWTs: A Departure from Conventional Work Design

Conventional approaches to job design divide work into narrow segments, thereby capitalizing on the benefits of specialization and division of labor as espoused by economist Adam Smith and management theorist Frederick W. Taylor. When stretched to the limit, however, these philosophies deliver the unanticipated side effects of low morale, poor quality, attendance problems, decreased commitment, and feelings of alienation.

The human relations movement helps alleviate these problems with the concepts of job rotation, horizontal job enlargement, and vertical job enrichment. Job rotation and horizontal enlargement can be viewed as extensions of the traditional approach to work design; rotation involves exchanging jobs, while horizontal enlargement expands the scope of work by adding extra duties. Both approaches attempt to reverse the damage done by overspecialization, but they are anemic attempts to inject some life into sterile job descriptions. Arguably, they are a step in the right direction, but such approaches do little to bring breadth or depth to jobs in any meaningful, permanent way. Continued detachment from the job, no sense of contribution, and job narrowness do little to eliminate apathy and alienation.

Vertical enrichment, by contrast, aims to fundamentally dismantle jobs with the intention of building motivational factors into the job package. As the job becomes more complex and challenging, the motivation to perform increases, particularly when the worker desires personal growth.[4]

SDWTs represent a practical application of vertical enrichment. Each member of a self-directed team performs many activities, and opportunities for personal growth are maximized through cross-training. Managers leave team members alone, staying out of their way as long as the team's output meets or exceeds established standards. Decisions typically within the domain of the supervisor are delegated to the team. A range of reward and recognition systems provides additional motivation.

SDWTs are a genuine departure from the usual way of organizing work. Rather than performing apparently meaningless job fragments that create distance between workers and the organization, members of SDWTs are part of a process: an inclusionary structure whereby decisions are shaped through the input of the people directly affected by them.

The Scope of SDWTs

A frequently asked question about SDWTs is how much autonomy teams should be given. The answer to this question is far from simple because it varies by organization. At Honeywell Canada, for instance, the scope of

Scope of Work for Team Members

Basic Responsibilities	Intermediate Responsibilities	Advanced Responsibilities
• Multiple work skills • Housekeeping • Movement to the point of need* • Safety • Customer satisfaction • Material replenishment • Quality at the source • Rework on the line • On-the-job training	• Total productive maintenance • Conflict resolution • Team meeting administration • Model changeover • Vacation scheduling	• Team member selection • Cost control • Performance appraisal

*Refers to reassigning personnel as required to clear bottlenecks in the operation.

Table 1. **Honeywell's Approach to Team Responsibilities**

SDWTs expanded over time as the teams matured and demonstrated a willingness to assume greater responsibility. (Honeywell's approach is instructive; see Table 1 for a summary.) The company anticipates that, in the future, team members' duties will expand yet again. But, after three years of experience, no team has progressed to the advanced stage.

By contrast, teams at Boart Longyear Canada, a mining bit manufacturer in Mississauga. ON, were already making decisions before being formally commissioned. Now, the Boart Longyear pilot team makes or participates in decisions about shift scheduling, training rotation, production prioritizing, and increasing crew requirements. In addition to holding these types of responsibilities, teams at Dofasco Steel make promotion decisions, and Motorola teams determine members' pay increases based on peer performance appraisals. Moving teams to these levels of responsibility takes time, and management must be satisfied that the teams can handle such duties.

Given these and other examples, it can be concluded that considerable variation exists when it comes to defining team scope. Deciding how quickly to escalate team responsibilities hinges on management's inclination, organizational culture variables, team members' needs and wants, and overall team maturity.

Implementing Self-Managed Teams

Several key variables must be carefully managed when implementing SDWTs. Inadequate planning will doom the initiative to failure, setting back the introduction of team concepts for many years as frustrated employees and protective unions fortify their resistance. Key factors to be considered include

A Case Study: Boart Longyear

To gain a clearer sense of how an organization goes from traditional management to self-directed work teams (SDWTs), consider Boart Longyear Inc., a manufacturer of mining machinery that has been experimenting with teams for over three years at its Mississauga, ON, operation.

The interest in teamwork at Boart Longyear evolved as a result of the company's intention to embrace world-class manufacturing techniques (WCM) to improve competitive position. Management recognized that the employees were key to WCM, and continuous improvement teams were established throughout all areas of the business. No crisis precipitated the move to SDWTs; instead, a planned-approach was adopted. SDWTs were viewed in a broad context and thus fit easily within the company's WCM program, which also included total predictive maintenance, quality management, just-in-time supply management, and employee involvement. The shift to teamwork was expected to deliver a number of benefits, including a greater sense of identification and ownership of the product for employees, improvements in productivity, and a higher standard of quality.

In preparation for the move to teamwork, Boart Longyear management anticipated that the union, while not vigorously resistant, would not publicly embrace SDWTs. In addition, because of the number of changes that had taken place at the company recently, management expected a certain amount of skepticism from employees—they might view SDWTs as just another fad and choose to ignore the movement. Management knew that communication and education would be necessary to deal with employee concerns. Fortunately, upper management and corporate support was evident.

A design team was formed composed of staff members from all levels and functions in the organization. This multifunctional group included representatives from engineering, operations, the shop floor, and the union. The group's task was to arrange site visits, gather information, analyze data, choose the pilot project, communicate with others in the company, and select consultants. Once the pilot project was under way, the design team disbanded.

To keep employees well informed, presentations were conducted both for senior managers and plant employees. Presentations to senior management were made by plant management, while consultants presented information to plant employees. These sessions were conducted not only to provide a general overview, but also to invite participants to buy into the teamwork concept.

Team preparation consisted of training by outside consultants, who provided 60 hours of instruction to members on topics such as interpersonal skills, leadership, group decision making, and running meetings. Training was achieved through a combination of lecture presentations, group discussions, field visits, videos, and self-assessments. Part of the training involved practice sessions in which fledgling teams made decisions about training schedules, work rotation, and team member selection.

A Case Study: Boart Longyear (continued)

One key stakeholder group that kept watch over employee interests was the union. Time was taken to assure the union of the company's intentions, discussions took place both informally and during contract negotiations. Management knew that it needed the union to understand that SDWTs were not in conflict with the collective-bargaining agreement. Parameters were set for how far team members could pursue various activities. For example, it was decided that team members would not be allowed to become involved with disciplining individuals, making strategic decisions, or addressing issues that required interpreting the collective-bargaining agreement. In many ways, these discussions set the groundwork for the future, when certain job functions might evolve into new roles.

In terms of startup problems, one of the greatest sources of frustration proved to be employees who were only comfortable being told what to do. Although the new culture called for more employee control and freedom, the first reaction of many was to ask their supervisor what to do next. There still appears to be a reluctance to step forward and assume responsibility, and no one person has emerged as a leader. This lack of leadership has resulted in unnecessary delays in decision making. (To avoid this, groups should begin to identify leaders early in the training process.)

One key factor contributing to the success of teams at Boart Longyear is the role played by plant manager Rick Langdon. Langdon has championed the SDWT effort in this organization, ensuring that training takes place, identifying prospective teams, and working closely with those supervisors who have since become coaches and mentors.

According to Langdon, the benefits of SDWTs to date include:

- Productivity has increased by an estimated 10% to 15%.
- Teams work harder because they do not want to appear to be underperforming.
- Individuals carry their weight, believing that they might otherwise be reprimanded by team members.
- Employees participate in regular meetings where quality- and production-related issues are discussed.
- Former supervisors work for the teams by obtaining information and securing resources.
- Employees like the concept of SDWTs because of the degree of control that they enjoy.

Langdon points to a number of fundamental concepts related to introducing teams, including doing preparatory homework, introducing a cross-functional design task force, and starting with a pilot approach. It should be emphasized that much of the work should be done up front, before start-up. The right work done at this stage saves headaches later. Failure to plan can doom implementation, cause frustration, displease the union, and make SDWTs look like just another program of the month.

- Management commitment
- Unions
- Training
- Communication
- Empowerment
- Rewards

Management Commitment

SDWTs can represent a radical departure from business as usual for most organizations. Benefits and gains from SDWTs can take many months or years to be realized. Organizations seeking a quick fix should look elsewhere for a remedy. Tangible results at Dofasco Steel, for example, were not realized until two or three years after SDWT implementation. Boart Longyear's pilot team, however, realized a 10% to 15% increase in productivity only a few months after installation (see the sidebar for details on Boart Longyear's implementation).

Clearly, an SDWT initiative must be delivered from the top down, as resources—and the green light for changes in operating practices—come from upper management. Patience and a willingness to stay the course are critical ingredients for success.

Management should devise a policy or statement of intention with respect to SDWTs. Articulating such a policy puts management on record—in the minds of employees, customers, and unions. Management might wish to include in the statement that there will be no layoffs due to the introduction of SDWTs and, where displacements do occur (for example, frontline supervisors and middle managers), retraining will be provided for alternate positions and income will be maintained at current levels for a specified period. Management should also outline the drivers for the introduction of SDWTs (for example, competitive pressures or the need for flexibility) and show a tie-in with previous improvement initiatives to provide a sense of continuity and integration.

The policy statement should be as timely and widely disseminated as possible. All members of management should also be formally advised of the organization's new direction. Making the assumption that everyone knows what is going on is a grave error.

A member of senior management should be selected as the champion and supporter of SDWTs. This will ensure that the initiative stays its course when progress is perceived to be slow, when disputes arise, and when key decisions must be made. In addition, this individual can allocate or lobby for funds to sustain training initiatives, special projects, equipment purchases, or customer visits. Such a champion can counsel organization members to be patient while SDWT results unfold. The SDWT vision will reside in this individual; he or she will be viewed as its keeper.

Management must seize every opportunity to speak out positively in favor of SDWTs. Walking the talk has become cliche, but the importance of behaving consistently with the new approach cannot be emphasized strongly enough. Lip service is not sufficient; walking the talk means making public statements in support of SDWTs, enthusiastically promoting this approach to employees and unions, and actively supporting team initiatives when requests are made for resources. Holding small, informal meetings with fledgling teams can do much to sustain the initiative. Detractors will be ever vigilant for any cracks in management commitment, and SDWTs can come to a grinding halt if there is a reluctance to spend a few dollars. Management must be either in or out for the long haul.

Unions

Seasoned management will anticipate that unions, in their effort to advance and defend employee interests, will be concerned about the implementation of SDWTs. Union resistance is to be expected and planned for accordingly. At best, management can hope for cooperation and enthusiasm; at worst, fierce opposition and sabotage. Most likely, a middle ground can be anticipated, in which the union does not speak out publicly against SDWTs. If the union does not denounce the initiative, moderate—albeit silent—support for SDWTs can be inferred. While it is preferable to have the active support of the union, passive acceptance greatly facilitates implementation.

The Canadian Auto Workers (CAW) is one union on public record as being philosophically opposed to programs involving risk sharing or participative management. The CAW has a standing committee that visits auto supplier operations where SDWTs and lean production techniques have been installed. Its major criticism of SDWTs is the lack of substantive decision making on the part of teams. Teams are viewed as a thinly disguised attempt on the part of management to exact greater output.[5]

One way to counter this type of skepticism is to include union representation on the SDWT design team. Organizations must seriously consider the attitudes of their unions before implementing SDWTs. Clearly, this responsibility rests with management. Securing union support should be an ongoing activity that begins prior to making the decision to proceed with SDWTs. The subject of a team-driven work environment can be broached during contract negotiations or at a regular problem-solving meeting.

Organizations must seriously consider the attitudes of their unions before implementing SDWTs.

Training

Richard Teerlink, Harley-Davidson's chief executive officer, has stated that if employees are empowered without the proper training, the only thing that will be certain is "you get bad decisions faster."[6] The definition of SDWTs used in this article suggests that these teams are composed of highly trained individuals. But announcing that a dozen individuals have been made into a team and turning them loose on production problems is a recipe for disaster because, for most employees, management decision making is unknown territory.

Training must address a spectrum of skills that vary in importance across employee groups. For instance, before proceeding to quality training or advanced technical skills, it is necessary to ensure minimum literacy and numeracy levels.[7] Studies consistently report illiteracy levels of 20% to 25% across North America, depending on the definition used; across Canada, figures range from 16% to 38%.[8] While management may believe that such illiteracy problems are challenges faced only by other organizations, it can be quite revealing to find the same level of inability resident within one's own company. In fact, prior to implementing its quality program, one Malcolm Baldrige National Quality Award winner found that the average numeracy and literacy skills of its work force were well below the high school level.

Training in technical, administrative, problem solving, quality management, productivity improvement, and interpersonal skills is necessary to varying degrees across teams and organizational levels. Table 2 suggests the type of training to be considered for various constituencies within a company. The numbers in the table represent the order of priority for training across groups in the organization.

Before pursuing the training identified in Table 2, it is recommended that all members of the organization receive awareness training. It should not be assumed that employees know what is happening simply because upper management has decided to proceed with SDWTs. Employees should be formally notified, and the implementation plan should be clearly explained to them.

Awareness training would cover such topics as the company's mission statement, structure and functions of SDWTs, the relationship between SDWTs and ongoing quality and productivity programs, reasons for the shift to teams, the transition to teams, new roles and responsibilities, team member selection, pilot projects, compensation and reward structures, team scope, and job security.

Consultants should be contracted to provide a substantial portion of the training. While lacking in company-specific experience, they bring a wealth of information and experience from their work in other organizations. Further, they are often more believable and competent teachers than insiders.

Communication

One of the most awkward adjustments for management is sharing information and data. Company information traditionally has been the exclusive domain of management, not to be divulged to the employees and certainly not to the union. If the organization is to work together as a team, however, full disclosure of relevant information is necessary. Employees must have access to all the information they need to do their jobs. Of course, certain confidential information, such as others' personnel files, or information that is of no direct relevance to the situation or decision at hand, should remain off limits to employees.

The group that will likely find this new policy disturbing is middle management, which regularly generates and manipulates data to produce information for decision making. Research has shown that middle managers can

If the organization is to work together as a team, full disclosure of relevant information is necessary.

Skill Group	Technical	Interpersonal	Administrative	Problem Solving	Quality Management	Productivity Improvement	Literacy and Numeracy
Team members	2	3	6	5	4	7	1
Support groups		2	3	4	1	5	
Coaches and facilitators	2	1	2	3	4	6	

The numbers in the table represent the order of training priority; the smallest number indicates the highest priority.
Support groups refer to technical support state in middle management; coaches and facilitators are those who assist with team functioning.

Table 2. **Training across constituencies**

be among the most significant contributors to the failure of employee participation programs.[9,10,11] Middle managers often feel threatened by such programs, which enable employees to propose—to upper management—solutions to problems deemed insoluble by middle managers.

Upper management must make it clear to managers and support staff that whatever information SDWTs need is to be provided without hesitation. Coaches and facilitators can be instrumental in securing data and information in the short term and, in the long term, educating team members in how to access this information for themselves.

Empowerment

The ability to make decisions and assume responsibility requires a suitable amount of authority to act autonomously. Richard S. Johnson, in his article "TQM: Leadership for the Quality Transformation," defined delegation as responsibility plus authority plus accountability.[12] The lesson here is that management must delegate increasing amounts of responsibility to teams as they show themselves capable of handling more. Where authority for spending is given, limits should be established in the same way that signing authorities and limits exist for various levels of management.

Traditionally, decision-making authority has rested firmly in the hands of the managers, who create systems and procedures, set performance standards, control and measure results, and take corrective action. With SDWTs, there will be a gradual transfer of operational decision making from managers to work teams. Such a transition can be traumatic for both parties—for managers, it means giving something up; for workers it means dealing with something new. Moving too quickly can spell disaster. A division of a major Canadian steel company discovered this when it made an aggressive movement toward teams and simply turned them loose on the organization. Declaring that teams are up and running hardly makes for a sound, full implementation.

There are no easy, cookbook answers as to how much authority to give employees and when. One possible approach is to monitor the amount of authority necessary to achieve particular goals and be ready to hand over more authority as the situation dictates. Teams must let their coaches or facilitators know, through their deeds and words, when they are ready to take on new responsibilities. The frontline supervisor, as coach, should stand ready to assist the team should it stumble, or expand its responsibility when it is ready.[13]

A note of caution: Too little authority hamstrings teams and is not much of a departure from the old way of doing things. But providing too much authority can confuse members who are unaccustomed to making decisions. Management must be ever vigilant for signs that it is time to extend or retract decision-making authority.

Rewards

A problematic area associated with SDWTs relates to reward systems. The issue is thorny for several reasons:

- Most organizations already have established compensation systems.
- Most employees want to know what's in it for them when they join an SDWT.
- Unions are typically not interested in changing or renegotiating compensation systems.

Unions will likely resist dismantling hard-won wage structures. Such structures represent certainty for employees in terms of take-home pay, and employees are not generally willing to expose their pay to potential loss. In the minds of many employees and unions, risk is management's problem.

For reward systems, management has a number of options, including:

- Adding no incentive to the present pay system
- Using a performance-based incentive system
- Switching to a profit-sharing system
- Introducing a one-time-pay incentive system for suggestions implemented, based on a percentage of savings to the organization
- Giving knowledge-based pay

Each of these options comes with a unique set of advantages and disadvantages. A detailed examination of each option, however, is beyond the scope of this article. Companies poised to proceed with SDWTs should carefully think through their policies, as the question surely will arise during awareness meetings.

Some Technical Details

A few final concerns should be addressed before embarking on the journey to SDWTs. Frontline supervisors are among the most concerned when it comes to SDWTs because their position might evaporate or evolve into a set of new responsibilities. While perhaps fewer supervisors will be required, it is clear that some amount of frontline supervision will always be needed, for the following reasons:

- Making decisions that the group is not allowed to make (such as disciplinary action, reviewing plant policies, and compensation)
- Resolving intragroup conflict
- Intervening when the interests of group members appear to conflict with those of the company

In an environment of self-managing teams, frontline supervisors can also anticipate facilitating the work of the team, assisting in problem solving, leading training programs, and negotiating for resources with upper management. Thus, while the traditional role of the supervisor might disappear, there will be enough meaningful work to keep the new coach and mentor more than busy.[14]

In unionized environments, job control and bumping rights are significant philosophical and operational concerns, and various questions will need to be addressed. For example, when positions open up in the team, must management respect seniority and job bidding arrangements, or can these be put aside for SDWTs? An associated issue is that of cross-training team members, where such training can result in individuals becoming eligible for higher-paying jobs. Union members outside the team may believe that they have been denied such training opportunities and decide to submit a grievance. When cutbacks are apparent, will the usual bumping into other departments prevail? The implication is that individuals who have not been trained in SDWT techniques will parachute into teams and disrupt team functioning. (Individuals who replace departing team members should attend training with other teams as they are assembled.) An analysis of the collective-bargaining agreement and a review of the workplace culture can be useful to head off such difficulties.

Management must decide whether team membership is to be voluntary or mandatory. It is preferable, of course, for individuals to join teams of their own accord. But, because of the realities of job classifications, contractual language, and so on, it is not always possible to have volunteers only. If an individual is opposed to joining a team, he or she should be excused from membership and reassigned as needed. Forcing employees to partake in team activities will merely poison the initiative.

It is advisable to establish a design or steering team representing all levels of the organization to assist the SDWT champion in planning and organizing the introduction of teams.[15] If applicable, union representation on the design team is also recommended.

Finally, job postings for positions within the team should state clearly that weekend, three-shift work, and other commitments could be part of team membership. Full, written disclosure will head off problems and eliminate surprises.

SDWTs Can Work

The benefits of SDWTs are real and many. The business literature provides evidence that organizations willing to move in this direction will be handsomely rewarded for their efforts. Visits to such companies will encourage management to proceed with this new form of organization. But courage is needed to undertake this bold initiative. Prosperity and marketplace survival will be the reward for those with the vision to empower employees to do the job they are capable of doing.

Notes

1. "The Celling Out of America," *The Economist*, 17 December 1994, 62.
2. Jim Clauson, "Cyberquality: Quality Resources on the Internet," *Quality Progress*, January 1995, 45.
3. R. Field and R. House, *Human Behaviour in Organizations: A Canadian Perspective* (Toronto: Prentice Hall Canada, 1995), 137.
4. R. Hackman, G. Oldham, R. Janson, and K. Purdy, "A New Strategy for Job Enrichment," *California Management Review* 17, no. 4 (1995): 62.
5. D. Robertson, J. Rinehart, C. Huxley, J. Wareham, H. Rosenfeld, A. McGough, and S. Benedict, "The CAMI Report: Lean Production in a Unionized Auto Plant," Canadian Auto Workers Union Research Department, Willowdale, Ontario, 1993.
6. "Now Hear This," *Fortune*, 22 August 1994, 20.
7. John Ryan, "Employees Speak on Quality in ASQC/Gallup Survey," *Quality Progress*, December 1993, 51.
8. "Adult Illiteracy in Canada: Results of a National Study," *Statistics Canada*, Ottawa, 1991: B. Perrin, "Literacy Counts," *National Literacy Secretariat*, Ottawa, Ontario, September 1990, and W. Fagan, "Adult Literacy Surveys: A Trans-Border Comparison," *Journal of Reading* (December–January 1995): 260–69.
9. R. Collard and B. Dale, "Quality Circles: Why They Break Down and Why They Hold Up," *Personnel Management* (February 1995): 82.
10. Michael W. Piczak, "Quality Circles Come Home," *Quality Progress*, December 1988, 37.
11. Charles N. Weaver, "How to Use Process Improvement Teams," *Quality Progress*, December 1993, 65.
12. Richard S. Johnson, "TQM: Leadership for the Quality Transformation," *Quality Progress*, March 1993, 91.
13. V. Hoevemeyer, "How Effective Is Your Team?" *Training and Development*, September 1993, 67.
14. S. Calminiti, "What Team Leaders Need to Know," *Fortune*, 20 February 1995, 93.
15. S. Phillips, "Teams Facilitate Change at Turbulent Plant," *Personnel Journal* (October 1994): 110.

Michael W. Piczak is a principal of Prime Consulting Limited and a professor of business administration at Mohawk College in Hamilton, Ontario. He received a master's degree in business administration from McMaster University in Hamilton, Ontario.

Reuben Z. Hauser is a principal of Prime Consulting Limited and a professor of industrial engineering at Mohawk College in Hamilton, Ontario. He received a master's degree in finance from McMaster University in Hamilton, Ontario.

Quality Transformation

SYSTEMS THINKING/ LEARNING ORGANIZATIONS

Self-Organization: The Irresistible Future of Organizing

Margaret J. Wheatley and Myron Kellner-Rogers

This article clearly explains how organizations as living systems are naturally self-organizing and the implications of this for effectively managing. The authors demonstrate that managing an organization as if it were a machine rather than a living system gets in the way of and distorts self-organization, but in no way thwarts it. The problem is that this distortion can threaten the organization's ability to adapt and quickly take advantage of opportunities. The ideas here are directly related to TQM's focus on systems, processes, and continuous improvement—another set of terms for focusing on interconnections, learning, evolution, and renewal to the benefit of all stakeholders.

Why do so many people in organizations feel discouraged and fearful about the future? Why does despair only increase as the fads fly by, shorter in duration, more costly in each attempt to improve? Why have the

---

This article is reprinted from *Strategy & Leadership* (formerly *Planning Review*), July–August 1996, with permission from the Strategic Leadership Forum (formerly The Planning Forum), The International Society for Strategic Management. Copyright © 1996 by The International Society for Strategic Management. All rights reserved.

best efforts to create significant and enduring organizational change resulted in so many failures? We, and our organizations, exist in a world of constant evolutionary activity. Why has change become so unnatural in human organizations?

We believe that the accumulating failures at organizational change can be traced to a fundamental but mistaken assumption that organizations are machines. Organizations-as-machines is a 17th century notion, from a time when scientists began to describe the universe as a great clock. Our modern belief in prediction and control originated in these clockwork images. Cause and effect were simple relationships; everything could be known; organizations and people could be engineered into efficient solutions. Three hundred years later, we still search for "tools and techniques" and "change levers"; we attempt to "drive" change through our organizations; we want to "build" solutions and "reengineer" for peak efficiencies.

But why would we want an organization to behave like a machine? Machines have no intelligence; they follow the instructions given to them. They only work in the specific conditions predicted by their engineers. Changes in their environment wreak havoc because they have no capacity to adapt.

These days, a different ideal for organizations is surfacing. We want organizations to be adaptive, flexible, self-renewing, resilient, learning, intelligent—attributes found only in living systems. The tension of our times is that we want our organizations to behave as living systems, but we only know how to treat them as machines.

It is time to change the way we think about organizations. Organizations are living systems. All living systems have the capacity to *self-organize*, to sustain themselves and move toward greater complexity and order as needed. They can respond intelligently to the need for change. They organize (and then reorganize) themselves into adaptive patterns and structures *without any externally imposed plan or direction*.

Self-organizing systems have what all leaders crave: the capacity to respond continuously to change. In these systems, change is the organizing force, not a problematic intrusion. Structures and solutions are temporary. Resources and people come together to create new initiatives, to respond to new regulations, to shift the organization's processes. Leaders emerge from the needs of the moment. There are far fewer levels of management. Experimentation is the norm. Local solutions predominate but are kept local, not elevated to models for the whole organization. Involvement and participation constantly deepen. These organizations are experts at the process of change. They understand their organization as a process of continuous organizing.

Self-organization offers hope for a simpler and more effective way to accomplish work. It challenges the most fundamental assumptions about how organization happens and the role of leaders. But it is not a new phenomenon. We have lived our entire lives in a self-organizing world. We watch self-organization on TV in the first hours after any disaster. People and resources organize without planning into coordinated, purposeful activity. Leaders emerge and recede based on who is available and who has information. Everything happens quickly and a little miraculously. These self-organized efforts create effective responses long before official relief agencies can even make it to the scene.

In the history of organizational theory, we have known about self-organization. Years ago, we called it the "informal organization." This was a description of what people did in order to accomplish their work. Often people ignored the formal structures, finding them ineffective and unresponsive. They reached out for the resources and relationships they needed; they followed leaders of their own choosing, those they knew they could rely on.

A more recent description of self-organization is found in a new term that describes organizations as "communities of practice." These "communities" are webs of connections woven by people to get their work done. People organize together based on their perception of needs and their desires to accomplish. The Xerox Corporation promotes this concept by stating that a successful company must acknowledge the power of community and adopt those "elegantly minimal processes" that allow communities to emerge.

And the Worldwide Web is probably the most potent and visible example of a self-organizing network forming around interests, the availability of information, and unbounded access to one another. It will be interesting to observe the Web's future now that control issues have become a paramount concern.

While there are many other examples of self-organization occurring in our midst, including well-documented experiences with self-managed teams, we will simply note that self-organization is not a new phenomenon. It has been difficult to observe only because we weren't interested in observing it. But as we describe organizations as living systems rather than as machines, self-organization becomes a primary concept, easily visible.

Order in Complex Systems

In the natural sciences, the search to understand self-organization derives from a very large question. How does life create greater order over time? Order is the unique ability of living systems to organize, reorganize, and grow more complex. But theoretical biologist Stuart Kauffman has demonstrated that the inevitable desire to organize is evident even in a non-living system of light bulbs. Kauffman constructed a network of 200 light bulbs, connecting one bulb to the behavior of only two others (using Boolean logic). For example, light bulb 23 could be instructed to go on if bulb 46 went on, and to go off if bulb 67 went on. The assigned connections were always random and limited to only two. Once the network was switched on, different configurations of on-and-off bulbs would illuminate. The number of possible on/off configurations is 10^{30}, a number of inconceivable possibilities. Given these numbers, we would expect chaos to rule. But it doesn't. The system settles instantly (on about the fourteenth iteration) into a pattern of on/off bulbs that it then continues to repeat.

How does life create greater order over time? Order is the unique ability of living systems to organize, reorganize, and grow more complex.

A few simple connections are sufficient to generate orderly patterns. Complex behavior originates from simple rules of connection. Order is not predesigned or engineered from the outside. The system organizes itself. We live in a universe, states Kauffman, where we get "order for free."

Emergence: The Surprise of Complexity

Social insects, bird flocks, fish schools, human traffic jams, all exhibit well-synchronized, highly ordered behaviors. Yet these sophisticated movements are not directed by any leader. Instead, a few rules focused at the local level lead to coordinated responses. Computer simulations that mimic flocking, swarming, or schooling behaviors program in only two or three rules for individuals to follow. There is never a rule about a leader or direction. The rules focus only on an individual's behavior in relation to that of its neighbors. Synchronized behavior emerges without orchestrated planning. (Recent commentators on the history of science note that scientists consistently avoided the conclusion that there was no leader. The belief in the need for planning and authority runs deep in Western thought.)

A startling example of complex and coordinated behavior emerging without leaders or plans is found in a species of termites. In Africa and Australia, certain termites build intricate towers 20 to 30 feet high; these are the largest structures on earth proportionate to the size of their builders. These towers are engineering marvels, filled with intricate chambers, tunnels, arches, and air-conditioning and humidifying capabilities. Termites accomplish this feat by following a bizarre job description. They wander at will, bump up against one another, and react. They observe what others are doing and coordinate their own activities with that information. Without blueprints or engineers, their arches meet in the middle.

Whether it be light bulbs, birds, termites, or humans, the conditions that create organization are the same. Individuals are similarly focused. Members develop connections with one another. Each determines its behavior based on information about what its neighbors are doing and what the collective purpose is. From such simple conditions, working communities emerge, self-organizing from local connections into global patterns and processes. Nothing is preplanned; patterns of behavior emerge that could not be predicted from observing individuals.

There is much to startle us in these scientific visions of how life organizes itself. Can human organizations be more intentionally self-organizing?

Three Conditions of Self-Organizing Organizations

If complex systems emerge from simple initial conditions, then human organizations similarly can be rooted in simplicity. During the past few years, our own search has focused on the simple conditions that support an organization's capacity to access its intelligence and to change as needed. We have seen evidence of these conditions in a wide variety of settings: in world-wide

> Individuals are similarly focused. Members develop connections with one another. Each determines its behavior based on information about what its neighbors are doing and what the collective purpose is.

manufacturers, in schools, in experiments with future battle strategy in the U.S. Army.

Organizations assume different forms, but they emerge from fundamentally similar conditions. A self gets organized. A world of shared meaning develops. Networks of relationships take form. Information is noticed, interpreted, transformed. From these simple dynamics emerge widely different expressions of organization. We have identified these essentials as three primary domains: identity, information, and relationships.

Identity—The Sense-Making Capacity of the Organization

How does an organization spin itself into existence? All organizing efforts begin with an intent, a belief that something more is possible now that the group is together. Organizing occurs around an identity—there is a "self" that gets organized. Once this identity is set in motion, it becomes the sense-making process of the organization. In deciding what to do, a system will refer back to its sense of self. We all interpret events and data according to who we think we are. We never simply "know" the world; we create worlds based on the meaning we invest in the information we choose to notice. Thus, everything we know is determined by who we think we are.

As we create perceptions of the world, we primarily use information that is *already in us* to make sense of something new. Biologist Francisco Varela explains that more than 80 percent of the information we use to create visual perceptions of the world comes from information already *inside* the brain. Less than 20 percent of the information we use to create a perception is external to the brain. Information from the outside only perturbs a system; it never functions as objective instructions. Varela describes this in an important maxim: "You can never direct a living system. You can only disturb it." This explains why organizations reject reports and data that others assume to be obvious and compelling. A system will be disturbed by information based on what's going on inside the organization—how the organization understands itself at that moment. This maxim also explains why organizations are never changed by assembling a new set of plans, by implementation directives or by organizational restructurings. You can never direct a living system, you can only disturb it.

The self the organization references includes its vision, mission, and values. But there is more. An organization's identity includes current interpretations of its history, present decisions and activities, and its sense of its future. Identity is both what we want to believe is true and what our actions show to be true about ourselves.

Because identity is the sense-making capacity of the organization, every organizing effort—whether it be the start-up of a team, a community project, or a nation—needs to begin by exploring and clarifying the intention and desires of its members. Why are we doing this? What's possible now that we've agreed to try this together? How does the purpose of this effort connect to my personal sense of purpose, and to the purposes of the larger system?

Think for a moment of your own experiences with the start-up activities of new projects or teams. Did the group spend much time discussing the

We never simply "know" the world; we create worlds based on the meaning we invest in the information we choose to notice.

deeper and often murkier realms of purpose and commitment? Or did people just want to know what their role was so they could get out of the meeting and get on with it? Did leaders spend more time on policies and procedures to coerce people into contributing rather than try to engage their desire to contribute to a worthy purpose?

Most organizing efforts don't begin with a commitment to creating a coherent sense of identity. Yet it is this clarity that frees people to contribute in creative and diverse ways. Clear alignment around principles and purposes allows for maximum autonomy. People use their shared sense of identity to organize their unique contributions. (This critical partnering of high alignment and high autonomy also appears in Information Technology discussions as design criteria for creating effective distributed data processing or client server systems.)

> Clear alignment around principles and purposes allows for maximum autonomy. People use their shared sense of identity to organize their unique contributions.

Organizations lose an enormous organizing advantage when they fail to create a clear and coherent identity. In a chaotic world, organizational identity needs to be the most stable aspect of the endeavor. Structures and programs come and go, but an organization with a coherent center is able to sustain itself through turbulence because of its clarity about who it is. Organizations that are coherent at their core move through the world with more confidence. Such clarity leads to expansionary behaviors; the organization expands to include those they had kept at bay—customers, suppliers, government regulators, and many others.

Information—The Medium of the Organization

Information lies at the heart of life. Life uses information to organize itself into material form. What is information? We like Gregory Bateson's definition, "Information is a difference which makes a difference," and Stafford Beer's explanation that "Information is that which changes us." When a system assigns meaning to data—"in-forms" it—data then becomes information.

Complex, living systems thrive in a zone of exquisitely sensitive information-processing, on a constantly changing edge between stability and chaos that has been dubbed "the edge of chaos." In this dynamic region, new information can enter, but the organization retains its identity. Contradicting most efforts to keep organizations at equilibrium, living systems seem to seek this far-from-equilibrium condition to stay alive. If a system has too much order, it atrophies and dies. Yet if it lives in chaos, it has no memory. Examples of both these behaviors abound in corporate America. The implosion of IBM and General Motors evidences how sophisticated information and measurement systems could create a sense of internal order while failing to allow for critical new information. And during the 1980s, many firms reached out chaotically without any sense of identity to markets and businesses they were incapable of managing.

Information that flows openly through an organization often looks chaotic. But it is the nutrient of self-organization. As one utility chief executive aptly put it: "In our organization, information has gone from being the *currency of exchange*—we traded it for power and status—to being the *medium* of our organization. We can't live without it; everyone feeds off of it. It has to be everywhere in the organization to sustain us."

Only when information belongs to everyone can people organize rapidly and effectively around shifts in customers, competitors, or environments. People need access to information that no one could predict they would want to know. They themselves didn't know they needed it until that very moment.

To say that information belongs to everyone doesn't mean that all decisions move to the most local units. When information is available everywhere, different people see different things. Those with a more strategic focus will see opportunities that others can't discern. Those on a production line similarly will pick up on information that others ignore. There is a need for many more eyes and ears, for many more members of the organization to "inform" the available data so that effective self-organization can occur. But it is information—unplanned, uncontrolled, abundant, superfluous—that creates the conditions for the emergence of fast, well-integrated, effective responses.

Relationships—The Pathways of Organization

Relationships are the pathways to the intelligence of the system. Through relationships, information is created and transformed, the organization's identity expands to include more stakeholders, and the enterprise becomes wiser. The more access people have to one another, the more possibilities there are. Without connections, nothing happens. Organizations held at equilibrium by well-designed organization charts die. In self-organizing systems, people need access to everyone; they need to be free to reach anywhere in the organization to accomplish work.

To respond with speed and effectiveness, people need access to the intelligence of the whole system. Who is available, what do they know, and how can they reach each other? People need opportunities to "bump up" against others in the system, making the unplanned connections that spawn new ventures or better integrated responses.

Where members of an organization have access to one another, the system expands to include more and more of them as stakeholders. It is astonishing to see how many of the behaviors we fear in one another dissipate in the presence of good relationships. Customers engaged in finding a solution become less insistent on perfection or detailed up-front specifications. Colleagues linked by a work project become more tolerant of one another's diverse lives. A community invited into a local chemical plant learns how a failure at the plant could create devastating environmental disasters, yet becomes more trusting of plant leadership.

The Dynamics of Self-Organization

The domains of identity, information, and relationships operate in a dynamic cycle so intertwined that it becomes difficult to distinguish among the three elements. New relationships connect more and more of the system, creating information that affects the organization's identity. Similarly, as information circulates freely it creates new business and propels people into new relationships. As the organization responds to new information and new relationships, its identity becomes clearer at the same time that it changes.

Earlier we stated that self-organization is not new in our experience of organizations, it just takes different eyes to see it. Self-organization has been going on all the time, but our attention has been diverted to perfecting the controls and mechanisms that we thought were making work happen. It is our belief that most people, whatever their organization, are using information, relationships, and identity to get work done. They work with whatever information is available, but it is usually insufficient and of poor quality. If they need more, they create misinformation and rumors. But always they are organizing around information. People also work with whatever relationships the system allows, often going around the system to make critical connections. Most people know which relationships would bolster their effectiveness, although this awareness may be voiced only as complaints. And as they do their work and make decisions, employees reference the organizational identity that they see and feel—the organization's norms, unspoken expectations, the values that are rewarded.

When errors or problems occur, the real work is to look into the domains of self-organization and determine what's going on at this subterranean level. In organizations, problems show up in behaviors, processes, or structures. Once we diagnose the problem, our collective practice has been to substitute new behaviors, new structures, new processes for the problematic elements. But this seldom works. *The problems that we see in organizations are artifacts of much deeper dynamics occurring in the domains of information, relationships, or identity.* If we can inquire at this deeper level, if we can inquire into the dynamic heart of organizing, both the problem and the solution will be discovered.

We observed the power of inquiring into these depths in a DuPont chemical plant in Belle, West Virginia. Safety had been a major focus for many years, addressed in many different ways. They had moved from 83 recordable injuries to none. But after more than a year with no recordable injuries, three minor personal accidents occurred within a few months. The leadership team knew from past experience that the solution to their safety problems did not lie in new regulations. Instead, they examined the organization in terms of these originating dynamics of identity, information, and relationships. What were they, as leaders, trying to accomplish? Did they still believe in their principles? How were their relationships with one another? Did everyone still have access to all information? These leaders could have responded in more traditional ways. They could have initiated disciplinary action, more regulations, safety training classes, or increased supervision. Instead, they questioned themselves more deeply and noted that because of several new members, they were no longer guided by the same shared clarity about safety. The re-creation of that clarity restored them to superior levels of safety performance.

If self-organization already exists in organizations—if people are naturally self-organizing—then the challenge for leaders is how to create the conditions that more effectively support this capacity. They do this by attending to what is available in the domains of information, relationships, and identity.

Leaders In Self-Organizing Organizations

What do leaders do in self-organizing organizations? As their organizations move towards a mode of operating that seems to exclude most traditional activities of planning and control, is there a role for leaders? Absolutely. Leaders are an essential requirement for the move toward self-organization. This is not laissez-faire management disguised as new biology. Given existing hierarchies, only leaders can commit their organizations to this path. But their focus shifts dramatically from what has occupied them in the past. In our work, we have observed many of the pleasures and perils of leaders on this path. We also are aware of some of the siren calls that seem to threaten the resolve of even the clearest of leaders.

The path of self-organization can never be known ahead of time. There are no prescribed stages or models. "The road is your footsteps, nothing else," as the South American poet Machados wrote. Therefore, leaders begin with a strong *intention*, not a set of action plans. (Plans do emerge, but locally, from responses to needs and contingencies.) Leaders also must have confidence in the organization's intelligence. The future is unknown, but they believe the system is talented enough to organize in whatever ways the future requires.

leaders begin with a strong intention, not a set of action plans.

This faith in the organization's ability and intelligence will be sorely tested. When there are failures, pressures from the outside, or employee resistance, it is easy to retreat to more traditional structures and solutions. As one manager describes it: "When things aren't going well, we've had to resist the temptation to fall back to the *perceived* safety of our old, rigid structures. But we know that the growth, the creativity, the opening up, the energy improves only if we hold ourselves at the edge of chaos."

The path of self-organization offers ample tests for leaders to discover how much they really trust their employees. Can employees make wise decisions? Can they deal with sensitive information? Can they talk to the community or government regulators? Employees earn trust, but leaders create the circumstances in which such trust can be earned.

Because dependency runs to deep in most organizations, employees often have to be encouraged to exercise initiative and explore new areas of competence. Not only do leaders have to let go and watch as employees figure out their own solutions, they also have to shore up their self-confidence and encourage them to do more. And leaders need to refrain from taking credit for their employees' good work—not always an easy task.

While self-organization calls us to very different ideas and forms of organizing, how else can we create the resilient, intelligent, fast, and flexible organizations that we require? How else can we succeed in organizing in the accelerating pace of our times except by realizing that organizations are living systems? This is not an easy shift—changing one's model of the way the world organizes. It is work that will occupy most of us for the rest of our careers. But the future pulls us toward these new understandings with an insistent and compelling call.

Meg Wheatley and Myron Kellner-Rogers are co-authors of A Simpler Way (Berrett-Koehler, September 1996) and partners in the consulting firm of Kellner-Rogers & Wheatley, Inc. Together they also lead the work of The Berkana Institute, a non-profit educational and research foundation seeking new organizational forms. Meg is author of the award-winning best-seller, Leadership and the New Science.

Strategic Process Improvement through Organizational Learning

John H. Grant and Devi R. Gnyawali

The principles of learning to improve processes, the quality of work life, and the quality of output are often not well practiced in organizations. Poor practice of these principles also compromises strategy development and implementation. The authors of this article suggest that "strategic processes and resulting economic performance can be improved through the effective integration of organizational learning practices at many levels."

Why do so many firms have difficulty improving their strategic management processes, even though many of their operating departments gain benefits from the "experience curve"? For example, an impressively successful midwestern bank improves the performance of acquired banks in contiguous markets but struggles with acquisitions in more remote parts of the country. A large diversified Asian manufacturer invests substantial sums of money annually documenting and distributing knowledge among top executives who are being transferred between countries, but the company cannot effectively use marketing data bases that are available across various divisions.

Why do such "learning problems" appear in the strategic management systems of so many organizations? What processes, systems, and conditions lead to effective and sustainable organizational learning (OL), and which ones prevent or subvert such learning in complex organizations? What can be done to help ensure that strategic performance will be better than the simple sum of individual performances?

With only a few notable exceptions, planning and related control systems have not been analyzed in terms of their contributions to organizational learning. At the same time, the "CLO" (chief learning officer) title and its variations that are appearing in some organizations do not automatically imply the

> The "CLO" (chief learning officer) title and its variations that are appearing in Some organizations do not automatically imply the involvement of these officers in key strategic systems.

This article is reprinted from *Strategy & Leadership* (formerly *Planning Review*), May–June 1996, with permission from the Strategic Leadership Forum (formerly The Planning Forum), The International Society for Strategic Management. Copyright © 1996 by The International Society for Strategic Management. All rights reserved.

involvement of these officers in key strategic systems. Further, as new organizational forms emerge, sophisticated OL systems will be needed for corporations to benefit from their investments in knowledge, much of which can "walk out the door" at a moment's notice.

Strategic processes and resulting economic performance in many organizations can be improved through the effective integration of organizational learning practices at many levels. By carefully scanning a variety of information sources for creative ways to enhance customer satisfaction and improve methods and processes, strategists can determine what their organizations need to learn in their various units in order to compete more effectively. At that point, specific steps can be taken to remove impediments to learning that are apt to exist and to implement complementary learning mechanisms that will facilitate continuing improvements.

The Need for Systematic Learning

Systematic attention to organizational learning processes is now urgent in most firms because of a combination of external and internal factors. The first factor relates to the fact that many markets are expecting more rapid and precise adjustments to technological developments, customer demands, investor aspirations, and regulatory expectations. Because of the increasing speed with which ideas can travel and products can be delivered, the need for organizations to convert either threats or opportunities into effective new offerings has never been greater.

Although some executives associated with recent successes in firms may be in positions of power and may not be interested in having the assumptions underlying the sources of their performance thoroughly critiqued, a second, even more pervasive, learning need arises because typical downsizing procedures have left many managers who previously might have registered contrasting views to stimulate learning now reluctant to do so.

Thirdly, various restructuring processes have moved many people into new positions so rapidly that they may not have had opportunities to effectively transfer their knowledge and key relationships to successors. Such mobility provides a stimulus for more systematic efforts to transfer the best ideas, while leaving open the possibilities for even better ones from the newly appointed person, but unfortunately, other factors serve to slow the responses to these demands.

Impediments to Learning

The "NIH" (not invented here) syndrome is a commonplace reminder of the pride that many individuals and units have in their unique procedures, regardless of their performance effects.

Conditions that create impediments to organizational learning exist at many hierarchical levels in most corporations. The "NIH" (not invented here) syndrome is a commonplace reminder of the pride that many individuals and units have in their unique procedures, regardless of their performance effects. At the individual level, employee turnover and relocation often lead to rapid loss of organizational memory unless systematic knowledge transfer processes are developed and used consistently.

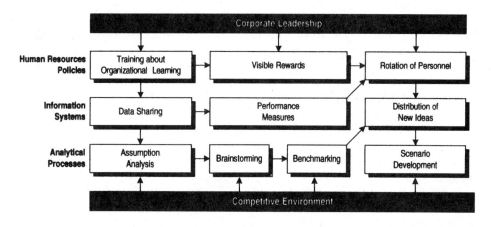

Exhibit 1. **Integrating components of organizational learning**

At the business-unit level, traditional concepts of performance measurement and perceived needs for commitment to existing assets and personnel have deterred many needed adjustments in organization strategies. The demands for consistency over time or among units make the process of timely adjustment to new insights very difficult.

At the departmental level, recent operating statistics provide strong evidence about what is working, so speculation or forecasts about why such actions won't continue to be successful in the future face major obstacles. Few organizations have conquered Professor March's challenge of developing the ability to "learn from samples of one or less," regardless of how good the one new procedure or product might be.

Unless the corporate-level actions of a publicly traded firm are sufficiently transparent that capital markets can judge its future prospects efficiently, there is a great danger that accounting-based historical data will be given more credence during internal decision processes than changes in stock price, which can often be effectively attributed to external factors or to "security analysts who don't understand our company."

The Components of Organizational Learning

While learning processes of individuals have been studied for decades, analyses at the organization level are much more recent. Some authors have begun analyzing the aggregation of learning in complex organizations; however, most such organizations need to progress much further in the integration of their "pockets of progress." Exhibit 1 describes the three areas in which strategic leaders can focus their integrating efforts.

Analytical Processes

These processes are designed to yield new insights into existing or changing conditions that should be considered during the strategic processes.

Assumption analyses can help executives determine which drivers of current performance are apt to remain predictable in the near future and which may be jolted by foreign competition, technological advances, or regulatory changes. Within business units there are often opportunities for learning from creative interaction across departments and with other units. Assumption analyses and other procedures stimulate the recognition that something can be done better—a key step in organizational learning.

Brainstorming exercises can be used to stimulate creative alternatives to current practices once assumption analyses have examined key internal and external relationships. Such unconstrained activities can help to identify new ways of serving customers, using technology, or entering new markets. With new possibilities identified, operating units are in a good position to consider just what types of activities ought to be benchmarked for "best practices" at the present time.

Benchmarking is a broadly applied and often costly practice, so the processes for gaining maximum leverage from creative applications is key to its economic benefits. Whether comparing minor activities or major processes, managers must think imaginatively about both what can be transferred and what can be improved in their particular setting. At the corporate level, capital market comparisons provide an aggregate benchmark that should stimulate questions to securities analysts if executives cannot explain price levels satisfactorily.

Scenario planning can provide an organizing framework for analyses linking variables of very different character into a coherent explanation describing a possible new condition or strategy for the department or corporation. Given the alternative scenarios generated, various decision analysis techniques can be used to help assess the risk-reward tradeoffs associated with varying rates of asset deployment or magnitudes of commitment. These and other useful analytical procedures can best aid organizational learning when they can interact with information systems that are efficient, robust, and timely.

Information Systems

The data held in information systems facilitate creativity as well as providing executives with enduring capacity to learn from prior experiences. Robust systems that can accommodate a broad array of both qualitative and quantitative data are crucial to developing alternative interpretations that will aid learning processes.

Accelerated data sharing permits knowledge gained in one part of the organization to be rapidly applied to related opportunities in distant locations or to unrelated customers. While the use of such systems in some professional service organizations may reduce the unique status of particular individuals, the opportunities to refine ideas quickly through a broad array of applications can enhance the overall capabilities of a firm.

Performance measures provide convenient linkages between information systems and human responses, so they offer unique opportunities to enhance learning. The preference in the development of measures should be for those that aid improved understanding of customer preferences, employee motivations, or investor expectations, rather than those that offer only precision, convenience, or low cost. When employees realize that they will be recognized or rewarded for something as nebulous as a good idea, one can infer that the measures are broad enough to stimulate and reinforce learning across the firm.

The distribution of new ideas may seem like such an obvious step that it is not worth mentioning, but it is amazing to note the varying qualities of distribution methods in some large firms. While some organizations post such improvements on an "intranet," others rely on slow, informal, word-of-mouth methods. In organizations operating across great distances and perhaps multiple cultures, the more casual approaches offer little chance of achieving competitive advantage.

Human Resources Policies

Carefully crafted human resources policies provide substantial opportunities for enhanced learning in most organizations. While various forms of training have received attention for years, most of it has been aimed at teaching current procedures, not analyzing methods to detect and communicate improvements derived from learning.

Training to support organizational learning involves many components, but problem solving and creativity skills are high among them. Encouragement of such skills in turn requires an understanding of the risk sharing (or "safe-failing" as some have described a similar idea) concepts surrounding experimentation within the organization. How are individuals treated when experiments fail? Are they criticized, or are they encouraged to communicate the negative results so their colleagues will not invest in similar experiments?

Visible rewards associated with both learning and teaching can facilitate communications about innovations that have succeeded. Whether there is a celebration for an effective small group effort or a major promotion for the delivery of a new product/service offering, the organization should signal that the OL component of the activity was important as well as the results, which may have been attributable to many factors.

The rotation or promotion of employees is an extremely important component of organizational learning, particularly when significant elements of the learning are quite tacit and thus difficult to document effectively. Such processes are particularly important across functional areas, business units, and different cultures, all of which can raise substantial barriers to effective communication.

Many of the significant interrelationships among these components of OL are identified in Exhibit 1, which emphasizes the internal corporate leadership and the external competitive environment as two primary stimuli for enhanced organizational learning. Leaders should be especially instrumental in the selection of training experiences, the implementation of reward systems, and the movement of key personnel into new assignments. At the same time,

> The rotation or promotion of employees is an extremely important component of organizational learning,

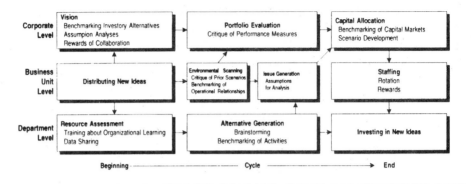

Exhibit 2. **Organizational learning throughout the strategic process**

the competitive environment will have an impact on the assumptions underlying a business, the ideas generated in brainstorming activities, the focus of benchmarking efforts, and the structure of scenarios to be developed.

Human resources policies, especially those related to training about the organizational learning expectations and establishing rewards, play a very important role in the overall effectiveness of the OL system. Performance measures are linked in several directions in order to calibrate the reward system and guide benchmarking activities.

Organizational Learning throughout the Strategic Process

The components of organizational learning can be implemented at various points in a typical strategic process. Exhibit 2 uses a common A-I-T-L (Activity-Involvement-Timing-Linkages) framework to summarize the strategic process. Many individual activities are designated within the boxes; the involvement of different people or units is suggested on the vertical axis; the timing of relationships is shown horizontally; and the arrows indicate the linkages among activities.

Typical key activities can be augmented with explicit organizational learning elements to help the firm become smarter faster. For example, at whatever time a leader chooses to articulate or refine the vision for a firm, such a statement should include some indication of the benchmarks against which progress will be measured, whether expressed in terms of financial results, marketing penetration, or quality of work life for employees. In addition, OL can be reinforced through the articulation of crucial assumptions surrounding the firm and the types of behaviors that will be rewarded over time.

Business unit personnel can become involved in environmental scanning, and critiques of prior scenarios can aid learning about previously expected results. Benchmarking at this level can involve the effects of various types of operational relationships from EDI (electronic data interchange) linkages with customers or suppliers to alliance partnerships.

At the department level, the alternative generation activity uses brainstorming in conjunction with the benchmarking of particular operating activities in other organizations. In developing an OL-focused strategic process, it is important to give explicit attention to the linkages—both human and mechanical/electronic—to become aware of which human linkages may atrophy from lack of nurture or reinforcement or be interrupted through personnel changes, and which electronic linkages may be destroyed when systems are upgraded or reconfigured.

In summary, the challenges facing the strategic and administrative systems of corporations have never been greater, even with all of the assistance of modern technology. Firms that adopt appropriate analytical procedures, robust information systems, and productive human resources policies to make OL a collective as well as an individual objective will be well positioned to prosper under the dynamic conditions that most will face in the years ahead.

John Grant is the Robert Kirby Professor of Strategic Management in the Katz Graduate School of Business, University of Pittsburgh.
Devi Gnyawali is a doctoral candidate working with Dr. Grant.

Quality Transformation

COMMUNICATION

TQM Can Be DOA Without a Proper Communications Plan

Kaat Exterbille

Everyone acknowledges the importance of communications no matter what your approach to management might be. However, many organizations, even those seeking to implement TQM, don't spend enough time developing a communications plan that will help ensure that TQM doesn't fail. This article addresses this issue and provides a chart of specific communication advice for each phase of TQM implementation.

Communications is the oil that lubricates the wheels and cogs of any change process. The introduction and maintenance of any quality and participation process is synonymous with change—changes in structure, working methods, corporate culture and so on down the line to the employee at the shipping dock or those who meet your customers on a daily basis.

Communication plans and change—Above all, the communications plan has to address the question of what the organization as a whole wishes to achieve. Employees need to know why they are being exposed to and expected to be a part of this change process. In short, the communications plan has to take into account:

- The potential contribution of every employee to the change process...
- The hoped-for attitudes and behaviors of all those concerned...
- The corporate culture one wishes to attain.

Reprinted with permission from the *Journal for Quality and Participation*, March 1996. Copyright © 1996 by the Association for Quality and Participation. All rights reserved.

"Here, now you do something with this!" doesn't work—Oftentimes, companies complete their TQM plans in the same manner as they used to complete their engineering plans—"When we're done with them, we just toss them over the wall to manufacturing" (read *communications* in this instance). So they say to their communications people "Go communicate the TQM message." The result is brochures, posters, training programs, etc. that may or may not make sense to the intended targets—the employees.

But what really matters is the impression the recipient gets of the TQM approach as a result of the communications effort—what is said, in what way and via which channel or media, is of crucial importance. Your real starting point should be based upon this foundation:

> *The viewpoint of the receiver of the message should be the starting point for your TQM communications process.*

Your total quality and participation approach must not only fit within the framework of your corporate strategy, it should be supported and strengthened by your communications activity.

If your objective is really to involve employees—What is the difference between real communications and the simple transmission of information? Communicating with employees in a manner that encourages and invites them to think and act differently, rather than just saying what we think they should think, say and do—that's the difference.

Your communications must start with your employees and should be adapted to their needs. The only criteria for the effectiveness of your communications strategy should be:

- Are employees familiar with the quality and participation philosophy?
- Are employees ready to accept change?
- Are employees involved?

Target/task oriented preparation—The key change issues and the target individuals and groups (those who will carry out the changes) need to be clearly identified. Your first step is to consider what your employees' needs are—before they will be prepared to commit themselves. To take that step, you need to answer these questions:

1. Who are our employees?
2. What do we expect from them?
3. What are their expectations of the organization?
4. What are their current attitudes?
5. What is the corporate culture?

The issues of quality and participation need to be evident in every communication, if the change is to be taken seriously. So after the preparation steps have been taken, we must look at integrating this change within the global corporate strategy. Since an organization's leaders are its chief integrators they are involved in the next step.

Leadership's role in the communications strategy—The task of top management is to regularly and consistently make the link between the quality and participation strategy and the operational results required by the organization's global strategy. In short, talking about how operational results flow from the quality and participation strategy is top management's chief role in supporting quality and participation as a strategic means to overall success. Such messages should not, repeat, should not be limited to conventional corporate media (videos, newsletters and memos, et cetera), but should also be evident in day-to-day communications.

Next to a clearly stated strategy and implementation plan on what to communicate, the most important success factor in the communications plan is the conviction and powers of persuasion of every member of management.

Communications and training links—Before developing quality and participation skills and techniques, it has to be made clear why the quality improvement effort is in and of itself necessary and why active commitment and involvement in it is needed from every employee. The firm must answer at least these two questions:

- What company results indicate that a TQM process is needed?
- How does our competitive situation, etc. tell us that TQM is needed?

Only when these questions are answered and in a manner that is understood by all should the approach to the quality issue be set out. Once the importance of the project is clear, then and only then will training in the different quality techniques be effective.

Measuring the communication process—Far from being an ad hoc or *nice-to-do* function, communications is one of management's key change process tools. The task of the communications strategy is to shape the process, so as to arrive at a change in attitudes and behavior.

As with any other strategic initiatives, being able to measure whether goals are being achieved is important. Thus, measurement techniques need to be envisaged to track the success of the communications process with benchmarks and key success indicators.

What's in it for me? The only effective way to get employees to buy into the quality philosophy is to make them fully aware of the personal advantages of this method. In other words, to communicate how their tasks will be more satisfying as they understand how their contribution leads to improved customer satisfaction and a more secure future.

On an ongoing basis, it is important to continuously communicate the results of the quality effort, as well as to give appropriate recognition to the individuals or teams who have achieved these results. Such recognition can take the form of small, not necessarily financial, awards as well as articles in the company magazine or even (what should have been normal) a simple *thank you!*

> Far from being an ad hoc or *nice-to-do* function, communications is one of management's key change process tools.

The Phases of a Complete Communication Plan for Your TQM Process

As our discussion so far already has indicated, there are distinct stages of phases in a TQM communications strategy. In communications, as well as change parlance, these are:

- A preparation phase...
- An awareness phase...
- An adaptation phase...
- An implementation phase...
- An adoption phase.

The awareness phase—This first phase involves explaining clearly why change is necessary. A mission statement setting out the values and aims of the company is an essential communications tool in this situation. Ideally the message will come directly and simultaneously from top management to everyone involved.

At the same time all managers need to receive background information and briefings in such a way that everyone tells the same story and that any bottom line explanations can be backed with facts and figures (see table 1).

An acclimation milepost... Employees need some period of time to identify with the new strategy. To ensure this, they need to be involved in the development of the strategy itself. Here effective *one-on-one* communications are essential. The message should not just come from the top down. Since, employees the world over watch a leader's feet more closely than their lips, a really key factor here is readiness by management to listen to the input from employees and to act accordingly. To facilitate this process, managers have to be coached and guided in developing personal communications skills. Management has to set the right example in its attitudes and behavior.

The adaptation or implementation phase—Once employees have identified with and are ready to commit themselves to the mission, they need to hear clearly and substantively exactly how they can contribute. Two communications techniques are relevant at this stage: task-oriented and personally-oriented messages.

Using both task and personally oriented examples, the mission has to be interpreted for every employee individually, while reflecting the corporate values and environment. Explanations of how strategic and operational applications of the mission benefit of day-to-day operations should be clear to both operational employees and human resources management and staff.

A final word on feedback—Feedback on the new organizational culture should be continuous. Opportunities should also be provided so that employees can give their views on the overall quality program and get clarification of their individual roles. Even those employees not directly involved in project groups should be associated with the program by receiving information on both the results and the colleagues who have helped achieve them.

> Once employees have identified with and are ready to commit themselves to the mission, they need to hear clearly and substantively exactly how they can contribute.

Table 1. **The Links between the TQM Plan and the Communications Planning Function**

Phase	Question	TQM action	Communications task

Preparation		• Top management meetings	• Establish a steering group
			• Employee research:
			  Needs of target groups
			  Perception of the organization
			• Desk research:
			  Corporate vision/mission
			  Corporate strategy
			• Provisional communications strategy and timetable

Awareness	Why change?	• Awareness sessions at all levels • Quality philosophy and methodology • Identification of process problem areas	• CEO message • Background briefings to line management • Identification of social and management problems areas
Process audit and adaptation	Why am I involved?	• Social Pareto • Social audit • Technical Pareto	• Vision, TQM mission • One-on-one communications motivational platforms • Management communications skills Leadership training • Support for middle management • Optimize communications methods and media

Implementation	How do I do it?	• Project groups
		• Method of implementation
		• Presentation to steering group
			• Task oriented communications
			  Skills training
			  Techniques
			  Information on techniques
			  Measurable tasks

	What do I have to do?		• Person/individual oriented communications
			  Communication about the desired new culture and its values on quality and participation
			  Feedback mechanisms and participation

		• Cross-functional coordination	• Cross-functional communications
Adoption	Are we doing it right?	• Adopt strategy • Strategic benchmarking • Re-engineering/work redesign • Quality function deployment • Networking	• Benchmark research • Customer and employee research • Communications network • Human resources evaluation of values and contribution to culture.

Kaat Exterbille is the managing director of Kate Thomas & Kleyn—a Brussels, Belgium, corporate communications and change management consulting firm.

The Partnership Facade

Jim Harris

We all know the phrase "walk the talk," but too often that is just another hollow phrase in the rhetoric of managers who are reluctant to give up their status and really communicate with employees. Still, such communication is vital; as this article points out, not sharing information and maintaining status-based relationsips is simply another symptom of organizations that do not have their act together. This article includes a useful short survey to help understand your company's "love quotient."

An easy way to gauge the strength of the internal partnerships in your company is to listen to the way your colleagues describe the company. People who feel a sense of partnership continually speak in terms of *"my* company," *"our* products," "it's all of *us* pulling together." By contrast, people who feel like hired hands say such things as *"the* company," "if *they* would only listen," "it's *their* problem."

Some companies change the words they use to describe each other in the hope of creating a stronger sense of internal partnership. Instead of calling each other *employees,* they decide to refer to each other as *associates, internal customers, teammates* and, yes, *partners.* But just changing a few words without embracing the core strategies that actually create vibrant partnerships only creates a partnership facade.

Real partnerships are made of more than words. They are made of actions, built up over time and proven through trust. The specifics may change from company to company, but the strategies for creating powerful partnerships are universal:

1. Squash Status Barriers

Status barriers are everywhere. I wish it were passé to discuss the negative impact status barriers have and how they erode the morale and commitment of employees. I also wish companies would recognize the often devastating psychological and emotional barriers that unnecessary status distinctions create between employees and the company. When employees live within a system of obvious "have and have-nots," morale is usually low and performance is often marginal.

> When employees live within a system of obvious "have and have-nots," morale is usually low and performance is often marginal.

Excerpted by permission of the publisher from *Getting Employees to Fall in Love with Your Company.* Copyright © 1996 by Jim Harris. Published by AMACOM, a division of the American Management Association. All rights reserved.

To create real partnerships, we must actively attack anything that artificially builds barriers between employees. As managers, we must ask ourselves some tough questions:

- How does reserved parking move the business forward?
- How does limiting performance incentives to the few inspire the commitment of the many?
- What is the impact of time clocks, pay docking, differential vacation or benefits plans and probationary-periods on our internal partnerships?
- How can we expect any employee to give his or her absolute world-class best if we promote and support a system that divides individuals into first-, second- and third-class status?

The number of management layers between the front line and top management is a status barrier we often overlook. There are only two layers of management between the CEO and the newly hired 18-year-olds at many companies, including Quad/Graphics, the large, Wisconsin-based magazine printer. There are only three layers of management between the CEO and the front line at Asea Brown Boveri (ABB), the $29 billion European engineering firm with 215,000 employees across 140 companies. There are only five layers of management between Pope John Paul II and all the parish priests in the 750 million-member worldwide Catholic Church.

The greater the distance between the CEO and the front line, the greater is the perceived status barrier. One headquarters accounting group for a major company has an unimaginable eight layers of management between the clerks and the chief financial officer—and they are all located on the same floor of the same building! An unbelievable eight layers of "status" within one 4,000-square-foot office.

2. Open the Company Books

Jack Stack, president of Springfield ReManufacturing, is often heralded as the leader of the open-book management craze. Open-book management, as the name implies, means opening the financial and operational statements to all employees. Stack believes that the best way to ensure the success of a company is by teaching everyone how to read the company's financial statements and learn how their function contributes to the company's profits. There is no better way, according to Stack, to help employees contribute to a profitable business than to teach them what a profitable business means to them and to their specific areas.

Too many companies continue to operate under the outdated philosophy of sharing the financial numbers with only a few select employees. This implies that:

- The company doesn't trust its employees;
- It believes employees are not capable of understanding the numbers; and
- It's not the employees' job to worry about the big picture.

Wrong, wrong, wrong. First, if you do not trust your employees, why did you hire them? Second, front-line employees who can raise families on $400 a week can certainly comprehend an income statement or balance sheet. Third, a supervisor cannot effectively contribute to overall company profitability if he or she is limited to understanding just the individual department's numbers. The maximum contribution to the company will occur only when the employee sees and learns how all aspects of the business contribute.

3. Pay for Performance, Not Titles

Pay has traditionally been a function of the position you held in the company hierarchy. It is usually based on such things as responsibility, the number of people who report directly to you, and the amount of budget you control. It represents an entitlement based on a title since the higher the position and the greater the span of control over people and budgets, the greater the pay. Therefore, we focus more energy on how to get promoted than on how to add value to a service or product.

We need to remember that a job title has never served a customer, repaired a machine or improved a manufacturing process. Dedicated people serve, repair and improve our company. In order to keep our best employees, we must refocus our pay efforts to regard actual performance rather than a mere job description.

Pay-for-performance programs are growing in popularity. They take many forms, including pay for knowledge, gain sharing programs, lump sum payments instead of raises, team productivity incentives, and employee stock option plans. The use of these systems will continue to expand, and for good reason: Employee commitment to company profitability and productivity increase when pay is directly tied to performance.

Progressive companies can create long-term partnerships with employees through pay-for-performance systems. It is common to find highly skilled, highly motivated and wonderfully productive 25- to 30-year veterans at pace-setting companies like Lincoln Electric in Cleveland and Nucor Steel in Charlotte.

4. Share the Bad Times as well as the Good Times

Consider some recent events:

- A major airline conducts a four-day board meeting in Paris while simultaneously negotiating huge labor concessions from its unions.
- A Fortune 500 company cuts the salaries for 120 executive secretaries while granting millions of dollars in bonuses to the secretaries' bosses.
- A board of directors grants huge stock blocks and bonuses to executives for slashing thousands of jobs and closing down dozens of operations.

> The maximum contribution to the company will occur only when the employee sees and learns how all aspects of the business contribute.

Are Employees Falling In Love With Your Organization?

Knowing where to start on the road to getting employees to fall in love with your company is very important. Here is a survey to help you decide on which of the five key priniciples you should begin. Circle one number for each statement. Add the numbers you circle for each principle and record the score in the space provided.

For each statement, circle the number that best describes your department or organization.
Circle 5 if you strongly agree with the statement.
Circle 4 if you somewhat agree with the statement.
Circle 3 if you are neutral on the statement.
Circle 2 if you somewhat disagree with the statement.
Circle 1 if you strongly disagree with the statement.

CAPTURE THE HEART
1. We have a written vision that is known to all and lived every day.   1 2 3 4 5
2. We seek a creative, low-cost way to balance work and family.   1 2 3 4 5
3. We love to celebrate and find ways to inject fun into the workplace.   1 2 3 4 5
                                        Capture the Heart Score_____

OPEN COMMUNICATION
1. It is obvious that management considers internal listening a priority.   1 2 3 4 5
2. Attention is given to using multiple communication channels—more than just using memos and e-mail.   1 2 3 4 5
3. Employees receive feedback in real time (immediate, direct, positive) rather than merely occasional performance appraisals.   1 2 3 4 5
                                        Open Communication Score_____

CREATE PARTNERSHIPS
1. There are few, if any, status barriers between employees (i.e., reserved parking, bonuses only for top management, special benefits).   1 2 3 4 5
2. We actively share financial numbers, ratios, and company performance measures with all employees.   1 2 3 4 5
3. Management visibly serves the front-line, customer-contact employee first (providing tools, resources, and training) before asking the front-line employee to serve us with reports, paperwork, etc.   1 2 3 4 5
                                        Create Partnerships Score_____

DRIVE LEARNING
1. We guarantee lifelong employability (rather than lifetime employment) through offering extensive training, cross-training, and work variety.   1 2 3 4 5
2. Special attention is given to creating visible, activity-filled programs that help drive learning through all levels of the organization—up, down, and laterally.   1 2 3 4 5
3. We actively support a philosophy of lifelong learning for our employees that goes beyond focusing only on today's job needs.   1 2 3 4 5
                                        Drive Learning Score_____

EMANCIPATE ACTION
1. We allow employees the freedom to fail and try again.   1 2 3 4 5
2. Constant attention is given to creating freedom from bureaucracy, unnecessary sign offs, outdated procedures, and office politics.   1 2 3 4 5
3. All employees are encouraged to openly challenge the status quo to help find better, faster, more profitable ways to serve our customers.   1 2 3 4 5
                                        Emancipate Action Score_____

Summary Score: The higher the score, the more you believe this principle is alive and well in your organization. The lower the score, the more your organization needs to address this principle.

Rare are the announcements that senior managers are sharing the burden. Yet far too often are the announcements that management boards profit in good times while greedily accepting huge bonuses during bad times. Such activities tear at the very soul of the vibrant partnerships necessary to win in today's marketplace. Even on Wall Street, the "share the wealth" mentality seems to have died.

Living on the front line is a scary proposition. Headlines endlessly proclaim the elimination of jobs through layoffs, downsizings, closings and restructurings. Most often the hardest hit is the front line. It is impossible to totally commit your heart, mind and soul to an organization when you know that another group usually gets the lion's share of the "good" while your group usually takes the brunt of the "bad." Any successful business partnership is built upon a foundation of all parties equitably sharing both the good and the bad times.

5. Serve the Front-Line Partners First

For close to 20 years, business experts have sung the praises of inverting the business pyramid. By placing managers at the bottom of the pyramid and the front-line staff on top, you demonstrate that a manager's key function is to support the front line. For many, the idea of managers serving the front line first is still an extremely foreign concept. The "power of inversion" continues to remain the philosophy of the few. Many companies continue to enforce policies and procedures mandating that front-line employees serve management first.

A powerful example of how management continues to demand being served first is the sacred operations manual. The overwhelming majority of operations manuals represent what the front line must do to make the job of managers easier. These manuals, filled with reports, forms, instructions, exceptions, returns, guidelines, lists and justifications procedures, often choke the front-line staff's ability to serve the customer. One large company has a checklist for all the checklists its in-store managers must complete! In this kind of environment, who is serving whom?

The philosophy of serving managers first would appall the military. Companies love to use military metaphors when referring to their operations. "It's a war," "we need to rally the troops," "let's dig in and win the battle," and "don't give up the ship" are common corporate proclamations. Every organization that views itself as being in a "war" and enjoys using military metaphors to describe its competitive environment ought to remember the one absolute and unbreakable law that all military commanders must obey: *Feed the troops first!*

Jim Harris is author of Getting Employees to Fall in Love With Your Company *(AMACOM, 1996), from which this is excerpted.*

> Rare are the announcements that senior managers are sharing the burden.

Quality Tools and Techniques

PROCESS REENGINEERING

Why Did Reengineering Die?

Oren Harari

Reengineering didn't really die, except as another quick fix, which like all other quick fixes, doesn't live up to the promise. Reengineering is, simply, taking radical actions to improve the efficiency and effectiveness of business processes to benefit company and customers. To succeed, however, reengineering requires hard work and thinking as unconventional as the new processes the company may be creating. This article by the plain-speaking and -writing Oren Harari explains such ideas in more detail.

Is reengineering really dead? Well, it depends. If it is revered, as it often is, as a guaranteed blueprint for organizational success, or as a magic vehicle that miraculously circumvents the vagaries of today's turbulent global economy, then it's been dead a long time. In fact, it was never alive. It came into the world of management hype stillborn, even as its apologists continued to worship it.

On the other hand, if reengineering is viewed as a set of commonsense ideas about revamping operations so as to improve efficiency, then it's as alive as any intervention that seeks to remove clutter and stupidity from organizational processes. But even with this more modest, realistic perspective,

Reprinted with permission from *Management Review*, June 1996. Copyright © 1996 by the American Management Association. All rights reserved.

most organizations that have launched reengineering have managed to kill it. More on this shortly.

First, a little history. In summer of 1990, Thomas Davenport and James Short published an article in *Sloan Management Review* titled "The New Industrial Engineering: Information Technology and Business Process Redesign." The article, as the title suggests, articulated means by which processes could be overhauled by capitalizing on advances in technology so as to, among other things, improve crossfunctional efforts. That same summer, Michael Hammer published a sexier piece in the *Harvard Business Review*, with the provocative title "Reengineering Work: Don't Automate, Obliterate." This article extended the thoughts of Davenport and Short by specifically calling for outright elimination of functional fiefdoms and, for good measure, many of the sacrosanct steps in conventional workflow. It promoted the idea of using information technology to link processes—like product development or order fulfillment—that sliced across functional groupings, so that workflow was organized around those integrated processes, not around typically discrete functional fiefdoms. Since the article was well-written and pragmatic, it was very persuasive. The requests for reprints were overwhelming.

But not as overwhelming as the response to the next tome, *Reengineering the Corporation*, by Hammer and his colleague James Champy. This book expanded the conceptual rationale for reengineering and provided the kinds of stories that made executives salivate. Like the story of IBM's credit issuance department, which somehow managed to reduce its turnaround cycle time from seven days to four hours (!!!) while increasing the volume of transactions by a factor of one hundred (that's one hundred times, not one hundred percent), all with fewer personnel.

Examples like these rolled off the pages of Hammer and Champy's book. Can you imagine the excitement they caused? The sheer possibilities blew manager's minds. Reading the book was a religious experience for some, an erotic experience for others. For finally, salvation had come. The magic bullet. The cure for all woes. The simple step-by-step to-do list that would transform any organization into a model of leanness, productivity and market leadership. It was a dream come true.

No, it was just a dream. As with TQM and other fads, the people who most benefited from what Davenport now derisively calls the "Reengineering Industrial Complex" were the external consultants and the in-house staff apostles. As Davenport asks: "How did a modest insight become the world's leading management fad? How did reengineering go from a decent idea to a $51 *billion* (my emphasis) industry?"

Reality Check

The answer is simple. As long as people are willing to pay for quick fixes and snake oil, someone will be out there selling it. Eventually, reality sinks in. Today, the $51 billion industry is downsizing dramatically, a delicious irony CSC Index, the original Hammer-Champy purveyor of consulting services, is losing people. Gemini Consulting, another reengineering guru-shop, is laying them off. The reason is that clients are getting hip to the fact that the underlying premises of reengineering (like eliminating counterproductive organizational barriers and process steps) are eminently sensible, but they've always been around. There's nothing secret or proprietary about them; it's old wine in a new bottle. There's no need to pay up the nose for a pretty package. In fact, the pretty package is seductive, for it may suggest that the changes which have to be made are relatively linear (one-two-three, follow the list) and relatively painless to execute.

> The premises of reengineering are easy to understand in principle, and damnably difficult to execute.

Nothing could be farther from the truth. In fact, the premises of reengineering are easy to understand in principle, and damnably difficult to execute in practice. That's why the research findings on the actual efficacy of reengineering interventions are sobering, to say the least. *The Economist* summarizes studies that indicate that 85 percent of interventions just plain and outright "fail." Even CSC's own follow-ups lead them to conclude that about two-thirds of interventions have yielded "mediocre, marginal or failed results." Wow.

Here's the reality flash: Reengineering died not because there's anything "wrong" with it. Quite the contrary; reengineering is one (though not the only) sensible way for leaders to rethink their business and overhaul their organization. The reason reengineering failed is the same reason that any change intervention fails: poor preparation, poor follow-through, poor foresight, poor business acumen, and a marked dearth of vision, courage and persistence from executives. In short, rotten execution.

> The reason reengineering failed is the same reason that any change intervention fails: poor preparation, poor follow-through, poor foresight, poor business acumen, and a marked dearth of vision, courage and persistence from executives. In short, rotten execution.

Rather than belabor this point any further, let's strike a more positive note by examining the flip side. Suppose you're enamored with the possibilities of reengineering, or suppose your organization has already committed to it. What can you do to help ensure its success? I propose four steps that are rarely followed with passion and commitment, but if they were, they would circumvent any premature death of reengineering. Indeed, these four steps would truly help you and your organization realize the genuine potential of the intervention.

1. **Link all reengineering efforts to the external market in general, and the customer in particular.** Many efforts fail because people focus their attention internally. They forget that their primary concern ought to be responding to what's "out there." They stolidly focus on the minutia of reengineering protocols even as consumer preferences shift and new competitors with new technologies enter the fray, thus rendering the reengineering activities useless. All the reengineering in the world won't revitalize a product line or technology that is becoming obsolete.

   In the world of consumer electronics, reengineering your organization won't help much if it's wedded to analog technology while the market leaders are moving toward digital interactivity. In the world of pop music distribution, reengineering your organization won't help in the long run if new no-name competitors are launching avenues for artists to record their own music, put it on the Net, and sell it there. In the retail banking arena, reengineering the organization will only buy you a little time if companies like software purveyor Oracle and credit card authorization king VeriFone are successful in jointly developing mediums for digital cash payments that will allow buyers and sellers to bypass the retail banking sector altogether. If you're in transportation, reengineering your centralized truck route system won't help you compete against folks like industrial packaging distributor Conifer Crent, which is in the process of eliminating centralized routes altogether; Conifer Crent visualizes a system where self-managed cells of truckers, sales and service people "own" a set of customers and are able to deliver any volume of customized industrial packaging to anyone at any time.

The point is, if you're going to reengineer, make sure your efforts aren't merely perpetuating a system that's dying. The most efficiently produced buggywhip is still a buggywhip. When leaders focus externally, they understand which internal interventions make sense to pursue and which don't. In the case of the examples above, they focus their reengineering efforts on ways to capitalize on digital technology, on Internet distribution, on electronic money, on self-managed transportation cells—because that's where the external marketplace is heading, and that's what it takes to delight increasingly fickle customers. In short, let the marketplace and the customer drive the direction of reengineering efforts, not a consultant with a set of standardized "to-do" protocols.

> *If you're going to reengineer, make sure your efforts aren't merely perpetuating a system that's dying.*

2. **Go 100 percent.** Your efforts are doomed to either mediocrity or failure if you apply reengineering to only one process, or you implement it in such a way that current processes are merely tweaked. Let me highlight a few selected comments that Michael Hammer voiced in a major conference in 1990:

"Reengineering is not incremental thinking. It is starting from scratch with 'green fields' design; rejecting conventional wisdom and thinking out of the box."

"Where quality programs focus on improving existing ways of working . . . reengineering questions the fundamental work mode. . . ."

"This is an order of magnitude change. . . . We are not going to 'fix' our organizations. We have to throw them away and start all over again."

Heady words. Whether or not you buy the Moses-on-the-Mount tone, Hammer's basic point is sound: Simply improving current processes will at best yield bland results. In fact, he suggests that the "theology of incrementalism makes it difficult to establish the appropriate climate for reengineering." That's why Ray Larkin, CEO of pulse oxymeter manufacturer Nellcor Puritan Bennett, argues that one would be wise to avoid goals which are attainable using current processes. He suggests that the trick is to shoot for goals which are clearly impossible using current processes, which will force people to scrap the old system and invent a new one if they want to be successful.

A couple years ago, Larkin proposed a series of so called impossible goals for Nellcor, like getting inventory turns from 3 percent to 20 percent, and reducing service turnaround time on a major product line from three weeks to one day. The key, says Larkin, is to first genially agree with everyone's inevitably furious reaction that the goals are impossible. Then comes the important part: Give people free reign to form project teams to dismantle the existing process and put one in that will meet those goals.

What happens? Well, as they mercilessly scrutinize operations, people first discover that there's no point in making more efficient what shouldn't be done at all. Then they discover that they can design entirely new ways of attacking work which will allow quantum leaps in performance and productivity. Larkin is so committed to this process that he asserts that giving people modest goals is both "insulting and demeaning."

> Giving people modest goals is both "insulting and demeaning."

3. **Concentrate on the human side of the intervention, not the technological.** Reengineering involves automation and technology, to be sure, but it is not equivalent to them. Reengineering may even require some downsizing, but that is not its primary goal—nor, argues Davenport, was it ever meant to be a primary goal.

To quote the *Economist:* "The 'soft' side of reengineering (winning over the workers) is even more important than the 'hard' side (such as installing new computers)."

Davenport is even more adamant. He asserts that reengineering is only valuable to the extent that ". . . it helps people do their work better and differently. Companies are still throwing money at technology—instead of working with the people in the organization to infuse technology."

Efforts that have a chance for success do not unilaterally impose reengineering on a confused, scared workforce. They do not hierarchically insist that people follow the sterile script of a consultant or an in-house technonerd. They do not expect to generate enthusiasm by terrifying people with insinuations of "inevitable" downsizing.

Rather, they include and involve people from Day One. They immerse people in training while providing them with the technology and authority to initiate reengineering. They encourage them to use the new tools and techniques to themselves overhaul processes and eliminate counterproductive steps, with the bigger purpose of enhancing revenues and reducing costs. In fact, successful interventions encourage people to obsolete their own jobs and figure out new ways of adding value—in effect, developing new skills and inventing new jobs for themselves in the organization.

I just finished reading a wonderful little story in *Fortune* magazine about a volunteer work group in Hewlett-Packard's North American distribution organization. In a short time frame of nine months, this work group managed to completely reengineer the creaky, slow process by which product reached customer. Among other things, the group reshaped a process that involved 70 separate computer systems and indeterminable byzantine steps into a single, unified database covering everything from customer order through credit check, manufacturing, warehousing, shipping and invoicing.

But this group was unique in that reengineering wasn't "done" to them. In fact, nothing was done to them. The team leaders went out of their way to step aside, not provide pat answers and—amazingly enough—not have any sort of "organization" at all. That is, team members did not have any formally defined responsibilities; it was up to them to figure out—individually and as a team—what to investigate, what to ask, what to work on, what to strive for. To quote Julie Anderson, one of the team leaders: "We took things away: no supervisors, no hierarchy, no titles on our business cards, no job descriptions, no plans, no step-by-step milestones of progress."

The team leaders provided training, tools, support and encouragement. The idea was to reduce the myopia and defensiveness that come from formally assigned responsibilities, and to create a sense of personal ownership. People used their own wits and visions to challenge the process and come up with a radically different alternative.

Now that's what reengineering is really about. How many organizations have tried to implement it that way? No wonder the track record is so dismal.

4. **Emphasize the vision, passion, courage, fortitude and dynamism parts of leadership.** If there's anything that requires leadership, reengineering does. When times are calm and stable, "management" will suffice. Whenever big change occurs, however, leadership becomes vital. Effective leaders demonstrate five fundamental attributes:

> If there's anything that requires leadership, reengineering does.

- Vision—they can paint a compelling, exciting picture of what reengineering is, what it can do for the company and the individual, and what it means for the organization's structure and culture.

- Passion—they are genuinely excited by the possibilities, they demonstrate that excitement every day, and they work hard to infuse everyone with the same feelings.

- Courage—they are not afraid to tackle sacred cows, they are not afraid to question, to challenge complacency, to ruffle feathers, to create discomfort with the status quo, and to provoke people to move beyond their comfort zone. Neither are they afraid to publicly apply the principles of reengineering to their own jobs.

- Fortitude—they are tenacious, and bulldog-persistent. They do not give up in the face of initial skepticism and reluctance from the troops. They push on, they slog, they demonstrate their day-in-day-out commitment in every meeting, every question, every report, every review. They set an example, because they understand that unless they demonstrate personal change in their own daily behavior, their credibility will be seriously jeopardized.

- Dynamism—they assume that reengineering is not an event, a "program" or a one-shot deal. Rather, they understand that nothing remains stable or static, and that the relentless questioning and reinventing of everything which is essential to reengineering will in fact be a permanent fact of life. Reinventing our work will never end, they say with relish.

The leader, thus, has a crucial role in visualizing a completely different organization, articulating that vision to others, doggedly pursuing that vision, and helping prepare others to jump in and join the effort. Without real leadership, reengineering is reduced to a soulless technique or a short-term fad, neither of which will yield anything other than marginal impact.

So there you have it: four reasons that reengineering can succeed, or the mirror perspective: four reasons that it died in the large majority of organizations which tried it. If you want to breathe life into the corpse, I sincerely hope that these four pathways will serve as good medicine for your organization.

The author is a professor at the University of San Francisco and a consultant with The Tom Peters Group in Palo Alto, California.

The Missing Piece in Reengineering

Nicholas F. Horney and Richard Koonce

Reengineering efforts completely change processes and jobs, but such efforts often don't take into account that people will need to do these new jobs and whether people have the skills and competencies needed. This article addresses this point and suggests a specific approach, the "competency alignment process," to match people with newly reengineered jobs.

Is your organization suffering from reorganization fatigue? Have you restructured, downsized, outplaced, and outsourced until all that remains is a skeleton staff of stressed-out employees and senior executives still waiting for the promised gains in productivity to materialize?

In recent years, hundreds of business books and management articles have been written about downsizing and restructuring: how to do it right, how to do it well, how to use it for sustainable change, and how to squeeze corporate vitality and productivity gains out of reengineered work processes and a slimmed-down workforce. Many of those articles spotlight the stage-setting importance of energetic CEO leadership and "envelope-pushing" missions and visions.

Leadership and organizational goals are important. But something is missing from the literature. Despite the media attention, the verdict on many reengineering efforts today is mixed at best. In fact, a recent business survey by a leading human resources consulting firm suggests that nearly two-thirds of all restructuring efforts are clear failures.

The reasons vary. Many restructuring efforts suffer from poor planning and have paid only scant attention to the importance of clear, consistent, and ongoing communication as part of restructuring initiatives.

Another problem is that workplaces have dealt inadequately with the "people variables" that are always at play in organizations in times of rapid change. Executives and managers need to pay more attention to the stress and anxiety that people feel during transition.

Still another reason for the failure of many restructuring and reengineering efforts is a lack of penetration to the deepest organizational levels. In essence, these initiatives ignore the issue of how people actually do their jobs

---

Reprinted with permission from *Training & Development*, December 1995. Copyright © 1995, American Society for Training & Development. All rights reserved.

each day. In other words, they fail to address one of the key ways for people to become engaged and energized as individual agents of change.

What we call competency alignment is a critical underpinning of successful business-process reengineering initiatives.

At its best, BPR involves a fundamental rethinking and radical redesign of "core business processes" within an organization. It necessarily implies taking a hard and systematic look not only at the organizational structures, management systems, beliefs, and values that are part of an organization's culture, but also at the jobs that people do on a daily basis and the systems that support and reinforce them.

Reengineering efforts should be targeted toward the specific goal of changing employee behaviors, processes, and systems at the "transactional" level in an organization (the level at which day-to-day business is actually done, according to change-management consultant and theorist W. Warner Burke).

Unless BPR examines the business at that level, all the CEO exhortations in the world aren't likely to bring about significant, long-term changes—either in organizational effectiveness or in a company's financial performance. Unless BPR pays attention to employees' day-to-day work, people are unlikely to fall enthusiastically into line to support new marketing goals, to work toward achieving the CEO's heartfelt desire to "go global," or to pursue more ambitious customer-service objectives.

Most organizations display an implicit (and sometimes explicit) systemic inertia. Systems and people resist change unless an organization addresses barriers methodically and systematically.

That's why competency alignment is critical. It gets you right down into the heart of an organization. It helps you focus time, energy, and attention on the details of how people work and interact on a day-to-day basis—with each other, with customers, with other stakeholders, with competitors, and with various human-resource processes and information systems in the organization.

By paying attention to how people interact "transactionally" in your company, organization, department, or work group, you put yourself in a powerful position to make changes that can reinforce reengineering goals or dramatic process improvement.

The Competency-Alignment Process

Coopers & Lybrand's competency-alignment process, or CAP involves the systematic study, analysis and assessment of job functions, tasks, and skills required by an organization that is reengineering one or more of its work processes. It focuses on analyzing, understanding, and optimally deploying people in the reengineered organization, ensuring the best job fit for everyone.

To do that, it methodically examines employee skill sets in order to determine where and when skill gaps exist and what can be done to remedy deficiencies—either through employee training, skill enhancement, redeployment, outplacement, outsourcing, or other efforts.

CAP provides a baseline methodology for retooling work processes at their most fundamental level—the level of the individual and the small work team.

CAP is an ideal mechanism for bringing employees into closer alignment with strategic organizational goals and objectives—a key success factor in creating a high-performing, improvement-driven organization, according to a recent C&L survey.

And it provides a means of refining and recalibrating that alignment over time—as job requirements change, as the structure of work within an organization changes, as production or manufacturing processes incorporate new technology, as employee skill sets age, and as external factors come into play.

A systematic and methodical approach can help you implement competency alignment in your own organization as a component of other reengineering efforts you are planning or implementing.

Where does competency alignment fit into the reengineering process? Think of it as a critical subset of larger-scale business-process redesign or reengineering efforts that are in the works or recently completed. It should be a part of any BPR initiative—whether the goal of the reengineering is to redraft your organization's entire mission or to overhaul one key business process such as research and development, marketing, manufacturing, or product distribution.

For many organizations, competency alignment has been the missing element or link in reengineering efforts. Even if it fell through the cracks in years past, it still may have influenced the outcome of productivity-improvement and change-management initiatives.

Nowadays, organizations can't afford to ignore it. The costs of employee recruitment, training, turnover, retraining, and poor job fit have become so high that they are clearly driving the need for organizations to get the most out of their BPR efforts, at a minimal cost.

Four Stages of CAP

An important outcome of completing competency alignment is the identification of current employees the organization can successfully place in new jobs or on new teams, as part of reengineering a key business process. Doing so takes great care, careful planning, and systematic implementation.

Most organizations should implement CAP in four stages: assess, deploy, learn, and align.

In stage 1, the assessment, it's important to conduct a task analysis of the reengineered process to determine the knowledge, skills, abilities, and competencies people will need in order to be effective contributors. You also will examine the suitability of current job holders to do that work.

A critical outcome of stage 2, deployment, is the identification of peoples' skill gaps. With this information, you can begin to make decisions about which employees to retain in their current functions and which to slate for outplacement or redeployment elsewhere in the organization.

Stage 3 of any CAP initiative deals with learning. This stage involves the development of skill-acquisition plans (such as training, outsourcing, or recruiting) to fill the skill gaps identified in stage 2.

> An important outcome of completing competency alignment is the identification of current employees the organization can successfully place in new jobs or on new teams, as part of reengineering a key business process.

> ### THE FOUR STAGES OF THE COMPETENCY-ALIGNMENT PROCESS
>
> **Stage 1: Assess**
> - Assess your process.
> - Assess your people.
> - Determine necessary tasks.
> - Determine necessary skills, abilities, and competencies.
> - Create a gap-analysis matrix.
>
> **Stage 2: Deploy**
> - Develop skill, ability, and competency profiles.
> - Use the profiles to deploy people into reengineered jobs, to redeploy them elsewhere in the organization, or to outplace them.
>
> **Stage 3: Learn**
> - Create training and career-development plans for employees.
> - Explore the use of different training approaches, formats, and methods.
> - Outsource non-core functions.
>
> **Stage 4: Align**
> - Align HR systems, including reward and recognition, compensation, and performance appraisal.
> - Conduct pilot tests.
> - Review, assess, and revise as appropriate.

You might, for example, decide to institute new training programs to help retained employees work more effectively in teams. You might also decide to import at least some new talent from outside the organization through targeted recruitment efforts. Or you might choose to outsource certain tasks that the organization no longer considers essential core functions.

Stage 4 of CAP is alignment. It focuses on developing and aligning an organization's human resource systems (such as the performance-appraisal system and the compensation system) to sustain the performance of people in the newly reengineered process.

Following this road map can help you ensure successful implementation of competency alignment in your organization. Now, let's take a detailed look at each of the four stages, which are summarized in the box.

Stage 1: Assessing the Process and the People

This stage is divided into two parts: assessing the competencies that the newly reengineered processes will require, and assessing the competency of existing employees to carry out the processes.

Process Competency Assessment

Let's say that your company or organization has decided to reengineer. You might be planning to reengineer your entire organization as part of a comprehensive change-management initiative (one that involves the systematic reengineering of all business processes, your business strategy, and your information-technology capabilities). Or you might plan to redesign only selected departments or work processes.

In any case, you'll need to get a clear bead on the kinds of tasks that will need to be done. And you'll need to know which skills and competencies people will need if they are to do the work in the future, after reengineering efforts are fully implemented.

In years past, you might have used job and task analysis to get at the heart of productivity problems or to understand better the different elements in a work process. You might have asked job holders to provide the following information:

- the core knowledge, skills, and abilities necessary for doing their jobs
- the amount of time spent each day or week on specific tasks
- ratings of tasks, in terms of relative importance.

That approach to job analysis was valuable in the past. But it becomes difficult to do when you are in the middle of reengineering a core business process—primarily because you don't yet have job holders or "incumbents" in the reengineered process. Instead, you'll need to use subject matter experts (SMEs) within the organization to help identify tasks and to describe the knowledge, skills, and abilities that are likely to be required of job holders once a map of the new process is fully developed.

SMEs can include current job holders, "process owners," key line managers, and others you deem to have broad knowledge of organizational goals as well as of specific processes and work content.

Now let's imagine that you are reengineering your company's order-management process. A key objective in stage 1 of CAP is to develop a process description, showing how work is performed now and how it will be done in the future. So you'll need to break the process down into individual tasks (for example, planning, order generating, scheduling, and shipping) and ask SMEs to identify the competencies people need for each task.

Initially, what you come up with may resemble a step-by-step view of the order-management process, with lists of specific skills tagged to each of the principle steps or tasks.

> ### HOW DO YOU SPELL SUCCESS IN A REENGINEERED WORKPLACE?
>
> So, you've been charged with leading the effort to assess current employees' suitability for working in a reengineered job context. Surveying those who work with the employees in question can give you a 360-degree view of worker competencies.
>
> Of course, many of the questions you'll ask are specific to the job you have in mind. But several factors tend to predict excellent performance in any job. To be successful in a job today—particularly in organizations that are undergoing incessant internal change and process improvement—a person typically must display the following traits:
>
> - the skills and abilities to do the actual work
> - the inclination or inherent ability to learn and adapt to a changing environment over time
> - motivation to do the work
> - compatibility with the organization's overall operating and management style
> - a sense of self-confidence about her or his ability to perform in the job over time.

Next, work with members of your reengineering team and with the SMEs to map out the way in which work will be done in the future. Ask such questions as these:

- What additional skills and competencies will people need to have?
- Will people work together differently than they do now? (For instance, will they spend more time in teams and on collaborative decision making?)
- What new technology will be integrated into the way work is done?
- What new skills will the technology require of workers?

To get answers to those questions and others, try conducting focus-group sessions, using groupware technology to catalog and organize peoples' responses. See the accompanying box, "A New Kind of Tool for Groups," for a discussion of groupware as a means for facilitating group sessions and collecting and analyzing data.

What typically emerges from an in-depth focus-group process is a detailed list of tasks, skills, and competencies that will be part of employees' work in the future.

A groupware session with subject matter experts will probably yield a sheet of formatted information that looks something like figure 1 (p. 439). This example was developed as the result of some work with a large financial institution to determine its employees' work tasks, knowledge and skill requirements, and competencies.

> ## A New Kind of Tool for Groups
>
> Groupware, or electronic meeting-support is a kind of software that organizations are using more and more often in brainstorming, data-gathering, and focus-group situations.
>
> Groupware technology can help you electronically capture and catalog large amounts of participant input, typically gathered in classroom sessions or through teleconferences. Usually, participants use laptop computers or keypads to input their answers to questions. The technology provides an accurate and quick way to capture data, compile statistics, set priorities for goals and objectives and build action plans.
>
> When using groupware with focus groups, you may find it helpful to ask participants to review an existing list of tasks, developed ahead of time by the reengineering team. Have focus-group members verify that the list is complete and that it accurately reflects all the transactions likely to be required as part of implementing a new work process.
>
> Once participants have signed off on the list, group the tasks and "subtasks" together in clusters. From those clusters the focus-group participants can determine what knowledge, skills and abilities people will need to have in order to perform future tasks in the organization.
>
> One common groupware feature, rank-order voting, may be especially useful in helping focus groups to determine the relative importance of various tasks, skills, and competencies.

What also frequently emerges from such data-gathering sessions are broad themes that suggest how much the nature and structure of work is changing.

For example, nowadays everyone from the boardroom to the loading dock needs hands-on familiarity—and preferably, a high comfort level—with computers. And today's workplace requires many people to have specific experience with such relatively new technology as local-area networks, "shareware," the Internet, and Windows applications. Such competencies will be even more essential in the future.

Another competency that people increasingly need in the workplace today is the ability to work effectively in groups. Since more and more work is team-based, you'll want to make sure that subject matter experts in your focus groups fully map the constellation of team skills and competencies that work will require in a newly restructured area of your organization.

By the time you've done all that, you'll have a clear handle on the competencies people will need for doing their jobs in the context of a reengineered work process. And by eliciting comments from process owners, supervisors, and others who are familiar with current processes, you create strong buy-in for the important employee-deployment decisions to come.

At the same time, you may acquire a sense of the work that lies ahead of you in actually implementing competency alignment in your organization, and of the tactics and strategies you'll need to use.

> ## MOVING PEOPLE INTO THE NEW WORKPLACE
>
> ### In the past, employees...
> - were familiar with mainframe computers and individual PCs.
> - worked as individual contributors, performing jobs defined by formal written job descriptions
> - dealt with very little change in their jobs
> - did what they were told, each person reporting to one boss who acted as a supervisor.
>
> ### Now they need training In. . . .
> - use of local-area networks, Windows, Lotus Notes, and other advanced computer technologies
> - how to work cooperatively on teams to perform project-driven work assignments, conflict-resolution and project management skills are a must
> - how to deal with constant technological and organizational change
> - serving many different "customers," both inside and outside the organization

For example, say you oversee training and development programs in a craft environment (such as a tool-and-die manufacturer) where the tradition and emphasis has long been on individual skill. Now, such factors as speed-to-market and concurrent engineering (the simultaneous development of a product and of the process for developing it) have emerged as critical to success.

You may face tough challenges if you intend to introduce team principles or large-scale, technology-assisted design into a manufacturing process in a traditional environment. Employees may be unfamiliar with (and even hostile to) new technology. They may lack an understanding of teamwork principles; they may have no interest at all in working on teams.

The people-assessment process that makes up the next part of stage 1 addresses such concerns by giving you tools for appropriate selection and retention of employees. It will also help you determine an individual employee's motivation to do new work, so that you can assess his or her suitability for working in an environment of changing norms and expectations.

Employee Competency Assessment

At this point, you've inventoried the skills and competencies people will need once a work process has been reengineered. So you have a road map with which to assess the suitability of current job holders to perform future jobs in your organization.

> **Job Tasks for Bank Employees, with Related Skills and Competencies**
>
> **Tasks**
> - Gather closing information.
> - Identify, read, review, and interpret loan documentation.
> - Identify legal issues.
> - Review and interpret loan authorization histories and amortization schedules.
>
> **Knowledge and Skill Requirements**
> - reading comprehension
> - knowledge of loan-servicing systems and loan documentation
> - knowledge of asset types and loan-classification schedules
> - knowledge of loan documentation, ranking, and legal issues, terminology, and definitions
>
> **Required Competencies**
> - detail orientation

Figure 1.

Your goal now is to assess the individual skills and backgrounds of current job holders.

Start by developing an assessment tool that looks at their interests and skills in the key areas you identified in the process of competency assessment.

A helpful tool at this stage is a 360-degree survey that lets supervisors, co-workers, and subordinates provide input on job holders. The responses will form accurate profiles of individual employees and their suitability to fill new jobs in the reengineered workplace.

Many of the questions to ask at this stage are specific to the process being reengineered. Others relate more generally to the work values and work styles of employees and to how well specific people are, likely to perform in a reengineered environment. Still others seek to assess peoples compatibility with, interest in, and motivation to do tasks in the reengineered job context.

For example, you might ask supervisors, co-workers, and subordinates to rate a person's ability to work with new technology, to think creatively, to deal with new situations, to handle stress, to solve problems, to work as part of a team, and to lead a team.

Some traits tend to predict success in almost any job—especially in organizations that are in states of constant reorganization. See the box, "How Do You Spell Success in a Reengineered Workplace?" for a rundown.

The outcome of assessing employees' backgrounds and skill levels is a gap-analysis matrix that includes each person who is involved in the work process as it stands before reengineering begins. The matrix covers a spectrum of skill areas that earlier steps have identified as important to the work process in question. For each competency, indicate whether each employee's skill level is weak, moderate, or strong.

Stage 2: Deploying People in a Reengineered Workplace

In essence, the matrix you created at the end of stage 1 enables you to assess the range of peoples' individual and aggregate abilities across a typical profile of what you need from an employee—both as an individual contributor and as part of a team.

You've now provided people with an overall "rating and ranking"—comparing their strengths and weaknesses with those of their co-workers, and taking into account the skills that are critical to the reengineered jobs. Armed with that information, it's possible to determine each person's suitability for training, for redeployment elsewhere in the organization, or for outplacement.

Determine people's scores on the matrix by taking the responses gathered from their supervisors, co-workers, and subordinates. Subject those responses to computer analysis that gives weighted averages to different skills and to the relative skills of one person compared with those of his or her co-workers.

That information will help you make the tough decisions about where and how each employee can best contribute in the reengineered work environment.

Stage 3: Creating the Means for Learning

You've determined the competencies people need for success in performing newly reengineered tasks. And you've profiled current job holders to assess their individual skill levels and their skill gaps.

Now you're in an ideal position to create training and career-development plans for employees, using the information you've collected. You also have the information you need for developing a plan to outsource specific tasks and functions that can now best be done outside the organization—for instance, benefits administration and payroll.

People from human resources, training and development, and various line operations should work in tandem to create training plans for employees. Those plans can be regularly updated and revised as needed. For instance, the introduction of new technology might necessitate additional training. So might the implementation of new work practices, whether they are specific to a single process or common across the organization.

This may also be the time to develop jointly a new learning philosophy for your organization—a philosophy that specifically supports job-redesign and process-reengineering priorities. For instance, you may want to inaugurate just-in-time training, computer-based training, distance learning (if you serve multiple geographic sites), or other training strategies to help support continuous-improvement efforts, whether they are process-specific or people-specific.

BPR presents an excellent time to develop, pilot, and roll out new training initiatives. They are another way to reinforce new work requirements and performance expectations in the reengineered environment.

What kinds of training do employees need to receive?

In addition to process and task-specific training, it's likely that your employees will need to develop better teamwork and communication skills. They may need updated management skills, or training in new technology.

In all likelihood, CAP will by this time have fully delineated the kinds of training you need to offer. Indeed, you may see a "before" and "after" picture emerging—one that gives a clear view of the skills that served employees well in the past, compared with the ones they now need to learn. That picture can point you in the direction you need to go in order to give employees the highest possible skill levels for performing reengineered jobs.

See the box "Moving People Into the New Workplace," for an example of changing skill and knowledge requirements in a reengineered workplace.

Stage 4: Aligning the Support Systems

Clearly, no amount of job reconfiguration is going to work unless you put systems in place to reinforce new behaviors and help support the design of new functions. Key systems include the reward and recognition system, the compensation system, and the performance-appraisal program.

That's why stage 4 of the CAP process must deal with building the right kind of infrastructure to support newly designed jobs.

You'll need to develop new philosophies and policies for performance appraisal, compensation, rewards, and incentives. Your reengineered environment probably includes more collaborative work, so the new systems should use measurements that are more team-based than in the past. You may need to retool your systems to reflect critical success factors such as customer-satisfaction levels, cycle time, quality improvement, and team performance.

But you'll also want to leave room for some measurements that key into individual contributions and effort on the job. For instance, what criteria will you build into your performance- appraisal process to recognize and acknowledge individual initiative?

In the Coopers & Lybrand survey of improvement-driven organizations, respondents from high-performing organizations in both the public and the private sector said their workplaces put a lot of stock in recognizing and rewarding individual as well as team efforts in the workplace.

In those organizations, quality-improvement accomplishments figure prominently in people's annual performance reviews. Job empowerment is a key operating philosophy. You might want to build such objectives into your own performance-appraisal process, as well.

> BPR presents an excellent time to develop, pilot, and roll out new training initiatives.

To undergird your competency-alignment, you'll need to field test the HR systems you are putting in place. Fully test each separate system (such as performance appraisal and measurement, compensation, or recognition) in a trial-period shakedown. Testing can help you ensure that each system is performing to expectations and is helping to reinforce the new work norms.

After you conduct separate tests of the different systems, evaluate the results and make revisions as necessary.

Making It All Come Together

How do you ensure that competency alignment becomes a highly effective component of your reengineering efforts?

Success begins with a realization that increasing corporate profitability or organizational effectiveness requires more than cutting costs or shedding staff. Instead, organizations must be purposeful in the ways in which they develop and leverage people as part of reengineering efforts.

Ultimately, the outcome of all this is to increase the bottom line or whatever other measurements your organization uses to gauge profitability or organizational vitality.

It is often easier, in the short term, to increase net income by reducing costs or head count. But true growth and vitality come from sensing new opportunities in the marketplace; building new competencies within the organization; and leveraging the skills, talents, and adaptiveness of employees to achieve organizational aims.

"Any company that is a bystander on the road to the future will watch as its structure, values, and skills become progressively less attuned to industry realities" and to the needs of the marketplace, note Gary Hamel and C.K. Prahalad in their book, *Competing for the Future*.

Pay conscious and purposeful attention to the importance of competency alignment as part of your reengineering efforts. Not only will it boost your organization's sustained vitality and profitability, but it also can enhance the resilience and resourcefulness of your organization and its employees in a climate of constant change.

Nicholas Horney is a managing associate with Coopers & Lybrand.

Richard Koonce is a career-planning consultant and the author of Career Power! 12 Winning Habits to Get You from Where You Are to Where You Want to Be *(AMACOM, 1994).*

Quality Tools and Techniques

BENCHMARKING

What Benchmarking Books Don't Tell You

Sarah Lincoln and Art Price

Benchmarking is another activity that seems to be a great idea, but when you sit down to do it, it turns out to be more complicated than you anticipated. As with most things, such complications come from misunderstanding what's involved and poor planning. This article offers five tips that can help benchmarking teams plan better and prevent mistakes that come from misunderstanding the benchmarking task.

When benchmarking is done well, it is a powerful competitive tool. When it is done poorly, it can be an incredible waste of money. A lot of good advice on how to effectively conduct benchmarking can be found in current literature. In fact, the number of books on benchmarking has increased dramatically over the past few years. These books can provide you with a wealth of information, but they don't tell you everything.

We have learned a number of interesting lessons while working with benchmarking teams at AT&T Global Business Communications Systems that we did not learn from the books. We would like to share some of them with you.

Tip 1: Do It Quickly or Don't Do It

Most benchmarking books do not discuss how long benchmarking studies can take. They warn you that teams can easily get bogged down in the technicalities of benchmarking, but they don't explain clearly that, more common than not, studies last from nine to 12 months. Why so long? Because benchmarking is still relatively new to most companies, which means that new

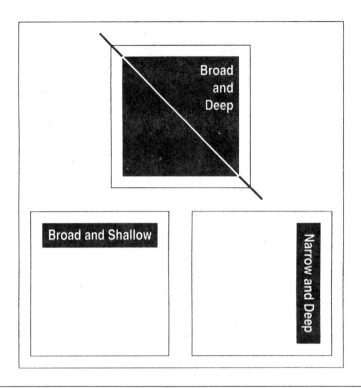

Figure 1. **Choosing the scope of benchmarking study**

teams are conducting the studies—and new teams don't usually know how to be expeditious.

While nine to 12 months is common, it's too long. Many circumstances can change in a company in that time. Team members might move to other job assignments, compromising the study's continuity. Or worse, the team's management could change and the study could be abandoned after months of hard work. So you need to get the benchmarking study done quickly—or you might not get it done at all.

Studies can drag out for several reasons:

- *The proper amount of resources is not applied.* Generally, team members are asked to conduct a benchmarking study in addition to their normal work. As a result, they devote a few hours a week (about 10% of their time) to the study. This breaks up the benchmarking activities, spreading them over months. Thus, momentum is difficult to maintain. If team members devote at least 20% of their time (about a day per week), teams can cut months off their studies' completion time.
- *Experts are not used.* Since benchmarking is difficult to do well, experts can increase the quality of a study while saving a lot of

time. While we do not advocate hiring a consultant to do the entire study for you, we do highly recommend engaging one for parts of it. At a minimum, pay for a literature search; experts can perform searches quicker and cheaper. You might consider hiring an experienced benchmarking facilitator to help the team become more efficient and avoid many common pitfalls. You might also consider paying to have the best-in-class companies identified and visits set up. You should make sure, however, that you get what you want by asking for proof that the companies are best in class and by screening the companies before visiting them. Using experts in these ways can save weeks, but be careful. Don't transfer the responsibility for the study's outcome to the consultant. Stay involved.

- *Groundwork has not been done.* Team members need to do groundwork—that is, collect customer, process, and performance information that will enable them to compare their company with others—before they can start benchmarking. Unless an organization is mature in its quality techniques, the benchmarking team will be doing this groundwork for the first time, which can take months. World-class companies continuously document their processes, analyze their customer needs, and compare their performance against those needs. If you adopt this as an essential part of your business activities, your benchmarking studies will get done faster.
- *Too large a scope is chosen.* Teams can easily bite off too much if they have not benchmarked before. Since they don't know what they are getting into, they don't realize the effect of their scope decision on the study's length. The next section gives some guidance on how to choose the proper scope.

Tip 2: Choose a Broad-and-Shallow or Narrow-and-Deep Scope

Choosing the scope of a benchmarking study is an art. Benchmarking books do a good job of helping teams choose relevant topics to benchmark.[1] They do not, however, give much advice on how to narrow the scope so that the study is achievable. They tend to just advise teams to avoid taking on too much. But what is "too much"?

If unguided, teams often choose to do a study that is broad and deep—that is, one that broadly covers a large process from beginning to end and goes into great depth in every aspect of that process. This is "too much." Instead, teams need to choose a scope that is either broad and shallow or narrow and deep.

Broad-and-shallow studies look across a process or function. They ask high-level, strategic questions, such as:

- What are the comparative costs of executing a similar process?
- What is a company's business strategy?

- What is the most effective organizational structure for a given function?

This type of study spans many functions and people and doesn't go into detail in any one area. It answers "What is done?" rather than "How is it done?" Broad-and-shallow studies are useful in developing strategies, setting goals, and reorganizing functions to be more effective.

Narrow-and-deep studies delve into one or two aspects of a process or function and look at how work is done. Operational-level questions are asked, such as:

- How are data automatically collected for the software development process?
- How does an organization exceed customer expectations in providing on-time delivery?
- How does a company decide what products to bring to market?

The kind of data collected in narrow-and-deep studies is very detailed. These studies dig deep to uncover the treasures within a process or organization. They are useful in changing how people do their work—namely, the processes they use to perform their jobs.

When teams try to answer "What is done?" and "How is it done?" simultaneously, they end up with hopelessly large, broad-and-deep studies. Thus, carefully choosing the scope of a benchmarking study is vital to success. There are many ways to control the scope. Some teams start with a broad-and-shallow scope and, after identifying a few areas of particular interest, go narrow and deep. Others are able to identify the narrow-and-deep target immediately, based on existing data or experience. It just depends on what a team is trying to accomplish and how much time it has. Just remember, if a team wants how-to information, it will eventually have to dig deep.

Tip 3: Integrate CSFs—They Are Critical

Critical success factors (CSFs) are "the few key areas where 'things must go right' for the business to flourish."[2] CSFs are derived from what is critical to a company's survival, whether that be its customers, competitive standing, financial stability, or business strategy. CSFs can differ between different businesses, organizations, and benchmarking teams.

Almost anything goes with CSFs. For example, a CSF for a package delivery service might be on-time delivery. A long-distance telephone service provider's CSFs are likely to be reliability and low cost. An internal mailroom's CSFs might be reliability, low cost, and accuracy. Keep in mind, however, that CSFs define the few *most critical* things, not just *important* things.

Some books discuss the importance of CSFs in relation to choosing what to benchmark, but teams need to go a step further.[3] CSFs need to influence not only the scope of the benchmarking process, but also its key measures, company selection criteria, questions, final analysis, and recommendations. This is important because no matter what you benchmark, you want to study,

WHAT BENCHMARKING BOOKS DON'T TELL YOU

measure, and collect information in the areas that are critical to your organization's success.

To ensure that CSFs influence the entire benchmarking process, use them when:

- *Choosing the benchmarking scope.* A benchmarking team can choose to benchmark a CSF. For example, if on-time delivery is a key to success, an effective use of benchmarking is to discover how other organizations do this well. If a team is interested in benchmarking a process or business strategy, it can include the top CSFs within the scope. By doing so, the team will collect information on how the process or strategy interplays with the CSFs and its recommendations will improve or maintain the company's performance in these critical areas.
- *Selecting key measures.* A benchmarking team should use CSFs as a means to select measures that will be used to indicate how the company performs in its critical areas. For example, with on-time delivery as a CSF, the team could track actual delivery times against promised delivery times or the customers' perception of the company's delivery reliability. If CSFs are used to guide the selection of key measures, the team is likely to discover that it is already collecting the internal data needed for the benchmarking study.
- *Identifying benchmarking partners.* CSFs and key measures form the basis for the criteria used to identify benchmarking partners. Good criteria can steer a team to the right partners. The right partners will share, at least partially, the team's view of what is critical to success. If the team and the benchmarking partners don't share common ground in critical areas, it's highly unlikely that the practices found will be relevant.
- *Developing benchmarking questions.* CSFs should influence the team's questions for the benchmarking partners. The team should include questions about the partners' ability to maintain or improve performance in CSFs. The level of detail depends on how near the CSFs are to the central focus of the study.
- *Preparing the final analysis and recommendations.* During final analysis, the benchmarking team should look for trends relating to how others achieve superior performance relative to key measures and, therefore, the CSFs. The team's recommendations will then, therefore, take CSFs into account and enhance the company's performance in its critical areas.

Keeping teams focused on CSFs throughout a benchmarking study increases the likelihood of a good return on investment because it guarantees that information is collected in the areas most critical to success.

Tip 4: Don't Fall for the Best-in-Class Fallacy

CSFs help benchmarking teams collect the right information, but from whom should they collect it? Teams should collect data from best-in-class companies, of course. While this practice is so clear-cut and simple in principle, its implementation is not.

Benchmarking books warn that finding best-in-class companies is one of the hardest steps in the benchmarking process. Despite these warnings, many benchmarking teams still believe the fallacy that, somewhere, there are best-in-class companies in the precise areas they are studying. This fallacy needs to be dispelled once and for all.

Finding a best-in-class company to benchmark is not an absolute. In other words, there is no preexisting magic list of best-in-class companies. Teams can't even count on the Malcolm Baldrige National Quality Award winners because they might not be best in the particular processes being studied.

In fact, a company that one team determines to be best in class can differ from another team's selection, even if both teams are conducting similar benchmarking studies. Best in class depends on a team's needs. Here is how a team can select the best-in-class companies that meet its needs:

> *Finding a best-in-class company to benchmark is not an absolute. There is no preexisting magic list of best-in-class companies.*

1. Formulate criteria that define a "class" of companies of interest. Base these criteria on fundamental attributes, such as the companies' customer base, global presence, technical focus, or quality maturity. In other words, ask: "What critical attributes must a company have to be a credible benchmarking partner?"
2. Define measures that can be used to compare companies to determine the "best" in class. These measures should be based on the team's CSFs. For example, "best" might be defined as the quickest response time, the most reliable products, the most maintainable products, or the highest productivity. Again, these depend on the team's CSFs and are specific to its situation.
3. Find companies that meet the team's class criteria and that appear to be the best performers relative to the defined measures. These are the team's best-in-class companies.

How hard a team searches to find the best performers in its class depends on what the team is trying to accomplish with the study.[4] If the team is reengineering a critical process, it might want to do a thorough search across the globe. If its budget is small, it might want to settle for "best-in-county" or "best-in-city" companies; a lot can be learned close to home. Whatever the search pattern, using this three-step approach can help teams find benchmarking partners from which they can learn.

Tip 5: Manage the Change from the Start

The purpose of benchmarking is to *change* a process or practice for the better. Unfortunately, many benchmarking studies never get beyond producing recommendations; they get bogged down when it comes time for implementation. If an organization does not properly manage change from the onset of the benchmarking study, the recommendations will sit on a shelf.

Most benchmarking books give valuable information on communicating benchmarking results and getting acceptance for a team's recommendations.[5] They also mention the value of having a sponsor for a study to increase its credibility in the organization. But this is not enough. Communicating the results of a study after the recommendations have been developed does not start the change process soon enough. Simply engaging a sponsor to help sell the team's recommendations might not work well either.

Benchmarking causes a shift in the team members' mind-set; the experience helps them accept change. This same shift has to occur not only in team members, but also in the entire organization. All of those who have a stake in the study—the managers, funders, process users, and customers—have to be appropriately informed before, not after, the benchmarking study and, if possible, be involved in it. By doing so, the stakeholders will likely accept the recommendations and help implement the necessary changes.

Using benchmarking facilitators who are trained in organizational change management techniques can also improve the effectiveness of benchmarking studies.[6] Applying change management techniques can increase the likelihood that the results of the team's efforts will actually be embraced by the organization—and that's what benchmarking is all about.

Wisdom and More Wisdom

The various benchmarking books currently available are a tremendous asset for benchmarking teams. From them, teams can learn much about the art of benchmarking. Based on our benchmarking experiences, we offer five more tips for success. We have found that getting studies done quickly, choosing a realistic scope, integrating critical success factors, avoiding the best-in-class fallacy, and managing the change from the start can make benchmarking studies more pleasurable and profitable.

Notes

1. For example, see R. C. Camp, *Benchmarking: The Search for Industry Best Practices That Lead to Superior Performance* (Milwaukee, Wis.: Quality Press, 1989); M. J. Spendolini, *The Benchmarking Book* (New York: AMACOM Publishing, 1992); and G. H. Watson, *The Benchmarking Workbook—Adapting Best Practices for Performance Improvement* (Cambridge, Mass.: Productivity Press, 1992).

2. J. F. Rockart, "Chief Executives Define Their Own Data Needs," *Harvard Business Review*, March–April 1979.

3. Spendolini, *The Benchmarking Book*, and Watson, *The Benchmarking Workbook*.

4. Spendolini, *The Benchmarking Book*.

5. For example, see Camp, *Benchmarking: The Search for Industry Best Practices*; Spendolini, *The Benchmarking Book*; Watson, *The Benchmarking Workbook*; and

G. J. Balm, *Benchmarking: A Practitioner's Guide for Becoming and Staying Best of the Test* (Schaumburg, Ill.: QPMA Press, 1993).

6. J. LaMarsh, *Changing the Way We Change—Gaining Control of Major Operational Change* (Reading, Mass.: Addison-Wesley, 1995).

Sarah Lincoln is the benchmarking manager at AT&T Global Business Communications Systems in Denver, Colorado.

Art Price is a DMTS (distinguished member of the technical staff) at AT&T Global Business Communications Systems in Denver, Colorado.

Benchmarking is Not an Instant Hit

Jim Morgan

This article, which appeared in Purchasing *magazine, provides a brief overview of the results of a survey on benchmarking practices in organizations today. As it has for many other such activities, the enthusiasm of many managers for this activity has waned. It seemed like a good thing to do, but then dramatic improvements didn't come. It seems that benchmarking, like other quality management practices, requires commitment, planning, and a long-term focus to get the maximum results.*

The life of the benchmarker can be a lonely one! Despite great strides made in benchmarking among the world's top competitors benchmarking for many in corporate America appear to be as much fun as a root canal. Where corporations have high-level management commitment and a willingness to back it up financially, benchmarking progress is, indeed, being made at a rapid pace. Best practices in a wide range of corporate processes and procurement procedures are being identified and appear to be on the way to implementation. On the other hand, the results of a recently completed mail survey and telephone poll of more than 500 randomly selected purchasing organizations are far from encouraging, as shown in figure 1.

In fact, our extensive polling and interviewing revealed a lack of enthusiasm, implementation, and, in many cases, commitment to benchmarking. There also appears to be a great deal of tension on the subject itself, which showed up in the unwillingness of buyers to go on record. More than in any recent major polls conducted by *Purchasing*, respondents invoked anonymity as a condition for comment.

The Sticking Points

Cause of this underwhelming reaction to benchmarking? For the most part the "unreaction" to benchmarking appears to be driven by these five factors:

■ **Lack of resources.** In interview after interview, purchasing executives stressed the fact that their operations and their companies, overall, are running lean and under pressure to cut costs further. "Right now we don't have enough time or staff to do what we need to do," says the buyer for a large

> **Benchmarking Survey Results**
>
> 1. Does your company have a formal benchmarking program?
>   No: 64.7% Yes: 21.9% Sometimes informal: 14.4%
>
> 2. If yes, does it use the benchmarking program to drive change?
>   Yes: 75.8% No: 15.3% Sometimes: 8.8%
>
> 3. How long has your company been using benchmarking?
>   1-3 years: 48.4% 3–5 years: 25.8% Over 5 years: 22.6% Under 1 year: 3.2%
>
> 4. Is your benchmarking program involved mainly in analyzing processes and determining gaps in performance vs. the benchmarked company? Or is the program more involved in comparative analysis?
>   Process analysis: 56.5% Competitive analysis: 21.2% Both: 18.2% Neither 3.1%
>
> 5. What company-wide processes has your company benchmarked or does it plan to benchmark? (on a scale of 1-10)
>   Order fulfillment: 7.5
>   Customer/supplier electronic interfaces: 5.9
>   Manufacturing setup: 4.1
>   New product development 4.4
>   Design change management 4.3

Figure 1.

packaging company. "We'll need more staff before we can even think about benchmarking."

- **Lack of commitment at the top.** Whether it's due to suspicion or lack of commitment, or other distractions, a fair number of respondents indicate that executive management is against benchmarking. Thus, a purchasing supervisor for a food company in California notes, "We are not permitted to benchmark because management feels competition within our industry is too great." In another case, the procurement manager for a computer storage equipment manufacturer reported that an early benchmarking effort on supplier assessment was "hopelessly muddled by management's lack of experience and commitment to benchmarking this process.... They were looking for some points that could be used in a marketing program and totally derailed our benchmarking effort."

- **Lack of suitable partners.** For many, especially in smaller companies, the problem revolves around where to find benchmarking partners. For instance, William A. Coakley, director of purchasing and materials management at Genzyme Corp., Framingham, Mass., notes that he has had "little trouble obtaining other biotech companies willing to share information." However, he still has a problem because "most companies in my industry lack the sophistication to be able to answer many of my questions."

He and may others surveyed meet this problem by going outside their immediate industries. The materials manager for a small chemicals company, for example, has solved the problem of finding suitable partners by "benchmarking on the basis of process. We ask who is best at this process and then try to set up a benchmarking study. It's a long process and requires much work so we have to restrict the amount of benchmarking we do." Another buyer for an electronics company in Texas simply finds "that it's too hard to determine who's world class or not." And Mike Kennedy, purchasing manager at Excel Corp., in Plainview, Texas, says that his industry (meat packing) is "playing catch-up" and indicates that many of his company's suppliers and competitors are "in the same situation . . . we're all just starting out and it's tough."

- **Worries about confidentiality.** "Up to now confidentially hasn't been a problem, but I'm afraid it could be in the future," says Steve Blose, Atlantic Electric, Pleasantville, N.J. Many other respondents are more immediately concerned. For instance, the purchasing director for a Florida food processing firm was quick to point out, "We're reluctant, our management is reluctant . . . to share information in benchmarking. We know we can use companies that we don't compete with, but there's still a danger that important competitive information will fall into the wrong hands."

- **Lack of understanding.** For a sizable number of respondents, benchmarking still represents an alien environment. Rather than seeking to compare their processes and procedures with the best available, they approach benchmarking in terms of averages—average inventories, average quality ratings, average number of orders processed per person, average leadtimes per class of commodity. Many other respondents candidly suggest that any benchmarking that needs to be done can be accomplished within the confines of their companies. And many fail and/or refuse to make any distinctions between processes and the functional parts of processes. In short, an alarming number of purchasing professionals have only a sketchy idea of what benchmarking is and how it works, some are suspicious of benchmarking and its proponents, but they constitute a tiny minority.

Fear of Getting Wet

The underlying truth that seems to emerge from the answers to this survey is most purchasing professionals—especially those in smaller companies—are tiptoeing along the shoreline. Most buyers surveyed, even when they're not entirely sure about the mechanics of benchmarking, are for it but not necessarily ready for it. A materials manager for an Indiana capital goods manufacturer probably best sums up the mood this way: "At the moment we don't have the staff or the resources to launch a major benchmarking program, but we can and are taking on one or two benchmarking projects each year. As we get better at it, we probably will do more. But right now we're just learning to walk."

Looked at from that vantage point (a few small programs at a time), many companies and their purchasing operations are making encouraging strides into benchmarking. Cathy Morrison, purchasing agent at Plastics Manufacturing, Concord, N.C., for instance, notes good success in benchmarking efforts at her company's metals facility. Another purchasing agent tells of significant downsizing of the company's truck/car fleet and the installation of a successful purchasing card program as the result of benchmarking projects.

Larry Bell at the John Deere Des Moines Works notes that there "usually are no major gains from any single benchmarking project" and what really counts is "the build-up of minor gains." Clarence Baker, senior EPO support services manager at Haliburton in Houston, Texas, agrees and notes that so far "our benchmarking has shown that we are on the right course on most procurement procedures and on the leading edge on some." Jack Ryel, purchasing and materials manager at Philips Petroleum, Borger, Texas, says benchmarking has helped "identify gaps—some in our favor, some not." And Genzyme's Coakley has been successful in finding six to eight "similar sized companies in my industry who have shared information with me on procurement cards, organizational issues, and purchasing volumes."

Jim Morgan is the editor of Purchasing *magazine.*

Quality Tools and Techniques

PROCESS MANAGEMENT AND MEASUREMENT

From Balanced Scorecard to Strategic Gauges: Is Measurement Worth It?

John H. Lingle and William A. Schiemann

Many managers believe that if you can't measure it, you can't manage it. Perhaps a better way to state this is that if you can't measure it, you can't improve it. This article amply demonstrates, on the basis on survey data, the importance of having clear and measurable strategic and tactical objectives. Besides making it easier to know where you're going and how you're doing, having such objectives also seems to facilitate better communication, shared information, teamwork, and a supportive corporate culture. The points in this article affirm what organizations practicing TQM know, that clearly defined and measured processes, teamwork, and continuous process and quality improvement all go together.

There is great commotion across the corporate landscape these days as executives attempt to master the three R's of the current management wisdom: reengineering, restructuring and renewal. With all the hubbub, how do executives know that their change efforts are producing results?

Reprinted with permission from *Management Review*, March 1996. Copyright © 1996 by the American Management Association, New York. All rights reserved.

A number of companies are answering the question by rediscovering the criticality of measurement as an important management tool. "You simply can't manage anything you can't measure," says Richard Quinn, vice president of quality at the Sears Merchandising Group.

The key questions, then, for managers on the firing line are: What are companies actually doing to measure results? Does measuring strategic performance make a difference? And is measurement being used to manage change?

To answer these questions, Wm. Schiemann & Associates Inc. conducted a national survey of a cross section of executives. The most significant conclusion from the research is that measurement plays a crucial role in translating business strategy into results. In fact, we have found that organizations which are tops in their industry, stellar financial performers and adept change leaders, distinguish themselves by the following characteristics: having agreed-upon measures that managers understand; balancing financial and nonfinancial measurement; linking strategic measures to operational ones; updating their strategic "scorecard" regularly; and clearly communicating measures and progress to all employees.

For those executives who have gone beyond the green eyeshades and stopwatches to assess the pivot points of their company's strategy—from how well customer expectations are met to the ability to manage relevant environmental and regulatory forces, to how adaptable the organization is—the measurement effort will yield ongoing results to the bottom line.

Measuring What, When?

In the dog-eat-dog world of real estate development, a misstep can be a deadly stumble, which makes information quality and performance measurement especially important. "Not all things are worth measuring all the time.... You have to determine what you want to measure and measure it properly," observes Ara Hovnanian, president of Hovnanian Enterprises Inc., the nation's 10th largest homebuilder.

Hovnanian's point is on target. In our research, we examined how executives measure six strategic performance areas that are crucial to long-term success: financial performance, operating efficiency, customer satisfaction, employee performance, innovation/change, and community/environmental issues. We asked our research sample these two questions: How highly do you value information in each strategic performance area? Would you bet your job on the quality of information in each of the areas?

Interestingly, information about customer satisfaction is highly valued by the largest percentage of executives.

Interestingly, information about customer satisfaction is highly valued by the largest percentage of executives, even more than the traditional management gauges of financial performance and operating efficiency. In a tough, competitive marketplace, knowledge about customers is power—and a competitive advantage. Two-thirds place a significant level of importance on employee performance, and nearly half of all managers place importance on innovation and change and community/environment.

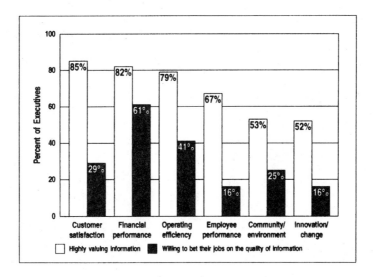

Figure 1. **Value vs. quality measurement**

Just how confident are executives about the quality of information upon which to base decisions in each strategic area? Not many would stake their jobs on the information that is available to them (as summarized in Figure 1). Even in the area of financial performance, only six-in-10 executives place confidence in the data that is available to them. There exists a wide gap between what is valued and what is treated as accurate. This trust gap cuts across internal and external environments. To the extent that an organization, like an individual, must "know thyself" to be effective, then executives face an urgent task of reexamining their measurement system to gain greater self-knowledge and self-confidence.

Why do executives feel so uncertain about the quality of information concerning customers, employees, innovation/change and other external stakeholders such as community groups or regulatory agencies? Our findings point to two factors that contribute to executive uncertainty: the clarity of measures in each strategic area of the business, and the frequency with which measurement is undertaken.

The responses in Figure 2 reflect a pattern similar to the confidence ratings in Figure 1. Apart from financial and operating efficiency, relatively few executives report that success measures in other areas are either clearly defined or updated at least semi-annually.

Most striking in these figures is how few managers feel that their organizations have been able to define in clear, unambiguous terms what the organization hopes to accomplish in the areas of employee performance, innovation/change and community/environment. Even in the area of customer satisfaction, the data indicate there is disagreement in many companies on what should be measured.

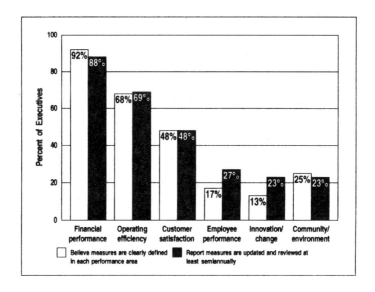

Figure 2. **Quality of measures**

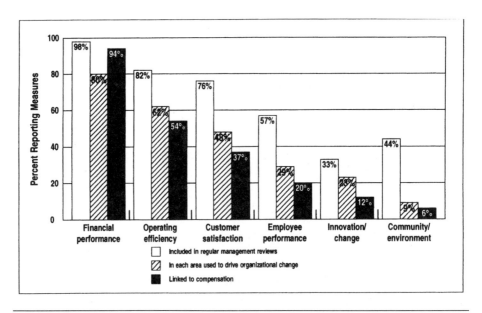

Figure 3. **Use of measures**

We probed deeper in our research to learn what measures were reviewed on a regular basis, which were linked to compensation and the extent to which different measures were used to drive organizational change (see Figure 3). Not surprisingly, what executives review and depend on to manage their organization parallels what they consider to be reliable.

While a substantial number of organizations have begun to examine performance measures beyond financial and operating efficiency at regular performance reviews, few have linked such measures to compensation or rely on them to drive organizational change. Without executives putting their money where their measures are, or without using measures to drive and evaluate strategic change, it is unlikely that these measures will become integral to how a business is directed and run.

If better measurement does not result in better performance, then why bother? Put differently, are "measurement-managed" companies more successful than those that downplay measurement? And, if so, how does measurement contribute to success?

If better measurement does not result in better performance, then why bother?

Measure by Measure

To answer these questions, we first identified those organizations characterized by respondents as being measurement-managed and compared them to a subset of organizations that did not appear to place much reliance on measurement. Measurement-managed organizations were those in which senior management was reported to be in agreement on measurable criteria for determining strategic success, and management updated and reviewed semiannual performance measures in three or more of the six primary performance areas. Fifty-eight organizations met both criteria for being measurement-managed.

We then identified as non-measurement-managed organizations those in which senior management reportedly did not agree on measurable criteria for determining strategic success, and performance measures in only one or two primary performance areas were reviewed on a semiannual basis. Sixty-four organizations met these criteria.

Once we had identified the two groups of contrasting organizations, we compared them on three success measures (see Figure 4). Measurement-managed companies fared better on each of the three measures.

A higher percentage of measurement-managed companies were identified as industry leaders, as being financially in the top third of their industry and as successfully managing their change effort. This last statistic implies that measurement-managed companies tend to anticipate the future and are likely to remain in a leadership position in a rapidly changing environment. "Don't spend your energy measuring what you can't change," advises J. Walter Kisling, chairman and CEO of Multiplex Co., a $27 million manufacturer of beverage-dispensing equipment. "Spend your energy keeping up with—and staying ahead of—change."

A higher percentage of measurement-managed companies were identified as industry leaders.

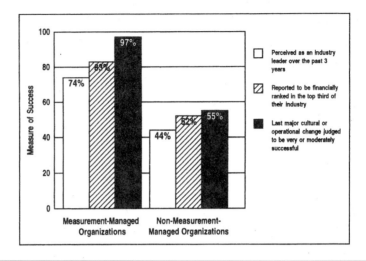

Figure 4. **The value of measurement management**

Ninety-seven percent of our measurement-managed companies reported success with a major change effort.

Given the number of change efforts that are buried in the graveyards of so many competitors, it is notable that 97 percent of our measurement-managed companies reported success with a major change effort. Even if we excuse this figure as overly self-congratulatory and lower the response value by 25 percent, the remaining figure would continue to be impressive.

One hypothesis for these findings is that successful industry and change leaders simply do a better job than nonleaders in measuring their workforce, which is where real change is won or lost. In fact, the survey findings reveal that employee measurement is the biggest single measurement area that separates successful from less successful firms, where success is defined as the three criteria in Figure 4. Knowing workforce values, morale, productivity and competencies appears to be a significant advantage for these successful firms.

In addition to employee measurement, another key distinction separating industry leaders from nonleaders is that industry leaders report reviewing on a more frequent basis a broader range of measures than do nonleaders, especially those related to customer satisfaction, employee performance, and community and environmental issues. These organizations keep performance measures continually in front of key executives and managers, knowing that if measures are out of sight, they're out of management.

Forget magic. Industry leaders we surveyed simply have a greater handle on the world around them.

This study supports the conclusion that good measurement is essential to good management. Forget magic. Industry leaders we surveyed simply have a greater handle on the world around them. While they carefully track financial performance and operating efficiency, they also apply rigor to the "soft side," such as customer satisfaction, human resources and innovativeness. These companies realize that their organizations are complex environments that require greater alignment and a balanced set of performance gauges.

Success Builders

Beyond balance, why is it that measurement-managed companies outperform those that are less disciplined? Our data point to four mechanisms that contribute to the success of measurement-managed organizations:

- *Agreement on strategy.* Only 7 percent of our measurement-managed companies reported a lack of agreement among top management on the business strategy of the organization. This compared to 63 percent of the non-measurement-managed organizations. The act of translating vision or strategy into measurable objectives forces specificity. It helps to surface and resolve those hidden disagreements that often get buried when the strategy remains abstract, only to return at some later date to haunt an organization.

- *Clarity of communication.* We asked managers how well their business strategy was communicated and understood from top to bottom at their organizations. Sixty percent of managers in measurement-managed organizations rated favorably how well the strategy was communicated throughout the organization, while only 8 percent of managers from non-measurement-managed organizations reported that their organization's strategy was well-communicated and understood.

  Good communication demands a clear message. If the strategy itself is unclear, insisting on measures for strategic goals can force clarity, as it has at Sears. "There was a real gap between the strategy and what it meant on a day-to-day, operational basis," explains Sears' Quinn. This left employees uncertain about how they could contribute. So Quinn and other senior executives added a number of "wraparound elements" to the strategy and then tied the key strategic goals to performance measures. "This helped to clarify our strategic thinking and bring to life what we want to become and how to get there," concludes Quinn.

  Measurement also provides a common language for communication. "People talk about how they're being measured," says Quinn. "It is almost the language in which communication occurs in an organization." Consistent with the notion of a common language, 71 percent of managers from measurement-managed organizations reported that information within their organization was shared openly and candidly, compared with only 30 percent in non-measurement-managed companies.

- *Focus and alignment efforts.* Effective organizations are organic, integrated entities in which different units, functions and levels support the company strategy—and one another. Not surprisingly, our measurement-managed organizations reported more frequently that unit performance measures were linked to strategic company measures (74 percent versus 16 percent for non-measurement-managed companies) and that individual performance measures were linked to unit measures (52 percent versus 11 percent, respectively).

> Seventy-one percent of managers from measurement-managed organizations reported that information within their organization was shared openly and candidly, compared with only 30 percent in non-measurement-managed companies.

For example, at Steelcase North America, a $2 billion manufacturer of office furniture, the company strategy is reviewed each year and modified as appropriate. Business unit and functional strategies are then developed. John Gruizenga, senior vice president for architectural products, points out that as strategy cascades down the organization, performance measures are established to link up with the strategic performance expectations of the entire company. "This alignment of effort ensures that the corporate strategy is carried throughout, right to tactics and individual performance measures," Gruizenga says.

Measurement-managed companies were also more likely to link multiple measures to compensation. Forty-seven percent of measurement-managed companies had measurements in at least three performance areas linked to compensation, while only 9 percent of the non-measurement-managed companies had measures in as many as three areas linked to compensation.

- *Organizational culture.* Call it culture, tone or style, but a number of mechanisms are at play in every organization that help provide a set of collective attitudes and behaviors that either sustain or impair competitiveness. Take teamwork. When compared to non-measurement-managed counterparts, managers of measurement-managed companies more frequently reported strong teamwork and cooperation among the management team (85 percent favorable ratings on teamwork versus 38 percent for non-measurement-managed companies).

> Managers of measurement-managed companies more frequently reported strong teamwork and cooperation among the management team.

We also asked respondents the extent to which employees in their organization self-monitored their own performance against agreed-upon standards. Forty-two percent of the measurement-managed companies reported excelling in this regard compared with only 16 percent of their non-measurement-managed counterparts.

Finally, increased willingness to take risks appears as an additional feature for building success in measurement-managed organizations. Fifty-two percent of managers from measurement-managed companies said employees in their organization generally were unafraid to take risks to accomplish their objectives, compared with only 22 percent of the non-measurement-managed companies.

While culture can be difficult to quantify—and change—measurement can play an important role in specifying values, thereby providing everyone in an organization with clear goals for focusing a culture change effort. At Gilbarco, a $350 million manufacturer of gasoline-dispensing equipment, a "hierarchy of measures" is in place that cascades down from corporate. "Measurement provides clear, visible targets throughout the organization," says Thomas Rosetta, Gilbarco's manager of U.S. operations. It also gives the team-based environment at Gilbarco a rallying point for galvanizing group effort.

Measurement Blockers

Few executives would disagree with the proposition that a good measurement system is an important management tool. Why, then, aren't more companies managing by deploying and reviewing a balanced set of measures, especially in the areas of customer satisfaction, employee performance, organization innovativeness and environmental/community influences?

Based on survey results, related telephone interviews and our consulting experience, here are the four most frequent barriers to effective measurement that we encountered:

- *Fuzzy objectives.* "If you wish to debate with me, define your terms." Aristotle's advice is equally valid for managing organizations. It's tough to run a business without clearly defined objectives. The development of measures requires that goals and objectives be defined with sufficient precision to be measurable. Typically, this precision exists in the financial and operational areas. Many companies do not invest the time needed to define with equal precision other areas of performance, such as customer satisfaction, employee performance and rate of change.

The development of measures requires that goals and objectives be defined with sufficient precision to be measurable.

A first step in achieving precision in hard-to-quantify areas is to translate "soft" objectives into clear statements of results and then ask, "How can this result be measured?" For example, Ara Hovnanian converted his company's guiding principles into specific "pledges" to customers, employees and stakeholders and then attached measurement criteria to each pledge.

Second, try specifying the behaviors that are implied by inherently imprecise terms such as "values," "culture" and the like. This allows managers to then measure where they hit or miss the mark. For example, at Sears a good deal of attention recently has been paid to defining the company's values. "The real issue is what behaviors look like every day, job by job, and whether they reflect the values we are espousing," Quinn says. "With this done, you can then set up a measurement system to determine where we're making progress and where we're not."

- *Unjustified trust in informal feedback systems.* There are many informal mechanisms by which companies receive feedback about their performance, such as complaints and criticisms from the salesforce about products and services. Companies can place unwarranted trust in these informal information channels as mechanisms for measuring performance.

While informal information channels can provide an idea of the range of issues an organization faces, they do not provide accurate information on the extent to which a problem exists across a larger customer or employee population. Non-measurement-managed companies often learn too late that an apparent problem which has absorbed resources is the concern of only a few squeaky wheels, while a more critical problem has gone unattended.

- *Entrenched measurement systems.* Most organizations have in place some type of measurement system by which they manage already. Employees can strongly resist new ways of defining success that are unfamiliar to them. They want to see the measures "work" for a while before they are willing to tie their financial future to them. Such resistance has defeated more than one attempt to increase management through measurement. Organizations often underestimate what is required not only to develop a sound system of measurement, but also to implement it in a way that results in active acceptance by the workforce.

Measurement-managed companies tend to involve the workforce in developing measures. For example, at Gilbarco, teams not only help determine specific performance measures, but determine how results will be tracked and reported. This ensures broad commitment to the measurement process.

- *The activity trap.* Measurement mania can be debilitating. Too many measures trivialize the effort. This occurs when companies focus on measuring activities, not results. "If you're measuring for measurement sake and do not use the data for fine-tuning the organization, or for spurring it on to achieve significant results, then you are going through wasted efforts," says Steven Fisher, CEO of Brown & Caldwell, a $25 million company specializing in environmental engineering.

With all the changes taking place, organizations are becoming more like molecules operating under Heisenberg's uncertainty principle than like the sure-footed leviathans that once stalked the marketplace just a decade ago. Establishing an effective measurement approach will not retard the rate of change or reduce the need for continuous improvement. But it should help organizations anticipate change and provide a balanced set of coordinates to monitor progress and manage through the chaos.

As Fisher puts it, "If you don't measure and track results, you won't know where you are and whether or not your strategies are working." Ultimately, that's the best justification for becoming a measurement-managed organization.

John H. Lingle is principal consultant with, and William A. Schiemann is president of, Wm. Schiemann & Associates Inc., a Somerville, New Jersey, management consulting firm specializing in strategic assessment and measurement-driven organizational change.

The Effects of SPC on the Target of Process Quality Improvement

Wen-Hsien Chen

This somewhat technical article suggests that there is a connection between using the data captured on control charts when coming up with process improvement targets. In many situations, the targets are chosen first, then the control charts are used to help meet these targets. This may not be the best approach. The author explains this with examples and various data tables.

Introduction

In the past decade, the globally competitive marketplace has exerted enormous pressure on firms to improve the quality of their products. In response, many firms have implemented various quality programs and made great improvements (see, for example, Reibstein, Washington, Levinson, and Shenitz (1992)). These improvements were typically due to application of more intensive statistical process control (SPC) and process improvement programs (see, for example, Blache, Stewart, Zimmerman, Shaull, Benner, and Humphrey (1988)).

The objective of SPC is to control a process in an ideal status with respect to a particular product specification. A natural vehicle to describe the process is the quality level of the items produced. A widely used process indicator is its output distribution characterized by the mean and variance. If the values of mean and variance are within prescribed limits, the process is operating in an in-control state. An assignable cause of variability may result in a shift in mean or variance or both to an out-of-control state, and thus lead to defective product, downtime, and cost of corrective action. SPC uses the process information from samples to identify process shifts and to initiate timely remedial actions. From a managerial perspective, SPC aims to maintain a process in its ideal status and to keep product quality loss at the minimum level during production. The application of $\bar{X}$ and R control charts are typical of this endeavor.

In addition to SPC, a major concern of quality management is to decrease production costs by improving process quality. Usually process quality can be improved by reducing output variability or the process failure rate or both.

From a managerial perspective, SPC aims to maintain a process in its ideal status and to keep product quality loss at the minimum level during production.

Reprinted with permission from the *Journal of Quality Technology*, April 1996. Copyright © 1996 by the American Society for Quality Control. All rights reserved.

For example, an improved coolant system for a machine tool can decrease the dimensional variability of finished parts; or a dust-free room can greatly decrease the process failure rate during the manufacture of semiconductors. Machine shop experience also reveals that better-trained workers, more stable raw materials, a cleaner environment, and higher quality cutting tools can contribute to either reducing variability of the process output or decreasing the frequency of the process shifting to an out-of-control state. (Hereafter, the shift of a process to an out-of-control state will be called the process failure rate.)

Certainly, process improvement brings benefit, but it also incurs cost. The higher the improvement target, generally the greater the cost. In practice, the improvement decision concerns primarily the tradeoff between the cost of the improvement effort and the direct benefits realized from an improved process. For example, a quality practitioner compares the potential improvement in product quality with the investment needed to reduce variability. Similar considerations can be observed in the decision to improve process failure rate. In either case, SPC data on output variability and process failure rate can be used to make decisions about process quality improvement. The determination of a process improvement target often makes little use of the information from SPC. The improvement effort is generally implemented independently by process engineers. Then, following process improvement, quality control personnel determine the values of control chart parameters (for example, sample size, inspection interval, and control limits of an $\bar{X}$ control chart). This sequential approach (SA) is widely used in practice. Much of the research on process control is also implicitly based on this sequential practice. Review papers by Montgomery (1980) and by Ho and Case (1994) on the economic design of control charts illustrate this approach.

Research on the design of control charts reveals that there may exist significant interactions between the process improvement target and the optimal design of control charts. This conjecture is based on computational results of Goel, Jain, and Wu (1968). Their work on economic design of the $\bar{X}$ control chart showed that a smaller process failure rate would result in a larger optimal value of the inspection interval and smaller total cost. A logical inference drawn from these observations is that the savings due to an improved failure rate can serve as an additional impetus for further process improvement. In addition, a significant effect of sampling cost on the optimal inspection interval revealed in the results of Goel, Jain, and Wu may be expected to exhibit a similar impact on decisions about process improvement. Similar reasoning may be applicable to the case of output variability improvement. This author contends that an improvement target obtained by the sequential approach may be far from optimum. In other words, ignoring the cost effect of process control can result in inappropriate decisions about the target of process quality improvement. An integrative approach to determining the target of process improvement and the values of control chart parameters may be more economically effective.

> Ignoring the cost effect of process control can result in inappropriate decisions about the target of process quality improvement.

The objective of this research was to investigate the effects of economic design of control charts on the target of process quality improvement. Specifically, this research explored the cost effectiveness of simultaneous decisions on the target of process improvement and the values of control chart parameters in comparison with that of the sequential approach. As described previously, there are two general approaches to process improvement—improved process failure rate and reduced output variability. Although the former is fully discussed, only some aspects of variability improvement are presented. A specific process was examined to illustrate this investigation. The process was monitored with an $\bar{X}$ control chart and its status was assumed to be in one of two states, either in control or out of control. We note that the $\bar{X}$ control chart is widely used in industry, and that it is the most popular scheme analyzed in the literature.

System Characteristics

The system characteristics related to process control and process improvement are now presented. A summary of notation used is shown in Table 1. The system examined in this research is a continuous process. The output characteristic is normally distributed with standard deviation σ. The process may be in one of two states, in control or out of control. In the in-control state, the process mean is at the value corresponding to the design target μ_0. A single assignable cause of variability results in a shift in the mean by a fixed magnitude $\delta\sigma$. The standard deviation of output in the latter state remains σ. Thus, the output mean in an out-of-control state is either $\mu_0 + \delta\sigma$ or $\mu_0 - \delta\sigma$. The elapsed period before the shift occurs is exponentially distributed with mean $1/\lambda$. Once the process shifts to an out-of-control state, it remains there until it is restored.

The process is monitored by an $\bar{X}$ control scheme with sample size n, sampling interval h, and control limits $\mu_0 \pm k\sigma/\sqrt{n}$. If the sample mean is outside the control limits, a search is initiated to determine the process status. If the process is confirmed to be out of control, the process is stopped, and a corrective action is taken. The time required to complete an inspection and record the result is proportional to the sample size. Therefore, for a sample of size n the time to complete an inspection is nE, where E is the sampling time per unit. During the search for an assignable cause (the time required is T_1), the process is allowed to continue operation. Once the process is confirmed to be out of control, the process is stopped, and the reset time T_2 and cost W are required to restore the process to the in-control state.

Cost Model for Process Failure Rate Improvement

The major cost components of the system are as follows:

1. Investment cost for process quality improvement;
2. Product quality losses due to variability; and
3. Operating costs of process control.

Table 1. **Notation for System Characteristics**

Symbol	Definition
h	Inspection interval
n	Sample size
k	Parameter defining control chart limits at $\mu_0 \pm k\sigma/\sqrt{n}$
λ	Process failure rate
λ_1	Process failure rate with the integrative approach
λ_2	Process failure rate with the sequential approach
σ	Standard deviation of process output
δ	Magnitude in standard deviation units of a shift in process mean
T_1	Time to confirm an out-of-control process
T_2	Time to restore an out-of-control process to an in-control state
E	Inspection time for one unit
nE	Inspection time for a sample of size n
τ	Elapsed time from the last inspection in an in-control period to the shift to an out-of-control state
α	Probability of a type I inspection error
β	Probability of a type II inspection error
W	Cost of downtime and repair of the assignable cause
a	Fixed sampling cost
b	Variable sampling cost per unit
Y	Cost of investigating a false alarm
C_0	Average product quality loss per unit time while in control
C_1	Average product quality loss per unit time while out of control
$g(\lambda)$	Cost per unit time to improve the process failure rate
y	Deviation from design target
$f(y)$	Product quality loss function
l, m	Constants in the product quality loss function
EPS	Measure of the process improvement cost
ECL	Expected production cycle time
ARL1	Average run length for an in-control process
ARL2	Average run length for an out-of-control process
$cost_1$	Total cost with the Integrative Approach
$cost_2$	Total cost with the Sequential Approach

Several assumptions were made to reflect practical situations and to make analyses feasible. They are discussed below.

First, the investment cost per unit time has a logarithmic function,

$$g(\lambda) = i - j \ln(\lambda)$$

with i and j specified constants. The logarithmic function specified is convex and decreasing in λ and is used to represent investment cost (for example, see Porteus (1986)). A notable property of this function is that it costs a fixed amount to reduce λ by a fixed fraction.

Second, product quality loss per unit time was assumed to be an exponential function, defined as $f(y) = l \times |y|^m$, where y is the deviation from the design target, and l and m are constants. This assumption is an extension of Taguchi's concept of quality losses. Taguchi (1986) advocated that any deviation from design target would result in quality loss and proposed a quadratic loss function to reflect this cost. When $m = 2$, $f(y)$ is equivalent to Taguchi's loss function. According to the definition of $f(y)$, the expected product quality loss per unit time of items produced in an in-control state, denoted C_0, is expressed as

$$C_0 = \int_{-\infty}^{\infty} \frac{\exp[-(y - \mu_0)^2/2\sigma^2]}{\sigma\sqrt{2\pi}} f(y) dy.$$

Similarly, the expected quality loss of items per unit time produced in an out-of-control state, denoted C_1, is expressed as

$$C_1 = \frac{1}{2} \int_{-\infty}^{\infty} \frac{\exp[-(y - \mu_0 - \delta\sigma)^2/2\sigma^2]}{\sigma\sqrt{2\pi}} f(y) dy$$
$$+ \frac{1}{2} \int_{-\infty}^{\infty} \frac{\exp[-(y - \mu_0 + \delta\sigma)^2/2\sigma^2]}{\sigma\sqrt{2\pi}} f(y) dy.$$

Third, as the system described above is a special case of the system studied by Lorenzen and Vance (1986) (hereafter referred to as LV), their cost model can be applied. According to LV, the operating costs for process control are related to inspection, false alarms, and downtime and repair of the process. Thus, the process control costs consist of the following components:

1. Inspection costs, $[(a + bn)/h][1/\lambda - \tau + nE + h(\text{ARL2}) + T_1]$, where a and b are the fixed and variable sampling costs, respectively, $\tau = [1 - (1 + \lambda h)e^{-\lambda h}]/[\lambda(1 - e^{-\lambda h})]$, $\text{ARL2} = 1/(1 - \beta)$, and β is the probability of type II inspection error;
2. The costs of investigating false alarms (type I inspection error), $sY/\text{ARL1}$, where $\text{ARL1} = 1/\alpha$, α is the probability of type I inspection error, $s = e^{-\lambda h}/(1 - e^{-\lambda h})$, and Y is the cost per false alarm; and

3. Costs of downtime and repair of the process, W.

The objective of economic decisions on the target of process improvement and on the values of control chart parameters is to achieve the minimum cost per unit time for the system. Incorporating the investment cost for process improvement, product quality losses, and the operating costs of process control, the total system cost per unit time, C, is obtained as

$$C = \frac{\frac{C_0}{\lambda} + C_1[-\tau + nE + h(ARL2) + T_1] + \frac{sY}{ARL1} + W}{ECL} + \frac{\left[\frac{a+bn}{h}\right]\left[\frac{1}{\lambda} - \tau + nE + h(ARL2) + T_1\right]}{ECL} + g(\lambda) \quad (1)$$

where C_0, C_1, and $g(\lambda)$ are as defined earlier, and

$$ECL = 1/\lambda - \tau + nE + h(ARL2) + T_1 + T_2$$

is the production cycle time.

The first two terms in (1) are costs considered by LV in the economic design of an $\bar{X}$ control chart, and the last term $g(\lambda)$ is the cost to improve the process failure rate. The decision variables of the cost model are n, k, h, and λ. This equation captures the costs of process control and process improvement.

The solution that minimizes the cost function (1) implicitly takes into account the interactions between process control and process improvement. This approach is called the *Integrative Approach* (IA), and the total cost obtained is denoted $cost_1$. In contrast, the traditional or *Sequential Approach* (SA) determines the process improvement target λ without considering the effect of SPC and specifies the values of control chart parameters n, k, and h for given process characteristics. While ignoring the cost of process control, decisions about the process improvement target with SA focus on the cost of downtime and corrective action, W, and the investment cost $g(\lambda)$. Thus, the target value of λ with SA can be obtained by minimizing the cost function C_λ

$$C_\lambda = W\lambda + g(\lambda). \quad (2)$$

Given the value of λ which minimizes (2), quality control personnel then determine values of n, k, and h for an $\bar{X}$ control chart by solving (1). The total cost obtained by this approach is denoted $cost_2$. Theoretically, there may exist discrepancies between the solutions (i.e., the values of n, k, h, and λ) of IA and SA. Furthermore, the total system cost (1) based on the solution of SA ($cost_2$) is expected to be greater than that of IA ($cost_1$). The cost penalty of SA is examined in the next section.

Computational Experiment

Algorithm to Minimize the Cost Function

Our objective was to examine the effect of the economic design of control charts on the process improvement target. The cost penalty of SA reflects the cost effectiveness of the proposed integrative approach to determining the target of process quality improvement. A computational experiment was designed to investigate this issue. Because (1) is a complicated function and an analytic solution is intractable, direct search was adopted to solve the cost function. The development of the solution procedure is based on the following observations.

First, as explained earlier, the improved process failure rate (λ) decreases the operating costs of process control, and in turn, the induced cost savings may stimulate further process improvement. Thus, the cost impact of SPC is expected to decrease the desired failure rate. Accordingly, the value of λ obtained from (2), designated λ_2, provides an upper bound on λ, which is also the optimal value of λ for SA.

Second, for a given λ, the cost function (1) is similar to that employed in the design of $\bar{X}$ control charts, and solution procedures developed in that context may be used to determine optimal values of n, k, and h. Therefore, an algorithm similar to that proposed by LV was used.

Third, preliminary experiment results showed that the cost function is smooth around the optimal value of λ and (1) may be convex in λ. Thus, the search for the optimal value of λ, denoted λ_1, can start from its upper bound with its value decremented until the total cost (1) exceeds the cost corresponding to the upper bound of λ (i.e., $cost_2$).

The solution procedures for IA are summarized in Figure 1. The solution procedures for SA is the same as that for IA, except that the value of λ is fixed at λ_2. Specifically, the optimal value of λ for SA is obtained by solving (2) instead of using an iterative search for λ_1 as shown in Figure 1.

Design of the Experiment

The choice of system parameters in this experiment was based primarily on the computations reported by Tagaras (1989). Values of these parameters (σ, δ, a, b, Y, E, T_1, T_2, l, m, W) are listed in Table 2. In order to reflect the cost of improvement efforts, a parameter EPS is defined as the ratio of investment cost to the corresponding reduction of cost from SA. Specifically,

$$\text{EPS} = \frac{g(\lambda_2) - g(\lambda')}{W(\lambda' - \lambda_2)}$$

where λ' is the original value of the process failure rate, and λ_2 is the improved value obtained with SA. Clearly, for the same values of λ' and λ_2, the greater the value of EPS, the more costly an improvement program. In this experiment, the values of EPS were set at 0.5, 0.8, and 1.0 to represent respectively

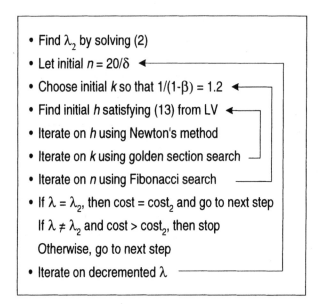

Figure 1. **Algorithm to minimize total system cost**

Table 2. **Parameters for the Computational Experiments**

Level	λ	σ	δ	a	b	Y	E	T_1	T_2	I	m	W	EPS
					Process Failure Rate Improvement								
Low	—	0.5	2.0	0.5	0.1	50	0.1	3	2	1	2	35	0.5
Middle	—	0.5	2.5	—	—	50	0.1	3	2	1	3	35	0.8
High	—	0.5	3.0	5.0	1.0	50	0.1	3	2	1	4	35	1.0
					Output Variability Improvement								
Low	0.005	—	2.0	0.5	0.1	50	0.1	3	2	1	2	35	0.5
Middle	0.005	—	2.5	—	—	50	0.1	3	2	1	3	35	0.8
High	0.005	—	3.0	5.0	1.0	50	0.1	3	2	1	4	35	1.0

low, moderate and high costs of improvement programs. Combining EPS and the parameters in Table 2, 108 test problems were designed.

The investment cost $g(\lambda) = i - j \ln(\lambda)$ exhibits a property such that it costs a fixed amount to improve λ by a fixed fraction. We also assumed that $g(\lambda)$ satisfies the following conditions (so that i and j can be specified for each test problem):

$$g(\lambda') = 0$$

and

$$C_\lambda(\lambda) = \text{EPS} \times W \times \lambda' + g(\lambda')$$
$$= \text{EPS} \times W \times \lambda'' + g(\lambda'')$$

with $\lambda' = 0.005$ cycle/year and $\lambda'' = 0.8\lambda'$. The first condition specifies that the investment cost is zero when λ is at the original level $\lambda' = 0.005$ cycle/year. The second condition indicates that the cost of downtime and corrective action are decreased by the amount $\text{EPS} \times W \times (\lambda' - \lambda'')$ as λ is improved to the value $0.8\lambda'$. For example, if $\text{EPS} = 0.5$ and $W = 35$ \$/cycle, then the two conditions reduce to

$$i - j \ln(0.005) = 0$$

and

$$0.5 \times 35 \times 0.005 + i - j \ln(0.005)$$
$$= 0.5 \times 35 \times 0.005 \times 0.8 + i - j \ln(0.005 \times 0.8).$$

Thus, $i = -0.415$ \$/year and $j = 0.0784$ \$/cycle are obtained by solving the two equations simultaneously.

Discussion of Experimental Results

Total Cost and Optimal Improvement Target The total cost of the system is based on (1). The computational results support the conjecture of significant cost effectiveness for the integrative approach (IA) in comparison with the sequential approach (SA). Some results describing the values of decision variables and total cost appear in Tables 3a and 3b. The results of IA and SA for problems defined by the parameter set $\sigma = 0.5$, $E = 0.1$, $T_1 = 3$, $T_2 = 2$, $W = 35$, and $Y = 50$ are presented for comparison. As explained earlier, cost_1 and cost_2 are total costs (1) based on values of n, k, h, and λ obtained with IA and SA respectively. A useful indicator to reflect the effectiveness of IA is the ratio of the cost penalty incurred from SA to the total cost of IA, or

$$\% \text{ Cost Penalty} = \frac{(\text{cost}_2 - \text{cost}_1)}{\text{cost}_1} \times 100\%.$$

Tables 3a and 3b reveal that the cost penalty may be substantial. For example, the penalty is 50.84% for case 7 of Table 3a. Table 3a shows that cost penalties exceed 15% for seven of the nine cases. Table 3b shows that cost penalties exceed 15% for ten of the twelve cases. The average cost penalties for various parameter choices appear in Table 5; the average for the 108 tests is 14.43%. The reported cost penalties indicate that the impact of economic design of control chart parameters on the determination of a process improvement target should not be neglected. Furthermore, the numerical results indicate that the optimal improvement target can be much greater than that obtained with SA. For example, $\lambda_1 = 0.00043$ versus $\lambda_2 = 0.00224$ for case 1 in Table 3a,

The reported cost penalties indicate that the impact of economic design of control chart parameters on the determination of a process improvement target should not be neglected.

Table 3a. Some Experimental Results for Process Failure Rate Improvement with $a = 5$, $b = 1$, and $\delta = 3$

No.	m	EPS	Integrative Approach					Sequential Approach					% Cost Penalty
			n	k	h	λ_1	$cost_1$	n	k	h	λ_2	$cost_2$	
1	2	0.5	2	2.67	120.09	0.00043	0.58	2	2.67	56.10	0.00224	0.68	16.04
2	2	0.8	2	2.67	83.04	0.00094	0.68	2	2.67	45.95	0.00359	0.78	13.98
3	2	1.0	2	2.67	69.76	0.00138	0.73	2	2.53	42.02	0.00448	0.82	12.71
4	3	0.5	2	2.67	110.49	0.00026	0.58	2	2.67	39.43	0.00224	0.75	30.17
5	3	0.8	2	2.67	75.64	0.00057	0.70	2	2.53	31.92	0.00359	0.89	27.13
6	3	1.0	2	2.67	63.08	0.00082	0.76	2	2.53	28.96	0.00448	0.95	25.66
7	4	0.5	2	2.67	106.37	0.00014	0.62	2	2.58	28.08	0.00224	0.93	50.84
8	4	0.8	2	2.67	70.65	0.00032	0.76	2	2.58	22.57	0.00359	1.12	47.31
9	4	1.0	2	2.67	57.25	0.00049	0.84	2	2.58	20.39	0.00448	1.22	45.57

Table 3b. **Some Experimental Results for Process Failure Rate Improvement with $m = 4$ and EPS $= 0.5$**

No.	a	b	δ	Integrative Approach					Sequential Approach					% Cost Penalty
				n	k	h	λ_1	$cost_1$	n	k	h	λ_2	$cost_2$	
1	0.5	0.1	2.0	5	3.11	25.89	0.00094	0.44	6	3.25	20.31	0.0022	0.46	6.04
2	0.5	1.0	2.0	3	2.53	58.98	0.00051	0.51	3	2.53	29.43	0.0022	0.59	15.54
3	5.0	0.1	2.0	7	3.11	97.16	0.00048	0.51	7	3.11	47.31	0.0022	0.59	16.46
4	5.0	1.0	2.0	4	2.53	137.05	0.00032	0.55	4	2.39	61.97	0.0022	0.68	23.45
5	0.5	0.1	2.5	4	3.47	19.45	0.00072	0.46	4	3.47	11.27	0.0022	0.51	10.67
6	0.5	1.0	2.5	2	2.67	47.58	0.00037	0.53	2	2.58	20.87	0.0022	0.65	22.38
7	5.0	0.1	2.5	6	3.47	87.60	0.00030	0.55	5	3.25	33.66	0.0022	0.70	27.44
8	5.0	1.0	2.5	3	2.67	124.60	0.00021	0.58	3	2.67	40.18	0.0022	0.79	35.77
9	0.5	0.1	3.0	3	3.47	17.91	0.00053	0.49	3	3.47	8.87	0.0022	0.57	17.24
10	0.5	1.0	3.0	2	2.89	40.95	0.00029	0.55	2	2.89	15.04	0.0022	0.73	31.32
11	5.0	0.1	3.0	4	3.47	80.51	0.00019	0.59	4	3.47	24.26	0.0022	0.83	41.91
12	5.0	1.0	3.0	2	2.67	106.37	0.00014	0.62	2	2.58	28.08	0.0022	0.93	50.84

where λ_1 and λ_2 are values of λ obtained with IA and SA respectively. These observations provide useful insight in the sense that one is strongly encouraged to set an aggressive target to improve process quality.

Effects of System Parameters Table 3a and 3b present some of the experimental results for the case of process failure rate improvement. Table 3a reveals a relatively large variation in cost penalty. First, it shows that the value of m in the loss function $f(y) = 1 \times |y|^m$ causes significant changes in cost penalty. Comparing the results of cases 1, 4, and 7, we observe that the cost penalty increases as the value of m increases. The value of m characterizes the behavior of product quality loss. For a given deviation from the design target, a larger value of m represents a greater quality loss. These numerical results indicate that it is economically crucial to apply IA in the case of larger product loss. Table 3a also indicates that a smaller value of EPS results in a larger penalty. For example, case 1 with EPS = 0.5 and a cost penalty of 16.04% versus case 3 with EPS = 1.0 and a cost penalty of 12.71%, as a smaller value of EPS represents a lower cost of improvement investment. This result indicates that simultaneous decisions on process improvement and process control may result in greater economic benefits when investment cost is lower.

Similar conclusions are observed for parameters a, b, and δ in Table 3b, where a, b, and δ are respectively the fixed inspection cost, the variable inspection cost, and the magnitude of shift in the process mean. Table 3b clearly indicates that larger values of these parameters incur larger cost penalties for SA relative to IA.

There are only slight differences for the solutions of n and k between IA and SA in Tables 3a and 3b. However, the values of the sampling intervals were greatly enhanced for IA. For example, in case 1 of Table 3a, the optimal sampling interval is 56.10 for SA versus 120.09 for IA, whereas $n = 2$ and $k = 2.67$ for both IA and SA. Obviously, the extra effort devoted to improving the process quality (the corresponding λ is 0.00224 versus 0.00043) results in decreased inspection effort.

Cost Model for Output Variability Improvement

Analyses for the case of output variability improvement are similar to those of failure rate improvement. The cost function is the same as (1) except that the investment cost is defined as

$$g(\sigma) = i - j \ln(\sigma)$$

where σ is the standard deviation of process output. The sequential approach to decisions about variability improvement focuses mainly on the product quality loss and the investment cost in determination of the target σ. Thus, this approach solves σ by minimizing

$$C_s(\sigma) = C_0(\sigma) + g(\sigma) \qquad (3)$$

where C_0 is defined in (1). Given the value of σ, the optimal values of n, k, and h with SA are then obtained by solving (1). The computational experiment for variability improvement is similar to that of failure rate improvement. The measure EPS is defined as

$$EPS = \frac{g(\sigma_2) - g(\sigma')}{C_0(\sigma') - C_0(\sigma_2)}$$

where $g(\sigma) = i - j \ln(\sigma)$, σ' is the original value of process variability, and σ^2 is the improved value obtained with SA. The system parameters used in the computational experiment are listed in Table 2. This resulted in 108 test problems. To enable values of constant i and j to be specified for each of test problem, the investment cost $g(\sigma)$ is assumed to satisfy the following conditions:

$$g(\sigma') = 0$$

and

$$C_s(\sigma) = EPS \times C_0(\sigma') + g(\sigma')$$
$$= EPS \times C_0(\sigma'') + g(\sigma'')$$

with $\sigma' = 0.5$ and $\sigma'' = 0.8\sigma'$. The first condition specifies that the investment cost is zero when σ is at the original level $\sigma' = 0.5$. The second condition defines that product quality loss is decreased by the amount EPS $\times$ $[C_0(\sigma') - C_0(\sigma'')]$ as σ'' becomes $0.8\sigma'$.

The computational results of 108 test problems are impressive. The cost penalty is 87.35% for SA as shown in case 12 of Table 4b. Like Tables 3a and 3b, Tables 4a and 4b reveal that larger values of a, b, δ, and m, and a smaller value of EPS result in larger cost penalties. Cost penalties for the 108 test problems averaged 23.1% (see Table 5). The large cost penalty indicates that IA is an economically effective approach for decisions on the improvement target of process quality and on the values of control chart parameters. Furthermore, the large differences between optimal values of σ for IA and SA indicates that SA is inadequate to determine a target for variability improvement. Hence, the determination of parameters for a process control chart has strong economic impact on decisions about a target for process improvement.

Conclusions

We have investigated the effect of economic design of the $\bar{X}$ control chart on the determination of a target for process improvement. Process quality improvement of two kinds—output variability and process failure rate—were discussed. By means of computational experiments, we compared the cost effectiveness of simultaneous determination of a target for process improvement and the values of control chart parameters. The computational results

Table 4a. Some Experimental Results for Output Variability Improvement with $a = 5$, $b = 1$, and $\delta = 3$

No.	m	EPS	Integrative Approach					Sequential Approach					% Cost Penalty
			n	k	h	σ_1	$cost_1$	n	k	h	σ_2	$cost_2$	
1	2	0.5	2	2.53	144.75	0.217	0.52	2	2.67	56.68	0.382	0.67	30.38
2	2	0.8	2	2.53	141.65	0.219	0.62	2	2.53	48.99	0.427	0.75	21.53
3	2	1.0	2	2.53	101.89	0.256	0.68	2	2.53	48.99	0.427	0.76	10.81
4	3	0.5	2	2.53	150.65	0.229	0.49	2	2.53	44.39	0.382	0.70	43.81
5	3	0.8	2	2.53	145.67	0.234	0.59	2	2.53	36.27	0.427	0.82	38.45
6	3	1.0	2	2.53	139.44	0.234	0.66	2	2.53	32.48	0.455	0.88	34.52
7	4	0.5	2	2.53	128.32	0.243	0.50	2	2.53	26.96	0.427	0.94	87.35
8	4	0.8	2	2.53	128.32	0.243	0.61	2	2.53	26.96	0.427	0.97	58.27
9	4	1.0	2	2.53	155.86	0.234	0.68	2	2.58	23.83	0.455	1.08	58.78

Table 4b. Some Experimental Results for Output Variability Improvement with $m = 4$ and $EPS = 0.5$

No.	a	b	δ	Integrative Approach					Sequential Approach					% Cost Penalty
				n	k	h	σ_1	$cost_1$	n	k	h	σ_2	$cost_2$	
1	0.5	0.1	2.0	5	3.47	125.43	0.269	0.36	6	3.25	20.46	0.427	0.45	23.44
2	0.5	1.0	2.0	3	2.53	103.99	0.298	0.40	3	2.53	30.87	0.427	0.57	41.12
3	5.0	0.1	2.0	7	3.11	146.14	0.314	0.42	7	3.11	49.69	0.427	0.58	36.95
4	5.0	1.0	2.0	3	2.17	151.13	0.331	0.47	4	2.39	66.74	0.427	0.65	38.68
5	0.5	0.1	2.5	4	3.47	38.73	0.269	0.39	4	3.47	10.94	0.427	0.50	28.18
6	0.5	1.0	2.5	2	2.53	90.66	0.256	0.42	2	2.53	19.74	0.427	0.65	52.67
7	5.0	0.1	2.5	5	3.47	175.68	0.256	0.44	5	3.25	33.48	0.427	0.70	60.04
8	5.0	1.0	2.5	3	2.53	134.70	0.283	0.49	3	2.53	40.47	0.427	0.79	61.16
9	0.5	0.1	3.0	3	3.47	27.49	0.256	0.42	3	3.47	8.43	0.427	0.57	37.33
10	0.5	1.0	3.0	1	2.53	91.48	0.208	0.44	2	2.89	14.46	0.427	0.73	66.49
11	5.0	0.1	3.0	4	3.47	140.63	0.231	0.47	4	3.47	23.53	0.427	0.85	79.43
12	5.0	1.0	3.0	2	2.52	128.32	0.243	0.50	2	2.53	26.96	0.427	0.94	87.35

Table 5. **Percent Average Cost Penalties from Experimental Results for Both Types of Process Quality Improvement**

Parameter	Process Failure Rate	Output Variability
$m = 2$	6.65	11.92
$m = 3$	13.24	20.43
$m = 4$	23.42	36.95
EPS = 0.5	15.76	30.81
EPS = 0.8	14.17	20.47
EPS = 1.0	13.37	18.04
$\delta = 2.0$	8.70	16.11
$\delta = 2.5$	14.01	23.01
$\delta = 3.0$	20.59	30.20
$a = 0.5$	9.80	17.17
$a = 5.0$	19.07	29.03
$b = 0.1$	11.54	19.11
$b = 1.0$	17.33	27.10
Grand Average	14.43%	23.10%

indicate that the impact of economic design of the control chart on the determination of the process improvement target should not be neglected. The effect of system parameters on the total costs of the sequential approach and the integrative approach were studied numerically. The results also show that the effort devoted to process improvement is compensated for by reduced inspection. In general, the integrative approach indicates that pursuit of a lofty process improvement target can realize benefits of decreased total system cost. The findings demonstrate the cost effectiveness of the integrative approach and encourage quality practitioners to set a lofty target to improve process quality.

References

Blache, K. M., K. C. Stewart, R. L. Zimmerman, J. E. Shaull, R. D. Benner, and G. P. Humphrey. 1988. Process Control and People at General Motors' Delta Engine Plant *Industrial Engineering* 20(3):24–30.

Goel, A. L., S. C. Jain, and S. M. Wu. 1968. An Algorithm for the Determination of the Economic Design of $\bar{X}$ Charts Based on Duncan's Model. *Journal of the American Statistical Association* 63:304–20.

Ho, C. and K. E. Case. 1994. Economic Design of Control Charts: A Literature Review for 1981–1991. *Journal of Quality Technology* 26:39–53.

Lorenzen, T. J., and L. C. Vance. 1986. The Economic Design of Control Charts: A Unified Approach. *Technometrics* 28:3–10.

Montgomery, D. C. 1980. The Economic Design of Control Charts: A Review and Literature Survey. *Journal of Quality Technology* 12:75–87.

Porteus, E. L. 1986. Optimal Lot Sizing, Process Quality Improvement and Setup Cost Reduction. *Operations Research* 34:137–44.

Reibstein, L., R. Washington, M. Levinson, and B. Shenitz. 1992. The Hardest Sell. *Newsweek*, 30 March, 50–52.

Tagaras, G. 1989. Power Approximation in the Economic Design of Control Charts. *Naval Research Logistics Quarterly* 36:639–54.

Taguchi, G. 1986. *Introduction to Quality Engineering.* White Plains, N.Y.: UNIPUB/Kraus International Publications.

Dr. Chen is an Associate Professor in the Department of Business Administration, National Cheng-Chi University, Taipei, Taiwan.

Simplify Your Approach to Performance Measurement

Philip Ricciardi

This article explains a method for measuring employee performance in terms of both productivity and quality. The author points out how many companies somehow separate these two items, but in fact they go together. Improvement in one area cannot come at the expense of the other. The performance "measurement" method presented here takes this into account and provides a numerical method to combine these two components in a meaningful way for managers and employees.

Overseeing business processes and measuring performance is more important now than ever before. In today's competitive environment, customers are always looking for better service, more knowledgeable support, faster response and lower prices. If any of your business processes have deteriorated, your customer base may dissolve before you can react.

You may already have established standards for measuring employee productivity. But a performance measurement program that doesn't encourage improvements in the volume and quality of output will not improve productivity or provide information to manage operations. It may even be detrimental to employee morale. And it certainly won't reduce operating costs or enhance profits.

Designing a Successful Program

To avoid these types of negative results, a performance measurement program must measure both productivity and quality, must provide incentives and, above all, must be fair.

Many companies fail to recognize the difference between productivity and quality of output, focusing instead on one or the other. Programs that concentrate on productivity create a nimble employee machine that processes work at a frenetic pace. Unfortunately, the machine makes mistakes that are passed on to customers. A focus on quality may generate great products, but also unacceptable lead times and missed delivery dates. The ideal program improves both productivity and quality by balancing the need for speed with doing the job right.

> A focus on quality may generate great products, but also unacceptable lead times and missed delivery dates.

Reprinted with permission from *HRMagazine*, March 1996. Copyright © 1996 by the Society for Human Resource Management. All rights reserved.

One of the toughest elements to get right in a performance measurement program is equity. Too many programs fail to assess each individual, department, or process in the same way. An accounts payable clerk should be measured in terms of checks processed per day, while a shipping dock worker is measured in terms of pallets loaded.

Even in the same department, inequities often exist. Suppose the customer service department standard is for each rep to handle a minimum of 20 calls per day. An employee who fields calls from newer customers, who tend to have more questions when they call, may resent this standard because these calls take longer than those of established customers. As a result, the employee consistently receives low marks and may leave the company. As an alternative, the employee may decide to defeat the standard by rushing through calls, delivering poor service and quite likely losing customers.

In this example, the performance measurement program not only fails to achieve its goals, it actually subverts quality and productivity. It is important that a program measure all employees accurately and generate comparable results, regardless of the type of work being performed.

Finally, the program must include incentives that recognize and reward outstanding performance. Incentive plans are not without risks, the introduction of any kind of performance measurement program has the potential for employee revolt if mishandled. If every employee has an opportunity to be rewarded for good performance, incentives will not cause problems. Quite the contrary, they may prove to be the most important element in ensuring a program's success.

Productivity and Quality: A Unique Relationship

Performance may be defined as productivity multiplied by quality—it is made up of both the amount of work completed and the value of the work to the customer. An overall performance rating, then, should measure the ability to deliver the right output in the right way, on time and in one effort.

Many organizations mistakenly focus on measuring and improving either productivity or quality, without understanding the intrinsic relationship between them. But the two affect each other as well as the organization's bottom line in an almost circular fashion.

Increased productivity reduces cost, since higher outputs per hour result in lower labor costs per unit. Increased productivity also increases service quality because faster delivery improves the timeliness of service, increasing quality to the customer. Increases in quality increase revenue since high-quality products increase client satisfaction and retention and sales. And increased quality improves productivity because performing tasks correctly the first time eliminates the need for inspection and rework, reducing costs per unit.

Measuring Components

Performance should be measured in terms of both its components: quality and productivity, which includes output and input. The examples below illustrate how to measure all of these components equitably.

> *Performance* may be defined as productivity multiplied by quality—it is made up of both the amount of work completed and the value of the work to the customer.

Productivity: output. *Productivity* is the relationship between process output and the resources used in that process, or output divided by input. If output increases while input remains constant, productivity increases.

The work required to produce any type of output can be measured in terms of *standard time*—the time a fully experienced employee needs to perform a given task and obtain its output. The concept of standard time helps measure all employees on a level playing field, regardless of the type of work they perform.

The first step in designing a performance measurement program is identifying the output of each process being measured and determining the standard time. Measuring all output in terms of standard time allows different types of work to be compared with one another. *Output* can then be thought of as volume completed (per process or task) multiplied by standard time per unit of output.

For example, a processed check may have a standard time of 30 minutes. If an accounts payable clerk processes 14 checks in a day, 7 hours of output has been created: $14 \times 30 = 420$ minutes or 7 hours. Similarly, a pallet loaded onto a truck may have a standard time of 5 minutes. If a shipping dock worker loads 84 pallets in a day, 7 hours of output has been created: $84 \times 5 = 420$ minutes or 7 hours. Although the accounts payable clerk and shipping dock worker performed very different jobs, both created the same amount of output in terms of standard time.

Productivity: input. The amount of time employees have available to perform tasks is *input*, the resource required to produce output. To determine input, subtract allowance time (for breaks, etc.), paid leave (for vacations, holidays, sick leave), and assignment-work time (work for which no standard time is established) from each employee's paid time.

For example, if a customer service rep paid for an 8-hour day received 15-minute breaks in the morning and afternoon, a 30-minute lunch break, and was given an unmeasured assignment for 1 hour, his or her input would be 6 hours:

```
  480 minutes
-  60 minutes (breaks and lunch)
-  60 minutes (assignment work)
  360 minutes, or 6 hours.
```

If this employee completes 20 phone calls today with a standard time of 20 minutes each, his or her *productivity score* would be 111, based on standard time.

Output: $20 \times 20 = 400$ minutes.
Input: $480 - 120 = 360$ minutes.
Productivity: $400 \div 360 \times 100 = 111$.

SIMPLIFY YOUR APPROACH TO PERFORMANCE MEASUREMENT

Because the productivity score is based on standard time, it can be compared with any other worker's productivity score on an even basis, regardless of the type of work performed. If 100 represents an average score, this employee's productivity is above-average.

Employees should spend their input time solely on *value-added* productive work. When productive work is not available, supervisors should not create "busy" work as assignment work. This practice artificially inflates productivity by decreasing input. The only way to maintain productivity when work is not available is to reduce paid time by encouraging employees to take vacation or stay home.

Quality. The output of every completed task must meet four criteria. Quality output is appropriate (the type required by the customer), accurate (correct), complete (every required aspect delivered) and on time. For example, if a customer orders a Jane Fonda workout tape in VHS format for Christmas, the warehouse must ship a tape that:

- Is a VHS tape, not a laser disc (appropriate).
- Is a Jane Fonda tape, not a Cindy Crawford tape (accurate).
- Contains accompanying brochures and coupons (complete).
- Arrives by December 24 (on-time).

> Employees should spend their input time solely on *value-added* productive work. When productive work is not available, supervisors should not create "busy" work as assignment work.

When output is not of acceptable quality, productivity suffers because work must be redone until correct. In terms of performance, *quality* may be viewed as the percentage of work done with acceptable quality, or total output minus output reworked, divided by total output.

Suppose the customer service rep in the example above completed 20 phone calls, but gave incorrect information to two callers. Realizing the mistakes, the employee calls the customers back with the correct information. Although 22 calls were made, we only want to give credit for the calls completed correctly in the first place. The employee's quality score would be calculated as total output minus output reworked divided by total output:

> In terms of performance, *quality* may be viewed as the percentage of work done with acceptable quality, or total output minus output reworked, divided by total output.

$$22 - 2 \div 22 = 91\%$$

As with the productivity score, the quality score can be compared with any other employee's quality score on an even basis, regardless of the type of work performed. Overall performance is determined by multiplying the employee's productivity score by his or her quality score: $111 \times 91\% = 101$.

Incentives

Employers can implement a monetary incentive or "gain-sharing" program using the existing performance measurement program as its basis. A simple shared-benefits concept drives the incentive calculation. Employees meeting work standards receive a performance score of 100. Employees attaining higher scores save money for the company by requiring fewer labor hours to

Table 1. **Measuring Productivity: Department Level Performance Report**

Department: Customer Service — Week 12

Employee	This Week Perf.	Six-Month Avg.	This Week Prod.	This Week Qual.	Output Time	Input Time	Incentive Payment
Allen Zaber	100.04	105.00	101.3	98.8%	40.5	40.0	$ 0.20
Barbara Young	95.50	100.50	95.5	100.0	38.2	40.0	
Charlie Xavier	87.25	85.30	87.3	100.0	34.9	40.0	
Donna Brown	105.27	106.20	107.8	97.7	43.1	40.0	29.52
Ernest Valdez	103.21	99.90	105.8	97.6	42.3	40.0	17.99
Freida Ulfson	93.48	96.40	98.5	94.9	39.4	40.0	
Greg Smith	94.99	94.00	115.0	82.6	46.0	40.0	
Hanna Smith	107.95	107.70	111.8	96.6	44.7	40.0	44.52
Irv Rozinski	91.03	84.20	97.3	93.6	38.9	40.0	
Jennifer Quinn	87.00	89.80	87.0	100.0	34.8	40.0	
Ken Porter	100.25	97.10	100.3	100.0	40.1	40.0	1.40
Linda O'Malley	93.99	93.60	96.5	97.4	38.6	40.0	
Mark Needham	105.50	103.80	105.5	100.0	42.2	40.0	30.80
Totals	**97.39**	**97.19**	**100.7**	**96.7%**	**523.7**	**520.0**	**$124.43**

complete work. To encourage high performance, the company shares some of these savings with their best performers.

For this example, assume the company has a policy of a 50% share of savings with employees.

Weekly pay rate = $560

Performance score for this week = 120

Full-time equivalent = 1.20

A full-time equivalent (FTE) rating simply means that the employee does the work of 1.20 average employees, so the company avoids having to pay an extra 0.20 average employees. Because the company is grateful for this good performance, it will return some of these cost sayings to the employee.

Company savings due to high performance: 0.20 FTE

This week's savings to the company:

$560 × 0.20 = $112.00

Incentive payment @ 50% = $56.00

This employee will receive an extra $56 this week in recognition of high performance.

To keep an incentive program under control, two policies must be strictly enforced. First, the basis for the incentive payments must be clearly documented in writing. Employees can get the mistaken impression that incentive payments are part of their normal salary or that they deserve payment every week. Employees must understand that incentive payments are a bonus for high performance, received only when performance warrants.

In addition, supervisors must not make "busy" work during slow periods to artificially reduce employees' available time (input). Incentive payments should be rare when work is slow for the company because employees have less opportunity to complete a high volume of work at high quality.

Keeping the Program Healthy

Performance can be calculated at various levels, from individual employee to company-wide, and across all categories of work. Maintain and analyze data over time to indicate trends and provide all employees with appropriate information for their level to track over time. Departmental performance reports with employee detail should also be maintained. Performance measurement programs provide data designed to drive management decisions and help identify and correct negative trends before they affect customers, so managers must be prepared to act on the data derived from the program.

A performance measurement system is not a set of calculations and reports that remains constant once established. The program is a dynamic entity that must be maintained by periodically reviewing the following factors:

- Performance standards: New technology may greatly improve the amount of work employees can process. Review standards after every technology change to make sure they are up to date.
- Daily data: Excessively high (over 140) or low (under 60) daily scores at the employee level may indicate a problem with input data. An employee may write down 4,000 units completed when he really meant 400, or a person given 3 hours of assignment work may forget to write it down. Produce an exception report for supervisors to review and reconcile on a daily basis.
- Equity: Review the type of tasks high and low performers work on. If everyone who works on a certain task tends to get a high or low score, the task's standard may need adjustment.

Employees will accept the program if it is explained in terms everyone can understand. Providing incentives for high performance is one way to generate enthusiasm at all levels. But most important, a solid program allows management to make objective assessments of performance and business processes, and provides an accurate source of information to manage and improve operations on a continual basis.

Philip Ricciardi, a supervisor in the operations management consulting practice at Richard A Eisner & Company, LLP, has designed and installed performance management systems in a wide variety of industries.

When Do I Recalculate My Limits?

Donald J. Wheeler

This brief article explains a question many quality practitioners have about when to recalculate the limits on a control chart. Wheeler, a well-known author and speaker, answers this question clearly and concisely.

Of all the questions about Shewhart's charts, this is perhaps the most frequently asked question. While there is no simple answer, there are some useful guidelines.

The first guideline for computing limits for Shewhart's charts is: You get no credit for computing the right number—only for taking the right action.

Without the follow-through of taking the right action, the computation of the right number is meaningless. Now, this is contrary to everyone's experience with arithmetic. Early on we are trained to "find the right number." Thus, when people are introduced to Shewhart's charts, this natural anxiety will surface in the form of questions about how to get the "right limits."

While there are definite rules for computing limits, and right and wrong ways of computing such limits, the real power of Shewhart's charts lies in the organization's ability to use them to understand and improve their processes. This use of Shewhart's charts—as an aid for making decisions—is the true focal point of the charts. But it is so easy to miss and so hard to teach.

The second guideline for computing limits for Shewhart's charts is: The purpose of the limits is to adequately reflect the voice of the process.

As long as the limits are computed in the correct way and reflect the voice of the process, then they are "correct limits." (Notice that the definite article is missing—they are just "correct limits," not "*the* correct limits.") Correct limits allow the user to separate probable noise from potential signals. Shewhart's charts are a tool for filtering out the probable noise. They have been proven to work in more than 70 years of practice.

Shewhart deliberately chose three-sigma limits. He wanted limits wide enough to filter out the bulk of the probable noise so that people wouldn't waste time interpreting noise as signals. He also wanted limits narrow enough to detect the probable signals so that people wouldn't miss signals of economic importance. In years of practice he found that three-sigma limits provided a satisfactory balance between these two mistakes.

Shewhart deliberately chose three-sigma limits. He wanted limits wide enough to filter out the bulk of the probable noise so that people wouldn't waste time interpreting noise as signals.

Therefore, in the spirit of striking a balance between the two mistakes above, the time to recompute the limits for Shewhart's charts comes when, in your best judgment, they no longer adequately reflect the voice of the process.

The third guideline for computing limits for Shewhart's charts is: Use the proper formulas for the computations. The proper formulas for the limits are well-known and widely published. Nevertheless, novices continually think that they know better and invent shortcuts that are wrong.

The proper formulas for average and range charts will always use an average or median dispersion statistic in the computations. No formula that uses a single measure of dispersion is correct. The proper formula for X-charts (charts for individual values) will always use an average moving range or a median moving range.

Within these three guidelines lies considerable latitude for computing limits. As Shewhart said, it is mostly a matter of "human judgment" about the way the process behaves, about the way the data are collected and about the chart's purpose. Computations and revisions of limits that heed these three guidelines will work. Calculations that ignore these guidelines won't.

So, in considering the recalculation of limits, ask yourself:

- Do the limits need to be revised in order for you to take the proper action on the process?
- Do the limits need to be revised to adequately reflect the voice of the process?
- Were the current limits computed using the proper formulas?

So, if the process shifts to a new location and you don't think there will be a change in dispersion, then you could use the former measure of dispersion, in conjunction with the new measure of location, to obtain limits in a timely manner. It is all a matter of judgment.

Remember, Shewhart's charts are intended as aids for making decisions, and as long as the limits appropriately reflect what the process can do, or can be made to do, then they are the right limits. This principle is seen in the questions used by Perry Regier of Dow Chemical Co.:

- Do the data display a distinctly different kind of behavior than in the past?
- Is the reason for this change in behavior known?
- Is the new process behavior desirable?
- Is it intended and expected that the new behavior will continue?

If the answer to all four questions is yes, then it is appropriate to revise the limits based on data collected since the change in the process.

If the answer to question 1 is no, then there should be no need for new limits.

If the answer to question 2 is no, then you should look for the assignable cause instead of tinkering with the limits.

If the answer to question 3 is no, then why aren't you working to remove the detrimental assignable cause instead of tinkering with the limits?

If the answer to question 4 is no, then you should again be looking for the assignable cause instead of tinkering with the limits. The objective is to discover what the process can do, or can be made to do.

Finally, how many data are needed to compute limits? Useful limits may be computed with small amounts of data. Shewhart suggested that as little as two subgroups of size four would be sufficient to start computing limits. The limits begin to solidify when 15 to 20 individual values are used in the computation. When fewer data are available, the limits should be considered "temporary limits." Such limits would be subject to revision as additional data become available. When more than 50 data are used in computing limits, there will be little point in further revisions of the limits.

So stop worrying about the details of computing limits for Shewhart's charts and get busy using them to understand and improve your processes.

Donald J. Wheeler is an internationally known consulting statistician and the author of Understanding Variation: The Key to Managing Chaos *and* Understanding Statistical Process Control, Second Edition. *© 1996 SPC Press Inc. Telephone (423) 584-5005.*

Functional Processes

SUPPLIERS AND PURCHASING

Purchasing and Quality

Eberhard E. Scheuing

This article by consulting editor in purchasing Eb Scheuing highlights the crucial role played by an organization's purchasing function in managing the quality of its outputs by managing the quality of its inputs. The article includes a case study on the Tennant Company and a review of the concept of "customer-driven quality."

Quality means consistently *meeting the requirements* of internal and external customers. Toward this end, it is necessary to know who the customers are, what their requirements are, and whether these requirements are being met.

Purchasing is the organizational function responsible for the quality of the products and services received from suppliers. The quality of these *inputs* determines the ability of internal customers to perform their responsibilities and meet the requirements of external customers.

As shown in figure 1, an organization's purchasing function is a crucial link in the *value chain*, which combines the efforts of several organizations to create value and satisfy the requirements of external customers:

- It manages upstream linkages to the organization's *supplier community*
    - selecting and monitoring suppliers
    - fostering continuous improvement of supplier quality, cost, and cycle time
    - soliciting supplier suggestions and involvement

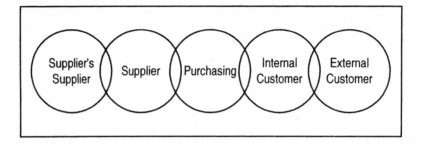

Figure 1. **Supplier-customer linkdates in a value chain**

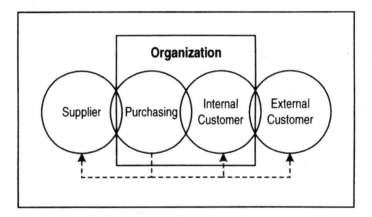

Figure 2. **Purchasing's role in managaing quality**

- It interacts with *internal customers*
  - understanding and serving their requirements
  - advising them
  - participating on cross-functional teams
- It needs to interact with *external customers*
  - understanding their operations and requirements
  - offering suggestions
  - providing the link to the supplier base

Managing Supplier Quality

Figure 2 illustrates the role of the purchasing function in managing quality: ensuring that the requirements of downstream customers are met by upstream suppliers. The figure further points out that a value chain is also a

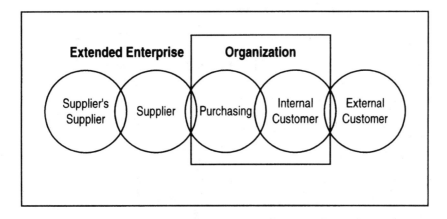

Figure 3. **Managing the extended enterprise**

chain of dependency. Downstream links depend on upstream players with regard to

- quality
- support and service
- timeliness
- cost

On all of these dimensions, downstream links can only be as good as their upstream counterparts. Inasmuch as suppliers are an organization's external resources, it is essential to extend the concept of the enterprise to include upstream value-chain links through

- upstream audits
- supplier training and assistance
- reengineering interactive processes

Figure 3 points out that to be effective in ensuring quality, purchasing professionals must extend their organization's enterprise to include multiple tiers of suppliers. It is axiomatic that any organization can only be as good as its suppliers. Accordingly, purchasers expend a great deal of effort on selecting and managing the organization's supplier base:

- They use quality audits to qualify a prospective supplier and periodically reexamine products, processes, and systems.
- They regularly evaluate supplier quality performance and communicate opportunities for improvement.
- They encourage, assist in, and reward continuous quality improvement by the organization's suppliers.
- They classify suppliers into different categories or levels based on quality performance.

> It is axiomatic that any organization can only be as good as its suppliers. Accordingly, purchasers expend a great deal of effort on selecting and managing the organization's supplier base.

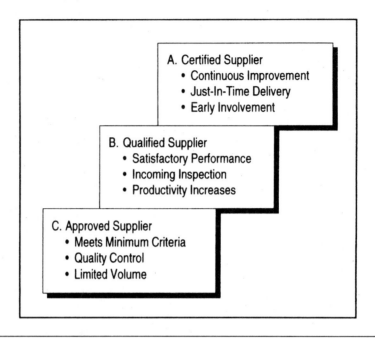

Figure 4. **Supplier quality evolution**

- They certify their top suppliers and may even publish their names in *The Wall Street Journal*—as Ford and Xerox do.

Supplier Certification

Supplier Certification involves the recognition of outstanding suppliers by specific customer organizations. To become *certified*, suppliers must have demonstrated

- commitment to customer satisfaction
- continuous quality improvement
- just-in-time delivery
- ideas for product and process enhancement
- productivity increases and cost savings

The progression of a supplier from the approved to the certified level is presented in figure 4. An approved supplier has been added to the list of approved sources after an initial quality audit determined that the supplier meets minimum criteria of good manufacturing practice and has a functioning quality-control system in place. Due to the lack of a track record, such a source is essentially on probation, in the form of a few trial orders, until a record of satisfactory performance has been established. At this point, the supplier becomes a serious contender by moving into the qualified category, which still involves incoming inspection but also expects productivity increases, speak: cost reductions.

Only a select few suppliers make it into the inner circle of certified suppliers. These exceptional sources demonstrate continuous quality improvement, provide just-in-time delivery of needed products, and participate proactively and early in new product development processes. These top suppliers are the organization's strategic partners who receive a large share of its purchase dollars.

In fact, on average, some 55 percent of the revenue stream received from external customers goes out the other door for purchases of goods and services from external suppliers. Organizations that purchase from thousands of suppliers are simply unable to evaluate and manage them all. Enlightened purchasers have accordingly been downsizing their supplier bases for years. When Xerox experienced serious quality problems in the early 1980s, it drastically shrunk its supplier base from 5,000 to 500 in two years. Although the pace and extent of this "weight loss" were exceptional, purchasing professionals across the country agree that concentrating efforts and dollars on high-performing suppliers is essential.

Understanding Cost of Quality

Xerox has identified three types of *cost of quality*, which can be described as the cost associated with meeting customer requirements:

Cost of conformance: Cost of "making sure," including statistical process control, education, problem prevention, measurement, and rewards

Cost of nonconformance: Cost of failing to meet or exceeding customer requirements

Cost of lost opportunity: Cost of customer defection due to lack of quality

The first type could be considered the "good" cost of quality because it is a necessary and appropriate cost. The other two must be labeled "bad" costs of quality because they are unnecessary and inappropriate. In a "zero defects" environment, these latter two types disappear.

Purchasers are responsible for ensuring zero defects in incoming products and services and thus eliminating the "bad" cost of quality for externally supplied inputs. Purchasers essentially serve as relationship and external resource managers by communicating the organization's requirements to suppliers, monitoring their performance, and highlighting improvement opportunities.

Supplier Quality Management at Work

The Tennant Company, a Midwestern manufacturer of motorized floor sweepers, exemplifies best practices in supplier quality management.[1] Tennant's supplier management teams, led by purchasers, thoroughly screen potential suppliers and evaluate current suppliers on an ongoing basis. They provide suppliers with continuous feedback on how well they meet Tennant's requirements and how suppliers can further improve their performance.

The company's relationships with its suppliers are driven by its quality policy and its adherence to seven *critical success factors*:

1. *Top-down emphasis on quality:* Quality must be driven from the top, and the CEO is firmly committed to the quality process.
2. *Sense of employee ownership—a corporate team approach:* The purchasing, design, production, marketing, and quality-control functions join forces in horizontally integrated teams.
3. *Customer/supplier partnership:* A relationship of trust and cooperation with suppliers yields powerful synergies.
4. *Preventing problems from reaching the customer:* Problems are avoided through frequent, open communication with suppliers and their early involvement in new product processes.
5. *Measuring and reporting quality:* Quality improves if it is measured, and it can be managed proactively only if it is measured.
6. *Qualifying suppliers:* Suppliers must understand and meet customer requirements.
7. *Annual business management sessions:* Annual high-level meetings with supplier teams enable open exchange of views and needs.

Tennant's *Supplier Satisfaction Rating Form* consists of three major sections. In sections I and II, suppliers can earn one of four ratings: excellent, good, average, or poor. Section I, *General Impression*, poses two questions:

- How well does this supplier meet Tennant's quality expectations?
- How well do they meet our expectations in the following areas? (Eleven areas, ranging from product reliability to top management involvement.)

Section II, *Customer Satisfaction*, contains four questions:

- How well does sales representation meet our expectations? (Seven areas, ranging from frequency of contact to accessibility.)
- How well do they meet our expectations on purchase orders and delivery? (Six areas, ranging from meet promise dates to packaging.)
- How well do they meet our engineering support expectations? (Seven areas, ranging from knowledge of our applications to overall responsiveness.)
- How well does this supplier work with us in containing costs? (Eight areas, ranging from controls price increases to performance vs. inflation.)

Section III presents the broad question:

- What other areas do we need to address to mutually improve our customer/supplier relationship?

A. Expectations	B. Relative Importance	C. Performance Ratings	D. Weighted Ratings
Totals	100%		

Figure 5. **Report card**

Quality in Purchasing

Enlightened purchasers do not just ride herd on suppliers to whip their quality into shape, they also practice what they preach. Purchasing professionals aim to deliver quality service to their internal customers. This effort of *customer-driven purchasing* involves understanding and satisfying customer requirements or expectations. It takes place as a four-step process:

1. Identify customers
2. Obtain customer input
    - Requirements/expectations
    - Relative importance (weights)
    - Perceived performance
    - Service quality gaps
3. Take improvement action
4. Obtain customer feedback
    - Performance ratings
    - Improvement opportunities

A powerful management tool for this process is the report card shown in figure 5. The report card can be used both to obtain customer input to understand their expectations and to obtain feedback in the form of performance

> Enlightened purchasers do not just ride herd on suppliers to whip their quality into shape, they also practice what they preach. Purchasing professionals aim to deliver quality service to their internal customers.

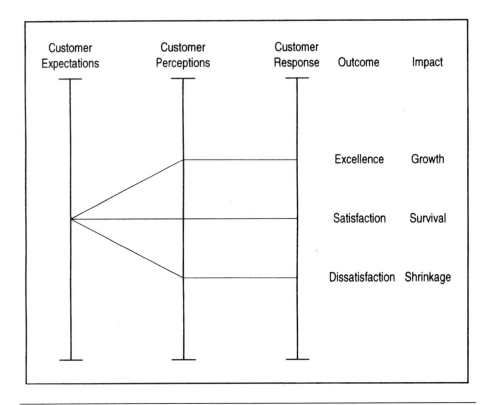

Figure 6. **Customer satisfaction and behavior**

ratings. Customers are asked to list their expectations, requirements, or evaluation criteria in column A and to indicate their relative importance by allocating percentage weights in column B. Customer perceived performance ratings are entered in column C according to the following scale:

5 = Excellent

4 = Very Good

3 = Good

2 = Fair

1 = Poor

The weighted ratings in column D are computed by multiplying the entries in columns B and C with each other. By summing up the weighted ratings, a single overall performance rating captures the level of internal customer satisfaction, supported by details that enable and focus corrective action.

Three potential patterns of perceived performance against customer expectations and their consequences in terms of customer response are depicted in figure 6, the scales of which can be defined as follows:

Customer expectations refer to the performance levels customers anticipate receiving from suppliers.

Customer perceptions are supplier performance assessments from the customers' perspective.

Customer response is the behavior customers carry out as a result of their comparison of expectations and perceptions.

Partnering for Quality

Applying the total quality management philosophy to their own actions as well as to those of their suppliers, purchasers need to partner both downstream with internal and external *customers* and upstream with *suppliers*.[2] In these partnering relationships, purchasers and their partners will benefit from a set of five proven power tools:

1. *Continuous improvement:* Inasmuch as quality is a journey rather than a destination, both sides must continually strive to enhance processes and products to keep them crisp and competitive. A vital approach to this task is benchmarking best-practice organizations.
2. *Long-term commitment:* Suppliers cannot and will not invest in process reengineering, new product development, and equipment upgrades unless suppliers are assured of a long-term relationship.
3. *Empowerment:* Individual players can do their best only if they have been granted the power to apply their talents and judgment and do what is right in a given situation, unencumbered by bureaucratic rules and procedures. Empowerment goes a long way toward making partnering for quality work by enabling each party's representatives to act in a way that serves both partners' long-term interests, regardless of short-term costs.
4. *Common values:* As core elements of corporate cultures, values are powerful forces that drive corporate and individual behavior. Any cooperative effort is greatly enhanced if the partners share commonly held values.
5. *Leadership:* Leadership involves formulating a strategic vision and rallying others in pursuit of this vision. It also requires providing adequate resources to enable the joint effort to succeed.

Partnering for quality produces a number of key *benefits*:

- Quality improves continuously
- Cost of quality decreases
- Lead time decreases

- Incoming inspection disappears
- Safety stocks disappear
- Just-in-time becomes possible
- Profits increase

Conclusion

Purchasing professionals are managers of resources and relationships. They hold the keys to an organization's marketplace success by ensuring the quality, timeliness, and cost competitiveness of inputs obtained from suppliers. In close cooperation with internal and external customers and their pool of suppliers, purchasers make essential contributions to the performance of their organizations as well as that of their supplier partners.

Notes

1. See Roger L. Hale, Ronald E. Kowal, Donald D. Carlton, and Tim K. Sehnert, *Managing Supplier Quality: How to Develop Customer/Supplier Partnerships That Work* (Exeter, N.H.: Monochrome, 1994).

2. See Eberhard E. Scheuing, *The Power of Strategic Partnering* (Portland, Oreg.: Productivity Press, 1994).

At CAT They're Driving Supplier Integration into the Design Process

Anne Millen Porter

Creating teamwork with their suppliers has allowed Caterpillar to develop higher-quality new products faster and cheaper. Deming's admonition to make partners of your suppliers is what this article is all about. It details how CAT has created teams of suppliers to work with the company in an integrated manner to design new products and supply the components to manufacture these products.

After 120 years in existence, it would seem there is little room for improving the design of an internal combustion engine. But leaders at Caterpillar's Engine Division say rising customer and environmental requirements make constant design improvements mandatory. To stay truly world competitive, says Don Western, vice president of operations, "we must produce million-mile engines with more features, more reliable performance, lower maintenance, outstanding product support, and progressively lower ownership and operating costs."

Market-driven mandates have caused the Engine Division to transform itself and its design process in three essential ways:

- **Internal reorganization.** Five years ago, Caterpillar changed from a highly functional organization to one of product-related, customer-oriented business units.
- **Internal supplier integration.** Under its Concurrent Product and Process Design (CPPD) approach, multifunctional commodity teams integrate suppliers into team driven design processes both for new product development and for improving existing products.
- **External supplier integration.** More recently, the division has begun to promote design integration among suppliers whose parts function together.

As George Benedetto, manager of Engine Division Supply Management (EDSM) sees if the reorganization begun in 1990 activated a radical change in

Five years ago, Caterpillar changed from a highly functional organization to one of product-related, customer-oriented business units.

Reprinted with permission from Purchasing *Magazine, 7 March 1996. Copyright © 1996 by Cahners Publishing Company. All rights reserved.*

EDSM's approach to design support. Instead of an engineering department we had seven product groups—each with its own set of designers. Roughly translated, this means that for each purchased commodity, EDSM has up to seven internal customers where it once had one or two. To complicate matters, Benedetto says EDSM had been downsized. "We had to do more facilitating between designers and suppliers with fewer people in our department."

Options were few. The supply management organization could either work harder or it could work smarter. According to Benedetto, EDSM chose the latter. Where once the organization was primarily concerned with purchasing activities, it has assumed the new proactive roles of "educator" and "enabler." And in assuming these roles, the supply management organization developed major initiatives to promote greater levels of integration with and among suppliers. These are:

- The Buying Game.
- Annual Quality Initiative (AQI) Awards.
- Supplier Technology Expo.

Defining the Endeavor

The Buying Game is a training program designed to facilitate interaction among supply managers, internal customers, and suppliers. It's a device aimed at developing a deep understanding of what's involved in balancing the issues of cooperation with those of control over supplier-customer relationships. "In the old days," observes Benedetto, "the buyer was a kind of communications funnel between Caterpillar and its suppliers." But this configuration served neither the supplier relationships that Caterpillar wished to develop nor the new Cat organization.

Deployment of multifunctional commodity teams rendered it necessary for a host of Caterpillar professionals from a variety of disciplines to deal directly with suppliers. "We needed to enlighten our internal users about the opportunities and pitfalls of dealing directly with suppliers," says Benedetto. "We had to create an environment in which our internal users were adept at conducting conversations with suppliers about potential working arrangements. We didn't, however, want to make buyers out of them."

The Buying Game, according to Pat McKee, project purchasing manager who helped develop the program, has been "a very effective tool." Hundreds of Caterpillar employees have undertaken the training, and as cooperation has flourished between Caterpillar and its suppliers, so too have the types of design improvements that ultimately upgrade the performance of Caterpillar engines.

Continuous Improvement

To recognize past improvements and to promote ongoing efforts, EDSM created an annual Quality Improvement (AQI) Award program. The annual awards program recognizes the most successful of supplier efforts undertaken on behalf of Caterpillar and the supplier.

"The AQI program is intended to provide incentive to suppliers to identify and manage successful projects that will provide benefits both to the supplier and to Caterpillar," says Terry Gramlich, commodity manager and chairman of the 1995 AQI committee.

At its inception five years ago, the AQI program was very small, says Gramlich. As he remembers it: "We held our first award ceremony in the dining room at our Mossville, Ill., facility," a room, that, according to Gramlich, accommodated roughly 40 people. That year, only AQI winners were present at the award ceremony. This past October, all 1995 AQI participants—roughly 400 supplier and Caterpillar personnel—were recognized at the Award ceremony in Peoria, Ill., and four winning projects were honored (see sidebar).

All AQI projects are implemented by teams comprising both Caterpillar and supplier personnel. "AQI projects," says Gramlich, "are generally process- and design-related. If a supplier comes forward with a project, many times it will require some type of design change and however large or small that change may be, it will always involve testing and validation."

But AQI is not limited to improving existing engine products, notes Gramlich. "More frequently, the AQI projects tend to occur at early stages of product development. More frequently, the effort is put in up front so that we are not correcting mistakes later on in production."

Ideas for AQI projects can be generated by either a supplier or a Caterpillar commodity team. Benedetto estimates that the breakdown is about 50:50. "Our experience is that suppliers have become very aggressive about initiating AQI projects."

Supplier Integration

"In the past," says Gramlich, "communications occurred primarily between one supplier and Caterpillar." But, in the most recent AQI competition Cat has begun to move suppliers into a more integrated design mode. Gramlich notes that EDSM has worked each year to improve and to expand the AQI Award process. Newest twist is the addition of a fourth category entitled "Special Achievement," which, in 1995, recognized the work of three related suppliers—Sadefa, Kolbenschmidt, and A. E. Goetze who became involved in a single project. The project was focused on combustion improvements and involved a piston supplier, a piston ring supplier, and a cylinder liner supplier.

Formerly, notes Gramlich, "our engineers would design the piston, the rings, and the liners and work separately and individually with each supplier." It's a story typical of many organizations where individual suppliers make parts to print, but the combined system often doesn't work or doesn't work optimally. By integrating the suppliers at the design stage, "all three suppliers now look at the total concept and they all have the simple objective of designing a system that will work."

Getting suppliers together is not a new concept, says Gramlich. "The innovation is in bringing three suppliers together and saying, 'Here's a challenge, now solve it.'" From a strategy perspective, integration among suppliers represents a new progression, suggests Gramlich.

> By integrating the suppliers at the design stage, "all three suppliers now look at the total concept and they all have the simple objective of designing a system that will work."

Having fostered better communication channels between internal customers and suppliers (with help from The Buying Game), and having implemented a program to promote continuous quality improvement between Caterpillar and suppliers (AQI), Caterpillar EDSM has taken an additional step toward driving technology deeper into the relationship. This step involves the staging of Cat's own biennial Supplier Technology Exposition. According to Benedetto, the "Expo brings together our many product people and supplier personnel in a cost- and time-effective way."

The first Expo was held in 1993 at the Peoria Civic Center in lieu of the division's annual supplier meeting. Last October, the 1995 Supplier Technology Expo featured nearly 150 exhibits and approximately 25 technical presentations by select Engine Division suppliers.

"The strength of this year's show," says Joel D. Feucht, supply Expo chairman, "is we built on what we learned in 1993." He notes, for example, that the Expo has two parts: (1) supplier booths where suppliers displayed components supplied to Caterpillar, and (2) technical presentations of leading-edge product and process technology developments that could have application to Cat products. For the 1995 Expo, the planners gave special attention to the presentation part of the program. To make presentations more useful to Expo attendees, program planners paid special attention to:

- **Subject material.** Particular care was taken to ensure that the subject material and technical scope of the presentations met the needs of Caterpillar's engineers and attending suppliers. "We chose our topics by actually going out and asking our engineers what they were interested in and needed to know."

- **Selecting the presenters.** Once expo planners got a good idea of the desired subjects and level of presentation, they took special pains to select from Caterpillar's preferred suppliers those most appropriate and selected topics.

- **Review.** Each supplier presentation also was previewed to see that it was appropriate and not just a high-powered sales pitch.

- **Technology.** Presentations often included information that goes well beyond descriptions of projects already in the works. In one case, for instance, a presenter showed how piston design affects the overall product design. (To insure that proprietary information was safeguarded, only Caterpillar suppliers were allowed to attend the presentations.)

In many ways, Feutch says, the Expo provides Caterpillar with an informal forum to facilitate supplier integration. Suppliers not only get a chance to talk with Cat engineers, they are encouraged to talk with one another. Often they also get a chance to share insights with suppliers and engineers on other Caterpillar products. Ideas that are hatched on one project often can be adopted and/or adapted to other Caterpillar products.

For some suppliers the Expo is a particularly good showcase of new technology. They can show new products and design features to a wide area of company—often to members of product teams with whom they ordinarily have little or no contact.

In some cases AQI projects have actually been spawned on the Expo floor when suppliers showing what they are doing in one area interest Cat engineers working on other products in other areas.

All Part of the Plan

In looking at EDSM's three major design initiatives, it's important to recognize that they're part of a well though out supply line management program. EDSM, for instance, operates with a very well established and refined supply base. Each commodity team in EDSM has a five-year plan, including a list of the core suppliers that the team plans to use.

Benedetto stresses that most decisions involving change of suppliers are made by a commodity team. "It's not just a purchasing decision, it's a commodity team decision." Commodity teams consist of representatives from the product groups and includes supplier quality, marketing, and logistics representatives.

He also emphasizes Cat's long-term commitment to established supplier relationships. "We try to facilitate (user-supplier interaction) by providing office space to our suppliers. We provide them with a working surface—a desk, files, a PC, telephone, etc. All of this is an attempt to improve the productivity of our suppliers as they interact with our people. Our ultimate objective is to come up with a reliable product, but the things we do also aim at getting the job accomplished in the most efficient manner."

Under Caterpillar's new approach to involving suppliers in design, supplier input is moved up to the original concept stage, notes Benedetto. "This is where we decide whether or not the product is feasible.

"Today when we have a concept study, we immediately call in those suppliers that we feel we probably will be working with." As a result, he notes, suppliers are involved with their process and product technology from the beginning.

In assessing the overall role of the Supplier Technology Expo, Benedetto looks on it as a unique tool in optimizing supplier participation in Caterpillar's product development. "There isn't anything we like better than to have our engineers go through the Expo saying, 'I didn't know they could do that!' or 'I need to talk to those guys!' If we hear enough of that, we know we've really hit the mark."

Anne Millen Porter is a contributor to Purchasing *magazine.*

Five Ways to Improve the Contracting Process

Pete Hybert

This article identifies five typical problems that regularly come up in the relationship between companies and those they contract with to supply solutions to their problems. It also suggests five ways to prevent such problems. As with many other process improvements, the emphasis here is on companies and their contractors working more closely together to take actions that prevent problems from ever occurring rather than on fixing problems once they occur. This approach means moving away from the lowest-bid method to one based on delivering the most value with least problems.

Companies that use a contracting process to provide customized, large-scale systems or products (e.g., information or telecommunication systems) have struggled to adapt the core quality principles to their businesses. While some have been successful, many have run into serious problems. The problems typically stem from trying to force-fit the use of quality tools. To improve the contracting process, companies need to focus on effectively achieving the mission of the contracting process rather than on figuring out ways in which to use the quality tools.

The contracting process, which is similar to many product development processes, can be found in a number of industries, including computer systems, telecommunication systems, aircraft and ship building, and construction. Contracting is the process by which customized systems are designed and delivered. Often, the term "solution" is used to describe a combination of customized products and services that are sold together as a solution to meet customer needs.

As Figure 1 shows, the contracting process is fairly generic. So, too, are its problems: The solution might not match the customer's real needs, schedules might be missed, or there could be so many change orders that they become obsolete before they are approved. The process often fails to yield customer satisfaction, produces unpredictable profits for the contractor, creates stress for the customer, and ties up more resources for more time than desired. Out of a number of potential problems like these, there are five primary ones that a quality-based approach can help solve:

> Contracting is the process by which customized systems are designed and delivered.

Reprinted with permission from *Quality Progress*, February 1996. Copyright © 1996 by the American Society for Quality Control. All rights reserved.

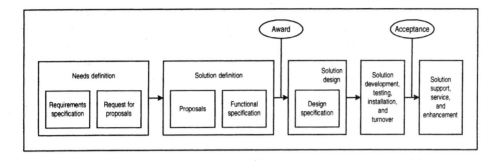

Figure 1. **A generic model of the contracting process**

- Poor up-front definition of customer needs
- Incomplete evaluation criteria for awarding the contract (an overemphasis on price)
- Poor planning of the project activities
- Poor assimilation of necessary midstream project changes (driven by problems or improvements discovered during the project)
- Metrics and rewards driving the wrong performance

Poor Up-Front Definition of Customer Needs

Since deficiencies in the needs definition process snowball as they roll further into the project, improving this process is a critical step toward more effective projects. The core of the needs definition process is the series of specifications that are developed by the customer and, possibly, by competing suppliers.

If contractors were to complete an application for the Malcolm Baldrige National Quality Award, they would probably view their specification process as a plus. After all, aren't specifications the result of working with the customer to define their requirements? Ideally, yes. In actuality, the specifications usually define only the technical aspects of the solution. They might not address quality requirements, customer satisfaction requirements, or cost considerations (such as development or life-cycle cost targets). So the specifications *do* describe customer requirements, but they are *incomplete* because they describe only one segment of the customer population: the technical evaluators. More important, because the specifications are developed up front rather than throughout the process, they are often inaccurate or incomplete in retrospect.

Specifications might also be written with the intent to *influence* the customer's requirements. Successful contractors try to minimize competitive bid pressure by working up front with customers to position themselves as the supplier of choice. This can result in specifications that don't indicate what the customers need but rather what the contractor wants to give them.

For example, one of the stated sales strategies for at least three contractors is to write the specifications for the customer. They train salespeople to position this idea as a time- and work-saver for the customer and provide salespeople with boilerplate specifications that favor their own products. In theory, this could be harmless if the requirements were closely matched to customer needs. In practice, however, salespeople tend to use the boilerplate specifications to shortcut the needs definition process. Initially, everyone is happy because the process moves more quickly. But the customers eventually discover the mismatches between their needs and what is being developed and have to make expensive midproject changes.

Specifications are typically not user-friendly documents; they do not *communicate* customer needs well, especially to those who were not involved in the process of developing them (such as new members on the project team or people who join the project downstream). Specifications are typically written in engineering- or legalese and rely on lengthy text with few diagrams, charts, or bullets to aid communication. Most important, sometimes the contents can be interpreted more than one way because the players involved in the project want it that way! Ambiguity prevents contractors from being held accountable for delivering solutions that don't quite solve customers' problems. It also prevents customers from being held to an early view of a requirement that might evolve as the project progresses. Where interpretations vary, the players must adopt hardball negotiating tactics, which certainly does not promote customer satisfaction with the project, regardless of the outcome.

Recommendation 1: Use a Quality Function Deployment Approach

Customers and contractors have a common interest in clearly defining needs in the early stages of a project. A quality function deployment (QFD) approach for defining functionality, quality, and cost requirements can reduce time and errors in this part of the process.

If the contractor's goal is to deliver a system or product that meets long-term customer requirements, its mind-set has to shift from deliverables and technical specifications (i.e., "What is the minimum I owe you?") to customer functionality (i.e., "What work will the customer be doing, and how will this system or product improve the case, cost, and quality of that work?"). For this to happen, the contractor must incorporate the needs of *all* stakeholders in the customer's organization.

For example, suppose the customer is a chemical processing plant looking for a control system for a new process. The plant's executives are concerned about return on investment, cash flow, and life-cycle costs. Its operations managers are interested in reliability, maintainability, and resources needed to support the system. The operators, or users, of the system are concerned with ease of learning, ease of use, and ease of access to reference information. (Too often, the users, including those who support and maintain the system, have the least input on its design.) Finally, technical experts are looking for system data for quality assurance tracking, the ability to adjust the system for varying situations, and so forth. There might even be regulatory needs as well. The needs of all these stakeholders must be met for the control system to improve the ease, cost, and quality of their work.

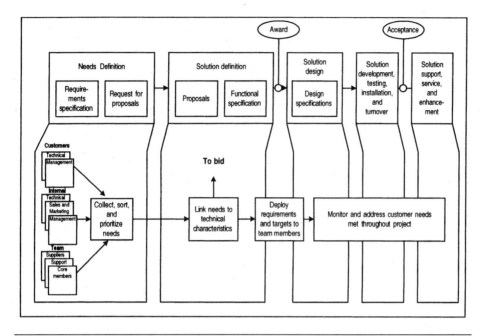

Figure 2. **How QFC can be integrated into the contracting process**

This is where the QFD matrix (often referred to as the house of quality) and the group consensus process can help. A team consisting of key personnel from both the customer and contractor organizations should use a consensus approach to develop the customer requirements portion of the QFD matrix (i.e., collect, sort, and prioritize customer requirements). Many approaches for gathering information can be used (e.g., focus groups or individual interviews), as long as the primary goal is to understand what the customer needs and expects—and not to create a perfect QFD matrix.

The value of the QFD approach is that the identified needs are clear, specific, and easily understood. They also truly reflect the customer's requirements because generating the matrix takes dialogue, clarification, and more than a single iteration (the needs can be reviewed and revised by all of the stakeholders before awarding the contract). Although documenting customer needs with the QFD approach will increase the time from initial project concept to bidding, it will result in time and dollar savings downstream because changes and conflicts will be reduced.

Figure 2 illustrates how other QFD elements can be integrated into the contracting process. Once the customer needs are defined, they can be translated into technical requirements and linked to elements of the solution, completing the QFD matrix. Upon award of the contract, the key result measures or solution elements can be deployed to various team members to ensure they are not later forgotten. Then, part of each team meeting can be spent focusing on progress toward meeting customer needs.

Incomplete Evaluation Criteria for Awarding the Contract

One of W. Edwards Deming's 14 points for management suggests that companies should end the practice of awarding business on the basis of price.[1] (Even when customers are not required to take the low bid, price is often the primary focal point of bid development and negotiation.) Yet, in practice if not in policy, it seems that price still drives many purchasing decisions. One reason is that customers are often unable to define their needs sufficiently to compare bids on an apples-to-apples basis.

Another problem is that although the users, technical experts, and managers need to review the various bids so that they have an opportunity to assess the proposed solutions against their own criteria, this practice isn't usually followed. Users usually have the most at stake in the decision, but they often have the least clout in the decision-making process.

Current bidding practices de-emphasize the importance of creating alignment between the contractor and customer so that they are both working toward the same end. Instead, these practices put the contractor and customer in an adversarial relationship and might even put one party in a position where it needs to take drastic measures to recover. For example, one contractor set change-order quotas for project team leaders to make up for how far it had cut into the profit margin during the bidding process.

> Current bidding practices de-emphasize the importance of creating alignment between the contractor and customer so that they are both working toward the same end.

Recommendation 2: Use a Teaming Process

Teaming with suppliers has been used in product development for years. The concept of teaming (or partnering in the nonlegal sense) with suppliers stresses having fewer suppliers and working closely with them so they understand the customer's needs well. This way, both the customer and the supplier have a stake in each other's success.

In the contracting process, the basic idea is for a customer to select a supplier (contractor) based on its ability to provide a solution. Then, the customer and contractor can develop the plan for the solution together, prior to or concurrent with pricing the project. (The plan for the solution is called a project plan, which covers everything after the award of the contract.)

For example, an industrial turbine manufacturer began the teaming process with a contractor that possessed a critical capability. The partnership evolved from negotiating volume sales agreements to sharing on-line ordering and quality data, sharing high-level business plans for new products and, finally, to including the supplier on a new product development team in the concept phase. The contractor and the manufacturer have both benefited from their partnership. The contractor has increased its business volume, improved its expertise in high-temperature metal coatings, and developed specific expertise in the turbine industry. The manufacturer now has higher-quality, lower-cost parts and a better end product.

Of course, there are also risks for both parties in the teaming approach. The contractor could engage in predatory pricing once it had a lock on the project; the customer could engage in heavy-duty arm twisting on price with the promise of future revenues once the supplier has committed resources to the project. Although the teaming approach requires mutual trust, it can reduce the need for costly risk management tactics, such as inflating prices, ducking difficult-to-address customer needs through loosely written specifications, or initiating change-order quotas.

Poor Planning of the Project Activities

Defects in a project plan can often be traced to faulty performance in early process steps. If the needs are poorly defined, the tasks required to meet those needs will be vague or wrong. Typically, project plans either lack the detail to prove that customer requirements are being met or are so detailed that nobody checks their validity or buys into them.

Defects in a project plan can often be traced to faulty performance in early process steps.

A worst-case example involved a project to develop a graphical interface for a controls system. The team leader asked the functional subteam leaders to put a plan together for their individual parts of the project. Once all of the plans were turned in, the team leader simply compiled them on his computer and distributed them to all key players. Unfortunately, everyone was too busy to review the compiled plan for the hand-offs (i.e., the passing of project information and deliverables from one subteam to another) that had to occur. The result was that the project was more than 12 months late, due to:

- The stacking of underestimated time frames for testing and approvals
- Misperceptions regarding the development of the project's design, prototype, training, and documentation
- Poor assumptions about when people would be available to work on the project

In a related case, a company decided to develop a computerized control system device. The project manager prepared a project plan based on subteam requirements. Management, however, had already promised a delivery date much earlier than that given in the team's plan. Not surprisingly, the device was delivered six months after the promised delivery date. (In fact, many of the previous product launches at this company were either late or on time but with reduced product functionality.)

Based on anecdotal evidence from many industries, these types of scenarios are much more common than one would expect.

Recommendation 3: Use a Group-Based Integrated Planning Approach

If the team members, including customer and contractor personnel from the key functions involved in the project, work as a group to develop the plan, they are more likely to develop an *integrated* project plan than if the project manager works independently. An integrated plan includes, at the minimum, sufficiently defined team deliverables and hand-offs to allow each subteam a reasonable comfort level.

Once the baseline project plan is developed, team members are in a better position to work the plan (i.e., make local decisions that fit the overall team direction) because they understand the details and trade-offs within the plan, they have committed to it in front of others, and they have developed the working relationships within the team that promote extra effort to make sure deadlines are met. This is probably the first point at which the plan should be entered into a computer program.

It could be argued that the group-based integrated planning approach cannot be used for complex, large-scale projects. But it could also be argued that, if the plan is so complicated that it cannot be comprehended, it cannot be effectively executed either. This approach will work on large-scale projects; they just might require several levels of planning.

Working together in meetings to determine the milestones and criteria for hand-offs will bring conflicts and difficult decisions, but the payoff is immense. For example, one contractor asked a group of five different suppliers (many of which had overlapping services and interests) to join the team to help prepare a bid for a project. Using the group-based integrated planning approach, a complex bid for more than $1 million of work was developed during a one-day meeting. In addition, the group accomplished some initial team building.

Poor Assimilation of Necessary Midstream Project Changes

When using a group-based integrated planning approach, the contracting process must allow for the extra dialogue and for modifications to the original plan. Rather than thinking that the defined and documented requirements are set in stone or forcing finalization earlier in the process than is reasonable, the project plan should include intermediate milestones and reassessments to better fit the reality of how people come to a common understanding of needs, contract deliverables, and so forth.

Most plans display the bias that change is an anomaly when, in fact, change is inevitable.

Most plans display the bias that change is an anomaly when, in fact, change is inevitable. Customers might change their business strategies during the project and, hence, their requirements for the solution. Technology might advance or the new technology planned for the solution might hit a snag in development. If the project plan has no room for adjustments to deal with unforeseen needs or problems, there is no room to work around those needs or problems.

Several words of caution, however, must be given in regard to making changes to the project plan:

- While the cost of small in-process changes can be absorbed, big ones can significantly increase the project cost. The team leader

- must be alert for all changes and not let big ones slip through unnoticed (many big ones look small at first).
- Although many contractors consider change orders to be "good business," they are a major source of potential customer dissatisfaction. From the customers' perspective, a change order means they have to go back to their managers to sell them on why more funds are needed for the project. Customers often feel taken advantage of because they know that margins are typically higher on change-order work than the original-project work. They might blame the contractors, thinking that the contractors didn't listen closely enough or didn't understand their businesses well enough to provide what they *needed*, instead of what they *asked for.*
- From the contractors' perspective, changes are very expensive (and frustrating) to incorporate in process. There is a real cost in productivity and energy associated with documenting a change, identifying its effect on other project areas, and communicating the information to everyone who needs to know. Every change is another opportunity to make a mistake: it introduces potential errors or discrepancies in the project information. Sometimes, the customer wants to change the solution back to what the contractor originally proposed, even though the customer initially refused it due to cost.

Recommendation 4: Change the Model

As Figure 1 showed, the typical model used to illustrate the contracting process is linear. A better representation, however, would be a spiral (see Figure 3). The spiral shows the incremental progress in all areas of the solution and ensures communication among all players in the process. In the product development world, this approach is sometimes called "concurrent engineering," "simultaneous engineering," or "integrated product development" because the people who produce, maintain, and support the solution begin developing their processes at the same time as the product is being designed.

Actually, the contracting process *can* be accurately represented using the typical linear model if a subtle point is well understood: The phases refer to activities, not individual functions, with team involvement in all phases. For example, when designing a software system, the design specifications should still be created before the final software code, but the software developers should be able to review and provide input on the specifications as they are written to ensure that the coding can be written efficiently.

The project plan should incorporate several features to minimize the pain of change and to account for in-process learning:

- "Cushions" should be allowed in schedules and budgets, but these cushions must be managed by the team and not buried in every task so that milestones are not taken seriously.

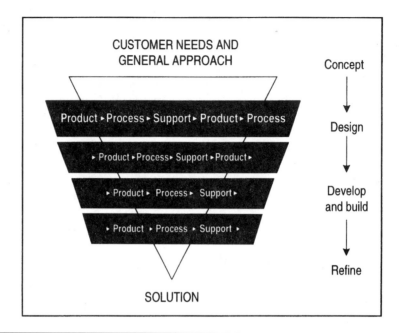

Figure 3. **The spiral model of the contracting process**

- Risks pertaining to potential changes should be systematically analyzed and managed.
- Accelerated prototypes or demos of critical elements should be prepared to get feedback on the solution earlier in the project (e.g., "rapid prototyping" as used in software development).

Of course, traditional change control and management procedures are also important.

Metrics and Rewards Driving the Wrong Performance

There is an old but true business adage, "You get what you measure." At a macro business level, contractors are rewarded by winning the contract if they underbid, exaggerate delivery capabilities, underestimate risk, or undersolve the customer's problem to get the price lower than their competitors. They are further rewarded with change orders for their ability to argue specification interpretation issues. At the project level, the same types of issues exist.

> At a macro business level, contractors are rewarded by winning the contract if they underbid, exaggerate delivery capabilities, underestimate risk, or undersolve the customer's problem to get the price lower than their competitors.

Recommendation 5: Focus Metrics on Customer Satisfaction and Quality

Measurement is a key part of the quality discipline. Project metrics can be determined up front if the project is geared toward a desired business result for the customer (such as reduced cost, more productivity, or improved quality). The people involved early in the sale of the solution are often involved in these types of issues, but the people doing its postsale implementation are not. This results in project metrics based on schedule and cost plan vs. actual. At its worst, this can result in a project manager completing a project on time at the planned margin but without it meeting the customer's needs.

To the customer, the most important consideration is the suitability of the end product over the long term. The customer must have a solution that serves its operational purposes and a system that provides the anticipated payback—or the project will be deemed a poor business decision. The right metrics need to be identified and made visible to the team, the customer, and the contractor.

Metrics for project outputs are typically assessed during acceptance testing. Metrics on the contracting process itself—such as customer satisfaction, disruption of customer operations, and decision-making cycle time—are more often overlooked. Yet, these types of data could serve as an early warning system for potential downstream problems. The QFD format can be used to help link metrics to requirements.

A Change in Mind-Set is Needed

The five recommendations described here are neither costly nor difficult to implement. An increasing number of contractors are already using QFD to define product and service features and quality requirements. The Department of Defense acquisition process includes a number of quality-related provisions (especially risk assessment and management provisions and provisions for evolving requirements) to improve the effectiveness of its contracting process.[2] Throughout the business world, experienced project managers are learning to create sound, well-integrated plans.

Improving the contracting process requires a change in mind-set. At issue is the trust between the customer, the contractor, and suppliers. Since contracting has historically been adversarial, it is unrealistic to expect companies to trust each other without first establishing a relationship. This relationship can be created by:

- Paying the contractor for the end deliverable *and* the consulting work needed to effectively define the customer's needs up front
- Creating a mutually beneficial partnership between the contractor and customer
- Developing well-integrated project plans that are aligned with the business goals of the contractor, customer, and suppliers
- Developing project plans that minimize the pain of change and account for in-process learning
- Setting up project and process metrics that focus on customer satisfaction and quality

Although establishing trust and adapting quality principles to fit a contracting environment are not easy tasks, there are certainly rewards to those that invest in the effort.

> Improving the contracting process requires a change in mind-set. At issue is the trust between the customer, the contractor, and suppliers.

Notes

1. W. Edwards Deming. *Out of the Crisis* (Cambridge, Mass.: Massachusetts Institute of Technology, Center for Advanced Engineering Study, 1986).

2. DOD 5000.2 "Defense Acquisition Management Policies and Procedures," 23 January 1991.

Pete Hybert is a senior associate at SWI, a management consulting services firm in Naperville, Illinois.

Functional Processes
LOGISTICS

Seven Trends of Highly Effective Warehouses

David R. Olson

TQM principles and practices have just as much relevance in warehousing and distribution as in any other business activity. With improved technology and better relationships among all members of the value chain, there are some dramatic changes going on in warehousing, resulting in more productivity and better customer service. This brief article profiles the trends that describe and drive these changes.

As we move from the 1990s and into the 21st century, we need to consider what trends are likely to have the staying power required to avoid becoming a fad. Looking back over the past several years, many have predicted the demise of warehousing—again and again—especially with the evolution of concepts such as just-in-time (JIT), quick response, efficient consumer response, direct store delivery, and continuous flow distribution. Common themes represented by these programs have caused a number of uninformed individuals to imagine a world without stockrooms, kitting operations, wholesalers, distributors, and distribution centers. Apparently, this world may use particle beam transmission, or possibly the Internet, to transmit goods and services at no cost from highly flexible, low cost, quick set-up production equipment all the way to the customer on time, all the time.

Reprinted from *IIE Solutions*, February 1996, with the permission of the Institute of Industrial Engineers, 25 Technology Park, Atlanta, GA 30092, (770) 449-0461. Copyright © 1996. All rights reserved.

Now, back to reality. We can be sure that warehouses will continue to play an important role in the logistics supply chain, just as they have done since Venice was the international hub of commerce in the "known world." We can be just as sure that warehousing will continue to be a dynamic function, driven by market forces toward continuous improvement. In keeping with this idea, the seven trends that will take industry into the next century include:

- Focusing on the customer;
- Consolidating operations;
- Continuous flow of material and information;
- Emphasis on value-added services;
- The application of information technology;
- Space compression; and
- Time compression.

1. Focusing on the Customer

The most successful, fastest growing, most profitable organizations have talked to and listened to their customers. They know that the customer wants value, at low cost with high functionality. They understand that quality must be a given. Their focus on quality goes beyond production, with consistently complete, accurate, and on-time shipments. They understand that the warehouse does add value—the utility of having the right product, at the right time, and making that product available to the customer.

They welcome special customer requirements as opportunities to differentiate themselves from their competition. They know how well they are doing because they have established relevant performance measures. These are customer-driven companies. As competitive pressures increase, customer service will be the competitive advantage in the 1990s. The most successful companies will be those that go beyond today's standard of customer focus, to develop true partnership relationships with suppliers and customers that transform the arm twisting of the 1990s to sharing of information, joint planning, and win/win agreements in the future. The warehouse will be in the middle of the action.

2. Consolidating Operations

Only the strong survive. The old song rings true. This law of the jungle prevails in all industry situations, including warehousing and distribution. Some organizations invest in themselves, others take profits. Some companies take their customers for granted, others look for ways to increase service and add value.

The 1980s were characterized by mergers and acquisitions. Customers realized it was to their competitive advantage to reduce their supplier base. An efficient deregulated transportation infrastructure has enabled higher and more reliable levels of customer service from fewer distribution points. This development allows economies of scale that can reduce operating costs including inventory, space, equipment, and labor. The result has been and will continue to be fewer distribution centers with greater individual mass. For some companies, the trend will be due to attrition, for others, consolidation will be due to strategic decision making. The end result will be the same: fewer operations of greater scope and efficiency.

3. Continuous Flow of Material and Information

World-class manufacturing practices developed in the late 1980s and early 1990s have had a spillover effect throughout the physical distribution arena. Manufacturers understood that once they established partnerships with suppliers, it simplified operations to move operations and inventory upstream. Store-direct shipments and supplier-maintained inventory are becoming increasingly common in the retail trade. Orders packed by store and shipped through distribution centers flow-through the facilities using cross-docking, or with only temporary storage. Some are drop-shipped from a secondary supplier to the customer. Items or products that were once picked and shipped monthly, or even quarterly, are now processed weekly and even biweekly. Will industry move to daily? In many cases, industry already have.

The trend toward more frequent shipments—and more receiving, putaway, picking, and shipping activities—will put greater demands on the material handling systems used in warehouses. This trend includes fork trucks, conveyors, and even carts. Demands on storage systems are sure to be different. The sizes of loads that are handled and stored will shrink. At the same time, stock keeping unit (SKU) proliferation will require more storage locations, not less. Because of the consolidation, there will be bigger distribution centers with more orders to process daily.

Just as material flow is becoming more continuous, so is the flow of information. On-line and even realtime information systems are replacing batch systems. As the number of material handling transactions increase within and between warehouses, so will information transactions. In this age of information, we are going to see perhaps the greatest change in warehousing with respect to information flow.

4. Emphasis on Value-Added Services

Perhaps one of the hottest topics and the most overused term in today's warehouse discourse is "value-added activities." The kind of activities that are often referred to as value added include special labeling, customized packaging, servicing vendor managed inventories, pre-ticketing retail merchandise, making retail merchandise "store ready" at the distribution point, and even electronic data interchange (EDI) transmission of information. There are two characteristics of these activities that qualify them as value added: They are not traditional warehouse functions (e.g., receive, store, pick, ship) and they have been moved upstream from the customer to the supplier. Therefore, they should be viewed as services that provide extra value to the customer, inasmuch as the supplier can perform the services effectively, and reduce the customers operating costs more than it increases the selling price. This is not to say that some of the traditional warehousing functions didn't add value of having the right product, in good condition, at the right place, at the right time. In fact, we may find that those products that are now referred to as value-added may be considered standard operating procedure in the future. Nevertheless, as the supply chain becomes more efficient, there will be new opportunities to move processes upstream.

> As the supply chain becomes more efficient, there will be new opportunities to move processes upstream.

5. The Application of Information Technology

Perhaps the most popular technology topic in warehousing today is that of bar coding. Of course, bar coding is a tool, like a fork truck, that when properly applied, performs a job efficiently and effectively. Coupled with radio frequency data communication, intelligent software, a dedicated computer platform, and maybe even electronic interfaces with host and external computers, creates a powerful system. Whenever material moves whether by fork truck, conveyor, or hand cart, it can be recorded automatically. The pwer of the computer is harnessed to direct the activities and select the resources that make the best use of those resources, while fully satisfying customer requirements.

Interest in this technology is well founded. As the warehouse management system industry matures, and understanding of its powerful capabilities grows, justification of information technology in the warehouse will become easier and the number of installations will grow dramatically.

6. Space Compression

Space as a scarce resource is nothing new to warehousing. If anything, it will be more of a problem in the future than it is today. There are many causal factors. The value of the warehousing function is less understood in traditional manufacturing organizations than elsewhere in industry. The result is that warehousing is constantly under pressure to give up space to production operations. Because of the continuing focus on customers and the expanding global marketplace, the SKU proliferation that has already taken place will continue. The same product will be packaged for more markets, with more languages, and even for specific customers. The more SKUs, the more storage locations and the more space that will be required. This is accentuated where consolidation takes place. Greater emphasis on throughput often results in more dock, staging, buffer, and sortation system space. Without adequate planning, the allocation of space can have adverse consequences on efficiency and service.

7. Time Compression

Who doesn't feel the pressure of time? Two-income families, quick gratification, down sizing and right sizing, and heightened competition all tend to give the impression that there just aren't as many hours in the day. This is perhaps more true in the warehouse than anywhere else in industry. Better, cheaper, and faster is the battle cry of the future. A wholesale supplier wants to simultaneously increase service levels with same day shipments and reduce SKUs and space requirements by packing to order. On the other hand, a retailer needs to simultaneously balance the warehouse workload and deliver store shipments on Friday ahead of the weekend sales rush. A wholesaler is convinced that sales are maximized with the traditional end of the month push. Another retail supplier ships 40 percent of its annual volume in the last month of the year. In this scenario, warehouse lives constantly on the edge from a time perspective, pushing the performance envelope in ways not understood by the purchasing, sales, or customer representatives that they work with.

Conclusion

Obviously, warehousing managers face some pretty steep challenges during the next century. But those same challenges offer a number of exciting opportunities. If companies can capitalize on these challenges and trends, prospects for improved logistical opportunities look excellent, especially for those companies seeking to achieve world-class status.

David R. Olson is Southwest Region General Manager for Tompkins Associates Inc., in Plano, Texas. He is an author and frequent lecturer on warehouse planning and operations subjects.

Build for Speed

Perry A. Trunick

Speed at delivering what customers want, when and where they want it, is key to gaining the competitive advantage today. Understanding this, manufacturers, warehousers, and distributors are becoming much more closely aligned. Such alignment allows for the teamwork and speed necessary to effectively and efficiently meet customer needs in today's dramatically more competitive and more global marketplace. This article goes into detail on this subject and provides some real-world examples of how various companies are doing this.

Some of the most important concepts in logistics have suffered from long development cycles, allowing them to dwell too long in the "theory" column. By the time they have made the leap to the "practice" side of the balance sheet, their names (or, more likely, the three-letter abbreviations) were passed in logistics circles. This is a dangerous attitude in today's hard-charging business environment. If your organization isn't implementing or using some of those concepts you may already be three or four years behind.

Leading companies are moving rapidly ahead in adopting concepts supporting supply chain management. They began refining some earlier methods and learned from mistakes made implementing partial solutions. And, probably most significant, many of these organizations aren't waiting for the market to drive them to change—spurred either by customer demand or competitive disadvantage.

One such company, Levi Strauss, decided to reengineer when it was at the top of its game. Sales and profits were strong, but the company recognized the need to change to meet the future.

After a rocky start with quick response (QR) and electronic data interchange (EDI), the clothier recognized it had not addressed its own infrastructure issues. Now, it is moving towards more regionalized distribution with suppliers linked to manufacturing and distribution facilities in clusters, according to Sandy Golden, business implementation manager, automatic replenishment systems. The goal is to regionalize inventory and deliver product within 72 hours.

> The current reality is that information moves faster than product, but the goal is to get these in synch.

The current reality is that information moves faster than product, but the goal is to get these in synch. The cycle consists of 24 hours for processing and 48 hours for delivery.

Reprinted with permission from *Transportation & Distribution*, February 1996. Copyright © 1996 by Penton Publishing, Inc., Cleveland, Ohio. All rights reserved.

"You're either integrated or you're not functional," says Golden. Levi Strauss is working towards weekly store orders based on agreed upon stocking terms and actual sales patterns. It will employ one-number forecasts, which experience has demonstrated is more consistent. Weekly sales forecasts extend out 60 weeks and provide greater visibility on future business. The forecasts also drive weekly inventory goals and weekly replenishment. Shortening the cycle time and employing vendor-managed replenishment means inventory and costs are reduced for each of the supply chain partners.

Not Just for Retail

General Motors' service parts division found it was able to reduce inventories and costs not only for its own operations but also for its dealer customers. Like Levi Strauss, the GM division started to change its supply chain processes three to five years ago. It had two objectives: service and cost. Its goal was what GM termed the one-visit repair.

For the one-visit repair to become a reality, the dealer had to have the part a customer needed and be able to complete the repair without requiring the customer to return. With 360,000 part numbers, this was quite an undertaking.

Darrel Manning, director of the material flow improvement team, General Motors Service Parts Operations, compared the operation in 1993 to its current status (as of late 1995):

- GM's service parts division reduced the number of suppliers by 20% to 3,200 actual suppliers in 1995.
- Its carrier base dropped 91% from 393 in 1993 to 35 target carriers in 1995.
- The parts tracer process was cut by 89% from 14 days to 1.5 days.

To make all of this work required close cooperation with a contract logistics supplier (Schneider Logistics) and a change to more dedicated routing and cross docking.

A Combination of Methods

Sunbeam Household Products was driven by a desire to reduce its time to market. Its solution employed many of the manufacturing and distribution concepts which have evolved over the last decade. All of this comes together in a new facility in Hattiesburg, MS. (The choice of location was part of the company's strategy, but for this discussion, let's look at the operations.)

The Sunbeam facility is 725,000 sq ft (73,627 sq meters) with a 160,000-sq ft (16,249-sq meter) high-cube, pick-to-light distribution center.

On the production floor, the manufacturing operation is set up with a focused factory approach. There are point-of-use receiving docks and a sophisticated vacuum delivery system connecting outdoor silos of raw resins directly to the molding machines. (This can be reconfigured so that each time

There are point-of-use receiving docks and a sophisticated vacuum delivery system connecting outdoor silos of raw resins directly to the molding machines.

a mold is changed, the resin supply line can be changed in a central control room to ensure the correct raw materials are delivered to the line.)

An automated material handling system removes finished parts from the line and transports them into storage or direct to the dock for shipment.

Part of the focused factory approach is the use of work teams, including a super crew. The super crew is cross trained in all functions and works three days of 12-hour shifts.

With quick-change capability and automated delivery systems—from the resin silos to the molds, then to the dock—the Sunbeam plant is designed for smooth flow and minimal inventory. Match that with vendor-managed inventory or replenishment and supplier-managed replenishment and the manufacturing line is more like a distribution cross-dock than a conventional factory—the goods merely pause to be reconfigured and move onto the next link in the supply chain.

The Future is Now

These examples show how a unity of purpose can drive change. But even the best intentions aren't enough to effect radical change in an organization—much less an entire supply chain.

"Six years ago, companies were split in the way they operated," observes John Williford, president and CEO of Menlo Logistics. He is addressing the first of four logistics trends he describes as:

- Globalization
- Conversion to demand pull
- Reverse logistics
- Cross-industry best practices.

Williford's discussion of globalization is the same as many organizations' efforts to put their domestic operations in order. Whether it is across divisions, business units, or continents, many organizations had different inventory management systems, different transportation systems, and different information systems. Now, says Williford, just about every company he sees is changing to a single world approach. They have, or will have, one inventory management system, one transportation system, and the information database will be centralized.

While Williford describes these trends in part as they apply to Menlo Logistics, a contract logistics provider, logistics practitioners describe their own organizations in similar terms.

"I'm responsible for coordinating materials through 30 manufacturing facilities," said one 16-year veteran. Only three of those facilities are owned by his company, he continues, concluding by saying it wasn't always that way. Another shipper agrees, "We literally had hundreds of warehouses in the U.S. Every business defined their own for their own customers." A third logistics professional talks about reducing the number of warehouses in his organization's network by 30%—but it took three years to do it.

Two factors are at work here. One is the effort to cut costs. The other is a restructuring of the logistics network. At first, the cost cutting was simple. Eliminate duplication, consolidate and streamline where possible, and give up ownership of costly overhead.

Inventory reductions and product line consolidation stripped out some of the need for warehouse space, so those warehouses could be sold and seasonal space could be put under contract to lower costs and reduce overhead. This may have led to some centralization or other structural changes, but another factor would prove a stronger force in restructuring—customer demand.

Centralized Hybrid

Lively discussion used to be unavoidable if you mentioned centralized vs. decentralized distribution or push vs. pull. Many logistics professionals now recognize their distribution systems are hybrids.

Information has to be centralized. "For realtime optimization, you're looking at integrating all of the [transactions] into a central repository so the sales force, customer service, and logistics can be reviewing the same information," said one manufacturing logistics manager. In his industry, he expects an evolution where manufacturing will concentrate on manufacturing and technology for the end product. All of the support functions—including accounting, purchasing, and transportation—will be handled through third parties. This will require tremendous integration of information systems, he says.

What will remain of the logistics function will be what he refers to as his supply chain services organization—basically a staff of internal consultants with the tools and techniques to go into a business unit and help define what the supply chain is.

Another logistics manager supports this view. He is responsible for three different divisions. Each one uses different parts of the services the central logistics function offers. So, while information will almost certainly be centralized, some functions will not. Managing freight rates, preaudit, and approval will remain central for one logistics manager. But, he says, divisions may source and use that central database to perform different functions like paying freight bills.

Demand Pull

The evolution of hybrid networks would occur much more slowly but for a development that is compelling change. That driving force is the customer.

Some of the best evidence of demand pulling the supply chain comes from the retail segment. Retailers are confronting a need to reduce inventories, says Alan Dabbiere, president of Manhattan Associates. But retailers aren't just reducing store inventory, they're stripping inventory out of the entire supply pipeline regardless of who owns it.

One of the goals of retailers is to have better selection in the stores with the same amount of space.

> ## Progress Isn't Easy
>
> Pioneers, as the saying goes, usually take a lot of arrows.
>
> When Ahold Nv. of The Netherlands elected to adopt cross docking in its U.S. operations, it probably expected some costs associated with the change and a little reluctance on the part of some employees. It certainly wasn't prepared for a contingent of Teamsters and others attacking it on its home ground.
>
> The International Brotherhood of Teamsters alleged Ahold's decision would "cost thousands of jobs and devastate local communities." The Teamsters enlisted an economic impact study by Cornell University which projected job losses in the Buffalo area (where Ahold's Tops chain operates its distribution center) at 5,300 and said it could cost area taxpayers $53 million in increased social services expenditures and lost tax revenues.
>
> The Teamsters made these and other claims through an open letter to Ahold management published as a full-page advertisement in leading Dutch newspapers.
>
> In other reports, the change in distribution operations was credited with more than 500 job losses. Further claims that the distribution methods would harm small, local shopkeepers were countered by Ahold. "Neighborhood shops will be better able to provide fresh produce through the faster distribution system," an Ahold spokesman was quoted as saying. Ahold also said the changes had been discussed by its U.S. subsidiaries and local union representatives.
>
> In an earlier financial analysis of the retail sector, Reuter news service quoted analysts who said that the Dutch retail sector would have to focus attention on cost control and efficiency to counter overcapacity. The retail food sector was plagued by discounting in 1995, though the analysts suggested this would level off in 1996 and should not seriously hurt earnings.
>
> Ahold realized 45% of its 1995 first-half sales from U.S. operations.

One way to accomplish this is to turn the floor inventory often and quickly. Short cycle replenishment is the term Dabbiere uses to describe the process. It's breaking down the barrier between industries, making the manufacturer and customer one virtual company, he continues. It's less an economic than a service issue, but turning up the velocity of the supply chain has financial benefits in cash flow as well as reduced inventories. The whole concept of electronic commerce makes short cycle ordering possible.

Point-of-sale and point-of-use messages pass up the supply chain via EDI and advance ship notices come back down. Manufacturing takes its cue on what must be made, and the distribution center looks at where the demand is occurring so it can receive, reconfigure, and cross dock product. For Levi Strauss, this last part is important. One retailer may require jeans to be

It's breaking down the barrier between industries, making the manufacturer and customer one virtual company.

shipped with a single fold, shelf ready. Another wants jeans double folded. Satisfying the customer requirement is paramount.

Increasingly, goods shipped to retail stores are preticketed and packaged floor ready. Distribution centers are changing to accomplish this, combining pick and pack with cross dock. It can mean pick, modify at pick face, merge with cross-docked product, pack, and deliver. The steps can be as simple as adding the appropriate price tag or hanging a garment on a store's own hanger, or it can involve a number of configuration steps. (An example of the latter is Motorola's cellular telephone plant in Scotland which postpones final configuration of telephones until it knows which market they are sold into and, therefore, which standard they use.)

While this sounds like a juggling act, it gets more complex. Major retailers don't necessarily want all of your product prepared the same way. One retailer has seven distribution changes, each with different labeling, packaging, routing, and EDI requirements, says Dabbiere. Which channel you use may depend on the size of the order, whether it's going to a general merchandise warehouse, cross dock, direct to store, or into the fashion channel. There's no single set of retail standards, and there may be more than one standard per retailer, so shippers have their hands full trying to stay in compliance. Many retailers have financial penalties for shipments that are not in compliance with their standards. And, ultimately, the entire account could be at risk if compliance is a recurring issue.

Changing Distribution

Velocity and a plethora of other issues are changing the face of distribution. Distribution on the side closest to the customer is becoming much more regionalized—not a warehouse across the street, but distribution centers within a next-day or second-day truck delivery. Manufacturing distribution tends to be more centralized and on a national scale. Add returns moving back up the supply chain (reverse logistics), and you find many companies reconsidering which elements they want to manage.

Regional retail distribution centers may handle store deliveries (or plant deliveries if this model is applied outside the retail sector), including delivery on their own trucks. They could handle a number of vendors across a high-velocity cross dock. Manufacturers would maintain their own distribution centers distributing to a network of retail distribution centers—some operated by the retailer, some by a number of retailers, and some by third parties. Suppliers to those manufacturers will manage manufacturing inventories of raw materials. And the whole thing will be driven by the end consumer taking a product off the shelf.

Henry Ford saw supply chain management clearly in 1926 when he said, "The time element in manufacturing stretches from the moment the raw material is separate from the earth to the moment the finished product is delivered to the ultimate customer." Unfortunately, he didn't have the technology to make his a demand-pull system.

Perry A. Trunick is editor of Transportation & Distribution.

> Distribution on the side closest to the customer is becoming much more regionalized—not a warehouse across the street, but distribution centers within a next-day or second-day truck delivery.

Functional Processes
INFORMATION TECHNOLOGY AND MANAGEMENT

Getting In the Pink

Anita Lienert

This is the story of how Owens-Corning Fiberglass, using information-management software from a German firm, turned itself around. This software allowed for centralization of information processing, and aligned information across the company. It made access to information across the company possible. This software actually forced the company to reduce its dependence on hierarchy to manage itself. From a quality perspective, the technology has made Owens-Corning processes operate much better and made the company more responsive to customers.

Hotdogging freestyle skiers leap into the air, executing high-flying scissor splits, 360-degree turns and other gravity-defying moves. To some, they may be the epitome of foolishness. To others, they are among the most courageous and talented athletes in the world.

You can see the freestylers on the slopes at the Olympic Games and in the most surprising of places: a dramatic poster display in the lobby of Owens-Corning Fiberglass Corp. in Toledo, Ohio. The $3 billion company; which most people know by its Pink Panther mascot, sponsors the daredevil sport.

Why not bowling or golf—something a little more down-to-earth and less risky?

Reprinted by permission of the publisher from *Management Review*, May 1996. Copyright © 1996 by the American Management Associaton. All rights reserved.

Because corporate officials say it wouldn't reflect how the company has pushed "the boundaries of the possible," breaking through such obstacles as an unfriendly takeover attempt in 1986, the death of a beloved chairman in 1991 and thousands of asbestos lawsuits that could have brought the 55-year-old maker of advanced glass and composite material to its knees.

But sponsoring freestyle skiing is only a small part of the story of how a rust-belt giant with 18,000 employees determined to reinvent itself—and now thinks of itself as an industrial visionary.

"Survival was the genesis of the vision," explains Glen H. Hiner, Owens-Corning chairman and CEO, who is overseeing the corporate reengineering dubbed "Advantage 2000," with its goal of turning Owens-Corning into a $5 billion company in four years. "We didn't have the capability to communicate with our customers in the way that they are demanding."

The vision includes how Owens-Corning embraced SAP, a tough German systems integrator that is not only remaking the company's information systems, but is also forcing radical changes in long-standing and often ineffective business practices. Owens-Corning is said to be the first U.S. corporation to go global with the SAP approach. And executives hope an end result will be an increase in Owens-Corning sales outside of the United States from 27 percent in 1995 to 40 percent by 2000, as well as getting a bigger slice of the $250 billion global home-improvement market.

Owens-Corning will invest $62 million between 1995 and 1997 on Advantage 2000, but says the program will generate at least $43 million per year in savings by 1999. Pink may be a recognizable color at Owens-Corning—there are plenty of life-size Pink Panther stuffed animals decorating executives' offices—but it's the last hue anyone here wants to see on a balance sheet.

Because SAP does not customize a system to a client's needs (it uses off-the-shelf technology), it forces business processes to be tightly integrated across applications and across departments, ensuring corporatewide consistency of information.

With the old computer systems, the sales representative's computer in Des Moines, Iowa, couldn't "talk" to the manufacturing plant's computer in Irving, Texas, in the same language. Crucial sales, inventory and production numbers often had to be entered by hand into separate computer systems.

Working this way is costly, time-consuming and likely to kill the most determined reengineering effort. SAP software not only allows Owens-Corning to standardize its information systems, but provides a road map for centralizing business activities.

To make SAP work, Owens-Corning had no choice but to remake its operation, including getting rid of all the plant accountants at more than 30 factories in North America. SAP also encourages busting traditional corporate hierarchies, which means that assignments are related to competency, not job title. In fact, a 22-year-old college intern named Molly Price had the honor of showing the chairman how to use the new system.

> Owens-Corning will invest $62 million between 1995 and 1997 on Advantage 2000, but says the program will generate at least $43 million per year in savings by 1999.

> SAP software not only allows Owens-Corning to standardize its information systems, but provides a road map for centralizing business activities.

Owens-Corning also took a hidebound salesforce—"old guys with sausage-fingers out of Peoria," as one executive put it—and yanked them into the '90s, with new variable pay-and-benefits programs and no more field offices. Learn the computers, work out of your homes and forget the luxury of having a secretary to help handle customers, they were told.

Bonuses are now being given on a quarterly—rather than annual—basis, making the company more responsive to good performance; but there are also risks involved. Employees will get only 35 cents on the dollar through the company's 401(k) pension plan, 15 cents less than the old days. But if Owens-Corning performs well, the company may match up to 75 cents.

Finally, there's the story of Toledo, a town that has suffered in the past from corporate defections and "low community self-esteem that fueled a negative image," according to a local economic development council report, yet somehow managed to prevent Owens-Corning from pulling up stakes.

Of course, a tempting tax-abatement plan helped. And so did the fact that many of the company's key executives are sons of steelworkers from places like Hicksville and Columbus, Ohio—men with a stake in the fate of the Midwest.

"It would have been much easier for them to go to the suburbs or let some Southern governor romance them," said Rick L. Weddle, president of the Toledo Regional Growth Partnership. "But they didn't. And it became a defining moment for the community."

Serious Second Looks

The Owens-Corning comeback is the biggest story in Toledo, a blue-collar town on the Michigan border that's famous for its Jeep plants and Tony Packos, a Coney Island restaurant that became a pop culture icon thanks to frequent mentions by Corporal Klinger on the TV-show M*A*S*H. That local hot spot lines its walls with plastic-encased hot dog buns autographed by everyone from Bill Clinton to Eva Gabor.

But now, both Toledo and Owens-Corning are getting more serious second looks. The company won the 1995 Overall Excellence category in *CFO* magazine's Reach Awards, beating out 250 other contenders, including Pepsi-Cola and AT&T.

"What the Reach Award winners illustrate is the ability to think outside the box, to break the rules, shatter assumptions and find radically new ways of doing things," wrote one judge.

Every interview with Owens-Corning employees and executives seems to lead to new insights and new tangents in the comeback story.

Gregory Thomson, senior vice president of human resources, gets settled in his office, ready to explain the variable pay-and-benefits system that began in January 1996, when he suddenly jumps out of his chair. "Follow me," he tells a visitor, leading the way across the hall and pulling open the drapes on a picture window to reveal a huge construction site hugging a curve in the Maumee River.

> ### SAP: REENGINEERING SPECIALIST
>
> Perhaps the best illustration of the success of German upstart SAP AG is the fact that four of its cofounders made the 1995 billionaires' list in *Forbes* magazine.
>
> "He who best fits a situation gets the biggest chunk of the cake," SAP Deputy Chairman Hasso Plattner told Forbes. "The traditional U.S. competition got caught totally wrong-footed."
>
> Headquartered in Walldorf, Germany, SAP (short for Systems, Applications and Products in Data Processing) is the undisputed leader in client/server software designed to manage everything from personnel to manufacturing under a single interface. What this means is that all computing functions of an enterprise are integrated in a seamless web. Example: when a company adds an employee, programs that handle payroll, personnel and profit-and-loss are automatically updated.
>
> Founded 24 years ago by four ex-IBMers, SAP now counts as clients such U.S. companies as Apple Computer, Dow Chemical and American Airlines, and generated $1.2 billion in revenues in 1994.
>
> The company's North American subsidiary, based in Wayne, Pennsylvania, charges $1 million for a typical installation.
>
> For more information, check out the SAP Web. site: http://www.sap.com.

"That's our new $100 million corporate headquarters that will be ready in September," Thomson says, pointing to the facility that Owens-Corning will lease from the city. "It's three football fields long and physically representative of the way we intend the corporation to be. It'll be more like an airport terminal than an office building—socially interactive, exciting and bright."

Good-Enough Reengineering

It's no surprise that the architect who designed Cleveland's Rock & Roll Hall of Fame is responsible for Owens-Corning's new home. But as company officials admit, the changes have sometimes struck a discordant note. Executives and managers are fine-tuning and rewriting the score as they go along. Even things like corporate terminology come under intense scrutiny. For example, employees selected to undergo SAP training were first called "power users." But since that was perceived as being "too aggressive," according to one executive, they've been rechristened "champions" and undergo "champion training."

And there's no predicting what the next few years will bring as the company continues a gradual rollout of the SAP system and Advantage 2000. But the corporate mantra seems best summed up by Michael D. Radcliff, vice president and chief information officer: "Avoid the perfection mentality. We

"Avoid the perfection mentality. We want 'good-enough reengineering.' So we are providing an environment where you make as many mistakes as quickly as you can."

want 'good-enough reengineering.' So we are providing an environment where you make as many mistakes as quickly as you can."

The first corporate guinea pig was the 400-member finance organization. Domenico Cecere, president of the roofing and asphalt division and former vice president and controller, says SAP is not just about bringing 5,000 new PCs into Owens-Corning.

"If you're taking your old processes and trying to use SAP, that's cramming your foot into the wrong shoe," he said. "I will tell you, the problem that you have is your foot is the wrong size. And you really have to take a look at what big feet you have. If we just went out and bought new hardware and software, we never would have gotten any savings. You can't get the efficiency if all you do is put new stuff in and do business the old way. If you put the old way we did things onto SAP, you'd probably bring it to its knees."

One of the first changes in finance: taking 30 different ways of doing payroll and payables out of what Cecere calls the "fiefdoms" of the North American factories and handling it with $7-an-hour help out of a central shared services location in Charleston, W.Va. The initial result: Owens-Corning was unable to pay its factory vendors for a time last summer because the system crashed.

"It was very disruptive," said Cecere. "To the point where the plant managers were calling, saying, 'You're going to prevent me from being able to manufacture because my vendors won't deliver because you haven't paid the bills.'"

Cecere credits top management with helping middle managers work through the snags.

"George Kiemle, the vice president of manufacturing, didn't say to me, 'Dom, you SOB, I'll never do this again. I'm going back to the old way,'" said Cecere. "He called and said, 'Dom, how can I help?' When you do these kinds of things, it will be disruptive and you'll have some problems. You have to have senior management behind you all the way.

Six months later, not only has that situation been resolved, Cecere reports positive feedback from U.S. plant managers.

"The system has moved us forward," Cecere said. "It used to take four accountants to close the books at one plant. We now have one accountant to close four plants."

Open Books, Open Goals

Charles Chambers, director of financial reporting and forecasting, says SAP and other corporate changes have made his life easier. For example, he says quarterly and annual filings with the U.S. Securities & Exchange Commission have changed from a "hernia-provoking experience that took two days to a nonevent that takes two hours." And now 14 people in his department do what it used to take 19 to accomplish.

Chambers is anxious for late 1997, when the whole corporation will be linked to a common computer system, accessed by all salaried personnel and 30 percent to 40 percent of assembly-line workers.

"When we all get on-line, I'll be able to walk into my office on a Monday or any other day and punch a few buttons and find out what sales are month-to-date on a worldwide basis," he said. "We can't do that now because our systems in Europe don't talk to systems in the United States. We'll also be able to know critical information like profitability by customer and the margin contribution by each salesperson. The goals are limitless."

Still, there are fears about security, especially in a new atmosphere of sharing information.

"The concept is that anybody can go in and run our financial schedule using SAP," Chambers said. "But I can lock people out, too. I don't want quarterly financial reports going to just anyone until we've released them. And they won't."

The highest expectation is for the company to become more sensitive and responsive to do-it-yourselfers and small contractors. Those buyers who comprise the backbone of Owens-Corning's business were often unwittingly shunted aside in the past—victims of incompatible computer systems and outmoded sales practices. Because many of the company's computers were not linked, it was sometimes impossible for salespeople to check on availability of products from the factory, make decisions on pricing or correct problems on an invoice.

With SAP and Advantage 2000, the Owens-Corning salesforce will not only have access to up-to-date information, but have pricing authority and more opportunities to make decisions rather than handing them off as well.

Explained Frank Glover, vice president of process development: "Under the old system, it would take two to three months for a customer to have a claim handled. It was absolutely terrible. The information wasn't documented or it was inaccurate. There were all kinds of problems with invoices. In the future, a sales guy can plug into the system and say immediately to a customer, 'You're right. There's a problem. I'll fix it right now.'"

Anita Lienert is a Detroit-based writer and a frequent contributor to Management Review.

Data Warehouses: Build Them for Decision-Making Power

John Teresko

This article points out that in most companies, information is not the problem. The problem is easy and quick access to this information to make effective decisions. To remedy this problem, companies are installing "data warehouses." These are essentially centralized storage of data from across the organization that enable executives to find out what they need to know about company operations. This article explains more about this emerging technology.

The surging popularity of the data-warehouse concept marks an important watershed in the way computers fit into the business world. Finally, 50 years after ENIAC, computer technology shows the achievable potential of fulfilling the strategic-knowledge needs of decision-makers. By expanding the computer's reach beyond such traditional automation tasks as accounting, design engineering, and production, the emerging data-warehouse technology may also finally redeem the troubled function of the chief information officer. By building data-warehousing concepts via available software technology, the CIO may finally be able to deliver what may be the most revolutionary change in information technology in the 1990s. No longer will the history of IT be marred by grandiose expectations reduced to disappointing realizations.

Although the desire to leverage operational data to support decision-making dates back to the early days of computers, that craving is now being reinforced by the confluence of strategic need—heightened competition—and a spectacular array of hardware and software technology that can make it possible through data warehousing.

An essential idea behind data warehousing is to bypass all the traditional problems that executives and other business analysts confront in trying to get useful knowledge from information storehouses that had been designed to support another purpose: day-to-day operations.

"The problem clearly has not been a lack of available data, but rather one of access," says Chris Erickson, president and CEO of Red Brick Systems Inc., Los Gatos, Calif., a data-warehouse solution vendor. "Ironically, while senior management is awash in data from the automation of corporate functions, too many corporate decisions are having to be made seat-of-the pants style. While the automation greatly increased the operating efficiency of businesses, the resulting accumulation of information did little to enhance any analysis of the business performance.

"In essence, computers have optimized all the business processes except the one of managing the business. The dirty little secret is that many senior managements in prior decades were—many still are—forced to manage using intuition drawn from the increasingly outmoded business models of prior experience. And if attempts were made to query the operational databases, the MIS [management information systems] effort took so long that the answers never arrived soon enough to be used as a competitive weapon." (He quips that the reason most MIS departments have been renamed IT is because they never really provided *management* information.)

So despite the massive amounts of data accumulated by the operational systems, the information potential is rarely realized, adds Van Symons, an IBM Corp. AS/400 client/server executive in Dallas. "Accessing information from the operational systems poses problems. First of all, there is the risk that business analysts could bring production databases to their knees just by the sheer volume of queries. Then consider the lack of tools—or their complexity. Also, data may not be stored in formats that are appropriate for analysts. Another problem is the timing of the data—it may not be appropriate for making business decisions. Since the parameters of the system were selected to support the operation, only limited historical data would be kept online—and possibly no trend analysis. Lastly, data intended to support the process may be stored inconsistently in multiple locations and in many different formats."

"The approach for a data warehouse," explains Randy Betancourt, program manager for data warehousing at SAS Institute Inc., Cary, N.C., "is to take the data from those operational environments and physically separate [those data] into a more meaningful and suitable format for decision-makers." He sees at least two factors contributing to warehousing's current prominence in the strategic use of IT:

"One is simply the competitive pressure businesses face in today's marketplace. It is driving the need to use every bit of available information to target customers better via database marketing. Another is affordability—the price points of computer hardware and software technology are being driven down to where organizations can seize the opportunity to buy effective solutions. Intensifying those factors is the increasing business literacy of IT professionals. In efforts to make their companies more competitive, they're transitioning their departments from cost centers to profit centers."

The idea of data warehousing is not new. Bill Inmon, the person generally credited with the concept, says it goes back to at least the mid-1980s when he began his own discovery that accessing information is a far more complex issue than merely getting at data. He explains it in terms of a process: "First, what data [are] the appropriate data to get your hands on? Then where [do] the data reside? How do I merge the files together to have a consolidated view of the data? Then, what about historical data?"

He says people who think that information is easy to get haven't done it before. "In order to have easily accessible information, you need to have the information designed and organized into a data warehouse, an integrated data architecture that is both detailed and summarized, with historical data and something called metadata [data about data] that allows easy discovery of the needed information—much as a card catalog in a library allows you to quickly find the books of your choice."

What led Inmon to formalize his thinking on data warehouses was a reaction to what he considered a simplistic view that all that was needed for data access was a relational database. Inmon did more than conceptualize the term. In addition to writing *Data Architecture: The Information Paradigm*, he is a founder and currently vice president of technology at Prism Solutions Inc., Sunnyvale, Calif.

The common characteristics of the early adopters of the concept seem to be centered on their need for making timely decisions amid rapidly changing business conditions. Although those conditions might suggest that companies heavily involved with logistics and retailing dominate the list, early adopters are spread across a broad spectrum of industries—aerospace, finance, high technology, insurance, manufacturing, pharmaceuticals, public utilities, and telecommunications. A report by International Data Corp. projects the overall growth of the data-warehouse market to be 23.9% between 1993 and 1998, driving revenues to $2.2 billion by 1998.

Consider Hughes Aircraft Co., a firm in transition from a traditional defense-based business to one that can successfully respond to competitive technological challenges in both defense and commercial environments. Incompatibility in a heterogeneous data environment was one of the primary obstacles to using data effectively, compounded by the myriad of platforms, database management systems, and applications at the company's various business units.

A seemingly simple request for information—such as identifying the top 10 purchased commodities and suppliers across three business units—could not be answered as quickly as the users wanted. The process of sourcing, extracting, and formatting the data into reports produced results that could not be easily validated. "The director of materiel needed the ability to find information about the acquisition, storage, or location of materiel in 10 minutes, not days," explains Robert Cuneo, director of Computer Sciences Corp. (CSC) applications services at the Hughes Support Center. (CSC is a computer service provider at the Hughes site.)

The goal was to allow Hughes to more effectively leverage its materiel requirements, negotiate corporate discounts with suppliers, and better manage purchased parts. The initial data-warehouse project was funded as part of an IT infrastructure project so that a pilot project could quickly be built and evaluated.

Prism Solutions provided the data-warehouse software tool, and Hughes used the Red Brick Warehouse as its target database-management system on a UNIX server. The implementation, serving the information needs of 400 users, paid for itself in its first year. Substantial savings were made in the cost of generating ad hoc reports and in data-management operating expenses. More important, Hughes expects a significantly higher ROI in the long term through the procurement-leveraging activity supported by the data warehouse.

Being able to easily access and leverage knowledge out of existing operational data has important strategic benefits to customer service, adds Russell Donovan, vice president-marketing at Praxis International Inc., Framingham, Mass., a provider of software tools for the data-warehouse market.

As a customer-service example, Donovan cites Whirlpool Corp. "The accumulation of customer information begins at the point of purchase, enabling Whirlpool personnel to access all aspects of the ownership and service record of the appliance during its life. Used that way, data warehousing can make a big corporation look more personable and knowledgeable about customers."

In a data-warehousing survey, IT analysts at the META Group Inc., Stamford, Conn., report that marketing is the only business area targeted more often than customer-information systems. Most companies are targeting multiple business areas, they report.

Two conditions signal an organization's suitability for data-warehousing—large amounts of data in operational systems and the absence of any IT strategy that would make it useful. "Often those factors are exacerbated by companies that have geographically separated operational data repositories," observes Steve Mann, director of product strategy for business applications at Computer Associates International Inc., Islandia, N.Y. He believes that "organizations as small as $150 million can benefit."

Regardless of size, data-warehouse solutions aren't bought—users or system integrators can buy only the software building blocks, the access tools, and sometimes the hardware with which to implement. To accommodate customers who want to buy solutions rather than technology, many vendors are forming solution-oriented alliances to facilitate implementations. For example, KPMG Peat Marwick LLP and Sun Microsystems Inc. are collaborating to provide solution-seekers with a laboratory environment in which they can perform hands-on testing of data-warehousing tools for performance, interoperability, and scalability.

In the chronology of organizational developments, the emergence of data warehouses seems to be but a prologue to a revised future of IT. "One immediate impact is that data warehousing is focusing attention on the importance of information to the corporation," says Martin B. Slagowitz, vice president of Information Builders Inc.'s Professional Services Div., New York. In the intermediate term he sees this understanding "putting pressure

The implementation, serving the information needs of 400 users, paid for itself in its first year.

on operational systems to better integrate with the needs of data warehousing."

An early indicator is the emergence of a category of corporate executive: the chief knowledge officer. Companies such as Hewlett-Packard, EDS, and Monsanto already have them in place.

John Teresko is a contributing editor to Industry Week.

Functional Processes
ACCOUNTING/ FINANCE

ABC and High Technology: A Story with a Moral

Frank H. Selto and Dale W. Jasinski

This is a case study of a high-tech company that lost an important customer and had to re-create itself with a new strategic direction. The article tells how decisions were made using traditional accounting data, and how those decisions would have been affected had the company been actively using activity-based costing (ABC), an approach that more realistically tracks costs by activities that generate costs.

Except in some large companies that are well-staffed, well-trained, and well-funded, there is not much evidence that ABC is understood well enough to be designed or implemented successfully as a stand-alone system, let alone one that is integrated with strategy. Even in large firms, widespread success of ABC is not obvious.

Because most job creation and innovative economic activity occur in small firms, we describe here the efforts of an accounting staff to design and implement an activity-based accounting system to support strategic changes in a relatively small, high-technology firm with approximately $100 million in annual sales. The experience has lessons for others who embark on similar tasks.

Reprinted with permission from *Managerial Accounting*, March 1996. Copyright © 1996 by the Institute for Management Accountants. All rights reserved.

The company, which chooses to remain anonymous and which we will call DataCom, is an entrepreneurial, privately held, high-technology concern that designs, assembles, and markets computer communication equipment. DataCom has faced challenges of rapid growth and equally rapid changes in both technology and customer needs that beset any high-technology company. Those that survive, and especially those that thrive, must be creative, agile, and quick to the market. DataCom encountered a dramatic change in its business environment that tested its abilities to survive.

Founded in 1985, DataCom quickly built its reputation as a premier supplier of computer communication links to mainframe computer original equipment manufacturers (OEMs) that incorporated DataCom's products without DataCom's name. DataCom enjoyed steady growth and added OEM partners. Founders of the company eagerly anticipated the right opportunity to take the firm public. In 1989, DataCom's primary OEM partner unexpectedly acquired its own mainframe communications technology and did not renew its contract. Sales dropped, but large cash reserves preserved the existence of DataCom.

DataCom's initial response was to redefine its marketing strategy as direct sellers to end users. Lack of name recognition and experience with direct sales, however, prevented significant sales growth. DataCom had no information that would measure the profitability of alternative markets or channels of distribution. DataCom downsized in 1990, primarily by reducing its direct sales force, and it focused on a few key markets. Cost cutting and focused marketing returned profitability in late 1990 and 1991. As a result, the company expanded its direct sales force to seek an increased market share, primarily in Europe where mainframe computing continued to be the technology of choice.

Economic recession in Europe in 1992 and rapid changes in U.S. computing technology to networked personal computers abruptly cut DataCom's already declining sales in half. DataCom had not foreseen the recession and had not expected domestic computing technology to switch so quickly away from mainframes. Although DataCom was not the only myopic firm at this time, in retrospect it was far more dependent on a dominant technology within a single market than was prudent. DataCom, however, never was blind to technological change and had continued to invest heavily in R&D even when sales first dropped in 1989.

By 1992 the combination of high expenditures for R&D, marketing, and administration had all but eliminated the company's sizable cash reserves, dropping its liquidity ratio from 5.0 to 1.0, with more than 50% of the 1992 numerator as receivables. DataCom found itself with a failing marketing strategy and a dependency on an apparently outmoded technology. It had neglected its creativity and agility and was in danger of being an also-ran in the market. During this threatening time, the accounting function continued to play a scorekeeping role—an increasingly unpleasant task.

Strategic Response

In 1992, DataCom's executive committee began urgent strategic planning. In the process, it affirmed that its comparative advantages were its abilities to:

1. Anticipate emerging communications technology accurately;
2. Create innovative, value-adding solutions to communications problems in the new technology;
3. Deliver outstanding functional quality.

The executive committee noted that continued investment in R&D had not resulted in significant new products. The causes were identified as diffused engineering efforts directed at putting out fires and making marginal changes to existing products to chase marginal sales. The executive committee sought a structural solution to refocus DataCom's considerable technological talent on its comparative advantages. The solution was to transform the highly centralized company into three distinct business units, each operating within technological boundaries. These business units were to be evaluated individually as profit centers that competed for scarce internal funds for growth.

The three business units were Harvest, Integration, and Network.

Harvest: The first unit was dedicated to prolonging the original line of business that connected mainframe computers to remote terminals. It would be operated as a harvest unit that generated cash to fund new product development. Technological risks were negligible, but the company expected that this market would disappear—but not, the company hoped, before new products were brought to market.

Integration: The second unit was designed to develop a complementary line of business that integrated communications among different computing technologies. This unit would extend DataCom's current expertise, but it followed previous R&D efforts. The integration problem is a significant barrier to computer communications, and DataCom believed it could develop innovative solutions in this large market. Growth in this market could prove to be steady but might level off within a few years.

Network: The third unit was organized to seek communication solutions to the emerging network computing technology. Though related to DataCom's previous efforts, this technology was sufficiently new to represent significant technological risk. If successful, however, solutions to network communication problems had the potential to repeat or outstrip DataCom's original level of success. This unit's performance would be the "home run" the executive committee sought, if its prediction was realized that networks would be the dominant computing technology of the foreseeable future.

> Common to all three divisions was a need to identify and measure the costs of support activities provided to each unit, including design and test engineering.

Activity-Based Information

The information needs to manage these three business units were somewhat different, requiring innovative solutions. Common to all three divisions was a need to identify and measure the costs of support activities provided to each

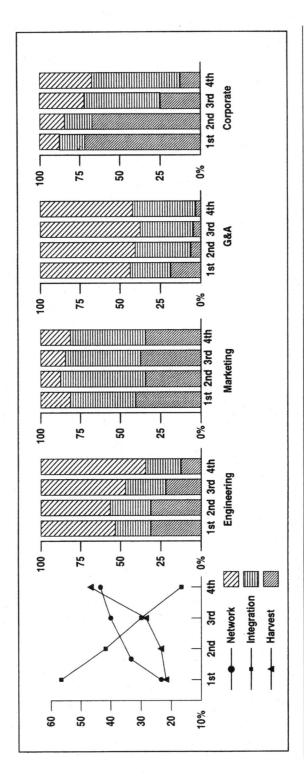

Figure 1. **Indirect expense allocation—1993**

unit, including design and test engineering. (Figure 1 shows the history of the shifting allocations over four quarters. As new directions were chosen, each unit was dramatically impacted.)

The Harvest business unit primarily required accurate scorekeeping to monitor profitability, cash flow, and marketing effectiveness—a familiar task for accounting. The new technology units, Integration and Network, weren't generating significant sales, but they did require expertise to evaluate the financial impacts of alternative technology and product development scenarios. The accounting staff was less prepared to fill this need.

Top management believed that an activity-based accounting system would enable them to better identify and cost DataCom's value-adding activities. The CEO, an engineer by training, saw that "as you get narrower (level of detail), you can look at the things that really impact your business. One of the things you find out as you do this is that your information systems may not be adequate to break down costs at the level that you need to make these decisions."

Accordingly, the CFO and the accounting department determined to develop an accounting system that would monitor the operations of the three business units as if they were separate entities and, at the same time, support managers' efforts to allocate resources efficiently. Some employees and activities were identified clearly with specific business units, including several financial staff who were to provide financial modeling assistance to unit managers. The majority of the company's employees still provided support activities from centralized service departments. These indirect expenses were and would continue to be a large percentage of total expenses.

The CFO began tracking service activity and costs to the business units. Interestingly, the CFO would not call this process ABC because of the connotation that ABC systems required special financial expertise and software, which at the time DataCom did not have and could not afford. At the same time, however, engineers in the centralized design unit were accustomed to recording their time, activities, and costs to projects and were well aware of the concepts and practices of ABC.

Despite the executive committee's desire to identify activities and to trace them to business units and products, the CFO never received sufficient resources to complete a thorough study. Because of the costs involved, the CFO never directly requested resources for that purpose. The engineers' comfort with ABC was not repeated in other service areas, and unassigned indirect costs averaged 38% of total expenses during 1993. The CFO allocated these expenses on the basis of revenues and hoped that business unit managers would reject these arbitrary allocations and demand more accurate, activity-based cost assignments—and then supply resources to generate the information. In 1993, based on revenue, the Harvest division received the great majority of the allocation, followed by the Integration division, which began to bring products to market. The Network division was shipping no products and was spared any revenue-based allocation.

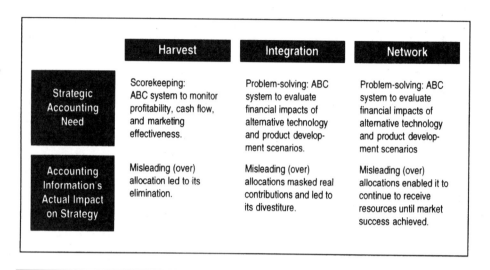

Table 1. **Business Unit—Information Needs**

Perhaps not surprisingly, the vice president of manufacturing and manager of the Integration business unit believed, "The new financial reporting that we have with the separation of the business units is clearly giving us a view of where the winners are and where the losers are. There have been some complaints from those who are spending time putting all these numbers together, saying, 'Where's the benefit? Where's the beef?', but for those of us who have become accustomed to it, it is a good decision-making tool."

Because the Harvest unit could "afford" the expense allocation and the other units saw no advantage to arguing for larger cost assignments, demand for improved activity-based information never materialized. Business unit managers did demand and receive extensive, centralized marketing services, which were traced only partially. Despite their increasing ABC skills, the accounting staff never got beyond their comfortable scorekeeping role, and they complained (somewhat incorrectly as it turned out): "The frustration from a planning point is—and we can do numbers as well as anybody else—that sometimes you still have to make a decision based upon those numbers, and we haven't made any decisions based on the numbers. We would view it [tracing activities and costs] as sort of a waste of time." (Table 1 displays the different information needs for each unit and the frustrating disparity between the needs of the units and what the accountants were able to provide.)

It was common knowledge that the revenue-based allocations were distorting business-unit profitability and that activity-based information could reveal a different picture: "If you use those [activity-based] numbers, you would discontinue the new products and continue to build the old, because they're making a bundle of money, and the new products are costing you a ton of money. But everyone thinks the future is Network.

Despite their increasing ABC skills, the accounting staff never got beyond their comfortable scorekeeping role.

"I'm wasting the company's dollars by spending my time tracking it now ... they're dumping all the dollars in revenue anyway ... and putting it into the Harvest business, which in reality has been taking substantially less of my time."

The accounting staff focused on scorekeeping of existing products rather than supporting the company's primary competencies—anticipating technological change, developing innovative solutions, and delivering high quality. The executive committee, however, did make critical decisions without accurate information about the activities necessary to support alternative product scenarios. Harvest was virtually eliminated, and Integration was divested, despite the facts that both generated positive cash flow and both probably were profitable.

Furthermore, the Integration market was growing more than had been anticipated and was receiving larger revenue-based allocations. Both units were judged to be insufficiently profitable to continue. (Integration is profitable under new ownership.)

The executive committee swung for the fence on Network and did hit the home run, as it turned out, while ignoring the high percentage hit that Integration delivered. This may be a defining difference between entrepreneurial, high-technology firms and more mature firms—the penchant for the home run and disdain for more conservative alternatives. Traditional management accounting's focus on scorekeeping may add relatively little value in this setting.

The CFO was sanguine about the prospects of developing improved, activity-based information, knowing that the future of the company rested on predicting the direction of computing technology accurately. The predisposition of the executive committee to go for the home run had been reinforced by misleading business unit profitability: "We have to go forward with the new development on our new product line [Network], which represents the future of business. Since we are trying to fund Network through the Harvest business unit, which happens to be the declining business, the decisions of how quickly to move forward would be the same [depending on cash flow]. So I am not convinced that we [would] have made different decisions [with better activity-based information]."

This comment probably was both rationalization and statement of fact. It was clear from the initial formation of the business units that Network was the executive committee favorite. Financial investigation of alternatives was limited because the CFO was the only member of the executive committee not given an explicit strategy formulation role. If aggregate scorekeeping was all that accounting could provide accurately, it could not bring much of value to the strategy table. Indeed, in the subsequent strategic planning process, the executive committee decided to solicit more input from throughout the organization and formed a planning committee of approximately 25 members. Any member of the executive committee could nominate an employee to serve on the committee. The only functional area not to have a representative was finance and accounting.

It seems likely that lack of accounting input will be repeated when DataCom eventually seeks a new home run technology to replace Network. After the sale of Integration, the accounting staff dropped efforts to trace activities and costs to any specific products or projects. An opportunity to add value to strategic decision making by developing activity-based information was missed, and aggregate scorekeeping is still accounting's primary function.

The Moral of the Story

Although ABC holds much promise for organizations, it cannot be implemented or evaluated independently of the organization's strategy. Unless ABC is integral to the company's strategy, it is unlikely that management accountants can break through the high-technology glass ceiling and become important members of the strategic planning team. Absent that input, it is clear that the job security of management accountants depends on providing relevant information to support the organization's decision making, and scorekeeping alone may be inadequate. Formulation and implementation of strategy will proceed with or without the accounting staff's contribution.

Frank H. Selto, Ph.D., teaches accounting and is chair of the Accounting and Information Systems Division at the College of Business and Administration, University of Colorado at Boulder.

Dale W. Jasinski will join the faculty at Idaho State University, Pocatello.

From Activity-Based Costing to Throughput Accounting

John B. MacArthur

This article tells of one company's search for an accounting system that would help the company accurately know its costs and profits on its highly varied product line. First it tried ABC but found it too complex and time consuming for the benefits offered. The company instead chose "throughput accounting," which has allowed the company for the first time to know the profitability on each item in its line.

Activity-based costing (ABC) has been widely supported as superior to conventional costing for providing cost information for a variety of uses such as product costing and long-run pricing. However, the additional data collection resources needed to obtain cost drivers and detailed activity costs may not be cost-beneficial, at least for some small- to medium-sized businesses. Traditional cost accounting information may measure resource consumption accurately enough for managers in companies with little product or process diversification.

Also, certain types of firms may adopt production techniques that lend themselves to simpler costing systems. For example:

Just-in-time (JIT) production and purchasing systems in general are well supported by backflush costing approaches. Backflush cost accounting systems eliminate job-cost records and some general ledger entries that are not required to support JIT production and purchasing systems, thus significantly reducing information processing costs.

Throughput accounting was designed to supply the very basic cost information required in synchronous manufacturing environments. As with backflush costing, throughput accounting eliminates a significant number of detailed costing entries, such as overhead cost allocation to products, that are typically associated with more traditional costing systems.

Management accountants should be flexible enough to adapt company costing systems to satisfy the changing information needs of managers in dynamic manufacturing and service environments. The accountants of one growing manufacturing company, Bertch Cabinet Mfg., Inc., recently exhibited such flexibility by deciding to change from ABC to throughput accounting to serve their internal customers better in a cost-beneficial way.

Cost Accounting at Bertch Cabinet Mfg.

Bertch Cabinet Mfg., Inc. (Bertch) is a fully integrated manufacturer of wood cabinets and their accessories (e.g., mirrors) located in the Midwest. The annual sales of Bertch place it in the top 10% of U.S. domestic cabinet manufacturers. Bertch has eight divisions: Bath Cabinet, Cultured Marble Top, Kitchen Cabinet, Mirror and Glass, Semi-Custom, Dimension, Transportation, and Administration.

From 1991 to 1993, Bertch experimented with ABC using a pilot study in two of its eight divisions, which was the first serious attempt to establish a formal costing system at Bertch. Previously cost information had been provided in an ad hoc manner. Bertch managers were pleased with the results from the initial ABC pilot study. ABC information was used to help:

- Price glass products;
- Assess the long-term profitability of mirror and cultured marble top products;
- Evaluate labor efficiency; and
- Manage the costs of nonvalue-added activities.

Bertch management originally planned to expand ABC into the remaining divisions and eventually to incorporate the ABC system into the general ledger. Subsequent efforts to further develop ABC at Bertch, however, were not successful for a number of reasons.

ABC Impact on Human Resources

Attempts to extend the implementation of ABC at Bertch beyond the initial pilot study convinced management that the required work was too time consuming and costly. The direct materials costs of the various wood and marble products were relatively easy to identify, but the assignment to products of direct labor and indirect costs of the Bertch divisions was difficult to determine in a timely manner with the available personnel resources. The manufacture of quality cabinets involves many complicated manufacturing processes.

For example, 21 separate machining operations are required to manufacture a single raised panel door. If each operation represents just one activity, it would be necessary to maintain costing information for 21 activity centers. Also, cost assignment would require 21 resource and 21 activity drivers. The costs of developing and running a complex ABC system were judged by Bertch management to be greater than the benefits it would receive from the costing system.

For instance, the benefit of ABC information for pricing decisions is limited because many of the prices for Bertch products are determined mainly by competitive market forces. Also, as a manufacturer of high-quality wood cabinets and accessories, the decision to in-source or out-source production items is determined largely by product quality and timeliness of delivery considerations rather than cost factors. As a result, all manufacturing processes currently are performed internally, and only raw materials (for example, wood and raw glass) are purchased from outside suppliers.

Another reason for the lack of success of ABC at Bertch was the use of temporary summer help as the primary personnel devoted to work on the ABC project. The person employed in 1991 to work on the initial ABC pilot study was very enthusiastic and competent and contributed significantly to the early success of the pilot study. Similar reliable help was found for 1993 for the follow-up work, but it is difficult for temporary help to become sufficiently competent in just a few months with the mechanics of a complicated concept such as ABC and the nuances of a new company to be able to make meaningful contributions in the development of a custom-made ABC system.

The significant production changes that took place between summers tended to make previous ABC work largely out of date. Using temporary accounting help to work on the ABC project for three summer months at a time meant that the continuity factor was missing.

Marketing Requirements

Possible delays in obtaining complex ABC product cost information could present problems to Bertch in obtaining new orders and new customers. For example, given the accounting labor hours available, it would not be possible to generate ABC cost information quickly enough to enable marketing managers to respond in a timely manner to purchase offers for semi-custom products from multiple outlet mass merchandisers with home centers. Such companies usually require swift acceptance of set prices that they offer as well as the provision of free displays from the manufacturers that supply their products. The semi-custom products are manufactured at Bertch by the relatively new Semi-Custom Division.

New Production Technologies and Processes

As a growing company, Bertch continually is upgrading its manufacturing technology. Also, major process reorganizations take place about every two years to improve production flows. It is not feasible for Bertch accountants constantly to update a complex ABC system to reflect the frequent technological and process changes.

Synchronous Manufacturing at Bertch

On February 1, 1993, synchronous manufacturing was introduced in the Dimension Division, and it became the catalyst for the new costing system adopted by Bertch management. The Dimension Division, the largest division of Bertch in terms of production volume, uses batch manufacturing for parts production and panel sizing. This division was the most obvious candidate for synchronous manufacturing and had the greatest potential for gain from its implementation.

A major education effort was undertaken to train the Dimension Division managers and employees in the new manufacturing approach. Managers and employees in general have become familiar with the synchronous manufacturing concept. More than 50% of Bertch personnel have read *The Goal*,[1] which, in story form, presents some of the basic principles of synchronous manufacturing.

The implementation of synchronous manufacturing proceeded more smoothly than anticipated and was operating with only a few minor problems after 60 days. Managers are pleased with the production improvements that included some of the following:

- Reduction in lead time from order to delivery of products. For example, the lead time was reduced from four to two weeks for Bath Cabinet Division products, which made them more competitive.

- Production batch sizes reduced from weekly to daily requirements. This improvement facilitates scheduling to exploit the system constraints, reduces the maximum number of items that can be spoiled through faulty production of a batch (wood products are easily spoiled), reduces the delay between spoilage and makeup work (each part is made daily instead of weekly), and reduces inventory levels (reduces clutter of the workplace and number of parts mislaid or buried in inventory piles). The use of CNC (computer numerical control) programmable router machines reduces setup time on intricate wood shaping and makes smaller batch sizes cost-effective. Also, workers became more efficient at setups through daily practice (learning effect). The longest setup is about 30 minutes. Further improvements in setup time to compensate for the considerable increase in the number of setups is expected in the future as synchronous manufacturing methods continue to be developed in the Dimension Division, as well as through the implementation of continuous improvement strategies.

- Thanks to the fact that now there are daily production schedules the materials flow is much more predictable and stable, plant emergencies are fewer in number and are handled more easily, and holiday scheduling is much simpler than previously was the case with weekly scheduling.

- Improved placement of inventory (buffers) that are placed strategically in front of constraints in the production process to ensure that they are kept busy. For many of the products manufactured by Bertch, the major constraint is deemed to be market demand.

- Improved morale of the workforce who are pleased to be able to ship products to customers more promptly and to work in a less cluttered plant.

- Improved customer satisfaction. A survey was sent by the Dimension Division to each of its internal customers (other divisions) to solicit input on any perceived improvements in service during the first six months of synchronous manufacturing in the division. The responses were predominantly positive.

Overall, synchronous manufacturing has helped to increase sales revenues and profits. In particular, sales volume has increased in the Bath Cabinet and Semi-Custom Divisions. To meet the increased demand, the number of shifts has increased from one to three in the Semi-Custom Division, and new facilities are in construction for this division. Also, new construction is in progress to release more space for the Dimension Division, which manufactures parts and panel sizing for three assembly divisions. These efforts to increase capacity to satisfy demand illustrate the theory of constraints concept of elevating the system's constraints.

The introduction of synchronous manufacturing techniques into the Dimension Division is not complete, and further developments are planned for the future. Synchronous manufacturing principles have not yet been formally introduced into the five assembly divisions. They already operate on a daily batch system with products assembled, boxed in cartons, and loaded on to trucks on the same day. The assembly divisions benefit from the improved service from the Dimension Division. It is planned to introduce some synchronous manufacturing principles into the assembly divisions in the future.

Bertch Accountants' Response

Given the inability of ABC to meet the costing needs of Bertch in a cost-beneficial way and the changed manufacturing environment, Bertch accountants sought a more appropriate costing system. As a first step, Bertch accountants attended a workshop to learn about throughput accounting. "Synchronous Flow Management: Managing Profit—The Flow of Money" was conducted by R.B. Vollum & Associates.

To support synchronous manufacturing, proponents of this technique recommend a simple, short-run costing concept called throughput accounting. Under throughput accounting, three "global operational measures" are advocated to replace current costing systems as follows:

- Throughput—The rate at which the system generates money through sales.
- Inventory—All the money the system invests in purchasing things the systems intends to sell.
- Operating Expense—All the money the system spends in turning inventory into throughput.

Throughput is defined as selling price less direct materials costs, which include payments to external parties for raw materials, components, subcontracted work, salespersons' commissions, transportation, and custom duties. Under throughput accounting the products to be emphasized are those with the highest throughput per unit of the scarce resources, such as machine hours at a machine that is a bottleneck.

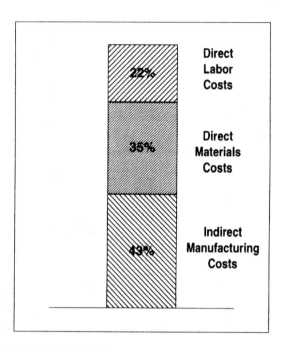

Figure 1. **Breakdown of Bertch manufacturing costs**

For example, if wood product X generates a throughput of $100 and requires 10 minutes at the bottleneck machine, the throughput per machine minute is ($100/10) $10. If wood product Y generates a higher throughput of $150 but requires 20 minutes at the bottleneck machine, the throughput per machine minute is ($150/20) $7.50. All other things being equal, wood product X should be emphasized because it produces the higher throughput per machine minute at the bottleneck resource.

As indicated above, throughput accounting theory requires that only direct materials costs be identified with individual products in order to calculate throughput margins, and this is a relatively easy task for Bertch accountants to accomplish expeditiously. As shown in Figure 1, materials costs are a significant portion of total production costs in the manufacture of wooden cabinets, especially since the 30%–40% price increases in 1993 because of wood shortages. The simpler throughput accounting concept that calls for the detailed costing of direct materials only is more appealing than ABC to Bertch management, and it was decided to implement it in place of ABC.

Direct labor workers at Bertch are paid an hourly rate, and in the main their wage costs are variable to units produced. Bertch management classify about 30% of direct labor as nonvariable to units produced, and about 70% is added to direct materials costs as unit-level variable costs. Cabinet manufacturing in general is labor intensive. As shown in Figure 1, this is particularly true in the case of Bertch, which serves the high-quality cabinet market and uses a high proportion of hand sanding, hand staining, and other labor intensive operations in its manufacturing operations. Since 1989, the number of Bertch personnel has more than doubled. The high percentage of Bertch total costs represented by labor justifies the inclusion of unit-level variable direct labor in its throughput model, in contrast with throughput theory which includes all manufacturing labor as part of operating expenses.

Beginning in January 1995, a full-time employee was devoted to the task of installing the throughput accounting system. During the previous six months, this work was conducted on a piecemeal basis. The speed of implementation was hindered by frequent changes to Bertch products such as new product lines, new door styles, and new product specifications (e.g., shelf thickness). Also, the implementation of throughput accounting was a big project for one person to undertake. The installation of the throughput accounting system was completed in September 1995. It took about 15 months to install.

The throughput accounting system is being used to calculate the throughput margin of the Bath Cabinet, Cultured Marble Top, Kitchen Cabinet, and Semi-Custom Divisions and their products. The other assembly division (Mirror and Glass) does not currently sell to outside customers but supplies the needs of the other divisions for glass and mirror items. This is the first time in Bertch's history that the profitability of all front-line divisions and their products has been determined. Management now can evaluate the profit performance of these divisions and their products in terms of throughput margin. Other uses for throughput accounting information at Bertch are input:

- In pricing decisions;
- In establishing amounts to bid on new contract opportunities; and
- In product emphasis decisions (based on throughput per unit of the constraint resource).

This is the first time in Bertch's history that the profitability of all front-line divisions and their products has been determined.

Meeting the Needs of Internal Customers

By definition, the primary purpose of a management accounting system is to satisfy the accounting needs of managers within the organization—the internal customers of management accounting information. The recent experience of Bertch documented in this article illustrates the efforts of management accountants in one company to satisfy the current needs of their internal customers.

In sum, Bertch management accountants:

- Changed course in the choice of a cost accounting system when this tactic was necessary to meet the changed needs of manufacturing, marketing, and other internal customers;
- Adopted a costing system that deviated from the standard throughput model to better suit the particular characteristics of their company.

To provide world-class service, management accountants should be proactive in pursuing design changes in costing systems and not delay taking corrective action until existing systems fail to meet the needs of their internal customers, especially in changing production and service environments. Also, management accounting information can be used in a proactive manner to stimulate operational changes that are desired by internal customers in higher levels of management.

For example, to encourage the design or redesign of products to spend the minimum amount of time at bottleneck resources, operating expenses could be assigned to bottleneck resources only and then to products in direct proportion to their processing time at bottleneck resources.[2] Under this approach, operational managers will be rewarded by a lower assignment of operating expenses if they are able to reduce the bottleneck processing time required by the products for which they are responsible.

In order to continue to provide value-added costing information, management accountants must be in touch with the current needs of their internal customers and in a timely manner provide the accounting information that the users require. Changes in underlying production and other processes likely will require corresponding changes in the costing system.

As the recent experience of Bertch demonstrates, an accounting change to meet user needs better can be a move to a simpler costing model. Greater complexity is not necessarily better because a more complicated costing system might be neither cost-beneficial nor timely. It is clear that costing systems should be dynamic and able to respond quickly to the changing needs of managers, subject to cost-benefit considerations. The Bertch experience is an important lesson for management accountants to put into practice.

Notes

1. Eliyahu M. Goldratt and Jeff Cox, *The Goal: A Process of Ongoing Improvement*, revised edition (Croton-on-Hudson, N.Y.: North River Press, 1992).

2. See Eliyahu M. Goldratt and Robert Fox, *The Race*, p. 171, and Robin Cooper and Robert S. Kaplan, *The Design of Cost Management Systems* (Englewood Cliffs, N.J.: Prentice-Hall, 1991), 171.

John B. MacArthur is associate professor of accounting, Department of Accounting and Finance, College of Business Administration, University of North Florida.

Functional Processes

PRODUCT DEVELOPMENT

Your Product Development Process Demands Ongoing Improvement

Preston G. Smith

Although managers routinely review their product development projects to ensure that each is meeting its objectives, managers seldom review the development process to identify and overcome its shortcomings. This article provides a twelve-step process for capturing the learning from each project. This is an elaboration on the continuous improvement appropriate for any organizational process, but with special emphasis on the needs of those in product development.

Most firms today have an established process for developing new products. However, managers typically view the development process as a business system that must be in place to ensure the quality of their products and control the use of resources, rather than viewing it as a competitive strength. Thus, they are unwilling to invest in continually sharpening the process to provide a competitive edge as the environment shifts.

In my experience working with many different companies to accelerate or otherwise improve how they develop new products, I have found that each firm can be put into one of four levels of product development maturity, depending on the attention management pays to its development process:

1. An *ad hoc* process.
2. An established process that is used on most or all projects but is not reviewed itself.
3. Some projects are reviewed to learn about the process, but there is no effective, ongoing way of using the review to improve the process.
4. Each project is viewed as a learning experience, and formal channels exist for modifying the process as a result of this learning.

Many companies have moved from the first stage to the second; fewer have advanced to the third stage, and very few have reached the powerful fourth stage. This article presents a 12-step process for helping companies reach these higher levels of maturity, in which product development is regarded as a continuously adaptive process that learns from all past projects.

The mindset underlying the higher levels of maturity is that each development project has two deliverables 1) The product that goes to market; 2) Improvements in the organization's development process. Any project that only produces an excellent product on time and within budget has only obtained half of the benefit. Leading companies invest in obtaining the second half.

Laying the Foundation

1. *Name your improvement process carefully* This first step is where many companies get off to a poor start. The most popular name is "postmortem," which suggests that something has died. Some companies place a lively twist on this situation by calling their process a postpartum.

Then there is the choice between "review" and "audit." Audit is an acceptable term in many companies, but it connotes unpleasant management heavy-handedness in others. The point is not that some terms are good and others are bad but that we should be particularly careful about the signals we send here, because any process that reviews project performance will get very close to evaluating individual performance too. Individuals will be unwilling to help the company improve if they suspect that their own career is at stake.

Senco Products (a Cincinnati, Ohio manufacturer of pneumatically powered hand tools) calls theirs a "process improvement review." Hewlett-Packard (Palo Alto, California) uses "retrospective analysis," and Farinon Division of Harris Corporation (a supplier of microwave products and systems in San Carlos, California) calls it "lessons learned." I will simply use "process review," but I suggest that readers choose a name that fits their culture and objectives.

2. *Piggyback on existing strengths.* After choosing your name, borrow from the best of your management processes. Perhaps you are using continuous improvement in another part of your business and can adopt this as a model for starting development process reviews. Perhaps you have redesigned one of your other business processes and can use this as a model. Many companies are training their employees in meeting and process analysis skills that will contribute to understanding the development process and how to improve it.
3. *Pick a reviewing pattern.* For process reviews to become routine, you will need a pattern for determining when they are to be conducted. There is no single answer here; it depends on objectives and company style. The most obvious pattern is simply to review each project as it is completed, as illustrated at the top of Figure 1. The major weakness of this approach is that it can take considerable time for a project to finish, and the learning is therefore delayed. Senco overcomes this problem by completing a minor review at the end of each phase and a more comprehensive review at project conclusion.

Another approach is to review the development process for all projects annually. This has two potential advantages. One is that several projects can be compared simultaneously for common issues, and the other is that an annual review can make it easier to obtain the senior management approval and attention crucial to making substantial changes.

The bottom diagram in Figure 1 illustrates a standing committee charged with continually upgrading the development process. The committee meets for perhaps a couple of hours every week and monitors every project regularly. This suggests a substantial commitment to improving the development process, and improvement can be rapid with such continual attention. Potential weaknesses are that the product developers may view the committee as "fluff," or it may have little authority to make major changes. Clearly, the committee must be well linked to senior management in order to benefit from the time devoted by the committee.

Figure 1 does not cover all of the possibilities. For example, Forma Scientific (a Marietta, Ohio producer of cell culture incubators and other laboratory equipment) set up a new product development team, not to improve the process but to actually build it. That is, rather than establish a development process in the abstract in advance, Forma's new product process development team interacted with the product development teams to develop its process as it developed the products.

To select a pattern for your process review process, consider your objectives and your current situation. Typical considerations include:

- The effort you are willing to commit to reviewing projects and implementing process changes.
- How much you are willing to burden development teams with this task.

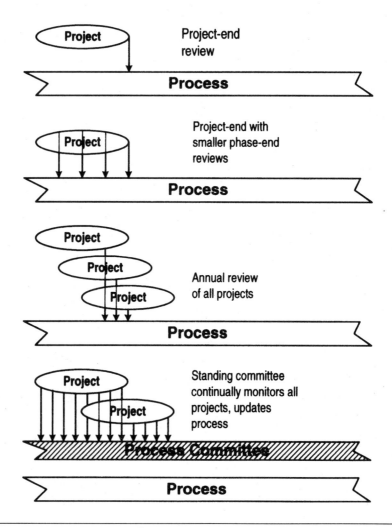

Figure 1. **From top to bottom, four of the many possible patterns for timing process reviews**

- How and when decisions on significant process changes are made in the company.
- How strongly you are wedded to project phase milestones, which will influence the review points.
- How you are going to compare across projects to generalize conclusions.
- How quickly you would like to correct specific weaknesses in your process.
- The relative importance of development time, product cost, development expense, and product quality in your business.

To achieve your specific objectives, a mix or hybrid may be appropriate. For example, Senco combines a standing quarterly review with the reviews of individual projects mentioned earlier. Says Senco's product development manager, Scott Allspaw: "Until we develop the habit of ongoing process improvement, we're meeting quarterly to ensure a continuous upgrading of our product development efforts."

The development team at Baker Oil Tools (a manufacturer of downhole safety valves in Broken Arrow, Oklahoma) is authorized to make minor process changes and try alternatives stemming from the team's weekly development process meetings. Then, at major project milestones, the project team meets with a senior management committee to recommend broader improvements.

Managers are familiar with development, design or project reviews in which accomplishments regarding time, cost and design quality are compared against plan and variances are noted. Such reviews are sufficient for managing projects but are inadequate to reach our goal here—learning from projects. To learn, we must go further, asking:

- How did this occur?
- How often does it occur?
- What could be done to prevent recurrence?
- Who will work on fixing the process?
- How will we know when it is fixed?

Conducting Reviews

4. *Assign a review.* After you know how you will be reviewing your process, consider who the reviewer will be. This is not an issue with relatively informal development team self-reviews, but it is critical for broader-based or more intensive reviews. Basically, the reviewer(s) must have an independent viewpoint, without a vested interest in the project under review. In addition, the reviewer should be able to engender real sharing and learning about the process, rather than an attitude of "auditing" it for mistakes. Three types of individuals may qualify:

- Product developers currently assigned to other projects.
- Those from "neutral" departments, such as corporate quality or human resources.
- Outsiders, such as consultants, facilitators, retirees, or benchmarking partners.

Reviewers should also be familiar with your products, processes and product development in general, so that they can place suggestions in context and spot a gem or red herring when they see it. In this regard, an outside product development expert or an employee who has recently transferred from a competitor can add value to the process by drawing from other experiences.

5. *Constructively balance positive and negative findings.* Strive for balance in conducting the review. Some companies tend to concentrate on the pleasant experiences and avoid conflict or failures. Consequently, there is the temptation at all levels to make the project look good, as though the people were writing a press release on it. The only problem with this is that it distracts us from what we are about, which is learning from our mistakes.

On the other hand, negative feedback that "poisons the well" is especially destructive. We must take time to understand what we did well and the reasons for the success, and to reinforce what we need to continue doing.

At Convex Computer (a Richardson, Texas-based producer of supercomputers and scalable servers) the review is structured to cover both successes, called the "gold buckets," and mistakes, called the "brown buckets" because of their dirtier nature. Reviews are structured to cover the brown buckets first in order to get people talking, then to close with the gold buckets to help people leave the review with a positive feeling.

6. *Focus on an improved process.* One problem with calling the reviews "audits" is that an audit suggests the purpose is to ensure that all rules were followed, whereas our objective is to learn how to do the job better, faster or more effectively. In fact, effectiveness may be enhanced by not following all steps every time, and a good review will uncover non-value-added activities that might be eliminated. Thus, "audit" suggests that the process is sacred, missing the whole point that our job instead is to streamline the process. An audit may stifle the very creativity we are trying to foster in learning about the process. For example, Fluke Corporation (an Everett, Washington producer of electronic test instruments) does conduct true process audits to inspect for conformance to process criteria, but they are careful to keep these audits totally independent of their process reviews.

Just as "audit" sends the wrong message as to what we are about, tying a process review to the company's reward system corrupts the objective and inhibits the free flow of information. You may wish to recognize a development team's accomplishments, but this should have nothing to do with learning about how the process might be improved. Specifically, maintain a clear separation between process reviews and personal performance appraisals.

> Tying a process review to the company's reward system corrupts the objective and inhibits the free flow of information.

As you modify your development process, remember that control systems, such as development procedures, tend to grow naturally with time. Consequently, maintain a conscious effort to eliminate as much from the process as you add to it. Otherwise, the process will grow imperceptibly slower and more bureaucratic as it is "improved."

Collecting Information

Process reviews are beneficial to the extent that they produce data supporting process change. Accordingly, reviewers collect two types of mutually supportive information about the project: qualitative information, in the form of participant interviews, and quantitative information, usually as metrics.

7. *Interview key participants* In selecting interviewees, consider four types of participants:

- Project participants (team members), who will have an inside view of the process's strengths and weaknesses.
- Those in supporting roles, who contribute to the product's development but are not considered members of the development team, including buyers, industrial engineers and cost accountants; to them, this project is just one of many they must service.
- Internal customers of the development team's efforts, usually production and sales; these people are in a good position to judge the quality of the development work, because they will have to deal with the resulting problems.
- The process beneficiary, who is the executive (usually of sales or marketing) who most directly benefits from having a timely, successful new product; this individual should have the clearest picture of the corporate value of an improved development process.

My normal approach is to interview the process beneficiary first to get the big picture as to how well the project and the product satisfied broader organizational objectives. This provides guidance on where emphasis should be placed for the remainder of the review. The next step is to work with the development team as a group to capture where this project advanced the development process and where the team encountered difficulties with the process. Finally, I interview selected internal customers and those in supporting roles to understand strengths and weaknesses from their position in the organization; their opinion on a certain point may vary from that of the development team.

8. *Back up the interviews with the data.* These interviews provide a great deal of relatively soft information. Quantitative measures, or metrics, can be used to focus on specific questions that have arisen in the interviews or to measure specific variables that relate to the firm's success. For example, manufacturing may observe that there were too many design changes late in the development cycle, so we collect data on engineering change notices issued by project phase and relate them to other comparable projects. This is a process of hypothesis testing, firming up the verbal data.

Generalizing on a practice often used in software development, Fluke Corporation monitors several types of defects, such as the number of parts changed per week, changes in specifications, and project plan changes. Fluke has also found complexity to be a revealing parameter, so they measure percent reuse of parts, percent reuse of integrated circuit design and the number of manufacturing processes used for the product.

Baker Oil Tools measures Suggested Dimensional Changes (from manufacturing), numerical control programming changes, and changes in manufacturing methods as indicators of how well their development process is working.

Randy Englund, a project manager at Hewlett-Packard, points out that metrics should be used to reinforce desired behavior. "People should know what they can do differently to change how the data reflect their performance," he says. "For example, a report could show progress on software defect resolution."

9. *Measure progress with ongoing metrics.* Although this use of metrics has the advantage of being responsive and adaptive, it provides little continuity with which to judge progress from project to project. Thus, other metrics are needed to provide an ongoing measure of performance relative to corporate goals. These may include cycle time for the entire development process or for critical phases or activities, such as prototyping cycles.

Cost and effectiveness measures are other common possibilities. Such broad metrics may not be powerful diagnostic tools, but they monitor how well the development process is achieving its objectives. Consequently, these trend metrics should be:

- Relatively permanent, as their value increases with the number of projects they can be used to compare.
- Well aligned with process objectives, to avoid misguiding people; for example, it would seem appropriate to track how well the project met its expense budget, but project economics analyses seldom suggest that project expense is a strong driver of new product profitability.[1]

Convex Computer uses the metric of manufacturing scrap dollars over the product introduction period to compare projects and monitor where improvements occur. This has the advantage of being "bottom line" oriented, and it is already tracked by the financial systems. Note that this measure aims at exposing root causes of the scrap: design process weaknesses early in the cycle and unanticipated redesign effort during the production startup phase, both of which lead to development expense overruns and schedule slippage.

> Eastman Chemical tracks net present value of their research projects as they pass various gates of Eastman's innovation process.

Eastman Chemical (1993 Malcolm Baldrige National Quality Award winner, which manufactures chemicals, plastics and fibers in Kingsport, Tennessee) tracks net present value of their research projects as they pass various gates of Eastman's innovation process. This ensures that they are concentrating effectively on those projects that add value to the company's bottom line.

Meyer provides additional guidance on measures, focusing particularly on ones that help a development team to monitor itself.[2] One of his points is that, to maintain focus, the number of metrics must be limited to fewer than 15. (See "Software Development: A Rich Source of Metrics Ideas," box.)

SOFTWARE DEVELOPMENT: A RICH SOURCE OF METRICS IDEAS

Although many managers are drawn to the concept of using metrics, they find it difficult to specify useful metrics. I have found it helpful to borrow heavily from what software developers have learned about constructing metrics.

Software is a valuable starting place, both because it lends itself to quantification and because it has shown itself to be notoriously difficult to manage, thus drawing considerable remedial attention. Grady and Caswell, for example, explain how the software development process is measured at Hewlett-Packard.[3,4] A pair of their charts below illustrates how metrics can help orient developers toward the reuse of existing code—or existing mechanical or electrical components, or existing chemical compounds, as the case may be.

Although it may appear obvious that reused product components reduce development labor and decrease risk dramatically, developers who pride themselves in inventing often need powerful evidence to adopt such changes in style. Leading software companies have learned how to use such metrics effectively to improve their product development.

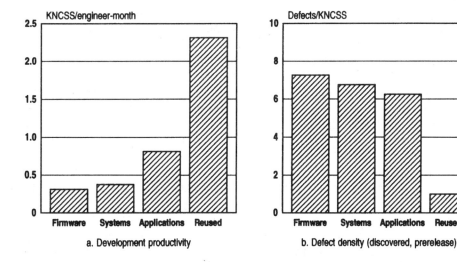

Figure 2. Hewlett-Packard divided its software development projects into the four categories represented by the four bars, demonstrating that projects making considerable use of existing code carried considerable advantage in both productivity and defect reduction. KNCSS = Thousand lines of noncomment source statements.

Source: Grady/Caswell, *Software Metrics: Establishing a Company-Wide Program,* copyright 1987, pp. 111–12. Reprinted by permission of Prentice-Hall, Inc., Englewood Cliffs, New Jersey.

Closing the All-Important Feedback Loop

The only reason we conduct process reviews is to improve the process. While this may seem obvious, this final step fails to occur in all too many cases. Analyzing and discussing improvement opportunities is far easier than committing to actually improve.

10. *Establish a closure mechanism.* In order to benefit from the effort expended on the review, you need a leak-proof system to convert findings into action. Those with the authority and resources to make changes in the process must initiate and monitor the change process. Although some changes can be handled relatively informally at a low level, others will require a massive effort led from the top of the company.

AT&T's Global Business Communications Systems (GBCS—Middletown, New Jersey) analyzes their process using several quality tools, such as affinity analysis, to identify candidate process improvements. Then, they charter quality improvement teams to develop and deploy process improvements. To ensure that the improvement concepts are incorporated into the GBCS process, they schedule activities allocate budgets, assign staffing, and act on the results of their quality improvement teams in the same way as they handle development projects.

Although each firm will have its own style for guaranteeing closure, fundamentally it comes down to the tools of effective management accountable individuals or teams, due dates and available resources.

The resource issue cannot be overemphasized. Conducting effective reviews will divert resources from activities, such as product development, with more immediate payoff, and taking action on certain findings will usually demand even more resources. Thus, an improved-development process must be viewed as a strategic investment, just like a new laboratory. How much you are willing to invest depends on how quickly you wish to improve. Viewing the benefits as being free is a prescription for failure.

Each company should design its own change management system in accordance with its culture and style, its needs, its position on the learning curve, and the resources it is willing to invest. In most cases, a process manager is an important ingredient of a successful system. This manager—who actually may range from one person working part time to a group of dozens in a large company—consolidates, updates and preserves the list of improvement options, and monitors both the review process and the current action projects. Generally process managers do not decide which improvement actions are undertaken or actually work on the improvement actions themselves.

Here are a few examples of how organizations have applied these principles to actually change the way they do business:

- Harris-Farinon has learned that for certain components with specific development needs, such as application-specific integrated circuits (ASICs) or microwave power amplifiers, they can improve

supplier responsiveness by tailoring subprocesses to their suppliers' individual process needs rather than imposing a Farinon process.
- Through process reviews, Convex Computer learned that technical risk assessments and planning early in the project were considered to be obstacles by the engineers. Yet, the same process reviews also demonstrated the need for such risk management. Consequently, resistance to risk management abated, which has led to early identification of latent problems and has been a key to reducing development cycle time by 50 percent while also reducing scrap costs.
- Through its process reviews, Baker Oil Tools has discovered the need for cross-functional job titles that reflect concurrent design of products and their associated manufacturing processes. For example, a senior tool engineer is now called a senior product development technologist.

Institutionalizing the Process

Companies that do well at learning from and improving their development processes do not regard process improvement as a one-time activity or a luxury reserved for times when they can afford it. Instead, they seek to make it a way of life, a normal part of business activity. Their first step is to appoint a permanent process manager and assign basic resources for reviews and improvement actions.

11. *Review every project.* Do not review an occasional project that proceeded especially well or poorly, but expect to learn from every project (above a certain threshold in size). There are practical reasons for this. Most companies do not have a large-enough sample of projects to see patterns clearly; they need all of the reviews they can get.

    More subtly, selecting certain projects for review makes them special. They turn out to be congratulatory exercises for the participants—or worse witch hunts—which erodes the learning and suggests to the participants that both they and the process are on trial. Selective reviews weaken the desired habit, which is that reviews are an expected and normal part of every project.

12. *Align the process with corporate objectives.* As with most cultural changes, a key part of building the review-is-normal environment is ascertaining that it is supported by corporate expectations and rewards. For example, those who participate in the process should be given time to participate and compensated adequately for their involvement. At a practical level, consider providing a charge number for those participating in process improvement, so that the time is not viewed as being stolen from other activities. Also, cultivate intrinsic motivators, such as

> Do not review an occasional project that proceeded especially well or poorly, but expect to learn from every project.

the satisfaction that comes from knowing one is making an important contribution to the company.

At Eastman Chemical, top management demonstrates its interest in improving its innovation process by regularly setting aside time to review project learnings. Also, the firm has changed its Quality Management Process to explicitly state that "checking and acting" are parts of a systematic management process.

In my final example, process improvement activities at AT&T's GBCS have evolved from what was viewed as "extra work" that got in the way of getting the job done, to what has become an integral part of product realization. It is now a way of life and no longer questioned.

The Payback

In working with dozens of companies to accelerate their development cycles, I have found that the ones that are winning with new products are those that are investing in their development process. The benefits they derive from these investments occur in several areas:

- Faster product development, as opportunities are found to adapt the development process to the specific needs of a project and to overlap activities.
- Higher-value products, because the link between the customer and the feature set is shortened and strengthened over time.
- Lower-cost products, as manufacturability decisions are made ever earlier in the product design phase.
- More products per dollar, because extraneous, non-value-adding activities are rooted out of the process.
- More responsiveness to turbulence in markets, technologies and the regulatory environment, due to a process that is built to continually adapt to change.

Notes

1. Preston G. Smith and Donald G. Reinertsen, *Developing Products in Half the Time* (New York: Van Nostrand Reinhold, 1995).

2. Christopher Meyer, "How the Right Measures Help Teams Excel," *Harvard Business Review* 72, no. 3 (1994): 95–103.

3. Robert B. Grady and Deborah L. Caswell, *Software Metrics* (Englewood Cliffs, N.J.: Prentice-Hall, 1987).

4. Grady, Robert B. *Practical Software Metrics for Project Management and Process Improvement* (Englewood Cliffs. N.J.: Prentice-Hall, 1992).

Preston Smith is a principal at New Product Dynamics, a Portland, Oregon, consulting firm specializing in managing the product development process. Smith is co-author of Developing Products in Half the Time *(Van Nostrand Reinhold, 1995) and holds a Ph.D. in engineering from Stanford University.*

Early Supplier Involvement: Leveraging Know-How for Better Product Development

Francis Bidault and Christina Butler

Getting trusted suppliers involved early in the product development process can make the process go faster and more efficiently. This article provides key guidelines for working with suppliers in the areas managing (1) the preparations phase, (2) the negotiations phase, and (3) the operations phase.

"ESI (Early Supplier Development) is often a matter of survival. Many customers say to us that if they do not capitalize on the resources of their supplier, they will be in trouble," states Gary Deaton, manager of marketing and manufacturing. The Minco Group, Dayton, Ohio.

Car manufacturers have reported tremendous cost savings on labor, material, overheads, cycle time, and quality improvements as a result of implementing ESI in the product development process. Hearing of these benefits, other manufacturers have begun to look closely at auto industry practices in this area to see if and how they might be applied in their own industries.

Following the apparent success of ESI in the automotive industry and the lack of research into the adoption of ESI by other sectors, a major M2000 (Manufacturing 2000) research project was launched to investigate the implementation of this practice outside the car industry. Between March 1994 and May 1995, we examined and contrasted the practice of ESI across companies and regions to identify "best practices" behind achievements to date and those required for challenges ahead. A key area we looked at was the challenge of moving to ESI.

M2000 research shows that manufacturers are offering more variety in their products and in the technologies employed in those products. At the same time, manufacturing operations are growing in complexity. Manufacturers are increasingly looking to the support of their suppliers for specialist

Manufacturers are increasingly looking to the support of their suppliers for specialist advice in areas where they are more knowledgeable and effective than the manufacturers.

Reprinted from *Target*, March–April 1996, with permission of the Association for Manufacturing Excellence, 380 W. Palatine Road, Wheeling, IL 60090-5863, (847) 520-3282. Copyright © 1996 by the Association for Manufacturing Excellence. All rights reserved.

advice in areas where they are more knowledgeable and effective than the manufacturers.

Understanding ESI

To successfully transfer responsibility in the area of innovation, manufacturers must work closely with the supplier. The investment required to do so is expensive in terms of time for the manufacturer. The company, therefore, needs to be selective, working closely with fewer, but stronger, suppliers. This leads to increased dependence on those suppliers. The first step to successful ESI is to select the right component or subsystem as the basis for a new supplier relationship.

ESI can be used most effectively in the design of custom-made parts. The Japanese automotive industry, where ESI originated, simplifies the identification of appropriate parts by segmenting all parts into three categories:

- Standard parts—for example, nuts and bolts, with no customization required:
- "Black Box" parts—for example, anti-lock braking systems, where development has always been under the responsibility of the supplier; and
- Custom-made parts—for example, bumpers, which manufacturers used to design with limited contribution from their suppliers, and thus where ESI has potential.

In contrast with the traditional development process where a supplier is selected by formal price competition in the engineering phase, ESI requires choosing a supplier well before the product design is complete—preferably at concept selection—and involving that supplier, to varying degrees, in the design of the product.

Extent of Information Sharing

ESI is not a "static state" that a company does or does not practice, but a continuum of involvement of suppliers in the development process. At the lowest level, "design supplied," the manufacturer takes full responsibility for the design, and the supplier simply shares information about its equipment and capabilities. Those companies that selected Level 1 in the survey questionnaire fall into this category.

A company can then move to the next level, "design shared," where the manufacturer still takes full responsibility for the product-design, but the supplier provides early feedback on the design, including suggesting cost, quality, and leadtime improvements. Companies responding that they are at Level 2 are in this category. An excellent example of a development handled at the "design shared" level would be the frame or chassis for the "Liberty" laser printer co-designed by Lexmark International, Inc., an American manufacturer of printers and other office supply equipment, and its long-time plastics supplier, the Minco Group.

Firms can then aim for an even higher level of involvement, "design sourced." Here the supplier takes full responsibility for a system or subassembly from concept to manufacture, incorporating one or more parts which the supplier also designs based on in-depth understanding of the manufacturer's requirements. Companies in the survey selecting Levels 3 to 5 are in this category. An example of a development managed at this level is a paper tray subassembly for a photocopier subcontracted by Xerox to a first-tier supplier.

Managing the ESI Process
Preparations Phase
Success in the initial phase relies on detailed preparation prior to approaching potential partners for the first time, although preparation is important for all subsequent projects. During preparations, three critical factors need to be aligned: (1) strategy: (2) human resources, and (3) operating policies.

Strategy
Top management, and the company more generally, must make ESI a priority. If an understanding of, and commitment to, the process does not exist at the highest level, there will inevitably be conflict between corporate and project level strategies. Individual project managers will have trouble succeeding, unnecessarily jeopardizing existing supplier relationships.

The Liberty team at Lexmark struggled due to a lack of top management and broad company support. This situation meant that the team had difficulty securing human and other resources as early as necessary, and the product was launched a year later than otherwise would have been the case, with a negative effect on market share.

Other companies in the sample suffered from an indecisive attitude toward ESI. In several instances, existing supplier relationships were even destroyed. Xerox, a company with a number of major ESI successes to its credit, has countered this issue by training all 3000 head office employees in its benefits and management. Kodak, with seven years of ESI projects behind it, believes that it is also necessary to empower individual project teams. And Motorola, another experienced ESI practitioner, says it is essential to treat ESI as an attitude or a frame of mind, and not a technique.

Human Resources
To avoid failure at the outset, it is critical that manufacturers select the right in-house people, especially for the first project where a crisis would make a subsequent attempt unlikely. Traditional thinking by two functional areas in particular—design/engineering and purchasing—block genuine attempts to change. M2000 research revealed that some designers were unable or unwilling to recognize the potential contribution of suppliers. While purchasers at Bosch Power Tools were pushing for the introduction of an ESI approach on some developments, certain engineers seemed reluctant to proceed, because they simply failed to see how small suppliers could possibly provide a company like Bosch, a clear leader in its industry, with information they did not already possess in-house.

Diebold, the American manufacturer of automated teller machines (ATMs), experienced the human resource issue from a different angle. With the best of intentions, a project team overwhelmed an ESI supplier on an ATM facia development which required more tools than that supplier could handle. The Diebold designers involved were unaware of the need to step out of their old roles in order to teach the supplier how to manage more complex projects successfully.

Some purchasers, on the other hand, were reluctant to relinquish the classic tendering approach to supplier selection. At Volvo Cars, a senior purchaser agreed to the need for suppliers to participate in product design, but only as long as the component or subassembly involved was eventually put up for tender. Insistence on this type of competition for an ESI development, particularly outside the car industry, are considered destructive to a manufacturer's long-term relationships with the "losing" suppliers. Other purchasers, who remained price-obsessed and were unable to focus on other selection criteria, made embarking on ESI projects problematic for colleagues. Lexmark countered potential problems of this nature on the Liberty project by selecting an open-minded core team composed of individuals who had good reputations with the prospective ESI supplier companies.

Xerox says that working with the suppliers to bring them up to speed saves money in the long run. Based on that company's ten years of experience with ESI, it takes an average of two to three projects to develop suppliers, and costs 25 percent less than doing the job in-house.

Operating Policies

Formal support for ESI is particularly needed in three functional areas: purchasing, design/engineering, and manufacturing, and the necessary mechanisms should ideally be put in place as early as possible in the company's ESI history.

Purchasing needs to work with design/engineering to select ESI suppliers. Instead of encouraging price/performance decisions, the company should reward purchasing on the information it provides to design/engineering on a wide range of industries and associated suppliers. Kodak accomplishes this by making purchasing a full member of product development teams early on.

Design/engineering must be rewarded for involving suppliers early in the development of customized parts, preferably at concept selection, instead of on in-house finished project drawings. This functional area must also work in the same way with its own manufacturing area. By doing so, Lexmark managed this angle of the Liberty project well. As a result, the program experienced no emergency engineering changes.

Lastly, manufacturing must be rewarded for avoiding production problems before they happen, while projects are still in the development stage. This means working early with both designers/engineers and suppliers, guiding the development of the design, instead of solving problems later on the line. As above, the Liberty project at Lexmark is a good example of this type of involvement.

The key is to really create cross-functional teams with membership from the three areas to select, initiate, and develop ESI relationships, including making partner suppliers full members of those project teams.

Negotiations Phase

The next phase is the selection of a partner and negotiation of a new kind of supply arrangement. The three top issues for a manufacturer to manage are (1) respect for confidentiality and intellectual property, (2) fair pricing, and (3) an emphasis on both through communication and mutual recognition.

Confidentiality

Confidentiality can be a real concern for both parties. Philips Japan, producer of kitchen and personal care appliances, realizes that it fails to benefit completely from its partnerships, as the company is reluctant to share a project's business case with ESI suppliers. The fear is that the supplier will share confidential information with a Philips competitor with whom the vendor also works. While this fear was often expressed by companies in our sample, suppliers told us that they did not, in fact, share business cases with their customer's competitors.

Intellectual property issues should be addressed early in conjunction with discussions on confidentiality. Braun, the German appliance manufacturer, has backed away from ESI as a result of a supplier using a co-development with Braun competitors. Needless to say, the relationship between the two partner companies was severed. While Braun argues, rightly, that patenting the idea involved was not practical, the break could probably have been avoided by talking about the issue up-front, prior to development, instead of assuming the co-development would be its sole property. Xerox emphasizes that confidentiality agreements should always be signed, but then must be put away. If an agreement needs to be referred to during the development, the partnership is in trouble.

Fair Pricing

The achievement of an equitable pricing arrangement is a major concern for both sides. It also encourages the partnership to develop by acknowledging the supplier's increased responsibilities and associated risks. The development of a win/win situation is often obstructed by a lack of understanding of how ESI works. As a result, companies remain reluctant to share financial information.

A large, technically excellent U.S. plastic molder complains that some manufacturers expect to keep all margin increases resulting from design improvements initiated by the supplier for themselves, despite talk of a "partnership" with the molder. Ironically, these manufacturers are the ones that know the least about plastics themselves. This attitude does nothing to encourage the supplier to make improvements in designs. Xerox points out that sometimes one of its own designers causes a problem for a supplier by asking for unnecessarily light tolerances, not appreciating that his request pushes up costs for the supplier. Diebold suggests that bringing a supplier in to participate in target costing helps both sides to understand and manage costs appropriately and fairly throughout the operations phase. Honda of America believes that cost savings resulting from collaboration should be split equitably, and thus reinforces the idea of true partnerships with its suppliers.

> Diebold suggests that bringing a supplier in to participate in target costing helps both sides to understand and manage costs appropriately and fairly throughout the operations phase.

Communication and Mutual Recognition

Manufacturers need to respect suppliers in their new role, or they risk jeopardizing the partnership at the outset. To avoid problems of this nature, companies must acknowledge suppliers' responsibilities, contributions and expectations, both up-front and throughout the development project. One high-tech company deplores the fact that certain manufacturers do not allow the supplier to make suggestions for improvements and refuse to listen to them when they do. These manufacturers feel that they are the experts, despite their claims of "partnership," and the supplier's evident technical superiority in the area of plastics. Lexmark encountered problems working with Minco on the Liberty project because it had not made its changed expectations of the supplier's role clear. Confusing signals from the manufacturer led Minco to believe nothing had changed in their relationship, despite the use of the term "partnership."

Kodak has found that the best way to counter these issues, and prevent lost opportunities and broken relationships, is to visit the supplier often, and talk to the company in-depth, and at all levels of the organization, prior to establishing such an intimate relationship. The process does take time, especially when a company first moves to ESI but, in the end, and on future projects, it is both worthwhile and necessary. Motorola suggests that setting up a supplier council to communicate their concerns as a group can pay dividends in understanding and managing the relationship successfully.

Operations Phase

Once the ESI project reached the operations stage, the partners need to keep talking to build trust and address potential stumbling blocks, of which the three most important are (1) changing behaviors, (2) understanding each other's processes, and (3) acknowledging altered time requirements.

Changing Behaviors

Change is difficult. Manufacturers often struggle to assume their new roles and relinquish control. Olivetti-Canon, the photocopier and printer joint venture, finds it tough to select the final producer of an ESI part at the time the design is commissioned, even though the supplier involved in the design usually gets the job. A company acting in this way risks creating the impression of a lack of commitment to the relationship and a lack of trust in the supplier. Even an experienced practitioner such as Kodak continues to wrestle with the concept after seven years. A number of its engineers still find it difficult to take the design risks required to work in parallel with suppliers. As a result, full benefits from the process are not obtained, and there is the possibility that suppliers will sense a lack of trust in their capabilities. One way around this issue is to conduct team-building exercises with each new project team, something Lexmark did for the Liberty printer. It works even better if suppliers are included in the process.

Honda of America believes it is important to measure the performance of both sides and does so regularly with all its partner suppliers. Performance improvement measures like cost reductions and engineering hours saved help to convince skeptics on both sides of the benefits of the ESI process. And Diebold says a final review session with the entire project team, including the suppliers, is vitally important to everyone's learning and the more efficient management of partnership projects in the future.

Understanding Processes and Time Requirements

Misunderstandings occur if mutual processes and/or time requirements are poorly understood by either company. Delays and potential breakdowns in partnership result from a lack of knowledge of each other's practices and from not investing sufficient time in constructive dialogue. Lexmark and Minco were slow to master this concept; the results were communication problems and delays both in the design process at Lexmark and the tooling process at Minco. Misunderstandings related to timing can also lead to unnecessary and potentially damaging circumstances.

ESI requires a substantial investment in time up-front in order to complete the design. During this initial period, nothing much appears to be happening. All subsequent phases, however, move much more quickly compared to the traditional approach, and the time overall will be reduced. Project teams at Kodak, Lexmark, and ETA have all had trouble convincing top management and, in one case, purchasing, that this preparation time is being well spent, although without it ESI cannot succeed. Diebold says that one solution to this issue is to have suppliers participate in project scheduling, which helps to eliminate potential problems. Bose and Kodak recommend inviting suppliers in as in-plants on the engineering and/or purchasing side to learn about the company's business and product development process.

Ultimately, it is necessary to educate top management and the company generally on a continuous basis throughout all phases. This necessity brings the company full-circle back to the need for thorough preparations. A change agent from either design/engineering or purchasing is needed to make the necessary shift in company attitude.

Framework for Success

Getting a supplier to invest in the design process demonstrates the vendor's commitment. The level of commitment required in ESI, however, often leaves the supplier exposed to a high degree of risk in its relationship with the manufacturer and differs from the way business is traditionally conducted. From the sample, 90 percent of suppliers interviewed experienced significant engineering costs related to ESI—people (free "guest" engineers or in-plants), assets (co-location of specific equipment/technology), and development of

> Delays and potential breakdowns in partnership result from a lack of knowledge of each other's practices.

new expertise—and 70 percent had no contract covering those expenditures. To get this commitment from a supplier, manufacturers need to demonstrate commitment both to the supplier and the development project, as well as convince the vendor of the company's excellence in the particular industry. Manufacturers must take the lead in demonstrating trustworthiness, given the change required from the old style manufacturing/supplier relationship. Mutual trust can then develop in all aspects of the relationship: ethical, technical, and strategic.

A review of our research results makes it clear that the balancing act that firms need to manage to succeed at ESI is indeed a fine one. At any step along the way, a company could make a slip from which it is potentially impossible to recover. In fact, there does seem to be a general mindset that leads to failure and another that leads to success.

Some manufacturers have a tendency to focus on achieving benefits for themselves at the expense of the other party. This tendency leads the company to emphasize a control mentality rather than generating a partnership one. The 'partner' supplier is discouraged from investing in the project, and sticks strictly to the terms of the contract. The end result is that the manufacturer "wins" only when the supplier loses. An atmosphere of mistrust is generated and the partnership is likely to fail. As a result, the project is also likely to fail and the investment in time and resources is lost.

A company that succeeds with ESI approaches the process from a different point of view. The focus is on benefits for both parties right from the outset. This attitude means that the firm also provides a supplier with the necessary support, and the supplier is thus encouraged to enhance its commitment to the project, via an increased investment in resources, which ultimately provides improved results. The situation is one of win/win in an environment of mutual trust. As a result, the project is more likely to succeed with real, and often substantial, benefits accruing to both sides. And the stage is set to grow the relationship and reap even greater rewards on future joint developments.

Francis Bidault is professor of strategy and technology management and program director, Strategic Alliances, and IMD in Lausanne, Switzerland.

Christina Buller is currently completing her Ph.D. at the London Business School. As a research associate from IMD's Manufacturing 2000 project, she studied ESI and new product development.

Functional Processes
HUMAN RESOURCES

Do Employee Involvement and TQM Programs Work?

Susan A. Mohrman, Edward E. Lawler III, and Gerald E. Ledford Jr.

This article is a kind of report on a research project undertaken to find out the relationship between employee involvement and the successful implementation of TQM. The results demonstrate that EI programs are more likely to succeed in companies that have implemented TQM, and that EI helps enhance the implementation of TQM. The only caution in all this is that the authors forgot, in titling this article, that TQM is not a program.

Perhaps the most interesting results from the current study are how EI and TQM work together to impact organizational performance. Although both stress employee involvement as well as training and skills development, there are some key differences between them:

- The TQM literature attends more to work process and customer outcomes.
- The employee involvement literature emphasizes design of the work and business units for fuller business involvement and employee motivation. In addition, employee involvement emphasizes making the employee a stakeholder in business performance through reward systems such as gainsharing and through business education.

In practice, these two management approaches may contribute to organizational effectiveness in a complementary and reinforcing manner such that their individual impact is weakened by the absence of the other. This is the major focus of this article, but before we look at the results of our study, we will briefly describe the study.

Use of and satisfaction with employee involvement and TQM programs—We asked companies whether they use self-managing teams, gainsharing programs, training programs, open-book management, and a variety of other practices that are supportive of employee involvement.* There was a clear general trend from 1987 to 1993 toward the increasing use of a wide variety of practices that move information, knowledge, power and rewards downward in the *Fortune 1000* companies. Self-managing teams and knowledge-based pay are the practices which show the greatest growth.

In the case of total quality management programs, we asked about the use of quality councils, cross-functional planning, re-engineering, customer satisfaction, monitoring and other so-called core TQM practices. We also asked about production oriented practices: self inspections, system process control, JIT, and work cells. From 1990 to 1993, we found an increase in the use of all quality management practices, and an increase in the number of companies with total quality management programs.

There is a strong correlation between the adoption of employee involvement approaches and the use of TQM practices . . . The extent to which companies employ the core TQM practices, the production-oriented practices, collaboration with customers, and cost-of-quality monitoring all have a highly significant relationship to our measure of overall employee involvement use.

The relationship to the development of knowledge and skills is particularly strong, demonstrating the strong emphasis in TQM programs on the development of skills. TQM's relationship to power sharing is also very strong, reflecting the focus in TQM on problem solving and decision making groups in general and on work cells and teams.

We asked the 1993 respondents in the *Fortune 1000* companies how positive their experience had been with their employee involvement and total quality management programs. The overwhelming majority of the companies believe that their overall employee involvement and total quality management efforts are successful. They have a slightly more positive feeling toward their TQM activities than their EI activities, but this difference is not statistically significant. This overall positive reaction fits with the continued growth of EI and TQM (table 1).

Combined impact on performance—We began our analysis of the impact of EI and TQM on organizational performance by looking at the degree to which TQM practices make EI programs more successful. The results are clear when we look at the success of EI programs. Overall, the more organizations use TQM practices, the more positive results they get from their EI efforts. This result provides evidence for the close and complementary relationship

> Overall, the more organizations use TQM practices, the more positive results they get from their EI efforts.

---

*Our definition of employee involvement includes a wide variety of practices that move information, knowledge, power and rewards downward in organizations.

> ## STUDY DESIGN
>
> In 1987, 1990 and 1993, we surveyed the *Fortune 1000* corporations. The survey has changed slightly from year to year, but it has always focused on the adoption of employee involvement and TQM practices as well as on their impact. In addition, we have followed the financial performance of the companies. The focus of the current article will be the 1993 survey which was commissioned by the AQP. It drew responses from 279 companies in the Fortune 1000. The complete results of the study as well as the questionnaire can be found in our recently-published book *Creating High Performance Organizations: Practices and Results of Employee Involvement and Total Quality Management in Fortune 1000 Companies*, published by Jossey-Bass in 1995.

Percent of companies	with EI	with TQM
Very negative	0%	0%
Negative	1%	1%
Neither negative nor positive	18%	16%
Positive	68%	66%
Very positive	13%	17%

Table 1. **Overall, How Positive Has Your Experience Been?**
The overwhelming majority of the companies believe that their overall EI and TQM efforts are successful. They have a slightly more positive feeling toward their TQM activities than their EI activities, but this difference is not statistically significant.

of the two initiatives. A similar pattern appeared when we looked at the relationship between impact of TQM programs and the use of EI efforts. The more organizations use EI practices, the more likely they are to have successful TQM programs.

The table above shows that some of the positive impact of TQM practices are actually the result of introducing EI practices (table 2). The partial correlations** between the use of core and production oriented TQM practices and the TQM outcomes all go down when the effect of EI is eliminated. The relationship between the use of core practices and employee outcomes loses

---

**To study more closely the intertwined impact of EI and TQM, we used a statistical technique called partial correlations. It allows a researcher to look separately at the impact of two related factors or in this case, programs such as EI and TQM.

	TQM outcome ratings		
Core TQM practices	Direct performance outcomes@	Profitability and competitiveness	Employee satisfaction and QWL
Without EI	.25***	.26***	.19
With EI	.31***	.35***	.25***
Production-oriented practices			
Without EI	.17	.18	.05
With EI	.23***	.27***	.11***

@Productivity, customer satisfaction, quality, and speed

Significant correlation coefficents key:
* = weak but significant (p ≤ .05)
** = moderate relationship (p ≤ .01)
*** = strong relationship (p ≤ .001)
Note: without = partial correlations.
With = zero order correlations.

Table 2. **Use of TQM Practices versus Outcomes.**
The relationship of the extent of use of TQM practices to outcomes of TQM programs with and without the effect of EI. Overall, the more organizations use TQM practices, the more positive results they get from their EI efforts. This result provides evidence for the close and complementary relationship of the two initiatives.

The impact of TQM programs that do not also include EI practices, will be less positive both for employee outcomes and for performance outcomes.

its statistical significance, indicating that it resulted in large part from employee involvement. The relationships of production-oriented practices to direct work performance outcomes and company performance also become nonsignificant without the EI impact.

The findings with respect to the combined impact of TQM and EI are not unexpected: most TQM proponents advocate high levels of employee involvement as part of their TQM efforts. However, the findings do make a very important point: the impact of TQM programs that *do not also include EI practices*, will be less positive both for employee outcomes and for performance outcomes.

The impact of the use of EI approaches on EI outcomes when the impact of TQM practices is eliminated—Again we see most of the relationships declining and many of them losing their significance. Without TQM practices, the overall use of EI and the amount of information that is shared remain significantly related to direct performance and company outcomes.

It appears that the use of TQM practices is an important part of successfully involving employees in processes that lead to improvements in organizational performance.

Knowledge and skills are weakly tied to company outcomes under the same conditions. Only power sharing is weakly related to employee outcomes when the impact of TQM is statistically partialled out. Thus, it appears that the use of TQM practices is an important part of successfully involving employees in processes that lead to improvements in organizational performance (table 3).

	EI outcome ratings		
	Direct performance outcomes@	Profitability and competitiveness	Employee satisfaction and QWL
EI Overall			
Without EI	.20**	.17*	.14
With EI	.33***	.28***	.22***
Information			
Without EI	.21**	.20**	.08
With EI	.29***	.27***	.14
Knowledge and skills			
Without EI	.13	.18*	.08
With EI	.31***	.31***	.21
Power sharing			
Without EI	.06	.10	.15
With EI	.22**	.23***	.23***
Rewards			
Without EI	.12	.09	.08
With EI	.17***	.13	.12

@Productivity, customer satisfaction, quality, and speed

Significant correlations coefficients key:
* = weak but significant (p ≤ .05)
** = moderate relationship (p ≤ .01)
*** = strong relationship (p ≤ .001)
Note: without = partial correlations.
With = zero order correlations.

Table 3.

Impact of How EI and TQM Are Managed

To further study the joint effects of EI and TQM programs, we asked companies how the relationship between EI and TQM is managed... The employee outcomes from TQM and the performance outcomes from EI are highest when the two initiatives are managed in an integrated fashion. Coordination helps somewhat, but both employee and company outcomes are highest if the two are integrated parts of a single improvement effort (table 4).

We believe this is a very important point. It argues strongly against organizations having separate officers and staff groups managing employee involvement and total quality management activities.

Indeed, it may argue against the whole idea of having EI and TQM programs. It suggests that it may be best to have an organizational improvement effort which combines and integrates EI, TQM and other practices which give the organization the type of performance capabilities it needs.

We also asked companies how their EI and TQM programs are positioned relative to each other... The performance outcomes from TQM and EI are perceived to be highest when EI is seen as an important part of TQM rather than the reverse. This result no doubt reflects the task oriented focus of TQM.

How EI and TQM are managed			
	Two separate programs	Two separate programs	One integrated program
Direct performance outcomes@			
EI	4.04	4.20	4.21
TQM	4.02	4.09	4.14
Profitability and competitiveness			
EI	3.78	3.90	4.00*
TQM	3.85	3.96	4.00
Employee satisfaction and QWL			
EI	3.92	3.98	3.96
TQM	3.70	3.76	3.89*

@Productivity, customer satisfaction, quality, and speed
Key: * = Significantly higher than the other forms of managing
Note: means are presented and compared

Table 4.

How EI and TQM are positioned relative to each other		
	EI is an important part of TQM	TQM is an important part of EI
Direct performance outcomes@		
EI	4.20**	3.98
TQM	4.14**	3.95
Profitability and competitiveness		
EI	3.95**	3.73
TQM	3.97	3.80
Employee satisfaction and QWL		
EI	3.96	3.90
TQM	3.86	3.67

@Productivity, customer satisfaction, quality, and speed

Key:
* = weak but significant ($p \leq .05$)
** = moderate relationship ($p \leq .01$)
*** = strong relationship ($p \leq .001$)
Note: means are presented and compared

Table 5.

Financial measures	Low use	Medium use	High use
Return on sales	6.3	8.3	10.3
Return on assets	4.7	5.8	6.9
Return on investment	9.0	11.8	14.6
Return on equity	16.6	19.7	22.8

Table 6.

For better or worse, it is often easier to get management commitment to these efforts if they see that the focus is directly on task and process, rather than on ways to involve employees in the business (table 5).

To complete our analysis of the combined impact of EI and TQM, we studied the relationship between the adoption of EI and TQM practices and corporate financial performance ... We shifted from looking at ratings of performance to publicly reported financial data. The results of this analysis generally confirm our earlier analysis:

- The more companies use EI and TQM practices, the better their financial performance.
- Companies that are high users of EI and TQM show the best return on sales, return on assets, return on investment and return on equity (table 6).

Table 6 shows the percent return for measures of financial performance as related to EI and TQM usage. We believe these results are particularly important because it is the first study to show the long term effects of EI and TQM on the actual performance of large corporations. It strongly suggests that EI and TQM pay off in terms of the financial performance of companies.

Summary and Implications

The evidence substantiates the close interrelationship and complementary nature of employee involvement and total quality management. They help make each other more successful:

- Companies with more extensive forms of employee involvement report higher outcomes from their TQM programs than companies with less employee involvement.
- Employee involvement programs are more successful when they are used in conjunction with TQM programs.

There is a particularly close link between the impact of production oriented TQM practices and the use of employee involvement approaches. The impact of production oriented TQM practices is positive only when they are used in a manner that is highly involving.

The study also provides some interesting data on how TQM and EI programs should be positioned relative to each other . . . Performance outcomes are highest if managers focus on TQM and see EI as part of it rather than the other way around. The highest impact is achieved if EI and TQM are managed as an integrated program. Finally, using them both leads to better financial performance.

The results of our study have some clear implications for how an organization should be managed . . . They strongly suggest employee involvement and TQM practices should be woven into an interrelated effort to improve organizational performance.

They don't allow us to specify exactly what this combination should look like, but that may be for the best. It is unlikely that there is a generally appropriate mix of practices that organizations should adopt. It is much more likely that organizations need to fit the use of EI and TQM practices to their particular technology, environment and need for improvements in organizational performance. It is clear, however, that the introduction of new practice goes best when employees are treated as major stakeholders in an organization and are involved in the business of the organization. This ensures that not only will the performance of the organization improve, but employees will develop a more positive relationship with the organization. In short, it has the greatest potential to produce a win/win situation.

Susan Mohrman is a senior research scientist at the Center for Effective Organizations.

Edward E. Lawler III is the founder and director of the University of Southern California's Center for Effective Organizations.

Gerald E. Ledford Jr. is a senior research scientist at the Center for Effective Organizations.

World Class Suggestion Systems Still Work Well

John Savageau

As will most initiatives in organizations, suggestion systems (yes, they are systems with processes like all aspects of organizational activities), if not well planned and executed, will fail. This article catalogs some specific reasons why suggestion systems in many companies have failed and describes how a world-class suggestion system works. Employee-initiated suggestions have the potential of cumulatively saving big costs for organizations, not to mention raising morale as employees find managers listening and then actually implementing their ideas.

And isn't our prejudice against suggestion systems, which in the words of noted author Pat Townsend, "... are at the bottom of the evolutionary ladder of employee involvement," enough to convince me that the topic is dead?

Well, yes it is 1996. And yes there are team-based employee involvement systems that could be classified as "up the evolutionary ladder," and most of us have been taught to be skeptics regarding the effectiveness of suggestion systems.

But despite all this, there are at least four good reasons why a business involved in TQM might consider using a suggestion system today:

1. World class suggestion systems are exceeding 40 ideas per person annually, with greater than 80 percent implementation rates, and high levels of participation.
2. Just as our understanding of teams has evolved dramatically in the past decade, so too has our understanding of suggestion systems. Today, a world class suggestion system need not be treated as a second class citizen of its team based employee involvement (EI) cousins. Rather, it can become a strong partner that helps enhance dialogue and learning throughout the business.
3. Organizations with excellent improvement teams often do not get *total* involvement. And even when they do, it is not likely that teams will surface and act upon all the good ideas that people generate. Inevitably some ideas will fall through the cracks.

Reprinted with permission from the *Journal for Quality and Participation*, March 1996. Copyright © 1996 by the Association for Quality and Participation. All rights reserved.

4. Suggestion systems can be used to focus on areas that teams do not. Improvement teams and self-managed teams are most frequently focused on process control and improvement. Suggestion systems are sometimes used to generate ideas for new products or services. When the focus is the same, suggestion systems can also be used as feeders to improvement team projects.

At Milliken and Toyota the passion they have for quality is reflected in their world class suggestion systems. Both of these companies project to be at or above 50 ideas per person this year. The implementation rate for ideas hovers around the 80th percentile. This is certainly performance far above the norm.

Debbie Hartung, president of the Employee Involvement Association (formerly the National Association of Suggestion Systems) says that over the past ten years, idea systems have been improving. Their 1992 report indicates participation rates at 8 percent, while the overall ideas contributed per person is about 2.4, and implementation of ideas contributed is 35 percent. Although these numbers seem to pale when contrasted with those of world class Milliken and Toyota, it is important to note that the reported total savings dollars was $2.2 billion.

> Yet today's numbers still indicate that most businesses have a lot of work to do to move their idea system up to world class standards.

Yet today's numbers still indicate that most businesses have a lot of work to do to move their idea system up to world class standards. Such performance with relatively low involvement, low idea submissions, and low implementation rate has contributed to the *bad reputation* suggestion systems have developed.

World class is not just an old suggestion box painted over— The poor reputation was well earned by what most of us have come to know (and despise) as the *suggestion box*, a traditional process dating back at least to the 1920s. This is the process which has proliferated throughout American industry, and is the starting point for most all suggestion systems.

Such systems usually operate as follows: ideas are submitted to a central location, a review board assesses and decides to approve or reject the suggestion; implementation is then directed by the review board. Additionally the review board or committee is tasked to promote and support the idea system, clarify legal issues, and define the reward and recognition process. Evaluation and improvement of the system itself are often totally neglected. Although historical performance data is not readily available, it is probably fair to assume that performance from the suggestion box type of system is typically at or below the EIA data. The following recent example supports this argument.

An automated suggestion process is not world class automatically . . . In 1990 a leading high-tech manufacturer implemented an automated suggestion system based on traditional methods. Ideas were sent via e-mail to a common address. A central group was formed to track and monitor ideas which were then forwarded to appropriate review boards. The review boards most often were comprised of senior management or their direct reports. Ideas were supposed to be reviewed and responded to within a *reasonable* time period. However, in practice, the reasonable response time operated at or above 8–12 weeks, and some ideas were never responded to. Despite having a state of the art computerized database, tracking software, and electronic mail network, the process results were lower than the EIA numbers for the same year. Total participation was just below 5 percent, with .8 ideas submitted per person/year and a 20 percent implementation rate.

The poor performance of the computerized suggestion system in this example may shed light on the reasons for the common belief that suggestion systems don't work, are archaic, and a waste of time. Worse, they may do more harm than good. So how is it that some firms get 20, 30, 40 and more ideas per person? What have they learned? To find the answer, let us look at the major flaws inherent to the design of the traditional suggestion box process.

Flaws in the Old System

Bottlenecks in the flow of ideas— The central review group(s) limits quantity of idea flow. This is not caused by any evil or planned intent. Just the opposite in fact. Most review groups see themselves as big promoters of idea generation processes. The problem is simply one of capacity. In a group with 200 people, 20 ideas per person would produce 4,000 ideas annually. A review group that meets once per week or month cannot review that many ideas. Thus the system is self limiting. This is why suggestion systems of this type are often plagued by backlogs, delays in response and implementation, lost interest, skepticism, and ultimately bad feelings on the part of the idea generator(s).

The disempowerment of management... Review groups are given the power and authority to implement ideas. The decisions from the group get delivered to a manager (sometimes across functions) instructing them to act. This essentially *disempowers* the manager by stepping on decisions that are usually hers or his to make. When this happens it may cause managers to feel left out of the process. Additionally when ideas are received from other functions they are frequently perceived as *bombs* lobbed over the fence. This resistance to cross-functional idea input is normal, but also destructive if the process is not designed to gradually develop trust and confidence in the overall intent of the ideas process.

Lack of management responsibility for making employee involvement happen as a day to day reality... By forwarding ideas off to a central location (often looked at as a black hole) managers are essentially *off the hook* for participating in this employee involvement activity. They are not responsible for giving OK's on ideas, for sponsoring an idea to a higher level, for coaching people on the quality of ideas, for encouraging ideas, for rewarding ideas, for empowering people to question their current paradigms of work, nor for tapping into the passion for work that is too often suppressed.

Lack of focus . . . When most idea systems are implemented only one broad goal is stated: "We want your ideas to improve things around here," is often what gets communicated. The obvious next question then is "What things?" and since no other goals exist, the answer provided to people who

> By forwarding ideas off to a central location (often looked at as a black hole) managers are essentially *off the hook* for participating in this employee involvement activity.

ask this question will vary dramatically by person, but will eventually become, *anything* goes. Unfortunately anything goes does not assure that ideas will be directed toward improvement.

In the case of the high-tech company mentioned earlier, suggestions came in to improve the exercise room, and to paint new lines in the driveway. These ideas came at a time when the business was in the process of losing hundreds of millions of dollars annually. In addition, most ideas submitted were targeted toward fixing some other department. The system did not encourage them to improve their own job function. In 1994 the firm discontinued its suggestion system and the opponents of participation can all say "See we tried it. It doesn't work!"

A World Class Solution: Design the System to Meet Desired Outcomes!

> In world class suggestion systems primary focus is toward improving one's own job or work area as it relates to the larger business goals.

In world class suggestion systems employees are aware of the business improvement goals. They are encouraged to submit ideas which are directed toward the goals. The primary focus is toward improving one's own job or work area as it relates to the larger business goals. *Reduce defects 50 percent reduce cycle time 30 percent improve customer service levels*, are examples of goals or key results from a strategic quality plan that a suggestion system can be focused on.

"The key is decentralization!" What Milliken, Toyota, and others have found is that the answer is simple in design and theory, but requires a great deal of effort and education to create. Best known practices are de-centralized. At Milliken the direct manager/supervisor is the first recipient of an idea. The manager is then tasked to respond and take action within a short time frame, typically 3 working days (figure 1).

The responding manager has four options from which to choose:

a. Say yes and give the immediate go ahead . . .
b. Suggest the formation of an improvement team if the idea appears workable but requires further analysis . . .
c. Sponsor the suggestion up a level if she or he doesn't have the authority to approve . . .
d. Say no and provide the business reasons why the idea is not viable or appropriate at the time.

And in each of these cases, thank the person for submitting an idea in the first place. In addition, the responding manager is obligated to provide coaching to the idea generator. Many people have good ideas but fail to communicate them well. Some lack focus, while others may be laden with blame and negativity.

Good coaching encourages future ideas and suggests how to word them more effectively.

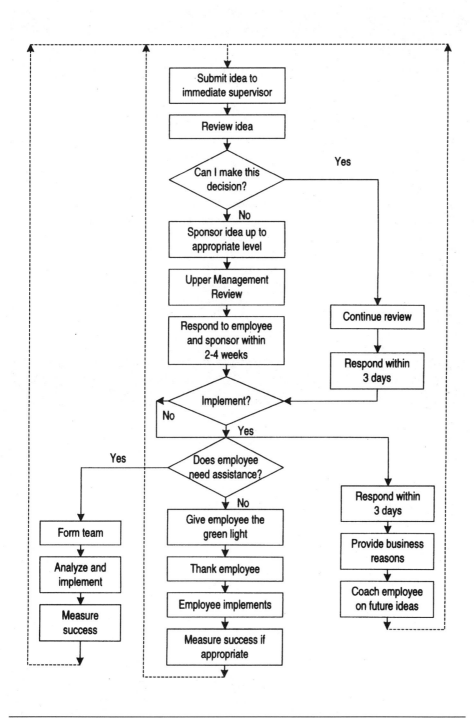

Figure 1. **A model of a world class suggestion system**

REQUIREMENTS TO MAKE YOUR SUGGESTION SYSTEM EFFECTIVE

Form an idea process support team to build and maintain the new process. The purpose of the idea support team as mentioned above is to:

- Design (or reengineer) the suggestion system
- Clarify the roles of supportive infrastructure
- Align the goals of the process to key business goals (toward TQM goals if they exist)
- Set participation targets
- Develop a reward and recognition process
- Establish a communication process
- Oversee the employee involvement and suggestion system training efforts
- Monitor, evaluate and improve the process
- Provide training to managers and individual contributors on the process and on employee involvement covering what it is, and how to do it.

SUGGESTED TOPICS FOR EMPLOYEE INVOLVEMENT TRAINING

- How the suggestion process works
- Annual goals of the suggestion process
- Definition of EI
- Benefits of EI
- Scenario of organization with no EI
- Management behaviors to promote EI
- Individual behaviors to promote EI
- Management responsibility in the new idea process
- Individual responsibility in the new idea process
- Coaching for focused high quality ideas
- Motivating for high involvement
- How to fill out an idea form
- How to help an individual contributor fill out the idea form.

Maintaining a balance— A delicate balance must be maintained between trusting managers and preventing out-of-hand no's (and yes's) caused by personality conflicts, or other manager/subordinate difficulties. Checks and balances should be maintained but the appearance of having *cops on the beat*

> **REVIEW CONTRIBUTOR HABITS AND BEHAVIORS THAT INHIBIT OR PROMOTE EI**
>
> This activity is very useful for management staffs during start-up (first 6–12 months).
>
> - Plan actions to eliminate undesirable behavior
> - Development of reward and recognition activities for all participants.
> - Define the recognition process
> - Define reward process
> - Oversee the implementation of recognition and reward processes
> - Assure equity between all levels of all functions
> - Ongoing continuous improvement of the process by the support team
> - Include input from the whole organization
> - Provide encouragement and energy . . . publicize successes
> - Customize the process to meet ever changing organization and environmental requirements

pushing ideas through should be avoided. A negative response to an idea submission, which does not include a sound business reason, may automatically prompt further review.

As the responsibility to review ideas moves to *all managers*, the role of the review committee switches from one of deciding which ideas to implement, to one of support by establishing, monitoring, and reviewing the idea process (see box).

The suggested name for this new committee is the *Idea Process (insert your term) Support Team*. The support team monitors how well the process is working, providing safeguards against abuse. It also works with top management to set the direction and context for ideas. It encourages, rewards, and works actively to gain full participation (ideas from every person in their business.)

Measuring success—Typical process metrics are in the areas of quantity, quality, turnaround time, and implementation of ideas. The support team is also charged with setting up checks and balances to make certain that good ideas aren't lost or rejected out of hand. Again, ideas should only be accepted and rejected for good practical business reasons.

Closing Thoughts

In the final analysis, suggestion systems reflect the intent of management. Is the intent to provide a mechanism for complaints and appeasement of hard feelings. Or is it to add another vehicle to tap into the collective brain power of the business? Those businesses with world class suggestion systems have opted for the latter. They see that this process fosters continuously expanding levels of dialogue between functional levels of authority. That dialogue is seen as a catalyst which fosters an open, learning environment.

Continuous improvement is synonymous with continual learning. When organizations achieve 20–50 ideas per person, continuous learning is well under way. And over time what starts off as a formal procedure, becomes an informal norm. People talk to, listen to, and learn from each other.

It seems that there is renewed energy for implementing suggestion systems. Critical to their future success is the adoption of a more capable process to unleash their real power. With an improved process, suggestion systems can regain their position as a management tool which helps to achieve a fuller utilization of human minds at work.

John Savageau is president and senior consultant of Pendleton Productivity Center, Inc. (Northborough, Massachusetts.) His firm is dedicated to achieving bottom line results through high involvement strategic quality management. Both a consultant and trainer, Savageau has focused his work on employee involvement and total quality since the early 1980s. He is a Massachusetts Quality Award Examiner, and a member of both the AQP and ASQC.

Gainsharing: A Lemon or Lemonade?

Woodruff Imberman

Gainsharing holds real promise to reinforce teamwork and operational success by sharing with employees the profits gleaned from higher performance. However, many companies have found gainsharing to be problematic and, after starting such programs, dropped them. This article points out, however, that the problem is not with gainsharing itself but with the way it has been carried out. When you don't combine gainsharing with quality-management initiatives, when you think of gainsharing as quick remedy for deep-seated and dysfunctional organization behavior patterns, it's bound not to work, just as TQM in general will not work under such circumstances. The author includes six recommendations for helping to ensure that gainsharing will enhance productivity if combined with sound management practices.

In 1981, gainsharing received attention from the federal government when the U.S. General Accounting Office issued a report entitled *Productivity Sharing Programs: Can They Contribute to Productivity Improvement?* The conclusion of that report was that pay-for-performance compensation systems—known as gainsharing—"resulted in labor cost savings averaging *17 percent*, along with improved work relations, reduced absenteeism, reduced turnover, and fewer grievances."

What sort of elixir is this gainsharing? It is *not* an incentive or bonus plan for individuals exceeding a standard or quota. That's the old piecework system. Gainsharing is a group bonus plan aimed at modifying employee behavior. The entire factory work force—as a unit—is involved in an effort to exceed past performance. If successful, the gain is translated into cash and *shared between the company and the employees.* Normally, the work force receives 50 percent of the gain in bonuses, and the company receives an equal share in cost savings. This is gainsharing in its simplest form, a form that often works very well.

Unfortunately, gainsharing programs that don't work well are rarely mentioned in the literature. Though gainsharing can be a great boon to a company and its work force, on occasion it turns out to be a lemon instead of lemonade. Under what circumstances does gainsharing benefit a company, and when does it become a detriment?

Gainsharing is a group bonus plan aimed at modifying employee behavior.

Academic Interest

Gainsharing is not new. It originated around 1935 in the form of the Scanlon plan. But the federal government's endorsement by the GAO in 1981 aroused the interest of the academic community, which followed up with a whole series of erudite studies. Professors R. J. Bullock and E. E. Lawler, for example, reviewed the experience of 33 companies with gainsharing and concluded in a 1984 report that in most cases "overall success was found in organizational effectiveness, quality, innovation, labor-management cooperation, and pay."

Thereafter, about 50 separate academic studies were published covering the use of gainsharing in various manufacturing and service industries, including banks, hospitals, and distribution. All studies were laudatory, finding that in various ways gainsharing helped boost productivity, quality, and company earnings. In 1989 Professor Paula B. Voos reported in the *Journal of Labor Research* on gainsharing programs at unionized Wisconsin firms. Her conclusion was that gainsharing achieved "greater positive effects on productivity, flexibility in utilizing labor, and company performance outdoing other labor management cooperation programs, including profit sharing, labor-management committees, and employee ESOPs." Professors T. L. Ross, L. Hatcher, and D. Collins found in 1992 that gainsharing showed better results than employee involvement, quality circles, quality of work life, suggestion boxes or suggestion systems, profit sharing, labor-management committees, employee stock ownership programs (ESOPS), job enrichment, information sharing, job rotation, survey feedback, information sharing, or Total Quality Management. The American Center on Productivity predicted that gainsharing "will become one of the fastest growing strategies in the U.S. in the 1990s and beyond."

It does not, however, seem to have grown very fast. Only about 2,500 companies are using some variety of gainsharing today, the American Center reports. It is used mainly by larger companies, such as Inland Container, General Signal, Eaton Corporation, TRW, General Electric, Cosco, Rexnord Fasteners, Federal Mogul, Cincinnati Milicron, Exide Electronics, Whirlpool, Rockwell International, Frigidaire, Ingersoll-Rand, and others. Why not more? And why not smaller companies? Despite the voluminous and overwhelmingly positive literature on gainsharing, many of these programs have failed. But these failures are usually mentioned only in passing in the many publications on the subject. And no one seems really interested in ascertaining the extent or causes of unsuccessful gainsharing programs. Has gainsharing indeed often been a flop? If so, why?

No Garden of Eden

A study done at the behest of the American Management Association and published in 1989 reported that gainsharing is not always a free ticket to an industrial Garden of Eden. Covering three years of experience with gainsharing in 83 companies, the AMA study reported that *only one-third* of the companies had success with gainsharing, boosting their productivity handsomely and radically reducing the cost of waste, spoilage, rejects, and rework. Two-thirds of the 83 company gainsharing plans were flops: they plateaued in a year or so, and were discontinued.

Three major causes of these failures were found. The first was the faulty payout formula by which the gains were to be measured and bonuses paid. Reasonable care was not taken to define the short-term performance objectives to be measured, and to ascertain that they were relevant to the company's long-term goals. Decisions on gainsharing plan objectives cannot be made by an auditor or plant manager with a spare hour or two to do the calculations. When the payout formula devised by management was too "tight"—that is, no matter how hard the work force tried, the payout was minuscule—failure was a sure thing. When the formula was based on the expectation that the work force would achieve zero defects overnight, or that a 100 percent boost in productivity would instantly emerge in the first months of the plan, failure ensued. Guidance in formula determination is almost mandatory.

Second, the plan itself was initially presented to the employees in an overly optimistic manner, "oversold" by a fervent management. Anxious to gain the cooperation of the work force, some managements made seductive sales talks to employees, leading them to believe that a gainsharing plan virtually guaranteed an automatic, substantial bonus with little or no special effort. The work force became disenchanted when no bonus or only a small bonus was forthcoming. Under those circumstances, employees concluded they had been purposely misled "to push the wheelbarrow faster" without any adequate recompense or management support. This sort of disenchantment occurs when a zealous management devises a plan and tries to "sell" it without input or suggestions from the work force. Explaining to employees what a gainsharing plan is, how it works, and what employees must do to make it successful and earn bonuses has to be done carefully, adroitly, and cautiously. Such presentations must be conservatively structured, and employees must be given enough time to digest the information and ask questions. Again, some guidance is helpful.

Third, lack of support by middle managers cropped up in most instances of failure. Middle managers resisted the need to spend more time on the factory floor—in short, to act as "internal consultants" to employees seeking to develop new problem-solving ideas for improving productivity or quality under gainsharing. When employees take a gainsharing plan seriously and begin to generate new ideas for improving plant performance, executives must enthusiastically and resolutely insist that prompt energetic action be taken on all reasonable, safe suggestions for improvement. This applies to ideas concerning process, methods, products, safety, inventory control, and scheduling. Not all ideas and suggestions will be earthshaking, but the cumulative effect of many small, suggested improvements often turns out to be quite large in total. This means that executives must pay more attention to the factory work force, which often requires some indoctrination and training. If senior management supports the training efforts and sends proper signals to middle management and supervisors to alter their style—to lead by example and persuasion rather than simply by giving orders—then most middle managers will make the transition gladly and successfully.

Update

Now the 1989 AMA study has been updated. From 1990 to 1993, I conducted a study of 147 companies that had gainsharing programs. Had anything changed from the 1989 findings that two-thirds of the gainsharing programs flopped? Not really. The numbers had changed slightly, but the basic findings were still the same. Of those 147 company plans, 63 (43 percent) failed to achieve the employee payouts pictured by management when the plans were first installed. Again, in most cases this was because management had set the productivity-quality goals too high; no matter how hard the work force scrambled, no reasonable effort could achieve those goals. So the payout was either miserly, minuscule, or nonexistent, and after a year or so the plans were discontinued.

Twenty-six percent of the companies failed mainly because the product mix or the marketing strategies changed considerably. In those cases the plans were not redesigned to take care of altered circumstances. For instance, in one surgical products company in western New York, the gainsharing emphasis was on reducing the cost of poor quality—curtailing rejects and rework while maintaining productivity. The company employed 170 people and produced a wide array of scalpels, needle holders, clamps, retractors, and other devices used by surgeons. The real challenge was not how to boost productivity, but how to improve quality performance. The gainsharing program worked extremely well until the company radically changed its product line by adding metal hospital beds, which became about 70 percent of the company's sales volume. Metal hospital beds called for no such quality precision as scalpels and other surgical instruments. The result: Interest in quality performance diminished as the hospital beds were produced in volume but the company made no change in its gainsharing plan. After about a year of this situation, the plan was dropped.

Nineteen (or 13 percent) of the 147 company gainsharing plans did not succeed because management changed its basic personnel policy. Usually, the company president, the vice president of operations, the plant manager, or some other key executive left the firm, and the new executive had other aims, perhaps thinking the factory personnel should perform at top speed without an incentive plan. For example, in an Ohio grey iron foundry, the object of the gainsharing plan was to reward employees for a cut in manhours per ton of good output (minus rejects, spoilage, and rework). During the base period of 1990, the number of manhours per *good* ton averaged 28.4 for the year; in 1993, the average was 22.58 manhours—a gain of 6.21 manhours per good ton, and a considerable savings in money with 15,000 tons produced per year. But new company officials torpedoed the gainsharing plan by maintaining there should be no additional bonuses for such improvements. "Top performance is what the work force is being paid for," the executives maintained, "so why should they be paid any extras?" Goals were boosted to unreasonable heights and payouts were cut; top performance was not maintained, and gainsharing was eventually eliminated.

Only 45 firms (31 percent) in the sample had successful gainsharing plans and results—almost identical in percentage to the previous 1989 study. The earlier study had found that when the gainsharing formula was well researched and the work force properly indoctrinated as to the goals and rewards, not only were productivity and quality improvements forthcoming, but collateral benefits accrued as well:

- On-time shipments rose 18 percent.
- Customer returns decreased 12 percent.
- Absenteeism dropped 84 percent.
- Lost-time accidents fell 69 percent.

The 1993 study found the same general collateral benefits among the companies that had successful gainsharing programs. Thus, the data for success and failure with gainsharing were remarkably consistent: *approximately one-third of the companies had smashing success, while two-thirds of them experienced failures.*

Rules for Success

Is there any way to change these proportions? Having examined scores of gainsharing plans and installed a variety of successful programs in 23 companies to date, I suggest six rules to ensure a successful gainsharing plan in any company:

1. The payout formula must be *reasonable* and *doable.* A formula that pays minuscule bonuses or is based on the illusion that large improvements will result instantaneously is doomed to failure. Some expert guidance is helpful in setting the formula. "Everything You Ever Wanted to Know About Gainsharing But Were Afraid to Ask" (1993) provides a detailed description of how formulas are devised in different industries.
2. The first six to nine months of a program are vital to long-term success. Key determinants of that success are employee involvement in devising the plan and employee understanding of the details. Consequently, a program of employee education must accompany the introduction of the gainsharing bonus formula. Here, too, guidance is helpful.
3. Employee participation in drafting the gainsharing plan has a stronger effect in motivating the work force to improve productivity and quality than even bonus share or bonus frequency. "Employee Participation: What Is It, How It Works" (1993) describes how to achieve that participation.

> A formula that pays minuscule bonuses or is based on the illusion that large improvements will result instantaneously is doomed to failure.

4. Though some executives like to pay gainsharing bonuses quarterly, this is usually too long an interval to maintain a high level of motivation. Monetary rewards should follow performance with minimal delay. Monthly payouts seem to function best: weekly payments are sometimes possible.
5. Unionism per se is neither a plus nor a minus in gainsharing. Of the 23 company gainsharing plans I have designed and installed, about half are in unionized plants and work as well as in nonunion plants. Preliminary education about the plan is vital. With gainsharing, union leadership and management can concentrate on problem-solving and rewarded improvements rather than sparring with each other.
6. As was emphasized in the 1989 American Management Association study, the greater the degree of expert guidance, the better the resultant gainsharing performance. All successful plans are tailor-made for particular plant situations.

Despite the rate of failure reported in the two AMA studies, gainsharing obviously has its phenomenal successes. Recently, *Crain's Chicago Business* (1994) published a chart summarizing the experience of 110 plant managers with their gainsharing programs. The chart showed that of the 110 managers studied, 93 reported highly favorable results in productivity improvement.

Examples abound. At Kaiser Aluminum in Jackson, Tennessee, a manufacturer of auto components, the use of gainsharing contributed to an 80 percent productivity boost over five years and a 70 percent decrease in poor quality costs—all with its effect on the bottom line. General Tire's 1,950-employee plant in Mount Vernon, Illinois, found that its gainsharing program generated $30 million in savings over a five-year period, $20 million of which was paid out to workers in the form of bonuses while the company profited by $10 million. "It is the basis of our operations now," said Floyd Brokman, coordinator of the program. Federal Mogul Corporation's bearings plant in Greensburg, Indiana realized productivity and quality savings worth nearly $400,000 in the first seven months of its gainsharing plan, which were shared 50/50 with the site's 500 employees. Maytag Corporation cited enhanced manufacturing efficiency in reporting an income jump of 90 percent while gaining market share as a result of its gainsharing program.

Wrought Washer Manufacturing Company of Milwaukee, Wisconsin, producer of a full line of washers and stampings, reported its first year's experience with gainsharing. According to *Tooling & Production* ("Investing in People Pays Off," 1994), the company's productivity grew by a cumulative 39 percent in 1993, earning the work force $165,737 in extra bonuses over their regular earnings and saving the company an additional $110,490. And productivity and quality continue to improve. At Whirlpool Corporation's Benton Harbor, Michigan plant, where metal rods are turned into parts for washers and dryers, productivity and quality improvement came as a result of a gainsharing program that reduced costs, bolstered profits, benefited customers, and raised blue-collar take-home pay by an average of about $3,000 a year.

Since the onset of the gainsharing plan in 1988, the plant has shown productivity gains of about 19 percent *annually* with each year building on the previous year. And quality? The number of parts rejected has sunk to 4 per million from 837 per million. Other firms that have learned how to successfully manage their gainsharing programs are Speed Queen, Merck & Co., Solar Turbines, Corning, Sony Electronics, Carrier, Dresser-Rand, Consolidated Diesel, Georgia-Pacific Paper, Federal Mogul, Colgate-Palmolive, Allstate, CIGNA Corp., Travelers Insurance, several financial institutions, and a number of other producers of durable and nondurable goods.

Gainsharing is used by many companies to achieve top quality output and bolster the bottom line. So why don't more of these programs produce lemonade rather than lemons? Gainsharing is not a procedure that can be done with the left hand while the right hand concentrates on other matters. It requires management's attention, encouragement, and support. The ultimate explanation for the number of failures may lie along the lines suggested by the noted management authority. Peter Drucker: "Inertia in management is responsible for more loss of market share, more loss of competitive position, and more loss of business growth than any other single factor."

References

Bullock, R. J., and Edward E. Lawler. "Gainsharing: A Few Questions, and Fewer Answers." *Human Resource Management* 23, no. 1 (1984): 23–40.

Crain's Chicago Business, April 21/May 5, 1994.

Drucker, Peter F. *Managing for the Future: The 1990s and Beyond.* New York: Harper's, 1992.

Imberman, Woodruff. "Employee Participation: What It Is, How It Works." *Target Magazine* (Association for Manufacturing Excellence), January–February 1993, 20–28.

———. "Everything You Ever Wanted to Know About Gainsharing But Were Afraid to Ask." *Target Magazine* (Association for Manufacturing Excellence), May–June 1993, 19–26.

Imberman, Woodruff, and Betty Flasch. "Gains and Losses from Gainsharing." *Industry Forum,* supplement to *Management Review* (American Management Association, New York, December 1989).

"Investing in People Pays Off." *Tooling and Production,* August 1994, 21–24.

Ross, T. L., L. Hatcher, and D. Collins. "Why Employees Support (and Oppose) Gainsharing Plans." *Compensation and Benefit Management* 8, no. 2 (1992): 17–27.

Voos, Paula. "Influence of Cooperative Programs on Union-Management Relations." *Journal of Labor Research* 10, no. 1 (1989): 104–17.

Woodruff Imberman is the president of Imberman and DeForest, Inc., a management consulting firm in Evanston, Illinois.

Standards and Assessments
ISO 9000

Now That You're Registered...

By Richard C. Randall

In this article, consulting editor Richard Randall discusses future directions for companies that have their initial ISO 9000 registration. Randall contrasts the approach taken by what he calls "reactive" companies with that of "proactive" companies. Attitude makes all the difference. Proactive companies are not standing still. They see ISO as an opportunity and are taking new actions, such as reengineering, implementing the ISO 14000 standard, and moving to something called QUENSH (QUality/ENvironmental/Safety and Health). Randall emphasizes that registration is good for the company and can make it more responsive to customers and all stakeholders and increase profits at the same time.

Since the introduction of ISO 9000 in 1987, thousands of companies the world over have implemented the standard and become registered. But after the initial implementation of the standard—and congratulations on having achieved registration status—many companies are asking what the future holds. In what direction can management most profitably and productively steer an ISO 9000–registered company? Traveling throughout the United States performing ISO 9000 surveillance audits for various registrars has afforded me the opportunity to observe the solution to this dilemma firsthand.

From my experience, companies tend to polarize themselves into one of two groups: those who take a *reactive* stance toward a minimal maintenance of the standard, and those who are *proactive* in utilizing it fully, even evolving beyond it.

Reactive Companies

Executive management of reactive companies takes a minimalist, hostile stance toward the requirements of ISO 9000, feeling that they have been coerced into compliance with the standard by their customers. But by investing time and energy into finding ways to "beat the system," reactive companies miss the valuable opportunity and vast rewards a well-honed, cost-cutting, time-saving quality system can provide.

When reactive companies complain about the expense associated with documenting their processes, for example, they do not fully realize the potential of their investment. Documenting one's processes requires examination, which reveals opportunities for improvement, including waste reduction and cost cutting. Furthermore, this documentation serves as an effective tool in communicating expectations to both employees and contractors. Documented processes help companies transition from being *people-dependent* organizations (whose success is contingent upon the expertise of indispensable individuals) to becoming *system-dependent* organizations (relying on more permanent and controllable processes).

Reactive companies wait for, rather than anticipate, revisions to ISO 9000 and will jump lamely through the new hoops of requirements during their next surveillance audit. Reactive companies are also terrified as to what ISO 14001 (environmental management systems-specification with guidance for use) might bring.

Oddly enough, many reactive companies claim to have had TQM (total quality management) systems in place prior to their forced compliance with ISO 9000. Although popular as a buzz term, TQM is characterized by its lack of defined requirements and typically manifests itself as "total quality mediocrity" in reactive companies.

> By investing time and energy into finding ways to "beat the system," reactive companies miss the valuable opportunity and vast rewards a well-honed, cost-cutting, time-saving quality system can provide.

Proactive Companies

Proactive companies are an inspiration to visit and a joy to audit. In contrast to reactive companies, proactive companies implement ISO 9000 in order to give themselves a marketing advantage over their competitors and enhance existing customer satisfaction. Proactive companies focus on measuring the financial impact their quality system has on the bottom line. Having met the requirements of the standard and tailored them to resolve real quality issues, proactive companies are genuinely excited by a vast array of positive results.

Case in Point

One company I enjoy visiting had not seen a profit for over ten years. After implementing ISO 9000 and achieving registration, *every division in the company realized a significant profit within six months*. The final company division to implement the standard went from a negative 4.6 percent ROS (return on sales) to a positive 14.7 percent ROS within that same time frame.

This particular company had never had a formal quality system prior to implementing ISO 9000. Furthermore, all of the strides made concurrent to implementing the standard were achieved while maintaining a consistent number of employees. Even though time was—and is—regularly taken away from production at the company's sites to provide a steady stream of quality-system training to its employees, the company continues to experience the highest overall production in its history.

The Value of Quality

What separates reactive companies from proactive companies? In a quest for the root cause of why reactive companies pursue minimal compliance to a standard that promotes customer satisfaction, one soon recognizes that quality is a "value." Like honesty and integrity, quality is—or is not—part of a company's culture.

In a reactive company, management's lack of commitment to the value of quality is reflected throughout the organization. Unfortunately, the organization is typically condemned to management's position, despite good intentions from other company factions, including the quality department. Such companies are often arrogant and motivated by little more than greed.

In other instances, the management of a potentially proactive company may have good intentions but lack the knowledge or skills necessary to nurture the value of quality. Upon recognizing this weakness, a cultural transformation can take place through education, training, and leadership. With such a metamorphosis, the organization inherently becomes proactive.

> In a quest for the root cause of why reactive companies pursue minimal compliance to a standard that promotes customer satisfaction, one soon recognizes that quality is a "value." Like honesty and integrity, quality is—or is not—part of a company's culture.

What the Future Holds . . .

Proactive companies who have experienced the greatest benefit from ISO 9000 are enthusiastically seeking ways in which their quality systems can evolve beyond compliance with the standard. Some companies pursue profitability and decreased waste by looking more closely at their processes, while others are investigating ways to translate the ideas and spirit of ISO 9000 to other areas and functions within their organization. The following sections examine some of the options that such companies employ.

Reengineering

Fundamentally changing how an organization operates, reengineering can be thought of as turbocharged continuous improvement.

The documentation process often serves as a rewarding cornerstone for the reengineering process. Formal, documented quality systems are foreign to many companies before they implement ISO 9000. During the documentation process, these companies often find that actual work practices haven't evolved since the company was formed. Old processes are often riddled with inefficiencies in light of current knowledge. The technological improvements instituted over the years may not have been planned, implemented, and integrated into the system as well as they might be, given hindsight. Whether as a response to necessity or to the dictates of the standard, the seeds for reengineering have been planted.

When visiting a company in the process of reengineering, an external quality auditor observes a ratchet effect taking place, in which a growing number of improvements are implemented throughout the organization at an accelerated pace. Within a year or two, the company can be operating in an entirely different manner than it was prior to implementing ISO 9000 and a formal, documented quality system.

Although companies that have well-established, mature quality systems have an advantage over those that do not, the rapid changes in available technology provide a variety of other options for evolution. Replacing or supplementing paper-documented quality systems with on-line versions, reexamining processes from the perspective of adding value to the customer, redesigning computer networks to take full advantage of shared information databases, and shifting from an emphasis on Management By Objective (MBO) to process-improvement teams are simple examples of aggressive and effective reengineering practices.

Marketing Quality Assurance

As a company begins implementing ISO 9000, it soon becomes apparent that marketing—the function that first introduces the company to potential customers—is completely neglected by both ISO 9001 and ISO 9002. Many proponents of the standard consider this to be a serious oversight that should have been corrected in the 1994 revision and see a revision that includes the marketing function as inevitable.

In most British companies, marketing remains a key area of weakness, according to research performed on behalf of the United Kingdom Department of Employment. In 1990, to address this need, Marketing Quality Assurance Ltd. (MQA) developed and published a marketing quality specification titled MQA 01 that applies the concepts contained in ISO 9000 to the area of marketing. A number of companies have used the MQA 01 specification to help them achieve successful ISO 9001 registration for their marketing, sales, and customer-service functions. Few U.S. companies have taken advantage of this opportunity, though it provides distinct benefits.

ISO 14000 and Environmental Management

Violating environmental regulations can be devastating for any company. Financial consequences including fines and cleanup costs can all but outweigh a tarnished reputation—and seriously impede a company's long-term success.

Waste, whether produced directly in the form of pollution, or indirectly in the form of disposable product packaging, translates into inevitable costs. Every cost incurred unnecessarily, in whatever form, depletes profits. Addressing and combating pollution-causing practices creates obvious win-win circumstances that cannot (and should not) be deferred.

The ISO 14000 series resulted from the desire to export the concepts found in ISO 9000 to an environmental management standard. Companies that put ISO 14001 to the test find themselves as proud of their improvements in efficiency as of their contribution to our environmental well-being.

> Companies that put ISO 14001 to the test find themselves as proud of their improvements in efficiency as of their contribution to our environmental well-being.

Many registrars will be offering ISO 14001 registration services in the form of a joint ISO 9000/14001 audit. Although the majority of ISO 9000–registered companies may never seek ISO 14001 registration, many appreciate the growing potential for "green" (environmentally friendly) marketing. As such, ISO 14001 implementation is an attractive option to a growing number of ISO 9000–registered companies.

QUENSH

Many quality professionals believe that the next step beyond ISO 9000 and ISO 14001 compliance is the establishment of a QUENSH management system. QUENSH is an acronym for QUality/ENvironmental/Safety and Health. QUENSH reduces non-value-added redundant systems by incorporating these three areas under a single management system.

> QUENSH reduces non-value-added redundant systems by incorporating these three areas under a single management system.

One ISO 9002–registered company, having only loosely implemented a QUENSH system in a single facility, saw medical costs associated with lost-time accidents drop from $324,000 (408 lost days) in 1992 to only $23,000 (15 lost days) in 1995. Clearly, even a modest QUENSH system can generate a significant cost benefit.

At the request of several member countries, ISO is currently considering development of an occupational health and safety management system (OHSMS) standard. The proposed standard is strongly supported by many countries who participate in the International Labor Organization (ILO). Proactive companies, however, need not wait for a formalized standard before realizing the appreciable cost savings a QUENSH system can provide.

Integrated Management Systems (IMS)

For some proactive companies, QUENSH doesn't go far enough. Expanding their formal controls to include financial management systems, these companies are implementing *integrated management systems* (IMS). An open-ended term, *integrated management systems* was adopted for its ability to encompass every aspect of an organization.

Although no specific practices or accounting systems are required for the financial function of an IMS, one must certainly wonder whether an ISO standard for financial management systems can be far away. Modern management often finds itself frustrated by the inability of traditional accounting methods to provide a basis for performance measurement. Easily specified costs are often hidden under generalized categories, distorting the true expenses for products and services while placing excessive emphasis on direct labor. Such a practice typically results in poor sourcing decisions, faulty investment analysis, and an overall lack of meaningful data to support management decision making.

In the late 1980s, a new concept in accounting methods called *activity based costing* (ABC) was introduced, which appears to resolve this issue. ABC methods uncover the true costs resulting from each activity associated with the manufacturing, handling, storing, and shipping of product. Management is thus afforded accurate data on which to base decision making. However, successful ABC methods are contingent upon a company's financial function being integrated throughout the entire organization, particularly in the quality function.

Ultimately, the concept of an integrated management system could evolve to the extent that functions such as quality, environmental, safety and health, finance, operations, and human resources are absorbed into executive management rather than being delegated to separate departments. This holistic management approach should be a natural progression for companies with high values, led by inspired and knowledgeable executive officers.

Conclusion

The future is bright for proactive companies in their quest to transcend the advantages of ISO 9000 by continuing to cut costs, boost production, and exceed customers' expectations. These companies actively seek methods for turbocharging their commitment to continuous improvement. They are aware of and address the relevance of reengineering, new accounting methods (such as ABC), quality assurance for their sales and marketing functions, environmental management options such as ISO 14000, and the cost advantages of new approaches to health and safety, including QUENSH and OHSMS.

Proactive companies view the value of quality as an indispensable part of their company culture and are becoming much more holistic in their management approach, seeking win-win solutions for all parties concerned. Consideration is given not only to the operational processes that make a company profitable, and to the customers through whom profits are earned, but to the health and safety of the employees who contribute to the company's success, as well as to the ecological environment that supports and sustains that success.

Proactive companies are enthusiastic regarding the potential for integrated management systems to bring the ideals we recognized in total quality management into being.

ISO 9000: A Practical Step-By-Step Approach

Roger S. Benson and Richard W. Sherman

This article succinctly describes the steps one company, Grace Specialty Polymers, went through to attain ISO 9000 registration. An important point the authors make early and reiterate at the end is the importance of having top management's commitment in undertaking the ISO registration process.

There are many articles and publications available that address the philosophy of, and reasons and benefits for, becoming ISO 9000 registered. The reasons for obtaining registration are numerous, and there are tangible benefits. But, what does an organization do once the scope of certification has been determined and the decision to proceed has been made?

Most companies that have been through the process recognize that obtaining ISO 9000 registration is a major accomplishment, and they are more than willing to share their experiences. Grace Specialty Polymers (GSP) is one of those companies, and the details of its registration efforts might inspire other companies to achieve ISO 9000 registration.

Preparing for ISO 9000 Registration

In early 1993, GSP management established a company goal to become ISO 9000 registered by the end of 1994. It also decided to pursue a multisite certification of four separate locations to ISO 9001-1987: headquarters, the research and development (R&D) facility, and two manufacturing locations. All four were within 30 miles of one another, so travel barriers were minimal.

The company had several points in its favor when it began the registration process:

- The vice president/general manager was a visionary who had recognized the benefits of total quality improvement (TQI) many years earlier.
- A strong TQI program was in place, providing a solid base for team activities and training.
- The company had a background in military standards (MIL-I-45208A) and had achieved automotive quality (Ford QI) requirements.

ISO 9000 registration seemed to be a natural fit and an extension of existing activities. To begin, an executive steering committee was chartered, consisting of the general manager and employees who reported directly to him. As top-level management, they provided the commitment, direction, and resources that are an absolute must in achieving registration.

Next, an ISO Implementation Team (IIT) was chartered. This seven-member cross-functional team was made up of mostly department managers or supervisors from four local sites and three major departments, including R&D, manufacturing, and quality.

The TQI director was designated by the steering committee as the team leader. Since he was also a member of the steering committee, he acted as a liaison between the committee and the ITT, reporting progress and relaying important resource needs and decisions. Each member was directly responsible for coordinating ISO 9000 activities at his or her site.

An in-house lead assessor training program was arranged by pooling resources and requirements with other W. R. Grace & Co. business units. Each team member attended the program and received a certificate for successful completion.

Immediately after the completion of lead assessor training, the team began meeting on a regular basis. Weekly meetings were scheduled well in advance so team members could arrange their schedules to prevent conflicts.

The team leader prepared a road map to ISO 9000 registration that was based on DuPont's stair-step approach and outlined the procedures GSP would follow to achieve registration by the end of 1994 (see Figure 1). This map was reviewed and approved by the IIT at the first meeting.

While team members were in lead assessor training, the steering committee attended a one-day management awareness training session that discussed basic ISO 9000 requirements and explained management responsibilities. Additionally, companywide awareness training was conducted at each site using commercially available videotape presentations and customized information pertinent to that site. Each functional manager instructed his or her staff. This showed upper management's commitment and helped ensure buy-in to the program at all levels.

Pre-Audit Conducted

Discussions with other registered companies indicated a lack of highly qualified consultants at that time. It wasn't uncommon for the client to know as much, or more, about the standard than the consultant. Based on this information, cost, and GSP's collective quality systems experience, it was decided to forgo consulting services and conduct pre-audit or gap analysis within the company.

The IIT was divided into three audit teams that systematically audited the sites to all sections of ISO 9001-1987. Noncompliance reports (NCRs) were written for all noncompliances, with the designation "major" or "minor." Audit teams presented their NCRs at weekly IIT meetings where team

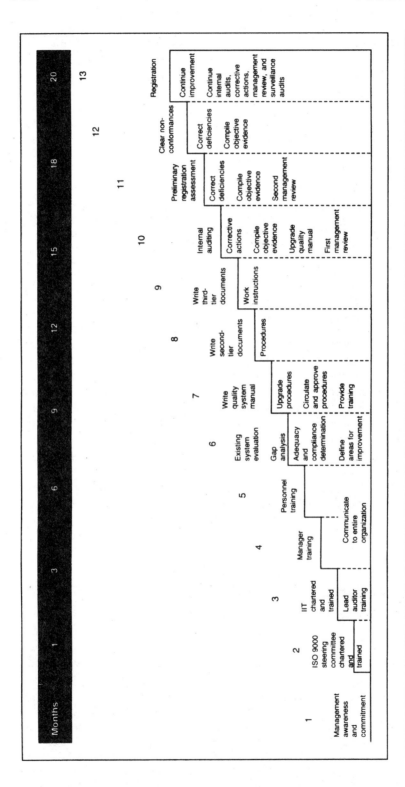

Figure 1. **Road map to ISO 9000 registration**

members had a chance to question and critique them. This helped ensure that all team members agreed with the noncompliance and understood its importance. Typically, the team also discussed possible solutions that could be used for corrective action.

The NCRs were used to drive corrective actions and push the overall quality system toward ISO 9000 compliance. For many of the minor NCRs, the responsibility for corrective action was assigned to the team member who had the greatest knowledge in that area. These NCRs usually required modifying existing procedures either to comply with the standard or reflect current practices.

After reviewing the NCRs that were assigned to them, the team members estimated time frames and the resources needed for completion. All of these estimates were then put into a master time line that was used to measure progress and estimate total project completion time.

Major noncompliances consisted of missing ISO 9000 elements and corrective action requirements that were beyond the ability of any single team member. Many of these cases had been documented but did not sufficiently meet ISO 9000 requirements. These corrective actions, which required a change in basic business practice or philosophy, had to be handled in a completely different manner.

For most major noncompliances, the team created subcommittees of two to four members that met outside the normal scheduled meetings. The subcommittees researched potential solutions and drew up a proposal listing possible solutions, projected implementation time, costs, benefits, advantages, and disadvantages. These proposals were then submitted to the appropriate functional manager.

The functional manager reviewed the various proposals and made a final decision based on the company's philosophy and availability of resources. Once a decision was made, it was reported to the IIT and resources were assigned. The IIT leader kept the steering committee informed of all proposals submitted.

From the beginning, benchmarking was used extensively to provide a baseline or reference point from which to begin the registration process and assist in setting direction. Other business units within W. R. Grace & Co. and suppliers and customers that had completed the registration process were contacted. The information obtained was very valuable, and in many cases served as a model for documentation format and content.

Documenting Control Procedures and Work Activities

Developing document control procedures was a major hurdle. Like most other companies, GSP settled on the three-tier approach (see Figure 2).

And, like most other companies, GSP struggled with the question of how many work activities needed to be documented, especially at level 3. Because GSP had only 225 employees and already had an existing quality manual, the team decided to centralize the control of all documents in the quality assurance department. One exception was made for design control. Product design

Developing document control procedures was a major hurdle.

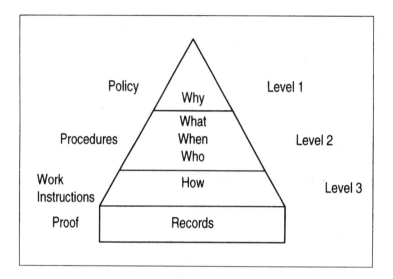

Figure 2. **Quality system document structure**

was entirely under the auspices of R&D; therefore, product design created and controlled its own procedures, which eventually evolved into the new-product development manual.

The existing quality assurance manual served as a foundation from which a new manual consistent with ISO 9000 requirements was built. A working copy of the manual was issued to the IIT members. Team members reviewed, commented on, and further revised these procedures. The procedure review process was conducted during and outside regular team meetings. Ultimately, this process of writing, reviewing, and modifying procedures transformed the existing quality assurance manual into the new ISO 9000 manual.

R&D used monthly staff meetings and ad hoc meetings facilitated by the R&D ISO coordinator to develop and refine the design control process, as well as the supporting procedures.

Implementing a Policies and Procedures Training Program

Upon completion of the new manual, a training program in policies and procedures—not a direct requirement of the standard—was created to ensure smooth implementation. This effort was facilitated by the creation of a training program development responsibility matrix, which assigned generic training modules to IIT members. The modules were developed by the IIT members and distributed to training administrators as required. In addition, a representative from each site developed a training plan that designated who needed training and what level of detail was required. IIT members were prepared to administer the training modules to the employees in group meetings.

In addition to procedures training, training guidelines for each employee's job function were developed. Unlike the policy and procedures training, this training is a requirement of the standard under section 4.18. Resident site training coordinators and administrators were designated. Because training records had not been formally maintained in the past, a training system was created that included a training procedure, a training requirement matrix for each employee, request forms for training, and training certificates. These documents were used to establish and record minimum job requirements, request training, certify that training has been provided, or certify that the employee already meets the minimum job requirements.

Based on benchmarking of several companies, employees who had been in the same job function at GSP for five years or more were considered to have met the minimum requirements of the job. Employees who had not been in the same job function at GSP for at least five years required evidence of certification, evidence that training was provided, or, at minimum, that a personal training plan was in place. A member of the steering committee administered a quality policy training module for all employees because the standard requires that all employees be aware of and understand their company's quality policy.

The Final Assessment

The IIT leader was responsible for determining which registrar to use for the assessment. This was done by benchmarking other organizations and contacting different registrars directly.

The original road map had planned for a pre-assessment with additional time for corrective action before the final assessment. The company felt comfortable that its systems were strong, so it decided to forgo the pre-assessment and proceed directly to the final assessment. This approach had several advantages. It saved an appreciable amount of time in meeting the company's goal. It saved money; a pre-assessment by a third party is expensive. Further, the company knew that if it didn't pass the actual assessment, it would be given a certain amount of time (usually 40 days) to correct noncompliances. During this time the company could focus all its efforts on these issues since they would be the only areas reaudited to confirm compliance. The assessment took place over three days: one-half day each at R&D and headquarters, and one day at each of the two manufacturing sites. Two auditors conducted the assessment; one of the auditors was designated as lead auditor. Two guides—IIT members who were selected for their experience and knowledge of the areas being audited—were assigned to the auditors at each site to help facilitate the audit process. The quality assurance manager, who was one of the guides, accompanied the auditors throughout the entire assessment.

Having two guides accompany the auditors was efficient because it gave them the opportunity to immediately address noncompliances as they arose without interrupting the assessment. This approach proved effective at R&D,

where one guide addressed design control issues while the other guide addressed calibration issues, correcting "observations" as they arose. (An observation is not considered a noncompliance, but it is an issue that could lead to a noncompliance and should be corrected in a timely manner.)

The IIT leader was one of the guides at headquarters. During the assessment of the other sites, he remained at headquarters, but attended the daily closing meetings to review the days' events at each site. From headquarters he was able to track the assessment's progress and relay important information ahead to the sites not yet audited. Additionally, he was able to plan and coordinate corrective action activities, addressing noncompliances as they were uncovered. Using this method, the company was able to solve seven of 11 minor noncompliances before the end of the assessment.

After the assessment of the final site, GSP had no major noncompliances and only four minor noncompliances. As a result, on Sept. 16, 1994, GSP was recommended for registration—a full three months ahead of schedule.

GSP's Recommendations for Achieving ISO 9000

Although there were not many ISO 9000 guidelines in place when GSP began the registration process, careful preparation, execution, documentation, and training enabled it to reach its goal. Other companies may wish to consider the following recommendations:

- Gain full support and commitment from the highest level of management. It must cascade down through the entire organization for successful implementation to take place.
- Form teams and committees to drive the ISO 9000 implementation process. Such teams include an executive steering committee, to support and guide the process from the highest level; an implementation team, to plan, assess, and execute actions as required; subcommittees, to assess and provide recommendations to management; and corrective action teams, to resolve specific noncompliances uncovered during internal assessments.
- Set a goal with milestones, conduct status reports, and pay close attention to timing.
- Enroll key individuals in a lead assessors' training program. This brings the knowledge and skills for ISO 9000 certification in-house, which may be more cost effective than hiring an outside consultant.
- Engage in benchmarking and networking activities. Typically, those who have been through the certification process are willing to share their experiences and might provide their policies and procedures for perusal.
- Attempt to capture what is being done today when writing procedures for an initial assessment. Make minor modifications and enhancements only where feasible and necessary.

- Structure quality manuals to follow the ISO 9000 format. A quality manual that is arranged according to standard format looks well planned and effective and is less confusing for the auditors. This helps the audit progress faster and more efficiently.
- Consider forgoing a formal pre-assessment. Pre-assessments, especially by a third party, can be costly and time consuming. Many minor noncompliances can be cleared as the regular assessment progresses, and efforts can be focused on areas that need attention.

Becoming ISO 9000 registered doesn't have to mean flying blind. These practical steps can make the process easier and can assist your company in reaching its ISO 9000 goals.

Roger S. Benson is senior process engineer at Grace Specialty Polymers in Lexington, Massachusetts. He has a bachelor's degree in biology from Northeastern University in Boston, Massachusetts.

Richard W. Sherman is a quality assurance manager at Grace Specialty Polymers in Woburn, Massachusetts. He has a bachelor's degree in chemistry from Elmhurst College in Illinois. Sherman is a senior member of ASQC, a certified quality engineer, and a certified quality auditor.

ISO 14001 Certification: Are You Ready?

Gregory J. Hale and Caroline G. Hemenway

ISO 14001 is the environmental standard that supplements ISO 9000 quality standards. This article describes this standard and the process for becoming certified. It also includes a good list of reasons why companies should consider becoming ISO 14001 certified.

Well, you did it. It took five years, but you finally convinced your company's senior managers that getting third-party certification to one of the ISO 9000 quality management system standards was the right thing to do.

While you may not be able to prove that the certification directly reflected your company's ability to procure contracts and/or customers, you and your senior managers are fairly certain that the internal benefits greatly exceed the price of having a yearly audit.

Armed with this information, you head into the executive corridor to obtain senior-level commitment for third-party certification to the newly developed environmental management standard—ISO 14001. But you can count on the following challenges to your proposal.

- Why should we seek certification?
- What benefits can our organization expect from certification?
- How much will certification cost?
- Who will do the work?
- How do we go about obtaining certification?

Regarding ISO management standards, the term certification is synonymous with registration in the United States. Registrars are also certification bodies. Because certification has legal connotations in the United States that are associated with products, the term registration prevails here. But it might be worthwhile to use certification because it is the term most accepted worldwide, where you may be conducting business.

Reprinted with permission from *Quality Digest*, July 1996. Copyright © 1996 by QCI International, Red Bluff, California. All rights reserved.

Who's Becoming Certified and Why?

You know that certification is becoming an important issue when companies such as IBM, 3M, Ford Motor, Sony, Canon, SGS-Thomson MicroElectronics and Philips Components acknowledge the benefits related to certification. These companies are among some 400 companies and sites around the world to become certified to ISO 14001, BS 7750—the British EMS standard—and/or the European Eco-Management and Audit Scheme—an ISO 14001 look-alike.

While it's too early in the process to determine what benefits a certified company will reap, it's fair to say that the potential is unlimited. An Apple Computer environmental health & safety division conducted a survey of 99 companies the latter part of 1995, revealing several key drivers of EMS implementation and certification. Respondents included representatives from the electronics, manufacturing, pharmaceutical and engineering/consulting fields. Many companies are now adopting an environmental management system for the following reasons:

- *Ease of trade*—International standards obviate the need for and proliferation of national and regional standards likely to hinder trade by erecting barriers, bureaucratic complexity and redundancy.
- *Improve compliance with legislative and regulatory requirements*—This includes requirements that make public certain information relating to environmental performance.
- *Credibility*—Third-party certification ensures a company's credibility and commitment to regulatory compliance and its continuous, institutional focus on environmental protection.
- *Reduce liability/risk.*
- *Regulatory incentives*—Organizations can take advantage of incentives that reward companies showing environmental leadership through certified compliance with an EMS.
- *Sentencing mitigation*—It is likely that federal and state sentencing guidelines will accept a corporate EMS as a way to levy individual and corporate fines.
- *Prevent pollution and reduce waste.*
- *Profit*—Customers—from consumers to governments—are preferring to purchase "green" products, but want assurances they are green.
- *Improved internal management methods*—and the improvements in efficiency and savings that result.
- *Pressure from shareholder groups*—They are more likely than ever to look for environmental responsibility in investments and financial reports.
- *Pressure from environmentalists (and stockholders)*—They bring a raft of legal precedents to bear on companies they consider poor environmental players.

- *Community goodwill.*
- *A high-quality work force*—They seek empowerment and involvement in addition to healthy and safe working conditions.
- *Insurance*—Insurance companies are less willing to issue coverage for incidents of pollution unless the firm requesting coverage has a proven environmental management system in place.
- *Sustainable development*—Management standards will become a stepping stone for less-developed countries to begin achieving an equivalent level of environmental protection found in neighboring countries. Because management standards require considerably fewer resources to implement, less-developed countries can reap the benefits of more focused and organized environmental protection activities. Over time, as their economies grow in concert with environmental regulations, they can acquire appropriate technologies to maximize environmental protection.
- *Preference in bank loans*—Some institutions, such as the World Bank, may view ISO 14000 as a test of a country's sincerity in its promotion of environmental protection and sustainable development.

Certification Scope

Many organizations that implemented the ISO 9001 quality management system specification standard at the departmental and sometimes individual process levels are rethinking this approach when it comes to ISO 14001 certification, according to Joseph Cascio, vice president of EMS for the Global Environment and Technology Foundation and former program director for Environment, Health and Safety Standardization at IBM. He says the ISO 9000 certification approach caused a proliferation of expensive and disruptive certifications brought on by multiple certifications at one site—which were fueled by multiple registrars performing individual audits. He vows that ISO 14001 certification is different.

Registration to ISO 14001 should be more refined because organizations learned their lesson and don't want to repeat their mistakes, explains Cascio, who is also chairman of the U.S. Technical Advisory Group to ISO TC 207, which developed the standard. Most organizations considering ISO 14001 certification are doing so at the "highest subunit level that fits the logical and physical realities of their firms," he notes.

"ISO 14001 allows total freedom in defining the 'organizational' unit seeking certification," explains Cascio. "As specified in ISO 14001, virtually any organized business activity, such as companies, departments, plants, divisions, construction sites, refineries, mines, branch offices, administration centers, schools, banks, restaurants, fire houses and mills, can be considered 'organizations' capable of implementing an environmental management system."

Experts agree that many considerations must be weighed in choosing the proper unit for certification, but most agree the logical option for larger companies should be an individual location.

However, the majority of small and medium-sized companies will want to pursue certification at the company level. A single company registration may make sense for multinational companies, but only if they are committed "to bringing all operations into line with the environmental ethic and to involve the corporate office in management reviews," says Cascio. "Individual processes and operations within a location will seldom qualify as organizational entities because environmental protection must normally be assured at a location basis for regulatory authorities."

Certification Process

Elizabeth Potts, president of ABS Evaluations, an ISO 9000/ISO 14000 registration organization, explains that the five basic steps to registration include:

- Application or contract
- Initial or preliminary assessment/document review
- Certification assessment
- Certification
- Certification maintenance

Each step depends on the success of the others. She says most certification bodies will require that an application or contract be completed. This document contains the rights and obligations of the certification body and audited organization.

"Although the organization's compliance management will be one of many subjects the audit will address, the certification body is not there to assess your compliance with individual regulatory documents and requirements," explains Potts. "The certification body will assess how your organization ensures that all applicable regulatory requirements are identified and incorporated into the EMS and how well the EMS is functioning."

The compliance-related segment of the audit focuses on the system and how it functions to satisfy the compliance commitment of its policy and compliance objectives and targets of your organization.

The compliance segment of the audit won't focus on whether each regulatory requirement complies fully. Any compliance issues noted during the audit will be brought to your organization's attention as part of the assessment. Regulatory compliance auditing responsibility remains with the organization being audited.

Initial or Preliminary Assessment

The next step in the certification process is the initial or preliminary assessment and document review. Most certification bodies will wish to conduct this phase of the certification process on-site, according to Potts.

The following documents are typically reviewed during this phase of the process:

- The EMS manual (if your organization has chosen to develop one).
- Analysis of environmental aspects and impacts.
- Applicable regulatory requirements.
- Audit reports.
- Organization charts.
- Training programs.
- Management review minutes.
- Your organization's continual improvement plans.

While this part of the certification process evaluates your organization's readiness to continue with a formal audit, it also helps the certification body plan the audit. In contrast to internal or second-party audits, the certification body cannot (and should not) provide substantive guidance on how to achieve conformity with ISO 14001.

"In other words, the certification body cannot be both a consultant and a certification body, as this clearly constitutes a conflict of interest," stresses Potts. "The certification body can and should, however, provide your organization with interpretive guidance by openly engaging in discussions about these concerns."

A certification assessment normally follows a successful initial assessment/document review. Usually, the primary difference between this assessment and internal audits is the "formality" of the assessment. "Your organization should expect the same independence, competence and professionalism of the lead auditor and audit team during a certification assessment as during an internal audit," notes Potts.

"To assess whether an EMS is in conformity with ISO 14001 and has been fully implemented and documented in manuals, supporting procedures and other records, all levels of personnel will be interviewed," explains Potts. "Your organization must be able to demonstrate that the EMS is designed (and implemented) to satisfy its policy commitments and the various ISO 14001 criteria. Most certification bodies conduct a daily debriefing with your organization to keep key individuals informed of progress and any apparent deficiencies noted."

A major difference between third-party ISO 14001 certification and internal EMS audits appears in the final phase of certification—surveillance. Teams from the certification body will return at prescribed intervals, usually every six or 12 months, to assess the continued conformity of your organization's EMS with ISO 14001.

"Corrections of deficiencies identified by previous assessments or surveillances will be verified, and selected elements of the standard will be reevaluated," notes Potts. "During the surveillance phase, emphasis is typically placed on internal EMS audits, management reviews, corrective/preventive action systems and continual improvement efforts."

No Shortage of Registrars

Finding a registrar to assess your organization shouldn't be difficult. In fact, it's possible to use the same registrar for both ISO 9000 and ISO 14001 services. What you should be more concerned with is how much registration will cost and how you can cut costs.

Some experts estimate that certification costs could range from $30,000 to $100,000 per site, depending on the organization's size and nature. Registration costs are expected to be similar to those associated with ISO 9000, with some monetary credit given to companies with certified quality management systems already in place. Steven Bold, manager of the environmental compliance group for Continental Circuits Corp., says his medium-sized company expects to initially invest close to $100,000 in the certification process—approximately $40,000 in software and $60,000 in labor costs associated with implementation and preassessments.

"We employ about 12,000 employees and have close to $120 million in annual sales, and are fortunate that we have both the financial and human resources to commit to our ISO 14001 certification process," explains Bold. "For the majority of electronic circuit manufacturers, the decision to become ISO 14001-certified is considerably more difficult.

"The growing pressure to become ISO 14001-certified threatens their bottom line, and some feel that it threatens their very existence. Some companies have stated that they may have to hire additional employees just to manage the ISO certification process."

For most small companies, the stacking of additional layers of environmental control on top of current regulatory requirements is seen as burdensome, according to Bold. "Once certified, organizations may be forced to pass certification costs on to customers, which may result in customers turning elsewhere, possibly overseas, to purchase their electronic interconnectors," he warns.

Cascio disagrees with Bold and says certification costs should be reasonable, noting that in March and April 1996, two IBM facilities in Germany and the United Kingdom received certification to ISO 14001 and EMAS at an estimated cost of $30,000 per facility.

SGS-Thomson Experience

In early January 1996, SGS-Thomson MicroElectronics Inc., located in San Diego, became the first U.S. site registered to the most recent draft of ISO 14001. The certification cost was $100,000, give or take 10 to 20 percent. While obtaining its ISO 14001 registration, the company also contracted to have the site assessed against the requirements of the EMAS regulation. As part of a corporatewide mandate, all 16 SGS-Thomson facilities located in the United States, Europe, Southeast Asia and Northern Africa will be validated for EMAS and/or certified to ISO 14001 by the end of 1997. Several SGS-Thomson facilities also are certified to ISO 9000 quality management systems and receive both corporate and third-party assessment regularly.

The facility's ISO 14001 certificate to the DIS will expire on the date the International Organization for Standardization prints ISO 14001 as an official ISO standard, which is expected to happen in late 1996. If no substantive changes are made to the ISO 14001 international standard, the facility's certificate will be updated automatically. However, if significant changes are made, the organization could be subject to another assessment.

Patrick Hoy, the site environmental manager for SGS-Thomson's Rancho Bernardo facility, was responsible for preparing the facility for the EMAS verification and ISO 14001 certification audits. Hoy used experience gained from the company's ISO 9001 quality management system to develop an implementation plan for both EMAS and ISO 14001.

"The key to any environmental management system is the mechanism employed for identifying aspects and impacts associated with a company's processes, products and services," explains Hoy. To prepare for the EMAS formal assessment, the Rancho Bernardo facility followed the failure-mode-and-effect analysis corporate standard developed for assessing the potential significance of each environmental effect identified.

As part of the corporate standard, Hoy and his team reviewed regulations that applied to the facility, pending environmental legislation, corporate requirements and media-specific permitting conditions.

"We examined potential impacts associated with water, air, hazardous waste, solid waste, soil and groundwater, energy, chemical management, external noise, raw materials, new product processes, product planning and emergency response planning," he reports.

Hoy ranked each impact according to its significance and where it appeared on the corporate list of overall environmental effects. He notes that the facility compiles a list or register of potential environmental impacts and makes it available to the public through its annual environmental statement.

What Does the Future Hold?

As the official publication date of the ISO 14001 standard draws closer, with some predicting middle to late October 1996, more facilities will "get off the fence" and seek certification. Lucent Technologies, Akzo Nobel and Toyota are among those expected to join the registration movement, certifying multiple sites around the world. Clearly, multinational companies will lead the ISO 14001-certification movement, with small and medium-sized companies following closely behind.

What next? We'll all just have to take a seat and see what happens with the market. It's possible, but highly unlikely, that ISO 14001 will simply be implemented as an internal tool. Most likely, state and federal regulators will begin using the certification process in regulatory incentives for companies tired of the command-and-control process.

Presently, key regulatory bodies in some two-dozen states, including Pennsylvania, California, Wisconsin, Minnesota and Massachusetts, are taking steps to use the ISO 14001-certification process to increase environmental protection.

While we could not predict all of your ISO 14001 implementation and certification questions, in this series we strove to anticipate key concerns. Now, the decision to pursue this subject is yours—we provided you with what we hope is a useful source of information to move forward. Armed with this information, you can feel confident that, when you turn that knob to the vice president's office, you won't be alone. Good luck.

Caroline G. Hemenway is publisher of CEEM Information Services in Fairfax, Virginia.

Gregory J. Hale is associate editor of International Environmental Systems Update, *a monthly newsletter on ISO 14000 developments and implications.*

Weighing Alternative ISO 9000 Registration: Not for All Companies

Paul Scicchitano

This article explores alternatives that some companies are proposing to ISO 9000 registration. The main proposed alternative is the Supplier Audit Confirmation. Scicchitano looks at the pros and cons of this idea, which some large companies such as Hewlett-Packard and Motorola are supporting, and suggests that it is still unclear whether such a registration will come into being.

With ISO 9000 registration costing thousands of dollars a whack, the concept of alternative registration has been gaining momentum in the United States and Europe. But whether it will be a viable option for most companies or just for the Fortune 500 set is anybody's guess.

The concept, initially proposed by the computer-printer giant Hewlett-Packard in November 1994, has been widely interpreted as a combination of stepped-up internal audits and reduced registrar oversight. The real savings for companies, say industry experts, will lie in shifting the burden of registrar follow-up visits from a third-party audit team back to the end users of the quality system.

Some experts, however, believe that few companies will be able to qualify for the new alternative, and they question whether those companies that do qualify will realize any savings as a result.

I would guess that the candidates for [alternative registration] in this country are numbered in the tens, not in the thousands," observes Ian Durand, a key U.S. delegate to the international committee that created the quality-assurance and quality-management standards, which were first published in 1987.

"By and large, the only companies that would find it useful are those that already have a very comprehensive quality system in place," explains Durand, president of Service Process Consulting, Inc., an ISO 9000 consulting firm and training organization. "Usually that means the large, often multisite, often multinational companies."

Reprinted with permission from *Compliance Engineering*, May–June 1996. Copyright © 1996 by *Compliance Engineering*. All rights reserved.

Typically, North American companies must undergo a full-blown ISO 9000 registration audit once every three years. This generally involves a documentation review followed by on-site auditing, which can take anywhere from one day to several weeks, depending on the size and complexity of the operation.

At the most basic level, third-party auditors are looking to ensure that the quality system is functioning as described in the company's quality manual, and that procedures have been put in place to cover key aspects of the system—that is, those aspects for which, in the auditors' judgment, an absence of documentation could adversely affect the quality of the product or service being produced. Once registered, companies are subject to less comprehensive follow-up audits, commonly referred to as surveillance visits, every six months or so.

Under the proposed alternative-registration scheme, now called Supplier Audit Confirmation (SAC), companies with effective quality systems would assume more responsibility for surveillance. Although specific details of the program are still being worked out, the idea has gained the attention of ISO 9000 registrars and accreditors worldwide.

The International Accreditation Forum (IAF), made up of representatives from about 20 of the most influential accreditation bodies, including the American National Accreditation Program for Registrars of Quality Systems, recently cleared the way for pilot testing of the concept.

The first tests are being conducted at a few sites in Europe and the United States, but the results probably will not be known for about a year as ISO 9000 registrars and accreditors attempt to determine whether the effectiveness of the quality system can be sustained over a period of time with a diminished level of third-party scrutiny.

The IAF's position is based on its belief that existing international guidance documents on registration and accreditation are sufficiently flexible to allow for recognition of a company's history of good performance where appropriate. The pilot tests are contingent upon the accreditation bodies' being able to ensure that their confidence is not misplaced.

In essence, the IAF "aims to establish a system that will enable a certificate to be issued regardless of the conformity assessment methodology used, without compromising confidence or credibility," according to an official statement.

In theory, alternative registration would be available only to those companies able to demonstrate the presence of an effective quality system, but so far, no one has defined exactly what that means. The idea may be gaining support because it offers an option for companies that do not require the level of oversight, necessary for traditional registration, suggests John Donaldson, Vice President of Conformity Assessment at the American National Standards Institute (ANSI), which, along with the Registrar Accreditation Board, operates the U.S. accreditation program for ISO 9000 registrars.

"If it loses credibility, it won't fly. If it can be seen to be credible, absolutely it will fly," predicts Donaldson. "They're going to see if [they] can do this and maintain the inherent credibility [of the third-party registration system]."

Donaldson sees the alternative-registration movement as a marketplace correction of the third-party system. "Some of their thinking is that there're a hell of a lot of steps in the registration process overall that are unnecessary," he says. "They're really saying, Let's go back and reappraise the registration process."

The concept has been refined over the past two years in an effort to make it more palatable to both the ISO 9000 industry and prospective end users. At first it drew strong objections from a number of ISO 9000 consultants, trainers, and registrars, who felt it was tantamount to self-certification, but the opposition softened after proponents, led by Hewlett-Packard and Motorola, took their case directly to accreditation bodies and standards writers all over the world. The proposal has since evolved into a much more moderate approach.

"Everybody knows it is not self-declaration. All the mystery is out of it now," maintains David Ling of Hewlett-Packard. The concept is well understood. The only thing now is to work out the criteria."

Ling himself says he, has no idea how many companies may ultimately be affected or how much they may be able to save through an alternative-registration arrangement. I can't say if [the plan will work] for 10 companies or for a thousand or for all companies that will be certified to ISO 9000." The key, he acknowledges, lies in how companies view third-party registration. Some may welcome the fresh eye of an outside audit team, while others view it as an intrusion into their corporate culture.

"We saw the emergence of management system standards and the role of third-party certification in place with the realities of those things, plus and minus. We wanted to have some input on a more business-focused way of demonstrating confidence," adds Ling. You can't think of savings in the traditional sense; it could be just in the philosophical sense."

The strongest reaction to the proposal by a major purchaser has come from the Big Three automakers—Chrysler, Ford, and General Motors—which incorporated ISO 9001 in their QS-9000 requirement and are relying on the third-party-registration industry to ensure compliance on the part of thousands of automotive suppliers. In a communiqué, the Big Three task force charged with overseeing QS-9000 indicated that under no circumstances would alternative registration be accepted from any supplier.

R. Dan Reid, General Motors' representative on the task force, said the automakers are concerned that alternative registration will not provide an equivalent level of assurance. Although they are "watching the impact of it," he says, "we don't support it." According to Reid, the Big Three would not be opposed to SAC's becoming a voluntary route to ISO 9000 conformity, provided that its use were noted on the registration certificate, something not envisioned by proponents.

Consultant Durand agrees that the alternative-registration option poses some risk. "I think the danger is that there won't be as much discipline and integrity in that process as in the regular third-party registration, [which itself is] not perfect either," he says. "It's introducing another layer where problems can creep in."

Elizabeth A. Potts, President of ABS Quality Evaluations, Inc. (ABS QE), an ISO 9000 registrar, said her firm will be among those participating in the pilot tests of the alternative-registration concept. She concedes that traditional surveillance visits may be of little value for companies that have been registered over an extended period of time.

"When you first go in and a company is implementing ISO 9000, you find holes in the system," Potts explains. "As the system matures and is internalized by the company, you find fewer and fewer."

In the United Kingdom, which claims nearly half of the world's total ISO 9000 registrations, Lloyd's Register Quality Assurance Ltd. (LRQA) employed an SAC-like model at British Nuclear Fuels two years before Hewlett-Packard's proposal surfaced in the United States, according to David Hadlet, LRQA's Director of Business Development.

In the case of British Nuclear Fuels, the savings were a matter of time more than money. "It costs time and effort to host an ISO 9000 audit, and it interrupts [a company's] ongoing audit programs," says Hadlet. "There's no question that for companies with mature, effective quality systems, the traditional type of surveillance visit does not add value." For such companies, the primary benefit of surveillance audits is to assure to outside parties that the system remains in compliance.

"Our concern is that it [may become] a one-size-fits-all approach," Hadlet adds. "Our view has always been that alternative approaches are acceptable when they've been designed to meet the specific needs of an individual supplier."

"The pilot programs that are now being undertaken have been worked out on a case-by-case basis. Right now each of the ones that are being handled is being done in an individualized negotiation among the registrar, the client, and the accreditation body," says Potts. "I think it's just something you are going to evolve into if you are truly a registrar that has entered into a long-standing partnership with your client. You are naturally going to to modify surveillance activities to meet evolving client needs."

Paul Scicchitano is managing editor of Quality Systems Update, *a monthly newsletter and information service devoted to ISO 9000 and ISO 14000 issues.*

Standards and Assessments
BALDRIGE CRITERIA/ QUALITY AUDITS

Is the Baldrige Still Meaningful?

Barbara Ettorre

This article answers the question its title asks. The answer is definitely yes. The award signifies that an organization is exceptional in managing its processes and in paying attention to its customers. Beyond the award, the criteria have become an important and very valuable tool for organizations of all sorts to evaluate where they are now and how to improve. Moreover, the criteria are constantly being upgraded and refined a sign that the Baldrige, in all its iterations, indeed has an important role to play in American, and international, organizations.

The Malcolm Baldrige National Quality Award is 9 years old—in the world of prizes, an exuberant preadolescent. But will the Baldrige even make it to its teens?

In recent months, a cost-conscious Congress has broached the idea that the Baldrige—the only award given by the U.S. government to the private sector for overall performance—should be turned over to a private sponsor.

Additionally, the number of applicants for the prize has steadily declined from a high of 106 in 1991 to only 47 last year. This suggests that, as corporate America's preoccupation with quality has faded, especially in the past two years, the award may have outlived its usefulness.

Reprinted by permission of the publisher from *Management Review*, March 1996. Copyright © 1996 by the American Management Association, New York. All rights reserved.

As corporate awards go, probably only the Deming Prize awarded in Japan is more prestigious worldwide than the Baldrige. Set up by an act of Congress in 1987 and first awarded in 1988, the Baldrige is managed by the Department of Commerce via the National Institute of Standards and Technology (NIST), an agency of the department's Technology Administration. The American Society for Quality Control helps administer the award program under contract to NIST. The award itself is named for Malcolm Baldrige, CEO of Scovill Inc. and later a respected secretary of commerce during the Reagan era. Baldrige, an avid rider, was killed in a rodeo accident in 1987.

Annual awarding of the Baldrige is attended by appropriate hoopla. Winners are given their awards by the President at a special White House ceremony. These organizations are universally recognized as world-class, particularly because a condition of winning the award requires them to share best practices. Indeed, this is the heart of the Baldrige. The elite group of 24 winners since 1988 has certainly turned itself inside out as role models for other organizations to emulate.

But there is something else that is emerging as more important. The Baldrige criteria are being disseminated far and wide to enterprises (including agencies of the U.S. government and the military) that have little intention of ever applying for the award. They use the criteria as a self-assessment tool.

"If they never award another Baldrige, it still will have a huge impact on many organizations," says Joseph P. O'Leary, partner in charge of Arthur Andersen's Customer Satisfaction practice. "It's the structure and the rigor. A new language has been created. The Baldrige has greatly increased awareness. As a result, there is significant success at many organizations the Baldrige will never, ever hear from."

Moreover, in a spate of imitative quality awards, the Baldrige model has been adopted at state and local levels, even abroad. Today, there are well over 40 awards, with new ones being created every year. Even Japan has been investigating a version of the Baldrige. All this means that there are fewer applicants for the Baldrige itself but, collectively, at least several hundred applicants for Baldrige progeny around the world. Clearly, the process is alive and well.

> "If they never award another Baldrige, it still will have a huge impact on many organizations."

> There are fewer applicants for the Baldrige itself but, collectively, at least several hundred applicants for Baldrige progeny around the world.

The Crux of the Matter

The criteria, which initially dealt with quality products and services, have evolved into wide-ranging standards addressing corporate processes from soup to nuts. There has been a steady shift in criteria from separate processes to overall business practices and excellence, including such areas as strategic planning, sophisticated customer service, advanced human resource management and employee satisfaction.

"Quality of products and service is only a part of the criteria today," says Curt W. Reimann, former director for quality programs at NIST and formerly director of the award. Reimann retired in December and a successor has not yet been determined.

"The set of Baldrige criteria is very sophisticated in its current incarnation," Reimann continues. "This has shocked some applicants who have only kept pace. The criteria are growing faster than companies can improve. We keep raising the bar."

So, is it fair to say that it is harder to win a Baldrige today? "Very definitely," says Reimann, "even as we have respect for our early winners."

Reimann says that the most rewarding part of his job as chief Baldrige administrator was helping to set evolving criteria every year. He stresses that the Baldrige *model* is the heart of the issue, noting that, from the inception of the prize, Congress intended the criteria to be widely circulated as an improvement tool.

Reimann notes that NIST has never copyrighted the criteria; nor has it prohibited authors, organizations, and state and local agencies from publishing them in countless books, periodicals and appendices and from reproducing them in any print or electronic form. The criteria are also on-line.

In fact, several years ago, when the Baldrige was the only quality game in town, NIST distributed an annual high of 240,000 sets of the criteria. Even today, with the criteria so readily available worldwide, NIST still disperses an average of 175,000 copies a year. Reimann estimates that states photocopy another 175,000, to cite just one source.

"The Baldrige hasn't moved far from its original mandate [of fostering self-assessment]," Reimann says. "But the criteria have moved. We've never been criticized on the criteria."

Many quality experts agree, stating that the Baldrige has served to educate organizations and governments to think critically about their processes and reasons for existence. "It is really out-of-date to call the Baldrige a quality process," says Donald C. Fisher, executive director of the Mid-South Quality Productivity Center and a former member of the Baldrige Board of Examiners. "Five years ago, that was okay. Now there is more sophistication. Organizations learned to cry, learned to walk—now they're learning to run."

Beyond striving for the prize itself, companies are finding that the real value of the Baldrige criteria is their continued use in strategic improvement, says Fisher, author of *Measuring Up to the Baldrige: A Quick & Easy Self-Assessment Guide for Organizations of All Sizes* (AMACOM, 1994). "When you're doing an award, you're trying to dress up an organization," he says. "When you are doing a self-assessment, you're trying to dress down an organization, to get into the core."

Fisher says that all kinds of public and private organizations—from governments, corporations, hotels and schools to religious affiliations and prisons—are using Baldrige criteria. Fisher has conducted Baldrige assessments for eight years. He says he has made more than 100 site visits, both as a Baldrige board member and as a private consultant. "I have 20 visits scheduled thus far in 1996," he says. "The Baldrige is very alive. It's not being touted because it has become grass-roots, a way of doing business."

Organizational change experts are finding that clients who use Baldrige criteria in a mode of continuous improvement confront world-class quality

> ## TO THINE OWN SELF BE TRUE
>
> Does your organization want to use the Baldrige criteria for self-assessment? Assemble a small core group of committed senior executives and apply these tips from David E. Setzer, quality director at AT&T Network Systems Group, a 1992 Baldrige winner.
>
> - Read, discuss and understand the criteria thoroughly.
> - Reach out to all constituents. Cast a wide net, finding people at all levels in the organization with, as Setzer says, "a fire in their gut."
> - Job titles, affiliations and levels are known only to one or two in the senior core group to eliminate hierarchical barriers and to deter stereotyped expectations.
> - Make it clear to all that regular job responsibilities will continue, even as this will be some of the hardest work the team will ever do. Those who decide they can't do it can leave gracefully.
> - The whole team, not a specialized staff group, must address and answer criteria questions. It doesn't pay to lock people in a room to talk about the criteria. They have to go where business is conducted. Examine operations. Solicit opinions. Ask questions. Listen.
> - Prepare the application as completely as you would for a Baldrige and get it analyzed by trained evaluators (within the company if you have them, or from the outside). Disseminate the evaluation. Publicize it throughout the organization.
> - Develop an improvement plan using the evaluation as a guide. Get input from the assessment team and other interested parties. Most importantly, repeatedly test the organization against the criteria for more feedback. This, says Setzer, is really what a continuous process of improvement looks like.

head-on. "That's the brilliance of the criteria," says Randall A. Lipton, president of the Lipton Group, based in Laguna Niguel, Calif. "The Baldrige got its job done and is continuing to be a very powerful tool."

It is also an exacting one, especially for organizations that haven't a clue about how rigorous Baldrige self-assessment actually is. Says Lipton, "For companies that are all lip service, the criteria say to them, 'Okay. You say you have world-class quality? Let's define what it is and bring it down to earth.'"

Will This Be on the Exam?

For Baldrige entrants themselves, helping to bring an organization down to earth is a network of volunteer examiners trained by NIST. These individuals, which NIST calls evaluators, make site visits and prepare exhaustive evaluations for the corporate entrants.

NIST has been training 260 unpaid evaluators a year almost since the award's inception. Prospective Baldrige evaluators must undergo a rigorous course in criteria management every year. They must reapply for fresh annual training because the criteria are constantly changing. Reimann estimates that about 5,000 volunteer evaluators are involved nationwide in site visits, evaluations and other paperwork for Baldrige imitator quality awards. Some of them were originally trained by NIST, others by state and local prize administrators.

Baldrige evaluators are unpaid volunteers (except for travel expenses). It costs from $6 million to $7 million a year, however, to administer the award. The Baldrige office at NIST has a staff of 25. Most of the money comes from entry fees and from a private Baldrige foundation that raises funds from a cross section of businesses to endow the award program. The budget-minded Republican Congress is perhaps a little too picky about the Baldrige: Of the total administration cost, the federal government's share is only $2.9 million under the Department of Commerce budget.

Reimann estimates that maintaining the family of independently run Baldrige imitators at state and local levels is a $100 million industry, comprising site visits, evaluator training, fees, administration, donations and ceremonies. "This is high leverage," he says, citing the high visibility and the learning process for all involved.

> Many corporations—especially those active in the Baldrige process and continuous self-assessment—maintain their own evaluators trained by the company itself or by NIST.

Many corporations—especially those active in the Baldrige process and continuous self-assessment—maintain their own evaluators trained by the company itself or by NIST. For example, AT&T, a three-time Baldrige winner (two in 1992, one in 1994), has about 300 voluntary examiners in all parts of the company, most of them trained in-house. Their work as examiners is in addition to their regular jobs.

"In my view, training to be an evaluator is a prerequisite to being a leader," says David E. Setzer, quality director at AT&T Network Systems Group, a 1992 winner. He notes that his operating group has been using the Baldrige process and criteria since 1988, when "we stumbled onto the guidelines and said, 'Wow. These are terrific questions. We ought to be asking ourselves these.'"

Setzer says his area continually invokes the criteria as newer, more refined guidelines are introduced by NIST. "We are finding flaws in our business by using this process, asking the questions over and over," he says. By the time his operating group won the Baldrige, Setzer estimates it had cycled through the criteria "eight or nine times."

"After awhile, [the improvement process] went beyond the questions," Setzer continues. "It took on a life of its own." He is a staunch defender of the process: "Quality carries a lot of baggage. TQM is not about quality, but about creating a great business using quality tools."

Nevertheless, some quality specialists who have questioned the relevance of the Baldrige cite the rise of ISO 9000, the certification standard for doing business internationally. They point out that ISO 9000 addresses real-world requirements, while the Baldrige deals with corporation-wide issues. ISO 9000 is about having metrics, processes and procedures in place; it concerns measurement organization and administration rather than world-class business practices and quality. ISO 9000 addresses the *administration* of an organization's quality program.

Are ISO 9000 and Baldrige mutually exclusive? Probably, in most areas they cover. It is possible that a Baldrige winner would not pass an ISO 9000 certification and vice versa.

"ISO 9000 and the Baldrige are really very different beasts," says Reimann. "ISO 9000 is a minimum standard; whereas the Baldrige is a stretch standard and much more comprehensive. The overlap is less than 10 percent. Companies can't do both ISO 9000 and the Baldrige at once."

The Baldrige was originally set up for manufacturing, service and small-business participants. NIST is looking to expand the prize into other areas. Currently, pilot programs are under way in healthcare and education to give NIST information on the practicality of awards in these fields. Reimann says, "We want to be careful and thorough."

Baldrige experts will evaluate readiness, interest and level of business process expertise. NIST hopes to launch Baldrige awards in healthcare and education next year if the secretary of commerce recommends them and if Congress grants approval.

And future Baldrige criteria, no matter the prize? "They will continue to get tougher," says Reimann.

Barbara Ettorre is a professional business writer whose articles are found in many top business publications.

> "ISO 9000 is a minimum standard; whereas the Baldrige is a stretch standard and much more comprehensive."

PART FOUR

Quality References

We have created *The Quality Yearbook* for you to use as an anthology and reference to the quality-management movement. Parts 1 to 3 are the anthology part of the book. Part 4 is the reference part. Several sections make up this reference, each with its own list of information sources on total quality management. Specifically, you'll find the following:

- **Comprehensive Annotated Bibliography.** We have ordered this in a similar way to the Yearbook's overall organization. The bibliography lists hundreds of articles and books on quality management from a large number of sources. This bibliography is current through fall 1995, and it should be the first place you turn to find more information on any aspect of quality management you're interested in.

- **On-Line Quality Services.** We started this section in the 1995 yearbook, expanded it in 1996, and now, with the explosion of the Internet, provide you with a comprehensive directory of discussion groups and World Wide Web sites. New here is an improved set of annotations and a rating system as to the usefulness of these sites to managers interested in TQM.

- **Directory of Magazines and Journals.** This section lists nearly seventy different publications that focus on or regularly include articles on quality management. This directory provides information on subscribing along with descriptions of content and rates the magazines according to their usefulness to managers who are interested in quality management.

- **Directory of Organizations and Companies to Help You Identify Best Practices.** This is a new section for *The Quality Yearbook*. What you'll find here are lists of industry associations, publications and directories, and companies that can help you identify best practices and with whom you might be able to set up benchmarking agreements. This directory is divided into two parts. The first part is organized by industry and includes industry associations followed by related publications and directories that can supply you with information about best practices in these industries. The second part is organized by business function and practice and lists many companies that have been identified as having best practices in these areas.

- **Quality Quotes.** In this continuing section of the yearbook, we have once more included a selection of approximately 200 quotations from many different people, including business writers, philosophers, poets, and practicing managers. These are food for thought, and can be helpful when planning a speech or writing a report.

- **The 1996 Baldrige Award Winners and the 1997 Baldrige Criteria.** Despite the fact that the winners and criteria are announced less than three months prior to the publication of this book, we feel that it is vital to be able to provide our readers with this information. By special arrangement with our production staff and the publisher, we are able to include this information for you.

- **Index.** We have prepared a detailed index to supplement the table of contents. The index should be valuable in helping you find more information on many topics covered in our anthology.

Annotated Bibliography of Quality Books and Articles

This bibliography follows the familiar organization of previous editions. We have included books and articles that appeared after we went to press for the previous edition, so you will see some material published in late 1995, but the majority of the publications included here are from 1996.

Publications in 1996 continued the same patterns evident in 1995—case studies, how-to articles, and a continuing outpouring of material on ISO 9000. Reengineering received less attention, but business transformation and corporate cultural topics were important areas of focus, as were tools and techniques. The variety of book publishers and journals carrying material on quality-management practices expanded. As in previous years, this bibliography is selective, not definitive. We are placing more emphasis on material that has "new news" and, as in the past, on tools and methods for successful applications. We noticed that articles are becoming increasingly more explicit and tactical in their advice, a very positive improvement in these publications.

MANUFACTURING SECTOR
General Trends

Anonymous. "Manufacturers Fail 'World Class' Test," *Journal of Management Development* 14, no. 9 (1995): 41–44.
Reports that three out of four European companies are confident that they can compete globally. Yet this article also points out that only one in fifty are world class. This is the major finding of Philip Hanson, of the IBM Consulting Group, and Professor Voss, of the London Business School, in a report titled "Made in Europe: A Four Nations Best Practices Study." Hanson and Voss used six variables against which to measure European manufacturing companies. Incidentally, this study confirms other studies done in previous years in the United States about the lack of competitiveness of European companies.

Dean, James W., Jr., and Snell, Scott A. "The Strategic Use of Integrated Manufacturing: An Empirical Examination," *Strategic Management Journal* 17, no. 6 (June 1996): 459–80.
What are the strategic uses of TQM, just-in-time methods, and IT? Demonstrates their wide use and profound influence on competitive effectiveness, particularly the use of TQM.

Dellana, Scott A., and Coffin, Mark A. "Quality Management Tactics in U.S. Manufacturing: Do They Rival the Japanese?" *Engineering Management Journal* 8, no. 2 (June 1996): 27–34.
Compares U.S. and Japanese firms that compete globally, presenting a framework for describing their quality-management tactics. Demonstrates that U.S. manufacturers have met the Japanese challenge in manufacturing but lag the Japanese in quality issues affecting manufacturing cycle time. This lag is found in product design, supplier quality, and preventive maintenance.

Golovin, Jonathan. "'Best Practices' Makes Perfect," *Manufacturing Systems* 14, no. 4 (April 1996): 74–75.
Argues that high-quality products have a greater impact on customer satisfaction than any other factor. Poor quality drives up manufacturing costs because products have to be modified, repaired, and redesigned, and test times extend. Inventory losses rise and customer demand declines. Manufacturers have to continually improve operations and quality of products to do well.

Montgomery, Joseph C., and Levine, Lawrence O. (eds.). *The Transition to Agile Manufacturing: Staying Flexible for Competitive Advantage.* Milwaukee: ASQC Press, 1995.

This is a nuts-and-bolts book on agile manufacturing concepts. It emphasizes how to implement this approach by applying quality-management practices. Includes measurements, turning the workplace into a learning organization, applying technology, and taking a systems approach.

Mullarkey, S., Jackson, P. R., and Parker, S. K. "Employee Reactions to JIT Manufacturing Practices: A Two Phase Investigation," *International Journal of Operations & Production Management* 15, no. 11 (1995): 62–79.

The authors looked at how employees reacted to the introduction of just-in-time manufacturing processes on the shop floor. Employees did achieve greater autonomy and a rise in job performance demands, but the social well-being of the employees did not change compared to the pre-JIT period. The authors concluded that strains on employees did not increase with this dramatic change in their work style.

Aerospace

Scott, William B. " 'Ownership' Key to Delta Quality," *Aviation and Space Technology* 143, no. 13 (September 25, 1995): 101–3.

McDonnell Douglas's assembly plan in Pueblo, Colorado, has deployed a strategy involving workforce flexibility, quality programs, and accountability. The plant improved productivity by 40 percent and quality by over 50 percent over the past four years. This articles describes how that was accomplished, a strong success story.

Tanner, John R., Heady, Ronald B., and Zhu, Zhiwei. "TQM Payback Experience in Manufacturing Firms," *Industrial Management and Data Systems* 95, no. 9 (1995): 38.

Reports on the results of a survey conducted with manufacturing companies to determine the payback of quality-management programs. Key findings: initial investments in TQM were recouped in one year in 42.3 percent of the companies, and two years or less for 80.8 percent. The literature on TQM is almost silent on the issue of payback. All companies eventually recouped their initial investments.

Process Industries

Cruz, Clarissa. "Quality Programs Soften Boundaries at Champion," *Purchasing* 120, no. 1 (January 11, 1996): 73–76.

Describes the quality-management practices at Champion International Corporation in Stamford, Connecticut. Quality is a key strategy, and Champion teaches all employees about quality practices and about how profitability grows out of continuous improvements. Selective use of suppliers is also described.

Guest, David. "Single Suppliers Find the Way Forward on Maintenance," *PPI: Pulp & Paper International* 37, no. 12 (December 1995): 21–25. Reports results of a benchmarking study conducted by *Pulp & Paper* and the Maintenance Association of the Paper Industry. They discovered an enormous variance in the cost of maintenance—from $26.04 to $130.32 per ton—with an average of $63.96 per ton. Today maintenance management is seeking ways to prevent waste, predict and correct problems, and improve processes in order to enable effective and efficient production. ISO 9000 practices are becoming more common, and mills are setting goals for downtime, speed of repairs, and other activities. Some mills are going further by adopting TQM approaches that strive for no downtime.

Vardy, Joel M. "Process Control Improvement via Discovery," *Chemical Engineering* 103, no. 1 (January 1996): 88–90.

Advances in information technology are making process systems more significant, making them a competitive tool in manufacturing. Vardy describes a twelve-step process widely used to take advantage of software tools in process control. This is a best practices article.

Various Manufacturing Industries

Anonymous. "Bosses Not Instrumental at Texas Instruments," *Journal of Management Development* 14, no. 9 (1995): 29–31.

Reviews activities at the Texas Instruments plant in Malaysia that is one of the world's best examples of how to integrate work teams into a total quality management environment. As a result, almost all management has been eliminated at the plant. This is the ultimate example of the empowered worker at work; a fascinating case study.

Bergstrom, Robin Yale. "TQM in a Hurry," *Automotive Production* 108, no. 1 (January 1996): 50–52.

Presents the case of Hydromat, Inc./Turmatic Systems, Inc., in implementing TQM within a year in order to conform to Ford Motor Company's quality specifications for suppliers. The company succeeded by doing its homework, developing and sticking to a plan of action, involving everyone, encouraging suggestions, empowering workers, and moving as quickly as possible. Concurrently it developed a quality manual, created quality procedures and work instructions, and defined all its key processes.

Gondhalekar, Shrinivas, Babu, A. Subash, and Godrej, N. B. "Towards TQM Using Kaizen Process Dynamics: A Case Study," *International Journal of Quality and Reliability Management* 12, no. 9 (1995): 192–209.

Describes the use of a kaizen system in a manufacturing operation over twenty-three months. The authors discovered that the kaizen was an autoregressive process that led to the formation of habits in making improvements and thus can lead to self-sustaining, controllable process management and continuous improvement.

Harrisson, Denis, and Laplante, Normand. "TQM, Trade Unions and Cooperation: Case Studies in Quebec," *Economic and Industrial Democracy* 17, no. 1 (February 1996): 99–129.

Presents the experiences of four unionized plants implementing TQM programs. Management and unions had to work together in partnership to implement these programs. Results varied from one plant to another, but all four were the subject of extensive negotiations between the two parties.

Larson, Melissa. "Quality Plays Plantwide Role," *Quality* 35, no. 1 (January 1996): 24–28.

Reports on a survey of 1,000 participants on a broad range of quality practices in manufacturing. Seventy-one percent implemented new technologies; slightly more also expanded existing quality practices. Training on quality issues remains spotty.

Litsikas, Mary. "Crown Cork & Seal Cuts Scrap," *Quality* 35, no. 4 (April 1996): 122–23.

The Crown Cork & Seal company provides a case study on TQM. This article describes how operators were trained in the use of statistical process control tools and how through the use of these tools, operations improved. The firm has been practicing TQM since 1990.

SERVICES SECTOR
Banking and Finance

Anonymous. "Do Banks Have an Image Problem? You Decide," *ABA Banking Journal, How Customers See You Supplement* (October 1995): S2–S31.

Presents the results of a survey conducted in late 1994 on how customers viewed the banking industry. Customers blasted the industry for not being flexible, delivering poor customer service, and not educating customers on how to use banking services and criticized banks' weak involvement in their communities. Concludes this industry has large image problems.

Asarnow, Elliot. "Best Practices in Loan Portfolio Management," *Journal of Lending and Credit Risk Management* 78, no. 7 (March 1996): 14–24.

Banks can learn a great deal about how to manage portfolios by looking not just at best practices in other banks but also at portfolio managers. Insights can be found in how to delegate authority, how to measure performance, and how to manage investment and disclosure policies. To a large extent, banks would have to take a process view of the business; continuous improvement in performance is crucial to survive.

Bednar, David A., Reeves, Carol A., and Lawrence, R. Cayce. "The Role of Technology in Banking: Listen to the Customer," *Journal of Retail Banking Services* 17, no. 3 (Autumn 1995): 35–41.

Presents results of a study that indicates that customer service is the most important factor in establishing the level of overall satisfaction in forty-one banks using technology. Customers rated ATMs very low in satisfying their service requirements. Customers preferred to deal with people, not machines. But they liked using technology to check account balances and to get cash. The older the customer, the more he or she liked working with people rather than machines.

Cowling, Alan, and Newman, Karzin. "Banking on People: TQM, Service Quality, and Human Resources," *Personnel Review* 24, no. 7 (1995): 25–40.

Reports on the experiences of two British banks in introducing quality-management practices, with emphasis on the reaction of employees. Lessons learned and advice to management are included, as well as comments about quality management in the U.K. in general.

Kotha, Shravan K., Barnum, Michael P., and Bowen, David A. "KeyCorp Service Excellence Management System," *Interfaces* 26, no. 1 (January/February 1996): 54–74.

This is a very detailed article on customer service in retail banking and about how to achieve productivity gains in this environment. Describes the work done since 1991 at KeyCorp, which developed a service excellence management system (SEMS) to deal with productivity and service issues in 1,300 branches. SEMS measures activities such as customer wait time, teller proficiency, and productivity levels. These data help management select what processes to reengineer, schedule staff to match customer arrivals, and improve productivity. As a result, customer-processing time has shrunk by 53 percent, and customer wait time has improved sharply (only 4 percent wait more than five minutes, representing a 124 percent improvement). Personnel expenses are expected to drop by $98 million over the next five years because of SEMS, generating an internal rate of return for SEMS of 3,500 percent. This is a must-read article for all retail bankers.

Malhotra, Vikram. "Balancing Customer Needs in Retail Banking Distribution," *McKinsey Quarterly* 1 (1996): 180–82.

Argues that banks can use carrot-and-the-stick strategies to encourage customers to use low-cost distribution channels. For example, banks can offer free service through automated means but charge for involving a bank teller. Recommends that banks operate a portfolio of services and channels of distribution. This article is a wonderful example of ignoring customer wants and needs and simply focusing on internal requirements for cost control. It thus is an example of why this industry has such a terrible reputation with customers!

Strischek, Dev, and Cross, Rob. "Reengineering the Credit Approval Process," *Journal of Lending & Credit Risk Management* 78, no. 5 (January 1996): 19–28.

The reengineering process will allow management to understand better how they should perform credit checks and approvals to improve customer service. Describes the normal basic steps of reengineering, from understanding the current process through total redesign. Focuses on the value of looking at handoffs and decision points. This is an excellent article, demonstrating how reengineering is done using a specific process as an example.

Distribution (Wholesale/Retail)

Anonymous. "Getting Started with CED," *Chain Store Age* (January 1996): 15–19.

Describes how to implement consumer enhancement and development (CED), offering a scorecard to help assess a company's performance and identify opportunities for improving operations. This is an excellent description of the tool and its techniques.

Anonymous. "Stocking the Knowledge Warehouse," *Chain Store Age* (January 1996): 20–21.

This is a brief missive on the need for retailers to build a CED knowledge warehouse prior to performing any analysis. Gives advice on the level of detail required in such a database.

Miller, Paul. "1996 Benchmark Report on Operations," *Catalog Age* 13, no. 4 (April 1996): 76–86.

Presents results of *Catalog Age's* second annual benchmarking report on operations. Focus is on standards that can be measured. Uncovered that some catalogers are changing practices, such as staffing phone banks for longer hours. Nearly half (47 percent) claim not to have problems with forecasting inventory. Of those that say they did have problems, 61 percent of consumer mailers, 70 percent of business-to-business, and 15 percent of hybrids reported having trouble both making initial merchandise plans and managing inventory.

Mittal, Banwari, and Lassar, Walfried M. "The Role of Personalization in Service Encounters," *Journal of Retailing* 72, no. 1 (Spring 1996): 95–109.

Personalization is the social content of interaction between customers and retail employees. How this occurs is critical (e.g., the idea of "moments of truth"). The authors use SERVQUAL data to discuss how important customers consider their relations with a retailer and how they assess the quality of the service provided. Reports on 233 customers. Personalization is one of the most, if not the most, important influencer on customer judgments concerning quality of service provided.

Narus, James A. and Anderson, James C. "Rethinking Distribution: Adaptive Channels," *Harvard Business Review* 74, no. 4 (July/August 1996): 112–20.

Describes new ways companies are sharing resources to improve distribution. Also reflects current trends in the redesign of logistics systems in business. The article, although neither original nor novel, is a useful summary of widely used current processes.

Hotels

Allison, Jack H., and Byron, Mary Ann. "Aligning Quality Improvement with Strategic Goals at ANA Hotel San Francisco," *National Productivity Review* 15, no. 2 (Spring 1996): 89–99.

Customers are selecting hotels increasingly on the basis of quality and subjective relationships. To deliver what customers want, the ANA Hotel San Francisco implemented quality training programs to balance the needs of customers, profits, and growth plans. This article is a useful how-to case study from the hotel industry.

Bernstein, Charles. "Whatever Happened to Customer Service?" *Restaurants & Institutions* 106, no. 6 (March 1, 1996): 50.

This short article says that service is declining as companies try to cut costs and boost profits. Cites the example in the hotel business of charging a penalty fee to guests who check out of their rooms earlier than planned. Predicts this will backfire as customers go elsewhere for their lodging. Criticizes the airlines for similar practices, such as the Saturday-night layover or pay a $50 fine. Presents other examples of bad quality practices.

Breiter, Deborah, et al. "Bergstrom Hotels: A Case Study in Quality," *International Journal of Contemporary Hospitality Management* 7, no. 6 (1995): 14–18.

Describes how the application of TQM in a Wisconsin hotel chain since 1989 has evolved from a quality commitment into a continuous improvement management system. Describes widely used tools, such as teaming and database decision making.

Insurance

Anonymous. "TQM Slashes Liability Losses by 31 Percent, J&H Finds," *National Underwriter Property & Casualty* (February 26, 1996): 9.

Reports that professional liability losses in large engineering and architectural firms declined by 31 percent if they had TQM programs. This is based on a study of 200 large design firms. Those with risk management functions also experienced 23 percent less loss. The implications for insurers are obvious: Push TQM where it is not being practiced.

Utilities

Asaithambi, Durrairaj. "Re-engineering Collections at Houston Lighting & Power," *TMA Journal* 16, no. 1 (January/February 1996): 28–34.

In 1992, Houston consolidated customer service centers from twenty-two to fourteen then in 1994 reengineered the collections process. The article describes a classic example of reengineering and then reports results. Results: customer satisfaction went up, capital spending dropped to zero, company reported savings of $2.5 million per year with another $3 million possible later on.

Hall, Jacqueline, and Frederic, Marjorie. "Walking the Talk: Here's a Real Example," *Journal for Quality & Participation* 19, no. 1 (January/February 1996): 30–34.

The finance department at Eatery introduced process management into its operations. Beginning in 1993, the CFO, Jerry McInvale, introduced his employees to the concept of natural work teams. He used management training and retreats to get managers on board, too. He also used his company's quality department to create training and to review his operations periodically.

Wiley, Tim, and Newton, John. "Best Business Practices and Technology as Strategy for County Utility," *Water Engineering and Management* 143, no. 3 (March 1996): 27–29.

Discusses Pinellas County Utilities, which provides water, sewer, and solid-waste services to 600,000 residents in Clearwater, Florida. The company intends to provide outstanding customer service and thus applies customer satisfaction practices across the entire company. In addition, to facilitate the effort, the company is installing a local area network and wide area network infrastructure and is developing a GIS system. The county is also committed to the application of TQM, which helps.

Other Services and Small Businesses

Ahire, Sanjay L., and Golhar, Damodar Y. "Quality Management in Large vs. Small Firms," *Journal of Small Business Management* 34, no. 2 (April 1996): 1–13.

The findings are that TQM leads to better-quality products, there are no operational differences in TQM based on size of firm, and big and small firms implement TQM with the same effectiveness. Although small firms have fewer resources and less market clout, small firms can exploit strengths such as flexibility and innovation in implementing TQM.

Ghobadian, A., and Gallear, D. N. "Total Quality Management in SMEs," *Omega* 24, no. 1 (February 1996): 83–106.

This article focuses on small and medium-sized businesses, defending their importance in the economies of the world. Then the authors look at the role of TQM in this environment, reporting that SMEs have been slower to implement quality-management practices than large companies. Compares and contrasts TQM implementation in large and small organizations, relying on inductive research.

Haksever, Cegiz. "Total Quality Management in the Small Business Environment," *Business Horizons* 39, no. 2 (March/April 1996): 33–40.

Reviews how TQM is implemented in small businesses and shows differences with what occurs in large organizations. Two basic problems often present in small companies are the lack of business experience on the part of managers and owners and insufficient financial and human resources to implement TQM. Reports that the U.S. Small Business Administration (SBA) can help train and provide financing for implementation.

Struebing, Laura. "TQM Makes a Difference with Food Service Distributors," *Quality Progress* 29, no. 3 (March 1996): 17–18.

Presents new results for five food distributors initially studied in 1991. All continued to use TQM practices as of 1995 because they worked in improving service and business.

PUBLIC SECTOR
Federal, State, and Local Government

Anonymous. "Making the Big U Turn," *Quality Progress* 29, no. 3 (March 1996): 59–62.

Describes the reinvention program launched by the Clinton administration in 1993 for the U.S. government. The central theme was to improve service to the American public, that is, listen to what they wanted and then deliver that to the extent possible. The 1995 National Performance Review report presented customer service standards for 214 federal agencies. Standards were improving all over the government.

Johnston, Brent E. "Renewing Canada's Federal Public Service," *International Journal of Public Administration* 19, no. 1 (January 1996): 103–20.

This describes the Canadian government's own transformation, begun in 1989 and called Public Service 2000. It has failed so far to alter fundamental behavior patterns, however. The author suggests the use of TQM at the agency level and a major shift in strategy from PS 2000.

Mani, Bonnie G. "Measuring Productivity in Federal Agencies: Does Total Quality Management Make a Difference?" *American Review of Public Administration* 26, no. 1 (March 1996): 19–39.

Using the Internal Revenue Service as a case study, the author describes the application of TQM in the U.S. government. She presents data from the Federal Productivity Measurement System (FPMS) to report on results of some 2,500 output indicators. Then she presents an analysis, based on those data, of the effects of TQM on government productivity. This is an important article for federal managers to read.

Rago, William V. "Struggles in Transformation: A Study in TQM, Leadership, and Organizational Culture in a Government Agency," *Public Administration Review* 56, no. 3 (May/June 1996): 227–34.

This is a case study based on the Texas Department of Mental Health and Mental Retardation, which has applied TQM for the past four years. Managers faced the most difficult task of all because they had to change their thinking about management practices. Their barriers included purpose, coordination, communication, and empowerment within the context of public-sector management practices.

Slavik, Larry. "Refuse Department Benefits from TQM," *American City and County* 110, no. 11 (October 1995): 70.

Describes how the Department of Public Works at Lakewood, Ohio, implemented TQM to handle rising costs and the need to recycle and to improve productivity as staff retired. The results were positive and are described.

Health Care

Barber, Ned. *Quality Assessment for Healthcare.* New York: Quality Resources, 1996.

Health care is an industry that extensively uses both quality- and regulatory-driven assessments. Thus the industry has an extensive literature, published on a continuing basis, on how to do that. This short book continues that tradition of providing application tools and advice.

Beckham, J. Daniel. "The Most Important Day," *Healthcare Forum* 39, no. 3 (May/June 1996): 84–89.

Changes in competition and the need to reduce costs are causing the health care industry to examine its practices. Reducing expense and downsizing simply demoralizes employees. Beckham recommends that before you do that, take a day to think through what work you do from the perspective of the customer. Then he presents a methodology for going through that day. The process is very much a classic mapping exercise, but something not frequently described in an article, and is thus a useful piece to read. (Included in this edition of the yearbook.)

Boerstler, Heidi, et al. "Implementation of Total Quality Management: Conventional Wisdom Versus Reality," *Hospital and Health Services Administration* 41, no. 2 (Summer 1996): 143–59.

Hospitals around the United States are implementing quality-management programs. Presents the results of a study of ten hospitals, using site visits. This is a useful insight into the implementation experiences of various health-care facilities.

Carmen, James M., et al. "Keys for Successful Implementation of Total Quality Management in Hospitals," *Health Care Management Review* 21, no. 1 (Winter 1996): 48–60.

Presents findings on implementing TQM over a two-year period in ten hospitals. This is an important article for hospital administrators on best practices in implementation.

Castaneda-Mendez, Kicab. *Value-Based Cost Management for Healthcare: Linking Costs to Quality and Delivery.* New York: Quality Resources, 1996.

This is a short, useful, industry-specific guide to sound management practices. It should be useful to all who deal with the health-care industry, not just hospitals, and thus should be useful to insurance and government organizations as well.

Francis, Stuart D., and Alle, Patrick G. "A 'Patient Focus Review' of Surgical Services: Business Process Re-engineering in Health Care," *Business Process Re-engineering and Management Journal* 2, no. 1 (1996): 48–62.

Presents a case study from Waitemata Health Limited, in Takapuna, Auckland, New Zealand. Demonstrates the steps taken, the root cause analysis done, roles played, and the methodology used to change and improve. Also references secondary literature relevant to the project.

Gates, Judy L. "TQM: Pressure Ulcer Prevention," *Nursing Management* 27, no. 4 (April 1996): 48E–48H.

This is a case study from Maricopa Medical Center, which used a team to create a pressure ulcer prevention program. They followed a seven-step process: collected baseline data; created a TQM team; prepared a cause-and-effect diagram; surveyed; planned and proposed actions; presented their ideas; and summarized their thoughts.

Gropper, Elise I. "The Malcolm Baldrige Health Care Pilot," *Nursing Management* 27, no. 4 (April 1996): 56–58.

Goals of the pilot are to deliver better service to patients and other stakeholders, and to improve health care's overall organizational effectiveness and efficiency. Criteria cover issues such as patient-focused quality and value, leadership, improvement, organizational learning, staff participation and development, and fact-based management.

Hardy, Valerie Stafford, and Forrer, Joan. "A Comprehensive Quality Management Approach," *Nursing Management* 27, no. 1 (January 1996): 35–39.

Discusses value analysis and shared governance as they might be applied in the health industry, arguing that in tandem they would constitute a comprehensive quality program. Then the authors describe a purchasing process to demonstrate the benefits of such an approach.

Johnson, David W. "Take Two Classes and Call Me in the Morning: The Case for Training Wellness," *Hospital Materiel Management Quarterly* 17, no. 3 (February 1996): 21–28.

Talks about the value of a needs assessment and well-thought-out training programs as ways of improving training. This helps both organizations and individuals, critical in all industries, including the health industry. Provides a variety of recommendations on how to proceed.

Kohli, Rajiv, Kerns, Barbara, and Forgionne, Giusseppi A. "The Application of TQM in a Hospital's Casualty and Pathology Departments," *International Journal of Quality and Reliability Management* 12, no. 9 (1995): 57–75.

Accreditation and market pressures are finally forcing hospitals to apply TQM. The article describes how TQM principles can be applied in accident, emergency, and pathology operations, using a case study of a regional, medium-sized hospital. Describes benefits and lessons learned.

Magnusson, Paul, and Hammonds, Keith H. "Health Care: The Quest for Quality," *Business Week* no. 3470 (April 8, 1996): 104–6.

There is more to running a health-care facility than simply reducing costs of operation. The article announces that the industry is rediscovering quality. Big employers are driving the industry to adopt quality-management practices because when things are done right, costs go down. Employers also know that investing up front in preventive programs for chronic ailments (e.g., asthma and diabetes) reduces emergency room visits and other complications.

Motwani, Jaideep, Sower, Victor E., and Brashier, Leon W. "Implementing TQM in the Health Care Sector," *Health Care Management Review* 21, no. 1 (Winter 1996): 73–82.

Puts the implementation of quality management into four research streams. Then suggests an implementation model based on the literature of quality implementation. Ends the article with research questions related to the health-care industry that will have to be studied in the future.

Sloan, M. Daniel. *Using Designed Experiments to Shrink Health Care Costs.* Milwaukee: ASQC Quality Press, 1996.

Sloan provides detailed discussion of how to design experiments within the health-care community, complete with descriptions of the approach, exercises, tools, and case studies. The audience for this book includes doctors, nurses, and health-care administrators. The article includes discussions of both factorial and fractional factorial experimentation.

Weeks, Brenda, Helms, Marilyn M., and Ettkin, Lawrence P. "A Physical Examination of Health Care's Readiness for a Total Quality Management Program: A Case Study," *Hospital Materiel Management Quarterly* 17, no. 2 (November 1995): 68–74.

Examining perceptions concerning quality practices in a hospital is crucial to successfully implementating cultural and business transformation. Describes important assessment criteria and includes a comparison of management and employee perceptions in one hospital and what they taught management.

Zander, Karen (ed.). *Managing Outcomes through Collaborative Care: The Application of Caremapping and Case Management.* Milwaukee: ASQC Press, 1995.

Describes caremapping and how to apply it to foster collaborative care practices for patients. Experts on case management come together to write a practical guide. It is heavily process-based, using cases, forms, and strategies. Concludes with some analysis of how these activities will change over the next several years. It is a solid, street-level view of how to manage care for patients as a process.

Education (K–12)

Anonymous. "The TQM Route to the Top of the Class," *Journal of Management Development* 14, no. 9 (1995): 38–40.

This article is very critical of K–12 in its use of total quality management. Reports from the American Society for Quality Control suggest that only 165 U.S. school districts are applying TQM. Says students and teachers have to have a clear understanding of their common purposes (the education version of customer focus). For example, after reforming the grading system, students could suggest what teaching methods work best for them.

Leonard, James F. *The New Philosophy for K-12 Education: A Deming Framework for Transforming America's Schools.* Milwaukee: ASQC Quality Press, 1996.

This book translates Deming's philosophy into terms that make it possible for teachers and school administrators to implement. To transform it to this approach, the author demonstrates the application of a systems perspective and the use of statistical methods and leadership to create an environment of continuous improvement.

Manley, Robert, and Manley, John. "Sharing the Wealth: TQM Spreads from Business to Education," *Quality Progress* 29, no. 6 (June 1996): 51–55.

This is another case study demonstrating the value of quality-management practices in schools; this one is about the experience of West Babylon School District in New York. Like so many other successful cases, this one is based on Deming's fourteen points, which have clearly been proven to work in scores of schools. This article is simply more evidence of that fact. (Included in the "Education K–12" section of the yearbook.)

Sharpless, Kathleen A., Slusher, Michael, and Swain, Mike. "How TQM Can Work in Education," *Quality Progress* 29, no. 5 (May 1996): 75–78.

Describes the application of TQM principles in the Independent School District of Conroe, Texas, beginning in 1992. Four phases of the process are described: commitment, training, application and practice, and standardization and recognition. Lessons learned included the following: obtain commitment from management, develop a three-year implementation strategy, train and train, implement a communications process, and implement a support and recognition system based on TQM principles. This is an excellent, if brief, case study on K–12 implementation of TQM.

Thornett, Trevor, and Viggiani, Rosemary. "Quality in Education: Creating a Learning Society: The Pen y Dre Experience," *TQM Magazine* 8, no. 4 (1996): 29–35.

Tells the story of how a school in South Wales doubled its examination results by applying quality methods.

The school also won the Wales Quality Award for Education. Key message: Techniques used in quality management in business can be applied successfully in schools.

Higher Education

Daniel Seymour & Associates. *High Performing Colleges: The Malcolm Baldrige National Quality Award as a Framework for Improving Higher Education.* 2 vols. Maryville, Mo.: Prescott Publishing Co., 1996.

This is the first substantive publication on the implementation of Baldrige practices in higher education published after the U.S. government introduced a quality award for this industry. It is detailed, specific, and chock-full of implementation ideas.

Holmes, George, and McElwee, Gerard. "Total Quality Management in Higher Education: How to Approach Human Resource Management," *TQM Magazine* 7, no. 6 (1995): 5–10.

Says that some implementations of TQM in higher education are not correct and attempts to change that by arguing that there needs to be true professionalization, not just some soft thinking about collegiate culture. Focuses on the situation in British universities.

Turner, Ronald E. "TQM in the College Classroom," *Quality Progress* 28, no. 10 (October 1995): 105–8.

Describes the benefits and operations of TQM in higher education teaching. This is a very practical, sober discussion of how TQM can be applied in teaching; one of the best articles ever published on the theme. (Included in the "Higher Education" section of the yearbook.)

Nonprofit Associations

Anonymous. "Quality Service: Your Staff Holds the Key," *Nonprofit World* 14, no. 2 (March/April 1996): 36–39.

Opportunity Village, a nonprofit organization, worries about helping handicapped workers while providing customer service in productive ways. Their clients are now viewed as customers and are treated as such; service is the watchword. Uses word of mouth to recruit excellent employees as well and provides them training on skills and on the philosophy of commitment to clients.

Katz, Ray. "What Nonprofits Should Know about TQM," *Nonprofit World* 13, no. 5 (September/October 1995): 4–5.

Describes basic components of TQM then discusses the implementation of quality programs in nonprofit organizations. Four factors to take into consideration: focus on constancy of purpose, get the board involved early, drive fear out of the workplace, and select pilot projects very carefully.

QUALITY TRANSFORMATION
Leadership and Management

Farkas, Charles M., and De Backer, Philippe. *Maximum Leadership: The World's Leading CEOs Share Their Five Strategies for Success.* New York: Holt, 1996.

Not so much a book about the process of quality management, it is a collection of case studies and insights into topics of great concern to companies implementing quality-management practices. It proposes a methodology for leadership and success based on this research of cases.

Feigenbaum, Armand V. "Managing for Tomorrow's Competitiveness Today," *Journal for Quality and Participation* 19, no. 2 (March 1996): 10–17.

This is an article by one of the fathers of the modern quality-management world. Feigenbaum says the five most important aspects of quality leadership are the ability of senior managers to lead the quality effort; their ability to focus on organizational improvement goals; their ability to create and support a world of empowered employees; their ability to self-discipline in order to satisfy customers; and their ability to improve individual and group leadership across the organization. (Included in this edition of the yearbook.)

Grant, Robert M., and Cibin, Renato. "The Chief Executive as Change Agent," *Planning Review* 24, no. 1 (January/February 1996): 9–11.

Describes activities in the petroleum industry to drive up returns to shareholders by increasing efficiencies and responding to customer requirements and to market opportunities. Industry has moved away from an administrative management model and toward a model more characterized by entrepreneurialism. The modern CEO in this industry accepts the concept that the shareholder's requirement for value is the CEO's primary focus. Describes actions being taken to achieve value: application of total quality management, use of breakthrough teams in process and problem work, and major efforts to change corporate cultures.

Hale, Guy. *The Leader's Edge: Mastering the Five Skills of Breakthrough Thinking.* Burr Ridge, Ill.: Irwin Professional Publishing, 1996.

This short book is filled with ideas on how to improve one's leadership and thinking habits and actions. The author argues that there are five actions one can take and describes each using examples.

Kidder, Pamela J., and Ryan, Bobbie. "How to Survive in a Downsizing World," *Journal for Quality and Participation* 18, no. 7 (December 1995): 58–66.

Discusses the relationship between downsizing and quality-management practices. Describes trends related to both and the effects on organizations and workforce planning.

Kotter, John P. *Leading Change.* Cambridge, Mass.: Harvard Business School Press, 1996.

This highly regarded writer on leadership distills twenty-five years of experience and wisdom on his subject. The book shows that without leadership, reengineering, TQM, and such practices just don't work, but with leadership, greatness emerges.

Lindsay, William M., and Patrick, Joseph A. *Total Quality and Organization Development.* Delray Beach, Fla.: St. Lucie Press, 1996.

Discusses the effects of TQM on organizational development, covering such topics as TQM in organization development, customer satisfaction, continuous improvement, the role of facts, performance and people, and implementation strategies.

Morrison, Ann M. *Leadership Diversity in America.* San Francisco: Jossey-Bass, 1996.

This continues a long tradition at Jossey-Bass of publishing books on leadership, corporate change management, and personnel practices. This is a substantial volume and, although not strictly a book on quality-management leadership, is nonetheless an important addition to the growing body of material on management leadership in general.

Nanus, Burt. *Leading the Way to Organization Renewal.* Portland, Oreg.: Productivity Press, 1996.

Nanus is best known as an expert on the role of vision in corporate management. With this book, he continues his research and work on corporate renewal. His publications are always important, and this one is no exception.

Pojidaeff, Dimitri. "The Core Principles of Participative Management," *Journal for Quality and Participation* 18, no. 7 (December 1995): 44–47.

The key message is that if you don't apply quality and participative management, your company may go out of business. Companies that do not apply these practices lack competitive advantage in the private sector; moreover, not applying TQM is a primary reason for such high costs and bloated bureaucracies in government.

Pyzdek, Thomas. "The Pursuit of Happiness," *Quality Progress* 29, no. 5 (May 1996): 92–96.

Argues that market-based management is useful in improving a company's performance. Pyzdek's approach relies on dynamic, free-market economy circumstances instead of a command-and-control approach to drive activities. The approach is rooted in the ideas of Friedrich A. Hayek, who demonstrated in 1944 that market conditions can result in people applying their talents to the creation of value. Six systems are required to do this: mission statements, supportive values and culture, roles and responsibilities to define property rights, compensation and motivation as market incentives, creation and use of information as the free flow of ideas, and internal markets as price systems.

Rifkin, Glenn. "Leadership: Can It Be Learned?" *Forbes ASAP Supplement* (April 8, 1996): 100–103.

Declares that leadership training is very fashionable at the moment, making a comeback after the end of the recession in the United States. In 1995, 72 percent of all companies with over one hundred employees sent managers to leadership training provided by consultants, trainers, and professors in what is now a $15 billion executive-training market. The trend is to bring in people to develop made-to-order training programs. Cites examples.

Sims Jr., Henry P., and Manz, Charles C. *Company of Heroes: Unleashing the Power of Self Leadership.* New York: John Wiley, 1996.

This book deals with the role of individuals in providing leadership. The book contains case studies and examples and includes suggestions for what works and why. It continues a long-standing tradition in business publishing of producing books on leadership and improvement.

Yearout, Stephen L. "The Secrets of Improvement Driven Organizations," *Quality Progress* 29, no. 1 (January 1996): 51–56.

A recent survey discovered that organizations considered as high performing had common behaviors. These in turn may represent some best practices or indicators of how organizations may perform. They are cascading leadership; consistent, powerful customer focus; extensive alignment between organizational goals and people; a strong commitment to continuous improvement; several ways to measure improvement; excellent change management; and an environment nurturing innovation.

Cultural Transformation

Anonymous. "Characteristics of TQM Adopting Companies," *IIE Solutions* 28, no. 1 (January 1996): 13.

Reports briefly on a study by Paul Osterman of MIT's Sloan School of Management. The key finding is that the majority of U.S. companies that adopt total quality management practices are younger or usually internationally competitive divisions of large corporations. This finding is borne out by other studies and cases from the late 1980s and 1990s reported in other articles in previous issues of *The Quality Yearbook*.

Frank, Cap. "The Continuing Quest for Excellence," *Quality Progress* 28, no. 12 (December 1995): 67–70.

Describes attendance at the Quest for Excellence Conference as proof of the continuing interest in quality-management practices. Describes how the conference is a major source of information about Baldrige Award recipients.

Fredendall, Lawrence D., and Robbins, Tina L. "Modeling the Role of Total Quality Management in the Customer Focused Organization," *Journal of Managerial Issues* 7, no. 4 (Winter 1995): 403–19.

Says TQM has not been linked before to managerial theories, which is then done in this article. Also discusses how TQM enhances customer satisfaction and how employee empowerment improves relations with customers.

Kankus, Richard F., and Cavalier, Robert P. "Combating Organizationally Induced Helplessness," *Quality Progress* 28, no. 12 (December 1995): 89–90.

The helplessness comes out of organizations applying strategies from the 1970s and 1980s, a time before the advent of participative and TQM-based management styles. The key problem is the reluctance of management to empower employees. Makes general recommendations about what can be done to fix the problem.

Kekale, Tauno, and Kekale, Jouni. "A Mismatch of Cultures: A Pitfall of Implementing a Total Quality Approach," *International Journal of Quality and Reliability Management* 12, no. 9 (1995): 210–20.

Presents a model of different kinds of assumptions about people, quality management, and learning held by management and is followed with advice on how to use these assumptions to create quality-management approaches. The model is then tested against case studies to see how they play out.

Pfeffer, Jeffrey. "When It Comes to 'Best Practices' Why Do Smart Organizations Occasionally Do Dumb Things?" *Organizational Dynamics* 25, no. 1 (Summer 1996): 33–44.

Because barriers such as financial systems, social and political pressures, and hierarchical structures get in the way. The ways to knock down these kinds of barriers include visiting high-performance companies, studying benchmark data from successful firms, changing physical and production workplaces and systems, aligning rewards with outstanding performance, and demonstrating results through pilot projects.

Raiborn, Cecily, and Payne, Dinah. "TQM: Just What the Ethicist Ordered," *Journal of Business Ethics* 15, no. 9 (September 1996): 963–72.

Says many companies have developed corporate codes of ethics in the same period when quality-management practices began to expand. Argues that the two trends are closely related. In short, TQM represents good ethics.

Wright, J. Nevan. "Creating a Quality Culture," *Journal of General Management* 21, no. 3 (Spring 1996): 19–29.

Defines success factors for a mission statement: CEO must have a vision; mission statement must be specific and fit the culture of the organization; statement must be honest; input from employees increases ownership and must be communicated and be available. Cultural change takes planning, execution, and time.

Customer Focus

Brown, James R. "Resolving Patient Complaints: A Step-by-Step Guide to Effective Services Recovery," *Healthcare Financial Management* 50, no. 5 (May 1996): 64.

This short article is applicable to any industry. Although it is a review of a book by the same title written by Liz Osborne, the article does provide some of the lessons she advocates.

Colletti, Jerome A., and Wood, Wally. "Hold On!" *Across the Board* 33, no. 6 (June 1996): 27–31.

Discusses the high turnover in customers (about 15 percent for many firms) each year. There are many reasons for this turnover, but it is not inevitable. Customers change their needs, and companies should change to respond to those needs. The authors call for companies to measure the churn, understand its causes, and then respond to them as a way of becoming more effective.

Fojt, Martin. "Becoming a Customer Driven Organization," *Journal of Services Marketing* 9, no. 3 (1995): 7–8.

Forum Corporation research on customer relations indicated what customers wanted from salespeople. Sellers needed to know more about customer needs and wants than in the past, not merely about products.

Fojt, Martin. "Can a Company Be Both Low Cost and Service Oriented?" *Journal of Services Marketing* 9, no. 3 (1995): 13–14.

Competitive pressures are forcing companies to do both. The author provides a framework for anticipating service requirements and for creating the capability of delivering services profitably. The strategy has five components: explicitly prioritizing and targeting customers, creating differentiated commercial organizations, managing product offerings, adopting various yet differentiated service processes, and establishing a differentiated pricing policy. Although conceptual and very high level, this nonetheless is an article with food for thought.

Jacobson, John E. "What the Customer Wants: (1) Service, (2) Quality, (3) Price," *Iron Age New Steel* 12, no. 4 (April 1996): 52–56.

Presents results of a survey of 1,741 steel buyers. Key finding: service has the greatest impact on customer satisfaction; quality and price were lower on the list. Price was the least effective in improving customer satisfaction. Buyers of hot-rolled sheet and special bar quality bar are the most disappointed with their suppliers' services. Depending on what part of the business the buyer was in, satisfaction levels varied. For example, 59 percent of coated-sheet buyers were unhappy, but only 16 percent of those buying structurals were unhappy.

Johnson, Rose L., Tsiros, Michael, and Lancioni, Richard A. "Measuring Service Quality: A Systems Approach," *Journal of Services Marketing* 9, no. 5 (1995): 6–19.

Says there are problems with measuring service qualities today, which are described and then dealt with using a general systems theory approach. Argues the case for considering customer perception of service and the service process as well. In defense of the proposed framework, presents results of two studies about six types of services. Ends with recommendations for implementing a systems approach.

Jones, Christopher R. "Customer Satisfaction Assessment for 'Internal' Suppliers," *Management Services* 40, no. 2 (February 1996): 16–18.

Argues the case for customer focus as a driver for survival, competitiveness, and growth for any organization and also for its internal operations. Critical to this is customer satisfaction assessment for understanding their needs and wants. Presents a customer satisfaction assessment process.

Kessler, Sheila. *Measuring and Managing Customer Satisfaction: Going for the Gold.* Milwaukee: ASQC Quality Press, 1996.

The author describes a process to measure and manage customer satisfaction. It is a hands-on, how-to book, complete with tools for getting the job done. Short and to the point, Kessler focuses on planning and implementation as ways of improving the feedback process.

Kingman Brundage, Jane, George, William R., and Bowen, David E. "'Service Logic': Achieving Service System Integration," *International Journal of Service Industry Management* 6, no. 4 (1995): 20–39.

This is a detailed presentation of a service logic model to function as a managerial tool for going after cross-functional issues involving service systems. Focuses on three key service management issues: marketing, operations, and human resources. Each is described, then the authors argue that the real challenge is integrating all three to provide a positive service experience. They offer a step-by-step set of recommendations for accomplishing this integration.

Lengnick Hall, Cynthia A. "Customer Contributions to Quality: A Different View of the Customer Oriented Firm," *Academy of Management Review* 21, no. 3 (July 1996): 791–824.

To apply management practices for competitive advantage effectively, the firm must become customer focused. Yet many quality programs provide for little or no customer participation in the efforts. The roles customers normally play are described as applied in both manufacturing and service firms. Offers insights from organization theory, services marketing, strategic management, and TQM in an integrated manner to provide a conceptual model and ten propositions on the role of customers. This is a serious and important contribution to the subject of the role of the customer.

Muffatto, Moreno, and Panizzolo, Roberto. "A Process Based View for Customer Satisfaction," *International Journal of Quality and Reliability Management* 12, no. 9 (1995): 154–69.

The authors offer a framework for analyzing organizational processes needed to achieve high customer satisfaction. The processes looked at are planning, design, and monitoring. Demonstrates the value of looking at all three, not just monitoring.

Murray, Marty, and Hines, Jeffrey D. "Building Stable Customer Relationships That Stand the Test of Time," *Hospital Materiel Management Quarterly* 17, no. 4 (May 1996): 14–22.

Uses the Baldor Electric Company as a positive case study of how an organization can take the customer's perspective in its operations. In the 1990s, building on a heritage of quality practices initiated in the 1980s, the company moved a QI approach to an internally developed process for value improvement. The company kept asking cutomers what they valued and then worked to deliver on that. The company also developed an employee-taught training program for its people. Most of the article focuses on this training program.

Naumann, Earl, and Giel, Kathleen. *Customer Satisfaction Measurement and Management.* Milwaukee: ASQC Press, 1995.

This is the most detailed publication available on the subject. It describes how to implement such a process step-by-step and includes software to facilitate the process. Covers all forms of customer surveys by form, telephone, and focus group. Describes how to do this in large and small firms; an excellent source of information on this process.

Noe, Jeffrey. "Maintaining Customers during a Merger," *America's Community Banker* 5, no. 4 (April 1996): 19–20.

This is less of an article on banking than on good practices during a merger. However, because banks are experiencing many mergers in the mid-1990s, using several banks as case studies is instructive. Customers feared longer lines, poorer service, and higher costs when mergers were announced. The successful banks spent months planning and executing a customer communications process before the actual mergers occurred. These banks also implemented customer retention strategies. In one of the case studies, the customer base declined by only 12 percent, lower than is normal in this industry.

Paxton, Ken. "Corrective Action in the Real World," *Quality Progress* 29, no. 5 (May 1996): 184.

This brief article argues that a gap often exists between a customer's desire and the vendor's intent in fixing a problem. Regardless of the cause of the gap, it can be minimized by conducting training and giving customers guidelines on corrective actions so that they collect information that can be used by vendors to help solve problems.

Richman, Tom. "Service Industries: Why Customers Leave," *Harvard Business Review* 74, no. 1 (January/February 1996): 9–10.

Explores why customers switch from one service provider to another. The number-one reason was core service failures; second was service encounter failures (e.g., between employees and customers). Presents a research methodology to understand such issues.

Stevens, Tim. "Service with Soul," *Industry Week* 245, no. 3 (February 5, 1996): 29–30.

Quotes Tom Peters as saying that everyone today has good products; the real differentiator, therefore, is service. You do that by developing a winning team of employees, encouraging entrepreneurial spirit among them, cultivating trust among customers, and creating simple and effective service delivery systems. Peters says the key is the employee.

Weber, Jared. "Securing Loyal Customers," *Cellular Business* 12, no. 10 (October 1995): 88–90.

Argues that the cellular phone business faces severe problems with customer turnover. Describes a new customer relationship marketing strategy designed to retain customers. The company employed an organized telemarketing campaign to build customer loyalty. The article describes how this is done in a customer-centric way.

Woodruff, Robert B., and Gardial, Sarah F. *Know Your Customer: New Approaches to Understanding Customer Value and Satisfaction.* Milwaukee: ASQC Quality Press, 1996.

The authors focus on ways to measure customer value determination as a way to understand current and future customer value better, even basing satisfaction measurements on values. It is a detailed book on methodology. It is also an excellent introduction to the rapidly accepted notion of measuring customer perceived value.

Quality Implementation Strategies and Planning

Bergstrom, Robin Yale. "Looking for a Break in the Quality Drive? Don't," *Automotive Production* 108, no. 6 (June 1996): 26–27.

Juran Institute's Joe DeFeo says that there are many questions to ask about ISO and QS audit processes. Quality management is broad today, covering areas such as financial performance, strategy management, human resources, manufacturing process control and improvement, and supplier relations. DeFeo argues that the Baldrige criteria help companies more than ISO or QS because the latter do not provide total quality organizations, only ideas on quality control. This is a very useful short essay on trends in implementation.

Brelin, Harvey, Davenport, Kimberly, Jennings, Lyell, and Murphy, Paul. "Bringing Quality into Focus," *Security Management* 40, no. 2 (February 1996): 23–24.

Says organizations fail to implement TQM for three general reasons: senior executives do not get involved, believing benefits will occur automatically; some TQM teams lack focus, failing to link corporate objectives with daily activities; and some organizations may lack discipline to implement. TQM must be directed at improving processes that promise to yield the greatest benefits to the corporation. Key factors affecting success or failure include key processes, goals, measures, capability gaps, employee preparation, and rewards.

Cassidy, Michael P. "Streamlining TQM," *TQM Magazine* 8, no. 4 (1996): 24–28.

Announces that many companies have dropped their TQM programs, often because of changes in management, financial pressures, or the impact of bureaucracy. Cites Lucent Technologies' Power Systems business as an example of how the retreat from TQM can be avoided. The company changed management, its name (it was part of AT&T), and mission but found good reason to keep applying quality-management practices.

Gyani, Girdhar J. "Small Groups Bring Big Results," *Quality Progress* 28, no. 12 (December 1995): 73–75.

Uses a fictitious oil refinery to describe how TQM can be implemented by deploying small groups, infusing continuous improvement across the entire organization. Fascinating way to discuss implementation strategies!

Harte, H. G., and Dale, B. G. "Total Quality Management in Professional Services: An Examination, Part 2," *Managing Service Quality* 5, no. 5 (1995): 43–48.

Studies how eight organizations introduced TQM to their employees. Discoveries included that professionals do not initially understand TQM, think it is patronizing, and are so dispersed that they are hard to teach. To be successful, organizations must have a committed senior management team. It is crucial to have professionals and managers understand what it takes to introduce and implement TQM.

Hawley, John K. "Where's the Q in TQM?" *Quality Progress* 28, no. 10 (October 1995): 63–64.

Six aspects of organizational change influence the success of TQM: change can be made but difficult, imposed change is resisted, full cooperation and participation by management is crucial, change takes time, positive results come later not earlier, and change may go in unintended directions. To increase the odds of success, set realistic expectations; emphasize cost benefits from the start; recognize that people change slower than processes; begin with small projects; set objectives, not dictates, on how to achieve them; and understand your organization.

Higgins, James M. "Putting the Bang Back in Your TQM Program," *Journal for Quality and Participation* 18, no. 6 (October/November 1995): 40–45.

Linear thinking—the core of many quality approaches—can be enhanced and zipped up with horizontal-thinking practices to integrate creative thinking into quality practices. Provides a variety of techniques for doing this.

Hubiak, William A., and O'Donnell, Susan Jones. "Do Americans Have Their Minds Set Against TQM?" *National Productivity Review* 15, no. 3 (Summer 1996): 19–20.

Offers five mental models for why Americans fail to implement quality management in their organizations: individualism, competitiveness, problem-solving orientation, linear thinking, and control orientation. These are culturally determined and influence the behavior of managers and employees.

Litsikas, Mary. "Cliches Aren't Just Quality Rhetoric," *Quality* 35, no. 4 (April 1996): 70–76.

Says TQM must be ingrained in all employees to be successful. This was the lesson learned at Teradyne Inc. and at Analog Devices Inc. Senior executives in these two companies report that they needed to remain constantly involved. Change is probably coming from the outside, for example, from customers. Describes their experience in implementing TQM.

Masters, Robert J. "Overcoming the Barriers to TQM's Success," *Quality Progress* 29, no. 5 (May 1996): 53–55.

Describes nine common barriers: lack of management commitment, inability of an organization to change its culture, poor planning, absence of continuous training, incompatible organizations that have isolated people and departments, inadequate attention to customers, limited use of empowerment and teaming, poor measurement techniques, and lack of access to data and feedback on results. Discusses each point.

Poirier, Charles C., and Tokarz, Steven J. *Avoiding the Pitfalls of Total Quality.* Milwaukee: ASQC Quality Press, 1996.

The authors provide a series of strategies and examples of how to implement quality-management practices. They cover a broad range of issues such as strategy, organization, personnel practice, leadership, and culture. (An excerpt from this book is included in part 1 of the yearbook.)

Communication

Day, Abby, and Wills, Gordon. "When the Customer Is the Supplier," *Total Quality Review* 5, no. 4 (September/October 1995): 18–22.

Describes quality improvement done by MCB University Press. Includes how this organization reacted to the growing importance of the Internet in communications and publishing.

Exterbille, Kaat. "TQM Can Be DOA without a Proper Communications Plan," *Journal for Quality and Participation* 19, no. 2 (March 1996): 32–35.

TQM will not stick in an organization without a solid communications plan being implemented at the same time. It must address what the enterprise wants to get done, the role of all employees in the process, and what has to happen. Phases of a good plan include participation, awareness, adaptation, implementation, and adoption. (Included in the "Communication" section of the yearbook.)

Irwin, Nancy K. "Wire It In—Don't Role It Out," *Total Quality Review* 5, no. 4 (September/October 1995): 11–17.

Answers the question, How can communications be wired into other core company systems and work practices? Describes how this was done at Duke Power Company then offers six techniques all organizations can use.

Lawrence, John J. "Math Programming's Potential to Aid TQM Implementation," *Quality Progress* 29, no. 1 (January 1996): 76–80.

Describes how math programming can help the communication process, the only article published so far on this application. Math programming can help define how different groups can have common goals and be part of a company-wide communications effort. Although the article is not as forceful as it needs to be, it is a creative contribution to the subject of corporate communications.

Sharman, G. K. "Spreading the Message," *Total Quality Review* 5, no. 4 (September/October 1995): 27–30.

Communications professionals are rarely part of a company's communications process. This article describes what it takes to have public relations people in an organization apply quality-management practices and fold their activities into the mainstream of the firm.

Wilson, Jeanne M., and George, Jill A. "Communication Secrets of Effective Team Leaders," *Total Quality Review* 5, no. 4 (September/October 1995): 23–26.

In an empowered world, management needs to change how it communicates, shifting to a more inclusive communications strategy. Using stories to illustrate the key learning points, this article describes games one should not play and basic communications traps to avoid.

Young, Barbara A., and Heinrich, Charles A. "Communication: Cornerstone of Collaboration," *Total Quality Review* 5, no. 4 (September/October 1995): 7–10.

The authors argue that intentional communications and leadership provide the basis for quality results. They demonstrate how a workshop can bring employees and management together in a common vision, goals, and actions to be taken.

Training

Anonymous. "Training, Like Quality Improvement, Is Continuous," *Health Industry Today* (February 1, 1996).

This brief article features the work of Coulter Electronics, which has been training its employees since 1992 in customer-focused themes. Managers did the training for the company, particularly for customer service personnel. Presents the case for why using managers as trainers makes sense.

Crom, Steven, and France, Herbert. "Teamwork Brings Breakthrough Improvements in Quality and Climate," *Quality Progress* 29, no. 3 (March 1996): 39–42.

Presents the success of a British cardboard manufacturer in integrating its improvement strategies by using team training, structured problem-solving methods, team meetings, ongoing management support, and leadership. It is an excellent case study of how to integrate training into quality improvement strategies. Results were significant: scrap rates were reduced by over 70 percent, yielding a savings of $980,000 on sales of $40 million. Teams learned to have confidence in their capabilities, trust their managers, and create opportunities to improve company operations.

Friedberg, Alan H., Milici, Paul, and Krouskos, Steven J. "Benchmark Survey: Internal Auditor Training," *Internal Auditing* 11, no. 4 (Spring 1996): 54–59.

Cites the example of the audit department at W.R. Grace & Company, which conducted a benchmark study to determine new training requirements for its auditors. One of the things the study looked at was how much staffing was required. The average for other companies was $292 million in sales for each internal auditor. On average it cost fifty-five hours and $3,500 to train an auditor. It was not clear if this was an annual or one-time cost.

Goodman, Paul S., and Darr, Eric D. "Exchanging Best Practices through Computer Aided Systems," *Academy of Management Executives* 10, no. 2 (May 1996): 7–19.

Says well-run companies can diffuse best practices very quickly and then argues that computer-aided systems can be the first step in that effort. Describes four problems that prevent more companies from doing that, such as inherent motivational problems in system use and complexity of demonstrating how such systems can help the performance of an organization. The authors then provide ways of getting past these dilemmas in order to increase the use of best practices using information technology. This article is about training, CAI, problem solving, and quality-management practices.

Gunter, Berton H. *Making Training Work: How to Achieve Bottom-Line Results and Lasting Success.* Milwaukee: ASQC Quality Press, 1996.

Describes why most training programs fail, then details a strategy to avoid these problems. Gunter focuses on results-generating training programs, (for example, using QFD techniques to remain customer focused) and provides many nuts-and-bolts suggestions and examples.

Hitchcock, Darcy. "Learning from Chaos," *Journal for Quality & Participation* 19, no. 1 (January/February 1996): 42–45.

Reports the result of a survey of about 100 organizations across many industries conducted between December 1993 and April 1996 about management development activities. An important finding was that there was no correlation between use and effectiveness. Second, some of the most effective approaches, such as providing a consultant on demand were never used.

Lengermann, Paul Adrian. "The Benefits and Costs of Training: A Comparison of Formal Company Training, Vendor Training, Outside Seminars, and School Based Training," *Human Resource Management* 35, no. 3 (Fall 1996): 361–81.

Based on a large sample of U.S. adults aged 25–35, this article spends a great deal of time arguing that they do not receive adequate amounts of training and why. The benefits of training are explained.

Prior-Smith, Karen, and Perrin, Mary. "Ideas on Motivating People, Addressing Complaints and Training (IMPACT): An Application of Benchmarking, Learning Best Practice from Hewlett-Packard," *Business Process Re-engineering and Management Journal* 2, no. 1 (1996): 7–25.

H-P is an excellent practitioner of benchmarking. This article demonstrates its use in an attempt to improve training and productivity of employees. Describes the approach taken in detail, illustrates the actions with the forms and charts, and describes the complaint and training processes in place; an excellent piece.

Theibert, Philip R. "Train and Degree Them, Anywhere," *Personnel Journal* 75, no. 2 (February 1996): 28–30.

Because college graduates are increasingly not coming to work with the skills businesses need, some corporations are working with local universities to develop training programs that are taught to employees for college credits. The article cites many examples, and in detail that of the Caribbean Hotel Management Services.

Around the World

Akan, Perran. "Dimensions of Service Quality: A Study in Istanbul," *Managing Service Quality* 5, no. 6 (1995): 39–43.

Describes quality practices in the hotel industry in Istanbul. Reports that the service of four- and five-star hotels is fine but not personalized.

Barad, Miryam. "Some Cultural/Geographical Styles in Quality Strategies and Quality Costs (P.R. China versus Australia)," *International Journal of Production Economics* 41, nos. 1–3 (October 1995): 81–92.

Looks at quality practices in these two different cultures. Some common traits between Japan and China were evident, but there were many differences between Chinese and Australian companies. Differences involved why quality programs were initiated, objectives for empowerment, scope of work, and the role of quality control circles and cost reporting.

Devos, Johan F., Guerrero-Cusumano, Jose L., and Selen, Willem J. "ISO 9000 in the Low Countries: Reaching for New Heights?" *Business Process Reengineering and Management Journal* 2, no. 1 (1996): 26–47.

Offers a useful overview of the history and current application of ISO 9000 in Europe before looking at the Low Countries. Demonstrates by industry what is going on in significant detail. Also provides citations of relevant literature.

Evans, Patricia, and Bellamy, Sheila. "Performance Evaluation in the Australian Public Sector: The Role of Management and Cost Accounting Control Systems," *International Journal of Public Sector Management* 8, no. 6 (1995): 30–38.

In the last several years, creative leadership in business practices has increasingly been emerging out of Australia. The subject of this article is additional proof of the business vitality in that part of the Pacific. The authors discuss the implementation of total quality management practices in the Australian public sector. Management accounting and planning practices can be very useful within a public-sector environment, say the authors, including the application of costing systems—in effect arguing a case for management practices that exceed what American public-sector agencies have been willing to adopt.

Ghosh, B. C., and Hua, Wee Han. "TQM in Practice: A Survey of Singapore's Manufacturing Companies on Their TQM Practices and Objectives," *The TQM Magazine* 8, no. 2 (1996): 52–54.

Presents results of a survey of twenty-seven companies coming from various industries and nationalities of management teams, representing about 5 percent of the city's manufacturing labor force. Reports TQM is well established, particularly among medium to large companies, but not across all functions.

Harrington, H. James. "National Traits in TQM Principles and Practices," *TQM Magazine* 8, no. 4 (1996): 49–54.

Harrington, a highly regarded expert on process and business transformation, argues in this important article that quality-management practices vary by country. He provides data concerning Japan and Germany, for example, based on research done by Ernst & Young and the American Quality Foundation. Also presents insights on the concept of best practices as applied internally. (See original article by Harrington as introduction to the yearbook.)

Koo, L. C. "The Practices of Quality Circles in Hong Kong," *Asia Pacific Journal of Quality Management* 4, no. 4 (1995): 17–32.

Offers a history of quality circles in Hong Kong. Argues that businesses there need significant improvements in training in practical tools, management concepts, and quality. Offerings of the Hong Kong Quality Management Associates are also presented.

Macedo-Soares, T. Diana L. v. A. de, and Lucas, Debora C. "Key Quality Management Practices of Leading Firms in Brazil: Findings of a Pilot-Study," *TQM Magazine* 8, no. 4 (1996): 55–70.

This is a major study on management practices of Brazilian firms—considered by many authorities to be some of the best managed in Latin America—and compares their practices with U.S. firms. Identifies sociocultural factors as the root cause of differences and recommends variations that take into account cultural differences within nations.

QUALITY TOOLS AND TECHNIQUES
Concepts

Anderson, John C., et al. "A Path Analytic Model of a Theory of Quality Management Underlying the Deming Management Method: Preliminary Empirical Findings," *Decision Sciences* 26, no. 5 (September/October 1995): 637–58.

Presents a theory of quality management underlying Deming's, operationalizing his ideas with measurement statements devised by the World Class Manufacturing research project team located at the University of Minnesota and at Iowa State University. They apply path analysis to explore the empirical strength of relationships advanced in their theory. They suggest that their data indicate new relationships and confirm existing ones.

ASQC Statistics Division. *Glossary and Tables for Statistical Quality Control, 3rd Edition.* Milwaukee: ASQC Quality Press, 1996.

This is a standard technical reference work used by many quality control experts in business and science. This edition contains new terms and corrections from the second edition and remains a short, useful introduction.

Crosby, Philip B. "Illusions about Quality," *Across the Board* 33, no. 6 (June 1996): 38–41.

Any article by Crosby, a pioneer in quality management, is worth reading. In this piece, he attacks the notion of management that quality can be implemented by systems such as TQM, Baldrige, or ISO 9000. He identifies five false assumptions widely in vogue: belief that quality means goodness, luxury, shininess, or weight; belief that quality is intangible, thus unmeasurable; belief that quality has its own economics; belief that quality problems originate with employees; and belief that quality starts in the quality department.

Dettmer, H. William. *Goldratt's Theory of Constraints: A Systems Approach to Continuous Improvement.* Milwaukee: ASQC Quality Press, 1996.

The author discusses Eliyahu M. Goldratt's (author of *The Goal*) theory of constraints, providing methods and cases of how to apply these analytics. Dettmer's book provides several chapters, for example, on the use of trees in analysis, and exercises for PRT, FRT, CRD, and CRT. Includes a glossary of terms.

Gomes, Helio. *Quality Quotes.* Milwaukee: ASQC Quality Press, 1996.

This is a collection of 1,500 quotes on various quality topics. They are organized by topic.

Hoare, Clive E. "Appreciation for a System," *TQM Magazine* 8, no. 4 (1996): 20–23.

This is a very useful introduction to basic systems theory, applied to a funny scenario. The author applies Deming's notion that you have to appreciate a system as an integral part of achieving profound knowledge. Ends with a challenge to the readers to apply the perspective.

Jacques, March Laree. "Fifty Years of Quality: An Anniversary Retrospective," *TQM Magazine* 8, no. 4 (1996): 5–16.

Looks at the past and future of quality management, using observations made by senior American experts and leaders in quality on the occasion of the fiftieth anniversary of the ASQC. Shows that quality management has a rich history and an optimistic future.

Kim, Kee Young, and Chang, Dae Ryun. "Global Quality Management: A Research Focus," *Decision Sciences* 26, no. 5 (September/October 1995): 561–68.

Defines global quality management (GQM) as the strategic planning and integration of products and processes to achieve high customer acceptance and low organizational dysfunctionality across country markets. Says GQM and TQM are similar, except that GQM is more global. In GQM, customer sets are wider thus making it more difficult to implement customer-focused strategies.

Lakhe, R. R., and Mohanty, R. P. "Understanding TQM in Service Systems," *International Journal of Quality and Reliability Management* 12, no. 9 (1995): 139–53.

Explains TQM in service systems. Attempts to synthesize various implementation strategies and theoretical constructs of the issue, ending with a case study of what was done at one bank. Describes a conceptual model for measuring the effectiveness of TQM in a service system.

Leach, Lawrence P. "TQM, Reengineering, and the Edge of Chaos," *Quality Progress* 29, no. 2 (February 1996): 85–90.

The science of complex systems offers insight into organizational behavior. Whereas TQM provides those elements needed to grow and change a business, reengineering makes it possible to design and build a business. TQM works because it enables a business to be adaptable and dynamic. TQM, the author suggests, will be more successful than reengineering because TQM is more comprehensive. Suggests managers view themselves as systems designers.

Reed, Richard, Lemak, David J., and Montgomery, Joseph C. "Beyond Process: TQM Content and Firm Performance," *Academy of Management Review* 21, no. 1 (January 1996): 173–202.

Argues that the literature on TQM has focused more on process than on content. Looks at TQM is relation to the orientation of companies and their market advantages, efficiency of product design, efficiency of processes, and product reliability. This is a theoretical piece attempting to put TQM into the context of traditional management concerns about firm successes.

Seawright, Kristie W., and Young, Scott T. "A Quality Definition Continuum," *Interfaces* 26, no. 3 (May/June 1996): 107–13.

Says there are many definitions of quality management. Presents a variety of definitions across a spectrum, illustrating the relationships of these definitions to one other. Provides some comments on the impact of one over the other. Offers the continuum as a way of understanding basic concepts of quality management.

Stuart, Michael, Mullins, Eamonn, and Drew, Eileen. "Statistical Quality Control and Improvement," *European Journal of Operational Research* 88, no. 2 (January 20, 1996): 203–14.

Describes the contributions made by statistical analysis to the successes of quality management. Discusses SPC and industrial experimentation in detail.

Watson, John G., and Korukonda, Appa Rao. "The TQM Jungle: A Dialectical Analysis," *International Journal of Quality and Reliability Management* 12, no. 9 (1995): 100–109.

Positions TQM in relation to established conceptual models in management theory. If not done, TQM will disappear. This is a highly theoretical piece.

Benchmarking

Andersen, Bjorn, and Camp, Robert C. "Current Position and Future Development of Benchmarking," *TQM Magazine* 7, no. 5 (1995): 21–25.

Using a survey on the future of benchmarking, the authors report that users see benchmarking as an important tool for transferring best practices to their own organizations. Most would develop formal benchmarking programs in the future. Computing in benchmarking is also expanding, although more traditional methods are preferred today, for example, direct contact with other companies.

Callahan, W. Terrence. "Improving Performance through Benchmarking," *Business Credit* 98, no. 1 (January 1996): 42.

This short article discusses benchmarking done by the Credit Research Foundation. It describes benchmarking credit and accounts receivable processes but does not report results.

Fitzgerald, Patrick. "Benchmarking Pays Off," *Chemical Marketing Reporter* 249, no. 15 (April 8, 1996): 16–17.

Fitzgerald argues the case for using benchmarking, with the chemical industry as his target audience. He also reviews the work in benchmarking of the American Productivity and Quality Center, which now has over 350 member organizations. Cites examples of effective use.

Hiebeler, Robert J. "Benchmarking: Knowledge Management," *Strategy and Leadership* 24, no. 2 (March/April 1996): 22–29.

Focuses on the importance of managing institutional knowledge as a competitive tool. Explains the advantages of leveraging intellectual capital then describes the Knowledge Management Assessment Tool used by the American Productivity and Quality Center. This tool helps organizations analyze how effective employees manage the knowledge process and compares that to the activities of other firms. Argues that companies are just starting to manage knowledge.

Morgan, Jim "Benchmarking Is Not an Instant Hit," *Purchasing* 120, no. 8 (May 23, 1996): 42–44.

Says many U.S. companies do not like benchmarking despite the strong case made for the use of this tool. A survey of 500 purchasing agents was the basis for that comment. Problems these people pointed out were lack of resources, lack of senior management commitment, lack of suitable partners, worries about confidentiality, lack of understanding, and fear of "getting wet." (Included in the benchmarking section of the yearbook.)

Roberts, Sally. "Fresh Approach to Benchmarking," *Business Insurance* 30, no. 19 (May 6, 1996): 3, 12.

Describes efforts by the Risk and Insurance Management Society's chapter in Pittsburgh, Pennsylvania, to identify best practices and benchmarks for various risk management activities. The chapter generated five reports: safety program assessments, broker selection process, how to choose a risk management information system, cost allocation, and communicating the risk management message.

Rosengard, Jeffrey S. "Benchmarking: Springboard for Change," *TMA Journal* 16, no. 2 (March/April 1996): 29–32.

Defines benchmarking as used by Westinghouse Electric Corporation and Cargill, Inc. Westinghouse, for example, uses benchmarking as a way to find best practices and to inform its improvement strategy. Cites the example of how its treasury department was reengineering and reorganized using benchmarking as a guide. Cargill uses benchmarking, run internally but validated externally, to track progress on a monthly basis. Cargill is interesting because it is doing global reengineering projects.

Reengineering and Improving Processes

Cook, Sarah. *Process Improvement: A Handbook for Managers.* Brookfield, Vt.: Gower Publishing, 1996.

This is a very short book on the subject, listing basic steps for a continuous improvement strategy, joining a long list of books previously published on the subject. Nothing extraordinary is presented in this volume, however.

Keats, J. Bert, and Montgomery, Douglas C. (eds.) *Statistical Applications in Process Control.* New York: Marcel Dekker, 1996.

This is a large collection of case studies and detailed examples of applying statistical process control techniques. It is filled with useful examples.

Kelada, Joseph N. *Integrating Reengineering with Total Quality.* Milwaukee: ASQC Quality Press, 1996.

This book blends discussion about total quality and reengineering, offering an integrated management approach. It is one of the first books to span quality-management practices and classical reengineering in a detailed manner.

Kleiner, Brian M., and Hertweck, Bryan. "By Which Method?: Total Quality Management, Reengineering, or Deengineering," *Engineering Management Journal* 8, no. 2 (June 1996): 13–18.

Provides a tight discussion and evaluation of the differences and similarities between all three approaches. Compares them to the thinking of Deming, Hammer, and Wheatley.

Martinsons, M. G. "Radical Process Innovation Using Information Technology: The Theory, the Practice and the Future of Reengineering," *International Journal of Information Management* 15, no. 4 (August 1995): 253–69.

Overviews business process reengineering then discusses how IT can facilitate rapid and radical change. Offers an example from a Hong Kong bank and a review of the relevant literature on the topic. Ends with a forecast of what can be expected.

McAdam, Rodney. "An Integrated Business Improvement Methodology to Refocus Business Improvement Efforts," *Business Process Re-engineering and Management Journal* 2, no. 1 (1996): 63–71.

This is a case study of reengineering as done at Short Bros PLC, a large industrial employer in Northern Ireland. Provides a detailed description of corporate change, business reengineering, and implementation strategies. Includes citations.

Measurements

Anonymous. "Measuring Efficiency and Effectiveness," *TMA Journal* 16, no. 2 (March/April 1996): 31.

Looks at the cost of processing a single accounts payable invoice as a snapshot of a company's processes when compared to those of the best. Uses Hackett Group's benchmark performance data, suggesting this approach for treasurers who want to do a quick self-diagnostic "sniff test" of their own operations' efficiency and effectiveness.

Capon, N., Kaye, M. M., and Wood, M. "Measuring the Success of a TQM Programme," *International Journal of Quality and Reliability Management* 12, no. 8 (1995): 8–22.

Using the example of Colt International, the authors describe the comprehensive measures put in place to define the TQM successes in this corporation. They applied the Baldrige framework, covered the activities of thirty-seven teams, and reported results of 10 percent to 15 percent improvement per month. Urges use of quantitative measures of performance.

Cortada, James W. *Performance Measurement Strategy.* 60-minute video. Syosset, N.Y.: Computer Channel, 1995.

Explains what comprehensive business measures look like today, how to implement these new measures, and how to sustain them. The video combines a presentation and a question-and-answer discussion and provides examples.

James, Jack. "Plant Efficiency: What Gets Measured, Gets Improved," *Metal Center News* 36, no. 6 (May 1996): 94–98.

Argues that productivity measurements begin with sampling work and do not require complex processes to get started. But measurement does require the commitment of time and resources. The author says that all activities should eventually be tracked, including warehouse operations. Says most service center managers know which operations are weak. In addition to work sampling, managers can benchmark best practices, implement a suggestion process, develop a series of motivational programs, and use consultants.

Shepherd, C. David, and Helms, Marilyn M. "TQM Measures: Reliability and Validity Issues," *Industrial Management* 37, no. 4 (July/August 1995): 16–21.

Describes three recognized ways to determine the reliability of a measure: test/retest, alternate form, and internal consistency. Reliability suggests precision in measurements but not validity or degree to which one is measuring what one wants to measure. Validity is measured by four types of techniques: face validity, content validity, criterion validity, and construct validity.

Tatikonda, Lakshmi U., and Tatikonda, J. "Measuring and Reporting the Cost of Quality," *Production & Inventory Management Journal* 37, no. 2 (Second Quarter 1996): 1–7.

Many companies fail to realize the benefits of TQM because, in part, of the lack of relevant productivity measures of the cost of quality. When they do exist, the cost of quality is frequently understated by a wide margin. Explores how the cost of quality impacts profits and expenditures.

Tools

Aiken, Milam, Hasan, Bassam, and Mahesh, Vanjani. "Total Quality Management: A GDSS Approach," *Information Systems Management* 13, no. 1 (Winter 196): 73–75.

Describes how group decision support systems are used to increase the productivity and effectiveness of teams. Describes the three most popular packages: GroupSystems, TeamFocus, and VisionQuest. Details what tasks are performed, for example, electronic brainstorming, idea organizing, ranking and rating, anonymity, parallel communication, and automated record keeping. Demonstrates that users reduce meeting times by 91 percent and labor costs by 71 percent. Attendees are also more satisfied with the quality of the meetings actually held.

American Society for Quality Control. *ASQC Quality Press Publications Catalog.* Milwaukee: ASQC Quality Press, 1996.

Published periodically during the year, this catalog has nearly 100 pages of listings of books and videos published by ASQC and other presses on quality topics. The catalog, unfortunately, is selective, emphasizing its products over those of other publishers, and is incomplete. Nonetheless, it is a major reference on current publications on the subject of quality.

American Society for Quality Control. *Quality Progress Collection.* Milwaukee: ASQC Quality Press, 1996.

This is a CD-ROM collection of all the articles published in *Quality Progress* from January 1990 through June 1995, some 4,000 pages of material. It has search and index capabilities, comes in both IBM-compatible and Macintosh-Apple formats; a major work of reference.

American Society for Quality Control Statistics Division. *Glossary and Tables for Statistical Quality Control, Third Division.* Milwaukee: ASQC Quality Press, 1996.

Quality engineers have been using earlier editions of this reference work since the mid-1970s. It is organized like earlier editions as a one-volume reference work.

Anjard, R. P. "Reengineering Basics: One of Three, New, Special Tools for Management, Quality and All Other Professionals," *Microelectronics Reliability* 36, no. 2 (February 1996): 213–22.

Describes reengineering and its features. This is a relatively basic introduction to the topic; nothing new in this presentation.

ASQC Energy and Environmental Division. *Definitions of Environmental Quality Assurance Terms.* Milwaukee: ASQC Quality Press, 1996.

This is a technical reference work, one of many published by the ASQC by industry to help quality control professionals. This is a thirty-four-page booklet.

Chang, Fred S., and Wiebe, Henry A. "The Ideal Culture Profile for Total Quality Management: A Competing Values Perspective," *Engineering Management Journal* 8, no. 2 (June 1996): 19–26.

There is great interest in organizational culture as an explanatory variable in TQM implementation. However, there are few effective assessment tools. The authors provide one to assess the firm's cultural characteristics, which they call the competing value framework of culture. They focus on the role of TQM within a culture through questions asked and reliance on secondary literature.

Clauson, Jim. "Relaying QFD Messages Around the World," *Quality Progress* 29, no. 5 (May 1996): 137–48.

Looks at QFD L mailing list, which runs on Majordomo's mailing list software. Describes the list's origins in Japan and how it can be used to incorporate information about needs and wants of customers into the design, manufacture, delivery, and service of products and services. Several information-gathering techniques, decision-making processes, and quantitative engineering practices are required. The mailing list can be used to gather information, serve as a forum, and link together customers and vendors. This list is managed by the Quality Function Deployment Institute (QFDI) and can be used by any company. A fascinating article about a new tool. (Clauson, by the way, is moderator of the Deming discussion group on the Internet.)

Cohen, Lou. *Quality Function Deployment: How to Make QFD Work for You.* Milwaukee: ASQC Press, 1995.

This is a detailed, comprehensive look at this important tool: what it is, how it is used, where it fits in an organization, effects, planning, and how it relates to the voice of the customer. This may be the best book published so far on the topic.

Griffith, Gary K. *Statistical Process Control Methods for Long and Short Runs, Second Edition.* Milwaukee: ASQC Quality Press, 1996.

This is a step-by-step guide for SPC. This edition adds material on control charts, capability analysis, and statistical problem solving and includes exercises with answers. This is an excellent one-volume reference work on the subject.

Harbour, Jerry L. *Cycle Time Reduction: Designing and Streamlining Work for High Performance.* New York: Quality Resources, 1996.

This is a short how-to book on improving and simplifying processes, complete with construction of measurements of performance.

He, Z., et al. "Fourteen Japanese Quality Tools in Software Process Improvement," *TQM Magazine* 8, no. 4 (1996): 40–44.

Argues that the seven old tools for measuring quality can be applied to about 95 percent of all situations related to quality. Then introduces seven new tools that are being received well in Japan. Offers a software improvement framework that incorporates the SEI capability maturity model as a self-assessment tool and maps the fourteen tools to this.

Lambrecht, Marc R., Chen, Shaoxiang, and Vandaele, Nico J. "A Lot Sizing Model with Queueing Delays: The Issue of Safety Time," *European Journal of Operational Research* 89, no. 2 (March 8, 1996): 269–76.

Introduces the idea of safety time in a "make to order" production environment. The production facility is the queuing model, including setup time. Offers an approach to quantify the safety time and to compute associated service levels based on the queuing delay. The main result is a convex relationship of expected wait time, with the variance of the wait time and quoted lead time as a function of the lot size with a concave relationship of the service level also as a function of lot size. Continuous arrival of work is assumed in the model, not just batch arrivals. To do that required deriving a new closed-form analytical expression for expected wait time.

Williams, Paul B. *Getting a Project Done on Time: Managing People, Time, and Results.* New York: AMACOM, 1996.

This is a short introduction to basic, well-known concepts of project management.

Teams and Teamwork

American Society for Quality Control. *Teams in America.* Milwaukee: ASQC Quality Press, 1996.

This is a ten-minute video, an outgrowth of the American Team Archetype research project sponsored by the ASQC, Eastman Kodak, GM, Kellogg Company, and Walt Disney Company. The video describes teaming practices in the United States.

Anonymous. "Team Incentives Prominent Among 'Best Practices' Companies," *Quality* 35, no. 4 (April 1996): 20.

Presents results of a survey by Hewitt Associates of eighty-six "best practices" firms, in size from 220 to over 100,000 employees concerning incentive plans. Forty-five percent have team incentives for salaried employees in work teams, and 36 percent for hourly employees.

Carson, Paula Phillip, et al. "Power in Organizations: A Look through the TQM Lens," *Quality Progress* 28, no. 11 (November 1995): 73–78.

The authors identify five social power bases at work in any organization functioning with empowered employees: coercive power, legitimate power, reward power, referent power, and expert power. Each results in different outcomes and varies in its attractiveness as a tool depending on what outcomes are desired.

Fisher, Bob, and Thomas, Bo. *Real Dream Teams: Seven Practices Used by World-Class Team Leaders to Achieve Extraordinary Results.* Delray Beach, Fla.: St. Lucie Press, 1996.

Most team books focus on teams, but this one takes the same material and applies it to team leaders. At under 200 pages, it is short and to the point, a useful introduction to the topic.

Harrington-Macklin, Deborah. *Keeping the Team Going: A Tool Kit to Renew and Refuel Your Workplace Teams.* New York: AMACOM, 1996.

The title captures the essence of this book. It is a how-to-do-it collection of techniques and methodologies. It is easy to use and relatively basic—only 240 pages. It is one of the more useful publications to appear recently on teams, reflecting an emerging set of best practices in this area.

Ittner, Christopher D., and Larcker, David F. "Total Quality Management and the Choice of Information and Reward Systems," *Journal of Accounting Research* 33 (Supplement) 1995: 1–34.

This article contains much useful information based on a study that looked at the relationship between the use of advanced manufacturing practices and the choice of information and reward systems, and their impact on organizational performance. Findings provided partial support for the notion that organizational performance is a function of the interaction between adopting quality practices and using nontraditional information and reward systems.

Kerr, Daryl L. "Team Building and TQM: An Experiential Exercise for Business Communication Students," *Business Communications Quarterly* 58, no. 1 (March 1995): 47–48.

Teams are becoming the most widely adopted new building blocks of organizations; TQM is becoming part of that new approach. Describes an exercise that links TQM and team building.

Parker, Glenn M. *Team Players and Teamwork.* San Francisco: Jossey-Bass, 1996.

Although there are many books on the role of teams, this one is a useful, up-to-date discussion of the topic. Like other publications from Jossey-Bass, this book is well written and of high quality.

Phillips, Sandra N. "Team Training Puts Fizz in Coke Plant's Future," *Personnel Journal* 75, no. 1 (January 196): 87–92.

Cites the case of Coca-Cola Fountain Manufacturing Baltimore Syrup Operations' turnaround through team training programs. Over the past five years, senior management fostered employee involvement and invested in employees. The plant is now performing in an effective manner.

Rayner, Steven R. *Team Traps: Survival Stories and Lessons from Team Disasters, Near-Misses, Mishaps, and Other Near-Death Experiences.* New York: Wiley Publishers, 1996.

Despite the attempt by this publisher to provide a cute title, this is a book about best practices in teaming and gives many examples of problems and how to fix them. After a decade of many organizations working with teams, we are now beginning to see publications emerge based on actual experience rather than just theory. This book is one example of these new publications; it reads very well, too.

Weekley, Thomas L., and Wilber, Jay. *United We Stand: The Unprecedented Story of the GM-UAW Quality Partnership.* New York: McGraw-Hill, 1996.

This is a lengthy antidote to a large body of literature that argues that unions in the United States and in Europe are anti-change and anti-quality. This book, on the contrary, demonstrates specific examples of the opposite at work with lessons learned. The book is a long but solid testimonial to the role unions can play in revitalizing companies, in this case, General Motors.

FUNCTIONAL PROCESSES
Planning

Cortada, James W. "Do You Take This Partner?" *Total Quality Review* 5, no. 5 (November/December 1995): 10–14.

Describes what partnerships are, the characteristics of good ones, and how to ensure their success. Draws the analogy that a good partnership is similar to a good marriage.

Hulsey, Neven. "Keys to Avoiding Industrial Future Shock," *Metal Center News* 36, no. 6 (May 1996): 5–6.

This short article describes prerequisites companies need to plan for before entering into global alliances. These include taking a positive attitude toward a potential partner, gaining commitment from top management, involving multiple parts of the company, and using a TQM approach. The Earle M. Jorgensen Company's actions are used as a case study to demonstrate the practicality of this advise. The company established goals, measures, targets, and action plans for teams as part of its effort to create and use alliances with other organizations, particularly with its suppliers, effectively.

Metz, Philip D. "Integrating Technology Planning with Business Planning," *Research Technology Management* 39, no. 3 (May/June 1996): 19–22.

Presents findings from a two-year study done by the Industrial Research Institute on the integration of R&D planning with business planning. The study uncovered five best practices: establish a structured process for technology planning, foster active involvement of R&D and other departments, gain top management commitment, organize for effective planning and buy-in across the organization, and hold business units and R&D accountable for measurable results.

Pruett, Mark, and Thomas, Howard. "Thinking about Quality and Its Links with Strategic Management," *European Management Journal* 14, no. 1 (February 1996): 37–46.

This is a general overview that emphasizes that companies in the West have now adopted quality practices.

Despite using various approaches, they all have the common feature of dealing with management as a holistic process, an integrative approach to the management of quality. That kind of approach raises key questions about how strategic management should be performed as a process, issues which are discussed in this article.

Sommer, Jeff. "Using Purchaser Market Research Data in Strategic Planning," *Healthcare Financial Management* 50, no. 5 (May 1996): 20–21.

Argues that health-care institutions serve three classes of customers: patients, doctors, and purchasers (the latter are payers and employers). Failing to understand their respective needs can have negative business consequences for these institutions and, given the move to managed-care environments, cannot be allowed to occur. Suggests that telephone interviews of targeted purchasers are an effective market research technique because interviews provide useful information while avoiding logistic problems.

Sales and Marketing

Anonymous. "Transforming the Customer Connection," *Business Communications Review, Network Transformation Supplement* (January 1996): 6–16.

Provides cases studies on the use of call centers to transform connections with customers, describing activities at Nortel Inc., Wells Fargo & Co., and American Express Canada, Inc. Wells Fargo is attempting to understand how customers want to access bank services; Nortel implemented a speech recognition process; American Express Canada identified what kinds of transactions could be automated and which should not.

Barnes, James G. "Establishing Relationships—Getting Closer to the Customer May Be More Difficult Than You Think," *Irish Marketing Review* 8 (1995): 107–16.

Argues that companies have not thought through what it means to get closer to customers. There is more to building a long-term relationship than just locking customers in with long-term contracts or having highly effective databases and retention campaigns. Rather, understanding characteristics that bond customers to companies along some spectrum is more effective in guiding the actions of a company. In short, this article reflects the growing acceptance of the idea that companies need to respond to customer buying characteristics.

DeCarlo, Neil J. "AT&T Paradyne Manages Customer Value," *Total Quality Review* 5, no. 4 (September/October 1995): 33–38.

Describes how AT&T has applied the customer value management (CVM) approach, in which one measures customer satisfaction as their perceived value. Describes its new measurement schemes in support of this. CVM approaches are rapidly becoming very popular with large marketing organizations; a good introduction to the concept.

Sperry, Joseph P. "Warning from the Front: Six Ways Suppliers Sacrifice Key Accounts," *Advertising Age's Business Marketing* 80, no. 11 (November 1995): 5.

They are short-term thinking, many-headed confusion, inattention and drift, dynamic shifts, misaligned delivery systems, and growth and downsizing.

Wright, Phillip C., and Grant, E. Stephen. "The Strategic Application of TQM Principles to Salesforce Management: A Human Resource Perspective," *Journal of Marketing Theory and Practice* 3, no. 3 (Summer 1995): 10–22.

Says sales management literature pays very little attention to TQM even though quality management is now very important. Describes the role of a sales manager in a quality environment, defining specific activities and their implications.

Logistics and Supplier Relations

Burt, David N., and Pinkerton, Richard L. *A Purchasing Manager's Guide to Strategic Proactive Procurement.* New York: AMACOM, 1996.

Purchasing processes have been some of the most reengineered ones in American industry in the past decade. The literature of best practices in this area is also quite substantial, in large part because of the continuing changes occurring in this field. This book summarizes best practices, circa early to mid-1990s, and is comprehensive enough to be a useful one-volume guide to the topic.

Genna, Albert. "Are All Small Suppliers Really Worth the Trouble?" *Purchasing* 120, no. 6 (April 25, 1996): 2–28.

This journal polled buyers, who reported advantages to using small suppliers, who sometimes provide better service than larger ones. Flexibility is the most important benefit. Communications tend to be better, as small supplier's work more closely with their customers because each customer is so much more important. Ninety-four percent of the buyers polled said they look closely at how well small suppliers can complete work on time. Many buyers seek references and probe for flexibility. Quality rather than price is more important when working with small suppliers.

Gooley, Toby B. "How Logistics Drives Customer Service," *Traffic Management* 35, no. 1 (January 1996): 45–47.

Argues that logistics are closely tied to customer service. To understand why requires appreciating what customer service is all about. Provides examples of how companies link the two together.

Hegji, Charles E. "On the Economics of Developing Input Supplier Relationships," *Mid-Atlantic Journal of Business* 31, no. 3 (December 1995): 247–57.

Describes the issues associated with single-vendor practices so touted by TQM. The two key questions are (1) what are the profit-maximizing circumstances for the company buying from only one vendor? and (2) what incentives does the supplier have to improve its quality via the relationship? Discusses both without giving definitive answers.

Lamming, Richard, and Hampson, Jon. "The Environment as a Supply Chain Management Issue," *British Journal of Management* 7 (special issue) (March 1996): S45–S62.

Explores purchasing and logistics issues within the context of environmentally sound management. Discusses customer views, legislative considerations, and sound management. Talks about the role of vendor assessment, TQM, and various lean supply and collaborative supply strategies. Draws parallels between well-established practices and new imperatives. The authors use five company case studies to demonstrate their ideas, arguing the benefits of sound environmental management in logistics and purchasing practices.

Lewis, Jordan D. "Practices for Joint Creativity," *Total Quality Review* 5, no. 5 (November/December 1995): 20–27.

The author demonstrates how customers and suppliers can apply a series of techniques to be creative together. He argues that a strategy that fosters creativity encourages effective partnerships.

Trunick, Perry A. "Boxmaker Stacks Up Logistics Savings," *Transportation and Distribution* 37, no. 5 (May 1996): 72–77.

This is a case study of Stone Container Corporation's return from financial difficulties. A key to its success was the installation of a logistics optimization system designed to improve customer service and to take costs out of the system. Stone also integrated the marketing and transportation departments under customer service and logistics so that the entire process is now in one part of the organization. Results: Stone reduced lead time from order processing to shipping.

Product Development and Manufacturing

Chiesa, Vittorio, Coughlan, Paul, and Voss, Chris A. "Development of a Technical Innovation Audit," *Journal of Product Innovation Management* 13, no. 2 (March 1996): 105–36.

Presents a framework for auditing technical innovation management, focusing on problems, needs, and measurements while generating information needed to develop action plans for improving performance. The framework is based on a process model for technical innovation and is comprehensive. This important article identifies four basic processes: generation of concepts, product development, innovation of processes, and acquisition of technology. The audit is done on the processes and on the outcomes of the four subprocesses. One can take this article and quickly begin to implement its techniques.

Hawley, John K. "Automation Doesn't Automatically Solve Problems," *Quality Progress* 29, no. 5 (May 1996): 59–63.

We focus too much on engineering issues when attempting to implement automated processes. We pay too little attention to the impact this would have on employees. Because people still play a part in automation, we cannot ignore their role, for example, what they do and how they are supervised. The challenge is how an operator can have and sustain interaction with the automated process without slowing down the work being done. The answer lies in looking at the role of operators in job design, operator/computer interfaces and training and maintenance of skill levels. Each issue is discussed in some detail.

Lee, Hau L. "Effective Inventory and Service Management through Product and Process Redesign," *Operations Research* 44, no. 1 (January/February 1996): 151–59.

An important problem is product proliferation because it negatively impacts one's ability to forecast demand and thus inventory levels remain too high and customer service is at risk. Global marketing is a source of concern, but product proliferation is also expanding as customer bases grow. Thus different variations of products are developed for different market niches around the world, complicating matters. The author suggests that inventory management and customer service can be enhanced by paying attention to these issues during product design phases. For example, simplifying logistics through product design can help in addition to the more traditional concerns regarding functionality, performance, and manufacturability. Presents models of how to factor logistical issues into design considerations, using real examples.

Martini, William J. "The Electronic Change Request: Applying TQM to the Creative Design Process," *Production and Inventory Management Journal* 37, no. 1 (First Quarter 1996): 1–6.

Argues that there can be a simple, practical, and effective approach to measure quality in any design process. Using well-defined statistics, errors in such a process can be identified, causes analyzed and located, and prevention implemented. Cost of redesign can be measured by the cost of rework needed to correct design errors. Called the electronic change request (ECR) approach, it is a software package used by Westinghouse and developed by the firm, a recognized best practices leader in the use of metrics. Argues that this company has improved design quality by over 30 percent using the tool. Furthermore, ECR has made it possible to improve operations continuously.

May, Douglas R., and Flannery, Brenda L. "Cutting Waste with Employee Involvement Teams," *Business Horizons* 38, no. 5 (September/October 1995): 28–38.

Argues that it is cheaper and easier to minimize waste than to recycle; therefore companies should focus on the former, not the latter, using team-based strategies. Explains how that can be done. This article, although listed in this section of the bibliography, is really applicable to all organizations, regardless of sector.

Mills, J., Platts, K., and Gregory, M. "A Framework for the Design of Manufacturing Strategy Processes: A Contingency Approach," *International Journal of Operations Product Management* 15, no. 4 (1995): 17–49.

The authors present a framework for taking into account the factors important in the design of manufacturing strategy processes. It incorporates elements from cellular manufacturing, TQM, and JIT and offers a holistic approach.

Montgomery, Joseph C., and Levine, Lawrence O. (eds.). *The Transition to Agile Manufacturing: Staying Flexible for Competitive Advantage.* Milwaukee: ASQC Quality Press, 1996.

This book is an introduction to agile manufacturing techniques, providing a solid discussion of what they are, why they are valuable, and how they are done. It is aimed at manufacturing management and engineers.

Pine, B. Joseph II. "Serve Each Customer Efficiently and Uniquely," *Business Communications Review, Network Transformation Supplement* (January 1996): 2–5.

The author describes the application of mass-customization techniques by Levi Strauss & Co., Andersen Corporation, and Ross Controls. Pine, a *Quality Yearbook* contributing editor, is a leading proponent of manufacturing goods and performing services tailored to the individual customer, not necessarily to specific markets. He describes his ideas briefly.

Pugh, Stuart. *Creating Innovative Products Using Total Design.* Reading, Mass.: Addison Wesley Longman, 1996.

This is a detailed, technical discussion of the topic. Innovation has become a fashionable term in the last two years, but this book builds on proven practices rather than on current trends. It is one of the more important books to appear recently on the subject, a real hands-on contribution.

Suri, Rajan, Veeramani, Raj, and Church, Joe. "Industry Teams Up with University to Drive Quick Response Manufacturing," *IIE Solutions* 27, no. 11 (November 1995): 26–30.

Describes how the University of Wisconsin–Madison Center for Quick Response Manufacturing assists companies to solve problems. Includes case studies of work done with Beloit Corporation, Ingersoll Cutting Tool Company, and Marathon Electric. Reports specific positive results. An excellent example of how a local university can help manufacturing companies apply technical knowledge in practical ways fast.

Swanson, Edward T., and Jambekar, Anil B. "A Technology Choice for Noise Reduction: A Quality Management Case Study," *International Journal of Quality and Reliability Management* 12, no. 9 (1995): 44–56.

Describes how noise was reduced in the operation of a clothes dryer, using systematic application of quality-management tools. The result was elimination of one type of irritating noise and led to new ways of improving the product.

Woodruff, Davis M. "Ten Essentials for Being the Low Cost, High Quality Producer," *Management Quarterly* 36, no. 4 (Winter 1995–1996): 2–7.

Says that there are ten focus areas all businesses must concentrate on if they are to change: Create an effective and credible leadership, concentrate on people, concentrate on customers, define "work to be done" (WTBD), implement a quality program that actually works, practice continuous improvement, get rid of waste, develop your own resources and assets, measure results, and use technology effectively.

Information Technology

Benjamin, Colin O., Lu, Chaoyuan, and de Neufville, Richard. "Classifying Risks on Information Technology Projects," *Engineering Management Journal* 7, no. 4 (December 195): 45–54.

Offers a qualitative model of the risks associated with implementing IT projects. The authors identified high-risk factors based on six case studies. They are offering a best-practices discussion of what project managers need to know when making decisions and evaluating risks based on the complexity of projects.

Braithwaite, Timothy. *The Power of IT: Maximizing Technology Investments.* Milwaukee: ASQC Press, 1996.

This short book—150 pages—discusses quality-management practices, conditions in the computer industry, and the role of automation. It is not a detailed best practices book but does offer a model on how users can communicate with IT professionals.

Hinton, John. "Quality Counts," *Accountancy* 117, no. 1232 (April 1996): 44–45.

This is a case study of quality-management practices at BNR Europe, Ltd. Demonstrates the value in precise terms for implementing ISO practices.

Longenmecker, Clinton O., Simonetti, Jack L., and Mulias, Mark. "Survival Skills for the IS Professional," *Information Systems Management* 13, no. 2 (Spring 1996): 26–31.

In addition to having a technical background, IS professionals today must also be business professionals and have a clear appreciation of customer expectations if they are to have successful careers. Presents the results of a study for seventy-five IS professionals to guide the development of skills needed today. Departments must have a customer focus, concentrate on business issues, provide feedback on what customers have to say, develop well-rounded professionals, and evaluate and reward performance.

Maglitta, Joseph. "Know How, Inc.," *Computerworld* 30, no. 3 (January 15, 1996): 73–75.

Describes what knowledge management is in processes, procedures, patents, reference works, best practices, formulas, forecasts, and fixes. Describes computer-based tools that can be used to manage information. Advice from the experts: Don't confuse knowledge management with decision support, adapt it to your situation, understand how people work, and use rapid application techniques.

Pearson, J. Michael, McCahon, Cynthia S., and Hightower, Ross T. "Total Quality Management: Are Information Systems Managers Ready?" *Information and Management* 29, no. 5 (November 1995): 251–63.

Presents the results of a study that showed that 41 percent of IS managers understood TQM concepts and believed they were useful in their operations. They identified the best benefits of TQM as improved customer service, enhanced quality of products and services, and increased flexibility, but stated that such improvements took three to five years to achieve.

Phan, Dien D., George, Joey F., and Vogel, Douglas R. "Managing Software Quality in a Very Large Development Project," *Information and Management* 29, no. 5 (November 1995): 277–83.

The biggest challenges in software project management concern quality control. This is also a subject about which little research has been done. The authors describe what thousands of programmers have done to improve quality control.

Pollalis, Yannis A. "A Systematic Approach to Change Management: Integrating IS Planning, BPR, and TQM," *Information Systems Management* 13, no. 2 (Spring 1996): 19–25.

Argues that IS planning should be linked to BPR and TQM. The author shows how to link the three, creating a platform for integrating them. The author sees four links into planning: alignment of IS and corporate goals, a focus on customers, IT-based process changes, and creating a learning organization. In effect, it is a holistic approach to understanding what IS must do.

Robertson, Maxine, Swan, Jacky, and Newell, Sue. "The Role of Networks in the Diffusion of Technological Innovation," *Journal of Management Studies* 33, no. 3 (May 196): 333–59.

Looks at the British experience, using three case studies of implementing computer-aided production management (CAPM). Although managers learned about this technology in many ways, the most widely influential resource were IT vendors who were promoting products and best practices.

Robbins, Donna. "IT Professionals Learn Customer Service," *Managing Office Technology* 41, no. 1 (January 1996): 58–59.

Today, IT professionals need to understand end users and customers. This article provides a case study of a company that created twenty-five guidelines for working with clients for the IT community to use at the end of a project to ensure that customer service had been delivered. Argues that end users were still not completely satisfied, accusing IT of still being too inwardly focused. Cautions against that.

Woodall, Jack, and Rebuck, Deborah K. *Total Quality in Information Systems and Technology.* Delray Beach, Fla.: St. Lucie Press, 1996.

Discusses the role of quality-management practices in IT, emphasizing quality concepts, planning, project teams, techniques, and electronic data interchange. Includes several case studies.

Research and Development

Sharman, G. K. "TQM and R&D: An Overview," *The TQM Magazine* 8, no. 2 (1996): 11–16.

Literature on this topic is very rare, so this is a welcome addition to the topic. The author argues that TQM applications to R&D are in their infancy. Few laboratories have tried TQM, so there are few data on the subject. He questions whether TQM principles can work in science; includes references.

Accounting, Finance, and Cost Justification

Ahire, Sanjay L. "TQM Age Versus Quality: An Empirical Investigation," *Production and Inventory Management Journal* 37, no. 1 (First Quarter 1996): 18–23.

Argues that a company does not have to wait for years to see if TQM efforts are paying off with benefits. The key is focusing on how one implements quality management. Presents data suggesting that benefits can accrue almost immediately, not years later. Yet higher operational results come in the second and third years, not dramatically in the first. Presents ten implementation recommendations that are proven to deliver benefits. The key is rigorous implementation of these ten initiatives, which are briefly described.

Anonymous. "Giving TQM a Try," *Institutional Investor* 29, no. 10 (October 1995): 321.

Reports the result of a survey of pension plan managers concerning the extent of their use of TQM principles. Results: less than 10 percent have any quality programs, another 14 percent said they were thinking about implementing one.

Babicky, Jacqueline. "TQM and the Role of CPAs in Industry," *CPA Journal* 66, no. 3 (March 1996): 69–70.

Summarizes Deming's philosophy and briefly defines TQM. Then demonstrates the role CPAs can play in applying these concepts because these professionals understand logical approaches to problem solving and how to identify and record key indicators. They can also conduct benchmarking.

Burke, John V., Luecke, Randall W., and Meeting, David. "Identifying Best Practices for Audit Committees," *Healthcare Financial Management* 50, no. 6 (June 1996): 76–82.

Says health-care providers all have audit committees. Then discusses how they can have best audit practices, presenting results of a recent study done by an accounting firm. Best practices include a written charter clearly outlining the committee's responsibilities, knowledge and proficiency enhanced through training, and access to industry-wide information.

Clarke, Paul, and Bellis-Jones, Robin. Activity-based Cost Management in the Management of Change," *The TQM Magazine* 8, no. 7 (1996): 43–48.

The authors describe the role of activity-based management in determining the direction of change. They argue the case for understanding costs in business instead of simply aspiring to change and improvement and offer good techniques for doing this.

Cover, Martha, Cooke, David, and Hunt, Matt. "Estimating the Cost of High-Quality Documentation," *Total Quality Review* 5, no. 4 (September/October 1995): 39–44.

Describes what it costs to prepare documents, then suggests how investing presales dollars in accurate and complete reports reduces costs of customer support. Provides a case study and examples of how to do this.

Glad, Ernest. "Zero Waste Accounting," *Chartered Accountants Journal of New Zealand* 75, no. 7 (August 1996): 26–30.

Waste is a hidden cost, and it exists in all organizations. The author demonstrates how to measure the price of nonconformance (PONC). Calls for accountants to develop a financial measure of waste quality, not simply of physical waste, such as parts destroyed. Argues that waste can be grouped into four types: wasted resources or inputs, process or production waste, output or product waste, and customer-related waste. This is a very original and groundbreaking article.

Gupta, Mahesh, and Campbell, Vickie S. "The Cost of Quality," *Production and Inventory Management Journal* 36, no. 3 (Third Quarter, 1995): 43–49.

Discusses the merits of cost of quality (COQ) programs. Links COQ programs to ABC accounting methods. Concludes with examples and discussion of effects on business.

Ito, Yoshihiro. "Strategic Goals of Quality Costing in Japanese Companies," *Management Accounting Research* 6, no. 4 (December 1995): 383–97.

Offers a brief history of the evolution of cost accounting in the United States and in Japan. Reviews the results of experiments in two Japanese companies—Honda Motor Company and Omron Corporation—to determine what would be required to develop a quality costing model that could be used as a strategic management tool. One of the problems faced was the lack of definition of quality cost by previous researchers. Then discusses the differences in management perspectives regarding quality costs in Japanese and American firms.

Miller, John A. *Implementing Activity-Based Management in Daily Operations.* New York: Wiley Publishers, 1996.

Activity-based management is a rapidly growing expanded application of ABC accounting. This book is a useful introduction to the topic as process and not simply accounting. It continues this publisher's interest in publishing accounting-related volumes, written for the business generalist, on a highly defined topic.

Sisaye, Seleshi, and Bodnar, George H. "Reengineering As a Process Innovation Approach to Internal Auditing," *Internal Auditing* 11, no. 3 (Winter 1996): 16–25.

Argues that reengineering internal auditing functions requires major changes in how this process is carried out, leading to a view of "control self-assessment collaborative relationships" between functional areas and auditors. Offers a five-step approach for auditing reengineering.

Sprague, David A. "Adding Value and Value Analysis to TQM," *Journal for Quality & Participation* 19, no. 1 (January/February 1996): 70–72.

Encourages the use of value analysis—a tool used by purchasing departments for a half century—in costing quality programs. Both focus on customer value and apply cross-functional approaches, problem solving, and documentation. However, continuous improvement—key to TQM—has not been a part of value analysis. The fundamental concept of value is the most critical reason for applying value analysis to TQM.

Struebing, Laura. "ASQC Sections Help Auditors Gain RAB Certification," *Quality Progress* 29, no. 6 (June 1996): 25–26.

The Registrar Accreditation Board (RAB) now requires applicants for lead auditor to perform ISO 9000 audits to be certified. Various branches of the ASQC in the United States are helping auditors gain the experience and training necessary to get certified. Describes some of the specific actions taking place around the country.

Wong, Peter. "Applying Total Quality Management: A Corporate Treasurer's Perspective," *Asia Money, Cash Management Guide Supplement* (October 1995): 18–19.

Wong describes three tools that can be used to manage cash effectively. These are a process control tool, a problem-solving tool, and another for planning.

Human Resources

Allender, Hans D. "Reengineering Employee Performance Appraisals the TQM Way," *Industrial Management* 37, no. 6 (November/December 1995): 10–12.

Reviews why organizations have personnel reviews, then argues that the application of TQM practices is imperative to fulfill these objectives, offering a brief framework to accomplish that. Concludes that if an organization really has implemented TQM, then there is no need for appraisals. Performance of employees grows as a result of profound knowledge, daily motivation, good coaching, and pride in work.

Blair, Earl H. "Achieving a Total Safety Paradigm through Authentic Caring and Quality," *Professional Safety* 41, no. 5 (May 196): 24–27.

Says most experts know that TQM can improve safety, but it is a different way to do so. Describes the differences, arguing that caring and process are very important to eliminating all safety problems, with primary focus on the human element.

Byham, William C. *The Selection Solution: Solving the Mystery of Matching People to Jobs.* Bridgeville, Pa.: Development Dimensions International, 1996.

This is one of many new books to appear in the past several years on the subject of hiring, placement, and deployment of personnel, in the vein of a best practices study.

Decker, Paul T., and Thornton, Craig V. "The Long Term Effects of Transitional Employment Services," *Social Security Bulletin* 58, no. 4 (Winter 1995): 71–81.

Presents the results of a six-year study of transitional employment services. The study reports that these services can dramatically increase employment and earnings of people who have mental retardation. This is a study of the Social Security Administration's Transitional Employment Training Demonstration. Gives examples of companies that have had good success; most also have reputations for quality-management practices and focus on customer satisfaction.

Flannery, Thomas P., Hofrichter, David A., and Platten, Paul E. *People, Performance, and Pay.* New York: Free Press, 1996.

In recent years there has been a growing interest in the business community over the need to align performance and compensation for meeting corporate objectives. This book joins many that have appeared in recent years demonstrating how to link corporate objectives with performance using recognition and compensation to do that.

Gilbert, John. " 'A Job for Life' into 'A Life of Jobs': The New Employment Contract," *The TQM Magazine* 8, no. 2 (196): 36–42.

The author compares old and new employment contracts, featuring the benefits of the newer ones and providing hands-on advice on how to implement these. Keys to success include good leadership, training, communications, and recognition. Gilbert concludes his good article with a list of actions to guide organizations implementing this new way of managing employees.

Jerome, Laurie, and Kleiner, Brian H. "Employee Morale and Its Impact on Service: What Companies Do to Create a Positive Service Experience," *Managing Service Quality* 5, no. 6 (1995): 21–25.

Examines the personnel practices of three amusement parks: Walt Disney, Knotts Berry Farm, and Universal Studios. They see investments in training, information technology, and personnel practices as critical to the overall morale of employees. Briefly, the article reviews activities in recruitment, training, customer service strategies, motivation, morale, and other employee programs.

Johnson, Joanne, Baldwin, John R., and Diverty, Brent. "The Implications of Innovation for Human Resource Strategies," *Futures* 28, no. 2 (March 1996): 103–19.

Looks at three issues: pervasiveness of using technology, relationship between technology and training in Canadian manufacturing firms, and how innovation is found to be an important driver behind training in all sectors.

Jonas, Paul. "The Missing Letter in TQM," *Occupational Health and Safety* 65, no. 3 (March 1996): 18–19.

Focuses on the commitment to safety that must be made by organizations and the role of employees. TQM principles are perfect for improving safety, leading to a sound safety program.

Knouse, Stephen B. (ed.). *Human Resources Management Perspectives on TQM: Concepts and Practices.* Milwaukee: ASQC Quality Press, 1996.

This book focuses on the behavioral side of TQM implementation. It covers all aspects of personnel practices and TQM; the most comprehensive guide currently available.

Kochanski, James T., and Ruse, Donald H. "Designing a Competency Based Human Resources Organization," *Human Resources Management* 35, no. 1 (Spring 1996): 19–33.

Provides background on the pressures facing human resource organizations today then argues that there is a move afoot to improve HR competencies, not simply to apply process improvement and reengineering strategies. The hunt for process and competency is influencing how HR functions are being organized; describes examples. This is an important HR management article.

McConnell, Charles R. "After Reduction in Force: Reinvigorating the Survivors," *Health Care Supervisor* 14, no. 4 (June 1996): 1–10.

Says insufficient attention is paid to how to deal with employees who remain after an organization has gone through a period of downsizing and layoffs. Describes the problems these people face and what should be done to reinvigorate them and help them become productive again. Management's tools to accomplish this are communications, education, and people-centered management practices.

McNerney, Donald J. "HR Practices: HR Adapts to Continuous Restructuring," *HR Focus* 73, no. 5 (May 1996): 1, 4.

HR professionals now live in a period of continuous corporate restructuring, adapting to new conditions. Employment security is now gone. HR professionals have developed programs that offer workers opportunities to improve skills and thus their attractiveness in the marketplace. Underlying this are new HR processes and programs.

Mohrman, Susan A., Lawler, Edward E. III, and Ledford, Gerald E. "Do Employee Involvement and TQM Programs Work?" *Journal for Quality & Participation* 19, no. 1 (January/February 1996): 6–10.

Presents results of surveys done of Fortune 1000 companies in 1987, 1990, and 1993 on effectiveness of employee involvement (EI) and total quality management. TQM tends to emphasize work processes and customer outcomes. EI tends to place more stress on design of work and business unit to achieve fuller business involvement and motivated employees. EI also places more emphasis on gain sharing, rewards, and training in order to have an employee act as a stakeholder of the business. Both approaches are complementary and do result in organizational improvements in productivity and effectiveness. The more both were used, the better the financial performance of the surveyed companies.

Morgan, Ronald B., and Smith, Jack E. *Staffing the New Workplace: Selecting and Promoting for Quality Improvement.* Milwaukee: ASQC Quality Press, 1996.

This research-based study describes ways to recruit, assess, select, and promote employees in quality-focused organizations. This is a serious nuts-and-bolts book that contains specifics on techniques, tools, and examples; massive in size.

Peters, Lee A., and Homer, John. "Learning to Lead, to Create Quality, to Influence Change in Projects," *Project Management Journal* 27, no. 1 (March 1996): 5–11.

Says new project managers learn by being coached and trying and practicing, and can additionally expand skills through the use of simulations. Argues that simulation is an excellent way to develop project management skills, especially because specific examples can be used in areas such as team formation, team building, effective planning, work breakdown structure, scheduling, estimating, scope definition, and project control techniques.

Ready, Kathryn J. "How Can Human Resource Expertise Support Reengineering Efforts?" *Industrial Management* 37, no. 6 (November/December 1995): 14–19.

Defines reengineering then argues that HR can help the effort by preparing people for the coming changes. HR can do this largely by redoing processes, practices, and policies to facilitate the reengineered activities. HR can help the business by providing information on existing and needed skills, qualifying individuals to help in reengineering, and tracking and evaluating results.

Savageau, John. "World Class Suggestion Systems Still Work Well," *Journal for Quality and Participation* 19, no. 2 (March 1996): 86–90.

Presents four reasons to have a suggestion process: the best systems generate over forty ideas per employee per year; not seen today as a second-class process in a team-focused world; helps get people involved because all people don't get involved through teams alone; can be used to focus on things that teams do not. Flaws of old suggestion systems include bottlenecks in the flow of ideas, lack of focus, disempowerment of management, and lack of management responsibility for making people involved in daily activities. (Included in the HR section of the yearbook.)

Scully, John P. "TQM and Human Nature: Getting Beyond Organizational Misconceptions," *Quality Progress* 29, no. 5 (May 1996): 45–48.

Argues that management systems concentrate on exploiting employees' intellectual and technical skills instead of viewing people as talented, diverse, and imperfect but capable of doing great things if they work together. Thus the prevailing approaches are counterproductive and prevent employees from doing the best with their skills and intellect. Empowerment, TQM, and teaming will not work unless management recognizes that it must go beyond exploiting skills and intellect and deal instead with the realities of human nature. Briefly discusses the virtues of empowerment, teaming, alignment of support systems, replacement of managers with leaders, measured implementation, community and hierarchy, and the end of management.

Shearer, Clive. "TQM Requires the Harnessing of Fear," *Quality Progress* 29, no. 4 (April 196): 97–100.

Argues that some people fear change—a basic feature of TQM—so fear must be harnessed as a positive force. Presents six common problems with TQM that lead to fear: discounting existing quality, lack of integration, totally empowering employees, mistaking means for ends, failing to measure processes, and failing to punish resisters. Management must acknowledge the real existence of fear in the workplace and face fear, working to drive it out.

von Dran, Gisela M., Kapplman, Leon A., and Prybutok, Victor R. "Empowerment and the Management of an Organizational Transformation Project," *Project Management Journal* 27, no. 1 (March 1996): 12–17.

Uses a case study to demonstrate how limited empowerment of front-line employees led to major benefits for a bank implementing a new information system. Bank tellers were divided into two groups and received the same training on the new system, but one group was told when to train, and the second scheduled its own training times. The group that determined when it would be trained was happier with the results and was measurably more productive than the control group.

STANDARDS AND ASSESSMENTS
ISO 9000

Aalund, Niels A. "ISO 9000: A Supplier's View," *Transportation and Distribution* 37, no. 5 (May 1996): 101.

Provides a brief overview of ISO 9000 targeted at the logistics community. Emphasizes the importance of quality and consistency of performance as crucial to logistics. Says this is a way of improving warehousing and distribution operations.

Anonymous. "Steel Centers Sense ISO Pressure Rising," *American Metal Market* (April 11, 1996): 3.

There is great interest in the steel industry in ISO 9000. Pressure from customers to certify is an important trend, and companies are concerned about the cost and time of doing that. Some argued that certification is not about quality but conformity with obvious performance benefits. Includes citations from various companies.

ASQC Chemical and Process Industries Division. *ISO 9000 Guidelines for the Chemical and Process Industries, Second Edition.* Milwaukee: ASQC Quality Press, 1996.

This reference work reflects the 1994 ISO 9000 revisions and how ISO 9001 works in these industries. All of the book's contents have been updated and expanded, with examples as well.

AT&T Corporate Quality Office. *Using ISO 9000 to Improve Business Processes.* Milwaukee: ASQC Press, 1995.

This offers a comprehensive framework for appreciating ISO 9000, how to get started with it, and how it is used by AT&T—a leading practitioner of quality management—and offers advice on its ongoing use. The book repeats much of the same sorts of material available in various introductions to the topic.

Bibby, Thomas. "ISO 9000, A Catalyst, Not a Solution," *Manufacturing Engineering* 116, no. 4 (April 1996): 14.

Says ISO 9000 offers you two things: (1) five accounting principles for documenting quality practices and (2) a framework for demonstrating to customers how products are tested, employees trained, documentation is in place, and defects are corrected. ISO does not provide solutions such as quality-management practices, and it is vague. In addition to ISO, one still needs to implement quality design, manufacturing, and delivery.

Chowdhury, Subir, and Zimmer, Ken. *QS-9000 Pioneers: Registered Companies Share Their Strategies for Success.* Milwaukee: ASQC Quality Press, 1996.

This is a wonderful collection of stories, case studies, best practices, and useful advice on the topic. It is organized much the way articles are on the topic rather than as the traditional book-length literature on ISO 9000. It is also substantial at over 300 pages in length. (See excerpt from this book in the "ISO 9000" section of the yearbook.)

Curkovic, Sime, and Handfield, Robert. "Use of ISO 9000 and Baldrige Award Criteria in Supplier Quality Evaluation," *International Journal of Purchasing and Materials Management* 32, no. 2 (Spring 1996): 2–11.

Purchasing and materials managers have increasingly been paying attention to ISO 9000 registration, both for the selection of suppliers and as a way for their companies to compete in the marketplace. It is a useful tool to screen potential suppliers. Baldrige provides a framework for assessing the quality systems of a supplier. Both approaches are compared and contrasted, using a survey of 314 companies to validate the observations. The study suggested that ISO 9000 registration criteria fail to measure key quality management practices such as strategic planning, employee involvement, quality results, and customer satisfaction.

Greenberg, Eric F. "ISO 14000 to Change Business Environment," *Packaging Digest* 33, no. 4 (March 1996): 20–22.

ISO 14000 does for environmental management what ISO 9000 did for manufacturing in general. New standards are currently being drafted but will consist of five elements: environmental management systems, environmental auditing, environmental labeling, life cycle assessments, and environmental considerations of product standards. The article urges readers to start taking these new standards seriously.

Guerin, John M., and Rice, Robert W. "Perceptions of Importers in the United Kingdom, Germany and the Netherlands regarding the Competitive Advantages of ISO 9000," *Forest Products Journal* 46, no. 4 (April 1996): 27–31.

Argues that ISO provides a way of defining quality assurance and management programs for any company. Forest products firms are discovering a competitive advantage in being able to define their standards to foreign customers via ISO 9000 norms. Presents results of a survey that indicates ISO standards have not yet resulted in competitive advantage.

Hill, Sidney. "ISO Certification Provides Long Term Payoff," *Quality* 35, no. 4 (April 1996): 66–68.

ISO followers are still very enthusiastic about their quality program and are recruiting others to their ISO world. Attacks the Baldrige Award because most who vie for it never win it, thus probably driving many firms to ISO. ISO has a beginning, middle, and end, whereas Baldrige takes longer. One weakness of ISO is the lack of a clear link between registration and higher-quality products because ISO has no specific provisions for judging the quality of products.

Jahnke, Bernd, Bachle, Michael, and Simoneit, Monika. "Modelling Sales Processes as Preparation for ISO 9001 Certification," *International Journal of Quality and Reliability Management* 12, no. 9 (1995): 76–99.

Looks at how to survey service processes using a case study of a sales department of a computer manufacturer, applying process modeling techniques. It was important to find the weak points and to fix them as part of the ISO 9001 certification audit.

Label, Wayne A., and Priester, Wilbur. "Expanding Your Role in ISO 9000," *CPA Journal* 66, no. 6 (June 1996): 40–44.

European requirements for certification have created enough momentum to have U.S. firms adopt ISO 9000 as well. Explains how CPA firms can provide services to firms adopting ISO 9000 and describes the normal certification process a firm must go through with ISO 9000.

Lamprecht, James L. *ISO 9000 Implementation for Small Business.* Milwaukee: ASQC Quality Press, 1996.

This is written in particular for small to medium-sized companies that have no experience with ISO 9000. Describes ISO 9000, and how to implement all its various aspects and provides examples. It is brief and to the point.

Lee, Tat Y. "The Experience of Implementing ISO 9000 in Hong Kong," *Asia Pacific Journal of Quality Management* 4, no. 4 (1995): 6–16.

The author reports on a survey of Hong Kong companies (service, manufacturing, and building and construction). All were certified regardless of their diversity of backgrounds and experience with quality management. Those with strong quality practices tended to benefit from ISO registration more than others. The author concludes that ISO 9000 is only a foundation for a long-term quality implementation effort. ISO 9000 offers no guarantees of quality success. In fact, many registrants still needed to improve their process capabilities and control over subcontractors.

Mallak, Larry A., Lyth, David M., and Bringelson, Liwana S. "Cultural Factors Impact Registration," *Total Quality Review* 5, no. 4 (September/October 1995): 53–57.

Argues that corporate culture must change to make ISO 9000 work. Presents a survey of what works and what does not work. The key success factors are being highly organized and paying attention to detail, while the elements hindering registration included ignoring the rules, being informal, and taking risks.

Milmo, Sean. "The Way to Uniformity," *Chemical Marketing Reporter* 249, no. 15 (April 8, 1996): 12–13.

Notes the wide adoption of ISO 9000 by European chemical companies. Uniformity of standards will be reached in Europe through ISO 9000, rather than by having various national standards. Cites the European Chemical Industry Council as driving standardization via ISO 9000. Companies are also expanding their use of benchmarking.

Rada, Roy. "ISO 9000 Reflects the Best in Standards," *Communications of the ACM* 39, no. 3 (March 1996): 17–20.

Declares ISO 9000 the preeminent standard for management. The tough thing to do is to figure out how to invest in and implement ISO practices. Provides a rationale for why ISO is so important and useful. This message appeared in one of the most respected information technology journals.

Schuler, Charles, Dunlap, Jesse, and Schuler, Katharine. *ISO 9000: Manufacturing, Software and Service.* Albany, N.Y.: Delmar Publishers, 1996.

Continues a long tradition of publications on ISO 9000 introductions. The book is designed to be a practical guide to the topic and is very current.

Stewart, William. "ISO 9000 Work Instruction—Subject: Fun," *TQM Magazine* 8, no. 4 (1996): 17–19.

Describes activities that one should take to create fun in the workplace. Fun can improve productivity. Offers humorous observations and practical advice. Stewart says the basis of his article was observation of the best and worst management practices he could find. Concludes that quality and fun are mutually reinforcing.

Van Houten, Gerry. "The ISO Document Tidal Wave," *Records Management Quarterly* 30, no. 2 (April 1996): 12–20.

Systematic document control is crucial in all quality-management programs. Thus documentation is a key business function that should, when possible, be automated. Provides reasons for automating: read-only access makes it possible for more than one person to see a document at the same time, there is no loss of documents, and retrieval is virtually instantaneous.

Vloeberghs, Daniel, and Bellens, Jan. "ISO 9000 in Belgium: Experience of Belgian Quality Managers and HRM," *European Management Journal* 14, no. 2 (April 1996): 207–11.

Presents survey results concerning all certified Belgian organizations. Shows what kinds of companies are certified, their motivations for using ISO 9000 norms, an evaluation of their experiences, and the relationship between HR management and quality management. Identifies future trends of ISO 9000 and quality assurance. This is an important study on ISO 9000.

Weston Jr., F. C. "What Do Managers Really Think of the ISO 9000 Registration Process?" *Quality Progress* 28, no. 10 (October 1995): 67–73.

Presents results of a survey of ISO 9000–registered companies in Colorado. The study is part of a longer-range project to understand the relationships between ISO, Baldrige, and other quality strategies.

Wilson, Lawrence A. *Eight-Step Process to Successful ISO 9000 Implementation: A Quality Management System Approach.* Milwaukee: ASQC Quality Press, 1996.

This highly regarded expert on ISO 9000 describes an eight-step process he uses to help companies implement ISO 9000. He has also developed software that can be used with the book to facilitate the implementation process.

Zaciewski, Robert D. "ISO 9000 Preparation: The First Crucial Steps," *Quality Progress* 28, no. 11 (November 1995): 81–83.

Argues the case for applying a structured approach to the initial introduction of ISO 9000 so that the organization will be ready for the harder steps that come later. Describes how to select which ISO standards to adopt.

Awards, Other Standards, and Issues

Anonymous. "The 1996 Presidential Award for Quality," *Journal for Quality & Participation* 19, no. 4 (July/August 1996): 32–41.

Describes the case of the U.S. Army Research Development and Engineering Center, which received this award from the American government. The center's achievement was the application of statistical theory to process management. Also describes the criteria for the awards, of which there are three: the Presidential Award for Quality, the Quality Improvement Prototype, and the Achievement Award.

Baila, D. L. "The Deming Prize," *Journal for Quality & Participation* 19, no. 4 (July/August 1996): 16–19.

Reports on the 1995 prizewinners and describes this Japanese quality award. Awards are given out in four categories: individual, companies and other operating organizations, factories, and companies located outside of Japan.

Bohoris, G. A. "A Comparative Assessment of Some Major Quality Awards," *International Journal of Quality and Reliability Management* 12, no. 9 (195): 30–43.

Reviews Japanese, European, and U.S. quality awards, comparing and contrasting their features. This article is the most complete and current comparison of such awards, including the obvious ones (e.g., Deming and Baldrige).

Brown, Mark Graham. *Baldrige Award Winning Quality: How to Interpret the Malcolm Baldrige Award Criteria, Sixth Edition.* Milwaukee: ASQC Quality Press, 1996.

This is the bible for those applying the Baldrige criteria to their business or applying for the award. This is the single most widely used reference book on the subject; revised annually.

Assessments

Brereton, Malcolm. "Introducing Self-Assessment: One of the Keys to Business Excellence," *Management Services* 40, no. 2 (February 1996): 22–23.

Defines self-assessments as a comprehensive, systematic, and routine review of any organization's activities and results. Self-assessments should then lead to an improvement plan. Using a recognized model for the assessments speeds up the process. Suggests that two models to consider are the U.K./European Quality Award Model and total quality management, both of which have been used by European companies for years. The focus is on processes and releasing the talents of employees.

Cahill, Lawrence B., and Michelin, Lori Benson. "Achieving Quality Environmental Audits: Twenty Tips for Success," *Total Quality Environmental Management* 5, no. 3 (Spring 1996): 61–70.

Some of the most important are don't ignore or underestimate the need to prepare for the audit, its execution, and logistics; develop and sustain an agenda throughout the audit; manage time effectively; balance records review, interviews, and observations; be flexible; be on time even if local staff is not; audits will always be seen as performance audits by site staff; still apply good judgment; learn and apply audit protocols but balance these with common sense and normal curiosity.

Sisaye, Seleshi. "Two Approaches to Internal Auditing and Control Systems: A Comparison of Reengineering and TQM," *Internal Auditing* 11, no. 4 (Spring 1996): 37–47.

Compares and contrasts both approaches then says that reengineering can make auditing functions better, for example, by linking a company's accounting system to organization-wide management units. Although the concepts are not new, directing them at auditors in their language is useful.

Yusof, Sha Ri M. "A Quality System Assessment of an Electrical Contracting Company Based on BS 5750," *International Journal of Quality and Reliability Management* 12, no. 8 (1995): 64–73.

Compares BS 5750 to ISO 9000, then describes BS 5750's application as a tool to improve performance and quality in one company. Argues that the implementation of BS 5750 is a foundation in any company's quest toward high-quality performance.

Quality Resources On-line '97
(or Quality@Online.TQY)

The Internet is no longer a novelty or a medium accessed only by the technically proficient. In the two years since we started this section of *The Quality Yearbook*, the amount of material on quality management has increased exponentially. It is not possible to provide a comprehensive list of materials available on-line. However, it is possible to provide a pretty good listing. And because most World Wide Web sites have links to other sites (listings that allow you to click and visit them), the list gets you not only to those we've included but also to all those links the developers of the listed sites have chosen to include.

What You'll Find Here

There are three categories of on-line information included in our directory:

- **Internet Discussion Groups**. These are subscription groups that allow electronic mail on a particular subject to be sent to a central location and then broadcast as E-mail to everyone who is a member of that group. We've sought to ferret out as many as we could find that would be relevant to people interested, even peripherally, in quality-management issues
- **World Wide Web Sites**. These are sites on the Internet that you access using a Web browser such Netscape Navigator or Microsoft Explorer. Web sites make up the large majority of the material we've included.
- **Private On-line Information Suppliers**. These are companies that supply information on a subscription basis and require proprietary software to access.

Rating Internet Sites

To assist you in evaluating these sites and deciding which ones you might want to check out, we've given them a rating of one, two, or three stars. Here's what the ratings mean:

★★★ This site has a lot of good information on its subject, is well laid out, and is relevant to most quality-management practitioners. Such sites also have links to many other sites.

★★ This is a more specialized site that is still well done, but not as inclusive, interesting, or useful as a three-star site.

★ This site will generally be one to visit if you have an interest in its subject matter or, if it covers quality management in general, does not have the depth or breadth that more highly rated sites have.

Organization of Listings

To make this directory still more useful, we have divided the Web listings into the following categories:

- **TQM in General**. These sites cover a variety of topics all related to TQM.
- **ISO-Related Topics**. These sites have to do directly or indirectly with standards and with ISO 9000, QS 9000, or ISO 14000.
- **Publishers of Quality Materials**. These are the sites of various book and magazine publishers that are especially involved with TQM.
- **Organizations**. These are the sites of organizations that quality practitioners may find useful.
- **Consultants in Quality Management**. These sites were put up by various consultants in quality management. They usually have information about their offerings, but some have a lot of useful information as well. You'll be able to tell those sites that are especially useful by their ratings. This is a very limited list of consultants, but it is representative of what you'll find if you go searching yourself.
- **Training/HR Sites**. Listed here are sites that would be of interest to those of you looking for training materials or trainers in quality management. Also listed here are the sites of major players in the human resource management area.
- **Other**. This section includes useful sites such as technical bookstores, a site describing the rules of copyright, an Internet glossary site, and others.

Now, let's go to the directory.

Discussion Groups

Joining a discussion group requires that you send an E-mail message to an address where subscriptions are processed. For each of the groups listed, we profile the E-mail address where you send a message to subscribe. Subscribing in this sense means formally joining the list. In doing this, you do not usually have a subject line but simply write in the body of the message "subscribe [name of list] first name, last name." Sometimes you don't include your name. Sometimes you put the word "subscribe" in the subject line of the message and that's all. We tell you exactly what to say for each list. When you subscribe, you will usually receive an introductory message that describes the service, what you can expect, and how to participate. This message will also tell you how to stop your subscription if you don't find the group to be of value. The name of the list describes its subject matter. Also, some of these are maintained in the U.K.—you can tell which ones because they include *uk* in the address.

ASQC-CSD (ASQC's customer-supplier division)
Send to: majordomo@quality.org
In body: subscribe asqc-csd

ASQC-HCD (ASQC's health-care division)
Send to: majordomo@quality.org
In body: subscribe asqc-hcd

ASQC-DCD (ASQC's design and construction division)
Send to: majordomo@quality.org
In body: subscribe asqc-dcd

Malcolm Baldrige National Quality Award (State, local, and national quality awards)
Send to: majordomo@quality.org
In body: subscribe baldrige

BPMI (Business process management and improvement)
Send to: majordomo@quality.org
In body: subscribe bpmi

BPR (Business process reengineering)
mailbase@mailbase.ac.uk
in body: join BPR firstname lastname

BPR-IAC (Business process reengineering in government)
majordomo@quality.org
in body: subscribe bpr-iac

BRKTHR-L (Breakthrough thinking discussion list)
listserv@vm.usc.edu
in body: subscribe brkthr-l firstname lastname

CHANGE (Change and leadership list)
Send to: majordomo@mindspring.com
In body: subscribe change firstname lastname

CONSULTING (For consultants in quality areas)
Send to: majordomo@quality.org
In body: subscribe consulting

CQEN (Community quality electronic network)
Message to: cqen.list-request@deming.eng.clemson.edu
In subject line: subscribe
In body: nothing

CQI-L (Continuous quality improvement—with special emphasis on this subject in universities)
Message to: listserv@mr.net
In body: subscribe CQI-L firstname lastname

CREA-CPS (Creative problem solving list)
Send to: listserv@hearn.bitnet
In body: subscribe crea-cps firstname lastname

DEN (Deming Electronic Network)
Message to: den.list-request@deming.eng.clemson.edu
In subject line: subscribe
In body: nothing

ISO 9000 (ISO 9000 standards discussion)
Send to: listserv@vm1.nodak.edu
In body: subscribe iso9000 firstname lastname

ISO 9000-3 (ISO 9000-3 standards discussion)
Send to: majordomo@quality.org
In body: info iso9000-3

ISO 14000 (ISO 14000 discussion)
Send to: listserv@vm1.nodak.edu
In body: subscribe iso14000 firstname lastname

LEARNING-ORG (Learning organization discussion list focuses on issues first articulated by Peter Senge and others at MIT)
Send to: majordomo@world.std.com
In body line 1: subscribe learning-org
line 2: end

MGTDEV-L (Management development list)
Send to: listserv@miamiu.acs.muohio.edu
in body: subscribe mgtdev-l firstname lastname

MIL-QUAL-D (Military quality discussion list)
Message to: majordomo@quality.org
In body: subscribe mil-qual-d

NWAC-L (Changing nature of work)
Send to: listserv@psuvm.psu.edu
In body: subscribe nwac-l firstname lastname

QUALITY (Total quality management in manufacturing and service industries)
Send to: listserv@pucc.princeton.edu
In body: subscribe quality

Quality Management (Based in the U.K.)
Message to: mailbase@mailbase.ac.uk
In body: join quality-management firstname lastname

QFD-L (Quality function deployment discussion list) Message to: majordomo@quality.org
In body: subscribe qfd-l

QP-HEALTH (Quality issues for professionals in health care)
Send to: majordomo@quality.org
In body: subscribe qp-health

REGOnet (Reinventing government)
Message to: listserv@pandora.sf.ca.us
In body: information REGO-L

The four following relate to the federal government:

REGO-QUAL (Creating quality leadership and management)

REGO-ORG (Transforming organizational structures)

REGO-DOD (Department of Defense)

REGO-EOP (Executive Office of the President)
Message to: listproc@gmu.edu
In body: subscribe REGO-XXX firstname lastname (where 'XXX' is QUAL, ORG, DOD, or EOP, taken from the list above).

TEAMNET-L (Research on teams)
Message to: roquemor@terrill.unt.edu
In body: join Teamnet firstname lastname

Total-Quality-EcoSys (Total quality in environmental systems)
Message to: Mailbase@mailbase.ac.uk
In body: Join Total-Quality-EcoSys firstname lastname

Total-Quality-Statistics (Statistics for continuous improvement)
Message to: Mailbase@mailbase.ac.uk
In body: Join Total-Quality-Statistics

TQM-D (Unmoderated version of quality in manufacturing and service industries)
Send to: majordomo@quality.org
In body: subscribe tqm-d

TQM-L (TQM in higher education)
Send to: listserv@ukancm.cc.ukans.edu
In body: subscribe TQM-l firstname lastname

TQMLIB (Implementing TQM in libraries)
Message to: listserv@cms.cc.wayne.edu
In body: subscribe TQMLIB firstname lastname

TRDEV-L (Training and development discussion list—very good participation)
Send to: listserv@psuvm.psu.edu
In body: subscribe trdev-l firstname lastname

World Wide Web (WWW) Sites

What follows is our directory of Web sites, organized alphabetically and by topic.

TQM In General

Agile and Advanced Manufacturing on the World Wide Web
http://www.sandia.gov/agil/home page.html
This site includes information on agile manufacturing and extensive links to many other sites dealing with agile manufacturing topics. Rating: ★★★

The Agility Forum
http://absu.amef.lehigh.edu/
This is the site of the Agile Manufacturing Enterprise Forum. It includes extensive information and resources on this important topic. It includes something called AgileNet, a for-pay service that provides case studies and profiles of agility and research projects. This is a very complete site on this subject. Rating: ★★★

Mr. Bill's Quality Bookmark Page
http://www.grfn.org/~mr bill/Quality.html
The stated purpose of this page is to provide a bookmark page for, and information on, Internet resources relating to quality and continuous improvement. Lots of links to sites dealing with benchmarking, theory of constraints, ISO 9000, and project management, plus essays on selected topics. Rating: ★★★

A Business Researcher's Interests
http://www.pitt.edu/~malhotra/interest.html
This site is loaded with links to and other information on a wide variety of topics of interest to businesspeople. The emphasis seems to be on topics dealing with information management, but there is an especially good listing of resources having to do with the learning organization. Rating: ★★★

Center for Quality and Productivity Improvement University of Wisconsin–Madison College of Engineering
http://www.engr.wisc.edu/centers/cqpi/

This page includes information about the center and its research and its offerings, including courses in design of experiments. Rating: ★★

Crazy about Constraints
http://www.rogo.com/cac/
This site is dedicated to the work of Eliyahu Goldratt as first laid out in his book *The Goal*. If you have an interest in this perspective on quality management, this site is worth visiting. Rating: ★★

Critical Linkages II
http://www.erinet.com/patterwc/CLIIN/
CLII is a newsletter that emphasizes short articles and quality in education. Includes articles and other material of interest to managers at all levels. Rating: ★

The Deming Electronic Network
http://deming.eng.clemson.edu/pub/den/
The site was established to foster understanding of the Deming System of Profound Knowledge to advance commerce, prosperity, and peace. The DEN is administered with the support of, but independently from, the W. Edwards Deming Institute (WEDI) and its board of directors. The DEN is a communications infrastructure for the WEDI and its board but does not represent or speak for them. The site includes lots of articles and essays related to Deming's approach to management. Rating: ★★

Deming Study Group of Dallas
http://rampages.onramp.net/~dumont/
This home page incorporates information of interest to study-group members as well as various articles dealing with Deming's management approach. Lots of good links to quality-related sites. Rating: ★★★

Department of Navy Total Quality Leadership Office
http://tql-navy.org/
This page includes a complete overview of how TQL is being implemented in the U.S. Navy. Lots of good information for people dealing with the navy or in the navy. Ideas and initiatives being taken by the navy are applicable in many nonmilitary operations as well. Worth visiting. Rating: ★★★

Euroqual
http://www.euroqual.org/
This site aims to provide "quality information on quality" on a Europe-wide basis, covering a broad range of quality-related issues in an easily accessible, user-friendly format. Comprehensive site on what's going with TQM in Europe. Rating: ★★★

Fifth Discipline Fieldbook Resources Connection Home Page
http://www.resourcesconnect.com/fieldbook.html
This site includes extensive information on ideas and practices involving the learning organization. For those interested, it's definitely worth visiting. Rating: ★★★

The Global Business Network
http://www.gbn.org/index.html
Global Business Network is a unique membership organization specializing in scenario thinking and collaborative learning about the future. The network provides various services, such as seminars. The best part of this site is its book club, which lists many titles related to systems thinking and quality-management issues. Rating: ★★

The Healthcare Quality Assessment Page
http://www.qserve.com/hcass/
This is a privately maintained page by a consultant, but it includes a lot of good information and links and has a special emphasis on the application of the Baldrige criteria to assessing the performance of health-care organizations. Rating: ★★

John Hunter's Webplace
http://pages.prodigy.com/john/hunter.html
This is a terrific source for links to many other quality-related sites. Also includes articles on TQM and lots of information on what's going on with TQM in many different contexts. Highly recommended. Rating: ★★★

International Organization for Standards
http://www.iso.ch/
This is the ISO home page and includes lots of information on ISO issues and links to other standards sites. If you're interested in the organization behind ISO 9000, this site is worth visiting. Rating: ★★★

Joint Commission on Accreditation of Healthcare Organizations
http://www.jcaho.org/jc home.html
This is the site of the JCAHO, the major health-care organization accrediting body. It emphasizes quality-management practices and provides a variety of materials to help hospitals and clinics implement TQM. This site describes the organization and their various resources. A must if you are in the health-care field. Rating: ★★★

Bob Willard's "LeaderAid!"
http://www.oise.on.ca/~bwillard/leadaid.htm
This is identified as "Internet resources for leadership/management development." This site has many very good resources of interest to managers, including information on leadership and management, facilitation techniques, teams and teamwork, the learning organization, and lots more. Rating: ★★★

Learning-Org Dialog on Learning Organizations
http://world.std.com/~lo

This site is dedicated to Peter Senge's learning organization ideas and contains the archives of the Learning-Org discussion group. Included here are all the postings, organized by month and accessible by subject and author as well. Includes good basic information on learning-organization concepts. Rating: ★★

Malcolm Baldrige National Quality Award
http://www.nist.gov/director/quality_program/

Everything you could want to know about the Malcolm Baldrige National Quality Award and where, if not covered here, you can find out more. Includes the criteria for awards in business, health care, and education, which can be used as excellent internal auditing documents. Definitely worth visiting. Rating: ★★★

The MIT Center for Advanced Educational Study Deming Site
http://morning-star.MIT.EDU/deming/

The MIT AES is the publisher of Deming's two books, *Out of the Crisis* and *The New Economics*, and this site provides brief reviews of these books, Deming videotapes, and allows you to order these items. Rating: ★

National Committee for Quality Assurance
http://www.ncqa.org/

The National Committee for Quality Assurance (NCQA) is an independent, not-for-profit organization dedicated to assessing and reporting on the quality of managed-care plans, including health maintenance organizations (HMOs). This is the NCQA's home page, and it includes detailed information about this organization, about the performance of HMOs, and other information. Worth visiting if you are in the health-care business or want to find out about how well your HMO is rated. Rating: ★★★

National Performance Review/Reinventing Government
http://www.npr.gov/

This site has lots of really great stuff about how Vice President Gore's initiative is progressing. The site includes tools, examples, and links to various other government sites and really affirms that quality initiatives work. The site is aimed at federal employees but is worthwhile for any interested person to visit. Rating: ★★★

On the Edge with Michael Finley
http://www.skypoint.com/~mfinley/

Finley is a business writer and commentator who, among other things, is co-author of the award-winning book *Why Teams Don't Work*. This site includes many of his articles, reviews of his books, and other information on his activities. Peripherally related to quality. Rating: ★

The Pennsylvania State University Center for Quality and Planning
http://www.psu.edu/president/cqi/

This is the information site for all quality initiatives taking place at Penn State. If you are interested in how TQM can be implemented at a major university, this site is definitely worth visiting. Rating: ★★★

Product Data Management Information Center
http://www.pdmic.com/home.html

Product data management is a subject and a process that is critical to the effective management of information to reduce cycle time as well as waste and rework in the manufacture of products. This site includes extensive resources on this subject and allows visitors to sign up to receive information regularly. Lots of good stuff, including bibliographies, articles on-line, and other resources. If you are in manufacturing, definitely worth visiting. Rating: ★★★

The QFD Institute
http://www.nauticom.net/www/qfdi/

This site has the descriptive subtitle "For the advancement of quality function deployment." It includes a reasonable amount of information about resources available from the institute, but if you want to find out what it is or how to use QFD by visiting the site, you will be disappointed. Rating: ★★

The Quality Network
http://www.quality.co.uk/quality/

This site is maintained in the U.K. and includes information on quality-management issues along with many links to other sites dealing with issues related to quality management. Worth visiting. Rating: ★★★

The Quality Observer
http://www.thequalityobserver.com/

This page is for *The Quality Observer* magazine. It includes lots of information on this magazine plus especially complete links to other sites related to quality issues. Rating: ★★

Quality Online
http://www.pi.net/~cbon/qlink.html#discus

This page includes information on discussion groups and links to various sites about quality management, ISO 9000, and related topics. Worth checking out. Rating: ★★

The Quality Wave
http://www.xnet.com/~creacon/Q4Q/

This page includes lots of links to many different sites dealing with issues of quality management. The links are divided into topics such as TQM, standards, general systems, manufacturing quality, and several others. Rating: ★★★

The Stanford Learning Organization Web
http://www-leland.stanford.edu/group/SLOW/
This site includes a variety of bibliographic information on this subject, including books, articles, and newsletters. Rating: ★

System Corp/Ernst & Young LLP
http://www.systemcorp.com/
The site for this collaboration of a software company and giant management consultant describes their ISO training and implementation CD-based product. This product was developed with H. James Harrington, a highly respected consultant who gives seminars around the world on process and performance improvement. (See Harrington's article that opens this edition of the yearbook.) Worth finding out about if you're thinking about ISO. Rating: ★★

Total Quality Management on the Internet
http://www.infi.net/pilot/extra/tqm/tqm/
The main thing about this site is that it includes links to about twenty or so sites dealing with many different issues of TQM. Rating: ★★

TQM: An Integrated Approach
http://www.dmu.ac.uk/dept/schools/business/corporate/tqmex/book
This site includes information on the book *TQM, An Integrated Approach: Implementing Total Quality through Japanese 5-S and ISO 9000*, written by Samuel K. Ho. It comes from Great Britain. Rating: ★

ISO-Related Topics

American National Standards Institute, ANSI Online
http://www.ansi.org/
Everything you might want to know about ANSI and its many services. Includes information on courses offered and many other resources ANSI offers to help organizations develop and implement standards for measuring and maintaining consistency in processes and quality in outputs. Includes links to other related sites. Rating: ★★★

Environmental Management Systems ISO 14000 Information
http://www.stoller.com/iso.htm
This is the page of a consulting company specializing in the area of ISO 14000 issues. A very good place to start to learn about this subject and how to gain certification in the ISO 14000 standards. Rating: ★★★

International Organization for Standardization, ISO Online
http://www.iso.ch/
If you want to know about the organization that administers ISO 9000 standards and certification, this is a good place to get information and find out about resources offered. Not as complete as the ANSI site. Rating: ★★

ISO 9000/QS-9000 Support Group
http://www.cris.com/~Isogroup/
This is an organized group to help companies understand and become certified in various ISO categories, including ISO 9000, QS 9000, and ISO 14000. There is a lot of good information here in the way of articles and links. This might be the first site to check out if you're in ISO certification. Rating: ★★★

National Institute of Standards and Technology (NIST)
http://www.nist.gov/
A very valuable site for any company seeking information on NIST resources for improving manufacturing processes. NIST sponsors frequent conferences and training programs on standards and technology-related matters, and these are listed and described here. The institute also administers the Malcolm Baldrige National Quality Award, and you can reach that site (described elsewhere in this directory) via a link here. Worth visiting. Rating: ★★★

Publishers of Quality Materials

Lakewood Publications
http://www.lakewoodpub.com/
This is the home page of Lakewood Publications, publisher of *Training* magazine and other training and management newsletters, magazines, and related materials. A good overview of what the company has available. Rating: ★★

Minitab, Inc.
http://www.minitab.com/
This is the site of the publisher of the widely used statistical software. Lots of stuff here of interest to users of this program or people who want to find out more about it. Rating: ★

Productivity Press
http://www.ppress.com/
Productivity Press is one of the leading publishers of books dealing with quality-management issues. You can browse their list of books by topic and order books through this site. Rating: ★

Quality Digest *Magazine*
http://www.tqm.com/digest/
This is the site for *Quality Digest* magazine and includes information on subscribing and editorial guidelines, plus access on-line to articles from previous issues. Also includes annotated links page. Well done. Rating: ★★★

Quality *Magazine*
http://www.access.digex.net/~quality/
This site includes lots of material from the magazine, including complete articles with figures. Good site and worth visiting. Rating: ★★★

The Quality Observer
http://www.thequalityobserver.com/
This page is for *The Quality Observer* magazine. It includes lots of information on this magazine plus especially complete links to other sites related to quality issues. Rating: ★★

Organizations

American Institute of Total Productive Maintenance
http://www.europa.com/aitpm/
This site is basically an overview of the association dealing with TPM tools and techniques. The site shows the benefits of joining the organization and includes a form for joining over the Internet. Rating: ★

American Quality and Productivity Center
http://www.apqc.org/
This center focuses on benchmarking and developing best practices but also deals with all aspects of quality management, including benchmarking, knowledge management, measurement, customer satisfaction, and productivity. The site offers training and other resources in all these areas. The site includes articles from the center's magazine *Continuous Journey*. This site has a great look and is definitely worth checking out. Rating: ★★★

American Society for Quality Control
http://www.asqc.org/
This site includes a listing of ASQC services as well as a chat area and a quality forum area. Good information on ASQC, but not as useful as you might expect. Rating: ★★

American Society for Training and Development
http://www.astd.org/astd.htm
This is a comprehensive site that has links to other sites, articles from the society's magazine, *Training & Development*, and other information related to performance and training issues. Rating: ★★★

APICS—The Performance Advantage
http://lionhrtpub.com/APICS.html
This is the site for *APICS—The Performance Advantage*, the magazine of the American Productivity and Inventory Control Society. It includes articles from the current edition of the magazine, links to other sites, and other useful information. Rating: ★★

Institute of Electrical and Electronics Engineers
http://www.ieee.org/
This is the world's largest technical professional society that has the goal of "promoting the development and application of electrotechnology and allied sciences for the benefit of humanity, the advancement of the profession, and

the well-being of our members." The site includes information on standards in this field that can be purchased. (Only of interest to people in this field.) Rating: ★★

Consultants in Quality Management

David Butler Associates
http://www2.zoom.com/dba/
David Butler is a consultant who has developed a complete do-it-yourself TQM implementation product. From the description at his site, this product looks well done. The site itself is very well done and has lots of interesting resources for training and implementation of TQM. The site includes a variety of informational articles written by Butler on topics related to TQM. More complete and useful than many consultant sites. Rating: ★★★

The Fenman Company
http://www.fenman.co.uk/
The site of a British consulting company. Its main value is links to many other sites in both the U.K. and elsewhere. Rating: ★

Integrated Quality Dynamics, Inc.
http://www.iqd.com/tqm.htm
This site promotes the consulting company Integrated Quality Dynamics, Inc. Includes essays on quality and links to other sites, with special emphasis on hoshin planning. Rating: ★

InterDynamics
http://www.ozemail.com.au/~interdyn/
This site of an Australian consulting company has some interesting material, including articles on understanding the holistic systems view of organizations. Rating: ★

Jensen Quality Associates
http://www.access.digex.net/~quality/
This consulting company's page includes a variety of good links to other quality-related sites and other documents and references of interest to quality-management practitioners. Rating: ★★

Joiner Associates
http://www.joiner.com/
Joiner Associates is a consulting and training firm whose products include *The Team Handbook* and other books and products. The company is known for its seminars on understanding variation and its importance for managing effectively. Rating: ★★

PC Engineering Inc.
http://www.pcengineering.com/
This site provides lots of information about and examples of the company's Improvit software product, a highly rated SPC tool. Worth visiting if you're interested in such materials. Rating: ★★

Phoenix Consulting Group
http://www.phoenixcg.com/
This page includes a variety of useful links and descriptive materials, including a good section on the Shingo Prize for Excellence in Manufacturing. Rating: ★★

Renaissance Business Associates
http://www.bizcenter.com/rbai/index.html
This center, run by Let Davidson, Ph.D., focuses on issues that relate to affirming the human spirit at work. Some interesting articles. Davidson runs the the wisdom_at_work Internet discussion group. Rating: ★

Total Quality Engineering
http://www.tqe.com/
This page is run by the consulting company Total Quality Engineering. The page describes the company's services and also includes good information on hoshin planning and links to other quality sites. Rating: ★

Training and HR Sites

Department of Labor Employment and Training Administration
http://www.ttrc.doleta.gov/
There is lots of good information here, including articles on a wide variety of training and employment issues, plus much other information and links. Rating: ★★★

Human Resource Management Resources on the Internet
http://www.nbs.ntu.ac.uk/staff/lyerj/hrm   link.htm
This is also known as Ray's List of HRM Connections. It includes several categories of links that can be useful, including publications, organizations, and lots of other materials. Though it comes from the U.K., the majority of links seem to be U.S. operations. Rating: ★★★

Instructional Technology
http://www.sci.csupomona.edu/seis/ist/ist.html
This site from California State Polytechnic University–Pomona provides a listing of links to universities around the world that offer programs in instructional technology. Very complete listing. Rating: ★★★

The Keirsey Temperament Sorter
http://sunsite.unc.edu/personality/keirsey.html
This site includes a personality type indicator similar to Myers-Briggs. You can take a seventy-question test here and have it scored. For those interested in personality types, this site is worth visiting. Rating: ★★★

The Masie Center: The Technology and Learning Thinktank
http://www.masie.com/
This is the home page of a private training company. It includes good information on computer-based training. Rating: ★★★

Society for Human Resource Management
http://www.shrm.org/
This site includes lots of information about the major HR society in the United States and *HRMagazine*, the society's official publication. Also includes an extensive listing of links to other sites. Rating: ★★★

TCM (Targeted Communication Management) Training and Development Resource Page
http://www.tcm.com/trdev/
This site includes links to all kinds of information and sites dealing with training issues, including a comprehensive list of Internet discussion groups related to training. Very well done and complete. Rating: ★★★

Training and Development via the Internet
http://cac.psu.edu/~cxl18/trdev/
This site includes a wide variety of articles, case studies, and links devoted to the topic of training and development using the Internet. A very good resource on this topic. Rating: ★★★

The Training Registry
http://www.tregistry.com/ttr/home.htm
The Training Registry is a directory of training courses and vendors listed by category and topic. The registry provides a quick and easy way to link up to many training sites and browse descriptions on a variety of courses from many different training vendors. They update their listings regularly. Rating: ★★

The Training and Seminar Locator
http://www.tasl.com/tasl/home.html
This site includes databases that allow you to search for events, products, and providers on many different business topics. Especially useful to people looking for seminars to specific topics, either public events or events at your company. Rating: ★★

Other

The Biz
http://www.thebiz.co.uk/
This site includes listings of and links to organizations throughout the U.K. Recommended as a good reference site for gathering information on British businesses.

Book Stacks Unlimited
http://www.books.com/scripts/place.exe
This is not a quality management–related site, but it includes links to most publishers and can be a valuable resource if you are looking for a particular title. You can search by subject or by publisher.

The Copyright Website
http://www.benedict.com/

This site provides real-world, practical, and relevant copyright information to anyone interested in copyrighting writing, music, or whatever. Lots of useful information on this subject.

The Dilbert Zone
http://www.unitedmedia.com/comics/dilbert/

OK, everybody else references it, so we will, too. Includes cartoons, information on Dilbert and his creator, and all kinds of stuff you can buy related to Dilbert. Fun.

Ecola
http://www.ecola.com/

This is a search engine/directory for locating thousands of newspapers and magazines that have Web sites. You can click on any listing and go to that site. Organized by topic and location. Also includes directories for hardware and software vendors, colleges, and bookstores. A valuable resource.

Goff's Cartoons for Newsletters and Magazines
http://www.fileshop.com/personal/tgoff/

This is cartoonist Ted Goff's page. It has a lot of sample cartoons related to business and safety on the job and provides information on gaining permission to use the cartoons in newsletters and other publications. Cartoons are funny and relevant.

Internet Glossary
http://www.jwtworks.com/hrlive/netresults/glossary.html

This is a useful site that has short definitions of many terms dealing with the Internet that you can print out and have on hand.

Maplewood Books
http://www.total-info.com/bookstre.htm

This is the site of Maplewood Books, an on-line technical bookstore. They carry about 8,000 titles dealing with wide variety of business, science, and engineering titles. If you're looking for a technical book, this is a good place to try.

The MCB University Press
http://www.mcb.co.uk/

This is a major U.K. publisher of journals related to quality, training, and engineering, among other areas. This site includes detailed information about MCB's journals, how to subscribe, and lots of other related information. Extensive and well done.

McGraw-Hill
http://www.books.mcgraw-hill.com/
This is the McGraw-Hill Web site, from which you read about and order all books published by this company. The site includes a powerful search engine for finding titles by author, title, subject, or subject matter.

Private On-Line Quality Information Suppliers

The Benchmarking Exchange
This is a private site accessible only through proprietary software. TBE is a comprehensive and user-friendly electronic communication and information system designed specifically for use by individuals and organizations involved in benchmarking and process improvement. TBE provides users with a comprehensive, centralized, and specialized forum for all phases of benchmarking. You can learn about this commercial on-line service by checking out their Web site at **http://www.benchnet.com/**. There are various subscription prices, but the standard is $295 a year for an individual. There are various other plans for companies, including site licenses. Rating: ★★★

The Quality Online Forum (QOF)
Quality Online Forum (QOF) is on-line service devoted to quality improvement and related issues. It contains a wide range of resources including bibliographies, book reviews, the full text of magazines and newsletters, video listings, demo software, catalogs, over thirty discussion groups, an extensive events calendar, and much more. Some of this information is not available anywhere else on the Internet. QOF is accessible by direct dial or through the Internet. Beginning in early 1997, Quality Online will be accessible via any Web browser. Like America Online, QOF requires proprietary software, which is available free of charge (E-mail info@qof.com or call 913-379-5590). A one-year subscription is only $60. A one-month free trial subscription is also available. Rating: ★★★

Magazines, Journals, and Newsletters That Cover Quality

This section includes a comprehensive (though not exhaustive) listing of publications that regularly cover issues on quality. Some of them are dedicated to quality issues, and some of them cover quality as part of their regular editorial policy. Many quality organizations, such as the American Society for Quality Control and others, publish a variety of newsletters of interest to members, depending on area of specialization. Most local quality organizations also publish their own newsletters. Those publications are not listed here, as they are available only to members. Nearly all industry associations also publish newsletters or magazines. You might want to explore those by contacting organizations of interest listed in the section on best-practices organizations in this edition of *The Quality Yearbook*.

We have organized this listing into three categories:

1. Magazines, journals, and newsletters dedicated to quality

2. General business magazines, journals, and newsletters that often include articles on quality

3. Industry and special-interest magazines, journals, and newsletters that sometimes cover quality issues

All magazines, journals, and newsletters included have something of value to anyone interested in TQM, depending on background and specialty. However, to help you sort through these publications in terms of their value in learning about and implementing quality, we have developed the following ratings:

★★★ Highest recommendation, regularly includes very useful information on quality management
★★ Highly recommended, frequently includes useful articles
★ Recommended, includes some useful articles

Note: These ratings do not apply to the overall quality of the magazines but to their coverage of quality-management topics.

Magazines, Journals, and Newsletters Dedicated to Quality

Critical Linkages II Newsletter

Published by Sager Educational Enterprises five times a year (Jan., Mar., May, Sept., and Nov.), 21 Wallis Road, Chestnut Hill, MA 02167, phone: (617) 469-9644, fax: (617) 469-9639.

Subscriptions: U.S. 1 year $30; outside U.S. 1 year $36.

This newsletter is designed to help managers understand the interconnections between people, departments, customers, suppliers, schools, and the community, and how that understanding can help them manager better. Special emphasis on education. Rating: ★

Customer Service Manager's Letter

Published monthly by Bureau of Business Practice (Division of Simon & Schuster), 24 Rope Ferry Road, Waterford, CT 06386, phone: (800) 876-9105.

Subscriptions: U.S. 1 year $144; 2 or more subscriptions 1 year $130 each. International: call or write.

This is an eight-page, full-color newsletter that has advice aimed at customer service managers. Emphasizes issues involved in implementing TQM both in terms of teamwork and empowerment and in terms of service excellence. Rating: ★★

Eye on Improvement

Published twenty-four times a year by the Institute for Healthcare Improvement, P.O. Box 38100, Cleveland, OH 44138-0100, phone: (800) 895-4951. (Note the IHI also publishes *Quality Connection* out of Boston).

Subscriptions: U.S. and Canada: 1 year $120; International: 1 year $140.

A newsletter designed to spread the word quickly to frontline health systems leadership about useful information on quality-management techniques and practices. Most of the articles are short digests of material that has appeared in many other publications that would be of interest to health-care professionals. Edited by Donald Berwick, M.D., a leading figure in implementing TQM in the delivery of health care. Rating: ★★★

Healthcare Quality Abstracts

Published monthly, except in July, by COR Healthcare Resources, P.O. Box 40959, Santa Barbara, CA 93140, phone: (805) 564-2177, fax: (805) 564-2146.

Subscriptions: U.S. and Canada: 1 year $98; all other countries: 1 year $110.

This is a newsletter format abstracting current articles from a variety of journals all dealing with quality management in health care. The listing is broken up by topic and the sources for all articles are listed on the back page with addresses and phone numbers of publishers. Rating: ★★

IEEE Engineering Management Review

Published quarterly by the Institute of Electrical and Electronic Engineers, 345 East 47th Street, New York, NY 10017-2394, phone: (212) 705-7900.

Subscriptions: For IEEE Engineering Society members, included in membership. Nonmember subscriptions available on request.

This is an interesting journal. Each issue takes a certain theme, often related to quality-management issues such as teamwork or the virtual enterprise, and then includes a variety of articles by leading writers, researchers, and practitioners reprinted from other journals over the past few years. An interesting concept and well done. Rating: ★★★

International Journal of Quality and Reliability Management

Published nine times a year by MCB University Press Ltd., 60/62 Toller Lane, Bradford, West Yorkshire, England BD8 9BY. For more information, see http://www.mcb.co.uk/.

Subscriptions: $2,629 per year.

This journal deals with all aspects of business and manufacturing improvements. It includes a combination of technical and nontechnical articles aimed at managers and engineers. MCB journals tend to have very high subscription rates, so keep that in mind in considering this journal. Rating: ★★

International Journal of Reliability, Quality, and Safety Engineering

Published quarterly by World Scientific Publishing Co., Farrer Road, P.O. Box 128, Singapore 912805, or 1060 Main Street, River Edge, NJ 07661.

Subscriptions: Contact publisher for information.

This is a technical journal of articles that cover the topics of the title. Typical articles have titles such as "Failure Mechanisms and Life Models" and "Statistical Methods for Management of Process Quality: A State-of-the-Art Review." Rating: ★

The Joint Commission Journal of Quality Improvement

Published monthly under the editorial direction of The Joint Commission by Mosby-Year Book, Inc., 11830 Westline Industrial Drive, St. Louis, MO 63146-3318, phone: (314) 453-4351 or (800) 453-4351.

Subscriptions: $115 per year for U.S.; all other countries, $125 per year.

This journal replaced the former journal published by The Joint Commission, the *Quality Review Bulletin*. The goal of this refereed journal is to publish articles that emphasize the improvement of health-care quality, which includes the measurement, assessment, and improvement of performance in health-care quality and delivery. The journal includes how-to articles and case studies. Rating: ★★★

Journal for Quality and Participation

Published seven times a year (Jan./Feb., March, June, July/Aug., Sept., Oct./Nov., and Dec.) by the Association for Quality and Participation, 801-B West 8th Street, Suite 501, Cincinnati, OH 45203, phone: (513) 381-1959, fax: (513) 381-0070.

Subscriptions: Available as part of membership in AQP. Nonmembers: $60 per year in North America and $75 outside of North America.

Includes a variety of practical, detailed articles on implementing quality in different industries and organizations. Often includes articles by well-known quality practitioners, authors, and consultants. Articles are nearly always intriguing, useful, and well written. One of the best magazines available on quality for managers. Rating: ★★★

Journal of Quality Technology

Published quarterly by the American Society for Quality Control, 611 E. Wisconsin Avenue, P.O. Box 3005, Milwaukee, WI 53201-3005, phone: (800) 248-1946, (414) 272-8575, fax: (414) 272-1734.

Subscriptions: For ASQC members, annual subscription is $20 in the U.S., $38.50 in Canada, and $36 for other international; for nonmembers, $30 annually in the U.S., $40 in for other international.

A technically oriented journal, that uses statistics heavily, emphasizing the practical applicability of new techniques and providing instructive examples of the operation of existing techniques and the results of historical researches. Useful only to those involved in quality control technology. Rating: ★★

Manufacturing Engineer

Published monthly by the Institution of Electrical Engineers, Savoy Place, London, WC2R 0BL, United Kingdom, phone: 011-171-240-1871, fax: 011-171-240-7735.

Subscriptions: Write or call for information.

This is a British publication and seems quite well done, mixing technical and management-oriented articles in a color magazine format. Includes a number of departments, such as news digest, software news, and book reviews. Rating: ★★

National Productivity Review

Published quarterly by John Wiley & Sons, 605 Third Avenue, New York, NY 10158-0012, phone: (212) 850-6470, E-mail: SUBINFO@jwiley.com.

Subscriptions: U.S. and Canada, 1 year $220; international, 1 year $244. Discounts available on multiple copy subscriptions.

A journal that includes practical articles focusing on the implementation of quality in all types of organizations. The articles are often written by the people who did the work. The writing is uneven, but the content is nearly always valuable. The articles tend to be long, so they require some time to read and reflect on. Divided into three sections: Ideas and Opinions, Features, and Reviews. Rating: ★★★

PI Quality

Published bimonthly by Hitchcock Publishing Co., 191 Gary Avenue, Carol Stream, IL 60188-2292, phone: (630) 665-1000, fax: (630) 462-2225.

Subscriptions: Free to qualified individuals in process industries or $40 per year in the U.S., and $100 per year for all other countries.

Because it is aimed at qualified circulation, this magazine includes a variety of articles dealing with the use of various technologies in the process industries (which include food, beverages, chemicals, pharmaceuticals, soaps, textiles, papers, petroleum, rubber, plastics, stone, clay, glass, leather, and primary metal processing). Articles tend to short and technology oriented. Rating: ★

Quality

Published monthly by Hitchcock Publishing Company, 191 S. Gary Avenue, Carol Stream, IL 60188, phone: (630) 665-1000, fax: (630) 462-2225.

Subscriptions: U.S., 1 year $70. Canada and Mexico, 1 year $85. International, 1 year $195.

Includes articles that focus mainly on the technical aspects of implementing total quality management in production and manufacturing environments. Includes reviews and event calendar. Especially of interest to engineers and quality technicians. Rating: ★★★

The Quality Connection

Published quarterly by the Institute for Healthcare Improvement, One Exeter Plaza, Ninth Floor, Boston, MA 02116, phone: (617) 424-4800, fax: (617) 424-4848.

Subscriptions: This newsletter is distributed free of charge to anyone interested in quality management in health care.

The newsletter includes interviews, commentaries, and other topics of interest on implementing quality management in health care. Donald Berwick, one of the leaders in this area, is the president and CEO of IHI, and this newsletter is part of their effort to spread the word. Because it is free, it won't get a rating, but it is definitely worth receiving if you have an interest in this area.

Quality Digest

Published monthly by QCI International, 1350 Vista Way, P.O. Box 882, Red Bluff, CA 96080, phone: (916) 893-4095, fax: (916) 893-4095, E-mail: Qualitydig@aol.com, Web page: http://www.tqm.com/digest/.

Subscriptions: Official rate $75 per year but discounted to $49 for United States; Canada and Mexico, $59 per year; all other countries, $69 per year.

Includes a variety of how-to articles and pieces on how various organizations implement quality in both technical and people-related areas. Regularly includes articles on ISO 9000 issues. Also includes monthly columnists such as Karl Albrecht, Ken Blanchard, A. Blanton Godfrey, and Paul Scicchitano plus software reviews, book reviews, and other information. Rating: ★★★

Quality Engineering

Published quarterly by the American Society for Quality Control and Marcel Dekker, Inc. Subscriptions available through Marcel Dekker Journals, P.O. Box 5017, Monticello, NY 12701-5176.

Subscriptions: 1 year $46.75; add $14 for surface mail and $22 for airmail to Europe and $26 for airmail to Asia (ASQC members: $30.25 U.S. and $46.75 international).

Dedicated to quality-management articles that deal with "What the problem was, how we solved it, and what the results were." Articles tend to be detailed and moderately technical but quite relevant to quality professionals. Rating: ★★

Quality and Reliability Engineering International
Published bimonthly by John Wiley & Sons Ltd., Baffins Lane, Chichester, Sussex PO19 1UD, England.

Subscriptions: 1 year $645.

A technical journal carrying articles designed to fill the gap between theoretical methods and scientific research on one hand and current industrial practices on the other. Highly specialized and mathematical. Recommended only for corporate or university libraries for use by engineers in this area. Rating: ★★

Quality in Higher Education
Published monthly by Magna Publications Inc., 2718 Dryden Drive, Madison, WI 53704-3086, phone: (608) 246-3580 or (800) 433-0499.

Subscriptions: $129 per year with discounts available for additional subscriptions to the same location.

A monthly eight-page, two-color newsletter covering ideas on applying TQM principles, case studies, information on TQM tools, and other practical articles on quality management in colleges and universities. Rating: ★★

Quality Management Journal
Published quarterly by the American Society for Quality Control, 611 E. Wisconsin Avenue, P.O. Box 3005, Milwaukee, WI 53201-3005, phone: (800) 248-1946, (414) 272-8575, fax: (414) 272-1734.

Subscriptions: Available to members of ASQC in the United States at $50 annually, to nonmembers at $60; in Canada at $74 to members and $84 to nonmembers; international: $74 to members and $84 to nonmembers.

A peer-reviewed journal designed to present academic research on quality management in a style that makes it accessible to managers in all fields. The quality of the articles has been mixed. Rating: ★★

Quality Progress
Published monthly by the American Society for Quality Control, Inc., 611 E. Wisconsin Avenue, P.O. Box 3005, Milwaukee, WI 53201-3005, phone: (800) 248-1946, (414) 272-8575, fax: (414) 272-1734,
Web page: http://qualityprogress.asqc.org.

Subscriptions: Available as part of membership in the ASQC. $50 per year for nonmembers in the U.S. and $85 for first class to Canada and international airmail.

This is the foremost magazine dealing with quality-management topics. Every issue is devoted to a theme but also includes several other articles dealing with a broad spectrum of issues. The articles are nearly always practical and provide perspectives that anyone interested in quality management will find valuable. It is the source of many articles in *The Quality Yearbook*. Includes event calendars, reviews, and many other regular features. The Web page has abstracts of articles going back to 1990 by keyword, title, and author. Rating: ★★★

Strategies for Healthcare Excellence

Published monthly by COR Research Inc., P.O. Box 40959, Santa Barbara, CA 93140-0959, phone: (805) 564-2177.

Subscriptions: U.S. and Canada: 1 year $197; all other countries: 1 year $210.

This is a twelve-page, two-color newsletter that includes detailed case studies on how specific health-care facilities and managers are implementing quality principles. Each issue usually includes one long piece followed by other short articles on this subject. Rating: ★★

Target

Published six times a year by the Association for Manufacturing Excellence, 380 W. Palatine Road, Wheeling, IL 60090-5863, phone: (847) 520-3282, fax: (847) 520-0163.

Subscriptions: Available to members of AME as part of membership. Cost of annual membership: $125.

Includes many practical articles on how various industries and companies are implementing TQM in manufacturing and management. Also includes reports from regional chapters, book reviews, and an event calendar. Practical, accessible, and well done. Rating: ★★★

The Quality Observer

Published monthly by The Quality Observer Corporation, 3970 Chain Bridge Road, P.O. Box 1111, Fairfax, VA 22030, phone: (703) 691-9496, fax: (703) 691-9399, Web page: http://www.thequalityobserver.com/.

Subscriptions: 1 year $99, 2 years $162, and 3 years $237; Overseas, 1 year $135, 2 years $221, and 3 years $323; Libraries: 1 year $168; Corporations: 1 year $293 (5 copies each issue sent to same address). Visit the Web site for discounted rates.

This was formerly a tabloid newspaper publication but has switched to a regular magazine format. It publishes a wide variety of articles dealing with quality management such as company profiles, management improvement tips, ISO 9000, teamwork, and so on. Articles tend to be a little less detailed than in other similar quality magazines. Rating: ★★

The Systems Thinker

Published ten times a year by Pegasus Communications, Inc., P.O. Box 120, Kendall Square, Cambridge, MA 02142, phone: (617) 576-1231, fax: (617) 576-3114.

Subscriptions: All subscriptions, one year: $167.

A thoughtful and important newsletter for anyone interested in using the systems view of organizations to better understand and improve organizational processes. The newsletter includes articles that help you recognize various patterns of behavior and case studies of systems thinking implementation. Systems thinking is the foundation for effectively implementing TQM. Rating: ★★★

Total Quality Environmental Manager

Published quarterly by John Wiley & Sons, Inc., 605 Third Avenue, New York, NY 10158-0012, phone: (212) 850-6479, E-mail: SUBINFO@jwiley.com.

Subscriptions: U.S., Canada, Mexico 1 year $159; other countries 1 year $209.

Although aimed at managers and, to some degree, engineers in the environmental area, its articles are practical and cover a broad spectrum in the application of TQM principles to this field. Worth having in the library of any company interested in this field. Rating: ★★★

Total Quality Management

Published six times a year by Carfax Publishing Company, P.O. Box 25, Abington, Oxfordshire OX14 3UE, United Kingdom.

Subscriptions: Information available by writing to publisher.

An academic journal with a practical bent, it includes articles and case studies of interest to managers and academics. It covers all aspects of quality management. Rating: ★★

The TQM Magazine

Published bimonthly by MCB University Press Ltd., 60/62 Toller Lane, Bradford, West Yorkshire, England BD8 9BY. For more information, see http://www.mcb.co.uk/.

Subscriptions: $679 per year.

Organized around different themes each month, this magazine includes a variety of how-to articles and descriptions of what is going on in various organizations. Subscription is mainly oriented to university and corporate libraries with a direct interest. Rating: ★★

General Business Magazines, Journals, and Newsletters That Include Coverage of Quality

Across the Board

Published ten times annually (July–August and November–December are combined issues) by The Conference Board, Inc., 845 Third Avenue, New York, NY 10022, phone: (212) 759-0900, fax: (212) 980-7014.

Subscriptions: $20 annually for Conference Board associates and $40 annually for nonassociates; outside the United States, add $7.

A thoughtfully edited magazine on issues of interest to all managers. Includes columns, commentaries, how-to articles, company profiles, and issue-related articles. Articles are occasionally directly related to quality, but nearly all are indirectly related. Rating: ★★★

Business Ethics
Published bimonthly by Mavis Publications, Inc., 52 S. 10th Street, #110, Minneapolis, MN 55403-4700, phone: (612) 962-4700.

Subscriptions: $49 per year in the U.S. and $59 for international subscriptions.

Includes articles on topics of ethical and social concern to business, all of which are arguably related to quality in one way or another. Occasionally includes articles specifically on quality management and its implementation in socially responsible organizations. Rating: ★★

Business Horizons
Published bimonthly by JAI Press, Inc., 55 Old Post Road, No. 2, P.O. Box 1678, Greenwich, CT 06836-1678, phone: (203) 661-7602, fax: (203) 661-0792.

Subscriptions: United States, $75 per year; outside the U.S., surface mail $95 and airmail $105 per year.

Published out of the Indiana University Graduate School of Business, this journal includes a diversity of articles of interest to managers, often shorter and less profound than those found in journals such as the *Harvard Business Review*. Articles included regularly relate in a direct or indirect way to quality management. Rating: ★★

Business Week
Published weekly by McGraw-Hill, Inc., 1221 Avenue of the Americas, New York, NY 10020, phone: (800) 635-1200, (212) 512-2000.

Subscriptions: Official subscription rate: 1 year $49.95, 2 years $79.95, and 3 years $99.95. Widely available at discounted subscription rates.

The leading business newsweekly in the United States, *Business Week* will occasionally include articles that deal with quality-management issues or profile companies that practice TQM. Rating: ★★

California Management Review
Published quarterly by the University of California, 2549 Haas School of Business, University of California, Berkeley, CA 94720-1900, phone: (510) 642-7159, fax: (510) 642-1318, E-mail: cmr@haas.berkeley.edu.

Subscriptions: United States, Canada, and Mexico, 1 year $50, 2 years $90, and 3 years $120; international subscriptions, 1 year $80, 2 years $150, and 3 years $210.

A serious journal, but with a practical orientation, often including in-depth articles on management that are related directly or indirectly to TQM. The tone of the journal is somewhat more academic than the *Harvard Business Review*, as most contributors are professors. Rating: ★★

Executive Excellence

Published monthly by Executive Excellence Publishing, 3507 North University, Suite 100, Provo, UT 84604-4479, phone: (800) 304-9782.

Subscriptions: U.S. and Canada: 1 year $129, 2 years $199, 3 years $297. Other countries: 1 year $169, 2 years $279, 3 years $407.

This is the newsletter from Stephen R. Covey's group. It bills itself as "The magazine of leadership development, managerial effectiveness, and organizational productivity." It is usually twenty pages long, beautifully designed, and printed in full color. The articles are usually around one to two pages each, contributed by well-known consultants and executives. All the articles are fundamentally based on quality-management principles and Covey's seven habits. Reasonably well done, but superficial. Rating: ★★

Fortune

Published biweekly by Time Warner, Inc., Time & Life Building, Rockefeller Center, New York, NY 10020, phone: (800) 621-8000.

Subscriptions: Official subscription rate: $57 1 year, U.S., and $65 1 year, Canada. Widely available at discounted subscription rates.

Provides in-depth articles on management and other business topics and profiles of executives and companies. Articles do not often focus on quality per se, but they are useful as benchmarks for understanding quality-management practices in relation to traditional perspectives on managing. Rating: ★★

Harvard Business Review

Published bimonthly by the Harvard Business School, 60 Harvard Way, Boston, MA 02163, phone: (800) 274-3214, (617) 495-6800.

Subscriptions: United States, $85 per year; Canada and Mexico, $95 per year; all other countries, $145 per year.

Includes in-depth articles on management techniques in all functional areas by highly regarded researchers and executives. Often includes articles of direct or indirect relevance to those interested in quality. Rating: ★★★

Inc.

Published monthly by *Inc.*, 38 Commercial Wharf, Boston, MA 02110, phone: (800) 234-0999, (617) 248-8000, fax: (617) 248-8277, E-mail: subscribe@incmag.com.

Subscriptions: Available at various rates, often discounted. Often offered at $19 for one year or $29 for 2 years.

The leading magazine covering issues of interest to small and medium-sized businesses. Includes lots of practical how-to techniques and articles and company and executive profiles. Sometimes covers issues directly related to quality management, though it only occasionally uses the word *quality* to describe this approach. Rating: ★★

Industry Week

Published biweekly by Penton Publishing Company, 1100 Superior Avenue, Cleveland, OH 44114-2543, phone: (800) 326-4146, (216) 696-7000, fax: (216) 696-6023.

Subscriptions: Distributed as a closed circulation magazine to qualified executives in administration, finance, production, engineering, purchasing, marketing, and sales. To those who do not qualify, it is available by subscription in the U.S. at $60 for 1 year, $95 for 2 years; Canada $80 for 1 year, $140 for 2 years; Mexico $90 for 1 year, $160 for 2 years; all other countries $100 for 1 year, $180 for 2 years.

Includes articles, columns, and reviews of timely interest to managers. Over the past two years, with a change of editor, the emphasis has been less on topics directly relating to quality management. It also tends to feature profiles of individual managers. Rating: ★★

Management Review

Published monthly by the American Management Association Publications Division, Box 408, Saranac, NY 12983-0408. For information, contact American Management Association, 1601 Broadway, New York, NY 10019, phone: (212) 903-8283, fax: (212) 903-8033.

Subscriptions: $45 per year in the U.S.; Non-U.S. subscriptions $60 per year.

This is the official magazine of the American Management Association. It includes a variety of articles, columns, and case studies of interest to managers and often includes pieces directly or indirectly about quality issues. Articles from *Management Review* are regularly included in *The Quality Yearbook*. Rating: ★★★

Nation's Business

Published by the U.S. Chamber of Commerce, 1615 H Street N.W., Washington, DC 20062-2000, phone: (800) 638-6582, (202) 463-5650.

Subscriptions: 1 year $22, 2 years $35, and 3 years $46; Canadian and international subscriptions add $20 to each rate. Often available at discounted rates.

Called the "small business adviser," this magazine includes a wide variety of articles, mostly short, on many different business topics. It sometimes includes articles related to quality management. Rating: ★

Organizational Dynamics

Published quarterly by the American Management Association, 1601 Broadway, New York, NY 10019, phone: (212) 903-8283, fax: (212) 903-8033.

Subscriptions: United States, $63 1 year, $107.10 2 years (AMA members get a ten percent discount); all other non-U.S. subscriptions add $35 per year for postage.

This journal usually includes five or six long articles per issue dealing with topics of people management and organizational management. There have been entire issues devoted to learning organizations. The articles often deal with topics related to quality management in one way or another. The articles are thoughtful and practical. Well worth looking at. Rating: ★★★

Sloan Management Review

Published quarterly by the MIT Sloan School of Management, 292 Main Street, E38-120, Cambridge, MA 02139-4307, phone: (617) 253-7170, fax: (617) 253-5584.

Subscriptions: $59 per year; $79 for Canada and Mexico, and $89 for all other international subscriptions.

Includes practical yet thoughtful articles on a variety of issues of direct interest to managers. Articles are often directly or indirectly related to quality issues. Usually has a theme for each issue. Includes good book review section. Rating: ★★★

Strategy & Business

Published quarterly by Booz-Allen & Hamilton, 101 Park Avenue, New York, NY 10178, phone: (888) 557-5550. Editorial offices: 67 Mount Street, Boston, MA 02108, phone: (617) 523-7047, fax: (617) 723-3989.

Subscriptions: United States and Canada, $38 1 year, $66 2 years, $97 3 years; elsewhere, $48 1 year, $86 2 years, and $127 3 years.

This is a new magazine that includes a variety of thoughtful articles in a nicely designed two-color journal format. The authors are prominent researchers, executives, and Booz-Allen consultants. Well-done and relevant to quality practitioners. Rating: ★★★

Strategy & Leadership

Published bimonthly by The Strategic Leadership Forum (formerly The Planning Forum), 435 N. Michigan Avenue, Suite 1700, Chicago, IL 60611-1700, phone: (800) 873-5995, (312) 644-0829, fax: (312) 644-8557, E-mail: 76105.1042@compuserve.com.

Subscriptions: United States and Canada, $95 per year, all other countries $115 per year.

This new version of what was *Planning Review* is a very well-done magazine that includes thoughtfully written articles by prominent consultants, professors, and executives. These articles often deal with issues of interest to those in quality management. This edition of the yearbook includes four articles from recent issues. Rating: ★★★

Supervisory Management

Published monthly by the American Management Association, 135 West 50th Street, New York, NY 10020, phone: (800) 759-8520, (212) 903-8075, fax: (212) 903-8083.

Subscriptions: U.S. 1 year $65, 2 years $110.50; all other countries add $10 per year.

This is a twelve-page, two-color newsletter oriented especially toward the needs of first line supervisors. It regularly deals with issues related to quality management and has special regular sections on teams and teamwork. Lots of short pieces; practical and down-to-earth. However, it is heavy on principles and light on examples from real companies. Rating: ★★

Tom Peters on Achieving Excellence

Newsletter published monthly by TPG Communications, P.O. Box 2189, Berkeley, CA 94702-0189, phone: (800) 959-1059.

Subscriptions: U.S. 1 year $99. Call for international rates.

A slick, readable, and practical twelve-page, two-color newsletter in the Peters style. Each issue has a theme and lots of short pieces on what different companies and people are doing to solve various business problems. Rating: ★★

Industry and Special-Interest Magazines, Journals, and Newsletters That Cover Quality

APICS—The Performance Advantage

Published quarterly by the American Production and Inventory Control Society, Inc., 500 West Annandale Road, Falls Church, VA 22046-4274, phone: (800) 392-7294 (subscriptions), (703) 237-8344, fax: (703) 237-4316.

Subscriptions: Included as part of membership package. Nonmembers subscriptions: United States $45 per year; Mexico and Canada, $55 per year; elsewhere, $70 per year.

Covers the latest manufacturing principles and practices, case studies, columns, and news. Articles are sometimes technical in nature, as is appropriate given the audience. Often includes articles on quality management in its field. Rating: ★★

Bank Management

Published bimonthly by Bank Administration Institute, One North Franklin, Chicago, IL 60606, phone: (800) 655-2706.

Subscriptions: U.S. 1 year $59; international 1 year $89.

Oriented toward strategic issues in bank management. Attractively produced magazine that occasionally covers quality-management topics. Rating: ★

Change

Published bimonthly by Heldref Publications, 1319 Eighteenth Street NW, Washington, DC 20036-1802, phone: (212) 296-6267, (800) 365-9753 for subscriptions.

Subscriptions: United States, $31 per year; outside the U.S., add $12 per year.

Billed as the magazine of higher learning, it addresses issues of interest to the administration of colleges and universities and occasionally carries articles on quality-management topics. Rating: ★

Distribution

Published monthly, but twice in April (thirteen issues total), by Chilton Company, One Chilton Way, Radnor, PA 19089, phone: (610) 964-4000.

Subscriptions: United States, $65 per year; Canada, $70 per year; all other countries, $95 per year.

This magazine deals with a variety of logistics and distribution issues and occasionally covers subjects dealing with the role of logistics in lowering costs and improving customer satisfaction. Includes a cover story, feature articles, and various departments. Rating: ★

Educational Leadership

Published monthly September through May, except for a bimonthly December–January issue, by the Association for Supervision and Curriculum Development, 1250 N. Pitt Street, Alexandria, VA 22314-1453, phone: (703) 549-9110, fax: (703) 549-3891.

Subscriptions: Part of membership package in ASCD. For nonmembers: 1 year $36. Call or write for international subscriptions.

Popular journal for educators and administrators. Occasionally includes articles covering quality principles in teaching and administration. Rating: ★

Enterprise Reengineering

Published monthly by Enterprise Reengineering, 7777 Leesburg Pike, Suite 315N, Falls Church, VA 22043, phone: (703) 761-0646, fax: 761-0766, E-mail: ER@reengineering.com, Web page: http://www.reengineering.com.

Subscriptions: Available free to qualified subscribers. Call to find out to get subscription questionnaire.

This is full-color, tabloid-style publication that offers a variety of articles dealing with reengineering efforts. It also includes lots of ads related to this subject. Worth finding out about. Rating: ★★

Healthcare Executive

Published bimonthly by the American College of Healthcare Executives, 840 North Lake Shore Drive, Chicago, IL 60611-9842.

Subscriptions: Write to Processing Center, 1951 Cornell Avenue, Melrose Park, IL 60160. 1 year $45, 2 years $80. For subscriptions outside the U.S., add $15 per year.

Aimed at health-care managers, this journal regularly covers quality-management issues. Most articles are relatively short and basic. Rating: ★★

Healthcare Forum Journal

Published bimonthly by The Healthcare Forum, 425 Market Street, San Francisco, CA 94105, phone: (415) 356-9312, fax: (415) 356-4312.

Subscriptions: Available as part of membership in The Healthcare Forum, or 1 year $45, 2 years $80, 3 years $115. Canada and Mexico add $15 per year; all other countries 1 year $90.

A well-done journal that regularly includes articles on issues related to quality management. Articles are often in-depth yet down-to-earth and readable. Includes many ads on resources for health-care managers who wish to implement TQM. Rating: ★★★

Health Care Management Review

Published quarterly by Aspen Publishers Inc., 7201 McKinney Circle, Frederick, MD 21701, phone: (800) 638-8437.

Subscriptions: $110 per year in the U.S. and Canada. For international subscriptions, contact Swets Publishing Service, P.O. Box 825, 2160 SZ Lisse, The Netherlands.

A refereed journal that regularly includes articles in the field of health-care management. Articles tend to be five to ten pages, including references, but nontechnical in nature. Rating: ★★

Hospitals

Published the fifth and twentieth of each month by American Hospital Publishing, Inc., part of the American Hospital Association, 840 N. Lake Shore Drive, Chicago, IL 60611, phone: (312) 280-6000.

Subscriptions: Included as part of membership. Nonmembers $60 per year; add $30 per year outside the U.S.

Focuses mainly on management issues for health-care executives. Occasionally includes articles on quality management. Rating: ★

HR Focus

Published by the American Management Association, 1601 Broadway, New York, NY 10019, phone: (212) 903-8283, fax: (212) 903-8033.

Subscriptions: United States 1 year $78.75 and 2 years $133.88. Add $20 annually for postage to Canada and $35 annually for all other countries.

Slick four-color newsletter-like publication, usually twenty-four pages long. Covers a variety of issues and news items of interest to HR managers. Sometimes includes articles on quality subjects. Rating: ★

HRMagazine

Published monthly by the Society for Human Resource Management, 606 N. Washington Street, Alexandria, VA 22314, phone: (703) 548-3440, fax: 836-0367, E-mail: SHRM@shrm.org, Web page: http://www.shrm.org/.

Subscriptions: Members of SHRM receive the magazine as part of their membership. Nonmember subscriptions available in North America, $60 per year; all other international, $105 per year. Available at discounted rates from postcards in magazine.

Dedicated to human resource issues, this magazine carries articles related to quality-management topics such as training and teamwork on a fairly regular basis. Rating: ★★

IIE Solutions

Published monthly by the Institute of Industrial Engineers, 25 Technology Park, Norcross, GA 30092, phone: (800) 494-0460, (770) 449-0461, fax: (770) 263-8532, Web page: http://www.iienet.org/.

Subscriptions: Included as part of the annual dues of IIE members. Nonmember subscriptions: United States, $54 1 year, $98 2 years, and $130 3 years; outside the U.S. $77 1 year, $124 2 years, and $185 3 years.

This is the official member publication of the IIE. Although aimed at the institute's membership, the magazine regularly includes both technical and nontechnical articles that deal with issues related to quality management. A useful publication. Rating: ★★★

Industrial Management
Published bimonthly by the Institute of Industrial Engineers, 25 Technology Park, Norcross, GA 30092, phone: (800) 494-0460, (770) 449-0461, fax: (770) 263-8532, Web page: http://www.iienet.org/.

Subscriptions: Member rate: $27 per year. Nonmembers: United States, $39 1 year, $71 2 years, and $94 3 years; international subscriptions: $50 1 year, $90 2 years, and $120 3 years.

A serious journal with an academic bent on management topics for engineers that sometimes includes quality-oriented articles. Rating: ★

Journal of Business Strategy
Published bimonthly by Faulkner & Gray, Inc., 11 Penn Plaza, New York, NY 10001, phone: (800) 535-8403.

Subscriptions: All locations: $98 per year, plus $4.95 for shipping and handling.

Each issue has a special focus, for example, outsourcing or small business, and includes several pieces on this subject. It also includes a special cover story and various other features.

The material covered in this magazine is generally related to TQM, but not usually directly. Rating: ★★

Journal of Health Care Marketing
Published quarterly by the American Marketing Association, 250 S. Wacker Drive, Chicago, IL 60606, phone: (800) 262-1150, (312) 648-0536, fax: (312) 993-7542, Web page: http://www.ama.org/.

Subscriptions: United States and Canada, $70 per year; other countries, $90 per year.

This journal covers in practical terms all issues that affect the marketing of health-care services. Regularly includes articles on developing quality relationships with customers. Rating: ★★

Management Accounting
Published monthly by the Institute of Management Accounting, 10 Paragon Drive, Montvale, NJ 07645-1760, phone: (201) 573-9000, fax: (201) 573-0639.

Subscriptions: Included as part of membership in IMA. Nonmembers: $130 per year, all locations.

Regularly includes articles of relevance to quality practitioners on subjects such as activity-based costing and other subjects that relate accounting practices to efficient management and customer satisfaction. This magazine is a frequent source of articles for *The Quality Yearbook* accounting and finance section. Rating: ★★

Manufacturing Engineering
Published monthly by the Society of Manufacturing Engineers, P.O. Box 930, Dearborn, MI 48121, phone: (313) 271-1500.

Subscriptions: Closed-circulation magazine sent free to members of the society and others involved with manufacturing. Personal subscription information available by contacting SME.

Covers articles on manufacturing, with some coverage of quality issues, including a monthly column, "The Quality Adviser." Rating: ★

Modern Healthcare
Published weekly by Crain Communications Inc., 965 E. Jefferson, Detroit, MI 48207-3185, phone: (800) 678-9595.

Subscriptions: 1 year $110 and 2 years $200; add $48 for all non-U.S. subscriptions.

A health-care industry weekly magazine that includes news articles and a variety of articles on managing health-care facilities. Occasionally includes articles on quality management. Rating: ★

Performance Improvement
Published monthly except for combined May–June and November–December issues by the International Society for Performance Improvement, ISPI Publications, 1300 L Street N.W., Suite 1250, Washington, DC 20005, phone: (202) 408-7969, fax: (202) 408-7972, E-mail: info@ispi.org.

Subscriptions: Available as part of membership; annual subscriptions available to nonmembers for $50.

Includes articles on a variety of topics related to training and performance improvement written by practitioners. Quality of writing varies, but articles are almost always practical and often directly or indirectly related to TQM. Rating: ★★

Sales & Marketing Management
Published monthly by Bill Communications, 355 Park Avenue South, New York, NY 10010, phone: (212) 592-6308, fax: (212) 592-6309, E-mail: 758 3876@mcimail.com.

Subscriptions: United States, $48 per year. Contact publisher for international rates.

This magazine includes articles on marketing and sales management issues, with emphasis on technique as well as on executive and company profiles. Includes various columns, including books reviews. Sometimes includes articles about developing and improving customer satisfaction. Rating: ★

The School Administrator
Published monthly by the American Association of School Administrators, 1801 N. Moore Street, Arlington, VA 22209, phone: (703) 875-7905.

Subscriptions: Included as part of membership.

Dedicated to the problems and opportunities of school administrators, occasionally including articles on quality in schools. Rating: ★

Training

Published monthly by Lakewood Publications, Inc., 50 S. Ninth Street, Minneapolis, MN 55402, phone: (800) 328-4329, (612) 333-0471, fax: (612) 333-6526.

Subscriptions: United States: 1 year $64, 2 years $108, and 3 years $138; Canada and Mexico, 1 year $74; international, 1 year $85.

Includes articles on issues of interest to corporate trainers and managers in general, as well as frequent articles on quality management and training. Articles are practical, well-written, and useful to all involved in quality management. Often includes articles on quality-management topics. Rating: ★★★

Training & Development

Published monthly by the American Society for Training and Development, Inc., 1640 King Street, Box 1443, Alexandria, VA 22313-2043, phone: (703) 683-8100, fax: (703) 683-9203.

Subscriptions: Members of ASTD receive this magazine as part of their membership. Nonmember subscriptions: United States, $75 per year; international, $165 per year.

Includes articles on training, human resources, performance, and management issues, with frequent coverage of quality-management topics. A good resource for all areas of HR and people management. Has lots of practical features, including book reviews and new learning tools sections. Rating: ★★★

Directory of Organizations and Companies to Help You Identify Best Practices

This is a new section for *The Quality Yearbook*. What you will find here are lists of industry associations, publications and directories, and companies that can help you identify best practices and with whom you might be able to set up benchmarking agreements. We have divided this directory into two parts. The first part is organized by industry and includes industry associations followed by related publications and directories that can supply you with information about best practices in these industries.

Industry associations are frequently overlooked as sources of benchmarking and best-practices data, yet these associations frequently conduct surveys and collect comparative data on the performance and activities of their members. In industries undergoing profound changes, associations often conduct surveys that focus on performance and best practices. Some of the organizations and government agencies listed also publish industry-focused studies that are sources of good information on best practices. These associations may be willing to share with you benchmark consortium studies, most of which are not well publicized but are becoming critical. Higher education and the utilities industry, for example, are extensive users of such consortiums. You can also find them in health, insurance, banking, and other areas focusing on industry-wide practices.

We should also note that best practices are not necessarily industry specific. You may find that the functional practices of a company in one industry are directly applicable to a company in another. Further, it may be easier to benchmark the practices of a noncompeting company in a different industry than those of companies in the same industry. In addition, companies that have received the Baldrige Award have the obligation to share their management practices with others, and most have a detailed packet of information available on request.

The second part of this directory is organized by business function and practice and lists many companies that have been identified as having best practices in these areas. You will find that companies are often listed in more than one category, and several of these companies are Baldrige Award recipients. However, because best practices keep changing, it is always wise to validate that a company listed as having a best practice still does. They also know who the other organizations with similar best practices are.

Last, do not hesitate to call leading consulting companies such as the Big Six, IBM Consulting, and others to see what best practices they are willing to share. No self-respecting large consulting firm today would be without such a repository.

You will note that there are whole industries and major processes for which we do not list any obvious candidates. That is because the whole notion of collecting databases of best practices is still in its infancy. In addition, some companies consider their best practices a competitive advantage and thus do not want to broadcast them. For that reason, the only companies listed are those that have been publicly disclosed.

This list was put together in the best traditions of best practices and benchmarking. We started with the pioneering list of organizations and best practices put together by Robert J. Boxwell Jr. and presented in his book, *Benchmarking for Competitive Advantage* (McGraw-Hill, 1994), and added to it from other published sources.

Part One. Industry Associations and Publications

Accounting

American Institute of Certified Public Accountants
1211 Ave. of the Americas
New York, NY 10036-8775
212-575-6200

National Society of Public Accountants
1010 North Fairfax St.
Alexandria, VA 22314-1547
703-549-6400

Advertising

American Advertising Federation
1400 K St., NW, Suite 1000
Washington, DC 20005
202-898-0089

American Association of Advertising Agencies (AAAA)
6666 3rd Ave.
New York, NY 10017
212-682-2500

Association of National Advertisers, Inc.
155 E. 44nd St.
New York, NY 10017
212-661-8057

Magazine Publishers of America
575 Lexington Ave.
New York, NY 10022
212-752-0055

Aerospace

Aerospace Education Foundation
1501 Lee Highway
Arlington, VA 22209
703-247-5839

Aerospace Electrical Society
Village Station
P.O. Box 24BB3
Los Angeles, CA 90024
714-778-1840

Aerospace Industries Association of America
1250 I St. NW
Washington, DC 20005
202-371-8400

IEEE Aerospace and Electronics Systems Society (AESS)
c/o Institute of Electrical and Electronics Engineers
345 E. 47th St.
New York, NY 10017
212-705-7867

National Aeronautic Association of the U.S.A. (NAA)
1815 N. Fort Myer Dr.
Arlington, VA 22209
703-527-0226

Publications and Directories

Aerospace Daily
Murdoch Magazines
1156 15th St. NW
Washington, DC 20005
202-822-4600
212-512-2000

Aerospace Database
Technical Information Service
American Institute of Aeronautics and Astronautics
555 W. 57th St.
New York, NY 10019
212-247-6500

Aerospace Facts and Figures
McGraw-Hill, Inc.
1221 Ave. of the Americas
New York, NY 10020
800-262-4729
212-512-2000

National Aeronautics and Space Administration (NASA)
Scientific and Technical Information Division
300 7th St., SW
Washington, DC 20546
202-755-1099

PTS F&S Index
Predicasts, Inc.
11001 Cedar Ave.
Cleveland, OH 44106
800-321-6388

Apparel

Administrative Board—Dress Industry (ABDI)
450 7th Ave., 7th Floor
New York, NY 10001
212-239-2011

Affiliated Dress Manufacturers (ADM)
1440 Broadway
New York, NY 10018
212-398-9797

Allied Underwear Association (AUA)
100 E. 42nd St.
New York, NY 10017
212-867-5720

Amalgamated Clothing and Textile Workers Union (ACTWU)
15 Union Sq. W.
New York, NY 10003
212-242-0700

American Apparel Manufacturers Association (AAMA)
2500 Wilson Blvd., Suite 301
Arlington, VA 22201
703-524-1864

American Fashion Association (AFA)
The Dallas Apparel Mart, Suite 5442
Dallas, TX 75258
214-631-0821

American Fur Industry (AFI)
363 7th Ave., 7th Floor
New York, NY 10001
212-564-5133

Apparel Guild (AG)
Gallery 34, Suite 407
147 W. 33rd St.
New York, NY 10001
212-279-4580

Bureau of Wholesale Sales Representatives
1819 Peachtree Rd., NE, Suite 210
Atlanta, GA 3039
404-351-7355

Chamber of Commerce of the Apparel Industry (CCAI)
570 7th Ave.
New York, NY 10018
212-354-0907

Fiber, Fabric and Apparel Coalition for Trade (FFACT)
1801 K St., NW, Suite 900
Washington, DC 20006
202-862-0517

International Association of Clothing Designers
240 Madison Ave., 12th Floor
New York, NY 100016
202-685-6602

National Outerwear and Sportswear Association
240 Madison Ave.
New York, NY 10016
212-686-3440

Publications and Directories

Apparel Industry Magazine
Shore Publishing Co.
180 Allen Rd., NE, Suite 300N
Atlanta, GA 30328
404-252-8831

Highlights of U.S. Export and Import Trade
Bureau of the Census
U.S. Department of Commerce
Government Printing Office
Washington, DC 20402
202-783-3238

Monthly Labor Review
Bureau of Labor Statistics (BLS)
U.S. Department of Labor
Government Printing Office
Washington, DC 20402
202-783-3238

Overseas Business Reports
Industry and Trade Administration
U.S. Department of Commerce
Government Printing Office
Washington, DC 20402
202-783-3238

Textile Highlights
American Textile Manufacturers Institute, Inc.
1801 K St., NW, Suite 900
Washington, DC 20006
202-862-0500

Beverages

Can Manufacturers Institute (CMI)
1625 Massachusetts Ave., NW
Washington, DC 20036
202-463-6745

Carbonated Beverage Institute (CBI)
1101 16th St., NW
Washington, DC 20036
202-463-6745

Master Brewers Association of the Americas (MBAA)
4513 Vernon Blvd.
Madison, WI 53705
608-231-3446

National Beverage Dispensing Equipment Association (NBDEA)
2011 I St., NW, 5th Floor
Washington, D.C. 20006
202-775-4885

National Beverage Packing Association (NBPA)
c/o Jerry Testa
200 Daingerfield Rd.
Alexandria, VA 22314-2800
703-684-1080

National Juice Products Association (NJPA)
P.O. Box 1531
215 Madison St.
Tampa, FL 33601
813-229-1089

National Licensed Beverage Association (NLBA)
4214 King St., W
Alexandria, VA 22302
703-671-7575

National Soft Drink Association (NSDA)
1101 16th St., NW
Washington, DC 20036
202-463-6732

National United Affiliated Beverage Association (NUABA)
P.O. Box 9308
Philadelphia, PA 19139
215-748-5670

National United Licensees Beverage Association (NULBA)
7141 Frankstown Ave.
Pittsburgh, PA 15208
412-241-9344

Publications and Directories

Beverage Media
Beverage Media Ltd.
161 6th Ave.
New York, NY 10013
212-620-0100

Beverages (Periodical)
International Beverage Publications, Inc.
P.O. Box 7406
Overland Park, KS 66207
913-341-0020

Beverage World
Keller International Publishing Corp.
150 Great Neck Rd.

Great Neck, NY 11021
516-829-9210

Foods Adlibra
Foods Adlibra Publications
9000 Plymouth Ave., N
Minneapolis, MN 55427
612-540-3463

Hereld's 5000: The Directory of Leading U.S. Food, Confectionary, and Beverage Manufacturers
S.I.C. Publishing Co.
Whitney Center, Suite 330
200 Leader Hill Dr.
Hamden, CT 06617
203-281-6766

Impact Beverage Trends in America
N. Shanken Communications, Inc.
387 Park Ave, S.
New York, NY 10016
212-684-4224

Industry Norms and Key Business Ratios
Dun and Bradstreet Credit Services
1 Diamond Hill Rd.
Murray Hill, NJ 07974
800-351-3446
201-665-5330

National Food Review (Periodical)
Economics, Statistics and Cooperatives Service
U.S. Department of Agriculture
Government Printing Office
Washington, DC 20402
202-783-3238

Building Materials

American Building Contractors Association (ABCA)
11100 Valley Blvd., Suite 120
El Monte, CA 91731
818-401-0071

Associated Builders and Contractors (ABC)
729 15th St., NW
Washington, DC 20005
202-637-8800

Associated Building Material Distributors of America (ABMDA)
7100 E. Lincoln Drive, Suite D-220
Scottsdale, AZ 85253
602-998-0696

Builders' Hardware Manufacturers Association (BHMA)
355 Lexington Ave., 17th Floor
New York, NY 10017
212-661-4261

Building and Construction Trades Department
AFL-CIO
815 16th St., NW, Suite 603
Washington, DC 20006
202-347-1461

Building Officials and Code Administrators International (BOCA)
4051 W. Flossmoor Rd.
Country Club Hills, IL 60477-5795
708-799-2300

Door and Hardware Institute (DHI)
7711 Old Springhouse Rd.
McLean, VA 22102
703-556-3990

International Conference of Building Officials (ICBO)
5360 S. Workman Mill Rd.
Whittier, CA 90601
213-699-0541

National Association of Home Builders of the U.S. (NAHB)
15th and M St., NW
Washington, DC 20005
202-822-0200

NAHB National Research Center
400 Prince George's Blvd.
Upper Marlboro, MD 20772
301-249-4000

National Building Material Distributors Association (NBMDA)
1417 Lake Cook Rd., Suite 130
Deerfield, IL 60015
708-945-7201

National Council of the Housing Industry (NCHI)
15th and M St., NW
Washington, DC 20005
202-822-0520

Small Homes Council—Building Research Council
University of Illinois
1 E. Saint Mary's Rd.
Champaign, IL 61820
217-333-1801

Publications and Directories

Construction Review
Industry and Trade Administration
U.S. Department of Commerce
Government Printing Office
Washington, DC 20402
202-783-3238

Construction Review—Directory of National Trade Associations, Professional Societies, and Labor Unions of the Construction and Building Products Industries
U.S. Department of Commerce
Building Materials and Construction Division
Room 4043
Washington, DC 20230
202-377-0132

North American Wholesale Lumber Association—Distribution Directory
North American Wholesale Lumber Association
2340 S. Arlington Heights Rd., No. 680
Arlington Heights, IL 60005
312-981-8630

Chemicals

American Chemical Society (ACS)
1155 16th St., NW
Washington, DC 20036
202-872-4600

Chemical Communications Association (CCA)
c/o Fleishman-Hillard, Inc.
40 W 57th St.
New York, NY 10019
212-265-9150

Chemical Manufacturers Association (CMA)
2501 M St., NW
Washington, DC 20037
202-887-1100

Chemical Marketing Research Association (CMRA)
60 Bay St., Suite 702
Staten Island, NY 10301
718-876-8800

Chemical Sources Association (CSA)
1620 I St., NW, Suite 925
Washington, DC 20006
202-293-5800

Chemical Specialties Manufacturers Association (CSMA)
1913 I St., NW
Washington, DC 20006
202-872-8110

Council for Chemical Research (CCR)
1620 L St., NW, Suite 825
Washington, DC 20036
202-429-3971

Eastman Chemical Company
Stone East Building
Kingsport, TN 37662
(615) 229-5264

National Association of Chemical Distributors (NACD)
1200 17th St., NW, Suite 400
Washington, DC 20036
202-296-9200

Institute for Chemical Education (ICE)
Department of Chemistry
University of Wisconsin
1101 University Ave.
Madison, WI 53706
608-262-3033

Sales Association of the Chemical Industry (SACI)
287 Lackawanna Ave.
P.O. Box 2148
West Paterson, NJ 07424
201-256-5547

Publications and Directories

Annual Review of the Chemical Industry
United Nations Publishing Division
2 United Nations Plaza
New York, NY 10017
800-553-3210
212-963-8297

CA Search
Chemical Abstracts Service
P.O. Box 3012
Columbus, OH 43210
614-447-3600

Chemical Sources—International
Directories Publishing Co., Inc.
P.O. Box 1824
Clemson, SC 29633
803-646-7840

Directory of Chemical Producers—United States
SRI International
333 Ravenswood Ave.
Menlo Park, CA 94025
415-859-3627

Quarterly Financial Report for Manufacturing, Mining, and Trade Corporations
U.S. Federal Trade Commission and U.S. Securities and Exchange Commission
Government Printing Office
Washington, DC 20402
301-783-3238

Computers and Information Processing

ABCD: The Micro Computer Industry Association
1515 E. Woodfield Rd., Suite 860
Schaumburg, IL 60173-5437
708-240-1818

ADAPSO—The Computer Software and Services Industry Association
1300 N. 17th St., Suite 300
Arlington, VA 22209
703-522-5055

Association for Computer Educators (ACE)
c/o Dr. Ben Bauman
College of Business—IDS
James Madison University
Harrisonburg, VA 22807
703-568-6189

Association of Computer Professionals (ACP)
230 Park Ave., Suite 460
New York, NY 10169
212-599-3019

Computer-Aided Manufacturing International (CAMI)
1250 E. Copeland Rd., No. 500
Arlington, TX 76011
817-860-1654

Computer and Automated Systems Association of Society of Manufacturing Engineers (CASA/SME)
Box 930
1 SME Dr.
Dearborn, MI 48121
313-271-1500

Computer and Communications Industry Association (CCIA)
666 11th St., NW, Suite 600
Washington, DC 20001
202-783-0070

Data Processing Management Association (DPMA)
505 Busse Highway
Park Ridge, IL 60068
709-825-8124

Digital Equipment Computer Users Society (DECUS)
333 South St.
Shrewsbury, MA 01545
508-467-5111

IEEE Computer Society
1730 Massachusetts Ave. NW
Washington, DC 20036
202-371-0101

Independent Computer Consultants Association (ICCA)
933 Gardenview Office Parkway
St. Louis, MO 63141
314-997-4633

Information Technology Association of America (ITAA)
1616 N. Ft. Myer Drive, Suite 1300
Arlington, VA 22209-3106
703-522-3106

International Council for Computer Communications (ICCC)
P.O. Box 9745
Washington, DC 20016
301-530-7628

The International Microcomputer Information Exchange (TIMIX)
P.O. Box 201897

Austin, TX 78720
512-250-7151

National Computer Association (NCA)
1485 E. Fremont Circle S.
Littleton, CO 80122
303-797-3559

National Computer Dealer Forum (NCDF)
c/o National Office Products Association
301 N. Fairfax St.
Alexandria, VA 22314
703-549-9040

National Computer Graphics Association (NCGA)
2722 Merrilee Dr., Suite 200
Fairfax, VA 22031
703-698-9600

North American Computer Service Association (NACSA)
100 Silver Beach, No. 918
Daytona Beach, FL 32118
904-255-9040

Software Industry Section of ADAPSO
c/o ADAPSO
1616 N. Fort Myer, Suite 1300
Arlington, VA 22209-9998
703-522-5055

Software Maintenance Association (SMA)
P.O. Box 12004, No. 297
Vallejo, CA 94590
707-643-4423

Special Interest Group for Computers and Society (SIGCAS)
c/o Association for Computing Machinery
11 W. 42nd St.
New York, NY 10036
212-869-7440

World Computer Graphics Association (WCGA)
2033 M St., NW, Suite 399
Washington, DC 20036
202-775-9556

Publications and Directories

Applied Science and Technology Index
H.W. Wilson Co.
950 University Ave.
Bronx, NY 10452
800-367-6770
212-588-8400

Computer Industry Forecasts: The Source for Market Information on Computers, Peripherals, and Software
Data Analysis Group
P.O. Box 4210
Georgetown, CA 95634
916-333-4001

Computer Review
GML Corp.
594 Marrett Rd.
Lexington, MA 02173
617-861-0515

Computers and Office and Accounting Machines
Government Printing Office
Washington, DC 20402
202-783-3238

Computer Shopper
Ziff Davis Publishing Co.
Computer Publications Division
1 Park Ave.
New York, NY 10017
212-503-5100

Computerworld
I.D.G. Communications, Inc.
375 Cochituate Rd.
Framingham, MA 01701
617-879-0700

IBM Journal of Research and Development
IBM Corp.
Armonk, NY 10504
914-765-1900

Electronics

Aircraft Electronics Association (AEA)
P.O. Box 1981
Independence, MO 64055
816-373-6565

American Component Dealers Association (ACDA)
5201 Great American Parkway
Santa Clara, CA 95054
408-987-4200

Association of Electronic Distributors (AED)
9363 Wilshire Blvd., Suite 217
Beverly Hills, CA 90210
213-278-0543

Association of High Tech Distributors (AHTD)
1900 Arch St.
Philadelphia, PA 19103
215-564-3484

Audio Engineering Society (AES)
60 E. 42nd St., Room 2520
New York, NY 10065
212-661-8528

Electronic Industries Association (EIA)
2001 Pennsylvania Ave., NW, Suite 1100
Washington, DC 20006-1813
202-457-4900

Electronics Representatives Association (ERA)
20 E. Huron
Chicago, ILL. 60611
312-649-1333

IEEE Circuits and Systems Society (CSS)
c/o Institute of Electrical and Electronics Engineers
345 E. 47th St.
New York, NY 10017
212-705-7867

IEEE Consumer Electronics Society (CES)
c/o Institute of Electrical and Electronics Engineers
345 E. 47th St.
New York, NY 10017
212-705-7867

IEEE Industrial Electronics Society (IES)
c/o Institute of Electrical and Electronics Engineers
345 E 47th St.
New York 10017
212-705-7867

Industry Coalition on Technology Transfer (ICOTT)
1400 L St., NW, 8th Floor
Washington, DC 20005-3502
202-371-5994

Institute of Electrical and Electronics Engineers (IEEE)
345 E 47th St.
New York, NY 10017
212-705-7900

National Engineering Consortium (NEC)
303 E. Wacker Dr., Suite 740
Chicago, IL 60601
312-938-3500

Semiconductor Equipment and Materials International (SEMI)
805 E. Middlefield Rd.
Mountain View, CA 94043
415-964-5111

Semiconductor Industry Association (SIA)
4300 Stevens Creek Blvd., No. 271
San Jose, CA 95129
408-246-2711

Surface Mount Equipment Manufacturers Association (SMEMA)
4113 Barberry Dr.
Lafayette Hill, PA 19444
215-825-1008

Publications and Directories

Electrical and Electronic Abstracts
Institution of Electrical Engineers
445 Hoes Lane
Piscataway, NJ 08854

Electronic Business
Cahners Publishing Co., Inc.
275 Washington St.
Newton, MA 02158
617-964-3030

Electronic Industry Telephone Directory
Harris Publishing Co.
2057-2 Aurora Rd.
Twinsburg, OH 44087
216-425-9143

Electronic Market Data Book
Electronic Industries Association
2001 Pennsylvania Ave, NW
Washington, DC 20006
202-457-4900

Electronic News
Fairchild Publications, Inc.
7 E 12th St.
New York, NY 10003
212-741-4230

INSPEC
Institution of Electrical Engineers
Station House
Nightingale Rd.
Hitchin, Herts, England SG5 IRJ

Survey of Electronics
Leland Schwartz
Merrill Publishing Co.
1300 Alum Creek Dr.
Columbus, OH 43216
800-848-1567
614-258-8441

Food

American Institute of Food Distribution (AIFD)
28-12 Broadway
Fair Lawn, NJ 07410
201-791-5570

Association of Food Industries (AFI)
Sravine Dr.
P.O. Box 776
Matawan, NJ 07747
201-583-8188

Educational Foundation of the National Restaurant Association (EFNRA)
250 S. Wacker Dr., No. 1400
Chicago, IL 60606
312-715-1010

Food and Allied Service Trades Department (FAST)
AFL-CIO
815 16th St., NW, Suite 408
Washington, DC 20006
202-737-7200

Food and Drug Law Institute (FDLI)
1000 Vermont Ave., 12th Floor
Washington, DC 20005
202-371-1420

Food and Nutrition Board (FNB)
Institute of Medicine
2101 Constitution Ave. NW
Washington, DC 10418
202-334-2238

Food Equipment Manufacturers Association (FEMA)
401 N. Michigan Ave.
Chicago, IL 60601
312-644-6610

Food Marketing Institute (FMI)
1750 K St., NW, Suite 700
Washington, DC 20006
202-452-8444

Food Research and Action Center (FRAC)
1319 F St., NW, Suite 500
Washington, DC 20004
202-393-5060

Food Research Institute
Stanford University
Stanford, CA 94305
415-723-3941

Food Service Equipment Distribution Association (FEDA)
332 S. Michigan Ave.
Chicago, IL 60604
312-427-9605

Food Service Marketing Institute (FSMI)
P.O. Box 1265
Lake Placid, NY 12946
518-523-2942

International Dairy-Deli Association (IDDA)
P.O. Box 5528
313 Price Pl., Suite 202
Madison, WI 53705
608-238-7908

International Federation of Grocery Manufacturers Association (IFGMA)
c/o Grocery Manufacturers of America
1010 Wisconsin Ave., NW, Suite 800
Washington, DC 20007
202-337-9400

International Food Service Distributors Association (IFDA)
201 Park Washington Ct.
Falls Church, VA 22046
703-532-9400

International Frozen Food Assocation (IFFA)
1764 Ikd Meadow Lane
McLean, VA 22102
703-821-0770

National Food and Energy Council (NFEC)
409 Vandiver W., Suite 202
Columbia, MO 65202
314-875-7155

National Food Distributors Association (NFDA)
401 N. Michigan Ave, Suite 2400
Chicago, IL 60611
312-644-6610

National Frozen Food Association (NFFA)
4755 Linglestown Rd., Suite 300
P.O. Box 6069
Harrisburg, PA 17112
717-657-8601

National Soft Serve and Fast Food Association (NSSFFA)
516 S. Front St.
Chesaning, MI 48616
517-845-3336

Snack Food Association (SFA)
1711 King St., Suite 1
Alexandria, VA 22314
703-836-4500

Publications and Directories

Food Industry Newsletter
Newsletters, Inc.
7600 Carter Ct.
Bethesda, MD 20817
703-631-2322

Hereld's 5000: The Directory of Leading U.S. Food, Confectionery, and Beverage Manufacturers
S.I.C. Publishing Co.
Whitney Center, Suite 330
200 Leader Hill Dr.
Hamden, CT 06617
203-281-6766

Nielsen Retail Index
Nielsen Marketing Research
Nielsen Plaza
Northbrook, IL 60062
312-498-6300

Forest Products

American Forest Council
1250 Connecticut Ave., NW, Suite 320
Washington, DC 20036
202-463-2455

American Hardwood Export Council (AHEC)
1250 Connecticut Ave., NW, Suite 200
Washington, DC 20036
202-463-2723

American Lumber Standards Committee (ALSC)
P.O. Box 210
Germantown, MD 20875
301-972-1700

American Pulpwood Association (APA)
1025 Vermont Ave., NW, Suite 1020
Washington, DC 20005
202-347-2900

Association of Consulting Foresters (ACF)
5410 Grosvenor Lane, Suite 205
Bethesda, MD 20814
301-530-6795

Association of Western Pulp and Paper Workers (AWPPW)
P.O. Box 4566
1430 SW Clay
Portland, OR 97208
503-228-7486

Forest Farmers Association (FFA)
P.O. Box 95385
Atlanta, GA 30347
404-325-2954

Forest Industries Council (FIC)
1250 Connecticut Ave., NW, Suite 320
Washington, DC 20036
202-463-2460

Forest Products Laboratory
1 Gifford Pinchot Dr.
Madison, WI 53705
608-264-5600

Forest Products Research Society
2801 Marshall St.
Madison, WI 53705
608-231-1361

Hardwood Distributors Association (HDA)
1279 N. McLean St.
P.O. Box 12802
Memphis, TN 38182
901-274-6887

Hardwood Research Council (HRC)
P.O. Box 34518
Memphis, TN 38184-0518
901-377-1824

DIRECTORY OF ORGANIZATIONS AND COMPANIES TO HELP YOU IDENTIFY BEST PRACTICES

National Association of State Foresters (NASF)
Hall of States
444 N. Capitol St., NW, Suite 526
Washington, DC 20001
202-624-5415

National Forest Products Association (NFPA)
1250 Connecticut Ave., NW, Suite 200
Washington, DC 20036
202-463-2700

North American Wholesale Lumber Association (NAWLA)
2340 S. Arlington Heights Rd., Suite 680
Arlington Heights, IL 60005
708-981-8630

Northwestern Lumbermens Association (NLA)
1405 N. Lical Dr., Suite 130
Golden Valley, MN 55422
612-544-6822

Northwest Forestry Association (NFA)
1500 SW 1st Ave., Suite 770
Portland, OR 97201
503-222-9505

Southeastern Lumber Manufacturers Association (SLMA)
P.O. Box 1788
Forest Park, GA 30051
404-361-1445

Timber Products Manufacturers (TPM)
951 E 3rd. Ave.
Spokane, WA 99202
509-535-4646

Western Forest Industries Association (WFIA)
1500 SW Taylor
Portland, OR 97205
503-224-5455

Wood Products Manufacturers Association (WPMA)
52 Racette Ave.
Gardner, MA 01440
508-632-3923

Publications and Directories
Demand and Price Situation for Forest Products
U.S. Forest Service
Government Printing Office
Washington, DC 20402
202-783-3238

Forest
Forest Products Research Society
2801 Marshall Ct.
Madison, WI 53705
608-231-1361

Forest Industries
Miller Freeman Publications
500 Howard St.
San Francisco, CA 94105
414-397-1881

U.S. Timber Production, Trade, Consumption, and Price Statistics
U.S. Forest Service
U.S. Department of Agriculture
14th St. and Independence Ave, SW
Washington, DC 20250
202-447-3760

Franchising

The American Association of Franchisees and Dealers (AAFD)
P.O. Box 81887
San Diego, CA 92138-1887
800-733-9858

International Franchise Association
1350 New York Ave., NW, Suite 900
Washington, DC 20005
202-628-8000

Furniture

American Furniture Manufacturers Association (AFMA)
P.O. Box HP-7
High Point, NC 27261
919-884-5000

American Society of Furniture Designers (ASFD)
P.O. Box 2688
High Point, NC 27161
919-884-4074

Contract Furnishings Council (CFC)
1190 Merchandise Mart
Chicago, IL 60654
312-321-0563

International Home Furnishings Marketing Association (IHFMA)
P.O. Box 5687
High Point, NC 27262
919-889-0203

National Home Furnishings Association (NHFA)
P.O. Box 2396
High Point, NC 27261
919-883-1650

National Office Products Association (NOPA)
301 N. Fairfax St.
Alexandria, VA 22314
703-549-9040

National Unfinished Furniture Institute (NUFI)
1850 Oak St.
Northfield, IL 60093
708-446-8434

National Wholesale Furniture Assocation (NWFA)
P.O. Box 2482
164 S. Main St., Suite 404
High Point, NC 27261
919-884-1566

Publications and Directories

Furniture/Today
Communications Today Publishing, Ltd.
200 S. Main St.
High Point, NC 27261
919-889-0113

Retail Trade, Annual Sales, Year-End Inventories, and Accounts Receivable by Kind of Retail Store
Bureau of the Census
U.S. Department of Commerce
Washington, DC 20233
202-763-4040

Industrial and Farm Equipment

American Boiler Manufacturers Association (ABMA)
950 N. Glebe Rd., Suite 160
Arlington, VA 22203
703-522-7350

American Gear Manufacturers Association (AGMA)
1500 King St., Suite 201
Alexandria, VA 22314
703-684-0211

American Machine Tool Distributors' Association (AMTDA)
1335 Rockville Pike
Rockville, MD 20852
301-738-1200

American Supply and Machinery Manufacturers Association (ASMMA)
Thomas Associates, Inc.
1230 Keith Building
Cleveland, OH 44115
216-241-7333

American Textile Machinery Association (ATMA)
7297 Lee Highway, Suite N
Falls Church, VA 22042
703-533-9251

Associated Equipment Distributors (AED)
615 W 22nd St.
Oak Brook, IL 60521
708-574-0650

Equipment Manufacturers Institute (EMI)
c/o Tom Metzger
10 S. Riverside Plaza, Suite 1220
Chicago, IL 60606
312-321-1470

Fabricators and Manufacturers Association, International (FMA)
5411 E. State St.
Rockford, IL 61108
815-399-8700

Farm Equipment Manufacturers Association (FEMA)
243 N. Lindbergh Blvd.
St. Louis, MO 63141
314-991-0702

Farm Equipment Wholesalers Association
P.O. Box 1347
Iowa City, IA 52240
319-354-5156

Hydraulic Tool Manufacturers Association (HTMA)
P.O. Box 1337
Milwaukee, WI 53201
414-633-3454

Industrial Distribution Association (IDA)
3 Corporate Sq., Suite 201
Atlanta, GA 30329
404-325-2776

Industrial Fasteners Institute (IFI)
1505 E. Ohio Building
Cleveland, OH 44114
216-241-1482

Manufacturers Alliance for Productivity and Innovation (MAPI)
1200 18th St. NW, Suite 400
Washington, DC 20036
202-331-8430

North American Equipment Dealers Association (NAEDA)
10877 Watson Road
St. Louis, MO 63127
314-821-7220

Surface Mount Technology Association (SMTA)
5200 Wilson Rd., Suite 100
Edina, MN 55424
612-920-SMTA

Publications and Directories

Agricola
U.S. National Agricultural Library
Beltsville, MD 20705
301-344-3813

Agricultural Engineering Abstracts
CAB International North America
845 N. Park Ave.
Tucson, AZ 85719
800-528-4841
602-621-7897

Annual Survey of Manufacturers
Bureau of the Census
U.S. Department of Commerce
Government Printing Office
Washington, DC 20402
202-783-3238

Farm Equipment
Johnson Hill Press, Inc.
1233 Janesville Ave.
Fort Atkinson, WI 53538
414-563-6388

Farm Equipment Wholesalers Association Directory
Farm Equipment Wholesalers Association
1927 Keokuk St.
Iowa City, IA 52240
319-354-5156

Industrial Equipment News
Thomas Publishing Co.
250 W. 34th St.
New York, NY 10119
212-868-5661

NTIS Bibliographic Database
U.S. Department of Commerce
National Technical Information Service
5285 Port Royal Rd.
Springfield, VA 22161
703-487-4630

Jewelry and Silverware

American Diamond Industry Association (ADIA)
71 West 47th St., Suite 705
New York, NY 10036
212-575-0525

American Gem and Mineral Suppliers Association (AGMSA)
P.O. Box 741
Patton, CA 92369
714-885-3918

American Gem Trade Association (AGTA)
181 World Trade Center
P.O. Box 581043
Dallas, TX 75258
214-742-GEMS

Fashion Jewelry Association of America (FJAA)
Box S-8, Regency East
1 Jackson Walkway
Providence, RI 02903
401-273-1515

Gemological Institute of America (GIA)
1660 Stewart St.
Santa Monica, CA 90404
213-829-2991

Jewelers Board of Trade (JBT)
P.O. Box 6928
Providence, RI 02940
401-438-0750

Jewelers of America
Time-Life Building
1271 Ave. of the Americas
New York, NY 10020
212-489-0023

Jewelers Shipping Association (JSA)
125 Carlsbad St.
Cranston, RI 02920
401-943-6020

Jewelry Industry Distributors Association (JIDA)
120 Light St.
Baltimore, MD 21230
301-752-3318

Jewelry Manufacturers Association (JMA)
475 5th Ave., Suite 1908
New York, NY 10017
212-725-5599

Jewelry Manufacturers Guild (JMG)
P.O. Box 46099
Los Angeles, CA 90046
714-769-1820

Manufacturing Jewelers and Silversmiths of America
100 India St.
Providence, RI 02903-4313
401-274-3840

Leather

Leathercraft Guild (LG)
P.O. Box 734
Artesia, CA 90701
213-864-2420

Leather Industries of America (LIA)
1000 Thomas Jefferson St., NW, Suite 515
Washington, DC 20007
202-342-8086

Leather Industry Statistics
Leather Industries of America
1000 Thomas Jefferson St., NW, Suite 515
Washington, DC 20007
202-342-8086

Leather Workers International Union
11 Peabody Square
P.O. Box 32
Peabody, MA 01960
508-531-6200

Luggage and Leather Goods Manufacturers of America (LLGMA)
350 5th Ave., Suite 2624
New York, NY 10118
212-695-2340

Metals and Metal Products

Aluminum Association (AA)
900 19th St. NW, Suite 300
Washington, DC 20006
202-862-5100

American Bureau of Metal Statistics (ABMS)
P.O. Box 1405
400 Plaza Dr.
Secaucus, NJ 07094
201-863-6900

American Copper Council (ACC)
333 Rector Pl., Suite 10P
New York, NY 10280
212-945-4990

American Institute for Imported Steel (AIIS)
11 W. 42nd St., Suite 3002
New York, NY 10036-8002
212-921-1765

ASM International
9639 Kinsman
Materials Park, OH 44073
216-338-5151

Association of Iron and Steel Engineers (AISE)
3 Gateway Center, Suite 2350
Pittsburgh, PA 15222
412-281-6323

Association of Steel Distributors (ASD)
401 N. Michigan Ave.
Chicago, IL 60611-4390
312-644-6610

Industry Council for Tangible Assets (ICTA)
25 E St., NW, 8th Floor
Washington, DC 20001
202-783-3500

Metal Building Manufacturers Association (MBMA)
c/o Charles M. Stockinger
Thomas Associates, Inc.
1230 Keith Building
Cleveland, OH 44115
216-241-7333

Metal Fabricating Institute (MFI)
P.O. Box 1178
Rockford, IL 61105
815-965-4031

Metal Finishing Suppliers' Association (MFSA)
801 N. Cass Ave., Suite 300
Westmount, IL 60555
708-887-0797

Metal Trades Department (of AFL-CIO) (MTD)
503 AFL-CIO Building
815 16th St., NW
Washington, DC 20006
202-347-7255

Specialty Steel Industry of the United States (SSIUS)
3050 K St., NW, 4th Floor
Washington, DC 20007
202-342-8400

Steel Manufacturers Association (SMA)
815 Connecticut Ave, NW, No. 304
Washington, DC 20006
202-342-1160

Publications and Directories

American Metal Market
7 E. 12th St.
New York, NY 10003
212-741-4140

Journal of Metals
Metallurgical Society, Inc.
420 Commonwealth Dr.
Warrendale, PA 15086
412-776-9070

Metals Statistics
Fairchild Publications, Inc.
7 E. 12th St.
New York, NY 10003
212-741-4140

Metals Week
McGraw-Hill, Inc.
1221 Ave. of the Americas
New York, NY 10020
212-512-2000

Mining and Crude Oil Production

American Institute of Mining, Metallurgical and Petroleum Engineers (AIME)
345 E. 47th St., 14th Floor
New York, NY 10017
212-705-7695

American Society for Surface Mining and Reclamation (ASSMR)
21 Grandview Dr.
Princeton, WV 24740
304-425-8332

Center for Alternative Mining Development Policy (CAMDP)
210 Avon St., Suite 9
La Crosse, WI 54603
608-784-4399

Colorado Mining Association (CMA)
1600 Broadway, Suite 1340
Denver, CO 80202
303-894-0536

Mining and Metallurgical Society of America (MMSA)
210 Post St., Suite 1102
San Francisco, CA 94108
415-398-6925

Northwest Mining Association (NWMA)
414 Peyton Building
Spokane, WA 99201
509-624-1158

Pennsylvania Grade Crude Oil Association (PGCOA)
c/o Pringle Powder Co.
Box 201
Bradford, PA 16701
814-368-8172

Pittsburgh Coal Mining Institute of America (PCMIA)
4800 Forbes Ave.
Pittsburgh, PA 15213
412-621-4500

Rocky Mountain Coal Mining Institute (RMCMI)
3000 Youngfield, no. 324
Lakewood, CO 80215-6545
303-238-9099

Society for Mining, Metallurgy, and Exploration (SME, Inc.)
P.O. Box 625005
Littleton, CO 80162
303-973-9550

U.S. Bureau of Mines
2401 E St., NW
Washington, DC 20241
202-634-1001

Publications and Directories

American Mining Congress Journal
American Mining Congress
1920 N. St., NW
Washington, DC 20035
202-861-2800

Census of Mineral Industries
U.S. Bureau of the Census
Washington, DC 20233
301-763-4100

Colorado School of Mines Quarterly
Colorado School of Mines Press
Golden, CO 80401
303-273-3600

Earth and Mineral Sciences
Penn State College of Earth and Mineral Sciences
Pennsylvania State University
116 Deike Building
University Park, PA 16802
814-863-4667

Minerals and Materials
U.S. Bureau of the Census
Washington, DC 20233
202-634-1001

Mining Machinery and Equipment
U.S. Bureau of the Census
Washington, DC 20233
301-763-4100

Motor Vehicles and Parts

American Automobile Association (AAA)
1000 AAA Dr.
Heathrow, FL 32746-5063
407-444-7000

American International Automobile Dealers Association (AIADA)
1128 16th St. NW
Washington, DC 20036
202-659-2561

Automotive Body Parts Association (ABPA)
2500 Wilcrest Dr., Suite 510
Houston, TX 77042-2752
713-977-5551

Automotive Cooling Systems Institute (ACSI)
300 Sylvan Ave.
P.O. Box 1638
Englewood Cliffs, NJ 07632
201-569-8500

Automotive Engine Rebuilders Association (AERA)
330 Lexington Dr.
Buffalo Grove, IL 60089-6998
708-541-6550

Automotive Exhaust Systems Manufactuers Council (AESMC)
300 Sylvan Ave.
P.O. Box 1638
Englewood Cliffs, NJ 07632
201-569-8500

Automotive Market Research Council (AMRC)
300 Sylvan Ave.
P.O. Box 1638
Englewood Cliffs, NJ 07632
607-257-6700

Automotive Parts and Accessories Association (APAA)
5100 Forbes Blvd.
Lanham, MD 20706
301-459-9110

Automotive Refrigeration Products Institute (ARPI)
4600 E. West Highway, Suite 300
Bethesda, MD 20814
301-657-2774

Brake Systems Parts Manufacturers Council (BSPMC)
300 Sylvan Ave.
P.O. Box 1638
Englewood Cliffs, NJ 07632
201-569-8500

Championship Association of Mechanics (CAM)
P.O. Box 7
Howard, CO 81233
719-942-3611

Filters Manufacturers Council (FMC)
300 Sylvan Ave.
P.O. Box 1638
Englewood Cliffs, NJ 07632
201-569-8500

Motor and Equipment Manufacturers Association
300 Sylvan Ave.
Englewood Cliffs, NJ 07632
201-569-8500

Motor Vehicle Manufacturers Association of the United States (MVMA)
7430 2nd Ave., Suite 300
Detroit, MI 48202
313-872-4311

National Automobile Dealers Association (NADA)
8400 Westpark Dr.
McClean, VA 22102
703-827-7407

National Automotive Parts Association (NAPA)
2999 Circle 75 Parkway
Atlanta, GA 30339
404-956-2200

Publications and Directories

Annual Survey of Manufacturers
U.S. Bureau of the Census
Washington, DC 20233
301-763-4100

Automotive Industries Statistical Issue
Chilton Co.
Chilton Way
Radnor, PA 19089
800-345-1214
215-964-4000

Automobile Quarterly
Automobile Quarterly, Inc.
420 N. Park Rd.
Wyomissing, PA 19610
215-325-8444

Automotive Aftermarket News
Stanley Publishing Co.
200 Madison Ave., Suite 2104
Chicago, IL 60601
312-332-0210

Automotive Executive
National Automobile Dealers Association
8400 Westpark Dr.
McLean, VA 22102
703-821-7150

Automotive Market Report
Automotive Auction Publishing, Inc.
1101 Fulton Building
Pittsburgh, PA 15222
412-281-2338

Dictionary of Automotive Technology
VCH Publications, Inc.
220 E. 23rd St.
New York, NY 10010
800-422-8824
212-683-8333

Motor News Analysis
News Analysis, Inc.
32068 Olde Franklin Dr.
Farmington Hills, MI 48018
313-851-1377

MVMA Motor Vehicle Facts and Figures
Motor Vehicle Manufacturers Association of the U.S., Inc.
7430 2nd Ave., Suite 300
Detroit, MI 48202
313-872-4311

Petroleum Refining

American Gas Association
1515 Wilson Blvd.
Arlington, VA 22209
703-841-8400

American Independent Refiners Association (AIRA)
649 S. Olive St., Suite 500
Los Angeles, CA 90014
213-624-8407

Association of Petroleum Re-refiners (APR)
P.O. Box 427
Buffalo, NY 14205
716-855-2212

National Petroleum Refiners Association (NAPRA)
1899 L St., NW, Suite 1000
Washington, DC 20036
202-457-0480

Publications and Directories

Basic Petroleum Data Book
American Petroleum Institute
275 7th Ave.
New York, NY 10001
212-366-4040

National Petroleum News
Hunter Publishing Co.
950 Lee St.
Des Plains, IL 60016
708-296-0770

The Oil and Gas Producing Industry in Your State
Independent Petroleum Association of America
Petroleum Independent Publishers, Inc.
1101 16th St., NW
Washington, DC 20036
202-857-4766

Oil and Gas Reporter
Matthew Bender & Co., Inc.
11 Penn Plaza
New York, NY 10017
800-223-1940
212-967-7707

Oil/Energy Statistics Bulletin
Oil Statistics Co.
P.O. Box 127
Babson Park, MA 02157
617-651-8126

Pharmaceuticals

Academy of Pharmaceutical Research and Science (APRS)
c/o Naomi U. Kaminsky
American Pharmaceutical Association
2215 Constitution Ave., NW
Washington, DC 20037
202-628-4410

American Council on Pharmaceutical Education (ACPE)
311 W. Superior St., Suite 512
Chicago, IL 60610
312-664-3575

American Pharmaceutical Association (APhA)
2215 Constitution Ave., NW
Washington, DC 20037
202-628-4410

Generic Pharmaceutical Industry Association (GPIA)
200 Madison Ave., Suite 2404
New York, NY 10016
212-683-1881

Medical Economics Co., Inc.
680 Kinderkamack Rd.
Oradell, NJ 07649
800-223-0581
201-262-3030

National Association of Pharmaceutical Manufacturers (NAPM)
747 3rd. Ave.
New York, NY 10017
212-838-3720

National Council of State Pharmaceutical Association Executives (NCSPAE)
c/o Paul Gacanti
Virginia Pharmaceutical Association
3119 W. Clay St.
Richmond, VA 23230
804-355-7941

National Pharmaceutical Association (NPhA)
College of Pharmacy and Pharmacal Sciences
Howard University
Washington, DC 20059
202-328-9229

National Pharmaceutical Council (NPC)
1894 Preston White Dr.
Reston, VA 22041
703-620-6390

National Wholesale Druggists' Association (NWDA)
105 Oronoco St.
P.O. Box 238
Alexandria, VA 22314
703-684-6400

Pharmaceutical Manufacturers Association (PMA)
1100 15th St., NW

Washington, DC 20005
202-835-3400

Publications and Directories

Food and Drug Administration
Shepard's/McGraw-Hill
420 N. Cascade Ave.
Colorado Springs, CO 80901
800-525-2474
303-577-7707

Pharmaceutical Marketers Directory
CPS Communications, Inc.
7200 W. Camino Real, Suite 215
Boca Raton, FL 33433
305-368-9301

Publishing and Printing

American Book Producers Association (ABPA)
41 Union Square W., Room 936
New York, NY 10003
212-645-2368

American Business Press (ABP)
675 3rd. Ave., Suite 400
New York, NY 10017
212-661-6360

American Newspaper Publishers Association (ANPA)
The Newspaper Center
Box 17407
Dulles International Airport
Washington, DC 20041
703-648-1000

International Financial Printers Association (IFPA)
100 Dangerfield Rd.
Alexandria, VA 22314
703-519-8122

International Newspaper Marketing Association (INMA)
P.O. Box 17422
Washington, DC 20041
703-648-1094

Label Printing Industries of America (LPIA)
100 Dangerfield Rd.
Alexandria, VA 22314
703-519-8122

Machine Printers and Engravers Association of the United States (MPEA)
690 Warren Ave.
East Providence, RI 02914
401-438-5849

Magazine Publishers of America (MPA)
575 Lexington Ave.
New York, NY 10022
212-752-0055

National Business Circulation Association (NBCA)
Act III Publishing
c/o Steve Wigginton
401 Park Ave. South
New York, NY 10016
212-545-5140

National Newspaper Publishers Association (NNPA)
948 National Press Building, Room 948
Washington, DC 20045
202-662-7324

National State Printing Association (NSPA)
c/o Council of State Governments
Iron Works Pike
P.O. Box 11910
Lexington, KY 40578
606-231-1874

Periodical and Book Association of America (PBAA)
120 E. 34th St., Suite 7-K
New York, NY 10016
712-689-4952

Periodicals Institute
P.O. Box 899
West Caldwell, NJ 07007
201-882-1130

Printing, Publishing, and Media Workers Sector, Communications Workers of America (CWA)
1925 K St., NW, Suite 400
Washington, DC 20006
202-728-2326

Screen Printing Association International (SPAI)
10015 Main St.
Fairfax, VA 22031
703-385-1335

Publications and Directories

American Printer
Maclean-Hunter Publishing Co.
29 N. Wacker Dr.
Chicago, IL 60606
800-621-9907
312-726-7907

Publisher's Weekly
Bowker Magazine Group
Cahners Magazine Division
249 W. 17th St.
New York, NY 10011
800-669-1002
212-645-9700

Real Estate

National Association of Realtors (NAR)
430 North Michigan Avenue
Chicago, IL 60611-4087
312-329-8200

Rubber and Plastic Products

American Laminators Association (ALA)
419 Norton Building
Seattle, WA 98104
206-622-0666

American Society of Electroplated Plastics (ASEP)
1101 14th St., NW, Suite 1100
Washington, DC 20005
202-371-1323

Ames Rubber Corporation
Ames Boulevard
Hamburg, NJ 07419
201-209-3200

Chemical Fabrics and Film Association (CFFA)
Thomas Associates, Inc.
1230 Keith Building
Cleveland, OH 44115
216-241-7323

Eastman Chemical Company
Stone East Building
Kingsport, TN 37662
615-229-5264

International Institute of Synthetic Rubber Producers (IISRP)
2077 S. Gessner Rd., Suite 133
Houston, TX 77063-1123
713-783-7511

National Association of Plastics Distributors (NAPD)
6333 Long St., Suite 340
Shawnee, KS 66216
913-268-6273

Plastic and Metal Products Manufacturers Association (PMPMA)
225 W. 34th. St., Suite 2002
New York, NY 10001
212-564-2500

Plastics Education Foundation (PEF)
c/o Society of Plastics Engineers
14 Fairfield Dr.
Brookfield, CT 06804
203-775-0471

Plastics Institute of America (PIA)
Stevens Institute of Technology
Castle Point Station
Hoboken, NJ 07030
201-420-5553

Plastics Recycling Foundation (PRF)
1275 S St., NW, Suite 500
Washington, DC 20005
202-371-5200

Polyurethane Division, Society of the Plastics Industry (PDSPI)
355 Lexington Ave.
New York, NY 10017
212-351-5425

Polyurethane Manufacturers Association (PMA)
Building C, Suite 20
800 Roosevelt Rd.
Glen Ellyn, IL 60137
708-858-2670

Rubber Manufacturers Association
1400 K St., NW
Washington, DC 20005
202-682-4800

Rubber Trade Association of New York (RTA)
17 Battery Place
New York, NY 10004
212-344-7776

Society of Plastics Engineers (SPE)
14 Fairfield Drive
Brookfield, CT 06804
203-775-0471

Publications and Directories

Plastics News: Crain's International Newspaper for the Plastics Industry
Crain Communications, Inc.
1725 Merriman Rd., Suite 300
Akron, OH 44313
216-836-9180

Plastics Technology Manufacturing Handbook and Buyers' Guide
Bill Communications, Inc.
633 3rd Ave.
New York, NY 10017
800-343-1732
212-986-4800

Plastics World—Plastics Directory
Cahners Publishing Co., Inc.
1350 E. Toujy Ave.
Des Plaines, IL 60018
312-635-8800

Rubber and Plastics News—Rubbicana Issue
Crain Communications, Inc.
740 N. Rush St.
Chicago, IL 60611
312-649-5200

Rubber: Production, Shipments, and Stocks
U.S. Bureau of the Census
Washington, DC 20233
301-763-4100

Rubber World
1867 W. Market St.
Akron, OH 44313
216-864-2122

Rubbicana: Directory of North American Rubber Product Manufacturers and Rubber Industry Suppliers
Crain Communications, Inc.
1725 Merriman Rd., Suite 300
Akron, OH 44313
216-836-9180

Scientific and Photographic Equipment

Association for Science, Technology and Innovation (ASTI)
P.O. Box 1242
Arlington, VA 22210
703-451-6948

Carnegie Commission on Science, Technology and Government (CCSTG)
10 Waverly Place
New York, NY 10003
212-998-2150

Commission on Professionals in Science and Technology (CPST)
1500 Massachusetts Ave., NW
Washington, DC 20005
202-223-6995

National Association for Science, Technology and Society (NASTS)
117 Willard Building
University Park, PA 16802
814-865-9951

National Association of Photo Equipment Technicians (NAPET)
3000 Picture Place
Jackson, MI 49201
517-788-8100

Photographic Society of America
3000 United Founders Blvd., Suite 103
Oklahoma City, OK 73112
405-843-1437

Scientific Apparatus Makers Association
1101 16th St., NW
Washington, DC 20036
202-223-1360

United Nations Centre for Science and Technology for Development (CSTD)
1 United Nations Plaza, DCI-10th Floor
New York, NY 10017
212-963-8435

Publications and Directories

Guide to Scientific Instruments
American Association for the Advancement of Science
1333 H St., NW, 8th Floor
Washington, DC 20005
202-326-6446

Journal of Imaging Technology
SPSE: The Society for Imaging Science and Technology
7003 Kilworth Lane
Springfield, VA 22151
703-642-9090

Review of Scientific Instruments
American Institute of Physics
335 E. 45th St.
New York, NY 10017
212-661-9404

Scisearch
Institute for Scientific Information
3501 Market St.
Philadelphia, PA 19104
800-523-1850
215-386-0100

Selected Instruments and Related Products
U.S. Bureau of the Census
Washington, DC 20233
301-763-4100

Standards and Practices for Instrumentation
Instrument Society of America
67 Alexander Dr.
Research Triangle Park, NC 27709
919-549-8411

Soaps and Cosmetics

Cosmetic Industry Buyers and Suppliers (CIBS)
c/o Joseph A. Palazzolo
36 Lakeville Rd.
New Hyde Park, NY 11040
516-775-0220

Cosmetic, Toiletry, and Fragrance Association (CTFA)
1110 Vermont Ave., NW, Suite 800
Washington, DC 20005
202-331-1770

Independent Cosmetic Distributors (ICMAD)
1230 W. Northest Hwy.
Palatine, IL 60067
708-991-4499

Soap and Detergent Association (SDA)
475 Park Ave. South
New York, NY 10016
212-725-1262

Publications and Directories
Cosmetics and Toiletries
Allured Publishing Corp.
214 W. Willow
Wheaton, IL 60189
312-653-2155

Drug and Cosmetic Industry
Edgell Communications, Inc.
7500 Old Oak Blvd.
Cleveland, OH 44130
216-243-8100

Soap/Cosmetics/Chemical Specialties
MacNair Publications, Inc.
101 W. 31st. St.
New York, NY 10001
212-279-4455

U.S. Toiletries and Cosmetics Industry
Off-the-Shelf Publications, Inc.
2171 Jericho Turnpike
Commack, NY 11725
516-462-2410

Telecommunications/Telephony

Competitive Telecommunications Association
120 Maryland Ave., NE
Washington, DC 20402
202-783-3238

Federal Communications Commission
1919 M St., NW
Washington, DC 20554
202-632-7000

National Telephone Cooperative Association
2626 Pennsylvania Ave., NW
Washington, DC 20037
202-298-2300

North American Telecommunications Association
2000 M St., NW, Suite 550
Washington, DC 20036
800-538-6286
202-296-9800

United States Telephone Association
900 19th St., NW, Suite 800
Washington, DC 20006
202-835-3100

Publications and Directories

Annual Statistical Reports of Independent Telephone Companies
Federal Communications Commission
1919 M St., NW
Washington, DC 20554
202-632-7000

Communications News
Communication News
111 E. Wacker Drive, 16th Floor
Chicago, IL 60601
312-938-2300

Communications Week
CMP Publications, Inc.
600 Community Drive
Manhasset, NY 11030
516-365-4600

Datapro Reports on Telecommunications
Datapro Research Corp.
1805 Underwood Blvd.
Delran, NJ 08075
609-764-0100

Independent Telephone Statistics
United States Telephone Association
900 19th St., NW, Suite 800
Washington, DC 20006
202-835-3100

Quarterly Operating Data of 68 Telephone Carriers
Federal Communications Commission
1919 M St., NW
Washington, DC 20554
202-632-7000

Telecommunications Reports
Business Research Publications, Inc.
1036 National Press Bldg.
Washington, DC 20045
202-347-2654

Telecommunications Week
Business Research Publications, Inc.
1036 National Press Bldg.
Washington, DC 20045
202-347-2654

Telephony
Telephony Publishing Corp.
55 E. Jackson Blvd.
Chicago, IL 60604
312-922-2435

Textiles

American Fiber Manufacturers Association (AFMA)
1150 17th St., NW
Washington, DC 20036
202-296-6508

American Fiber, Textile, Apparel Coalition (AFTAC)
1801 K St., NW, Suite 900
Washington, DC 20006
202-862-0500

American Reuseable Textile Association (ARTA)
P.O. Box 1073
Largo, FL 34294
314-889-1360

American Textile Manufacturers Institute (ATMI)
1801 K St., NW, Suite 900
Washington, DC 20006
202-862-0500

Institute of Textile Technology (ITT)
P.O. Box 391
Charlottesville, VA 22902
804-296-5511

Institutional and Service Textile Distributors Association (ISTDA)
93 Standish Rd.
Hillsdale, NJ 07642
201-664-4600

Knitted Textile Association (KTA)
386 Park Ave., South
New York, NY 10016
212-689-3808

National Textile Processors Guild (NTPG)
75 Livingston St.
Brooklyn, NY 11201
718-875-2300

Northern Textile Association (NTA)
230 Congress St.
Boston, MA 02110
617-542-8220

Southern Textile Association (STA)
509 Francisca Lane
P.O. Box 190
Cary, NC 27512
919-467-1655

Textile Converters Association (TCA)
100 E. 42nd St.
New York, NY 10017
212-867-5720

Textile Distributors Association (TDA)
45 West 36th St., 3rd Floor
New York, NY: 10018
212-563-0400

Textile Fibers and By-Products Association (TFBPA)
P.O. Box 11065
Charlotte, NC 28220
704-527-5593

Textile Information Users Council (TIUC)
c/o Trudy Craven
Milliken Research Corp.
P.O. Box 5521
Spartanburg, SC 29304
803-573-1589

Textile Quality Control Association (TQCA)
P.O. Box 76501
Atlanta, GA 30328
404-252-9037

Textile Research Institute (TRI)
P.O. Box 625
Princeton, NJ 08542
609-924-3150

United Textile Workers of America (UTWA)
P.O. Box 749
Voorhees, NJ 08043
609-772-9699

Publications and Directories

America's Textiles International Directory
Billian Publishing Co.
2100 Powers Ferry Rd.
Atlanta, GA 30339
404-955-5656

Textile Research Journal
Textile Research Institute
601 Prospect Ave.
Princeton, NJ 08542
609-924-3150

Textile World
Maclean Publishing Co.
Textile Publications
4170 Ashford-Dunwood Rd., Suite 420
Atlanta, GA 30319
404-252-0626

World Textile Abstracts
Shirley Institute
Charlton St.
Manchester, UK M1 3FH

Toys and Sporting Goods

American Toy Export Association (ATEA)
c/o Kraemer Mercantile Corp.
200 5th Ave., Room 1303
New York, NY 10010
212-255-1772

Athletic Goods Team Distributors (AGTD)
1699 Wall St.
Mt. Prospect, IL 60056
708-439-4000

International Committee of Toy Industries
c/o David Hawtin
British Toy and Manufacturers Association
80 Camberwell Road
London, UK SE5 0EG

International Union of Allied Novelty and Production Workers
181 S. Franklin Ave.
Valley Stream, NY 11581
212-889-1212

National Association of Doll and Stuffed Toy Manufacturers, Inc. (NADSTM)
200 E. Post Road
White Plains, NY 10601
914-682-8900

National Association of Sporting Goods Wholesalers (NASGW)
P.O. Box 11344
Chicago, IL 60611
312-565-0233

Sporting Goods Manufacturers Association (SGMA)
200 Castlewood Drive

North Palm Beach, FL 33408
407-842-4100

Toy and Hobby Wholesalers Association of America (TWA)
P.O. Box 955
Marlton, NJ 08053
609-985-2878

Toy Manufacturers of America (TMA)
200 5th Ave., Room 740
New York, NY 10010
212-675-1141

Toy Wholesalers Association of America
66 E. Main St.
Moorestown, NJ 08507
609-234-9155

USA Toy Library Association
2719 Broadway Ave.
Evanston, IL 60201
312-864-8240

Transportation Equipment

American Public Works Association (APWA)
1313 E. 60th St.
Chicago, IL 60637
312-667-2200

High Speed Rail Association (HSRA)
206 Valley Ct., Suite 800
Pittsburgh, PA 15237
412-366-6887

International Mass Transit Association (IMTA)
P.O. Box 40247
Washington, DC 20016-0247
202-362-7960

The Maintenance Council of the American Trucking Association (TMC)
2200 Mill Rd.
Alexandria, VA 22314
703-838-1763

National Industrial Transportation League (NITL)
1090 Vermont Ave., NW, Suite 410
Washington, DC 20005
202-842-3870

Transportation Institute
5201 Auth Way
Camp Springs, MD 20746
301-423-3335

Transportation Research Board (TRB)
2101 Constitution Ave., NW
Washington, DC 20418
202-334-2934

Transportation Research Forum (TRF)
1600 Wilson Blvd., Suite 905
Arlington, VA 22209
703-525-1191

Transportation Safety Equipment Institute (TSEI)
300 Sylvan Ave.
P.O. Box 1638
Englewood Cliffs, NJ 07632-0638
201-569-8500

Publications and Directories

Census of Transportation
Government Printing Office
Washington, DC 20402
202-783-3238

Journal of Advanced Transportation
Institute for Transportation, Inc.
Duke Station
P.O. Box 4670
Durham, NC 27706
919-684-8834

National Transportation Statistics
Government Printing Office
Washington, DC 20402
202-783-3238

Transportation Journal
P.O. Box 33095
Louisville, KY 40232
502-451-8150

Transportation Research Information Service (TRIS)
Transportation Research Board
National Research Council
2101 Constitution Ave., NW
Washington, DC 20418
202-334-3250

Urban Transport News
Business Publishers, Inc.
951 Pershing Drive
Silver Spring, MD 20910
301-587-6300

Part Two. Organizations Known for Best Practices

Benchmarking

American Productivity and Quality Center
123 N. Post Oak Lane, Suite 300
Houston, TX 77024-7797
800-324-4673

Armstrong World Industries, Inc.
313 West Liberty Street
Lancaster, PA 17603
717-396-2766

AT&T Universal Card Services
8787 Baypine Rd.
Jacksonville, FL 32256
904-443-7500

Churchill & Company
P.O. Box 425214
San Francisco, CA 94142
415-981-7700

DEC
146 Main St.
Maynard, MA 01754
508-493-5111

Florida Power & Light
9250 W. Flagler St.
Miami, FL 33174
305-552-3552

Ford Motor Company
The American Road
Dearborn, MI 48121

IBM Corporation
Highway 52 & 37th St., NW
Rochester, MN 55901
507-253-9000

Motorola, Inc.
1303 E. Algonquin Road
Schaumburg, IL 60196

Xerox Corporation
800 Long Ridge Road
Stamford, CT 06904
203-968-3000

Billing and Collections

American Express
World Financial Center
New York, NY 10285
212-640-2000

AT&T Universal Card Services
8787 Baypine Road
Jacksonville, FL 32256
904-443-7500

MCI
1133 19th St., NW
Washington, DC 20036
202-872-1600

Compensation

American Compensation Association (ACA)
P.O. Box 29312
Phoenix, AZ 85038-9312
602-951-9191

Corning Telecommunications Division
Business Systems Corning, Inc.
Corning, NY 14831
800-525-2524

Competitive Analysis

The Society of Competitor Intelligence Professionals
1818 18th St., NW, No. 225
Washington, DC 20006
202-223-5885

Concurrent Engineering

Boeing Company
7755 E. Marginal Way South
Seattle, WA 98108
206-655-2121

3M Corporation
3M Center
St. Paul, MN 55144-1000
612-733-1110

Customer Feedback

AT&T Consumer Communications Services
295 North Maple Ave.

Basking Ridge, NJ 07920
800-473-5047

AT&T Universal Card Services
8787 Baypine Road
Jacksonville, FL 32256
904-443-7500

GTE Directories Corporation
West Airfield Drive
P.O. Box 609810
Fallas/Ft. Worth, TX 75261-9810
214-453-7751

Marriott Corporation
1 Marriott Drive
Bethesda, MD 20814
301-380-9000

Motorola, Inc.
1303 E. Algonquin Road
Schaumburg, IL 60196
708-576-5000

Zytec Corp.
7575 Market Place Drive
Eden Prairie, MN 55344
612-941-1100

Customer Focus

GE (Plastics)
260 Long Ridge Road
Stamford, CT 06927
203-357-4000

Wallace Company, Inc.
P.O. Box 1492
Houston, TX 77251-1492
713-237-5900

Westinghouse Electric Co.
Westinghouse Building, Gateway Center
Pittsburgh, PA 15222
412-244-2000

Xerox Corp.
800 Long Ridge Road
Stamford, CT 06904
203-968-3000

Customer Service

American Express
World Financial Center
New York, NY 10285
212-640-2000

AT&T Network Systems Group
Transmission Systems Business Unit
475 South St.
Morristown, NJ 07962
201-606-2000

AT&T Universal Card Services
8787 Baypine Road
Jacksonville, FL 32256

Banc One Corp.
100 E. Broad St.
Columbus, OH 43271

Cadillac Motor Car Company
2860 Clark St.
Detroit, MI 48232
313-492-7151

Florida Power & Light
9250 W. Flagler St.
Miami, FL 33174
305-552-3552

General Electric (GE)
260 Long Ridge Road
Stamford, CT 06927
203-357-4000

Globe Metallurgical, Inc.
6450 Rockside Woods Blvd. South, No. 3
Cleveland, OH 44131
216-328-0145

Granite Rock Company
P.O. Box 50001
Watsonville, CA 95077

Hewlett-Packard Company
300 Hanover Street
Palo Alto, CA 94304

IBM Corporation
Highway 52 & 37th St., NW
Rochester, MN 55901
507-253-9000

L.L. Bean, Inc.
Casco St.
Freeport, ME 04033
207-865-4761

Marlow Industries
10451 Vista Park Rd.
Dallas, TX 75238-2645
214-340-4900

Milliken & Company
920 Milliken Road
Spartanburg, SC 29303
803-573-2020

Motorola, Inc.
1303 E. Algonquin Road
Schamburg, IL 60196

Nordstrom Inc.
1501 5th Ave.
Seattle, WA 98101-1603

Procter & Gamble Co.
1 Procter & Gamble Plaza
Cincinnati, OH 45202
513-983-1100

Ritz Carlton Hotel Company
3414 Peachtree Road, NE, Suite 300
Atlanta, GA
404-237-5500

Solectron Group
2001 Fortune Drive
San Jose, CA 95131

Texas Instruments Inc.
13500 N. Central Expressway
Dallas, TX 75265
214-995-3333

Wallace Company, Inc.
P.O. Box 1492
Houston, TX 77251-1492
713-237-5900

Westinghouse Commercial Nuclear Fuel Division
Westinghouse Building, Gateway Center
Pittsburgh, PA 15222
412-244-2000

Xerox Corporation
800 Long Ridge Road
Stamford, CT 06904
203-968-3000

Zytec Corporation
7575 Market Place Drive
Eden Prairie, MN 55344
612-941-1100

Data Transfer and Check Clearing

First National Bank of Chicago
1 First National Plaza, Suite 0518
Chicago, IL 60670
312-732-4000

Design for Manufacturing Assembly

Digital Equipment Corporation (DEC)
146 Main St.
Maynard, MA 01754
508-493-5111

Motorola, Inc.
1303 E. Algonquin Road
Schaumburg, IL 60196
708-576-5000

NCR Division of AT&T
1700 S. Patterson Blvd.
Dayton, OH 45479
513-445-5000

Document Processing

Bell Atlantic Corp.
1717 Arch St.
Philadelphia, PA 19103
215-963-6000

Citicorp
399 Park Ave.
New York, NY 10043
212-559-1000

Employee Recognition

AT&T Universal Card Services
8787 Baypine Road
Jacksonville, FL 32256
904-443-7500

Milliken and Co.
920 Milliken Road
Spartanburg, SC 29303
803-573-2020

Employee Suggestions

Dow Chemical Co.
2030 Willard H. Dow Center
Midland, MI 48674
517-636-1000

Milliken and Co.
920 Milliken Road
Spartanburg, SC 29303
803-573-2020

Procter & Gamble Co.
1 Procter & Gamble Plaza
Cincinnati, OH 45202
513-983-1100

Toyota Motor Manufacturing USA, Inc.
1001 Cherry Blossom Way
Georgetown, KY 40324
502-868-2000

Wainwright Industries, Inc.
17 Cermak Boulevard
P.O. Box 640
St. Peters, MO 63376
314-278-5850

Employee Surveys

Mayflower Group, Inc.
9998 N. Michigan Road
Carmel, IN 46032
317-875-1469

Empowerment

Honda of America Manufacturing Inc.
Honda Parkway
Marysville, OH 43040
513-642-5000

Milliken and Co.
920 Milliken Road
Spartanburg, SC 29303
803-573-2020

Environmental Management

Ben & Jerry's Homemade Inc.
Junction of Routes 2 and 100
N. Moretown, VT 05676
802-244-5641

Coalition for Environmentally Responsible Economics (CERES)
711 Atlantic Ave.
Boston, MA 02111
617-451-0927

Council on Economic Priorities (CEP)
30 Irving Place
New York, NY 10003
212-420-1133

Dow Chemical Co.
2030 Willard H. Dow Center
Midland, MI 48674
517-636-1000

Environmental Law Institute
1616 P St., NW, Suite 200
Washington, DC 20036
202-328-5150

Environmental Protection Agency (U.S. EPA)
401 M Street SW
Washington, DC 20460
202-260-2090

Global Environmental Management Initiative
1828 L. Street NW, Suite 711
Washington, DC 20036
202-296-7449

National Technical Information Service (NTIS)
5285 Port Royal Road
Springfield, VA 22161

3M Corporation
3M Center
St. Paul, MN 55144-1000
612-733-1110

Executive Development

General Electric
260 Long Ridge Road
Stamford, CT 06927
203-357-4000

Facilities Management

Walt Disney World Co.
1675 Buena Vista Drive
Buena Vista, FL 32830
407-824-2222

Flexible Manufacturing

Allen-Bradley Company, Inc.
1201 S. 2nd St.
Milwaukee, WI 53204
414-382-2000

Baldor Electric Co.
5711 S. 7th St.
Fort Smith, AK 72902
501-646-4711

Motorola/Boynton Beach
1303 E. Algonquin Road
Schaumburg, IL 60196
708-576-5000

Health-Care Management

Adolph Coors Co.
12th St. and Ford St.
Golden, CO 80401
303-279-6565

Allied-Signal Aerospace Co.
2525 W. 190th St.
Torrance, CA 90504
213-321-5000

Southern California Edison Co.
2244 Walnut Grove
Rosemead, CA 91770
818-302-1212

Human Resources Management

American Society for Training and Development (ASTD)
1640 King Street
Alexandria, VA 22313
703-683-8100

Baxter International Inc.
1 Baxter Parkway
Deerfield, IL 60015
708-948-2000

The Conference Board
845 3rd Avenue
New York, NY 10022
212-759-0900
800-872-6273

IBM Corporation
Old Orchard Road
Armonk, NY 10504
914-765-1900

Institute for International Human Resources
606 North Washington Street
Alexandria, VA 22314
703-548-3440

International Foundation of Employee Benefits Plans
18700 West Bluemound Rd.
Box 69
Brookfield, WI 53008
414-786-6700

L.L. Bean Inc.
Casco St.
Freeport, ME 04033
207-865-4761

Ritz Carlton Hotel Co.
3414 Peachtree Road, NE, Suite 300
Atlanta, GA 30326
404-237-5500

3M Corporation
3M Center
St. Paul, MN 55144-1000
612-733-1110

Walt Disney World Co.
1675 Buena Vista Drive
Buena Vista, FL 32830
407-824-2222

Industrial Design

Black & Decker Corp. (Household Products)
701 E. Joppa Ave.
Towson, MD 21204
410-583-3900

Braun Corp.
1014 S. Monticello St.
Winamac, IN 46996
219-946-6153

Herman Miller Inc.
8500 Byron Road
Zeeland, MI 49464
616-772-3300

Inventory Control

American Hospital Supply/Circon Corp.
460 Ward Drive
Santa Barbara, CA 93111-2310
805-967-0404

Apple Computer, Inc.
20525 Mariani Ave.
Cupertino, CA 95014
408-996-1010

Federal Express Corp.
2005 Corporate Ave.
Memphis, TN 38132
901-369-3600

L.L. Bean, Inc.
Casco St.
Freeport, ME 04033
207-865-4761

Northern Telecom Inc.
200 Athens Eay
Nashville, TN 37228
615-734-4000

Westinghouse Electric Company
Westinghouse Building, Gateway Center
Pittsburgh, PA 15222
412-244-2000

Make-Versus-Buy Decisions

Boeing Company
7755 E. Marginal Way South
Seattle, WA 98108
206-655-2121

Manufacturing

A.L. Philpott Manufacturing Center
Martinsville, VA 24112
703-666-8890

Agile Manufacturing Enterprise Forum (Agility Forum)
P.O. Box 1582
Bethlehem, PA 18015-1582
610-758-5510

American Production and Inventory Control Society (APICS)
P.O. Box 4274
Falls Church, VA 22046-4274
703-237-8344

American Society for Quality Control (ASQC)
P.O. Box 3066
Milwaukee, WI 53201-3066
800-248-1946
414-272-8575

Association for Manufacturing Excellence (AME)
380 W. Palatine Road
Wheeling, IL 60090-5863
708-520-3282

AT&T Network Systems Group
Transmission Systems Business Unit
475 South St.
Morristown, NJ 07962
201-606-2000

Corning Inc.
Houghton Park
Corning, NY 14831
607-974-9000

Eastman Chemical Company
Stone East Building
Kingsport, TN 37662
615-229-5264

Hewlett-Packard Co.
300 Hanover St.
Palo Alto, CA 94304
415-857-1501

IBM Corporation
Old Orchard Road
Armonk, NY 10504
914-765-1900

Milliken & Company
920 Milliken Road
Spartanburg, SC 29303
803-573-2020

Motorola, Inc.
1303 E. Algonquin Road
Schaumburg, IL 60196
708-576-5000

National Association of Manufacturers (NAM)
1331 Pennsylvania Ave., NW, Suite 1500 North Lobby
Washington, DC 20004-1703
202-637-3000

Phillip Morris Companies, Inc.
120 Park Ave.
New York, NY 10017
212-880-5000

Solectron Group
2001 Fortune Drive
San Jose, CA 95131
408-957-8500

Texas Instruments Inc.
13500 N. Central Expressway
Dallas, TX 75265
214-995-3333

Toyota Motor Manufacturing USA, Inc.
1001 Cherry Blossom Way

Georgetown, KY 40324
502-868-2000

Wainwright Industries, Inc.
17 Cermak Boulevard
P.O. Box 640
St. Peters, MO 63376
314-278-5850

Westinghouse Commercial Nuclear Fuel Division
Westinghouse Building, Gateway Center
Pittsburgh, PA 15222
414-244-2000

Xerox Corp.
800 Long Ridge Road
Stamford, CT 06904
203-968-3000

Zytec Corp.
7575 Market Place Drive
Eden Prairie, MN 55344
612-941-1100

Marketing

American Marketing Association
250 S. Wacker Drive, Suite 200
Chicago, IL 60606
312-648-0536

The Conference Board
845 3rd Ave.
New York, NY 10022
212-759-0900

Direct Marketing Association, Inc.
11 West 42nd St.
New York, NY 10036-8096
212-768-7277

Helene Curtis Industries Inc.
325 N. Wells St.
Chicago, IL 60610
312-661-0222

IBM Corporation
Old Orchard Road
Armonk, NY 10504
914-765-1900

International Trade Administration (ITA)
U.S. Department of Commerce
Herbert C. Hoover Building
14th and Constitution NW
Washington, DC 20230
202-482-2000

Microsoft Corporation
1 Microsoft Way
Redmond, WA 98052-6399
206-882-8080

Procter & Gamble Company
1 Procter & Gamble Plaza
Cincinnati, Ohio 45202
513-983-1100

3M Corporation
3M Center
St. Paul, MN 55144-1000
612-733-1110

Xerox Corporation
800 Long Ridge Road
Stamford, CT 06904
203-968-3000

Plant Layout and Design

Cummins Engine Company, Inc.
500 Jackson St.
Columbus, IN 47202-3005
812-377-5000

General Electric
260 Long Ridge Rd.
Stamford, CT 06927
203-357-4000

Policy Deployment

Florida Power & Light
9250 W. Flagler St.
Miami, FL 33174
305-552-3552

Ford Motor Company
The American Road
Dearborn, MI 48121
313-322-3000

Process Improvement

Armstrong World Industries, Inc.
313 West Liberty Street
Lancaster, PA 17603
717-396-2766

AT&T Consumer Communications Services
295 North Maple Avenue
Basking Ridge, NJ 07920
800-473-5047

CSX Corp.
901 E. Cary St.
Richmond, VA 23219
804-782-1400

Federal Express Corp.
2005 Corporate Ave.
Memphis, TN 38132
901-369-3600

GTE Directories Corp.
West Airfield Drive
P.O. Box 609810
Dallas/Ft. Worth, TX 75261-9810
214-453-7751

Motorola Inc.
1303 E. Algonquin Rd.
Schaumburg, IL 60196
708-576-5000

Solectron Group
2001 Fortune Drive
San Jose, CA 95131
408-957-8500

Wainwright Industries, Inc.
17 Cermak Boulevard
P.O. Box 640
St. Peters, MO 63376

Xerox Corporation
800 Long Ridge Road
Stamford, CT 06904
203-968-3000

Product Development

American Express
World Financial Center
New York, NY 10285
212-640-2000

Digital Equipment Corporation
146 Main St.
Maynard, MA 01754
508-493-5111

Federal Express Corp.
2005 Corporate Ave.
Memphis, TN 38132
901-369-3600

Hewlett-Packard Co.
300 Hanover St.
Palo Alto, CA 94304
415-857-1501

Honda of America Manufacturing Inc.
Honda Parkway
Marysville, OH 43040
513-642-5000

Intel Corp.
2200 Mission College Blvd.
Santa Clara, CA 95052
408-765-8080

Motorola Corp.
1303 E. Algonquin Rd.
Schaumburg, IL 60196
708-576-5000

Sony Corporation of America
Sony Drive
Park Ridge, NJ 07656
201-930-1000

3M Corporaton
3M Center
St. Paul, MN 55144-1000
612-733-1110

Product Improvement

Motorola Corp.
1303 E. Algonquin Road
Schaumburg, IL 60196
708-576-5000

Northern Telecom Inc.
200 Athens Way
Nashville, TN 37228
615-734-4000

Project Team Management

Chaparral Steel Co.
300 Ward Road
Midlothian, TX 76065-9651
214-775-8241

PURCHASING

AMP Inc.
471 Friendship Road
Harrisburg, PA 17105
717-564-0100

Federal Express Corp.
2005 Corporate Ave.
Memphis, TN 38132
901-369-3600

Florida Power & Light
9250 W. Flagler St.
Miami, FL 33174
305-552-3552

Honda of America Manufacturing Inc.
Honda Parkway
Marysville, OH 43040
513-642-5000

NCR Division, AT&T Corp.
1700 S. Pattern Blvd.
Dayton, OH 45479
513-445-5000

Xerox Corporation
800 Long Ridge Road
Stamford, CT 06904
203-968-3000

Quality Management

American Society for Quality Control (ASQC)
P.O. Box 3005
Milwaukee, WI 53201-3005
800-248-1946
414-272-8575

Ames Rubber Corporation
Ames Boulevard
Hamburg, NJ 07419
201-209-3200

Armstrong World Industries, Inc.
313 West Liberty Street
Lancaster, PA 17603
717-396-2766

Association for Quality and Participation
801-B West 8th Street
Cincinnati, OH 45203-1607
513-381-1959

AT&T Consumer Communications Services
295 North Maple Avenue
Basking Ridge, NJ 07920
800-473-5047

AT&T Network Systems Group
Transmission Systems Business Unit
475 South St.
Morristown, NJ 07962
201-606-2000

AT&T Universal Card Services
8787 Baypine Road
Jacksonville, FL 32256
904-443-7500

Cadillac Motor Car Company
2860 Clark St.
Detroit, MI 48232
313-492-7151

Corning Telecommunications Products Division
Business Systems Corning, Inc.
Corning, NY 14831
800-525-2524

Digital Equipment Corporation
146 Main St.
Maynard, MA 01754
508-493-5111

Eastman Chemical Company
Stone East Building
Kingsport, TN 37662
615-229-5264

Federal Express Corp.
2005 Corporate Ave.
Memphis, TN 38132
901-369-3600

Florida Power & Light
9250 W. Flagler St.
Miami, FL 33174
305-552-3552

Globe Metallurgical, Inc.
6450 Rockside Woods Blvd. South, No. 3
Cleveland, OH 44131
216-328-0145

Granite Rock Company
P.O. Box 50001
Watsonville, CA 95077
408-724-5611

GTE Directories, Inc.
West Airfield Drive
P.O. Box 609810
Dallas/Ft. Worth, TX 75261-9810
214-453-7751

IBM Corporation
Highway 52 and 37th St., NW
Rochester, MN 55901
507-253-9000

Marlow Industries
10451 Vista Park Road
Dallas, TX 75238-2645
214-340-4900

Milliken & Company
920 Milliken Road
Spartanburg, SC 29303
803-573-2020

Motorola, Inc.
1303 E. Algonquin Rd.
Schaumburg, IL 60196
708-576-5000

Ritz Carlton Hotel Co.
3414 Peachtree Road, NE, Suite 300
Atlanta, GA 30326
404-237-5500

Solectron Group
2001 Fortune Drive
San Jose, CA 95131
408-957-8500

Texas Instruments Inc.
13500 N. Central Expressway
Dallas, TX 75265
214-995-3333

Toyota Motor Manufacturing USA, Inc.
1001 Cherry Blossom Way
Georgetown, KY 40324
502-868-2000

Wainwright Industries, Inc.
17 Cermak Boulevard
P.O. Box 640
St. Peters, MO 63376
314-278-5850

Westinghouse Commercial Nuclear Fuel Division
Westinghouse Center
Pittsburgh, PA 15222
412-244-2000

Westinghouse Company, Inc.
P.O. Box 1492
Houston, TX 77251-1492
713-237-5900

Xerox Corporation
800 Long Ridge Road
Stamford, CT 06904
203-968-3000

Zytec Corp.
7575 Market Place Drive
Eden Prairie, MN 55344
612-941-1100

Research and Development

AT&T Co.
32 Ave. of the Americas
New York, NY 10013-2412
212-605-5500

Hewlett-Packard Co.
300 Hanover St.
Palo Alto, CA 94304
415-857-1501

Shell Oil Co.
1 Shell Plaza
Houston, TX 77252
713-241-6161

Sales Management

IBM Corporation
Old Orchard Road
Armonk, NY 10504
914-765-1900

Merck and Company, Inc.
P.O. Box 2000
Rahway, NJ 07065-0909
908-594-4000

Procter & Gamble Co.
1 Procter & Gamble Plaza
Cincinnati, OH 45202
513-983-1100

Xerox Corp.
800 Long Ridge Road
Stamford, CT 06904
203-968-3000

Service Parts Logistics

Deere and Co.
John Deere Rd.
Moline, IL 61265-8098
309-765-8000

Strategic Planning

Eastman Chemical Company
Stone East Building
Kingsport, TN 37622
615-229-5264

The Planning Forum
5500 College Corner Pike
P.O. Box 70
Oxford, OH 45056-0070
513-523-4185

Supplier Relations and Support

Ames Rubber Corporation
Ames Boulevard
Hamburg, NJ 07419
201-209-3200

Bose Corp.
The Mountain
Framingham, MA 01701
508-879-7330

Corning Inc.
Houghton Park
Corning, NY 14831
607-974-9000

Ford Motor Company
The American Road
Dearborn, MI 48121
313-322-3000

Globe Metallurgical Inc.
6450 Rockside Woods Blvd. S., No. 3
Cleveland, OH 44131
216-328-0145

Levi Strauss and Co.
1155 Battery St.
San Francisco, CA 94120
415-544-6000

Motorola, Inc.
1303 E. Algonquin Rd.
Schaumburg, IL 60196
708-576-5000

3M Corporation
3M Center
St. Paul, MN 55144-1000
612-733-1110

Xerox Corp.
800 Long Ridge Rd.
Stamford, CT 06904
203-968-3000

Teaming and Self-Directed Teams

Corning Inc. (SCC plant)
Houghton Park
Corning, NY 14831
607-974-9000

Physio-Control Corp.
11811 Willows Rd. NE
Redmond, WA 98052
206-867-4000

Toledo Scale/Mettler Instrument Corp.
Princeton-Hightstown Road
Hightstown, NJ 08520
609-448-3000

Technology Transfer

Dow Chemical Co.
2030 Willard H. Dow Center
Midland, MI 48674
517-636-1000

Square D Co.
1415 S. Roselle
Palatine, IL 60067
708-397-2600

3M Corporation
3M Center
St. Paul, MN 55144-1000
612-733-1110

Telephone Customer Service

AT&T Consumer Communications Services
295 North Maple Avenue
Basking Ridge, NJ 07920
800-473-5047

AT&T Universal Card Services
8787 Baypine Road
Jacksonville, FL 32256
904-443-7500

Baxter International Inc.
1 Baxter Parkway
Deerfield, IL 60015
708-948-2000

Training

Ames Rubber Corporation
Ames Boulevard
Hamburg, NJ 07419
201-209-3200

AT&T Consumer Communications Services
295 North Maple Avenue
Basking Ridge, NJ 07920
800-473-5047

AT&T Universal Card Services
8787 Baypine Road
Jacksonville, FL 32256
904-443-7500

Ford Motor Company
The American Road
Dearborn, MI 48121
313-322-3000

General Electric
260 Long Ridge Road
Stamford, CT 06927
203-357-4000

Motorola, Inc.
1303 Algonquin Road
Schaumburg, IL 60196
708-576-5000

Polaroid Corp.
549 Technology Sq.
Cambridge, MA 02139
617-577-2000

Square D Co.
1415 S. Roselle
Palatine, IL 60067
708-397-2600

USAA
USAA Building
San Antonio, TX 78228
210-498-2211

Vendor Certification

Boeing Co.
7755 E. Marginal Way S.
Seattle, WA 98108
206-655-2121

Cummins Engine Company, Inc.
500 Jackson St.
Columbus, IN 47202-3005
812-377-5000

Dow Chemical Co.
2030 Willard H. Dow Center
Midland, MI 48674
517-636-1000

Motorola, Inc.
1303 E. Algonquin Road
Schaumburg, IL 60196
708-576-5000

Warehousing and Distribution

Citicorp
399 Park Ave.
New York, NY 10043
212-559-1000

Federal Express Corp.
2005 Corporate Ave.
Memphis, TN 38132
901-369-3600

Hershey Foods Corp.
100 Crystal A Dr.
Hershey, PA 17033-0810
717-534-4001

L.L. Bean Inc.
Casco St.
Freeport, ME 04033
207-865-4761

Mary Kay Cosmetics, Inc.
8787 Stemmons Freeway
Dallas, TX 75247
214-630-8787

Warranties

Ames Rubber Corporation
Ames Boulevard
Hamburg, NJ 07419
201-209-3200

Eastman Chemical Company
Stone East Building
Kingsport, TN 37662
615-229-5264

Xerox Corporation
800 Long Ridge Road
Stamford, CT 06904
203-968-3000

Waste Minimization

Dow Chemical Co.
2030 Willard H. Dow Center
Midland, MI 48674
517-636-1000

3M Corporation
3M Center
St. Paul, MN 55144-1000
612-733-1110

Quotes on Quality

The purpose of these quotes is to provide inspirational insight into what total quality management is all about, to get you thinking about this approach in new ways, to offer material useful in speeches and presentations, and to have a little fun as well. We like quotes and hope you do as well. We have organized the following quotes loosely by topic. Several of these quotes might have fit in several categories. We tried to put them in the one we felt was most logical. But in looking for thoughts on a particular topic, we suggest you still scan the entire section.

The quotes come from a variety of sources, including philosophers, poets, and scientists as well as business writers and managers. There is a reason for this. The ideas of quality management are not unique to this field. They naturally emerge in a variety of forms and contexts whenever people contemplate excellence and high performance in any human endeavor. We may come to see that quality management is ultimately about bringing out the best in all of us to facilitate our mutual success and prosperity. Wise people have appreciated such notions from time immemorial. Consider that as you read these quotes.

The categories we have chosen include Quality Defined, Managing for Quality, Quality Leadership, Culture of Quality, Customers, Training, Process Management and Reengineering, Measurement, Continuous Improvement, Communication, Vision, The Learning Organization, and The Wrong Stuff. Some categories include many more quotes than others. It just seems that people have said more quotable statements about leadership, for example, than about process management.

Quality Defined

Includes thoughts on what quality means to different people.

> Quality is the degree of excellence at an acceptable price and the control of variability at an acceptable cost.
> —Robert A. Broh

> Quality isn't asserted by the supplier; it's perceived by the customer.
> —John Guaspari

> Quality refers to the amount of the unpriced attribute contained in each unity of the priced attribute.
> —Keith B. Leffler

A business should quickly stand on its own based on the service it provides society. Profits should be a reflection not of corporate greed but a vote of confidence from society that what is offered by the firm is valued.

—Konosuke Matshushita

Managing for Quality

Includes thoughts on the concerns of people seeking to implement quality management.

A man is known by the company he organizes.

—Ambrose Bierce

He who would do good to another must do it in the minute particulars. General good is the plea of the hypocrite, scoundrel, and flatterer.

—Robert Blake

Those who would administer wisely must, indeed, be wise, for one of the serious obstacles to the improvement of our race is indiscriminate charity.

—Andrew Carnegie

Like a good marriage, a good partnership can bring profound benefits to the couple, but it requires a lot of work to be successful. Just as nearly half the marriages end in divorce, so, too, many partnerships fail and for the same reasons.

—James W. Cortada

Nothing astonishes men so much as common sense and plain dealing.

—Ralph Waldo Emerson

Quality control is applicable to any kind of enterprise. In fact, it *must* be applied in every enterprise.

—Kaoru Ishikawa

Criticism, as it was first instituted by Aristotle, was meant as a standard for judging well.

—Samuel Johnson

If you ask managers what they do, they will most likely tell you that they plan, organize, coordinate, and control. Then watch what they do. Don't be surprised if you can't relate what you see to those four words.

—Henry Mintzberg

People are always blaming their circumstances for what they are. I don't believe in circumstances. The people who get on in the world are the people who get up and look for the circumstances they want and if they can't find them, make them.

—George Bernard Shaw

Quality Leadership

Includes thoughts on individual leadership behaviors that bring about the success of all.

The true teacher defends his pupils against his own personal influence. He inspires self-trust. He guides their eyes from himself to the spirit that quickens him. He will have no disciple.
—Bronson Alcott

Predicting the future isn't hard, especially for those who create it.
—Anatole France

The last of the human freedoms—to choose one's attitude in any given set of circumstances, to choose one's own way.
—Viktor Frankl

We must be the world we want to create.
—Gandhi

Those who honestly mean to be true contradict themselves more rarely than those who try to be consistent.
—Oliver Wendell Holmes Jr.

The real distinction is between those who adapt their purposes to reality and those who seek to mould reality in the light of their purposes.
—Henry Kissinger

Our deepest fear is not that we are inadequate. Our deepest fear is that we are powerful beyond measure. It is our light, not our darkness, that most frightens us. We ask ourselves, "Who am I to be brilliant, gorgeous, talented and fabulous?" Actually, who are you not to be? You are a child of God. Your playing small doesn't serve the world. There's nothing enlightened about shrinking so that other people won't feel insecure around you. We were born to make manifest the glory of God that is within us. It's not just in some of us; it's in everyone. And as we let our own light shine, we unconsciously give other people permission to do the same. As we are liberated from our own fear, our presence automatically liberates others.
—Nelson Mandela, Inaugural Address, 1994

Rule #1: Don't sweat the small stuff. Rule #2: It's all small stuff.
—Dr. Michael Mantell

The superior man will not manifest either narrow-mindedness or the want of self-respect.
—Mencius

Every act of creation is first an act of destruction because the new idea will destroy what a lot of people believe is essential to the survival of their intellectual world.
—Pablo Picasso

Example is not the main thing in influencing others; it's the only thing.
—Albert Schweitzer

Most powerful is he who has himself in his own power.
—Lucius Annaeus Seneca (the younger)

Live with men as if God saw you; converse with God as if men heard you.
—Lucius Annaeus Seneca (the younger)

Courage is resistance to fear, mastery of fear—not absence of fear.
—Mark Twain

Laughing at someone else is an excellent way of learning how to laugh at oneself; and questioning what seem to be the absurd beliefs of another group is a good way of recognizing the potential absurdity of many of one's own cherished beliefs.
—Gore Vidal

If a man is in health, he doesn't need to take anybody else's temperature to know where he is going.
—E. B. White

In reality power is *not* something individuals have except as it is accorded to them by those for whom they work and who work for them. And these accorders of power only do this when it seems to be in their best interest to do so.
—John Woods

Culture of Quality

Includes thoughts that suggest the attributes of a culture that promotes quality attitudes and behaviors among all employees.

God grant me the serenity to prioritize the things I cannot delegate, the courage to say no when I need to, and the wisdom to know when to go home!
—Anonymous

Never esteem anything as of advantage to you that will make you break your word or lose your self-respect.
—Marcus Antoninus

If we had no winter, the spring would not be so pleasant: if we did not sometimes taste adversity, prosperity would not be so welcome.
—Anne Bradstreet

Nobody could make a greater mistake than to do nothing because he could do only a little.
—Edmund Burke

"If everybody minded their own business," said the Duchess in a hoarse growl, "the world would go round a deal faster than it does."
—Lewis Carroll

Trustworthy people require no worthy enemies as they go about their labors.
—Philip Crosby

A fool can learn from his own experience; the wise learn from the experience of others.
—Democritus

No institution can possibly survive if it needs geniuses or supermen to manage it. It must be organized in such a way as to be able to get along under a leadership composed of average human beings.
—Peter Drucker

Opportunity is missed by most people because it is dressed in overalls and looks like work.
—Thomas Edison

The reward of a thing well done, is to have done it.
—Ralph Waldo Emerson

It is not the employer who pays the wages—he only handles the money. It is the product that pays the wages.
—Henry Ford

Dare to be naive.
—R. Buckminster Fuller

I do not feel obliged to believe that the same God who has endowed us with sense, reason, and intellect has intended us to forgo their use.
—Galileo Galilei

The deed is everything, the glory nothing.
—Goethe

Adversity reveals genius; prosperity conceals it.
—Horace, Roman Poet

Do your work with your whole heart, and you will succeed—there's so little competition.
—Elbert Hubbard

Good timber does not grow with ease. The stronger the wind the stronger the trees.
—Willard Marriott

If we face a recession, we should not lay off employees; the company should sacrifice a profit. It's management's risk and management's responsibility. Employees are not guilty; why should they suffer?
—Akio Morita

When you make a world tolerable for yourself, you make a world tolerable for others.
—Anaïs Nin

The best preparation for tomorrow is to do today's work superbly well.
—Sir William Osler

The way to get good ideas is to get lots of ideas and throw the bad ones away.
—Linus Pauling

TQM does not and will not bring results overnight. The essence of TQM is a change of culture.
—Edward Sallis

It is not because things are difficult that we do not dare, it is because we do not dare that they are difficult.
—Seneca

If you don't do it excellently, don't do it at all. Because if it's not excellent, it would be profitable or fun, and if you're not in business for fun and profit, what the hell are you doing here?
—Robert Townsend

While an unchanging dominant majority is perpetually rehearsing its own defeat, fresh challenges are evoking fresh creative responses from newly recruited minorities, which proclaim their own creative power by rising, each time, to the occasion.
—Arnold Toynbee

If you are a middle manager avid to begin a quality initiative in a company ruled by an executive from the old school, look elsewhere for a job.
—Mary Walton

Organizations and the people who make them up are the same thing. You cannot look out for the organization without looking out for those people.
—John Woods

I don't believe evolution is about survival of the fittest. I believe it is about survival of the most useful.
—John Woods

Customers

Includes thoughts on the idea of serving customers.

If you're not serving the customer, then you better be serving someone who is.
—Karl Albrecht and Ron Zemke

If things are not going well with you, begin your effort at correcting the situation by carefully examining the service you are rendering, and especially the spirit in which you are rendering it.
—Roger Babson

Customers are the most important asset any company has, even though they don't show up on the balance sheet.
—Thomas Berry

Don't just anticipate your customer's future needs. Create them.
—Daniel Burrus

A market is never saturated with a good product, but it is very quickly saturated with a bad one.
—Henry Ford

He that plants trees loves others besides himself.
—Thomas Fuller

High quality means pleasing customers, not just protecting them from annoyance.
—David Garvin

All of management's efforts for Kaizen boil down to two words: customer satisfaction.
—Masaaki Imai

The business process starts with the customer. In fact, if it is not started with the customer, it all too many times abruptly ends with the customer.
—William Scherkenbach

I don't know what your destiny will be, but one thing I know: the only ones among you who will be really happy are those who will have sought and found out how to serve.
—Albert Schweitzer

Many a man would rather you heard his story than granted his request.
—Phillip Stanhope, Earl of Chesterfield

Profits are the organization's measure of the benefits and quality it has delivered to customers. So are losses.
—John Woods

When you understand that your success is directly tied to the world of which you are a part, then service to others is just what makes sense.
—John Woods

Training

Includes thoughts on the value of training.

They know enough who know how to learn.
—Henry Adams

The first object of any act of learning, over and beyond the pleasure it may give, is that it should serve us in the future. Learning should not only take us somewhere, it should allow us later to go further more easily.
—Jerome Bruner

Knowledge is the only instrument of production that is not subject to diminishing returns.
—J. M. Clarke

Like a good parent, the good mentor allows the person they are coaching to make mistakes, recognizing that the best learning often occurs from making errors.
—Francis Gouillart and James Kelly

Quality is the target: education gives people the tools with which to take aim.
—David L. Muthler and Lucy N. Lytle

Management will recognize the need for education and retraining when they realize that people are an asset and not an expense.
—William Scherkenbach

The great aim of education is not knowledge but action.
—Herbert Spencer

Give me the young man who has brains enough to make a fool of himself.
—Robert Louis Stevenson

Learning carries within itself certain dangers because out of necessity one has to learn from one's enemies.
—Leon Trotsky

Human history becomes more and more a race between education and catastrophe.
—H. G. Wells

Process Management and Reengineering

Includes thoughts on the idea of processes and their management and process renewal and reengineering.

Forget goals. Value the process.
—Jim Bouton

Probabilities direct the conduct of the wise man.
—Cicero

Everyone doing their best is not sufficient.
—W. Edwards Deming

It is not employees who cause the majority of errors; they are just unwilling pawns who operate in the environment often controlled by obsolete and cumbersome operating systems.
—H. James Harrington

If I had six hours to chop down a tree, I'd spend the first four sharpening the axe.
—Abraham Lincoln

Creativity can solve almost any problem. The creative act, the defeat of habit by originality, overcomes everything.
—George Lois

Creativity is merely a plus name for regular activity. . . . Any activity becomes creative when the doer cares about doing it right or better.
—John Updike

Some rules of life: (1) Keep your head down and swing easy. (2) Change your oil every three thousand miles.
—John Woods

Measurement

Includes thoughts on standards, measuring, and fact-based, intelligent action.

It is the theory which decides what we can observe.
—Albert Einstein

It has been said that figures rule the world. Maybe. But I am not sure that figures show us whether it is being ruled well or badly.
—Goethe

Measurement is the first step that leads to control and eventually to improvement. If you can't measure something, you can't understand it. If you can't understand it, you can't control it. If you can't control it, you can't improve it.
—H. James Harrington

In quality control, we try as far as possible to make our various judgments based on the facts, not on guesswork. Our slogan is "Speak with facts."
—Katsuya Hosotani

Without a standard there is no logical basis for making a decision or taking action.
—Joseph M. Juran

Perfect standards do not exist at any company. Conditions always change, and standards must follow suit.
—Shigeru Nakamura

The most savage controversies are about those matters as to which there is no good evidence either way.
—Bertrand Russell

You cannot compel people to comply with standards; compliance must be a voluntary decision. Compunction and coercion are anathema in successful TQM programs.
—Allan Sayle

Continuous Improvement

Includes thoughts on change and improvement.

Obstacles are things people see when they lose sight of their goals.
—Anonymous

He that will not apply new remedies must expect new evils; for time is the greatest innovator.
—Francis Bacon

Beaver: Gee, there is something wrong with just about everything, isn't there Dad?
Ward: Just about, Beav.
—*Leave It to Beaver*

Only the provisional endures.
—French Proverb

If I had a formula for bypassing trouble, I would not pass it round. Trouble creates a capacity to handle it. I don't embrace trouble; that's as bad as treating it as an enemy. But I do say meet it as a friend, for you'll see a lot of it and had better be on speaking terms with it.
—Oliver Wendell Holmes

The ideas of control and improvement are often confused with one another. This is because the quality control and quality improvement are inseparable.
—Kaoru Ishikawa

We must welcome the future, remembering that soon it will be the past; and we must respect the past, remembering that it was once all that was humanly possible.
—George Santayana

The only way to even *approach* doing something perfectly is through experience, and experience is the name everyone gives to their mistakes.
—Oscar Wilde

Without deviation, progress is not possible.
—Frank Zappa

Communication

Includes thoughts on the importance of communcation and its nature.

The truth can never be told so as to be understood and not be believ'd.
—William Blake

What is conceived well is expressed clearly, and the words to say it arrive with ease.
—Nicolas Boileau

To think justly, we must understand what others mean: to know the value of our thoughts, we must try their effect on other minds.
—William Hazlitt

Most communications problems can be solved with proximity.
—Richard A. Moran

A matter that becomes clear ceases to concern us.
—Friedrich Nietzsche

A recent publication claimed that jumping off a building could lead to "sudden deceleration trauma."
Reported by Mark Patinkin

Help fight truth decay.
—Sign in the office of polygrapher, Baltimore City Police Department

I distrust the incommunicable; it is the source of all violence.
—Jean-Paul Sartre

If you tell the truth, you don't have to remember anything.
—Mark Twain

Our words do not describe things but our relationship to things. In this lies the heart of misunderstanding.
—John Woods

Vision

Includes thoughts on what vision means and its effect on people.

Don't be afraid to take a big step if one is indicated. You can't cross a chasm in two small jumps.
—David Lloyd George

Genius means little more than the faculty of perceiving in an unhabitual way.
—William James

A rock pile ceases to be a rock pile the moment a single man contemplates it, bearing with him the image of a cathedral.
—Antoine de Saint Exupéry

If people don't have their own vision, all they can do is sign up for someone else's. This result is compliance, never commitment.
—Peter M. Senge

If there is genuine potential for growth, build capacity in advance of demand, as a strategy for creating demand. Hold the vision, especially as regards assessing key performance and evaluating whether capacity to meet potential demand is adequate.
—Peter M. Senge

Only the curious will learn, and only the resolute overcome the obstacles to learning. The quest quotient has always excited me more than the intelligence quotient.
—Eugene S. Wilson

The Learning Organization

Includes thoughts on what a learning organization is like, the nature of learning, and its place in life.

Chaos often breeds life, while order breeds habit.
—Henry Adams

The important thing about science is not so much to obtain new facts as to discover new ways of thinking about them.
—William Bragg

A state without the means of some change is without the means of its conservation.
—Edmund Burke

For a man to attain to an eminent degree in learning costs him time, watching, hunger, nakedness, dizziness in the head, weakness in the stomach, and other inconveniences.
—Miguel de Cervantes

Only the hand that erases can write the true thing.
—Meister Eckhart

Perhaps the mission of those who love mankind is to make people laugh at the truth, to make truth laugh, because the only truth lies in learning to free ourselves from insane passion for the truth.
—Umberto Eco

The very essence of human progress is applying new knowledge in new ways, continuously making the old way of doing things obsolete. Organizations don't make progress, *people* do. The organization is just a vehicle for human cooperation. Its form adapts as the needs of the people adapt, and peoples' needs change with the progressive acquisition and application of new knowledge.
—Francis Gouillart and James Kelly

Experience is that marvelous thing that enables you to recognize a mistake when you make it again.
—F. P. Jones

Mistakes are the portals of discovery.
—James Joyce

Where all think alike, no one thinks very much.
—Walter Lippman

The best way to view a present problem is to give *it* all you've got, to study it and its nature, to perceive *within* it the intrinsic interrelationships, to discover (rather than to invent) the answer to the problem within the problem itself.
—Abraham Maslow

An era can be said to end when its basic illusions are exhausted.
—Arthur Miller

Confusion is a word we have invented for an order which is not yet understood.
—Henry Miller

I like to think of my behavior in the sixties as a "learning experience." Then again, I like to think of anything stupid I've done as a "learning experience." It makes me feel less stupid.
—P. J. O'Rourke

In a sluggish system, aggressiveness produces instability. Either be patient or make the system more responsive.
—Peter M. Senge

A man should never be ashamed to own up when he has been in the wrong, which is but saying, in other words, that he is wiser today than he was yesterday.
—Jonathan Swift

The "silly question" is the first intimation of a some totally new development.
—Alfred North Whitehead

It's not whether we have learned from history—we have—but our awareness of what we have learned.

—John Woods

The Wrong Stuff

Includes thoughts on foolish behaviors and attitudes that lead to problems for individuals and organizations.

Voice 1: Please divert your course 15 degrees to the north to avoid a collision.

Voice 2: Recommend you divert *your* course 15 degrees to south to avoid a collision.

Voice 1: This is the captain of a U.S. Navy ship. I say again, divert *your* course.

Voice 2: No. I say again, you divert *your* course.

Voice 1: THIS IS THE AIRCRAFT CARRIER ENTERPRISE. WE ARE A LARGE WARSHIP OF THE U.S. NAVY. DIVERT YOUR COURSE NOW!

Voice 2: This is a lighthouse. Your call.

—Actual radio conversation released by the Chief of Naval Operations, October 10, 1975

A disaster is a good idea taken to extremes.

—Anonymous

It's a poor workman who blames his tools.

—Anonymous

Who is all-powerful should fear everything.

—Corneille

Whatever was required to be done, the Circumlocution Office was beforehand with all the public departments in the art of perceiving—*how not to do it.*

—Charles Dickens, *Little Dorrit*

How much easier it is to be critical than to be correct.

—Benjamin Disraeli

Most of what we call management consists of making it difficult for people to get their work done.

—Peter Drucker

I could never think well of a man's intellectual or moral character, if he was habitually unfaithful to his appointments.

—Nathaniel Emmons

Only the mediocre are always at their best.

—Jean Giraudoux

Our business world has accepted errors as a way of life. We live with them, we plan for them, and we make excuses for them. They have become part of the personality of our business. Our employees quickly recognize our standards and create errors so that they will not disappoint us.

—H. James Harrington

"To err is human, to forgive divine." (Well if that's true, we certainly have a lot of divine managers.)
—H. James Harrington

Well, if you've got work to do, Wallace, I don't want to interfere. I was reading an article in the paper the other day where a certain amount of responsibility around the home was good character training. Good bye, Mr. and Mrs. Cleaver.
—Eddie Haskell (not a role model)

Stability itself is nothing else than a more sluggish motion.
—Michel de Montaigne

Many are stubborn in pursuit of the path they have chosen, few in pursuit of the goals.
—Friedrich Nietzsche

The goods come back, but not the customer.
—Robert W. Peach

"Company policy" means there's no understandable reason for this action.
—Herbert V. Prochnow

To be feared is to fear: no one has been able to strike terror into others and at the same time enjoy peace of mind himself.
—Lucius Annaeus Seneca (the younger)

What is wrong with priests and popes is that instead of being apostles and saints, they are nothing but empirics who say "I know" instead of "I am learning," and pray for credulity and inertia as wise men pray for scepticism and activity.
—George Bernard Shaw

It is difficult to get a man to understand something when his salary depends on his not understanding it.
—Upton Sinclair

People who long to be rich are a prey to temptation; they get trapped into all sorts of foolishness and dangerous ambitions which eventually plunge them into ruin and destruction.
—1 Timothy 6:9–10 (Jerusalem Bible)

I know that most men, including those at ease with problems of great complexity, can seldom accept even the simplest and most obvious truth if it be such as would oblige them to admit the falsity of their conclusions which they have delighted in explaining to colleagues, which they have proudly taught to others, and which they have woven, thread by thread, into the fabric of their lives.
—Leo Tolstoy

And whoever begins by being a dupe ends by becoming a scoundrel.
—Voltaire

1996 Malcolm Baldrige National Quality Award Winners

In 1996, four companies received Baldrige Awards recognizing the companies' outstanding efforts to improve processes and satisfy customers. The recipients are

- **ADAC Laboratories,** located in Milpitas, California, in the manufacturing category. This company designs and manufactures products used in the health-care industry for nuclear medicine, radiation therapy, and health-care information systems.
- **Dana Commercial Credit Corporation,** headquartered in Toledo, Ohio, in the service industries category. Dana Commercial Credit is an operation of Dana Corporation and delivers a variety of financial services with a special emphasis on leasing programs.
- **Custom Research Inc.,** located in Minneapolis, Minnesota, in the small business category. CRI is a full-service national marketing research firm serving consumer, business-to-business, services, and medical markets.
- **Trident Precision Manufacturing Inc.,** located in Webster, New York, in the small business category. Serving several customers in different industries, Trident is a contract manufacturer of precision metal components, electromechanical assemblies, and custom products.

In 1996, there were twenty-nine applicants for the Baldrige Award—thirteen in manufacturing, six in services, and ten in small business. The smaller number of applications does not reflect on the importance of the award or the esteem in which it is held in American business. There are now many state awards (see the 1995 yearbook for a review of these), and many companies are investing effort in becoming ISO 9000 certified. The fact that there were four recipients out of the twenty-nine applicants attests to the high level of performance of those organizations that applied.

The Real Value of the Baldrige Award

Although receiving the Baldrige Award is a visible and public payoff for companies actively implementing total quality management as a philosophy and mode of operating, the award is only an acknowledgment that the firm is on the right path. This is a path that brings out the best in employees, helps the company get better and better in executing its processes, delivers ever higher quality outputs to customers, and very positively affects growth and profitability.

An important part of the Baldrige program and its continuously evolving award criteria is its use as guidelines for transforming a company. Thousands of organizations of all types use the Baldrige criteria to audit their current performance and help them proceed down the road toward creating ever more customer-oriented, efficient, and effective operations. The small investment of the government and others in the Baldrige program delivers great dividends to our country. The criteria provide specific categories for focusing efforts on improving and provide ways to measure how well a company is now doing and what it needs to work on to improve.

Although some business observers feel that the Baldrige Award is passé or not relevant anymore, you can be sure this is not the case. Generally, these people are the same observers and practitioners who feel that TQM was just another management fad—they tried it, and it didn't work. They have failed to appreciate that TQM is not just a set of techniques that a company might adopt. TQM requires a deep attitude and culture change as well. It requires that a company move away from an internal focus and toward a focus on and commitment to understanding and improving itself in relation to its customers and society as a whole.

TQM provides tools and techniques you can use to systematically improve. But without the change in attitude and culture, such efforts will not bring much improvement. But if such changes are made, the improvements will be dramatic. This is evident in Baldrige award winners—without exception, they have cultures that support customer satisfaction and foster a commitment to excellence in all aspects of their operations. Is such a culture necessary for a company to win the Baldrige? From our perspective, the answer is an unqualified yes. A firm does not achieve the recognition symbolized by the Baldrige award without highly motivated employees who go out of their way to do their jobs well and identify their personal welfare with doing a good job for the company. The good news is that the type of culture that brings out the best in people to improve processes and serve customers is also key to becoming a growing, vibrant, and profitable company.

Because Baldrige recipients become benchmarks of best practices, we present the following overviews of the 1996 recipients to give you a sense of what the Baldrige judges determined were outstanding examples of companies that practice managing for continuous process improvement and total quality. In doing this, we reproduce the announcements from the National Institute of Standards and Technology citing its reasons for selecting each recipient. We also give you names you can contact to get additional information on these companies and their individual approaches to managing for total quality.

You can learn more about this year's recipients, the award criteria, and other relevant information at a two Web sites maintained by NIST: http://www.quality.nist.gov and http://www.nist.gov/director/quality_program/.

1996 Baldrige Award Winner, Manufacturing Category: ADAC Laboratories

The following is an edited version of the press release from the National Institute of Standards and Technology announcing that ADAC Laboratories has received a Baldrige Award for 1996 in the manufacturing industries category.

Founded in 1970, ADAC Laboratories designs, manufactures, markets, and supports products for health-care customers in nuclear medicine, radiation therapy planning, and health-care information systems. These products and services are sold to hospitals, universities, and clinics throughout the world. ADAC's 710 employees work primarily at its headquarters facility in Milpitas, California, and at facilities in Houston, Texas, and Washington, Missouri. The following list highlights some of the reasons ADAC was chosen as a recipient of the 1996 Baldrige award:

- Customer focus at ADAC Laboratories is revealed by its core value, "Customers Come First." All executives are expected to spend 25 percent of their time with customers, personally take customer calls, and invite customers to attend weekly quality meetings. Customer satisfaction results have shown positive and improving trends over a five-year period for post-sales technical support (10 percent increase), customer retention (from 70 percent to 90 percent), and service contract renewals (from 85 percent to 95 percent).

- One of ADAC's measures of service quality is service cycle time, which determines the total time for getting a system back in operation. Service cycle time is critical to customers because they often cannot treat patients until a problem is fixed. Since the company began tracking this measure in 1990, the average cycle time has declined from 56 hours to 17 hours.

- ADAC's nuclear medicine market share has grown over the past five years from 10 percent to approximately 52 percent in the United States and from 5 percent to approximately 28 percent in Europe. Also, its revenue has tripled since 1990 compared to a 50 percent increase for the industry as a whole.

- ADAC consistently brings products to market faster than its competitors. Time-to-market leadership is evidenced by three product releases that averaged just over one-half the development time of competitors for similar products.

- Current levels and trends in key business measures demonstrate positive trends and performance. For example, revenue per employee has risen from about $200,000 in 1990 to almost $330,000 in 1995. On this overall measure of productivity, ADAC has achieved a 65 percent greater efficiency than its best competitor. Another measure is the number of direct labor dollars required to build cameras used to detect and diagnose health problems. Through more efficient processes and technology improvements, labor dollars per camera have decreased 40 percent since 1994.

- ADAC has a strong focus on process that is standardized through training and educating all employees. Over 100 customer and operational measures are reported in semiweekly quality meetings, which are open to all employees, customers, and suppliers.

- ADAC's business planning process, known as DASH, measures financial, customer, and operational performance and regulatory compliance. At quarterly DASH meetings, the company assesses its progress and, if needed, makes mid-course changes emphasizing recent performance compared with plans and the vision of the future.

- ADAC started an Advanced Clinical Research Program in 1992 to fund research at leading hospitals to improve the quality and efficiency of health care. ADAC donates approximately $350,000 annually to the program.

David L. Lowe is CEO of ADAC Laboratories. For more information, contact Kathy Call, Corporate Communications, ADAC Laboratories, 540 Alder Drive, Milpitas, CA 95035, (408) 321-9100.

1996 Baldrige Award Winner, Service Category: Dana Commercial Credit Corporation

The following is an edited version of the press release from the National Institute of Standards and Technology announcing that Dana Commercial Credit Corporation had received a Baldrige Award for 1996 in the service industry category.

Dana Commercial Credit Corporation, an operation of Dana Corporation, provides leasing and financing services to a broad range of business customers in selected market niches. Its primary offices are located in Toledo and Maumee, Ohio; Troy, Michigan; Oakville, Ontario, Canada; and Weybridge, Surrey, United Kingdom. Activities range from leveraged leases for power generation facilities and real estate properties with values up to $150 million, to customized programs assisting vendor-manufacturers in selling products such as in-store photo processing laboratories, to customized private label leasing programs that aid computer manufacturers, distributors, and dealers in selling systems that average $10,000 each. Dana Commercial Credit has assets of approximately $1.5 billion, and the company employs 547 people. The following list highlights some of DCC's quality-management initiatives:

- DCC consistently meets or exceeds key customer requirements, including completing transactions that competitors cannot, closing transactions on time, getting transactions done as agreed, and providing customized lease products before the competition. For example, DCC's Capital Markets Group (CMG), which accounts for 50 percent of new transactions, has closed all of its transactions on time for the past five years. DCC's Dealer Products Group (DPG) U.S., accounting for 20 percent of new volume, has reduced the time it takes to approve a transaction from about seven hours in 1992 to an hour or less in 1996.

- CMG and DPG U.S. have consistently achieved excellent performance levels with their primary customers. CMG maintains 50 percent and DPG U.S. has 20 percent of DCC's lease portfolio. Since 1994, CMG's customers have ranked them between 4 and 5 on a 5-point scale (1 being not satisfied, 5 very satisfied). DPG U.S. ranks 8 to 9 on a 10-point scale (0 being poor, 10 superior). The industry average is about 6.

- Each DCC employee receives an average of forty-eight hours of education, exceeding the industry average and the average of key competitors. DCC develops and teaches its employees more than forty classes in areas such as accounting, finance, and law, as well as interpersonal communications, quality, and marketing.

- DCC has a policy of promoting from within and fills 100 percent of senior leadership and 95 percent of supervisory and management positions internally, providing all employees with significant opportunities for advancement and growth.

- Most financial performance measures show sustained improvement and very good comparative performance. Return on equity and return on assets have increased more than 45 percent since 1991. Return on equity has been at or above 20 percent since 1992, compared to industry averages of 15 percent to 18 percent and exceeding DCC's two largest competitors by 2 percent and 10 percent.

- With more than 2,000 competitors, DCC ranks as number eleven with 0.6 percent of the market.

- DCC committed 45 percent of the money it saved as a result of a tax incentive from the City of Toledo to the Toledo School Board. As a result, the school board will receive one and one-half times more revenue than it would have received from DCC through a normal tax distribution. Toledo has adopted this approach as the standard for future tax incentives. This practice also has received positive recognition throughout the state of Michigan as a model of excellence.

Edward Shultz is chairman and CEO. The company headquarters are at 1801 Richards Road, Toledo, Ohio 43607. For more information, contact Tricia Akins, Director of Corporation Communications, at 201 W. Big Beaver Road, Suite 800, Troy Michigan 48089, (810) 680-4341.

1996 Baldrige Award Winner, Small Business Category: Custom Research Inc.

The following is an edited version of the press release from the National Institute of Standards and Technology announcing that Custom Research Inc. had received a Baldrige Award for 1996 in the small business industry category.

Custom Research Inc. (CRI) is a full-service national marketing research firm that has clients in consumer, business-to-business, services, and medical markets. The company works with large multinational companies to design and conduct projects that provide information to help make better business decisions. A privately owned corporation, CRI ranks thirty-sixth in size in a $4 billion industry with 3,000 competitors. CRI has 105 full-time employees at its headquarters in Minneapolis and in offices in San Francisco and Ridgewood, New Jersy (New York metropolitan area). CRI also has telephone interviewing centers in St. Paul, Minnesota, and Madison, Wisconsin. The following list highlights some of CRI's quality-management initiatives:

- Since 1988, feedback from clients on each of CRI's projects shows steadily improved overall project performance. CRI is now meeting or exceeding clients' expectations on 97 percent of its projects. Seventy percent of CRI's clients say the company exceeds expectations. CRI is rated by 92 percent of its clients as "better than competition" on the key dimension of "overall level of service."

- "Managing work through technology-driven processes" is one of CRI's key business drivers. CRI uses a standardized nine-step process to deliver customized proposals and projects to clients. Within this process, CRI's integrated software system links database questionnaires to the coding, tabulating, and reporting functions so that no rekeying is necessary after survey data are received from the field. As a result, cycle time for data tabulation has dropped from two weeks to one day.

- Trends for on-time delivery of final reports and data tables to clients have been favorable since 1993. 1995 results showed 99 percent of final reports and 96 percent of data tables being delivered on time.

- Revenue per full-time employee has risen 70 percent since 1988, when CRI reorganized into cross-functional, goal-directed teams. Since then, CRI has been substantially higher than its key competitors on this productivity measure.

- CRI reduced its overall client base from 138 in 1988 to 67 in 1995 as part of the company's strategy to better serve its biggest clients and build partnerships with them. Since then, the number of larger clients has increased from 25 to 34, and revenue has continued to grow.

- Ninety-four percent of CRI's employees agree that "All things considered, this is a good place to work." This is significantly above the norm of 76 percent nationally for business service companies. An important part of the company's employee development is a focus on training, which currently runs at over 120 hours per employee per year.

- CRI celebrates company successes and creates an atmosphere that encourages recognition. This includes monthly "Good News Meetings" and company-wide trips to celebrate achieving major business goals.

- CRI has played a key role within its industry in formulating the Council of Marketing and Opinion Research. CMOR speaks as one voice for the industry on key issues such as increasing the public's cooperation in marketing research studies.

CRI is managed by Judith Corson and Jeffrey Pope, Partners. For more information contact Beth Rounds, Senior Vice President, 10301 Wayzata Boulevard, P.O. Box 26695, Minneapolis, MN 55426-0864, (612) 542-0882.

1996 Baldrige Award Winner, Small Business Category: Trident Precision Manufacturing Inc.

The following is an edited version of the press release from the National Institute of Standards and Technology announcing that Trident Precision Manufacturing Inc. had received a Baldrige award for 1996 in the small business industry category.

Founded in 1979, Trident Precision Manufacturing Inc. is a privately held contract manufacturer of precision sheet-metal components, electromechanical assemblies, and custom products. The company develops tooling and processes to manufacture components and assemblies designed by its customers in a variety of industries, including office equipment, medical supplies, banking, computers, and defense. The company's 167 employees are based in a single manufacturing facility in Webster, New York. The following list highlights some of Trident's quality-management initiatives:

- On-time delivery indicates a positive trend over time, rising from 87 percent in 1990 to 99.94 percent in 1995.

- Trident monitors custom product reliability through defects per hundred machines. (Custom products go directly from Trident to the customer's distribution center for shipping to their customers.) For the past two years, Trident's custom products have had zero defects. As a result, Trident has been able to give its customers a full guarantee against defects.

- Results in achieving full employee involvement are indicated by 100 percent participation on departmental work teams since 1992, 97 percent of recommendations for process improvements being accepted in 1995, and over 95 percent accepted since 1991. Reward and recognition of employees has climbed steadily, from just 9 incidents in 1988 to 1,201 in 1995.

- Percent of direct labor hours spent on rework declined from 8.7 percent in 1990 to 1.1 percent in 1995.

- Financial indicators are positive: sales per employee rose from $67,000 in 1988 to $116,000 per employee in 1995. Return on assets rose from 7.9 percent in 1992 to 10 percent in 1995, compared to similar companies' return on assets of 4.7 percent in 1992 and 7.8 percent in 1995.

- Trident's quality rating (based on customer reports and customer rejects) for its major customers shows performance results consistent over time and above 99.8 percent.

- Customer satisfaction performance is strong: Trident has never lost a customer to a competitor, and sales volume has increased steadily from $4.4 million in 1988 to $14.3 million in 1995. Due to its strong customer focus, Trident has been able to maintain its status as a key supplier to major customers, even after those customers reduced their suppliers by 65 percent to 75 percent. Trident is the only supplier of General Dynamics to receive its Supplier Excellence Award.

- To ensure that new work being considered by the company will not introduce toxic or hazardous materials into Trident's manufacturing environment, the company works with potential customers before bids are placed on a project in an effort to ensure that no materials required for the project would pose a health or safety risk. If this cannot be done, Trident will not bid on the job.

- Employee turnover has shown dramatic improvement, falling from 41 percent in 1988 to 5 percent in 1994 and 1995. Trident's goal is to have less than 2 percent turnover.

- Trident's investment in training and education over the last several years is 4.6 percent of payroll, impressive for a small company. Expenditures in this area have consistently been two to three times the national average for the past seven years.

- Through customer participation in continuous improvement meetings, Trident shares manufacturing plans, methods, times, and costs with its customers in the quotation or concurrent engineering stages. Customer participation facilitates process modifications and plan changes and allows agreement on performance metrics.

Nicholas Juskiw is president and CEO of Trident Precision Manufacturing. For more information, contact Joseph Conchelos, Vice President of Quality, at 734 Sale Road, Webster, NY 14580-9796, (716) 265-1009.

1997 Malcolm Baldrige National Quality Award Criteria

This section provides you with a summary of the 1997 Baldrige Award criteria, including changes from the 1996 criteria and points given for each category. In reviewing this, you should note that for 1997, the number of criteria and point values for each criteria have been substantially revised. The most significant change is the heavy orientation toward business results; 450 out of 1,000 points are devoted to this category. The business results section is where the criteria explore whether good management intentions have paid off in the marketplace. You can receive a complete copy of the 1997 criteria and the 1997 application forms and instructions (criteria and application are two separate documents) by writing to

Malcolm Baldrige National Quality Award
National Institute of Standards and Technology
Route 270 and Quince Orchard Road
Administration Building, Room A537
Gaithersburg, MD 20899-0001
Phone: (301) 975-2036
Fax: (301) 948-3716
Internet: oqp@nist.gov
Web page: http://www.quality.nist.gov

1997 Award Criteria: Values, Concepts, and Framework

Award Criteria Purposes

The Malcolm Baldrige National Quality Award Criteria are the basis for making Awards and for giving feedback to applicants. In addition, the Criteria have three other important roles in strengthening U.S. competitiveness:

- to help improve performance practices and capabilities;
- to facilitate communication and sharing of best practices information among U.S. organizations of all types; and
- to serve as a working tool for understanding and managing performance, planning, training, and assessment.

Award Criteria Goals

The Criteria are designed to help companies enhance their competitiveness through focus on dual, results-oriented goals:

- delivery of ever-improving value to customers, resulting in marketplace success; and
- improvement of overall company performance and capabilities.

Core Values and Concepts

The Award Criteria are built upon a set of core values and concepts. These values and concepts are the foundation for integrating key business requirements within a results-oriented framework. These core values and concepts are:

Customer-Driven Quality

Quality is judged by customers. Thus, quality must take into account all product and service features and characteristics that contribute value to customers and lead to customer satisfaction, preference, and retention.

Value and satisfaction may be influenced by many factors throughout the customer's overall purchase, ownership, and service experiences. These factors include the company's relationship with customers that helps build trust, confidence, and loyalty.

Customer-driven quality addresses not only the product and service characteristics that meet basic customer requirements. It also includes those features and characteristics that differentiate them from competing offerings. Such differentiation may be based upon new or modified offerings, combinations of product and service offerings, customization of offerings, rapid response, or special relationships.

Customer-driven quality is thus a strategic concept. It is directed toward customer retention, market share gain, and growth. It demands constant sensitivity to changing and emerging customer and market requirements, and the factors that drive customer satisfaction and retention.

It also demands awareness of developments in technology and of competitors' offerings, and rapid and flexible response to customer and market requirements.

Customer-driven quality means much more than defect and error reduction, merely meeting specifications, or reducing complaints. Nevertheless, defect and error reduction and elimination of causes of dissatisfaction contribute to the customers' view of quality and are thus also important parts of customer-driven quality. In addition, the company's success in recovering from defects and mistakes ("making things right for the cutstomer") is crucial to building customer relationships and to customer retention.

Leadership

A company's senior leaders need to set directions and create a customer orientation, clear and visible values, and high expectations. The values, directions, and expectations need to address all stakeholders. The leaders need to ensure the creation of strategies, systems, and methods for achieving excellence and building knowledge and capabilities. The strategies and values should help guide all activities and decisions of the company. The senior leaders need to commit to the development of the entire work force and should encourage participation, learning, and creativity by all employees. Through their personal roles in planning, communications, review of company performance, and employee recognition, the senior leaders serve as role models, reinforcing the values and expectations and building leadership and initiative throughout the company.

Continuous Improvement and Learning

Achieving the highest levels of performance requires a well-executed approach to continuous improvement and learning. The term "continuous improvement" refers to both incremental and "breakthrough" improvement. The term "learning" refers to adaptation to change, leading to new goals and/or approaches. Improvement and learning need to be "embedded" in the way the company operates. Embedded means improvement and learning: (1) are a regular part of daily work; (2) seek to eliminate problems at their source; and (3) are driven by opportunities to do better, as well as by problems that must be corrected. Sources of improvement and learning include: employee ideas; R&D; customer imput; and benchmarking.

Improvement and learning include: (1) enhancing value to customers through new and improved products and services; (2) developing new business opportunities; (3) reducing errors, defects, waste, and related costs; (4) responsiveness and cycle time performance; (5) productivity and effectiveness in the use of all resources; and (6) the company's performance in fulfilling its public responsibilities and service as a good citizen. Thus, improvement and learning are directed not only toward better products and services but also toward being more responsive, adaptive, and efficient—giving the company additional marketplace and performance advantages.

Employee Participation and Development

A company's success depends increasingly on the knowledge, skills, and motivation of its work force. Employee success depends increasingly on having opportunities to learn and to practice new skills. Companies need to invest in the development of the work force through education, training, and opportunities for continuing growth. Opportunitites might include classroom and on-the-job training, job rotation, and pay for demonstrated knowledge and skills. On-the-job training offers a cost effective way to train and to better link training to work processes. Work force education and training programs may need to utilize advanced technologies, such as computer-based learning and satelite broadcasts. Increasingly, training, development, and work units need to be tailored to a diverse work force and to more flexible, high performance work practices.

Major challenges in the area of work force development include: (1) integration of human resource practices—selection, performance, recognition, training, and career advancement; and (2) alignment of human resource management with strategic change processes. Addressing these challenges requires use of employee-related data on knowledge, skills, satisfaction, motivation, safety, and well-being. Such data need to be tied to indicators of company or unit performance, such as customer satisfaction, customer retention, and productivity. Through this approach, human resource management may be better integrated and aligned with business directions.

Fast Response
Success in competitive markets demands ever-shorter cycles for new or improved product and service introduction. Also, faster and more flexible response to customers is now a more critical requirement. Major improvement in response time often requires simplification of work units and processes. To accomplish this, the time performance of work processes should be among the key process measures. There are other important benefits derived from this time focus: time improvements often drive simultaneous improvements in organization, quality, and productivity. Hence it is beneficial to integrate response time, quality, and productivity objectives.

Design Quality and Prevention
Companies need to emphasize design quality—problem and waste prevention achieved through building quality into products and services and efficiency into production and delivery processes. Costs of preventing problems at the design stage are usually much lower than cost of correcting problems that occur "downstream." Design quality includes the creation of fault-tolerant (robust) or failure-resistant processes and products.

A major success factor in competition is the design-to-introduction ("product generation") cycle time. To meet the demands of rapidly changing markets, companies need to carry out stage-to-stage integration ("concurrent engineering") of activities from basic research to commercialization. Increasingly, design quality also depends upon the ability to use information from diverse sources and data bases, that combine customer preference, competitive offerings, price, marketplace changes, and external research findings. More emphasis should also be placed on capturing learning from other design projects.

From the point of view of public responsibility, the design stage is critical. In manufacturing, design decisions determine process wastes and the content of municipal and industrial wastes. The growing environmental demands mean that design strategies need to anticipate environmental factors.

Consistent with the theme of design quality and prevention, improvement needs to emphasize interventions "upstream"—at early stages in processes. This approach yields the maximum cost and other benefits of improvements and corrections. Such upstream intervention also needs to take into account the company's suppliers.

Long-Range View of the Future
Pursuit of market leadership requires a strong future orientation and a willingness to make long-term commitments to key stakeholders—customers,

employees, suppliers, stockholders, the public, and the community. Planning needs to anticipate many changes, such as customers' expectations, new business opportunities, technological developments, new customer segments, evolving regulatory requirements, community/societal expectations, and thrusts by competitors. Plans, strategies, and resource allocations need to reflect these commitments and changes. A major part of the long-term commitment is developing employees and suppliers and fulfilling public responsibilities.

Management by Fact

Modern businesses depend upon measurement and analysis of performance. Measurements must derive from the company's strategy and provide critical data and information about key processes, outputs, and results. Data and information needed for performance measurement and improvement are of many types, including: customer, product and service performance, operations, market, competitive comparisons, supplier, employee-related, and cost and financial. Analysis refers to extracting larger meaning from data and information to support evaluation and decision making at all levels within the company. Analysis entails using data to determine trends, projections, and cause and effect—that might not be evident without analysis. Data and analysis support a variety of company purposes, such as planning, reviewing company performance, improving operations, and comparing company performance with competitors' or with "best practices" benchmarks.

A major consideration in performance improvement involves the creation and use of performance measures or indicators. Performance measures or indicators are measurable characteristics of products, services, processes, and operations the company uses to track and improve performance. *The measures or indicators should be selected to best represent the factors that lead to improved customer, operational, and financial performance. A comprehensive set of measures or indicators tied to customer and/or company performance requirements represents a clear basis for aligning all activities with the company's goals.* Through the analysis of data from the tracking processes, the measures or indicators themselves may be evaluated and changed to better support such goals.

Partnership Development

Companies need to build internal and external partnerships to better accomplish their overall goals.

Internal partnerships might include labor-management cooperation, such as agreements with unions. Agreements might entail employee development, cross-training, or new work organizations, such as high performance work teams. Internal partnerships might also involve creating network relationships among company units to improve flexibility, responsiveness, and knowledge sharing.

External partnerships might be with customers, suppliers, and education organizations for a variety of purposes, including education and training. An increasingly important kind of external partnership is the strategic partnership or alliance. Such partnerships might offer a company entry into new markets or a basis for new products or services. A partnership might also permit the blending of a company's core competencies or leadership capabilities with complementary strengths and capabilities of partners, thereby enhancing overall capability, including speed and flexibility. Internal and external partnerships should develop longer-term objectives, thereby creating a basis for mutual investments. Partners should address the key requirements for success, means of regular communication, approaches to evaluating progress, and means for adapting to changing conditions. In some cases, joint education and training could offer a cost-effective means to help ensure success.

Company Responsibility and Citizenship

A company's leadership needs to stress its responsibilities to the public and practice good citizenship. This responsibility refers to basic expectations of the company—business ethics and protection of public health, safety, and the environment. Health, safety, and the environment include the company's operations as well as the life cycles of its products and services. Companies need to emphasize resource conservation and waste reduction at their source. Company planning should anticipate adverse impacts from facilities, production, distribution, transportation, use, and disposal of products. Plans should seek to prevent problems, to provide a forthright company response if problems occur, and to make available information and support needed to maintain public awareness, safety, and confidence. Companies should not only meet all local, state, and federal laws and regulatory requirements. They should treat these and related requirements as areas for continuous improvement "beyond mere compliance." This requires use of appropriate measures in managing performance.

Practicing good citizenship refers to leadership and support—within limits of a company's resources—of publicly important purposes, including areas of public responsibility. Such purposes might include education improvement, improving health care in the community, environmental excellence, resource conservation, community services, improving industry and business practices, and sharing of nonproprietary information. Company leadership as a corporate citizen also entails influencing other organizations, private and public, to partner for these purposes. For example, individual companies could lead efforts to help define the obligations of their industry to its communities.

Results Focus

A company's performance measurements need to focus on key results. Results should be guided by and balanced by the interests of all stakeholders—customers, employees, stockholders, suppliers and partners, the public, and the community. To meet the sometimes conflicting and changing aims that balance implies, company strategy needs to explicitly include all stakeholder requirements. This will help to ensure that actions and plans meet differing stakeholder needs and avoid adverse impact on any stakeholders. The use of

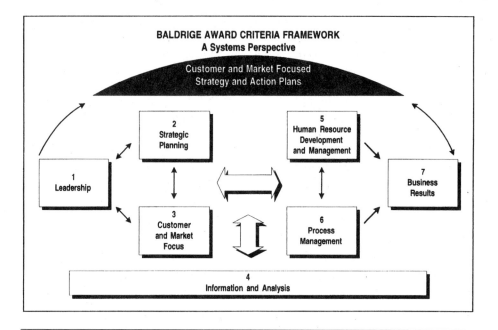

Figure 1. **Baldrige award criteria framework: a systems perspective**

a balanced composite of performance measures offers an effective means to communicate short- and longer-term priorities, to monitor actual performance, and to marshal support for improving results.

Award Criteria Framework

The core values and concepts are embodied in seven Categories, as follows:

1. Leadership
2. Strategic Planning
3. Customer and Market Focus
4. Information and Analysis
5. Human Resource Development and Management
6. Process Management
7. Business Results

The framework connecting and integrating the Categories is given in figure 1.

The framework has three basic elements, from top to bottom:

Strategy and Action Plans

Strategy and Action Plans are the set of company-level requirements, derived from short- and long-term strategic planning, that must be done well for the company's strategy to succeed. Strategy and Action Plans guide overall resource decisions and drive the alignment of measures for all work units to ensure customer satisfaction and market success.

System
The system is comprised of the six Baldrige Categories in the center of the figure that define the organization, its operations, and its results.

All company actions point toward Business Results—a composite of customer, financial, and non-financial performance results, including human resource development and public responsibility.

Information and Analysis
Information and Analysis (Category 4) are critical to the effective management of the company and to a fact-based system for improving company performance and competitiveness.

Award Criteria Organization
The seven Criteria Categories shown in the figure are subdivided into Items and Areas to Address:

Items
There are 20 Items, each focusing on a major requirement. Item titles and point values are given on page 779.

Areas to Address
Items consist of one or more Areas to Address (Areas). Information is submitted by applicants in response to the specific requirements of these Areas.

Changes from the 1997 Award Criteria

The Criteria continue to evolve toward comprehensive coverage of strategy-driven performance, addressing the needs and expectations of all stakeholders—customers, employees, stockholders, suppliers, and the public. The Criteria for 1997 strengthen the systems view of performance management, and place a greater focus on company strategy, organizational learning, and better integration of business results. The composite of business results (now Category 7) has been increased in point value (to 450 points) to indicate the degree of importance and to reflect the change in content. Specifically, Category 7 now includes customer satisfaction, financial and market indicators of performance, human resource results, and company-specific operational results key to achieving business success.

The most significant changes made in the Criteria and in the Criteria booklet are summarized as follows:

- The Criteria framework has been revised. The major changes that led to this revision are: (1) greater importance of company strategy to an effective performance management system; and (2) the need to treat all business results, including customer satisfaction results, in a parallel and integrated manner. In addition, some Items or parts of Items have been combined to better integrate requirements and to reduce the number of responses applicants need to make. As a result of these changes:

- All Categories from 1996 except Category 1 (Leadership) have been renumbered. A schematic diagram of the new Category framework is given on page 772.
- The number of Items has been reduced from 24 to 20.
- The number of Areas to Address has been reduced from 52 to 30. Each Area has been given a title to highlight its purpose within its Item. For Items having only one Area, the Item title is repeated as the Area title.
- The application page limit has been reduced from 70 to 50.
- The number of Item Notes has been reduced from 114 to 45. Information from Item Notes in the 1996 Criteria that provided examples is included in a revised Item Descriptions section, now titled 1997 Award Criteria: Item Descriptions and Comments starting on page 20. This section also takes into account the changed Category content and relationships.
- Greater emphasis is placed on organizational learning to underscore the importance of learning and change as well as continuous improvement.

Changes, by Category, are:

Leadership

- Items 1.1 and 1.2 from 1996 have been combined into a new Item 1.1, Leadership System. This combination provides the opportunity for coordinated presentation of the role of senior leaders and the company's leadership system. Additional emphasis has been placed on the senior leaders' role in organizational learning. The company performance review function from Item 1.2 (1996) is now added to Item 4.3, Analysis and Review of Company Performance.

- Item 1.2 replaces Item 1.3 from the 1996 Criteria. Area 1.2b (1.3b in 1996) addresses more clearly the company's community involvement and leadership, rather than leadership as a corporate citizen, to accommodate better companies of all sizes and also to emphasize employee involvement in key communities.

Strategic Planning

- This Category contains two Items with the same basic requirements as the corresponding Items in 1996. However, the titles of the Items have been modified to better communicate their main purposes and to sharpen the important distinction between these items.

- Item 2.1 is now Strategy Development Process instead of Strategy Development. This change is intended to emphasize that the Item examines primarily the overall *strategic planning process*—how it is done, what it considers, and how it leads to action plans.

- Item 2.2 is now Company Strategy instead of Strategy Deployment. This change is intended to emphasize that the Item calls for information on *the company's actual strategy* and how it is deployed, not on a general description of the deployment process. This Item now includes coverage of human resource plans, which in 1996 were addressed in the Human Resource Development and Management Category. This change is made to emphasize the need for better integration of human resource planning with strategic planning.

Customer and Market Focus

- This Category title is changed from Customer Focus and Satisfaction to reflect two important changes made in the Criteria: (1) inclusion of Customer Satisfaction Results with Business Results; and, (2) recognition that Criteria evolution has placed growing emphasis not only on customers, but also on markets.

- The Category now contains two Items, reduced by two Items from 1996. In addition to the movement of Item 7.4 (1996) to the Business Results Category, two Items from 1996 [Customer Relationship Management (7.2) and Customer Satisfaction Determination (7.3)] have been combined into Customer Satisfaction and Relationship Enhancement (Item 3.2). This change is intended to provide better integration of the different ways that companies use to understand customers, the factors that determine satisfaction, and how they stay close to customers. The change is intended to accommodate better the approaches of excellent smaller companies, many of which use regular customer contact more than survey methods both to determine satisfaction and to enhance relationships.

Information and Analysis

- This Category contains three Items with the same basic requirements as the corresponding Items in 1996. However, the titles of the Items have been modified to better communicate their main purposes.

- Items 4.1 and 4.2 now include Use in their titles to better reflect the central purpose of these Items.

- Item 4.3 is now Analysis and Review of Company Performance, making this Item the point within the Criteria for supporting a critical requirement in performance management—understanding the meaning of performance information to help guide the company's decisions and actions. The major change is the coverage of the company performance review function which was part of Item 1.2 in 1996.

Human Resource Development and Management

- This Category now contains three Items, reduced by one Item from 1996.

- Item 4.1 from 1996, Human Resource Planning and Evaluation, has been eliminated. The important planning included in this Item is now integrated within overall company planning as mentioned above under Strategic Planning.

- Item 4.2 from 1996, High Performance Work Systems, now becomes Item 5.1 and it titled Work Systems. Although this item retains its focus on high performance, its title is changed to avoid the appearance that its purpose is narrower—high performance work teams.

Process Management

- Items 5.1 and 5.2 from 1996 have been combined into a new Item 6.1, Management of Product and Service Processes. This combination permits a company to place appropriate emphasis on design processes and production/delivery processes, as dictated by the company's business. Area 6.1b includes a new requirement on the transfer of learning to other company units and projects. This is intended to foster better internal communication and to aid overall organizational improvement.

- Item 6.3, Management of Supplier and Partnering Processes, places specific attention on preferred suppliers and partnering arrangements in recognition of their growing importance and the attention they require. The Item requires information on the company's actions and plans to improve suppliers' and partners' abilities to contribute to achieving the company's goals. This change is intended to focus responses on considerations most important to the supplier/customer relationship.

Business Results

- This Category now contains five Items, including Customer Satisfaction Results (Item 7.4 in 1996), thus integrating all results Items into one Category.

- There are three 130-point Items in the Category, in recognition of the importance of customer satisfaction, financial and marketplace performance, and company-specific performance measures (see discussion below) to the success of the company.

- Item 7.2 has been created to provide a single Item focusing on Financial and Market Results. This Item stresses the importance of a set of key financial and market results tracked by senior leadership to gauge overall company performance. In this and other results Items, applicants are offered greater latitude in recognition of the fact that not all results are quantitative or can be trended in a meaningful manner (e.g., business growth into new geographic regions or new markets entered through adaptations of existing products).

- Item 7.4, Supplier and Partner Results, has been expanded to include company costs and/or performance improvements due to supplier and partner performance. This is an added measure of the effectiveness of the relationship and the linkage to important results.

- Item 7.5, Company-Specific Results, is a new Item, including some results called for in Items 6.1 and 6.2 in 1996, but offering applicants much greater latitude in showing beneficial changes and improvements in internal operations and operational results relevant to their businesses and strategies. Results appropriate for Item 7.5 include improvements in and performances of products, services, and processes; productivity; cycle time; regulatory/legal compliance and related performance; and new product and/or service introductions. Applicants are encouraged to report unique and innovative results that directly relate to their types of businesses or strategies.

1997 Baldrige Award Changes: Questions and Answers

The National Institute for Standards and Technology has prepared a question-and-answer document to explain the many changes in the 1997 criteria. Because we feel this will be valuable to you, we reproduce it here.

Q. *Why is the booklet previously titled "Award Criteria" now titled "Criteria for Performance Excellence"?*
A. The name change reflects the fact that the criteria are much more than a set of "rules" for an award contest. They are now widely accepted as *the* standard for performance and business excellence. In addition to serving as the basis for applying for the award, the criteria are used by thousands of organizations of all kinds for self-assessment, planning, training, and other purposes. More than a million copies have been distributed since 1989.

Q. *Why did you create a new framework for the Baldrige criteria?*
A. The framework was changed for the same reason that the category order was changed—to provide a better systems view of performance management. The framework has been redesigned to improve the focus on customer and market-driven strategy, the role of information and analysis, and the goal of improved business results.

Q. *Have the 1997 Baldrige criteria strayed away from their strong focus on customers?*
A. The emphasis on the customer is just as strong, if not stronger, in the 1997 criteria. All of the criteria rest on two primary building blocks: (1) quality is customer driven and (2) companies must continually improve their performance capabilities. These concepts are woven throughout all the criteria.

In particular for 1997, the customer and market focus category—tied closely to the leadership category—examines how the company determines its customers' requirements and expectations. In the business results category, the first item addresses the main customer-related results—customer satisfaction, customer dissatisfaction, and customer satisfaction relative to competitors. The framework depicts customer- and market-focused strategy as an overall umbrella for the Baldrige systems perspective on performance management.

Q. *Have you reduced the emphasis on product and service quality?*

A. No. The product and service quality results item from 1996 has been moved to the new business results category as part of company-specific results. This new item includes improvements in and performance of products, services, and processes; productivity; cycle time; regulatory or legal compliance and related performance; and new product or service introductions. The new category allows much greater flexibility in how a company shows results, changes, and improvements.

Q. *What hasn't changed?*

A. Although some criteria categories have been renamed and reordered, their content remains basically the same as it was in 1988. As in the past, the award criteria are built on the same basic core values and principles:

- Customer-driven quality
- Leadership
- Continuous improvement and learning
- Employee participation and development
- Fast response
- Design quality and prevention
- Long-range view of the future
- Management by fact
- Partnership development
- Company responsibility and citizenship
- Results focus

Q. *Have the eligibility rules changed?*

A. Yes. The rule that previously required a company to have more than 50 percent of its personnel or physical assets in the United States or its territories has been dropped. To be successful in a global economy, many American companies find they must have substantial overseas operations. The company must retain its headquarters in the United States. Also, if an applicant reaches the site visit stage, the applicant must ensure that the appropriate people and materials are available in the United States for examiners to evaluate the company. A company that receives the award must be willing and able to participate in the Quest for Excellence conference and be able to share information on the seven Baldrige categories in the company's U.S. facilities.

Q. *Does NIST consult with any other organizations before making changes?*

A. Yes. Every two years, NIST, in conjunction with members of the boards of overseers and examiners, past winners, and other organizations, performs an extensive evaluation of the criteria to determine whether they remain relevant and reflect current thinking regarding performance excellence. However, the ultimate decision regarding changes rests with NIST, the manager of the award.

1997 Award Criteria—Item Listing

1997 Categories/Items	Point Values
1.0 Leadership	**110**
1.1 Leadership System	80
1.2 Company Responsibility and Citizenship	30
2.0 Strategic Planning	**80**
2.1 Strategy Development Process	40
2.2 Company Strategy	40
3.0 Customer and Market Focus	**80**
3.1 Customer and Market Knowledge	40
3.2 Customer Satisfaction and Relationship Enhancement	40
4.0 Information and Analysis	**75**
4.1 Selection and Use of Information and Data	25
4.2 Selection and Use of Comparative Information and Data	15
4.3 Analysis and Review of Company Performance	40
5.0 Human Resource Development and Management	**100**
5.1 Work Systems	40
5.2 Employee Education, Training, and Development	30
5.3 Employee Well-Being and Satisfaction	30
6.0 Process Management	**100**
6.1 Management of Product and Service Processes	20
6.2 Management of Support Processes	20
6.3 Management of Supplier and Partnering Processes	20
7.0 Business Results	**450**
7.1 Customer Satisfaction Results	130
7.2 Financial and Market Results	130
7.3 Human Resource Results	35
7.4 Supplier and Partner Results	25
7.5 Company-Specific Results	130
TOTAL POINTS	**1,000**

Index

ABC, case studies, 539–554
Accounting, bibliography, 660–661; cases and practices, 539–554
ADAC Laboratories, Baldrige winner, 730
Aerospace, bibliography, 634
Aetna Life & Casualty, training at, 352–353
AFGE, 117
AFSCME, 117
AFT, 117
Ahold Nv., case of cross docking, 526
Akins, Tricia, Dana Commercial Credit, 762
Alfred, Richard, on community colleges, 175–187
Allspaw, Scott, quoted, 559
Allstate Insurance Company, skills process at, 364
American Assembly of Collegiate Schools of Business (AACSB), requirements of, 192
American Management Association (AMA), gainsharing studies of, 593–598
American National Standards Institute (AOSI), views of ISO, 622–623
American Society for Quality Control (ASQC), organizational changes in, 118
Assessments, bibliography, 664–667; cases and issues, 599–630; effect on best practices, 8; in higher education, 188–193; Xerox's, 309–313
Association for Quality and Participation, role of, 120
Assumption analysis, when to use and value of, 410
AT&T Global Business Communications Systems, analyzes processes, 564, 565; benchmarking at, 443–450
AT&T Universal Card, 104
Awards, bibliography, 666

Babson College, Baldrige and, 198–203
Baker, Clarence, quoted, 454
Baker Oil Tools, product development process at, 561
Baldrige Award, 1996 award winners, 758–765; address of, 203; issues, 625–630
Baldrige criteria, changes for 1997, 766–779; applied in K-12, 194–204
Banking, bibliography, 635–636
Banks, customer survey case, 340–345; value propositions in, 99–107
Barnes, Bill, quoted, 337, 338
Bateson, Gregory, quoted on definition of information, 402
Beck, John, leadership and teams, 376–383

Beckham, J. Daniel, on customer value in hospitals, 146–151
Beer, Stafford, definition of information, 402
Bell, Larry, quoted, 454
Belmont University, Baldrige and, 198–203
Benchmarking, best practices, 730; bibliography, 651; current wisdom about, 443–490; value of, 8–9, 410
Benedetto, George, views of, 501–502
Benson, George, training success stories, 352–359
Benson, Roger S., case of ISO certification, 605–612
Bertch Cabinet Manufacturing. Inc., ABC case study, 548–554
Best practices, what they are, 3–12
Betancourt, Randy, quoted, 535
Betton, John, quoted on business education, 188–193
Bibliography, quality practices, 633–667
BIC Corporation, strategic planning at, 235–236
Bidault, Francis, suppliers and product development, 567–574
Big Q, concept described, 50–57
Billing, best practices, 730
Blake, Norman, quoted, 260
Blose, Steve, quoted, 453
Blyth, Pamela L., on reengineering in health care, 152–161
Boart Lonyear Canada, self-directed teams at, 387–389
Boeing Company, training at, 361–362
Bold, Steven, investing in ISO certification, 618
Bolouki, Shawn, 154
Boyd, John, on Harley-Davidson, 298
Brand management, improves customer satisfaction, 315–321
Brainstorming, 273–274; value of, 410
British Nuclear Fuels, 624
Business schools, role of quality practices in, 188–193
Business, small, bibliography, 637–638
Butler, Christina, suppliers and product development, 567–574
Buzzell, Robert D., PIMS and, 48–49

Calek, Anne, on customer loyalty, 322–333
Call centers, applying TQM in, 108–114
Call, Kathy, ADAC Laboratories, 761
Camillus, John C., on strategic planning, 231–242
Canada, Deming principles in government, 122–129
Carlzon, Jan, moments of truth, 328–329
Carson, Judith, Custom Research, Inc., 764

Carson, Kerry D., use of social power, 254–266
Carson, Paula Phillips, use of social power, 254–266
Carter, Patricia, on community colleges, 175–187
Caterpillar Corporation, case study of teaming at, 501–505
Cecere, Domenico, quoted, 532
Cellular manufacturing, case study of, 91–98
Central DuPage Health System, customer focus at, 146–151
Chambers, Charles, quoted, 532
Charles Medical Center, teams and quality practices at, 158–161
Check clearing, best practices, 732
Chen, Wen-Hsien, SPC and processes, 465–481
Communication, bibliography, 647; quality practices, 414–423; teams and, 392–393
Community colleges, how to be competitive, 175–187
Compensation, best practices, 730
Competency-alignment process, described, 431–442
Competition, Feigenbaum on, 243–253
Competitive analysis, best practices, 730
Complaints, value of, 327–328
Conchelos, Joseph, Trident Precision Manufacturing, 765
Concurrent engineering, best practices, 730
Connellan, Tom, 327; on knowing what customers want, 323–324
Contracts, ways to improve process of, 506–516
Control, basis of, 17–21; nature of, 15–16
Control charts, recalculating limits in, 488–490
Convex Computer, product development process review, 560, 562
Cost justification, bibliography, 660–661; of training, 346–351
Cowen, Scott S., on higher education, 163–167
Critical success factors (CSFs), role in benchmarking, 446–447
Crosby, Philip, quoted on people's rights and rewards, 256, 260, 263
Cross docking, case of, 526
Cultural transformation, bibliography, 643; practices of, 285–314
Custom Research, Inc., Baldrige winner, 762–764
Customer feedback, best practices, 730–731
Customer focus, best practices, 731; bibliography, 643–645
Customer service, best practices, 731–732
Customer value, concept of applied, 99–107
Customer, cases of voice of, 315–345; value from hospitals, 146–151

Dabbiere, Alan, need to reduce inventories, 525
Dahl, Roger, quoted on power, 264
Dasch, Martha L., on hospital quality practices, 138–145

Dana Commercial Credit Corp., Baldrige winner, 761–762
Data transfer, best practices, 732
Data warehouses, use of, 534–538
Davis, Scott, product brands and customer satisfaction, 315–321
Deaton, Gary, quoted on supplier role, 567
Delegation, defined, 393
Deming, W. Edwards, drive out fear, 256; ending performance appraisals, 172; 14 Points in education, 218–228; ideas applied to government, 122–129
Deutsche Telecom, use of I/T, 65
Devia, Napoleon, on manufacturing competitiveness, 61–68
Diebold, ESI at, 570, 572
Digital Equipment Corporation, personnel assessments at, 365; training at, 362–363
DiJulio, Dave, quoted on Harley-Davidson's culture, 296
Distribution, best practices, 741
Distribution industry, bibliography, 636
Document processing, best practices, 732
Dolence, Michael G., on higher education, 163–167
Dow Chemical Company, customer surveying by, 338–339
Downsizing, popularity of, 54; problems with, 277–284
Drucker, Peter, quoted on change, 167; quoted on quality, 48
DuPont, 404

Early Supplier Involvement (ESI), described, 567–574
Eastman Chemical, customer surveying by, 337–338; tracks value of R&D, 563, 565
Education (K-12), bibliography, 640–641; quality practices in, 194–228
Electronic Hardware (EHC), cellular manufacturing at, 91–98
Employee Involvement Association, 585
Employee performance, measuring, 482–487
Employee recognition, best practices, 732
Employees, how to communicate with, 419–423; role in customer satisfaction, 330–332
Empowerment, best practices, 733; role in teams, 393–394; strategies for, 277–284
Englund, Randy, quoted on metrics, 562
Environment management, 602–603; best practices, 733; ISO and, 613–620
Erickson, Chris, quoted, 535
Ettore, Barbara, on relevance of Baldrige Award, 625–630
Executive development, best practices, 733
Exterbille, Kaat, how to develop communications program, 414–418

Facilities management, best practices, 733
Federal government, bibliography, 638

INDEX

Feigenbaum, Armand V., on leadership, 243–253
Filipczak, Bob, on Harley-Davidson's corporate culture, 293–300
Finance, functional area, bibliography, 660–661
Finance industry, bibliography, 635–636; cases and practices, 539–554
Fishboning, at Shasta Industries, 70–71
Fisher, Donald C., quoted on Baldrige, 627
Fisher, Steven, quoted on measuring, 463
Ford Motor Company, 67, 130, 134
French, J.R.P., on social power, 254–255
Functional processes, cases and best practices, 491–598

Gale, Bradley T., PIMS and, 48–49
Galpin, Timothy, on culture and organizational change, 285–292
Gainsharing, described, 592–598
Gemla, concept defined, 362–365
General Electric (GE), 176
General Motors (GM), delivery process of, 523; effects of cost reduction at, 147; rewards at, 261
George Westinghouse Vocational and Technical High School, TQM case study, 210–217
Ginnodo, Bill, quoted on quality, 49
Glover, Frank, quoted, 533
Gnyawali, Devi R., learning to improve, 407–413
Godfrey, Blanton, quoted, 322–323, 324, 330
Goldberg, Paul, quoted on quality potential in public sector, 117–118
Government, quality management in, 115–137
Grace Specialty Polymers, case of ISO certification, 605–612
Grant, John H., on learning to improve, 407–413
Gregg, Laura, on customer loyalty, 323; quoted, 325, 332
Greenberg, Sheldon, on Shasta Industries practices, 69–80
Gretzky, Wayne, quoted on how he became famous, 107
Growth, barriers to, 88–90

Hadlet, David, quoted, 624
Hale, Gregory J., on ISO 14001 certification, 613–620
Hamel, Gary, quoted, 442
Harari, Oren, on how to do effective process reengineering, 424–430
Harlan, Robert, on Shasta Industries, 69–80
Harley-Davidson, corporate culture of, 293–300; employee rewards at, 261
Harrington, H. James, on best practices, 3–12
Harris, Jim, communications and employees, 419–423
Hartung, Debbie, on idea systems, 585
Harwell, Chris, quoted, 201
Hashampour, Reza, quoted, 208

Hauzer, Reuben Z., on implementing self-directed teams, 389–396
Hayes, Bob, quoted, 330
Health care, best practices, 734; bibliography, 638–640; quality management in, 138–161
Heckscher, Charles, on empowerment strategies, 277–284
Heilmann, Ronald L., on higher education's future, 162–167
Hemenway, Caroline G., on ISO 14001 certification, 613–620
Hewlett-Packard, 562, 563, 623
Higher education, bibliography, 641; quality management in, 162–193
Hiner, Glen H., quoted, 529
Holland, Dennis S., on using TQM in call centers, 108–114
Holten, John, on customer surveys, 340–345
Honeywell, approach to teams, 387
Horine, Julie, on quality in K-12 education, 194–204
Horney, Nicholas F., people issues in reengineering, 431–442
Hotels, bibliography, 636–637
Hovnanian, Ara, quoted, 456
Hughes Aircraft Company, product development at, 64–65; uses data warehousing tools, 536–537
Human resources, bibliography, 661–664; cases and practices, 575–598; See also, Personnel
Hybert, Pete, improving contracting process, 506–516

IBM Rochester, quality at, 51
Imberman, Woodruff, on gainsharing, 592–598
Industrial design, best practices, 734
Industry, associations directory, 705–729
Information, defined, 402
Information Age, effect on higher education, 162–167
Information systems, value for learning, 410–411
Information technology (I/T), bibliography, 659–660; cases and practices, 528–538; on getting closer to customers with, 67
Inmon, Bill, father of data warehousing quoted, 535–536
Insurance, bibliography, 637
Internal Revenue Service (IRS), innovations at, 131, 133
International Accreditation Forum (IAF), defined, 622
International Quality Study, results of, 3–12; See also, Best practices
Inventory control, best practices, 734–735
Ishikawa, Kaoru, quoted on quality, 168–169
ISO 9000, bibliography, 664–666; cases and practices, 599–624; internet sources, 678–679
ISO 14000, 602–603
ISO 14001, certification described, 613–620

J. Walter Thompson, 67
Jasinski, Dale W., on ABC, 539–546
Johnson, Elaine, quoted on TQM and students, 208
Johnson, Richard S., defines delegation, 393
Journals, directory of, 686–703
Juskiw, Nicholas, CEO Trident Precision Manufacturing, 765

Kanban, defined and case study of, 95–96
Kauffman, Stuart, biological view of organizations, 399
Keenan, Jr., William, on customer surveys, 334–339
Kellogg Foundation, quality programs of, 120–121
Kellner-Rogers, Myron, on self-organizing, 397–406
Kennedy, Mike, quoted, 453
Kessler, Laura, on customer loyalty, 323
Kinnvison, Dianne, on classroom mission statement, 207
Kirkpatrick, Donald, his training model lacking ROI, 347
Kisling, J. Walter, quoted on measurements, 459
Klebar, Bob, quoted on Harley-Davidson, 296–299
Knight, E. Leon, use of social power, 254–266
Knowledge factories, described, 81–91
Kolind, Lars, quoted, 86
Kodak, ESI at, 569, 573
Konopnicki, Patrick, quality in classrooms, 205–209
Koonce, Richard, people issues in reengineering, 431–442
Kotler, Philip, on marketing, 32–46
Kuczmarski & Associates, brand management study of, 315–321

Landis and Gear, customer strategy of, 67
Lareau, William, behavior, 267–276
Law of Mortality, defined, 17–18
Lawler III, Edward E., on TQM and employee involvement, 575–583
Leaders, role in self-organizing organizations, 405–406
Leadership, best practices, 243–284; bibliography, 641–643; in context of teams, 376–383
Learning, making possible, 440–441
Learning organizations, practices in, 397–413
Ledford, Jr., Gerald E., on TQM and employee involvement, 575–583
Leo, Richard J., on Xerox's culture, 301–314
Levin, Wayne J., on Deming principles in government, 122–129
Levi Strauss, delivery process of, 522–523
Lexmark, ESI at, 570, 572, 573
Lienert, Anita, use of I/T, 528–533
Lincoln, Sarah, on improving benchmarking, 443–450
Lingle, John H., on measurements, 455–464
Lipton, Randall A., defends Baldrige, 627
Lloyd's Register Quality Assurance Ltd. (LRQA), 624

Local government, bibliography, 638
Logistics, best practices, 739; bibliography, 656–657; cases and practices, 517–527
Lowe, David L., CEO of ADAC Laboratories, 761
Loyalty, customer, 322–333; problems with, 280–282

MacArthur, John B., ABC case study, 547–554
Mack Truck, 65
Magazines, directory of, 686–703
Magnavox Electronics Systems Company, cost justifies training, 346–351
Make versus buy decisions, 735
Malcolm Baldrige National Quality Award, criteria for 1997, 766–779; point values 1997, 779; recipients, 1996, 758–765
Management, bibliography, 641–643; participatory, 277–284
Manley, John, on 14 Points in education, 218–228
Manley, Robert, on 14 Points in education, 218–228
Manufacturing, best practices, 735–736; bibliography, 633–635; flexible, 733–734; planning in, 235–236; quality practices in, 61–98; synchronous (case study), 549–554
Manufacturing design, best practices, 732
Marketing, best practices, 736; bibliography, 656; Kotler's concept of, 32–46; tasks of management, 43–44
MBNA bank, customer loyalty at, 323
McKee, Pat, 502
Measurements, 455–490, 562; bibliography, 652–653; of cultural changes, 290–292; value of, 11–12, 410–411
Meetings, effect on best practices, 7–8
Milliken and Company, quality practices in, 49; suggestion process at, 585
Minco Group, 567
Minnesota Academic Excellence Foundation, address, 203
Minnesota Partners for Quality Education, 202, 203
Mohrman, Susan A., on TQM and employee involvement, 575–583
Moments of truth, described, 328–329
Morgan, Jim, on benchmarking survey, 451–454
Morrison, Cathy, quoted, 454
Motorola, 65, 67, 176, quality at, 49; supplier council at, 572

Nagel, Roger N., on manufacturing competitiveness, 61–68
National Bicycle Industrial Company, how it makes bikes, 85
National Council on Educational Standard and Testing (NCTE), goals of, 188
National Performance Review, update on, 130–137
National Westminster Bank USA, supports TQM education, 216

INDEX

Naval Hospital Orlando, quality practices at, 138–145
Newsletters, directory of, 686–703
Nonprofit associations, bibliography, 641
Norris, Donald M., on higher education, 163–167
Norwest Mortgage, customer surveying by, 340–345

Objectives, on measuring, 455–464
Olson, David R., warehouse management trends, 517–521
Organizations, culture and change, 285–292; directory of, 704–742; learning, 397–413; noted for best practices, 730–742; Scholtes, Peter R., discusses teams in organizations, 360–375; on internet 680–681;
Oticon, jobs and roles at, 86
Otis Elevator, use of I/T by, 65–66
Otisline, described, 65–66
Overmyer-Day, Leslie, training success stories, 352–359
Owens-Corning Fiberglass, use of I/T by, 528–533

Pati, Niranjan, quality in business education, 188–193
PDCA, at a hospital, 138–145
Pearl River School District, Baldrige and, 198–203
People, bibliography, 661–664; reengineering issues and, 431–442
Performance, individual vs. team, 376–383
Personnel, best practices, 734
Petronius, quoted on reorganizations, 360
Philips Japan, 572
Phillips, Jack J., cost justifying training, 346–351
Piczak, Michael W., on implementing self-directed teams, 389–396
PIMS, described, 48–49
Pinellas County School District, Baldrige and, 198–203
Planning, best practices, 740; bibliography, 655–656; in contracting process, 511–512; quality practices, 231–242
Plant layout and design, best practices, 736
Poirier, Charles C., why TQM fails, 47–57
Policy deployment, best practices, 736
Pope, Jeffrey, Custom Research, Inc., 764
Porter, Anne Millen, on teaming with suppliers, 501–505
Potts, Elizabeth A., on steps for ISO certification, 616–617, 624
Prahalad, C.K., quoted, 442
Price, Art, trips on improving benchmarking, 443–450
Primates, how humans behave like, 267–276
Process improvement, best practices, 736
Process industries, bibliography, 634
Process management, bibliography, 651–652; design of, 236–242; Shasta Industries and, 69–80

Process reengineering, issues and practices, 424–442
Processes, role of organizational learning on, 407–413
Product development, best practices, 737; bibliography, 657–659; cases and practices, 10–11, 555–565; speeded with partnering, 501–505; strategies for, 64–66
Product improvement, best practices, 737
Products, analysis case of, 92–93; best practices in identifying new, 9–10
Productivity, measuring, 482–487
Project teams, best practices, 737
Propinquity, role of, 274–275
Public sector, bibliography, 638–641; quality practices in, 115–137
Publishers, quality materials (on internet), 670–680
Purchasing, benchmarking survey of, 451–454; best practices, 737–738; process descriptions, 491–516

Quality, implementation practices, 229–630; transformation cases and practices, 231–630
Quality and Productivity Management Association (QPMA), 49
Quality Function Deployment (QFD), use in contracting process, 508–510
Quality implementation, bibliography, 645–647
Quality management, best practices, 738–739
Quality services, online, described, 668–685
Quality tools, bibliography, 649–655
Quality transformation, bibliography, 641–649
QUENSH, defined, 603
Quinn, Richard, 461; quoted on value of measuring, 455
Quotes, on quality, 743–757

R&D, at Eastman Chemical, 563
Radcliff, Michael D., quoted on reengineering, 531
Randall, Richard C., proactive use of ISO 9000, 599–604
Raven, B., social power, 254–255
Reengineering, 424–442, 601–602; bibliography, 651–652; in health care, 152–161
Reid, R. Dan, pro ISO, 623
Reindenbach, Erick, on value propositions, 99–107
Reimann, Curt W., quoted on Baldrige, 626–627, 630
Reis, Dayr, quality in business education, 188–193
Research and development, best practices, 739; bibliography, 660
Resource optimization (RO), defined, 56–57
Retail industry, bibliography, 636
Rewards, 411; problems and issues, 394
Ricciardi, Philip, on measuring employee performance, 482–487
Ricoh Corporation, helped schools with TQM, 216
Roe, C. William, use of social power, 254–266
ROI, as applied to training, 346–351

Rooney, James, quoted, 325
Rosenthal, Gail, role at Harley-Davidson, 298–299
Ross Operating Valves, case study of, 67
RossFlex Process, described, 67
Roth, Aleda V., role of knowledge factories, 81–91
Rounds, Beth, Custom Research, Inc., 764
Ryel, Jack, quoted, 454

Sales, best practices, 739; bibliography, 656
SAP, use by Owens-Corning Fiberglass, 528–533
Savageau, John, on suggestion systems, 584–591
Sawyer, Ernie, quoted on TQM in classrooms, 207–208
scenario planning, value of, 410
Schargel, Franklin P., on TQM in education, 210–217
Scheuing, Eberhard E., purchasing's role managing quality, 491–500
Schiemann, William A., on measurements, 455–464
Scholtes Peter R. on value of teams, 360–375
Schultz, Edward, CEO Dana Commercial Credit, 762
Scicchitano, Paul, on alternatives to ISO certification, 621–624
SEIU, 117
Selto, Frank H., on ABC, 539–546
Senior citizens, power of, 148
Sensenbrenner, Joe, on public sector quality in 1986, 115–121
Service guarantees, effect on customers, 325–327
Services sector, bibliography, 635–638; quality practices in, 99–114
Setzer, David E., on using Baldrige criteria, 628–629
Severns, Michael B., 158
SGS-Thomson MicroElectronics, Inc., ISO certification case study, 618–619
Shasta Industries, how it transformed its manufacturing, 69–80
Sherman, Richard W., case of ISO certification, 605–612
Shewhart, Walter A., charts of, 488–490; on statistical quality control, 13–31
Shultz, Edward, CEO, Dana Commercial Credit Corp.,
Siebenaller, Jeff, mentioned, 49
Slagowitz, Martin B., quoted, 537–538
Sloan, Alfred, on letting people work, 278
Smith, Ken, quoted on change in public sector, 116
Smith, Preston G., product development process, 555–565
Society of Incentive Travel Executives (SITE), survey on employee incentives, 332
Sosville, Dick, quoted, 338, 339
Speed, role of, 57
Sprint, performance assessments at, 365
Standards, bibliography, 664–667; issues and cases, 599–630
State government, bibliography, 638

Statistical Process Control (SPC), effect on universal best practices, 7; role in process improvements, 465–481; Shewhart on, 13–31
Steelcase, marketing strategy of, 67
Stone, Stanley D., describes cellular manufacturing at Electronic Hardware, 91–98
Struebing, Laura, on customer loyalty, 322–333
Suggestion systems, best practices, 732–733; value of, 584–591
Sunbeam Household Products, delivery process of, 523–524
Supplier Audit Confirmation (SAC), defined, 622
Suppliers, best practices, 740; involving in product development, 567–574; process descriptions, 491–516
Supplier relations, bibliography, 656–657
Surveys, of customers, 334–339, 340–345; of employees,
Symons, Van, quoted, 535
Systems thinking, practices in, 397–413; role of teams in, 360–375

Teams, at Caterpillar Corp., 501–505; at Charles Medical Center, 158–161; at Thornton Hospital, 154–158; best practices, 740; bibliography, 654–655; effect on best practices, 7; in contracting process, 510–511; optimal size, 269–270; quality practices with, 360–396; Scholtes on, 360–375; work of, 279–280
Technology, best practices, 11
Technology transfer, best practices, 740
Teerlink, Richard, quoted on value of training, 391
Tektronic Education Consortium, described, 363
Telephone customer services, best practices, 740
Tennant Company, supplier quality management case, 495–496
Tenneco, quality practices at, 51
Texas Instruments, performance assessments at, 365–366; training at, 363–364
Thornton Hospital, teaming and reengineering at, 152–161
Throughput accounting, described with case, 547–554
Tokarz, Steven J., why TQM fails, 47–57
Tools, bibliography, 653–654
Tools and techniques, best practices, 424–490
Townsend, Pat, quoted, 584
Toyota, suggestion process at, 585
TQM, what it is, 1–2; why it fails and what to do about it, 47–57
Training, 411; best practices, 741; best practices and, 9; bibliography, 647–648; case of Shasta Industries, 78; internet sites, 682–683; on manufacturing processes, 97; quality practices in, 346–349; teams and, 391–392
Tresko, John, on data warehouses, 534–538

INDEX

Trident Precision Manufacturing, Inc., Baldrige recipient; 764–765
Trunick, Perry A., on speed of delivery, 522–527
Tuckman, Bruce, model of teams by, 378
Turner, Ronald E., on TQM in college classrooms, 168–174

Unions, role in public sector, 116–118; teams and, 391
Universal Card (AT&T), why successful, 104
U.S. Department of Veterans Affairs (VA), innovations at, 133–134
U.S. Government, status of reinvention initiative, 130–137
U.S. Postal Services, performance indicators at, 135
Utilities, bibliography, 637–638

Value propositions, on driving business, 99–107
Variation, Shewhart on, 13–31
Veleris, John, on Shasta Industries, 69–80
Vendor certification, best practices, 742
Virginia Beach Public Schools, quality in classrooms, 205–209

Walton, Sam, quoted on employee loyalty, 262
Warehouses, trends in managing, 517–521
Warehousing, best practices, 741

Warranties, best practices, 741
Waste minimization, best practices, 742
Watson, Carey, on value of complaints, 327–328
West Babylon School District, applied Deming's 14 points, 218–228
Westinghouse Electric Corporation, supports TQM education, 216
Wheatley, Margaret J., on self-organizations, 397–406
Wheeler, Donald J., recalculating limits on control charts, 488–490
Wholesale distribution, bibliography, 636
Williams Elementary School, benchmarking at, 207
Wilson, Terry C., on value propositions, 99–107
World Center for Community Excellence, founded, 120
World, quality, bibliography, 648–649

Xerox Corporation, communications at, 399; culture of described, 301–314; customer surveying by, 335–337; ESI at, 569, 570; measuring cost of quality, 495; quality results at, 49, 51
Xerox 2000, described, 301–314

Yeager, Neil, leadership and teams, 376–383

Zent, Max, introduced quality practices to Tenneco, 51